DATE DUE			

Isozymes

I

Molecular Structure

Organizing Committee and Editorial Board

Isozymes

I

Molecular Structure

EDITED BY

Clement L. Markert

Department of Biology
Yale University

ACADEMIC PRESS   New York   San Francisco   London   1975

A Subsidiary of Harcourt Brace Jovanovich, Publishers

ACADEMIC PRESS RAPID MANUSCRIPT REPRODUCTION

ACADEMIC PRESS, INC.
111 Fifth Avenue, New York, New York 10003

United Kingdom Edition published by
ACADEMIC PRESS, INC. (LONDON) LTD.
24/28 Oval Road, London NW1

Library of Congress Cataloging in Publication Data

International Conference on Isozymes, 3d, Yale University, 1974
    Isozymes.

    Bibliography: v. 1, p.
    Includes index.
    CONTENTS: v. 1. Molecular structure.
    1.   Isoenzymes–Congresses.   I.   Markert, Clement
Lawrence, (date)   ed.   II.   Title.
QP601.148 1974   574.1'925   74-31288
ISBN 0–12–472701–8 (v. 1)

Contents

CONTENTS

CONTENTS

CONTENTS

Contributors

H. Aebi Medizinisch-chemisches Institut der Universität Berne, CH — 3000, Berne 9, Switzerland

Klaus Apel The Biological Laboratories, Harvard University, 16 Divinity Avenue, Cambridge, Massachusetts 02138

William J. Arnold Division of Rheumatic and Genetic Diseases, Departments of Medicine and Biochemistry, Duke University Medical Center, Durham, North Carolina 27710

Isabelle Audit Unité 91 de l'N.S.E.R.M. — Hôpital Henri Mondor, 94010 Creteil, France

William M. Awad, Jr. Departments of Medicine and Biochemistry, University of Miami School of Medicine, P.O. Box 520875 Biscayne Annex, Miami, Florida 33152

David P. Baccanari Burroughs Wellcome Co., 3030 Cornwallis Road, Research Triangle Park, North Carolina 27709

John A. Badwey Biochemistry Department, University of Massachusetts, Amherst, Massachusetts 01002

Bohdan Bakay Department of Pediatrics, School of Medicine, University of California at San Diego, La Jolla, California 92037

Joann Basabe Department of Biological Chemistry, California College of Medicine, University of California, Irvine, California 92664

Kirstan Bazzell Oklahoma Medical Research Foundation, Oklahoma City, Oklahoma 73100

Kathy B. Benveniste Laboratory of Developmental Biology and Anomalies, National Institute of Dental Research, National Institutes of Health, Bethesda, Maryland 20014

Will Bloch Department of Chemistry, Reed College, Portland, Oregon 97202

Lawrence Bogorad The Biological Laboratories, Harvard University, 16 Divinity Avenue, Cambridge, Massachusetts 02138

Emory H. Braswell Biological Sciences Group, University of Connecticut, Storrs, Connecticut 06268

John R. Cann Department of Biophysics and Genetics, University of Colorado Medical Center, Denver, Colorado 80220

J. M. Cardenas Department of Biochemistry and Biophysics, Oregon State University, Corvallis, Oregon 97331

F. J. Castellino Department of Chemistry, University of Notre Dame, Notre Dame, Indiana 46556

Stephen D. Cederbaum Division of Medical Genetics, Mental Retardation Unit, UCLA Neuropsychiatric Institute, Los Angeles, California 90024

Sungman Cha Division of Biological and Medical Sciences, Brown University, Providence, Rhode Island 02912

Lee Chao Genetics and Cell Biology Section, Biological Sciences Group, The University of Connecticut, Storrs, Connecticut 06268

John Cheney Department of Botany and Forest Pathology, State University New York, College of Environmental Science and Forestry, Syracuse, New York 13210

Soo II Chung Laboratory of Biochemistry, National Institute of Dental Research, National Institutes of Health, Bethesda, Maryland 20014

Major L. Cohn Department of Anesthesiology, School of Medicine, University of Pittsburgh, Pittsburgh, Pennsylvania 15261

J. A. Cortner Department of Pediatrics and Internal Medicine, State University of New York at Buffalo, and the Children's Hospital, Buffalo, New York 14222

Rody P. Cox Division of Human Genetics, Departments of Medicine and Pharmacology, New York University Medical Center, New York, New York 10016

David M. Dawson Department of Neurology, Children's Hospital Medical Center and Harvard Medical School, Boston, Massachusetts 02115

George J. Doellgast Tufts Cancer Research Center and the Department of Pathology (Oncology), Tufts University School of Medicine, 136 Harrison Avenue, Boston, Massachusetts 02111

John A. Duley Department of Biochemistry, La Trobe University, Bundoora, 3083, Victoria, Australia

R. D. Dyson Department of Biochemistry and Biophysics, Oregon State University, Corvallis, Oregon 97331

D. J. Ecobichon Department of Pharmacology, Faculty of Medicine, Dalhousie University, Halifax, N. S., Canada

Masaharu Eguchi Laboratory of Genetics, Kyoto Technical University, Matsugasaki, Kyoto, Japan

W. Eventoff Department of Biological Sciences, Purdue University, West Lafayette, Indiana 47907

William N. Fishbein Biochemistry Branch, Armed Forces Institute of Pathology, Washington, D.C. 20306

Rachel A. Fisher MRC Human Biochemical Genetics Unit, University College, London, England

William H. Fishman Tufts Cancer Research Center and the Department of Pathology (Oncology), Tufts University School of Medicine, 136 Harrison Avenue, Boston, Massachusetts 02111

Michael Flashner, SUNY College of Environmental Science and Forestry, Syracuse, New York 13210

Tova Francus Department of Microbiology, Columbia University, New York, New York 10032

Thomas D. Gelehrter, M.D. Department of Human Genetics, University of Michigan Medical School, Ann Arbor, Michigan 48104

Larry Gerace Rockefeller University, New York, New York 10021

Nimai K. Ghosh Division of Human Genetics, Departments of Medicine and Pharmacology, New York University Medical Center, New York, New York 10016

Robert W. Gracy Department of Chemistry, North Texas State University, Denton, Texas 76203

Jerry M. Greene Department of Neurology, Children's Hospital Medical Center and Harvard Medical School, Boston, Massachusetts 02115

David H. Griffin Department of Botany and Forest Pathology, State University of New York, College of Environmental Science and Forestry, Syracuse, New York 13210

Martin J. Griffin Oklahoma Medical Research Foundation, Oklahoma City, Oklahoma 73100

Albert Grossman Department of Pharmacology, New York University School of Medicine, New York, New York 10016

M. L. Hackert Department of Biological Sciences, Purdue University, West Lafayette, Indiana 47907

James W. Hardin The Biological Laboratories, Harvard University, 16 Divinity Avenue, Cambridge, Massachusetts 02138

Harry Harris MRC Human Biochemical Genetics Unit, University College, London, England

Norman B. Hecht   Department of Biology, Tufts University, Medford, Massachusetts 02155

Roger S. Holmes   School of Science, Griffith University, Nathan, Brisbane, Queensland. 4111., Australia

D. A. Hopkinson   MRC Human Biochemical Genetics Unit, Galton Laboratory, University College, London, England

B. L. Horecker   Roche Institute of Molecular Biology, Nutley, New Jersey 07110

Paul A. Horgen   Department of Botany, University of Toronto, Erindale Campus, Mississauga, Ontario, Canada

Kenneth H. Ibsen   Department of Biological Chemistry, California College of Medicine, University of California, Irvine, California 92664

Ronald W. Johnson   Department of Pharmacology, New York University School of Medicine, New York, New York 10016

K. K. Kannan   Department of Molecular Biology, The Wallenberg Laboratory, Uppsala University, Uppsala, Sweden

Cecil J. Kelly   Division of Biological and Medical Sciences, Brown University, Providence, Rhode Island 02912

Philip M. Kelley   Department of Microbiology, Washington University School of Medicine, St. Louis, Missouri 63110

William N. Kelley   Division of Rheumatic and Genetic Diseases, Departments of Medicine and Biochemistry, Duke University Medical Center, Durham, North Carolina 27710

William C. Kenney   Department of Biochemistry and Biophysics, University of California, San Francisco, California, 94143 and Division of Molecular Biology, Veterans Administration Hospital, San Francisco, California 94121

Chi-Wei Lin   Tufts Cancer Research Center and Department of Pathology, Tufts University School of Medicine, Boston, Massachusetts 02111

Helga Lundgren   SUNY College of Environmental Science and Forestry, Syracuse, New York 13210

Clement L. Markert   Department of Biology, Yale University, New Haven, Connecticut 06520

Ivy R. McManus   Department of Biochemistry, Faculty of Arts and Sciences, University of Pittsburgh, Pittsburgh, Pennsylvania 15261

E. A. Meighen   Department of Biochemistry, McGill University, Montreal, Quebec, Canada

David B. Millar   Laboratory of Physical Biochemistry, Naval Medical Research Institute, National Naval Medical Center, Bethesda, Maryland 20014

Kenneth D. Munkres   Laboratory of Molecular Biology, University of Wisconsin, Madison, Wisconsin 53704

K. Nagarajan   Biochemistry Branch, Armed Forces Institute of Pathology, Washington, D.C. 20306

Ernst A. Noltmann   Department of Biochemistry, University of California, Riverside, California 92502

B. Notstrand   Department of Molecular Biology, The Wallenberg Laboratory, Uppsala University, Uppsala, Sweden

Maria S. Ochoa   Departments of Medicine and Biochemistry, University of Miami School of Medicine, P.O. Box 520875, Biscayne Annex, Miami, Florida 33152

Janis O'Donnell   Department of Biology, The Johns Hopkins University, Baltimore, Maryland 21218

Regina Pietruszko   Center of Alcohol Studies, Rutgers University, New Brunswick, New Jersey 08903

Jean Rosa   Unité 91 de l'N.S.E.R.M. — Hopital Henri Mondor, 94010 Creteil, France

Raymonde Rosa   Unité 91 de l'N.S.E.R.M. — Hopital Henri Mondor, 94010 Creteil, France

M. G. Rossmann   Department of Biological Sciences, Purdue University, West Lafayette, Indiana 47907

Milton J. Schlesinger   Department of Microbiology, Washington University School of Medicine, St. Louis, Missouri 63110

B. Scherz   Medizinisch-chemisches Institut der Universität Berne, CH-3000, Berne. 9. Switzerland

J. D. Schnatz   Department of Pediatrics and Internal Medicine, State University of New York at Buffalo, and the Children's Hospital, Buffalo, New York 14222

Marcia Schwartz   Department of Biology, The Johns Hopkins University, Baltimore, Maryland 21218

Warren Scurzi   Biochemistry Branch, Armed Forces Institute of Pathology, Washington, D.C. 20306

G. F. Sensabaugh   Forensic Science Program, School of Criminology, University of California, Berkeley, California 94720

James B. Shaklee   Provisional Department of Genetics and Development, University of Illinois at Urbana-Champaign, Urbana, Illinois 61801

G. E. Siefring, Jr.   Department of Chemistry, University of Notre Dame, Notre Dame, Indiana 46556

Steven Siegel   Departments of Medicine and Biochemistry, University of Miami School of Medicine, P.O. Box 520875 Biscayne Annex, Miami, Florida 33152

Jane Smith   The Biological Laboratories, Harvard University, 16 Divinity Avenue, Cambridge, Massachusetts 02138

J. M. Sodetz   Department of Chemistry, University of Notre Dame, Notre Dame, Indiana 46556

William Sofer   Department of Biology, The Johns Hopkins University, Baltimore, Maryland 21218

Carolyn J. Spencer   Department of Human Genetics, University of Michigan Medical School, Ann Arbor, Michigan 48104

Joseph F. Speyer   Genetics and Cell Biology Section, Biological Sciences Group, The University of Connecticut, Storrs, Connecticut 06268

S. J. Steindel   Department of Biological Sciences, Purdue University, West Lafayette, Indiana 47907

J. J. Strandholm   Department of Biochemistry and Biophysics, Oregon State University, Corvallis, Oregon 97331

Stuart W. Tannenbaum   SUNY College of Environmental Science and Forestry, Syracuse, New York 13210

William Timberlake   Department of Botany and Forest Pathology, State University of New York, College of Environmental Science and Forestry, Syracuse, New York 13210

Patricia Trippet   Department of Biological Chemistry, California College of Medicine, University of California, Irvine, California 92664

Orestes Tsolas   Roche Institute of Molecular Biology, Nutley, New Jersey 07110

Bruce J. Turner   The Rockefeller University, New York, New York 10021

Bryan M. Turner   Division of Medical Genetics, Department of Pediatrics, Mount Sinai School of Medicine of the City University of New York, New York, New York 10029

I. Vaara   Department of Molecular Biology, The Wallenberg Laboratory, Uppsala University, Uppsala, Sweden

Richard B. Vallee Laboratory of Molecular Biology, The University of Wisconsin, Madison, Wisconsin 53706

Klaus D. Vosbeck Departments of Medicine and Biochemistry, University of Miami School of Medicine, P.O. Box 520875 Biscayne Annex, Miami, Florida 33152

E. W. Westhead Biochemistry Department, University of Massachusetts, Amherst, Massachusetts 01002

Beulah M. Woodfin Department of Biochemistry, School of Medicine, University of New Mexico, Albuquerque, New Mexico 87131

S. R. Wyss Medizinisch-chemisches Institut der Universität Berne, CH-3000, Berne 9, Switzerland

Preface

Isozymes are now recognized, investigated, and used throughout many areas of biological investigation. They have taken their place as an essential feature of the biochemical organization of living things. Like many developments in the biomedical sciences, the field of isozymes began with occasional, perplexing observations that generated questions which led to more investigation and, finally, with the application of new techniques, to a clear recognition and appreciation of a new dimension of enzymology.

The area of isozyme research is only about 15 years old but has been characterized by an exponential growth. Since the recognition in 1959 that isozymic systems are a fundamental and significant aspect of biological organization, many thousands of papers have been published on isozymes. Several hundred enzymes have already been resolved into multiple molecular forms, and many more will doubtless be added to the list. In any event, it is now the responsibility of enzymologists to examine every enzyme system for possible isozyme multiplicity.

Two previous international conferences have been held on the subject of isozymes, both under the sponsorship of the New York Academy of Sciences—the first in February 1961, and the second in December 1966. And now, after a somewhat longer interval, the Third International Conference was convened in April 1974, at Yale University. For many years, a small group of investigators has met annually to discuss recent advances in research on isozymes. They have published a bulletin each year and have generally helped to shape the field; in effect, they have been a standing committee for this area of research. From this group emerged the decision to hold a third international conference, and an organizing committee of five was appointed to carry out the mandate for convening a third conference. This Third International Conference was by far the largest of the three so far held with 224 speakers representing 21 countries, and organized into nine simultaneous sessions for three days on April 18, 19, and 20, 1974. Virtually every speaker submitted a manuscript for publication, and these total almost 4,000 pages. The manuscripts have been collected into four volumes entitled, *I. Molecular Structure; II. Physiological Function; III. Developmental Biology; and IV. Genetics and Evolution.* The oral reports at the Conference and the submitted manuscripts cover a vast area of biological research. Not every manuscript fits precisely into one or another of the four volumes, but the most appropriate assignment has been made wherever possible. The quality of the volumes and the success of the Conference must be credited to the participants and to the organizing committee. The scientific community owes much to them.

Acknowledgments

I would like to acknowledge the help of my students and my laboratory staff in organizing the Conference and in preparing the volumes for publication. I am grateful to my wife, Margaret Markert, for volunteering her time and talent in helping to organize the Conference and in copy editing the manuscripts.

Financial help for the Conference was provided by the National Science Foundation, the National Institutes of Health, International Union of Biochemistry, Yale University, and a number of private contributors:

Private Contributors

American Instrument Company
Silver Spring, Maryland 20910

Canalco
Rockville, Maryland 20852

CIBA-GEIGY Corporation
Ardsley, New York

Gelman Instrument Company
Ann Arbor, Michigan 48106

Gilford Instrument Laboratories, Inc.
Oberlin, Ohio

Hamilton Company
Reno, Nevada 89502

Kontes Glass Company
Vineland, New Hampshire 08360

The Lilly Research Laboratories
Indianapolis, Indiana

Merck Sharp & Dohme
West Point, Pennsylvania

Miles Laboratories, Inc.
Elkhart, Indiana

New England Nuclear
Worcester, Massachusetts 01608

Schering Corporation
Bloomfield, New Jersey

Smith Kline & French Laboratories
Philadelphia, Pennsylvania

Warner-Lambert Company
Morris Plains, New Jersey

BIOLOGY OF ISOZYMES

CLEMENT L. MARKERT
Department of Biology
Yale University
New Haven, Connecticut 06520

HISTORY

Isozymes are now a common part of the scientific vocabulary but their recognition is relatively recent, being first announced in 1959 (Markert and Møller). Prior to that time, molecular heterogeneity had often been noted in enzyme preparations, but such heterogeneity was usually attributed to contaminants or to partially denatured or degraded enzyme molecules (e.g., cf. Meister, 1950). During the 1950s there were occasional suggestions that this heterogeneity might not all be artifactual but might indeed reflect reality within the cell (Neilands, 1952a, 1952b; Vesell and Bearn, 1957; Wieland and Pfleiderer, 1957). Most of these investigators used zone electrophoresis to separate enzyme preparations into several different active components, but their procedures were all laborious and had poor resolving power for separating the multiple forms of enzymes. Consequently, their reports on heterogeneity had little impact on investigators until a simple, easy method for assessing enzyme heterogeneity was developed. By coupling the starch gel electrophoresis technique of Smithies (1955) with histochemical staining procedures to identify separated enzymes, Hunter and Markert (1957) developed just such a technique -- the zymogram technique, which is characterized by great resolving power for the identification of the isozymes of many enzymes. This technique was applied with dramatic success first to esterases and then to several other enzymes including lactate dehydrogenase (Markert and Møller, 1959). These enzymes were fruitful choices because their isozymes were numerous, distinct, and easily resolved. Moreover, this technique was direct, simple, easy to use, and applicable to enzymes in crude homogenates. Perhaps as important as the technique was the recognition that isozymes are a natural and important aspect of cellular biochemistry and are to be found in nearly all organisms. As a consequence of these technological and conceptual innovations, thousands of investigations have focused on isozymes and the literature is replete with descriptions and analyses of isozymes from many points of view.

Once the zymogram technique became a standard part of the laboratory armamentarium, then many modifications and improvements were made and other techniques were also used to resolve enzyme preparations into multiple molecular forms. Chromatography, especially its refined version of affinity chromatography, various immunochemical techniques, differential denaturation, salt precipitation, and so forth, have all been used to resolve enzyme systems into distinguishable molecular forms -- that is, into isozymes. Like many advances in science, the development of the essential techniques of isozymology proved crucial not only for obtaining data but also for generating a deeper understanding. Clearly isozymes are now an important dimension of enzymology, biochemistry, cell biology, and genetics -- in short, of many of the major fields of biology.

UBIQUITY

The fact that many enzymes have already been resolved into two or more isozymic forms makes every enzyme a candidate for investigation of molecular multiplicity. In fact, with the acceptance of the concept that every enzyme is encoded by a gene and that every gene may mutate to produce functioning alleles, we are forced to the conclusion that every enzyme will eventually be found to exist in isozymic forms. Clearly, our recognition of isozymes generates a large and rich area of investigation and places a sharp responsibility on enzymologists. All enzyme preparations should be tested for isozymic heterogeneity before the properties of the enzyme are reported in the literature. Moreover, each enzyme from each species can now be examined for possible isozymes. If isozymes are found, then their physical-chemical nature, genetic or epigenetic origin, cellular distribution, physiological role, development and evolution, all become subjects for investigation. Clearly, enzymologists have much to do.

FORMATION OF ISOZYMES

The original definition of the term 'isozyme' was designed to have operational utility only and not to specify the molecular basis for any particular multiplicity. The vast literature on isozymes accumulated during the past 15 years has not only demonstrated the usefulness of the term but, perhaps more important, attested to the fruitfulness of the concept that organisms do commonly synthesize many of their enzymes in several different molecular forms, presumably to fulfill specialized metabolic requirements.

From the point of view of the living cell, a variety of mechanisms is available for generating molecular multiplicity of any enzyme (cf. Markert, 1968; Markert and Whitt, 1968).

2

The cell may begin by duplicating a gene or even by making
several copies, each of which can then diverge by mutation to
produce a modified version of the original enzyme. These
isozymes and their controlling genes are then subject to
evolutionary selection that will result in tailoring the
structure of the molecule to fit the specialized metabolic
requirements of the cell. Genetically specified isozymes are
obviously fundamental to cellular biochemistry, but the cell
is not restricted to this method for producing isozymes. Post-
translational or epigenetic modifications of enzymes can also
be affected, particularly through the reactive amino, carboxyl,
or hydroxyl groups of several of its amino acid residues.
Disulfide bridging, hydrogen bonding, and various types of
interaction between polypeptide chains can all modify the three-
dimensional structure of enzymes to create distinctly different
isozymes. The individual polypeptide chains may function singly
or as polymers of various sizes, or as parts of larger aggregates to achieve maximal functional utility for the cell.
Moreover, proteins may be conjugated with other molecules such
as carbohydrates to alter enzymatic function and to determine
the specific location of the protein molecule within the cell.
Also, the initial polypeptide chain can be shortened by proteolysis to produce functionally different molecules. All of these
methods, and probably others, have been exploited by cells to
produce numerous isozymic systems.

NOMENCLATURE

Although genetic and epigenetic isozymes have quite different
origins, they are so closely related in biological significance
that they should be described by a single general nomenclature.
All living systems apparently require multiple molecular forms
of certain enzymes in order to maximize biological capacity. A
variety of biochemical mechanisms have been developed for producing these isozymes. Therefore, despite occasional suggestions for altering or restricting the definition of isozymes,
the original definition still seems best: once the accepted
criteria for defining a collection of molecules from a single
individual or from members of a single species as an enzyme
have been successfully applied, then if these molecules can
be separated into distinguishable types by any method, then
these types represent isozymic forms of the enzyme. Such a
broad operational definition embraces many different kinds of
molecular multiplicity. However, once the detailed molecular
nature of any isozymic system is known, then the nomenclature
can become more precise and specific in accord with the information available. The general term 'isozyme' might then

3

appropriately be modified by such terms as allelic, non-allelic, homopolymeric, heteropolymeric, epigenetic, conjugated, conformational, and so forth.

FUNCTION OF ISOZYMES

Isozymes are numerous and characteristic of many cells, tissues, and organs, but what is their biological utility? Why are they such a ubiquitous aspect of the biochemical organization of cells? At one time it was believed that one gene coded for one enzyme, which was totally responsible for a single biochemical reaction. This simple molecular equation, one gene – one enzyme – one catalytic reaction, was a useful and stimulating generalization in the early days of biochemical genetics. But now we know that multiple varieties of an enzyme are needed to catalyze the same reaction, but under different metabolic conditions, or in different places in the same cell, or in different cells, or in the same cell at successive stages in differentiation. Apparently, isozymes have been tailored by evolutionary pressures to fit the fastidious requirements of the cell's metabolic machinery.

Isozymes in a cell are like singers in a choir. Together thay all sing the same melody, but there is an array of voices, each having its own tonal quality, range, timbre, resonance, and power. Each singer must occupy a designated position with reference to all the others in order to achieve an integrated pattern of performance. Moreover, the quality of each singer depends upon genetic endowment and epigenetic conditioning. As the program changes from time to time, the relative roles of the individual performers may also change and those with the most suitable characteristics, genetic or not, will be retained and the others dismissed. The ultimate unit of judgment, of selection, is the entire choir and not the individual performers. And so it is with cells and their constituent isozymes.

After the identification and molecular description of isozymes has been completed, then the principal problem is understanding the specific function of each isozyme. In this area progress has been very slow. Very few isozymic systems are well understood today. One of the earliest contributions was the recognition that isozymes could exist in different cell organelles and in different metabolic compartments within a single cell. The location of isozymes in cell organelles is well illustrated by mitochondria. Mitochondrial enzymes, malate dehydrogenase, e.g., are frequently different from the homologous enzymes in the cytosol even though both kinds of isozymes are encoded in nuclear genes. The distinction between

organelle and metabolic compartments is not always sharp. RNA
polymerases may fall in this ambiguous category; two isozymes
are associated with the nucleolus and one at least with mitochondria, but the topographic location of the others is uncertain. All of them must have access to the DNA of the chromosomes or of the mitochondria.

Metabolic compartmentation is perhaps best illustrated by
those isozymes that catalyze common steps in a branched biosynthetic pathway leading to the formation of two or more end
products. A common regulatory feature of such pathways is the
presence of isozymes with different susceptibilities to end-
product control. Such isozymic systems permit an effective
and independent feedback regulation of different synthetic
(or degradative) pathways involving the same catalytic activity
(cf. Umbarger, 1961; Stadtman, 1968). The isozymic forms of
an enzyme commonly have somewhat different kinetic properties,
although the significance of these differences, except for
feedback regulation, is seldom clear. This has been especially
true for the much studied LDH-A_4 and B_4 isozymic systems, but
the role of the different kinetic properties of these isozymes
in the economy of the cell is now becoming evident (Vesell,
1975; Everse and Kaplan, 1975; Nadal-Ginard and Markert, 1975).

Although the metabolism of all cells is basically similar,
nevertheless an enormous variety of cells is found in each
metazoan organism and among single-celled organisms as well.
Diversity in metabolic architecture must have appeared early
in cell evolution and is now characteristic of all contemporary
organisms. Many different molecular structures have been used
by cells to generate isozymic systems. These have ranged from
molecules distinguished by a single amino acid substitution to
molecules differing in many amino acids and produced by genes
long separated in evolution, to molecules so different that
one suspects that they arose by convergent evolution rather
than by divergence from a common ancestral gene, to many
epigenetically modified molecules with altered structures
and properties.

Isozymes are frequently identified by electrophoresis and
thus recognized by virtue of differences in net charge. However it should be remembered that most amino acid substitutions
do not alter the net charge on an enzyme, and thus most isozymes remain undetected by electrophoresis and probably have
so far escaped recognition altogether. The sequence of amino
acids at the active enzymatic site is usually highly conserved
and remains the same for the different isozymes of a single
enzyme. Charge differences among isozymes must be attributed
to other parts of the molecule. The physiological role of the
differences in net charge is a moot question but the differences

in charge distribution over the surface of the isozyme surely affect the topographic location of the molecule within the cell. The positioning of isozymes in different metabolic compartments is probably made possible by the specific distribution of charges over the surface of the molecule. A change in charge distribution should produce different macromolecular associations and thus alter the very structure of the cell. Mutations -- neutral, beneficial, or deleterious -- all affect molecular structure to some degree. Enzymes of altered structure may function equally well in vitro but may not be equally well integrated into the molecular ecology of the cell. Moreover, the stability of the enzyme within the cell and its rate of degradation may depend significantly upon stable associations with other molecules -- associations controlled by surface configurations including charge distributions. Very few aspects of the molecular structure of isozymes could be more important than those that determine intracellular location.

Isozymes are useful tools, perhaps the best, for probing the detailed metabolic activities of cells in discharging their physiological functions. This is so because alternative isozymes, like alternative alleles, allow us to examine the "normal" condition by contrast with a variant condition. In fact, alleles are commonly expressed as isozymes, thus making possible both a genetic and a biochemical comparison. On the level of molecular structure, isozymes should help to reveal the structural requirements for enzymatic activity, for stability, and for specifying associations among molecules. Eventually we should gain truly useful insights into the metabolic structure of cells through research on isozymes.

ISOZYMES AS GENETIC MARKERS IN DEVELOPMENT AND EVOLUTION

Isozymes provide rich material for investigating the structure and function of enzymes and for examining their role in cellular metabolism, but they also can facilitate studies in cell differentiation and in population genetics and evolution because they serve as excellent markers of gene function. In the field of evolution, isozymes are now extensively used to measure the frequencies of alleles in populations and thus to allow an assessment of selection pressures in evolutionary movement. They have also made possible the development of the provocative concept of neutral mutation and have stimulated considerable investigation on the significance in evolution of alternative protein structures as exemplified by different isozymes. Alternative protein structures may be equally advantageous to a cell, provided complementary changes have occurred

in associated macromolecules. In other words, the advantage
of a given enzyme structure is only relative and depends upon
the molecular environment in which it must function.

One area in the study of evolution in which isozymes may
prove of critical importance relates to the acquisition of new
genetic information during evolution. The possibility of a
completely new gene arising *de novo* today seems vanishingly
small. New information probably arises through the duplication
of genes and their subsequent divergence through mutation.
This procedure can lead to the evolution of one enzyme into
another and probably most enzymes have arisen this way. It
seems clear that isozymes can be generated by the duplication
of loci. Thus it should be possible to study the retention
of homologous properties in enzymes, encoded in duplicated loci,
as these enzymes diverge by mutation of the controlling genes.
We cannot today study the sequence of molecular events that
has already occurred during evolution, but we can probably
discover each type of change that has occurred by a detailed
molecular analysis of contemporary groups of related enzymes.

The use of isozymes in the study of molecular events in
developmental biology has also proved fruitful. Isozymic
patterns commonly change during embryonic development and
are highly characteristic of each type of cell at each stage
in its differentiation. The ontogeny of such isozyme patterns
reflects the changing patterns of gene activation and repression. New patterns of isozymes can also be generated as fully
differentiated cells are further transformed into cancer cells.
These neoplastic patterns are frequently reminiscent of an
earlier embryonic state of cell differentiation. Thus it is
obvious that cell-specific isozyme patterns must reflect a
corresponding specificity of gene function.

The molecular mechanisms by which genes are regulated during
the course of differentiation still completely elude us. However, studies of isozymes may provide some insight into this
central problem of experimental biology. For example, the
sensitivity and precision of the regulatory controls on the
expression of genes can be examined by following the behavior
of duplicated loci in producing their corresponding products,
either in organisms at different stages in evolution or in
cells at different stages in differentiation. Eventually,
duplicated loci probably become completely independent, respond-
ing to different regulatory signals and even encoding proteins
with distinctly different enzymatic properties. Thus, the
evolution of specific patterns of isozymes reflects a corresponding evolution of regulatory controls on the responsible
genes. By mapping the genes and the regulatory DNA, the
effects of specific chromosomal locations on the function

of genes may be determined. The packaging of genes in particular chromosomes -- that is, the karyotype, is highly specific and apparently unchanging in each species. Any abnormality in packaging the genes into different chromosomes almost always leads to severe developmental abnormalities, probably because the regulatory DNA has been altered. In time, even the significance of this karyotypic rigidity for gene behavior may be understood through research based on isozymes.

In summary, although isozymology is not a distinct discipline itself, it nevertheless constitutes a vital new dimension and significant refinement of many of the established disciplines of biology -- physical biochemistry, enzymology, physiology, cell biology, developmental biology, genetics, and evolution.

REFERENCES

Everse, J. and N.O. Kaplan 1975. Mechanisms of action and biological functions of various dehydrogenase isozymes. *II. Isozymes: Physiology and Function.* C.L. Markert, editor, Academic Press, New York.

Hunter, R.L. and C.L. Markert 1957. Histochemical demonstration of enzymes separated by zone electrophoresis in starch gels. *Science.* 125: 1294-1295.

Markert, C.L. 1968. The molecular basis for isozymes. *Ann. N.Y. Acad. Sci.* 151: 14-40.

Markert, C.L. and F. Møller 1959. Multiple forms of enzymes: tissue, ontogenetic, and species specific patterns. *Proc. Natl. Acad. Sci. USA.* 45: 753-763.

Markert, C.L. and G.S. Whitt 1968. Molecular varieties of isozymes. *Experientia.* 24: 977-991.

Meister, A. 1950. Reduction of α, γ-diketo and α-keto acids catalyzed by muscle preparations and by crystalline lactic dehydrogenase. *J. Biol. Chem.* 184: 117-129.

Nadal-Ginard, B. and C.L. Markert 1975. Use of affinity chromatography for purification of lactate dehydrogenase and for assessing the homology and function of the A and B subunits. *II. Isozymes: Physiology and Function.* C.L. Markert, editor, Academic Press, New York.

Neilands, J.B. 1952a. Studies on lactic dehydrogenase of heart I. Purify, kinetics, and equilibria. *J. Biol. Chem.* 199: 373-381.

Neilands, J.B. 1952b. The purity of crystalline lactic dehydrogenase. *Science.* 115: 143-144.

Smithies, O. 1955. Zone electrophoresis in starch gels: group variation in the serum proteins of normal human adults. *Biochem. J.* 61: 629-641.

Stadtman, E.R. 1968. The role of multiple enzymes in the regulation of branched metabolic pathways. *Ann. N.Y. Acad. Sci.* 151: 516-530.

Umbarger, H.E. 1961. In: *Control Mechanisms in Cellular Processes*. D.M. Bonner, editor. The Ronald Press Co., New York. pp. 67-86.

Vesell, E.S. 1975. Medical uses of isozymes. *II. Isozymes: Physiology and Function*. C.L. Markert, editor, Academic Press, New York.

Vesell, E.S. and A.G. Bearn 1957. Localization of lactic acid dehydrogenase activity in serum fractions. *Proc. Soc. Exp. Biol. and Med.* 94: 96-99.

Wieland, T. and G. Pfleiderer 1957. Nachweis der Heterogenität von Milchsäure-dehydrogenasen verschiedenen Ursprungs durch Trägerelektrophorese. *Biochem. Z.* 329: 112-116.

BIOCHEMISTRY OF ISOZYMES

B. L. HORECKER
Roche Institute of Molecular Biology
Nutley, New Jersey 07110

ABSTRACT. Biochemical studies of isozymes are important
for an understanding of their evolutionary and genetic origin and their role in the physiology of differentiated
cells. They can also provide valuable information on the
control of gene expression during the development of differentiated cells and on the nature of metabolic control in
these cells.

   Knowledge of the structure and kinetic properties of
isozymes holds the key to many of these questions. In this
presentation structural studies on aldolases A and C from
rabbit muscle and brain, and on the isozymes of fructose
bisphosphatase from rabbit liver, kidney, and muscle will
be reviewed as examples of the biochemical approach. Evidence for epigenetic modification and the relation of these
changes to enzyme regulation and turnover will also be discussed.

INTRODUCTION

   In his opening remarks, Dr. Markert emphasized the importance of isozymes for an understanding of a number of major
biological problems, including the evolution of populations,
the transformation of one gene into another, and the regulation of gene expression. Isozymes may also hold the key to
an understanding of metabolic regulation and the function of
metabolic pathways in differentiated tissues. As will be discussed by Dr. Vesell, they can also help the clinician to
understand metabolic and genetic disorders.

   Essential to all of these problems is a better understanding
of the structure of isozymes and their functions. I will con-
centrate on these aspects, using as examples two enzymes,
aldolase and fructose bisphosphatase, that have been studied
intensively in our laboratory. I will also consider certain
epigenetic modifications of these enzymes and their physiological implications. Moreover, it was suggested by Markert
five years ago (1969) at a meeting on phosphohydrolases organized by William H. Fishman and sponsored by the New York Academy of Sciences that more attention be paid to those mechanisms
whereby enzymes are degraded and removed from the cell.
Obviously, selective degradation can play an important role
in determining the intracellular steady state levels of enzymes. In addition, selective modification may give rise to

11

new forms of enzymes with altered catalytic and regulatory properties. This is a well-known phenomenon in the formation of proteolytic enzymes from pro-enzymes, but it has not yet been frequently encountered for intracellular enzymes. Selective degradation may indeed play an important role in regulation and differentiation; the loss of reticulum in the erythrocyte is an obvious example. The isozymes of RNA polymerase and aldolase found in the spores and vegetative cells of *Bacillus subtilis* have been identified as products of the enzyme found in the vegetative cell formed by the action of a proteolytic enzyme that appears to be specifically induced under sporulating conditions (Doi, 1972). Evidence for regulation of a mammalian enzyme by proteolytic modification will also be discussed.

Structure of Isozymes

Identification. In order to study isozymes, they must first be identified, and a number of techniques have been developed for this purpose. Rapid and convenient methods employing zone electrophoresis on semi-solid supports with staining for catalytic activity (Hunter and Markert, 1957), so-called "zymograms", have been widely adopted. Ion exchange chromatography (Peterson and Sober, 1962) has also proved valuable both for analysis and for isolation. Another powerful technique is isoelectric focusing (Vesterberg, 1967); this will often resolve isozyme systems where other methods fail.

Isozymes may also be distinguished by their kinetic properties, which are particularly useful in combination with the separation methods described. Examples are the use of high and low concentrations of glucose to distinguish the various forms of hexokinase in gel electrophoresis (Katzen et al, 1965) or high and low levels of pyruvate to identify the isozymes of lactate dehydrogenases (Rose and Wilson, 1966).

Separation and Isolation. In addition to the conventional methods for protein purification, the techniques that have served for the identification of isozymes can also be utilized for their isolation. Ion exchange chromatography and isoelectric focusing have been very useful, although there have been difficulties in the use of the latter for preparative purposes. These problems appear to have been solved by the use of an apparatus for discontinuous electrofocusing (Denckla, unpublished). The development of fluorescamine has increased the sensitivity of amino acid analysis to the point where this can now be carried out with the amounts of protein present in individual bands separated in conventional disc gel electrophoresis (Stein et al, 1974). Affinity chromatography

(Cuatrecasas and Anfinsen, 1971) promises to be a powerful
tool for the separation of isozymes, taking advantage of differences in ligand binding. Another novel development is ion
filtration chromatography, recently introduced by Kirkegaard
et al (1972). This method has the advantage of speed and high
capacity; I will later illustrate its application to the isolation of the isozymes of aldolase from rabbit brain.

Chemical Characterization. Analyses of structure-function
relationships usually begin with comparisons of similarities
and differences in primary structure, including fingerprint
analysis, chromatography of proteolytic digests and cyanogen
bromide cleavage products leading ultimately to determinations
of amino acid sequences, and crystallographic analysis, which
provide information on fine details of genetic and evolutionary
relationships. Isozymes may prove to be identical in primary
structure, but differ in other respects; for example, if they
are glycoproteins they may differ in the nature of the carbohydrate side chains. Modification of amino acid side chains
by phosphorylation or adenylylation are important in regulation
of activity, while deamidation and phosphorylation may influence secretion, uptake, and degradation.

Origin of Isozymes. Biochemical studies of the kind
described above may help to establish whether one or more genes
is involved in coding for isozymes, and can also provide clues
to the origin of these genes and to the extent of their homology. If the enzyme variants are formed by epigenetic processes, then structural studies are essential for an understanding of the nature of these processes and their regulation.
Many enzymes exist as interconvertible forms that might be
classified as isozymes, although in many instances these cannot
be separated by the usual electrophoretic or chromatographic
techniques. One might ask why the addition of the highly
charged phosphate group to an enzyme such as phosphorylase does
not change its electrophoretic mobility, when the replacement
of only a single carboxyl group in hemoglobin causes distinct
changes in mobility (Ingram, 1959). Has the protein undergone
a specific change in structure that compensates for the added
electric charge?

Function of Isozymes. This was also mentioned by Dr.
Markert in his discussion of the biology of isozymes. Obviously, an understanding of the biochemical role of the enzyme
is essential to an understanding of the need for isozymic
forms. In many cases the presence of an isozyme in a specialized tissue, or subcellular structure, suggests a specialized
role in metabolism. Some isozymes are designed to respond to
specific signals that regulate their specialized function in
the cell. Often the discovery of the existence of an isozyme

13

preceded an understanding of its role in the regulation of metabolism.

I have already alluded to the existence of modified forms of enzymes generated by the action of intracellular proteases. It will be of interest to explore the role of these proteases in enzyme turnover and possibly in the regulation of metabolic pathways.

Aldolase Isozymes

Occurrence. Fructose 1,6-diphosphate aldolase was discovered by Meyerhof et al (1936). The discovery of a functionally distinct aldolase in mammalian liver was made more than 20 years ago (Leuthardt et al, 1952; Hers and Kusaka, 1953). A third isozyme was identified in rabbit brain (Penhoet et al, 1966). All three isozymes are readily identified in mammalian tissues by their mobility in cellulose acetate electrophoresis (Fig. 1). Aldolase A, the form characteristic of skeletal muscle and most other adult tissues, moves toward the cathode, but more slowly than aldolase B, the predominant form in liver. Kidney contains aldolases A and B and also three hybrid forms with intermediate mobility, consistent with the tetrameric structure of aldolases. Aldolase C, the anodic form, occurs in brain, together with aldolase A and three hybrid forms yielding the characteristic 5 isozyme set.

Structure and Kinetic Properties. These are summarized in Table I. Aldolases A and B are similar, if not identical, in molecular weight, but C is significantly smaller. The evidence for this will be discussed later. The three aldolases are immunologically distinct, and also differ in their catalytic properties. Aldolase A, whose primary function is in glycolysis, shows higher activity toward fructose 1,6-diphosphate than toward fructose 1-phosphate. Aldolase B is more suited for utilization of fructose and catalyzes the cleavage of fructose 1-phosphate and fructose diphosphate at nearly equal rates. Aldolase B also shows higher affinity for triose phosphates, consistent with the role of liver in gluconeogenesis. The properties of aldolase C are intermediate, but the function of this isozyme in brain remains unknown. Clarke and Masters (1973) have reported that hybrids containing aldolase A subunits in brain tend to be associated with a particulate fraction, and this may provide a clue to the function of the two isozymes in this tissue. The three aldolase isozymes have been shown to occur in all vertebrate species studied, although their proportions may vary (Lebherz and Rutter, 1969).

Aldolases A and B are very similar in amino acid composition, the most significant difference being the number of methionine residues, three per subunit for aldolase A, and six

14

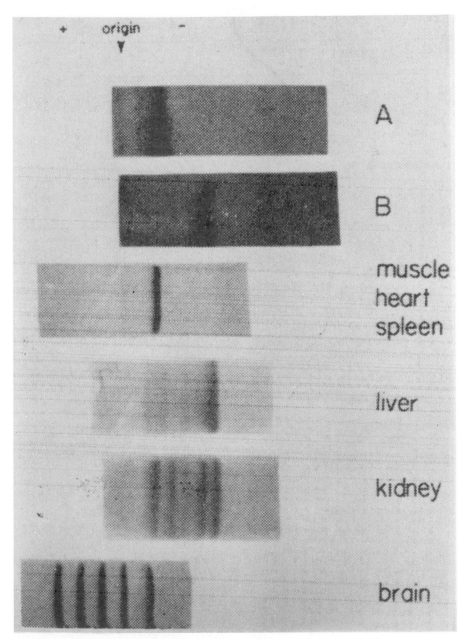

Fig. 1. Isozymes of fructose bisphosphate aldolase in adult rat tissues (From Penhoet et al, 1966). Similar zymograms have been obtained with extracts of adult rabbit tissues (Lebherz et al, 1969).

TABLE I

Properties of Aldolase Isozymes

Property	A	B	C
Molecular weight	160,000	158,000	(148,000)
Subunit weight	40,000	39,000	37,000
Activity ratio: Fru-P$_2$/Fru-1-P	50	1	25
% Inhibition by anti-A	100	0	0
% Inhibition by anti-B	0	100	0
% Inhibition by anti-C	0	0	95
K_m, Fru-P$_2$	3×10^{-6}	1×10^{-6}	3×10^{-6}
K_m, Fru-1-P	1×10^{-2}	3×10^{-4}	4×10^{-3}
K_m, Ga3P	1×10^{-3}	3×10^{-4}	8×10^{-4}

Data from Kawahara and Tanford (1966), Sia and Horecker (1968), Christen et al (1965), Lee and Horecker (1974), Penhoet et al (1969), Penhoet and Rutter (1971).

TABLE II

Amino Acid Composition of Aldolase Subunits

Amino Acid	A	B	C
Lysine	27	28	*19*
Histidine	11	9	*7*
Arginine	14	15	14
Cm-Cysteine	8	8	*6*
Aspartic Acid	*30*	35	34
Threonine	20	21	17
Serine	19	17	20
Glutamic Acid	44	41	*37*
Proline	19	17	18
Glycine	31	29	29
Alanine	*44*	38	39
Valine	21	25	25
Methionine	3	*6*	3
Isoleucine	19	20	17
Leucine	35	36	31
Tyrosine	11	10	9
Phenylalanine	7	11	9
Tryptophan	3	*4*	3

Data from Lai (1968), Lai et al (1974), Gracy et al (1969), Lee and Horecker (1974).

per subunit for aldolase B (Table II). Aldolase C contains
fewer lysine and histidine residues, which may account for its
anodic migration in electrophoresis. It also contains fewer
cysteine residues per subunit. The differences in methionine
content are interesting, since cleavage of the peptide chains
at methionyl linkages by the technique of Gross and Witkop
(1962) has proved a powerful tool in structure and sequence
analysis. It might be pointed out that aldolases vary enor-
mously in their content of methionine (Table III). Thus the
enzyme from codfish muscle contains 32 residues, or 8 per subunit, while frog muscle aldolase contains only 4 residues, or
1 per peptide chain. No phylogenetic basis is discernible for
this difference.

  The Structure of Aldolase C. Dr. Y. Lee in our laboratory
has recently carried out some studies on the structure of
aldolase C which have provided some interesting insights into
its structural and thus into its genetic relations to aldolase
A (Lee and Horecker, 1974). In order to accomplish these
studies, it was necessary to isolate pure aldolase C subunits
in quantities much greater than were previously available.
Her procedure illustrates the power of the ion filtration
technique of Kirkegaard (1972) mentioned earlier (Fig. 2).
The method is basically gel filtration through a charged resin
that will retard certain proteins by electrostatic interactions. By careful selection of pH and ion strength, conditions can be obtained in which the bulk of the protein, including aldolase A_4, emerges with the void volume, while the
three hybrids and aldolase C_4 homopolymer are retarded, emer-
ging as four distinct peaks. The advantage of the method is
its speed and high capacity. In the experiment shown, 10 g of
protein was applied to a 5000 ml column and the procedure,
carried out at room temperature, was complete within three
hours.

  For separation of the A and C subunits the hybrids containing C subunits were collected, carboxymethylated with [^{14}C]io-
doacetate and separated into the A and C components by ion filtration chromatography on Sephadex C-25.

  The cyanogen bromide cleavage patterns yielded important
information on the comparative structure of aldolases C and A
(Fig. 3). Each subunit contains three methionine residues and
yields four peptides that can be separated on Sephadex G-75.
The four peptides of aldolase A have been identified by Lai
(1968) as originating from the NH_2-terminus (N), the COOH-terminus (C), and the interior of the chain (A and B). The number
of cysteine residues is three for peptide N, two for peptides
C and A, and one for peptide B, accounting for all eight
cysteine residues in the molecule. Aldolase C yielded radio-

TABLE III
Methionine Content of Aldolase Subunits

Species	A	B	C
Codfish[a]	8		
Shark[b]	7		
Sturgeon[c]	6		
Lobster[d]	5		
Spinach leaves[e]	5		
Chicken[f,g]	4		2
Rabbit[h,i,j]	3	6	3
Boa constrictor[k]	3		
Frog[l]	1		

[a] Lai and Chen (1971)
[b] Caban and Hass (1971)
[c] Anderson et al (1969)
[d] Guha et al (1971)
[e] Fluri et al (1967)
[f] Marquardt (1969)
[g] Marquardt (1970)
[h] Lee and Horecker (1974)
[i] Lai (1968)
[j] Gracy et al (1969)
[k] Schwartz and Horecker (1966)
[l] Ting et al (1969)

active peptides corresponding to N, C, and B, but peptide A
was not radioactive and was detected only after hydrolysis in
alkali and ninhydrin analysis. Unlike peptide A from aldolase
A, which contains two cysteine residues, this peptide contains
no cysteine. Peptides C and A also emerged later than the
corresponding peptides from aldolase A, suggesting that they
are smaller in size and also that the aldolase C subunit might
be significantly shorter than the subunit of aldolase A. The
molecular weight of the aldolase C subunit, determined by
equilibrium sedimentation and SDS-disc gel electrophoresis,
was indeed found to be 37,000, compared with 40,000 for
aldolase A. Thus aldolase C is about 30 residues smaller than
aldolase A. The COOH-terminal and A-peptides must each contain
at least one deletion, including deletion of the two cysteine
residues in peptide A.

Dr. Lee's results provide some interesting information on
the structure of the two aldolases. Lai and his coworkers
(1974) have recently completed the sequence of aldolase A, and
by assembling the known functional groups at the active center,
have gained some idea of the 3-dimensional folding of the
molecule (Fig. 4). Lysines 107 and 227, cysteines 72 and 336,
and histidine 361 have all been identified as functional groups
at the active site. In addition, cysteines 237 and 287 have

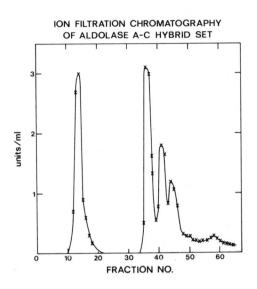

ION FILTRATION CHROMATOGRAPHY
OF ALDOLASE A-C HYBRID SET

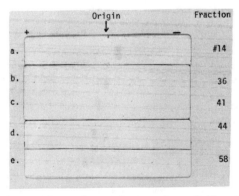

Fig. 2. Ion filtration chromatography of the AC-hybrid set
from rabbit brain. The upper portion shows the results of
analyses of fractions from the column for aldolase activity,
the lower part the results of cellulose acetate electrophoresis
of the peak fractions.

been shown to be exposed, on the surface of the molecule,
while cysteines 134, 149, 177, and 199 are buried and do not
react with sulfhydryl reagents in the native enzyme (Steinman
and Richards, 1970; Lai and Horecker, 1972). Two of these
buried cysteine residues, 177 and 199, are absent from aldolase
C. In aldolase A they must thus be buried in the subunit

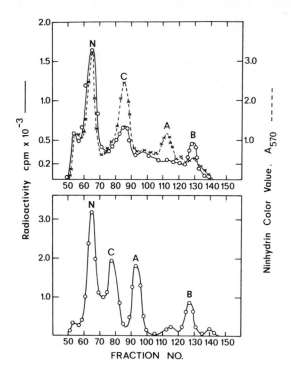

Fig. 3. Separation of BrCN peptides of aldolases C (upper
section) and A (lower section) on Sephadex G-75. (From Lee
and Horecker, 1974).

folding, and not between the subunits in the tetramer. Two of
the four missing histidine residues are also from this region
of the molecule. What is remarkable is that the aldolase A
and C subunits still assemble readily to form the A-C hybrids,
in spite of these differences in primary structure and molecular weight. Evidently the structure of the active site and
the domains of subunit interaction must still be similar in
the two isozymes. Recently, Dr. Lee has obtained evidence that
the active site is located in peptide N, rather than peptide A,
of aldolase C. It will be of interest to determine just how
extensive are the deletions and rearrangements in the primary
sequence as compared with that of aldolase A.

Microheterogeneity of Aldolase A. Aldolase A isolated from
adult rabbits has been shown to contain two different subunits
that can be separated by chromatography in the presence of 8 M
urea (Chan et al, 1967). The enzyme, which migrates as a single species in disc gel electrophoresis (see Fig. 1), can be

```
                                                                    20
Pro-His-Ser-His-Pro-Ala-Leu-Thr-Pro-Glu-Gln-Lys-Lys-Glu-Leu-Ser-Asp-Ile-Ala-His-
                                                                    40
Arg-Ile-Val-Ala-Pro-Gly-Lys-Gly-Ile-Leu-Ala-Ala-Asp-Gln-Ser-Thr-Gly-Ser-Ile-Ala-
                                                                    60
Lys-Arg-Leu-Gln-Ser-Ile-Gly-Thr-Glu-Asn-Thr-Glu-Glu-Asn-Arg-Arg-Phe-Tyr-Arg-Gln-
                                        72                          80
Leu-Leu-Leu-Thr-Ala-Asp-Asp-Arg-Val-Asn-Pro-Cys-Ile-Gly-Gly-Val-Ile-Leu-Phe-His-
                                                                    100
Thr-Glu-Leu-Tyr-Gln-Lys-Ala-Asp-Asp-Gly-Arg-Pro-Phe-Pro-Gln-Val-Ile-Lys-Ser-Lys-
                         107                                        120
Gly-Gly-Val-Val-Gly-Ile-Lys-Val-Asp-Lys-Gly-Val-Val-Pro-Leu-Ala-Gly-Thr-Asp-Gly-
                              134                                   140
Glu-Thr-Thr-Thr-Gln-Gly-Leu-Asp-Gly-Leu-Ser-Glu-Arg-Cys-Ala-Gln-Tyr-Lys-Lys-Asp-
                         149                                        160
Gly-Ala-Asp-Phe-Ala-Lys-Trp-Arg-Cys-Val-Leu-Lys-Ile-Gly-Gln-His-Thr-Pro-Ser-Ala-
                                        177                         180
Leu-Ala-Ile-Met-Glu-Asn-Ala-Asn-Val-Leu-Ala-Arg-Tyr-Ala-Ser-Ile-Cys-Gln-Gln-Asn-
                                             199 200
Gly-Pro-Ile-Glu-Val-Pro-Glu-Ile-Leu-Pro-Asp-Gly-Asp-His-Asp-Leu-Lys-Arg-Cys-Gln-
                                                                    220
Tyr-Val-Thr-Gln-Lys-Val-Leu-Ala-Ala-Val-Tyr-Lys-Ala-Leu-Ser-Asn-His-His-Ile-Tyr-
         227                                         237            240
Leu-Gln-Gly-Thr-Leu-Leu-Lys-Pro-Asn-Met-Val-Thr-Pro-Gly-His-Ala-Cys-Thr-Gln-Lys-
                                                                    260
Tyr-Ser-His-Gln-Gln-Ile-Ala-Met-Ala-Thr-Val-Thr-Ala-Leu-Arg-Arg-Thr-Val-Pro-Pro-
                                                                    280
Ala-Val-Thr-Gly-Val-Thr-Phe-Leu-Ser-Gly-Ser-Glu-Glu-Glu-Glu-Gly-Ala-Ser-Ile-Asn-
                    287                                             300
Leu-Asn-Ala-Ile-Asn-Lys-Cys-Pro-Leu-Leu-Trp-Pro-Lys-Ala-Leu-Thr-Phe-Ser-Tyr-Gly-
                                                                    320
Arg-Ala-Leu-Gln-Ala-Ser-Ala-Leu-Lys-Ala-Trp-Gly-Gly-Lys-Lys-Glu-Asn-Leu-Lys-Ala-
                                        336            340
Ala-Gln-Glu-Glu-Tyr-Val-Lys-Arg-Ala-Leu-Ala-Asn-Ser-Leu-Ala-Cys-Gln-Gly-Lys-Tyr-
                                                  359 360
Thr-Pro-Ser-Gly-Gln-Ala-Gly-Ala-Ala-Ala-Ser-Glu-Ser-Leu-Phe-Ile-Ser-Asn-His-Ala-
361
Tyr
```

Fig. 4. Primary sequence and functional groups of rabbit muscle aldolase. (From Lai et al, 1974).

---

resolved into five components by isoelectric focusing (Susor et al, 1969). Similar microheterogeneity was observed with a number of other crystalline enzymes (Fig. 5). Only muscle lactate dehydrogenase behaved as a single species in this system.

In the case of rabbit muscle aldolase the microheterogeneity has been shown to be due to deamidation of a single asparagine residue (Lai et al, 1970), Asn 258, leading to a random distribution of tetramers containing from zero to four deamidated subunits. This change has been shown to be age-dependent. Young animals contain only the native chains, while adult rabbits (5 mo. and older) contain approximately equal quantities of native and modified chains (Koida et al, 1969) and the half-time of conversion in vivo was shown to be about eight days (Midelfort and Mehler, 1972).

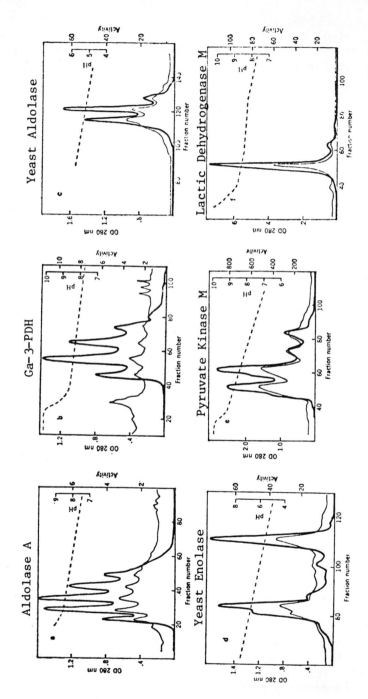

Fig. 5. Microheterogeneity of crystalline enzymes in electrofocusing. (From Susor et al, 1969.)

22

The observations on the deamidation of Asn 258 in rabbit muscle aldolase suggest that this is a particularly labile residue. Is deamidation catalyzed by a specific enzyme or is it due to factors intrinsic to the structure of the protein? Recently Robinson and Rudd (1974) have reviewed their studies on the deamidation of asparagine and glutamine residues in proteins and reached the surprising conclusion that deamidation is largely dependent on the primary sequence (Table IV). McKerrow and Robinson (1974) synthesized a number of peptides resembling that containing Asn 258 and found that a pentapeptide containing the Ser-Asn-His sequence was spontaneously deamidated at pH 7.5 with a half-life of six days, very similar to the half-life of Asn 258 of aldolase A in vivo. When histidine was replaced by alanine the half-life was increased by 10-fold. Thus deamidation was determined by the primary sequence and was little dependent on the 3-dimensional structure; if anything the Asn group in aldolase was more stable than in the simple pentapeptide.

TABLE IV
Deamidation Half-Times of Asn and Gln at pH 7.4, 37°

Peptide	$t_{\frac{1}{2}}$ (days)
-Ile-Ser-Asn-His-Ala-Tyr (α chain of aldolase A)	8 (in vivo)[a]
Gly-Ser-Asn-His-Gly	6[b]
Gly-Ser-Asn-Ala-Gly	52[b]
Gly-Ser-Gln-Ala-Gly	~900[b]

[a] From Midelfort and Mehler (1972)

[b] From Robinson and Rudd (1974)

Robinson and Rudd (1974) have proposed that deamidation is a common phenomenon and is specifically designed to regulate the turnover of intracellular proteins. This suggestion derives strong support from the studies of Flatmark and his associates (1968) who demonstrated that the isozymes of cytochrome C arose by deamidation in vivo. Robinson's group has synthesized peptides containing the sequences around all eight asparagine residues in cytochrome C (Table V) and found that

TABLE V
Deamidation Half-Times of Analogues
of the Asn and Gln in Cytochrome c

Peptide	$t_{\frac{1}{2}}$ (days)
Gly-Thr-Asn-Glu	16
Gly-Ala-Asn-Lys-Gly	54
Gly-Glu-Asn-Pro-Gly	80
Gly-Lys-Asn-Lys-Gly	94
Gly-Ala-Gln-Cys-Gly	113
Gly-Pro-Asn-Leu-Gly	277
Gly-Gly-Gln-Ala-Gly	418
Gly-Val-Gln-Lys-Gly	421

From Robinson and Rudd (1974)

their half-life varied from 16 to 421 days. They concluded
that the first deamidation is under sequence control, but the
second deamidation was faster than expected and is affected
by the tertiary structure (Robinson and Rudd, 1974). Deamidation is also accelerated by the presence of ascorbic acid and
O_2 (Robinson et al, 1973).

It will be of interest to establish whether the microheterogeneity reported by Susor et al (1969) for a number of enzymes
is also due to in vivo deamidation, as has been shown for muscle aldolase. It will also be of interest to determine whether
the reported microheterogeneity in aldolases B (Gürtler and
Leuthardt, 1970) and C (Kochman et al, 1968) are also due to
deamidation; it should be noted that these enzymes have COOH-
terminal tyrosine residues, but the rest of the COOH-terminal
sequence differs from that of aldolase A.

Isozymes of Fructose Bisphosphatase (Fru-P$_2$ase)

Comparative Properties and Structures. Fructose 1,6-biphos-
phatases (Fru-P$_2$ases) were originally described in liver and
kidney by Gomori (1943), and the enzyme has recently also been
shown to be present in white skeletal muscle (Krebs and
Woodford, 1965; Salas et al, 1965). I will review some of our
present knowledge of the structures and functions of these
enzymes. These studies have been carried out with crystalline
enzymes prepared in our laboratory from rabbit liver, kidney,
and muscle by Drs. T. Sasaki, B. Abrams, and A. Datta, respectively.

The enzymes from liver and kidney appear to be identical on the basis of their electrophoretic and immunological behavior (Fig. 6). They yield identical precipitin lines with anti-liver Fru-P_2ase (Enser et al, 1969). On the other hand the muscle enzyme migrates rapidly toward the opposite pole and does not yield a precipitin line with anti-liver Fru-P_2ase. The results of amino acid analysis (Traniello et al, 1972; Tashima et al, 1972) confirm the non-identity of the muscle enzyme but are inconclusive with respect to the identity or non-identity of the liver and kidney enzymes (Table VI). An interesting difference is the presence of tryptophan in the liver and kidney enzymes; this amino acid has not been detected in muscle Fru-P_2ase.

TABLE VI

Amino Acid Compositions of Fructose Bisphosphatases
from Rabbit Tissues

Amino Acid residue	Liver[a]	Kidney[b]	Muscle[c]
Lysine	120	114	95
Histidine	24	24	12
Arginine	54	43	48
CM-cysteine	22	16	17
Aspartic acid	135	148	118
Threonine	72	70	78
Serine	76	79	73
Glutamic acid	104	114	130
Proline	59	61	60
Glycine	102	111	119
Alanine	117	119	128
Valine	100	94	116
Methionine	33	30	31
Isoleucine	75	73	64
Leucine	111	113	125
Tyrosine	46	45	65
Phenylalanine	36	40	37
Tryptophan	4	4	0

[a] From Traniello et al (1972)

[b] From Tashima et al (1972)

[c] From Black et al (1972)

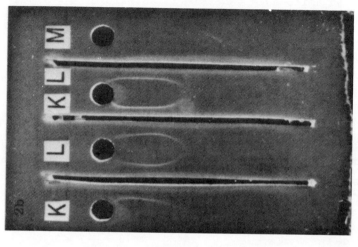

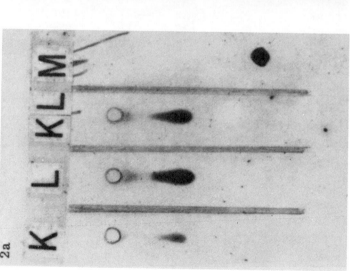

Fig. 6. Immunoelectrophoresis at pH 6.5 of Fru-P_2ases purified from rabbit kidney (K), liver (L), and muscle (M). The enzymes migrated toward the cathode, except for muscle Fru-P_2ase, which moved toward the anode. The cell containing the muscle enzyme was placed at the lower end of the plate. Antiserum to purified liver Fru-P_2ase was placed in the troughs. (a) Activity stain. (b) Immunoelectrophoresis pattern. (From Enser et al, 1969).

26

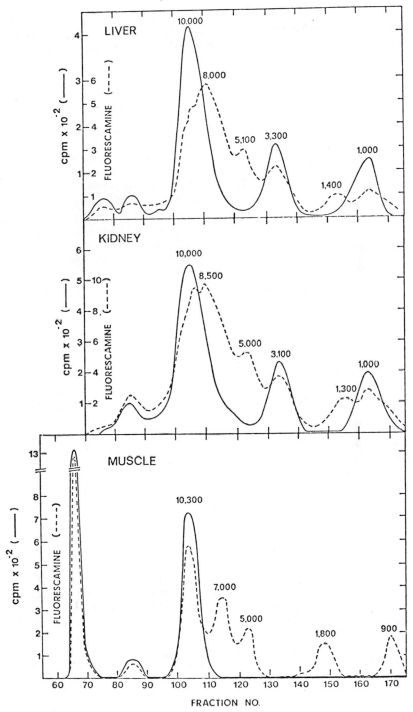

Fig. 7

27

(*Fig. 7 legend.*) Separation of BrCN peptides of liver, kidney, and muscle Fru-P$_2$ases on Sephadex G-75. The cysteine residues were labeled with [^{14}C]iodoacetate before cleavage. Solid line: radioactivity. Dashed line: fluorescence intensity after alkaline hydrolysis and reaction with fluorescamine (Sasaki, Abrams and Datta, unpublished experiments).

Cleavage with cyanogen bromide of the three isozymes and filtration of the peptides formed through Sephadex G-75 yielded the patterns shown in Fig. 7. Before cleavage each protein was carboxymethylated with radioactive iodoacetate. The liver and kidney enzymes each yielded three radioactive peptides, containing 3, 1, and 1 cysteine residues. The muscle enzyme yielded only two radioactive peaks, one of which emerged with the void volume. The dotted lines show the peptide patterns obtained after alkaline hydrolysis and analysis with fluorescamine. Clearly the patterns are very similar for the liver and kidney enzymes and quite different for the muscle enzyme. Pending the results of sequence analysis of some of the well-separated peptides, we may tentatively conclude that the same gene codes for the enzyme from liver and kidney, and that the muscle enzyme is determined by a different gene.

Function of the Fru-P$_2$ase Isozymes. Fru-P$_2$ases in liver and kidney catalyze one of the irreversible steps essential for gluconeogenesis, and if there should prove to be any differences between these enzymes these should reflect differences in the regulation of gluconeogenesis in these tissues. In muscle, however, this enzyme must have a different function, and several theories have been proposed (Krebs and Woodford, 1965; Newsholme and Crabtree, 1970; Newsholme et al, 1972). The most attractive proposes that the enzyme is involved in a futile cycle with phosphofructokinase for heat production (Newsholme et al, 1972). This hypothesis has been elegantly supported by the experiments of Lardy and his coworkers (Bloxham et al, 1973; Clark et al, 1973a) with the intact bumble bee, which must maintain a thoracic temperature of 30° in order to fly. They showed that the futile cycle did indeed function in the resting bee, but was shut off during flight, when sufficient heat would be provided as a by-product of contraction. They also implicated the PFK-Fru-P$_2$ase futile cycle as the source of excess heat production in malignant hyperthermia, a condition encountered on administration of certain anesthetics, such as halothane, in sensitive pigs (Clarke et al, 1973b) and also in man. However, it has not yet been established that the futile cycle is the source of heat in normal non-shivering thermogenesis in mammals.

Seasonal Changes in Fru-P$_2$ases and Proteolytic Modification.
In this final section, I will discuss modifications of liver
Fru-P$_2$ase by proteolytic enzymes, some of which appear to
occur in vivo and may be related to the regulation of gluconeogenesis. Two forms of Fru-P$_2$ase have been isolated from rabbit liver. One, apparently the native form, has a neutral pH
optimum; the other, which appears to rise by proteolytic
modification during purification, has an optimum above pH 9.
The neutral Fru-P$_2$ase can be converted to the alkaline form
by exposure to papain (Pontremoli et al, 1971) (Fig. 8) or by

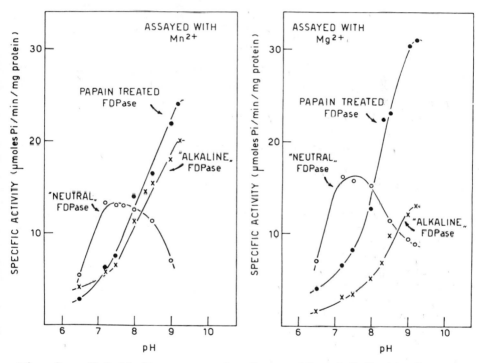

Fig. 8. pH Activity curves for "neutral" and "alkaline" Fru-
P$_2$ases and the "neutral" enzyme treated with papain. (From
Pontremoli et al, 1971).

exposure to subtilisin (Traniello et al, 1972). However, the
"native" neutral enzyme also exists in two forms which can be
distinguished by differences in their chromatographic properties (Pontremoli and Grazi, 1968), and also by the effects of
subtilisin (Pontremoli et al, 1973a) (Fig. 9). On exposure to
subtilisin both enzymes show the characteristic shift from the
neutral to alkaline pH optimum, due largely to the increase in

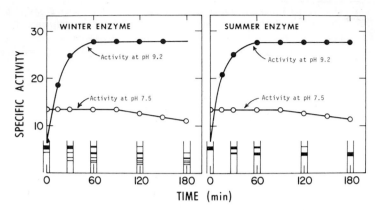

Fig. 9. Changes in catalytic properties and molecular structure of Fru-P$_2$ases during digestion with subtilisin. The winter enzyme was isolated in January and the summer enzyme in
June, from livers of the same commercial strain of rabbits.
(From Pontremoli et al, 1973a).

catalytic activity at pH 9.2. However, in Na dodecyl sulfate
gel electrophoresis the summer enzyme treated with subtilisin
shows evidence of only a single smaller subunit (MW = 29,000),
in contrast to the winter enzyme which is cleaved to yield
many smaller pieces. Evidently the winter enzyme has a looser
structure, with a number of bonds susceptible to subtilisin,
in contrast to the summer enzyme, which is attacked at only
a single site. Despite this difference, the catalytic properties of the two enzymes, before and after exposure to subtilisin, are identical.

A clue to the nature of this difference has been obtained
by studying the enzymes purified from the livers of animals
exposed to cold, or fasted. Although the catalytic properties
of the two forms are indistinguishable, and the subunit size
is the same, only the enzyme from fed animals contains tryptophan (Table VII). It has been established that the tryptophan
residue is situated near the NH$_2$-terminus (Pontremoli et al,
1973b; Pontremoli et al, 1974), and the results suggest that
in the fasted animals a small peptide containing this tryptophan residue has been removed. The COOH-terminus is not
affected (Table VIII).

The properties of the various forms of liver Fru-P$_2$ase are
summarized in Table VIII. The native enzyme has a subunit
molecular weight of 36,000 and contains tryptophan. No amino
acids are released with leucine aminopeptidase, suggesting

TABLE VII
Molecular Properties of Fru-P_2ase Purified from
Livers of Fed and Fasted Rabbits

Conditions	Subunits[a]		COOH-terminal sequence			Tryptophan Content[b]
	% heavy	% light	Ala[b]	His[b]	Gly[b]	
Fed	100	-	3.28	2.15	0.15	4.1
Fasted, 36 hr	90	10	3.54	1.98	0.18	0.7
Fasted, 96 hr	87	13	3.38	2.32	0.31	0.4

[a] Corresponding to molecular weights of 36,000 and 29,000,
respectively, as estimated by disc gel electrophoresis in
Na dodecyl sulfate. (From Pontremoli et al, 1974).

[b] moles/mole enzyme

that the NH_2-terminus may be blocked. The COOH-terminus is
alanine. The winter enzyme differs in its sensitivity to
subtilisin, but whether or not it contains tryptophan has not
been established. However, the fasted enzyme, with similar
properties, has been shown to lack tryptophan. The first
three forms listed in Table VIII are known to occur in vivo.
An additional form can be generated from the native enzyme by
exposure to subtilisin (Traniello et al, 1972; Pontremoli et
al, 1973b). It has an alkaline pH optimum, a smaller molecular weight, and a new unique NH_2-terminus. This form can also
be generated by exposure to lysosomes in vitro (Pontremoli et
al, 1973a). Finally, the form isolated from acetone powders,
and presumably modified by lysosomal proteases, contains two
non-identical subunits, no tryptophan, and has also been modified at the COOH-terminus, so that it contains nearly equal
amounts of COOH-terminal alanine and glycine (Sia et al, 1969).
    The role of these modified enzymes in gluconeogenesis con-
trol remains to be determined. They appear not to be simply
intermediates in the degradation of Fru-P_2ase, since the modified enzyme in fasted animals reaches levels several-fold
higher than those found in fed animals, and the modification
appears to be related to the increased rate of gluconeogenesis.
An interesting suggestion has come from the work of Benkovic
et al (1974) who found that the Trp-less enzyme is readily
desensitized to inhibition by AMP on exposure to subtilisin.

TABLE VIII

Properties of the Native and Modified Forms of Rabbit Liver Fru-P$_2$ase

Enzyme species	Subunit	NH$_2$-terminus	Tryptophan	COOH-terminus	pH optimum
Native[a]	36,000	blocked	present	Ala	neutral
Winter[b]	36,000	undetermined	undetermined	Ala	neutral
Fasted[c]	36,000	undetermined	absent	Ala	neutral
Modified by subtilisin[d]	29,000	Leu	absent	Ala	alkaline
Alkaline[e]	36,000 and 29,000	unknown	absent	Ala Gly	alkaline

a Isolated from fresh livers homogenized in isotonic sucrose, during the summer months, (Traniello et al, 1971).

b Isolated as in footnote a, from rabbits collected in January, (Pontremoli et al, 1973a).

c Isolated from livers of animals fasted for 36-96 hours, (Pontremoli et al, 1974).

d Native enzyme after treatment with subtilisin, (Pontremoli et al, 1973b).

e Isolated from rabbit liver acetone powders, (Pontremoli et al, 1965).

32

Loss of AMP sensitivity may well be the key event in regulation of $Fru-P_2ase$, and the complex pattern of proteolytic modification of $Fru-P_2ase$ may be designed to accomplish this end.

REFERENCES

Anderson, P. J., Ian Gibbons, and Richard N. Perham 1969. A comparative study of the structure of muscle fructose 1,6-diphosphate aldolases. *Eur. J. Biochem.* 11: 503-509.

Benkovic, S. J., W. A. Frey, C. B. Libby, and J. J. Villafranca 1974. The role of tryptophan in neutral fructose diphosphatase. *Biochem. Biophys. Res. Commun.* 57: 196-203.

Black, W. J., A. Van Tol, J. Fernando, and B. L. Horecker 1972. Isolation of a highly active fructose diphosphatase from rabbit muscle: its subunit structure and activation by monovalent cations. *Arch. Biochem. Biophys.* 151: 576-590.

Bloxham, David P., Michael G. Clark, Paul C. Holland, and Henry A. Lardy 1973. A model study of the fructose diphosphatase-phosphofructokinase substrate cycle. *Biochem. J.* 134: 581-587.

Caban, C. E., and L. E. Hass 1971. Studies on the structure and function of muscle aldolase V. Molecular characterization of the enzyme from shark (*Mustelus Canis*). *J. Biol. Chem.* 246: 6807-6813.

Chan, W., D. E. Morse, and B. L. Horecker 1967. Nonidentity of subunits of rabbit muscle aldolase. *Proc. Natl. Acad. Sci.* 57: 1013-1020.

Christen, Ph., H. Göschke F. Leuthardt and A. Schmid 1965. Über die aldolase der kaninchenleber: Molekulargewicht, dissoziation in untereinheiten. *Helvetica Chim. Acta.* 48: 1050-1056.

Clark, Michael G., David P. Bloxham, Paul C. Holland, and Henry A. Lardy 1973a. Estimation of the fructose diphosphatase-phosphofructokinase substrate cycle in the flight muscle of *Bombus affinis*. *Biochem. J.* 134: 589-597.

Clark, M. G., C. H. Williams, W. F. Pfeifer, D. P. Bloxham, P. C. Holland, C. A. Taylor, and H. A. Lardy 1973. Accelerated substrate cycling of fructose-6-phosphate in the muscle of malignant hyperthermic pigs. *Nature* 245: 99-101.

Clarke, F. M. and C. J. Masters 1973. On the distribution of aldolase isoenzymes in subcellular fractions from rat brain. *Arch. Biochem. Biophys.* 156: 673-683.

Cuatrecasas, Pedro, and Christian B. Anfinsen 1971. Affinity chromatography. in *Methods in Enzymology* (Colowick and Kaplan, eds.). Academic Press, Inc., New York, 22: 345-378.

Doi, Roy H. 1972. Role of proteases in sporulation. in *Current Topics in Cellular Regulation*. (Horecker and Stadtman, eds.) New York, Academic Press, 6: 1-20.

Enser, M., S. Shapiro, and B. L. Horecker 1969. Immunological studies of liver, kidney, and muscle fructose 1,6-diphosphatases. *Arch. Biochem. Biophys.* 129: 377-383.

Flatmark, Torgeir, and Knut Kletten 1968. Multiple forms of cytochrome c in the rat. *J. Biol. Chem.* 243: 1623-1629.

Fluri, R., T. Ramasarma, and B. L. Horecker 1967. Purification and properties of fructose diphosphate aldolase from spinach leaves. *Eur. J. Biochem.* 1: 117-124.

Gomori, G. 1943. Hexosediphosphatase. *J. Biol. Chem.* 148: 139-149.

Gracy, Robert W., Andras G. Lacko, and B. L. Horecker 1969. Subunit structure and chemical properties of rabbit liver aldolase. *J. Biol. Chem.* 244: 3913-3919.

Gross, Erhard, and Bernhard Witkop 1962. Nonenzymatic cleavage of peptide bonds: the methionine residues in bovine pancreatic ribonuclease. *J. Biol. Chem.* 237: 1856-1860.

Guha, Arabinda, C. Y. Lai, and B. L. Horecker 1971. Lobster muscle aldolase: isolation, properties and primary structure at the substrate-binding site. *Arch. Biochem. Biophys.* 147: 692-706.

Gürtler, von B., and F. Leuthardt 1970. Über die heterogenität der aldolasen. *Helvetica Chimica Acta.* 53: 654-658.

Hers, H. G., and T. Kusaka 1952. Le metabolisme du fructose-1-phosphate dans le foie. *Biochim. Biophys. Acta.* 11: 427-437.

Hunter, R. L., and C. L. Markert 1957. Histochemical demonstration of enzymes separated by zone electrophoresis in starch gels. *Science* 125: 1294-1295.

Ingram, V. M. 1959. Abnormal human haemoglobins. III. The chemical difference between normal and sickle cell haemoglobins. *Biochim. Biophys. Acta.* 36: 402-411.

Katzen, Howard M., Denis D. Soderman, and Harold M. Nitowsky 1965. Kinetic and electrophoretic evidence for multiple forms of glucose-ATP phosphotransferase activity from human cell cultures and rat liver. *Biochem. Biophys. Res. Commun.* 19: 377-382.

Kawahara, Kazuo, and Charles Tanford 1966. The number of polypeptide chains in rabbit muscle aldolase. *Biochem.* 5: 1578-1584.

Kirkegaard, Leslie H. 1972. Ion filtration chromatography: a powerful new technique for enzyme purification applied to *E. coli* alkaline phosphatase. *Anal. Biochem.* 50: 122-138.

Kochman, M., E. Penhoet, and W. J. Rutter 1968. Characterization of the subunits of aldolases A and C. *Fed. Proc.* 27: 590, Abs.

Koida, M., C. Y. Lai, and B. L. Horecker 1969. Subunit structure of rabbit muscle aldolase: extent of homology of the α and β subunits and age-dependent changes in their ratio. *Arch. Biochem. Biophys.* 134: 623-631.

Krebs, H. A., and Muriel Woodford 1965. Fructose 1,6-diphosphatase in striated muscle. *Biochem. J.* 94: 436-445.

Lai, C. Y. 1968. Studies on the structure of rabbit muscle aldolase I. Cleavage with cyanogen bromide: an approach to the determination of the total primary structure. *Arch. Biochem. Biophys.* 128: 202-211.

Lai, C. Y., and C. Chen 1971. Codfish muscle aldolase: purification, properties, and primary structure around the substrate-binding site. *Arch. Biochem. Biophys.* 144: 467-475.

Lai, C. Y., C. Chen, and B. L. Horecker 1970. Primary structure of two COOH-terminal hexapeptides from rabbit muscle aldolase: a difference in the structure of the α and β subunits. *Biochem. Biophys. Res. Commun.* 40: 461-468.

Lai, C. Y., and B. L. Horecker 1972. Aldolase: a model for enzyme structure-function relationships. in *Essays in Biochemistry.* (Campbell and Dickens, eds.) New York, Academic Press, Inc., 8: 149-178.

Lai, C. Y., N. Nakai, and D. Chang 1974. Amino acid sequence of rabbit muscle aldolase and the structure of the active center. *Science* 183: 1204-1205.

Lebherz, H. G., and W. J. Rutter 1969. Distribution of fructose diphosphate aldolase variants in biological systems. *Biochem.* 8: 109-121.

Lee, Y., and B. L. Horecker 1974. Subunit structure of rabbit brain aldolase. *Arch. Biochem. Biophys.* (in press).

Leuthardt, F., E. Testa, and H. P. Wolf 1952. Über den stoffwechsel des fructose-1-phosphäts in der leber. *Helv. Physiol. Pharmacol. Acta* 10: c57-c59.

Markert, Clement L. 1969. Summation of the session on development and physiology. in *The Phosphohydrolases: Their Biology, Biochemistry and Clinical Enzymology.* Ann. *N. Y. Acad. Sci.* 166: 523-524.

Marquardt, Ronald R. 1969. Multiple molecular forms of avian aldolases. I. Crystallization and physical properties of chicken (*Gallus domesticus*) breast muscle aldolase. *Can. J. Biochem.* 47: 517-526.

Marquardt, Ronald R. 1970. Multiple molecular forms of avian aldolases. IV. Purification and properties of chicken (*Gallus domesticus*) brain aldolase. *Can. J. Biochem.* 48: 322-333.

McKerrow, James H. and Arthur B. Robinson 1974. Primary sequence dependence of the deamidation of rabbit muscle aldolase. *Science* 182: 85.

Meyerhof, O., K. Lohmann, and Ph. Schuster 1936. Über die aldolase, ein kohlenstoff-verknupfendes ferment. II. Mitteilung: aldolkondensation von dioxyacetonphosphorsäure mit acetaldehyd. *Biochem. Z.* 286: 201-318.

Midelfort, C. F., and A. H. Mehler 1972. Deamidation in vivo of an asparagine residue of rabbit muscle aldolase. *Proc. Natl. Acad. Sci.* 69: 1816-1819.

Newsholme, E. A., and B. Crabtree 1970. The role of fructose-1, 6-diphosphatase in the regulation of glycolysis in skeletal muscle. *FEBS Letters* 7: 195-198.

Newsholme, E. A., B. Crabtree, S. J. Higgins, S. D. Thornton, and Carole Start 1972. The activities of fructose diphosphatase in flight muscles from the bumble-bee and the role of this enzyme in heat generation. *Biochem. J.* 128: 89-97.

Penhoet, Edward E., Marian Kochman, and William J. Rutter 1969. Molecular forms of fructose diphosphate aldolase in mammalian tissues. *Biochem.* 8: 4396-4402.

Penhoet, E., T. Rajkumar, and W. J. Rutter 1966. Multiple forms of fructose diphosphate aldolase in mammalian tissues. *Proc. Natl. Acad. Sci.* 56: 1275-1282.

Penhoet, Edward E., and William J. Rutter 1971. Catalytic and immunochemical properties of homomeric and heteromeric combinations of aldolase subunits. *J. Biol. Chem.* 246: 318-323.

Peterson, Elbert A., and Herbert A. Sober 1962. Column chromatography of proteins: substituted celluloses. in *Methods in Enzymology*. (Colowick and Kaplan, eds.) New York, Academic Press, Inc. 5: 3-27.

Pontremoli, Sandro, and Enrico Grazi 1968. Gluconeogenesis. in *Carbohydrate Metabolism and its Disorders*. (Dickens, Randle, and Whelan, eds.) New York, Academic Press, Inc., 1: 259-295.

Pontremoli, S., E. Melloni, F. Balestrero, A. T. Franzi, A. De Flora, and B. L. Horecker 1973a. Fructose 1,6-bisphosphatase: The role of lysosomal enzymes in the modification of catalytic and structural properties. *Proc. Natl. Acad. Sci.* 70: 303-305.

Pontremoli, S., E. Melloni, A. De Flora, and B. L. Horecker 1973b. Conversion of neutral to alkaline liver fructose 1,6-bisphosphatase: Changes in molecular properties of the enzyme. *Proc. Natl. Acad. Sci.* 70: 661-664.

Pontremoli, S., E. Melloni, F. Salamino, A. De Flora, and B. L. Horecker 1974. Changes in activity and molecular properties of fructose 1,6-bisphosphatase during fasting and refeeding. *Proc. Natl. Acad. Sci.* (in press).

Pontremoli, Sandro, Edon Melloni, and Serena Traniello 1971. Conversion of "neutral" to "alkaline" fructose 1,6-diphosphatase by controlled digestion with papain. *Arch. Biochem. Biophys.* 147: 762-766.

Pontremoli, S., S. Traniello, B. Luppis, and W. A. Wood 1965. Fructose diphosphatase from rabbit liver I. Purification and properties. *J. Biol. Chem.* 240: 3459-3463.

Robinson, Arthur B., Karen Irving, and Mary McCrea 1973. Acceleration of the rate of deamidation of GlyArgAsnArgGly and of human transferrin by addition of l-ascorbic acid. *Proc. Natl. Acad. Sci.* 70: 2122-2123.

Robinson, Arthur B., and Colette J. Rudd 1974. Deamidation of glutaminyl and asparaginyl residues in peptides and proteins. in *Current Topics in Cellular Regulation.* (Horecker and Stadtman, eds.) New York, Academic Press, Inc., 8: (in press).

Rose, Richard G., and Allan C. Wilson 1966. Peafowl lactate dehydrogenase: Problem of isoenzyme identification. *Science* 153: 1411-1413.

Salas, Margarita, E. Viñuela, J. Salas, and A. Sols 1964. Muscle fructose-1,6-diphosphatase. *Biochem. Biophys. Res. Commun.* 17: 150-155.

Schwartz, E., and B. L. Horecker 1966. Purification and properties of fructose diphosphate aldolase from Boa constrictors. *Arch. Biochem. Biophys.* 115: 407-416.

Sia, C. L., and B. L. Horecker 1968. Dissociation of protein subunits by maleylation. *Biochem. Biophys. Res. Commun.* 31: 731-737.

Sia, C. L., S. Traniello, S. Pontremoli, and B. L. Horecker 1969. Studies of the subunit structure of rabbit liver fructose diphosphatase. *Arch. Biochem. Biophys.* 132: 325-330.

Stein, S., C. H. Chang, P. Böhlen, K. Imai, and S. Udenfriend 1974. Amino acid analysis with fluorescamine of stained protein bands from polyacrylamide gels. *Anal. Biochem.* (in press).

Steinman, Howard M., and Frederic M. Richards 1970. Participation of cysteinyl residues in the structure and function of muscle aldolase. Characterization of mixed disulfide derivatives. *Biochem.* 9: 4360-4372.

Susor, Walter A., Marion Kochman, and William J. Rutter 1969. Heterogeneity of presumably homogeneous protein preparations. *Science.* 165: 1260-1262.

Tashima, Y., G. Tholey, G. Drummond, H. Bertrand, J. S. Rosenberg, and B. L. Horecker 1972. Purification and properties of a rabbit kidney fructose diphosphatase with neutral pH optimum. *Arch. Biochem. Biophys.* 149: 118-126.

Ting, Shu-Mei, C. L. Sia, C. Y. Lai, and B. L. Horecker 1971. Frog muscle aldolase: Purification of the enzyme and structure of the active site. *Arch. Biochem. Biophys.* 144: 485-490.

Traniello, S., E. Melloni, S. Pontremoli, C. L. Sia, and B. L. Horecker 1972. Rabbit liver fructose 1,6-diphosphatase. Properties of the native enzyme and their modification by subtilisin. *Arch. Biochem. Biophys.* 149: 222-231.

Traniello, S., S. Pontremoli, Y. Tashima, and B. L. Horecker 1971. Fructose 1,6-diphosphatase from liver: Isolation of the native form with optimal activity at neutral pH. *Arch. Biochem. Biophys.* 146: 161-166.

Vesterberg, Olof 1967. Isoelectric fractionation, analysis, and characterization of ampholytes in natural pH gradients. V. Separation of myoglobins and studies on their electro-chemical differences. *Acta. Chem. Scand.* 21: 206-216.

MULTIPLE FORMS OF MAMMALIAN DNA-DEPENDENT DNA POLYMERASES

NORMAN B. HECHT
Department of Biology, Tufts University
Medford, Massachusetts   02155

ABSTRACT.   Aqueous subcellular fractionation experiments
with murine testis, liver, and regenerating liver have revealed that the nuclear fraction contains a low molecular
weight (3.5S) DNA-dependent DNA polymerase while two high
molecular weight (6-8S) DNA-dependent DNA polymerases are
present in the cytosol.   The predominant cytoplasmic DNA
polymerase from testis and liver is convertible by 0.25M
KCl to a form that appears similar to the nuclear 3.5S activity by the criteria of sedimentation coefficient, molecular weight, elution from phosphocellulose, and diverse
enzymic properties.   The conversion is reversible by dialysis and gel filtration.   Reduced concentrations of salt
(0.125M KCl) produce an active 4.7S DNA polymerase activity
suggesting a monomer trimer relationship between one of the
cytoplasmic DNA polymerases and the nuclear activity.   The
second high molecular weight cytoplasmic DNA polymerase
does not change sedimentation coefficient in the presence
of 0.25M KCl and is found at greatly increased levels in
regenerating liver.

INTRODUCTION

The synthesis of DNA during spermatogenesis has been extensively studied by autoradiographic methods but virtually
nothing is known about the factors that regulate this DNA
metabolism.   One of the possible sites for the regulation of
DNA synthesis is DNA polymerase.

The existence of multiple forms of DNA-dependent DNA polymerases has been reported by investigators using a wide
variety of eukaryotic cells (Weissbach et al., 1971; Baril et
al., 1971; Chang and Bollum, 1971; 1972; Sedwick et al., 1972;
Adams et al., 1973; Hecht, 1973; Lazarus and Kitron, 1973).
When aqueous subcellular fractionation methods are used, there
is general agreement that a low molecular weight (3.5S) DNA
polymerase is found predominantly in the nucleus, an apparently distinct DNA polymerase activity is isolated from a highly
purified mitochondrial fraction (Meyer and Simpson, 1968; 1970),
and a heterogeneous high molecular weight (6-8S) DNA polymerase
activity, or group of activities, is recoverable from the or-
ganelle-free cytoplasmic fraction (Chang and Bollum, 1971;
Sherman and Kang, 1973; Hecht, 1973; Lazarus and Kitron, 1973).

Studies of cells under differing physiological conditions have demonstrated that the 6-8S cytoplasmic activities increase significantly with cell growth (Chang and Bollum, 1972; Chang et al., 1973; Baril et al., 1973) and several investigators have reported the existence of a high molecular weight activity in the nucleus (Weissbach et al., 1971; Sedwick et al., 1972; Loeb, 1969). However, due to the many different procedures used by investigators to isolate subcellular fractions and the difficulty in relating "in vivo" location to where the enzyme is found after cellular fractionation, one cannot safely designate the true cellular location of these enzymes. It will be the intention of this author when describing an activity as a cytoplasmic DNA polymerase to refer only to the location of the activity after cellular fractionation.

We have previously reported the isolation from mouse testis of a chromatin-bound low molecular weight (3.5S) DNA polymerase, a heterogenous high molecular weight (6-8S) activity from the cytosol, and an activity specific to the mitochondrial fraction (Hecht, 1973). This communication compares the distributions and activities of several DNA-dependent DNA polymerases in testis, liver, and regenerating liver. Evidence will be presented for two distinct cytoplasmic high molecular weight DNA polymerases and for conversion studies whereby a low molecular weight activity is derived from one of the high molecular weight DNA polymerases.

MATERIALS AND METHODS

Preparation of Organelle Extracts

Nuclei and cytosol were isolated from testis and liver of 8-12 week old $B6AF_1/J$ mice by a sucrose method (Hecht, 1973). Enzyme extracts were prepared by sonication of each subcellular fraction at 4°C for 60 seconds in low salt extraction buffer (0.02M Tris, pH 8.1, 0.05M NaCl, 0.003M β-mercaptoethanol, 0.001M EDTA), followed by incubation for 30 min at 25°C in DNase (40 µg/ml) and $MgCl_2$ (0.01M), and centrifugation at 100,000 Xg for 60 min. The supernatant was removed and dialyzed overnight against low salt extraction buffer. An alternative high salt extraction method involved suspension of each subcellular fraction in high salt extraction buffer (1 M KCl dissolved in low salt extraction buffer) for 60 min and a subsequent centrifugation at 100,000 Xg for 60 min. The supernatant was dialyzed overnight against the low salt extraction buffer.

DNA Polymerase Assays

DNA polymerase activity was measured by a modification of a method previously described (Hecht, 1973). Incubation was at 37°C for 60 min. The standard incubation mixture (0.3ml) contained 5 μmol of Tris-HCl buffer at pH 8.8, 2.5 μmol $MgCl_2$, 3 μmol β-mercaptoethanol, 5 μmol KCl, 160 nmol activated calf thymus DNA, 10 nmol each of dCTP, dATP, and dGTP, 1 nmol [³H]TTP (1000 cpm/pmol), and enzyme. The calf thymus DNA was activated by the method of Aposhian and Kornberg (1962). The reaction was linear for at least 2 hr (not measured longer). After incubating the assay tubes were placed in ice and the reaction was stopped by the addition of 0.5 ml cold lN perchloric acid containing 0.02M sodium pyrophosphate. The precipitates were collected on glass fiber filters (Whatman GF/A) and washed with 10 ml of 5% perchloric acid containing 0.02M sodium pyrophosphate and with 10 ml of 95% ethanol. The filters were dried and the bound radioactivity was counted in a liquid scintillation spectrometer.

Activity was measured at DNA polymerase levels where nucleotide incorporation was proportional to enzyme concentration. Therefore, in terms of the total amount of nucleotide incorporated the values for the calf thymus DNA-primed assay are approximately four times the amount of [³H]-TMP incorporated into acid insoluble form.

Sucrose Gradient Centrifugation

Sedimentation coefficients were determined using the method of Martin and Ames (1961). Centrifugation conditions and the procedure for gradient fractionation are described in the figure captions.

Gel Filtration

Sephadex G-100 (Pharmacia) was equilibrated with either low salt extraction buffer containing 20% ethylene glycol in the presence or absence of 0.25M KCl. Columns (1.0 cm in diameter and 50 cm in length) were prepared and protein was eluted with the equilibration buffer.

Phosphocellulose Chromatography

Whatman phosphocellulose (P-11, purchased from Reeve Angel) was washed by the method of Burgess (1969). After washing, it was equilibrated with 0.02M Tris pH 8.1, 0.05M NaCl, 0.003M β-mercaptoethanol, 0.001M EDTA, and 20% ethylene glycol. Phosphocellulose columns (1 cm X 15 cm) were poured and washed overnight by gravity. After the crude enzyme extract was loaded, the columns were washed with equilibration buffer and eluted with a continuous gradient from 0 to 1.0M KCl in equilibration buffer.

RESULTS

Enzyme extracts were prepared from the nuclear and cytosol fractions of the testis by the low salt extraction method. Aliquots of each crude supernatant were layered over 5 to 20% sucrose gradients (sucrose dissolved in low salt extraction buffer) and centrifuged (Figure 1A and B). The DNA polymerase from the nuclear fraction (A) had a sedimentation coefficient of 3.5S while the cytoplasmic fraction (B) exhibited a broad, often split, peak of activity between 6 and 8S. When the same preparations were analyzed under conditions of higher ionic strength (sucrose dissolved in low-salt extraction buffer plus 0.25M KCl), no change was observed in the sedimentation of the nuclear fraction (C), but the majority of the cytoplasmic activity sedimented at 3.5S (D). A small reproducible fraction with a sedimentation coefficient between 6 and 7S was always observed with testicular cytoplasmic extracts at higher ionic strength (D). Higher concentrations of salt in the gradient gave identical results. When nuclear and cytoplasmic extracts were mixed and analyzed at higher ionic strength, greater than 95% of the initial activity was recovered with a sedimentation coefficient of 3.5S.

Gel permeation provided a second means to assess the molecular changes occurring in the presence of higher ionic strength. DNA polymerase activity from nuclear extracts analyzed in the presence or absence of 0.25M KCl had an estimated molecular weight of 50,000 (Figure 2A). In contrast the molecular weight of the testicular cytoplasmic activity varied with the ionic strength of the elution buffer. The activity was excluded from the gel in the presence of low salt extraction buffer (Figure 2B) but when higher ionic strength buffer was used (Figure 2C), most (peak II) had an estimated molecular weight of 50,000. In the presence of 0.25M KCl, a small amount (peak I) of the total cytoplasmic DNA polymerase continued to elute in the exclusion volume. Other experiments using Sephadex G-200 indicate that peaks I and II of the testicular cytoplasmic DNA polymerases have molecular weights of 170,000 and 150,000, respectively. These estimates are preliminary due to poor recoveries of the loaded activity from the slow running columns.

The difference between the estimated molecular weights of the predominant testicular cytoplasmic DNA polymerase in the

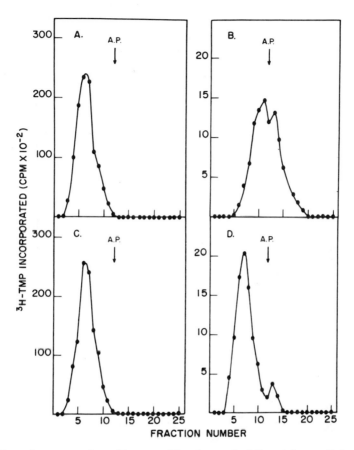

Fig. 1. Sucrose density gradient centrifugation of testis DNA polymerase at low and high ionic strength. Linear 5 to 20% sucrose gradients were prepared in low salt extraction buffer, with or without 0.25M KCl. Aliquots (200 µl) of organelle extracts were centrifuged in a Beckman SW50.1 rotor for 16 hr at 4°C at 40,000 rev/min. Sedimentation was to the right. Fractions (0.2 ml) were collected from the top of the gradient and aliquots (0.1 ml) were assayed for DNA polymerase activity. The arrow marked A.P. indicates the sedimentation position of alkaline phosphatase ($S_{20,w}$ = 6.3). A. Nuclear extract in low salt extraction buffer. B. Cytoplasmic extract in low salt extraction buffer. C. High ionic strength gradient (0.25M KCl) with same load as (A). D. High ionic strength gradient (0.25M KCl) with same load as (B).

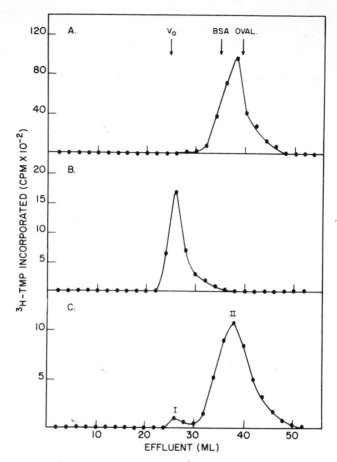

Fig. 2. Gel filtration of testis DNA polymerase at low and high ionic strength. Nuclear (1 mg) and cytoplasmic (8 mg) extracts were applied to Sephadex G-100 (1 cm X 50 cm) equilibrated with low salt extraction buffer (Fig. 2A, 2B) or low salt extraction buffer containing 0.25M KCl (Fig. 2C). V_0 is at 25 ml. Fractions of 2 ml were collected and 0.1 ml aliquots were assayed for DNA polymerase. The columns were calibrated with dextran blue dye, lactic dehydrogenase, *E. coli* alkaline phosphatase, bovine serum albumin, and ovalbumin.

presence and absence of 0.25M KCl led us to consider the possibility of a trimer to monomer conversion. If this were to occur, one would predict that a reduced ionic strength could lead to the formation of a molecule of intermediate size.

Figure 3 shows the results of three sucrose gradients run in
(a) low salt buffer (b) low salt buffer + 0.125M KCl and (c)
low salt buffer + 0.25M KCl. In the presence of 0.125M KCl
three DNA polymerase activities were evident -- unconverted
cytoplasmic enzyme, an intermediate activity of sedimentation
coefficient 4.7S and an activity of sedimentation coefficient
3.5S (Figure 3B). At the salt concentration of 0.125M KCl the
4.7S constitutes approximately 50% of the total activity with
the unconverted 6-8S and the totally converted 3.5S sharing
the remainder. Raising the ionic strength converts the 4.7S
activity to a 3.5S sedimenting activity. The use of 0.25M KCl
converts most of the original cytoplasmic 6-8S activity to the
3.5S sedimenting species (Figure 3C).

Although KCl has been used to effect conversion between the
polymerase forms throughout the studies reported here, the
conversion is not KCl specific. NaCl, NH_4Cl, Na_2SO_4, and
$(NH_4)_2SO_4$ work equally well. Lower concentrations (0.125M)
of the latter two are capable of effecting complete conversion,
while the presence of 0.0625M Na_2SO_4 or $(NH_4)_2SO_4$ will produce
a 4.7S intermediate.

To further characterize the nuclear and cytoplasmic DNA
polymerases, crude supernatants were chromatographed on phosphocellulose. The activities derived from the nuclear and
cytoplasmic fractions both eluted at 0.78M KCl (Figure 4).
Often, a small amount of DNA polymerase eluted between 0.50
and 0.60M KCl (peak I) (Figure 4B). This activity was not
seen in every preparation. Peak fractions from these columns
were dialyzed and used for all further investigations.

The nuclear and cytoplasmic DNA polymerase activities that
elute from phosphocellulose at 0.78M KCl have many similar
properties. They both show maximal activity in the presence
of all four deoxyribonucleoside triphosphates, require Mg^{++},
are not inactivated by 0.3mM N-ethylmaleimide, and are similarly inhibited by ethidium bromide, ethanol, and single stranded
DNA (Table 1). Furthermore, both DNA polymerases have pH
optima of 8.6-9.0, have broad Mg^{++} optima of 4-10mM in the
presence of activated (DNase digested) DNA, and have sedimentation coefficients of 3.5S in the presence of 0.25M KCl. These
properties suggest that one of the testicular cytoplasmic
activities could be a polymeric form of the 3.5S nuclear
enzyme.

Since the DNA polymerase activities found in testis may not
be representative of all mouse tissues, the cytoplasmic and
nuclear DNA polymerases of liver and regenerating liver were
characterized by phosphocellulose chromatography. Nuclear and
cytoplasmic extracts from liver gave similar results to those
obtained from testis, most of the activities eluting at 0.78M
KCl with occasional small peaks between 0.50 and 0.60M. In

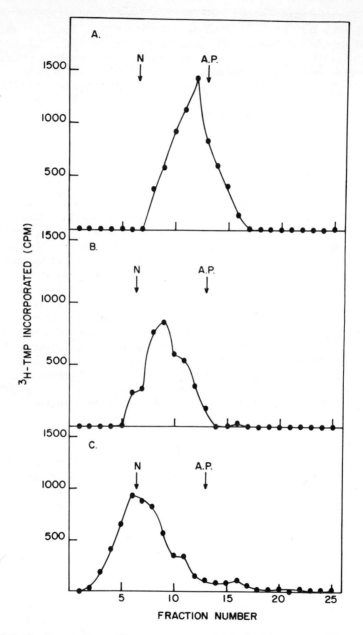

Fig. 3. Sucrose density gradient centrifugation of the cytosol DNA polymerases from testis in the presence of increasing ionic strength. The procedure followed was as described in Fig. 1. The arrow N indicates the sedimentation position of a

(Fig. 3 legend continued)
marker 3.5S activity isolated from the nucleus and run in a
separate tube.   A.   Low salt extraction buffer gradient.   B.
Low salt extraction buffer + 0.125M KCl gradient.   C.   Low
salt extraction buffer + 0.25M KCl gradient.

TESTIS

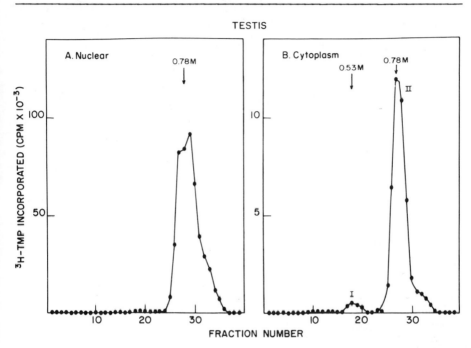

Fig. 4.   Phosphocellulose chromatography of nuclear and cytosol
extracts from testis.   Ten ml of nuclear extract (0.4 mg/ml)
or 20 ml of cytoplasmic extract (1.7 mg/ml) were loaded onto
columns (1 cm X 15 cm) at a flow rate of 4 ml/hr.   The columns
were washed with 40 ml of equilibration buffer (low salt extraction buffer containing 20% ethylene glycol) and eluted
with continuous gradients from 0 to 1.0M KCl in equilibration
buffer.   Fractions (0.7 ml) were collected at a flow rate of
4 ml/hr directly into bovine serum albumin (final concentration 1 mg/ml) and aliquots (0.02 ml) were assayed for DNA
polymerase activity.

contrast, regenerating liver forty-eight hours after hepatectomy, had two distinct cytoplasmic activities eluting at 0.53M
KCl (peak I) and 0.80M KCl (peak II) (Figure 5B).   Sucrose
gradient studies performed at higher ionic strength (0.25M KCl)
indicate peaks I and II have sedimentation coefficients of 7.2S

TABLE 1

ENZYMATIC PROPERTIES OF NUCLEAR AND CYTOPLASMIC
DNA POLYMERASES FROM TESTIS

Reaction conditions	[^{3}H] TMP Incorporated (pmoles/hr)	
	Nuclear DNA polymerase	Cytoplasmic DNA polymerase
Complete	117	95
-DNA	0	0
-Mg^{++}	0	0
-Enzyme	0	0
-activated DNA + native DNA	44	25
-activated DNA + single stranded DNA	5	7
-dCTP, dGTP, dATP	12	10
+KCl (100 mM)	170	101
+KCl (200 mM)	152	78
+Ethidium bromide (8 μM)	98	80
+Ethidium bromide (32 μM)	95	81
+Ethanol (10%)	110	86
+Ethanol (20%)	101	77
+Single stranded DNA (25 μg)	56	51
+Single stranded DNA (50 μg)	47	32

Note: The assay conditions were as described in MATERIALS
AND METHODS except for the modifications listed.

and 3.5S, respectively. The nuclear fraction from regenerating
liver elutes at 0.78M KCl and when analyzed by sucrose gradient
centrifugation has an S value of 3.5. Occasionally an earlier
peak of activity was seen eluting before the 0.78M KCl eluting
peak (Figure 5A). The relationship between this activity and
any others is unknown but this phenomenon may be similar to
that reported by McCaffery et al. (1973). Preliminary studies
indicate the peak I activity isolated from regenerating liver
cytoplasm is similar to the peak I from testis or liver by the
criteria of salt sensitivity, sedimentation coefficient,
molecular weight, and inhibition by N-ethylmaleimide.

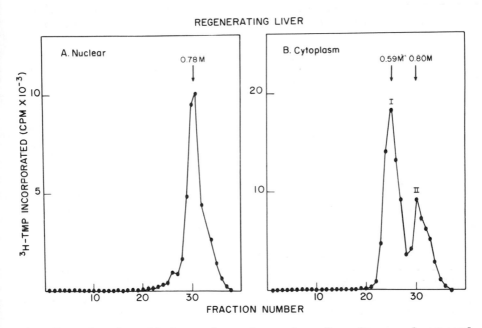

Fig. 5. Phosphocellulose chromatography of nuclear and cytosol
extracts from regenerating liver. Mice were hepatectomized by
the method of Higgins and Anderson (1931). Forty-eight hours
later, the mice were sacrificed and nuclear and cytosol extracts
were isolated from the liver. Twenty ml of nuclear extract
(0.2 mg/ml) or 28 ml of cytoplasmic extract (2.1 mg/ml) were
loaded onto columns (1 cm X 15 cm) and analyzed as described
in Fig. 4

DISCUSSION

As part of a study of the regulation of DNA synthesis during
spermatogenesis, we have been characterizing the DNA-dependent

DNA polymerase activities from mouse testis. Using aqueous methods to isolate subcellular fractions, we have found a low molecular weight nuclear DNA polymerase activity, two high molecular weight cytoplasmic DNA polymerase activities, and a mitochondrial DNA polymerase (Hecht and Davidson, 1973; Hecht, 1973). The predominant DNA polymerase in the cytoplasm of mature mouse testis is convertible by 0.25M KCl to a molecule with many properties similar to the low molecular weight nuclear DNA polymerase. A second high molecular weight activity that does not change sedimentation coefficient in the presence of 0.25M KCl is also present. This form of DNA polymerase may be the one that has been postulated to be involved in DNA replication since a greatly increased amount of a similar activity is found in the cytosol of regenerating liver.

Since the cellular roles of the multiple DNA polymerases are unknown and it is conceptually difficult to relate a DNA polymerase in the cytoplasm to DNA replication, reasons for the existence of a purported replicative enzyme in the cytosol must be considered. Cytoplasmic DNA polymerases may be 1) artifacts due to nuclear leakage during the aqueous subcellular fractionation, 2) enzymes involved in cytoplasmic DNA synthesis or 3) part of a regulatory system involving a shuttling of DNA polymerases between the nucleus and cytoplasm.

Although the predominant testicular cytoplasmic DNA polymerase is readily distinguishable from the nuclear activity by the criteria of size and elution from DNA-cellulose and from DEAE-cellulose (Hecht, 1973), the conversion studies and the many similar enzymatic properties they exhibit suggest a monomer-polymer relationship. The molecular weight determinations and the two-stage salt-induced conversion are consistent with a monomer-trimer model. However, definitive proof that the low molecular weight nuclear DNA polymerase and the 3.5S dissociated component from the cytoplasmic DNA polymerase are identical awaits purification of the enzymes to homogeneity.

Although a monomer-trimer model is inviting, aspects of the experimental evidence indicate that this idea may be an oversimplification. The first and most compelling observation that suggests the involvement of a cytoplasmic factor is the reversible effect of salt in the conversion of only the cyto-plasmic activity. Removal of the higher ionic strength (0.25M KCl) from the cytoplasmic DNA polymerase by gel filtration or by dialysis allows reconstitution of the high molecular weight activity. A similar result is not obtained when the nuclear 3.5S polymerase is treated identically. Furthermore, if a cytoplasmic extract is raised to higher ionic strength (0.25M KCl) and analyzed by sucrose gradient centrifugation in a low salt buffer, only 3.5S sedimenting activity is seen. This

50

argues for the presence of a cytoplasmic factor that is
capable of exerting a reassociating influence after dialysis
or gel filtration but is prevented from doing this by centrifugation. Perhaps a cytoplasmic peptide of size vastly different from the active DNA polymerase components may be involved in the polymerization phenomenon.

The existence of a polymeric cytoplasmic DNA polymerase is
not dependent upon the method of extraction. Nuclear and cytoplasmic activities can be solubilized by either a low salt procedure involving sonication, treatment with DNase I and dialysis or by a high salt procedure involving extraction with
1M KCl at 4°C followed by centrifugation and dialysis. Although the former procedure solubilizes between 60 and 90% of
the total testicular activities and the latter solubilizes
over 95%, the properties of the subcellular DNA polymerases
are identical. The low molecular weight activity is recovered
from the nuclei and a polymeric high molecular weight activity
is recovered from the cytosol suggesting the influence of
cytoplasmic factors on production or maintenance of the polymeric state.

The phenomenon of enzymic variation due to differences in
polymeric state of apparently identical subunits has been reported before for serum choline esterases, glutamate dehydrogenase, β-galactosidase, and skeletal muscle phosphorylase
(La Motta et al., 1965; Bitensky et al., 1965; and Alpers et
al., 1968). Often the equilibrium between monomeric and
polymeric enzyme forms is shifted by effector molecules thereby allowing enzyme activation or inhibition. Since the physiological role in DNA biosynthesis of either the 3.5S DNA polymerase or the cytoplasmic polymeric activity is unknown, one
can only speculate a function. I propose that the monomer-
polymer forms of this DNA polymerase could serve a regulatory
role. The polymerization would be controlled by cytoplasmic
factor(s) that cause several 3.5S components to aggregate.
Assuming that transport into the nucleus is specific for the
low molecular weight form of the DNA polymerase this polymerization would be an effective means to regulate the amount of
nuclear 3.5S DNA polymerase. Conditions equivalent to the "in
vitro" salt dissociation would be required for transport of
this DNA polymerase into the nucleus. Alternatively, it is
equally possible that the 3.5S DNA polymerase is an artifact;
the "in vivo" enzyme being a polymer of the 3.5S activity
either associated with chromatin or in the cytoplasm. The
nuclear low-molecular-weight DNA polymerase that is isolated
may be a dissociated monomer unit due to the rigorous extraction conditions needed to solubilize it from the chromatin.

LITERATURE CITED

Adams, R.L.P., M.A.L. Henderson, W. Wood, and J.G. Lindsay. 1973. Multiple forms of nuclear deoxyribonucleic acid polymerases and their relationship with soluble enzyme. *Biochem. J.* 131: 237-246.

Alpers, D.H., E. Steers, S. Shifrin, and G. Tomkins. 1968. Isozymes of the lactose operon of *Escherichia coli*. *Ann. N.Y. Acad. Sci.* 151: 545-555.

Aposhian, H.V., and A. Kornberg. 1962. Enzymatic synthesis of deoxyribonucleic acid. IX. The polymerase formed after T2 infection of *E. coli*. *J. Biol. Chem.* 237: 519-525.

Baril, E.F., O.E. Brown, M.D. Jenkins, and J. Laszlo. 1971. DNA polymerase with rat liver ribosomes and smooth membranes: purification and properties of the enzymes. *Biochem.* 10: 1981-1992.

Baril, E.F., M.D. Jenkins, O.E. Brown, J. Laszlo, and H.P. Morris. 1973. DNA polymerases I and II in regenerating rat liver and Morris hepatomas. *Cancer Res.* 33: 1187-1193.

Berger, H., R.C.C. Huang, and J.L. Irvin. 1971. Purification and characterization of a deoxyribonucleic acid polymerase from rat liver. *J. Biol. Chem.* 246: 7275-7283.

Bitensky, M.W., K.L. Yielding, and G.M. Tomkins. 1965. The effect of allosteric modifiers on the rate of denaturation of glutamate dehydrogenase. *J. Biol. Chem.* 240: 1077-1082.

Burgess, R.R. 1969. A new method for the large scale purification of *E. coli* DNA-dependent RNA polymerase. *J. Biol. Chem.* 244: 6160-6170.

Chang, L.M.S., and F.J. Bollum. 1971. Low molecular weight DNA polymerase in mammalian cells. *J. Biol. Chem.* 246: 5835-5837.

Chang, L.M.S., and F.J. Bollum. 1972. Low molecular weight deoxyribonucleic acid polymerase from rabbit bone marrow. *Biochem.* 11: 1264-1272.

Chang, L.M.S., and F.J. Bollum. 1972. Variation of deoxyribonucleic acid polymerase activities during rat liver regeneration. *J. Biol. Chem.* 247: 7948-7950.

Chang, L.M.S., M. Brown, and F.J. Bollum. 1973. Induction of DNA polymerase in mouse L cells. *J. Mol. Biol.* 74: 1-8.

Haines, M.E., R.G. Wickremasinghe, and I.R. Johnston. 1972. Purification and partial characterization of rat liver nuclear DNA polymerase. *Eur. J. Biochem.* 31: 119-129.

Hecht, N.B., and D. Davidson. 1973. The presence of a common active subunit in low and high molecular weight murine DNA polymerases. *Biochem. Biophys. Res. Comm.* 51: 299-305.

Hecht, N.B. 1973. Interconvertibility of mouse DNA polymerase activities derived from the nucleus and cytoplasm.
Biochem. Biophys. Acta 312: 471-483.

Hecht, N.B. 1974. In manuscript.

Higgins, G.M., and R.M. Anderson. 1931. Restoration of the
liver of the white rat following partial surgical removal.
Arch. Path., Chicago 12: 186-202.

La Motta, R.V., R.B. McComb, and H.J. Wetstone. 1965. Isozymes of serum cholinesterase: a new polymerization sequence. *Can. J. Physiol. Pharmacol.* 43: 313-318.

Lazarus, L.H., and N. Kitron. 1973. Cytoplasmic DNA polymerase: polymeric forms and their conversion into an
active monomer resembling nuclear DNA polymerase. *J.
Mol. Biol.* 81: 529-534.

Loeb, L. 1969. Purification and properties of deoxyribonucleic acid polymerase from nuclei of sea urchin embryos.
J. Biol. Chem. 244: 1672-1681.

Martin, R.G., and B.M. Ames. 1961. A method for determining
the sedimentation behavior of enzymes. *J. Biol. Chem.*
236: 1372-1379.

McCaffery, R., D.F. Smoler, and D. Baltimore. 1973. Terminal
deoxyribonucleotidyl transferase in a case of childhood
acute lymphoblastic leukemia. *Proc. Natl. Acad. Sci.,
U.S.A.* 70: 521-525.

Meyer, R.R., and M.V. Simpson. 1968. DNA biosynthesis in
mitochondria. *Proc. Natl. Acad. Sci., U.S.A.* 61: 130-
137.

Meyer, R.R., and M.V. Simpson. 1970. Deoxyribonucleic acid
biosynthesis in mitochondria. *J. Biol. Chem.* 245: 3426-
3435.

Sedwick, W.D., T.S. Wang, and D. Korn. 1972. Purification
and properties of nuclear and cytoplasmic DNA polymerases
from human KB cells. *J. Biol. Chem.* 247: 5026-5033

Sherman, M.I., and H.S. Kang. 1973. DNA polymerases in mid-
gestation mouse embryo, trophoblast, and decidua. *Dev.
Biol.* 34: 200-210.

Weissbach, A., A. Shelabach, B. Fridlender, and A. Bolden.
1971. DNA polymerases from human cells. *Nature New
Biology* 231: 167-170.

THE RNA POLYMERASES OF THE NUCLEUS AND CHLOROPLASTS OF MAIZE

JAMES W. HARDIN
KLAUS APEL
JANE SMITH
LAWRENCE BOGORAD
The Biological Laboratories
Harvard University
16 Divinity Avenue
Cambridge, Massachusetts 02138

ABSTRACT. Three forms of RNA polymerase have been isolated from leaves of maize. Two of these enzymes are of
nuclear origin. The third appears to be unique to chloroplasts. Nuclear RNA polymerase II and the chloroplast
polymerase have been highly purified and their properties
examined. While differing in sensitivity to inhibitors
and in chromatographic properties, they are similar in
response to monovalent and divalent cations and in that
each has a subunit of 180,000 daltons.

The activity of the plastid DNA-dependent RNA polymerase has been studied at various points in the development of greening plastids. Mechanisms which might account
for the dramatic increase in plastid polymerase activity
during the course of greening have been examined.

Template specificity of the nuclear II enzyme has been
studied, and the presence of a polypeptide associated with
an increase in activity with native DNA template has been
described.

INTRODUCTION

It has been well established in plant cells that chloroplasts as well as nuclei contain DNA from which RNA is tran-
scribed; and that, in addition to the nuclear-dependent protein synthesizing system, the cell contains at least one
other protein-synthesizing system within the chloroplasts
(Kirk and Tilney-Bassett, 1967). The morphogenesis of the
chloroplast depends on the interaction of both the nuclear
and chloroplast genomes (Mets and Bogorad, 1971; Mets and
Bogorad, 1972; Davidson, Hanson, and Bogorad, 1974; Apel and
Schweiger, 1972). It is not known however, how this interaction is regulated. One possible means of regulation may
be exerted at the level of transcription involving both the
nuclear and chloroplast DNA-dependent RNA polymerases.
DNA-dependent RNA polymerases associated with nuclei and

chloroplasts have been described for several species (Kirk, 1964; Bottomley et al, 1971a; Polya and Jagendorf, 1971); from one species, maize, both chloroplast-specific and nuclear DNA-dependent RNA polymerases have been isolated and highly purified (Bottomley et al, 1971b; Strain et al, 1971).

In order to investigate the possible involvement of the different RNA polymerases in the regulation of chloroplast development we have begun a study of polymerases of the greening maize plant. It is well documented that after illumination of dark-grown plants, a number of biochemical and morphological changes occur in plastids (Bogorad, 1967). One of these changes is a rapid and specific increase in RNA synthesis (Smith et al.,1970).

In this paper we present results which suggest that the chloroplast RNA polymerase is involved in the regulation of this light-dependent RNA synthesis. Furthermore, we have characterized the properties and the structure of the nuclear and the chloroplast RNA polymerases. Based on these data, we propose some possible ways in which the different enzymes may be linked.

MATERIALS AND METHODS

Growth of Plants

Maize seeds (*Zea mays* WF9 TMS x B37 or FR9^{C}MS x FR37) were soaked overnight in water, sown in moist vermiculite and grown in the dark at 28°C for 5-7 days. Plants were removed from the dark and illuminated for periods from 2 hr to several days depending upon the particular experiment.

Isolation of Chloroplasts

Chloroplasts were isolated and purified as described by Bottomley et al.(1971b). The washed plastids were either assayed for RNA polymerase activity or stored in liquid nitrogen until needed for RNA polymerase purification.

Estimation of Chlorophyll

Chlorophyll was estimated by the procedure of Arnon (1949).

Solubilization and Purification of Plastid RNA Polymerase

RNA polymerase was solubilized from plastids by incubation of the washed plastids with EDTA at elevated temperatures as

previously described (Bottomley et al, 1971b). The enzyme
was further purified by glycerol gradient centrifugation in
high and/or low salt, and by phosphocellulose chromatography.

Isolation of Nuclear RNA Polymerases

Nuclear RNA polymerases were isolated by a modification
of the procedures of Strain et al.(1971) and Mullinix et al.
(1973). For purification of the nuclear enzyme II the following procedure was employed as shown in Fig. 1: Sepharose
4B chromatography, polyethylene glycol-Dextran phase parti-
tion and chromatography on DEAE cellulose, DNA cellulose and
phosphocellulose.

ISOLATION OF MAIZE RNA POLYMERASES

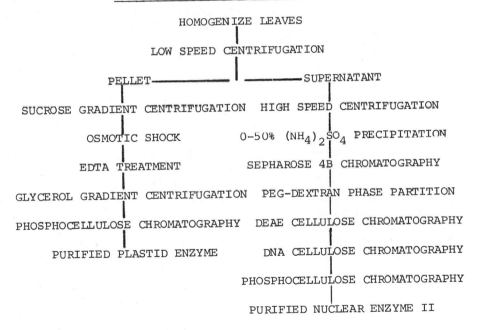

Fig. 1. Purification of the chloroplast and nuclear II DNA-
dependent RNA polymerase of maize.

Assay for RNA Polymerase

The activity of the plastid and nuclear DNA-dependent RNA
polymerases was assayed by the methods of Bottomley et al.
(1971b) and Strain et al.(1971) respectively.

Polyacrylamide Gel Electrophoresis

Sodium dodecyl sulfate gels were run as described by Weber and Osborn (1969). Gels were stained as described by Fairbanks et al (1971).

RESULTS

1. *Purification and characterization of maize RNA polymerases*

Three different forms of DNA-dependent RNA polymerases have been isolated from maize leaves. One of these RNA polymerases has been found in chloroplasts, while the two other enzymes have been isolated from a soluble (presumably nuclear) fraction. The enzymes have been purified and characterized as described by Bottomley et al (1971b), Strain et al (1971), and Mullinix et al (1973).

The purification of the nuclear enzyme II has been modified in such a way that a more highly purified, stable, and active enzyme has been obtained. This modified purification includes chromatography of the enzyme on phosphocellulose. Since nuclear enzyme I is unstable we have concentrated our efforts on isolating nuclear enzyme II and comparing it to the chloroplast enzyme.

Figure 1 presents an outline of the purification procedures employed for the plastid and nuclear polymerase II. As can be seen, the purification procedures differ for the nuclear and chloroplast enzymes after the initial low speed centrifugation. It should be emphasized that the plastid enzyme is tightly bound to the DNA-membrane matrix of the purified and osmotically-shocked plastids and is not released until solubilized by incubation with EDTA at elevated temperatures.

The plastid enzyme has several properties which are distinct from those of the nuclear enzyme. The two enzymes exhibit different characteristics at some purification steps. The nuclear enzyme elutes at a lower salt concentration from DEAE cellulose and phosphocellulose than does the chloroplast enzyme (Strain et al, 1971). These two enzymes also differ in their response to the drug α-amanitin. The nuclear enzyme II is similar to other eucaryotic enzymes II in that it is inhibited by α-amanitin; the chloroplast enzyme is completely insensitive to the drug.

The two enzymes are similar with regard to several other properties. They are both large, multimeric proteins of approximately 500,000 daltons. Both enzymes have a temperature optimum which is considerably higher than those reported for other eucaryotic polymerases (48°C for the plastid enzyme and 44°C for the nuclear enzyme). Both exhibit maximal activity with Mg^{2+}, and both are inhibited by even low concentrations of monovalent cations.

Analysis of the two enzymes by sodium dodecyl sulfate polyacrylamide gel electrophoresis (SDS-PAGE) reveals both similarities and differences in their subunit structure (Fig. 2) (Mullinix et al, 1973; Smith and Bogorad, in preparation). While the complete structure of both enzymes is yet to be elucidated, it has been shown that each enzyme contains two large subunits which are present at all stages of purification. The molecular weights of these polypeptides is 180,000 and 160,000 daltons for the nuclear enzyme (Fig. 2a) and 180,000 and 140,000 daltons for the chloroplast enzyme (Fig. 2b). When a sample containing both the plastid and nuclear enzymes is applied to a single gel and analyzed by SDS-PAGE, a single 180,000 dalton band, a 160,000 and a 140,000 dalton band can be seen on the gel (Fig. 2c). This analysis suggests that the 180,000 dalton polypeptides of the nuclear and plastid enzymes are of similar, if not identical, molecular weights.

2. *Light-induced changes in chloroplast RNA synthesis*

Synthesis of chloroplast RNA in etiolated plants is rapidly increased following brief exposure to light. It is possible that this increase in RNA synthesis is regulated by the chloroplast RNA polymerase.

We illuminated etiolated maize plants for different lengths of time, isolated the plastids and estimated their RNA polymerase activities. For the first 6 hr of illumination there is no difference in activity between the plastids from illuminated plants and from the dark-grown controls. Within the next 6-10 hr of illumination the activity increases rapidly until a point at about 16 hr at which the activity is increased 5 to 6-fold times that of the dark-grown controls (Fig. 3).

We have eliminated the possibilities that this increase is caused by differences in optimal assay conditions, or in concentration of nucleases and proteases in samples from illuminated and dark-grown plants (Apel and Bogorad, in preparation). The increase of RNA polymerase activity in the crude plastid extracts following exposure to light might be explained by one of the following mechanisms:

(a) Light may induce a rapid *de novo* synthesis of RNA polymerase molecules in etioplasts.

(b) The specific activity of the enzyme may change upon

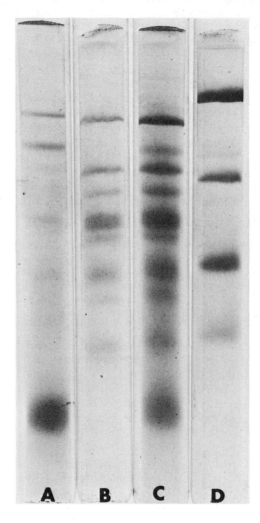

Fig. 2. Sodium dodecyl sulfate polyacrylamide gel electrophoretic analysis of (A) Maize nuclear RNA polymerase II,
purified through DEAE cellulose.
    (B) Maize chloroplast RNA polymerase purified through phos-
phocellulose.
    (C) Co-electrophoresis of maize nuclear II RNA polymerase
and maize chloroplast RNA polymerase.
    (D) Standard proteins: myosin, 212,000; β-galactosidase,
130,000; bovine serum albumin, 68,000; ovalbumin, 43,000
daltons.

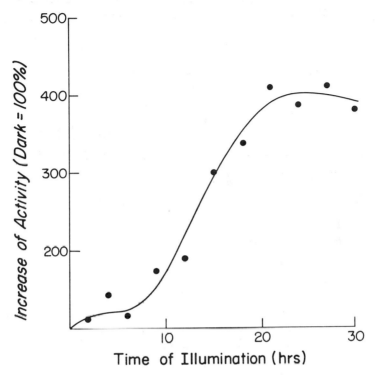

Fig. 3. RNA polymerase activity in suspensions of maize plastids isolated from etiolated plants illuminated for various periods of time.

illumination.

(c) Illumination may affect some change(s) in the template or release the enzyme from an inhibitor.

The chloroplast enzyme was solubilized and purified through the phosphocellulose step as described in Materials and Methods. SDS-PAGE analysis of purified chloroplast enzyme isolated from plants which were illuminated 2 and 16 hr reveals no differences in the polypeptide composition of the two samples. The specific activity, as well as total amount of purified enzyme per gram of leaf tissue, is the same for the two samples. These experiments indicate that neither a change in number of enzyme molecules nor a change in specific activity is responsible for the change in polymerase activity in the plastid suspensions following illumination. The fact that the differences in activity seen in the two plastid preparations (Fig. 3) disappears after the solubilized enzyme preparations are subjected to high speed centrifugation and separated from the bulk of DNA and membranes (Fig. 4) suggests that the third

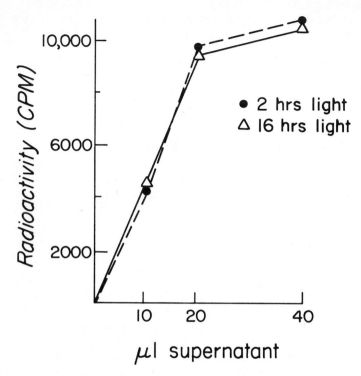

Fig. 4.   Solubilized plastid RNA polymerase activity from eti-olated maize plants illuminated for 2 hr or 16 hr. The RNA polymerase was separated from the membrane-DNA complex by centrifugation and assayed as described in Materials and Methods.

possibility may be involved in the phenomenon of the increased RNA synthesis following illumination. This is currently under investigation.

3. *Regulation of template specificity of nuclear enzyme II*

Two forms of nuclear enzyme II have been described. Form IIa prefers a denatured DNA template for maximum activity while form IIb prefers a native DNA template. It appears that IIb can be converted to IIa (Strain et al, 1971). With the procedure currently being employed for purification of enzyme II, it was observed that enzyme purified through DNA cellulose

exhibited some activity on native DNA, indicating that after
this purification step we still have a mixture of both forms
of enzyme II. After the next purification step, phosphocellulose chromatography, a large fraction of the native DNA-
preferring activity disappears (Table 1), suggesting that the
native preferring form IIb has been converted to IIa. The
ability of the phosphocellulose purified enzyme II to utilize
native DNA could be restored by the addition of a thermolabile
fraction eluted from phosphocellulose at a higher salt concentration than that which eluted the enzyme (Table 1). SDS-PAGE
analysis of samples of the DNA cellulose and phosphocellulose
enzymes, and the late phosphocellulose fraction reveals that
the late phosphocellulose fraction contains one major polypeptide - a polypeptide present in the DNA cellulose enzyme,
but greatly diminished in amount in the enzyme eluted from
phosphocellulose (Hardin and Bogorad, in preparation).
Studies are currently in progress on the nature of this late
phosphocellulose fraction and its function in the template
specificity of the nuclear enzyme II.

TABLE 1

EFFECT OF THE PHOSPHOCELLULOSE FACTOR ON NUCLEAR II RNA
POLYMERASE ACTIVITY* WITH DENATURED AND NATIVE CALF THYMUS DNA

	DNA cellulose-purified enzyme	Phosphocellulose-purified enzyme
Native DNA	776	334
+ Phosphocellulose factor	828	890
Denatured DNA	816	878
+ Phosphocellulose factor	960	1046

*Nuclear II RNA polymerase has been purified as described in
Materials and Methods. Activities of the enzyme eluted from
DNA-cellulose and phosphocellulose in the presence or absence
of the phosphocellulose protein factor were compared.
Activity is expressed in cpm per 10 μl of enzyme, assayed for
20 min under standard assay conditions.

DISCUSSION

The results which we have presented indicate that there are
at least three RNA polymerases in maize leaves. Two of these

have been isolated from the soluble fraction of leaf homogenates; the third is tightly bound to chloroplast membranes
and only solubilized at high temperature in the presence of
EDTA. That these enzymes are distinct is further supported
by differences in purification properties, response to the
drug α-amanitin and subunit structure.

More detailed discussions of the properties of these enzymes are found in Bottomley et al (1971b), Strain et al
(1971), and Mullinix et al (1973).

Since enzymes from mitochondria have been found in other
eucaryotic cells we expect that such an enzyme will be found
in maize leaves. Preliminary work on the maize mitochondrial
RNA polymerase is in progress. Thus, it may be possible to
use the maize system to examine the properties and relationships of the several DNA dependent RNA polymerases.

Chloroplast development depends upon the interaction of
the nuclear and plastid genomes. For example, two genes for
Chlamydomonas chloroplast ribosomes have been identified in
the nuclear genome; a gene which specifies a third ribosomal
protein is in the chloroplast genome (Mets and Bogorad, 1971,
1972; Davidson, Hanson, and Bogorad, 1974). Both nuclear
and plastid genes are involved in thylakoid membrane formation in *Acetabularia* (Apel and Schweiger, 1972). Another
example of this interaction is the formation of the chloroplast specific enzyme ribulose diphosphate carboxylase. This
enzyme is composed of two types of subunits. There is evidence that information for one subunit is transmitted uni-
parentally and is thus likely to be contained in the chloroplast DNA (Chan and Wildman, 1972) and that synthesis of this
polypeptide takes place on chloroplast ribosomes (Gooding et
al, 1973), while the information for the other subunit is
transmitted in a Mendelian manner and is thus probably coded
for by nuclear DNA (Kawashima and Wildman, 1972) and synthesized in the cytosol on 80S ribosomes (Gray and Kekwick,
1973).

The problem is how are such interactions controlled? One
possible means of control might be affected at the transcriptional level involving RNA polymerases in both nuclei and
chloroplasts. For a number of systems there is evidence that
changes in RNA polymerase control or influence development.
For instance, changes in activity of RNA polymerase accompanying changes in development have been associated (1) with
the appearance of new polypeptides which bind to the polymerase molecule (Horvitz, 1973), or (2) with the appearance
and/or disappearance of a species or form of RNA polymerase
(Chamberlin et al, 1970; Wilhelm et al, 1974).

We have examined the possible role of chloroplast RNA

polymerase as a control in the chloroplast development by studying changes in the enzyme during the light-induced greening process in maize leaves. We have shown that the activity of RNA polymerase is increased dramatically after illumination of dark-grown plants. There is no evidence that increased activity is caused by a stimulating factor or by the appearance or disappearance of a new enzyme species. Furthermore, we have ruled out the possibility that this increase is due to a change in specific activity or increased synthesis of the enzyme following illumination. Although we cannot completely rule out the possibility that some modification of the enzyme has occurred, our results from analysis of the enzyme by SDS-PAGE demonstrate that the enzymes isolated at different stages of plastid development have identical subunit structure.

The finding that separation of soluble RNA polymerase isolated from plastids at different developmental stages no longer exhibit differences in activity after the solubilized enzymes have been separated from the plastid DNA-membrane-complex suggests that the light stimulation is caused by something which remains associated with the DNA-membrane complex. At present we are examining this phenomenon.

We have no information about changes in properties of the nuclear RNA polymerases after illumination of the dark-grown plants. However, it is well known that products of nuclear transcription are critical to the development of chloroplasts. Thus, there must be interaction and integration of the products of nuclear and plastid transcription.

We believe that our work on the comparison of subunit structure of RNA polymerases suggests some possibilities for interaction of the two genomes at the level of these enzymes in the nucleus and chloroplast.

The presence of a 180,000 dalton polypeptide in both enzymes suggests that the chloroplast and nuclear enzymes may contain a common polypeptide subunit. If such identity is demonstrated, it would be interesting to know where the information for the polypeptide is located - whether in the nuclear or plastid genome or whether in both genomes - and whether this polypeptide is synthesized on 70S or 80S ribosomes or on both types of ribosomes. However attractive these speculations are in terms of interaction of cellular functions, it is still necessary to demonstrate unequivocally that the 180,000 dalton polypeptide of the plastid and the nuclear enzymes are identical. Additional characterization of these polypeptides by fingerprints of tryptic digests and amino acid analysis will be necessary in order to establish the identity (or lack of identity) of the 180,000 dalton polypeptides in the two organelles. A project has been initiated for the isolation of sufficient enzymes to pursue these studies.

We have described a protein which may interact with the nuclear enzyme II and facilitate its transcription on native DNA in vitro. However, as yet we have no evidence that this factor modifies the function of this nuclear RNA polymerase in vivo.

We believe that future work on the maize DNA-dependent RNA polymerases may provide us with valuable hints on the interaction of the different genomes of the plant cell and their possible roles in cellular development.

LITERATURE CITED

Apel, K. and H. G. Schweiger 1972. Nuclear dependency of chloroplast proteins in *Acetabularia*. *Eur. J. Biochem.* 25: 229-238.

Arnon, D. I. 1949. Copper enzymes of isolated chloroplasts. Polyphenoloxidase in *Beta vulgaris*. *Plant Physiol*. 24: 1-15.

Bogorad, L. 1967. Control mechanisms in plastid development. in *Developmental Biology, Supplement 1, The 26th Symposium of the Society for Developmental Biology*, "Control mechanisms in developmental processes." 1-31.

Bottomley, W., D. Spencer, A. M. Wheeler, and P. R. Whitfeld 1971a. The effect of a range of RNA polymerase inhibitors on RNA synthesis in higher plant chloroplasts and nuclei. *Arch. Biochem. Biophys*. 143: 269-275.

Bottomley, W. B., H. J. Smith, and L. Bogorad 1971b. RNA polymerases of maize. Partial purification and properties of the chloroplast enzyme. *Proc. Natl. Acad. Sci.* (U.S.) 68: 2412-2416.

Chamberlin, M., J. McGrath, and L. Waskell 1970. New RNA polymerase from *Escherichia coli* infected with bacteriophage T_7. *Nature* 228: 227-231.

Chan, P. H., and S. G. Wildman 1972. Chloroplast DNA codes for the primary structure of the large subunit of fraction I protein. *Biochim. Biophys. Acta* 227: 677-680.

Davidson, J. N., M. R. Hanson, and L. Bogorad 1974. An altered chloroplast ribosomal protein in ery-M1 mutants of *Chlamydomonas reinhardi*. *Molecular and Gen. Genet.* in press.

Fairbanks, G., T. L. Steck, and D. F. H. Wallach 1971. Electrophoretic analysis of the major polypeptides of the human erythrocyte membrane. *Biochem*. 10: 2606-2617.

Gooding, L. R., H. Roy, and A. T. Jagendorf 1973. Immunological identification of nascent subunits of wheat ribulose diphosphate carboxylase on ribosomes of both chloroplast and cytoplasmic origin. *Arch. Biochem. Biophys.* 159: 324–335.

Gray, J. C. and R. G. O. Kekwick 1973. Synthesis of the small subunit of ribulose 1,5-diphosphate carboxylase on cytoplasmic ribosomes from greening bean leaves. *FEBS Lett.* 38: 67–69.

Kawashima, N. and S. G. Wildman 1972. Studies on Fraction I protein. Mode of inheritance of primary structure in relation to whether chloroplast or nuclear DNA codes for a chloroplast protein. *Biochim. Biophys. Acta* 262: 42–49.

Kirk, J. T. O., and R. A. E. Tilney-Bassett 1967. *The Plastids.* W. H. Freeman and Co., San Francisco.

Horvitz, H. R. 1973. Polypeptide bound to the host RNA polymerase is specified by T_4 control gene 33. *Nature New Biol.* 244: 137–140.

Mets, L. and L. Bogorad 1971. Mendelian and uniparental alterations in erythromycin binding by plastid ribosomes. *Science* 174: 707–709.

Mets, L. and L. Bogorad 1972. Altered chloroplast ribosomal proteins associated with erythromycin-resistant mutants in two genetic systems of *Chlamydomonas reinhardi. Proc. Natl. Acad. Sci.* 69: 3779–3783.

Mullinix, K. P., G. C. Strain, and L. Bogorad 1973. RNA polymerases of maize. Purification and molecular structure of DNA-dependent RNA polymerase II. *Proc. Natl. Acad. Sci.* 70: 2386–2390.

Polya, G. M. and A. T. Jagendorf 1971. Wheat leaf RNA polymerases. I. Partial purification and characterization of nuclear, chloroplast, and soluble DNA-dependent enzymes. *Arch. Biochem. Biophys.* 146: 635–648.

Smith, H., G. R. Stewart, and R. Berry 1970. The effects of light on plastid ribosomal-RNA and enzymes at different stages of barley etioplast development. *Phytochemistry* 9: 977–983.

Strain, G. C., K. P. Mullinix, and L. Bogorad 1971. RNA polymerases of maize: nuclear RNA polymerases. *Proc. Natl. Acad. Sci.* (U.S.) 68: 2647–2651.

Weber, K. and M. Osborn 1969. The reliability of molecular weight determinations by dodecyl sulfate-polyacrylamide gel electrophoresis. *J. Biol. Chem.* 244: 4406–4412.

Wilhelm, J., D. Dina, and M. Crippa 1974. A special form of deoxyribonucleic acid dependent ribonucleic acid polymerase from oocytes of *Xenopus laevis*. Isolation and characterization. *Biochem.* 13: 1200–1208.

Wildman, S. G., N. Kawashima, D. P. Bourque, F. Wong, S. Singh, P. H. Chan, S. Y. Kwok, K. Sakano, S. D. Kung, and J. P. Thornber 1972. Location of DNAs coding for various kinds of chloroplast protein. in *Proc. Intl. Conf. "Biochemistry of gene expression in higher organisms"* (Pollak and Lee, eds). Australian and New Zealand Book Co., Sidney, 443-456.

THE RNA POLYMERASES
OF FUNGI.
GLUTARIMIDES AND THE REGULATION OF RNA POLYMERASE I.

DAVID H. GRIFFIN, WILLIAM TIMBERLAKE,
JOHN CHENEY AND PAUL A. HORGEN

Department of Botany and Forest Pathology
State University of New York,
College of Environmental Science and Forestry
Syracuse, New York 13210 and
Department of Botany
University of Toronto
Erindale Campus
Mississauga, Ontario
Canada

ABSTRACT. Multiple forms of RNA polymerase have been reported from representatives of six classes of fungi. The
number of forms reported varies from two to four. The
general properties of these forms of RNA polymerase suggest
that they are very similar to the multiple forms of RNA
polymerase reported from higher plants and animals. All
species studied have at least one polymerase sensitive to
α-amanitin. All forms are more active in the presence of
manganese than in the presence of magnesium. Where studied,
the fungal enzymes have high molecular weight and multiple
subunits.
    Very little is known about control functions. Template
specificity studies have yielded contradictory results and
critical information on the quality of the templates used
is unavailable. A stimulatory factor from yeast has been
reported. Attempts to relate levels of polymerase activity to growth rates and developmental stages have been made.
However, lack of understanding of the nature of the holoenzyme makes interpretation of these experiments difficult.
    Reports of inhibition of RNA polymerase I by cyclohex-
imide in vitro are inconsistent. However, in vivo evidence
is most easily interpreted as direct inhibition of the
enzyme by cycloheximide. The effects of other inhibitors
of protein synthesis and various glutarimide analogs of
cycloheximide suggest the presence of a cellular regulator
of RNA polymerase which glutarimides mimic.

As early as 1964, Widnell and Tata (1964) suggested that
two RNA synthesizing enzymes existed in the nucleus of eukaryotic cells, one stimulated by Mg^{++} synthesizing G-C rich RNA,

and the other stimulated by Mn^{++} synthesizing DNA-like RNA (Widnell, 1965). It wasn't until late 1969 that Roeder and Rutter isolated two forms of RNA polymerase from rat liver and three forms from developing sea urchin embryo. They suggested that each of the forms may have specialized functions within the cell (Roeder and Rutter, 1969). During 1970, reports appeared confirming Roeder and Rutter's initial discovery in higher animal cells (Jacob, et al., 1970; Kedinger, et al., 1970). Since the initial reports of multiple enzymes in animal cells, more than one enzyme has been reported from a variety of higher plant cells (Mondal, et al., 1970; Strain, et al., 1971; Horgen and Key, 1973). At least one enzyme (polymerase I or polymerase A) is compartmentalized within the nucleolus (Jacob, et al., 1970; Roeder and Rutter, 1970a; Roeder and Rutter, 1970b; Tocchini-Valentini, et al., 1970) and is presumably involved in the synthesis of ribosomal RNA in vivo. There also exists at least one enzyme (polymerase II or polymerase B) in the non-nucleolar nucleoplasm (Jacob, et al., 1970; Kedinger, et al., 1970; Lindell, et al., 1970) which is presumably involved in the synthesis of DNA-like RNA (mRNA) in vivo. This enzyme was specifically inhibited by α-amanitin (Jacob, et al., 1970; Kedinger, et al., 1970; Lindell, et al., 1970). A third nuclear enzyme (polymerase III or polymerase A III) is reported from a large number of eukaryotic tissues and is also thought to be nucleoplasmic (Roeder and Rutter, 1969; Lindell, et al., 1970). This form may function in the synthesis of 4S and 5S RNA (Roeder, these proceedings). Organelle-specific RNA polymerases are present in mitochondria (Horgen and Griffin, 1971b; Reid and Parsons, 1971; Küntzel and Schäfer, 1971; Tsai, et al., 1971; Scragg, 1971) and in chloroplasts (Bottomley, et al., 1971).

Horgen and Griffin (1971a) reported that the water mold *Blastocladiella emersonii* possessed multiple RNA polymerases. Vegetative (OC) thalli of *B. emersonii* possessed three chromatographically distinct activities of RNA polymerase eluted from DEAE cellulose (Fig. 1). Anion exchange on DEAE cellulose or DEAE Sephadex has been a standard technique for the initial separation of the eukaryotic soluble RNA polymerases. A linear salt gradient (ammonium sulfate, potassium chloride, ammonium chloride) is used to elute the enzymes from the column. RNA polymerase I (nucleolar enzyme) elutes at the lowest salt concentration, RNA polymerase II (nucleoplasmic enzyme) at the intermediary concentration and RNA polymerase III at the highest concentration of salt (Fig. 1). Form III in *B. emersonii* was shown to be mitochondrial (Horgen and Griffin, 1971). Rechromatography of the eluted enzymes indicated no dissociation of the individual activities or interconvertibility among

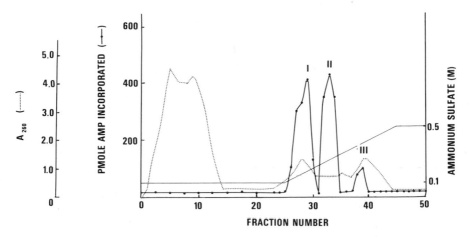

Fig. 1. DEAE-cellulose chromatography of the RNA polymerases of *Blastocladiella emersonii* solubilized and assayed as per Horgen and Griffin (1971a). Peaks I and II are nuclear enzymes, peak III is mitochondrial (Horgen and Griffin, 1971b).

the three forms. Stimulation by exogenously added DNA and dependency upon the presence of the four nucleoside triphosphates suggest that the activities are in fact, DNA-directed RNA synthesizing enzymes (Horgen and Griffin, 1971a). RNA polymerase II of *B. emersonii* was sensitive to α-amanitin (Horgen and Griffin, 1971b) suggesting that this activity was very similar to the RNA polymerase II reported in higher plants (Strain, et al., 1971; Horgen and Key, 1973) and animal cells (Jacob, et al., 1970; Kedinger et al., 1970; Lindell, et al., 1970). All evidence in fact suggests that the primitive eukaryote *Blastocladiella* possesses RNA synthesizing enzymes very similar to the multiple RNA polymerases reported from higher plants and animals.

Since the initial observation by Horgen and Griffin (1971a), multiple RNA polymerases have been demonstrated in *Saccharomyces cerevisae* (Ponta, et al., 1971; Brogt and Planta, 1972; Dezélée, et al. 1972; Adman, et al., 1972; Sebastian et al., 1973), *Achlya bisexualis* (Timberlake, et al., 1972), *Rhizopus stolonifer* (Gong and Van Etten, 1972), *Allomyces arbuscula* (Cain and Nester, 1973), *Physarum polycephalum* (Hildebrandt and Sauer, 1973; Gornicki et al., 1974), *Dictyostelium discoideum* (Pong and Loomis, 1973) and *Neurospora crassa* (Tellez de Iñon, Leoni and Torres, 1974). Table 1 compares some of the general properties of fungal RNA polymerases.

TABLE I

PROPERTIES OF FUNGAL DNA-DEPENDENT RNA POLYMERASES

ORGANISM	FORM[1]	SALT OPTIMA (mM)			ACTIVITY RATIOS	TEMPLATE PREFERENCES		REFERENCES
		NH$_4^+$	Mn^{2+}	Mg^{2+}	Mn/Mg	Homologous or Heterologous	Native or Denatured	
Blasto-cladiella	I	10	1.0	75	2	-	-	Horgen (1972)
	II	100	2.0	50	2	-	-	
Allo-myces	I	50	2.5	10	1.3	None	Denat.	Cain and Nester (1973)
	II	100	2.5	25	1.4	None	Denat.	
	III	150	5.0	25	0.9	None	Denat.	
Achlya	I	50	1.0	10	1.1	-	None	Cheney (unpubl.)
Saccharo-myces	I (A)	100 (KCl)	1.0	16	1	Heter.	Native	Brogt and Planta (1972)
	II (B)	100 (KCl)	3.0	8	10	Heter.	Denat.	
	III (C)	-	-	-	-	Heter.	Denat.	
Saccharo-myces	IB	50	-	-	-	Heter.	Native	Adam et al. (1972)
	II	75	-	-	-	Heter.	Denat.	
	III	250	-	-	-	Heter.	Native	

TABLE I (CON'T)

				N.A.[2]				
Saccharo-myces	I (A)	20	2	N.A.	Infinite	Heter.	Denat.	Ponta et al. (1972)
	II (B)	50	2	N.A.	Infinite	Heter.	Denat.	
	III (C)	75	2	5	1.7	Heter.	Denat.	
Rhizopus	I	20	1	5	1.4	--	Denat.	Gong and van Etten (1972)
	II	90	1	10	1.3	--	Denat.	
	III	110	1	6	1.0	--	Denat.	
Dictyo-stelium	I	50	--	--		Homol.	None	Pong and Loomis (1973)
	II	100	--	--		Homol.	None	
Neuro-spora	IA (I)	--	--	4	--	--	None	Tellez de Iñon, et al. (1974)
	IB (II)	--	--	-	--	--	-	
	IIA (III)	--	--	4	--	--	None	
	IIB (IV)	--	--	2.5	--	--	Denat.	
Physarum	II (S)	80	1	5	1.5	--	Denat.	Gorniki, et al. (1974)
	III? (R)	20	2	5	2.5	--	Denat.	
Physarum	II (B)	--	2.5	10	6	Heter.	Denat.	Hildebrandt and Sauer (1973)
	III? (A)	--	2.5	10	2	Heter.	Denat.	

[1]Probable designation according to Roeder and Rutter (1969); authors terminology in ( ) when different.
[2]N.A. = Negligible Activity

In all fungi that have been examined, at least two and as many as four nuclear enzymes have been eluted from DEAE columns. In *A bisexualis* (Timberlake, 1972) and in *S. cerevisae* (Adman, et al., 1972), polymerase I can be fractionated into at least two activities. The nucleoplasmic enzyme (polymerase II) from fungi has been shown to be sensitive to α-amanitin. That the enzymes are distinct from one another has also been demonstrated in yeast by immunological techniques (antiserum for polymerase II did not effectively cross react with other nuclear enzymes) (Ponta, et al., 1971).

Although the optimal concentrations vary somewhat, both polymerase I and polymerase II in fungi are most active utilizing manganese ion (Table I). This is similar to what has been reported for animal polymerases (Roeder and Rutter, 1969; Chambon, et al., 1970), but differs from divalent cation responses in higher plants (Strain, et al., 1971; Mondal, et al., 1972; Horgen and Key, 1973).

Template specificity studies indicated that the fungal polymerases varied in response to exogenously added DNA (Table 1). All three polymerases from Allomyces preferred denatured to native DNA and showed no preference for the homologous template (Cain and Nester, 1973). The multiple polymerases of Rhizopus also were more effective utilizing denatured templates (Gong and Van Etten, 1972). Synthetic poly d(A-T) was effective as a template for Rhizopus enzymes. In Physarum, denatured calf thymus DNA was twice as effective as denatured homologous DNA for both polymerase I and II, but native homologous DNA was more active by a factor of 7 than native calf DNA (Hildebrandt and Sauer, 1973). Denatured calf thymus DNA was the most effective template for both enzyme forms (Hildebrandt and Sauer, 1973; Gornicki et al., 1974). Poly d(A-T) was also a very effective template for the true slime mold polymerases. Dictyostelium polymerase I and II showed no differences from one another in the utilization of templates (Pong and Loomis, 1973). Both preferred homologous DNA and appeared to be quite effective on various native templates (Pong and Loomis, 1973). No denatured templates were examined. The Neurospora polymerase IIB (IV) preferred denatured calf thymus DNA. The other forms showed no preference (Tellez de Iñon, Leoni and Torres, 1974). Homologous template was not examined for its effectiveness with the Neurospora enzymes.

The most widely studied fungus is *S. cerevisiae*, therefore the reports on template specificity vary considerably. Ponta et al., (1971) found that all three polymerases were most effective with denatured calf thymus DNA. Furthermore, polymerases I and II showed very little activity with native yeast DNA, whereas polymerase III showed relatively high activity with the "natural template" (Ponta, et al., 1971).

Brogt and Planta (1972) found that native DNA (both calf thymus and yeast) was a better template for polymerase I than denatured DNA, whereas polymerase II and III exhibited the highest activity with a denatured template. All three enzymes showed lower activity with homologous template than with heterologous template (Brogt and Planta, 1972). Adman, et al., (1972) found that polymerase Ib and III were 2-3 times more active on native than on denatured DNA, whereas polymerase II preferred denatured template. All three enzymes utilized poly d(A-T) very effectively, but not preferentially to natural templates. Furthermore, it was demonstrated that template preference could be altered by changing the salt conditions (Adman, et al., 1972). Sebastian, et al., (1973) reported that native yeast RNA polymerase I and II are equally active with yeast native DNA, but that polymerase II was 4 times as active with denatured yeast DNA. In addition, they demonstrated that poly d(A-T) was more effective than any natural template. Dezélée, et al., (1972) reported that polymerase II of yeast was 10 times more effective on denatured templates than native. Interpretation of these apparently contradictory results is difficult without knowledge of the integrity of the DNA used (Hossenlopp, Outdet and Chambon, 1974; Meilhac and Chambon, 1973; Flint, de Pomerae, Chesterton and Butterworth, 1974). The differences reported may be the result of either template quality or the properties of the enzymes as isolated.

Molecular weight and subunit composition have been reported for some of the fungal polymerases. Generally, the molecular weight appears to be around 400-550,000 daltons for both polymerase I and II (Dezélée et al., 1972; Pong and Loomis, 1973; Ponta, et al., 1972; Adman, et al., 1972; Gornicki et al., 1974). The enzymes seem to have two large subunits and possibly three smaller subunits very similar to the RNA polymerase of *E. coli*. Furthermore, a "factor" stimulating transcription of all eukaryotic and prokaryotic polymerases examined was described from yeast (diMauro, et al., 1972). No one is certain what the holoenzyme is like in vivo, or if it even has been isolated in vitro.

An interesting question that has been posed concerning RNA polymerases is whether the levels of the enzymes or their activities change during growth and development. Horgen (1971) examined the zoospores of *B. emersonii* for the presence of multiple polymerases. The spores, which lack the ability to synthesize RNA in vivo (Murphy and Lovett, 1966), were found to possess all three multiple RNA polymerases (Horgen, 1971). Furthermore, no general or specific inhibitor was detected in the protoplasm of the spores (Horgen, 1971). RNA synthesis

in vivo may therefore be under some other kind of internal
restraint mechanism. Conversely, Gong and van Etten (1972)
examined a similar condition in spores of *R. stolonifer*. They
found that the dormant spores were deficient in RNA polymerase
II. RNA polymerase II was first observed between two and three
hours after germination began. Furthermore, they found that
polymerase I responded differently to divalent cations in un-
germinated spores than it did during the vegetative growth
stages and germination (Gong and van Etten, 1972). They concluded that both qualitative and quantitative changes in RNA
polymerase occur during the germination of Rhizopus spores.
More recently, it has been reported that no changes in the
multiple RNA polymerases occur during development of *D. dis-
coide m* (Pong and Loomis, 1973), or of *A. arbuscula* (Cain and
Nester, 1973). The authors felt that in these fungi, RNA
polymerase probably does not play a role in controlling development. However, in *S. cereusiae*, it was found that the level
of RNA polymerase I increased as the growth rate of the yeast
was increased (Sebastian, Mian and Halvorson, 1973). Gross
and Pogo (1974) report that the stringent response of yeast
spheroplasts to amino acid withdrawal did not cause a change
in extractable RNA polymerase activities.

Significant functional differences in the multiple forms
are suggested by their differential responses to divalent ions,
ammonium sulfate and template, and by their intranuclear localization. However, since template quality (Hossenlopp, Oudet
and Chambon, 1974; Meilhac and Chambon, 1973; Flint, de Pomerae,
Chesterton and Butterworth, 1974) and the presence of protein
factors (Stein and Hausen, 1970; diMauro, Hollenberg and Hall,
1972; Seifart, Juhasz and Benecke, 1973) and other poorly
understood variables may have large effects on in vitro activity, it is not clear how many of these differences represent
the complete enzyme as it functions in vivo and how many are
artifactual.

Other functional differences are suggested by the responses
to inhibitors. The fungal toxin, α-amanitin, was thought to
be a specific inhibitor of polymerase II, in vitro. However,
other forms of polymerase are now reported to be sensitive to
this toxin (Roeder, these proceedings; Benecke et. al., these proceedings). The *Streptomyces* antibiotic, cycloheximide, has also
been reported as an inhibitor for RNA polymerase I, but reports
from various investigators are inconsistent. Horgen and Griffin,
1971a; Timberlake, McDowell and Griffin, 1972; Timberlake,
Hagen and Griffin, 1972; Rudick and Weisman, 1973 report inhibition of polymerase I by cycloheximide whereas others (e.g. Cain
and Nester, 1973; Gross and Pogo, 1973; Gong and van Etten,
1972) report no inhibition.

Although it is generally accepted that cycloheximide is a specific inhibitor of eucaryotic protein synthesis in vivo, there have been several reports that it also inhibits RNA synthesis in vivo in animals, plants, and fungi with preferential inhibition of ribosomal RNA synthesis (Gottlieb and Shaw, 1967; Fiala and Davis, 1965; Fukuhara, 1965; deKloet, 1966; Willems, Penman and Penman, 1969; Muramatsu, Shimada, and Higashinakagawa, 1970). In general, the conclusion from these studies is that continued protein synthesis is required for the continued synthesis of ribosomal RNA in the nucleolus, and that the observed effect of cycloheximide on RNA synthesis is due to its direct effect on ribosomes. However, there are several reports of inhibition of ribosomal RNA synthesis in vivo which cannot be explained on the basis on protein synthesis inhibition (Farber and Farmar, 1973; Timerlake and Griffin, 1974a and 1974b; Benecke, Ferencz, and Seifart, these proceedings).

The results of studies on RNA polymerase from fungi and from animals indicate that while there may be some differences in the sensitivity of RNA polymerase I to cycloheximide, it is quite probable that the extraction, purification, and assay procedures are most important in explaining the differences. This is particularly true if interactions between RNA polymerase and other cellular constituents are involved in the inhibition, since minor differences in procedures could result in the differential loss of such factors.

Various extraction procedures have been used for isolating RNA polymerase I. Enzyme extracted from nuclear pellets of Blastocladiella by low salt (0.01 M ammonium sulfate) followed by high salt (0.15 M) at 37° C yielded RNA polymerase I sensitive to 10^{-4} M cycloheximide (Horgen and Griffin, 1971a). Enzyme extracted by 0.3 M ammonium sulfate with sonication from nuclear fragments of Acanthamoeba also yielded enzyme sensitive to 10^{-4} M cycloheximide (Rudick and Weisman, 1973). Use of the bile salt detergent, deoxycholate, for solubilization resulted in enzyme from Achlya sensitive to 10^{-6} M cycloheximide and enzyme from rat liver sensitive to 10^{-9} M cycloheximide (Timberlake, McDowell, and Griffin, 1972; Timberlake, Hagen and Griffin, 1972).

We can now say that the solubilization method is very important in determining whether the enzyme is sensitive to cycloheximide. Other properties of the enzyme are also drastically affected by the extraction procedure. Solubilization of the enzyme from Achlya nuclei by the Roeder and Rutter technique (Cheney, unpublished), sonication in 0.3 M ammonium sulfate, and by extraction with 0.3 M ammonium sulfate without sonication, yielded enzyme that was insensitive to cycloheximide.

Enzyme preparations which showed sensitivity to cycloheximide in our hands had two outstanding characteristics: (1) a short time course, linear for one to two minutes and (2) enzyme dilution curves which suggest the presence of an inhibitor in the preparation (Timberlake, McDowell and Griffin, 1972; Timberlake, Hagen and Griffin, 1972). Enzyme extracted from nuclear pellets by 0.3 M ammonium sulfate with or without sonication exhibited time courses linear for 20 minutes or more and had linear dilution curves (Cheney, unpublished).

Many artifacts may be generated in the course of isolating an enzyme as complex as RNA polymerase. As was pointed out by Jacob (1973), the solubilization techniques used are quite harsh, with sonication being especially vigorous. We have had considerable difficulty in purifying RNA polymerase from Achlya, since further chromatography of the DEAE fraction of both the deoxycholate solubilized enzyme and the sonicated enzyme resulted in multiple activity peaks that were very unstable (Timberlake, 1972). Enzyme solubilized by 0.3 M ammonium sulfate without sonication has been purified nearly to homogeneity as shown by acrylamide gel electrophoresis, Fig. 2 (Cheney, unpublished). This enzyme rechromatographs on DEAE sephadex as a single peak, Fig. 3B, but when sonicated as in the Roeder and Rutter (1970a) solubilization technique, this enzyme yields multiple activity peaks that are unstable, Fig. 3A (Cheney, unpublished). Furthermore, these unstable peaks of activity are three to four times as active initially as the unsonicated activity. It is apparent that sonication illicits some significant changes in the enzyme activities. The nature of these changes caused by sonication and other solubilization procedures are not known. Perhaps they alter the structure of the RNA polymerases. It is not known what effects sonication and other solubilization procedures have on the structure of the RNA polymerases.

Since the short time course and the dilution properties of the deoxycholate solubilized enzyme seems artifactual, we have attempted to determine whether the cycloheximide sensitivity is also an artifact. Our approach to this problem was to examine the effects of several inhibitors of protein synthesis and a variety of glutarimide antibiotics on both protein synthesis and ribosomal RNA synthesis in vivo (Timberlake and Griffin, 1974a and b). The rationale is that if the effect of cycloheximide on ribosomal RNA synthesis is indirect, the result of inhibiting the synthesis of some rapidly turning-over moiety of RNA polymerase (cf. Muramatsu, Shimada and Higashinakagawa, 1970), then inhibition of protein synthesis by other drugs should show comparable inhibition.

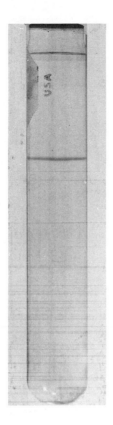

Fig. 2. Polyacrylamide gel electrophoresis of RNA polymerase
I isolated from *Achlya bisexualis* (Cheney, unpublished).
Electrophoresis of 57 ug of protein according to Brewer and
Ashworth (1969) in 5% polyacrylamide gel. The top 5 mm of the
gel is the stacking gel.

Three inhibitors, puromycin, blasticidin-S and DL-p-
fluorophenylalanine were used at the lowest concentration which
gave maximal inhibition of growth (Timberlake, 1974b). Puromycin and blasticidin-S caused rapid and complete inhibition
of the accumulation of ^{3}H-proline into proteins, as did cycloheximide (Timberlake and Griffin, 1974b). While DL-p-
fluorophenylalanine did not inhibit the accumulation of label
into protein, it did inhibit the accumulation of alkaline
phosphatase activity (Timberlake and Griffin, 1973). None
of the drugs affected the uptake of uridine into the cells.
Cycloheximide caused a complete cessation of the accumulation
of uridine into ribosomal RNA within 2.5 minutes, whereas the
other inhibitors caused a delayed and gradual inhibition over

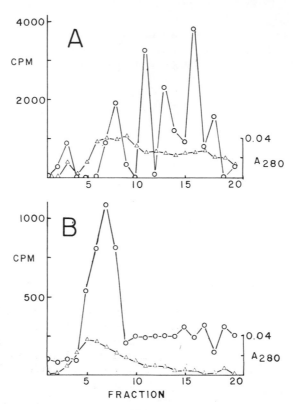

Fig. 3. DEAE-Sephadex rechromatography of RNA polymerase I
from *Achlya bisexualis* (Cheney, unpublished). Two equal
aliquots of the enzyme preparation shown in Fig. 2 were
rechromatographed on DEAE-Sephadex with a linear ammonium
sulfate gradient (0.1 M to 0.3 M) after sonication (Roeder
and Rutter, 1969), A; and with no further treatment, B.

30 to 40 minutes. In the presence of each of the inhibitors
except cycloheximide, the capacity of the cells to accumulate
label into ribosomal RNA decayed rapidly following pseudo-first
order kinetics with half-lives between 10 and 23 minutes (Timberlake and Griffin, 1974b). In the presence of cycloheximide
this capacity decays with a half-life of 1.4 minutes. Such
rapid inhibition in contrast to the slower inhibition by the
other drugs could be explained in terms of a direct inhibition
of the RNA polymerase by cyclohexmide. (Benecke et. al. These proceedings) reports that in vivo treatment with
does not effect the amount of RNA polymerase I activity extractable from rat liver.

Comparison of nine different glutarimide analogs of

cycloheximide also allowed us to separate the effect of protein synthesis inhibition from the effect on RNA synthesis (Timberlake and Griffin, 1974a). Of particular interest were the effects of three glutarimides modified on the hydroxyl: anhydrocycloheximide, dehydrocycloheximide and cycloheximide acetate. These affected the accumulation of labelled proline into protein variously, having respectively: no effect, delayed and incomplete inhibition, and rapid and complete inhibition. However, all three caused a stimulation of incorporation of uridine into RNA, Fig. 4 (Timberlake and Griffin, 1974a). In addition, streptovitacin A and streptimidone both inhibited accumulation of label into protein while having no effect on uridine incorporation into RNA. Both of these are modified on the cyclohexanone ring.

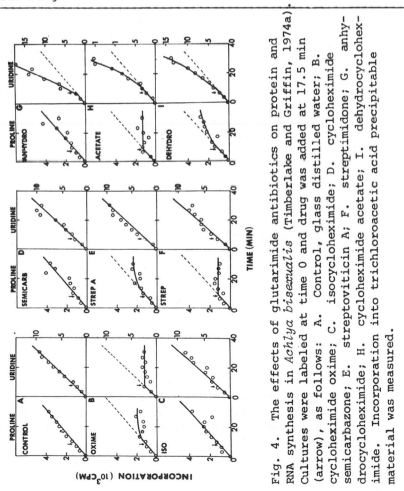

Fig. 4. The effects of glutarimide antibiotics on protein and RNA synthesis in *Achlya bisexualis* (Timberlake and Griffin, 1974a). Cultures were labeled at time 0 and drug was added at 17.5 min (arrow), as follows: A. Control, glass distilled water; B. cycloheximide oxime; C. isocycloheximide; D. cycloheximide semicarbazone; E. streptoviticin A; F. streptimidone; G. anhydrocycloheximide; H. cycloheximide acetate; I. dehydrocyclohex-imide. Incorporation into trichloroacetic acid precipitable material was measured.

These experiments indicate that ribosomal RNA synthesis requires continued protein synthesis, but that cycloheximide has a direct effect on RNA polymerase I activity in addition to its effect on the ribosomes. A model for the control of RNA polymerase I activity which accounts for these observations, and for the observation that certain cycloheximide analogs apparently stimulate RNA synthesis regardless of their effect on protein synthesis has been proposed (Timberlake and Griffin, 1974a). This model, Fig. 5, supposes that there is a compound A* required for activity of RNA polymerase I. The cellular concentration of this compound is dependent on its rate of synthesis and degradation. The synthetic enzyme is rapidly turning over, so that continued protein synthesis is required to maintain A*. The glutarimide antibiotics are structurally similar to A*, and depending on the details of their structure, may interact with RNA polymerase in such a way to replace A* and either inhibit or activate the enzyme. The site of interaction on the polymerase may be a loosely bound subunit which is easily lost or altered during purification. This model would also explain the release from stringent control of RNA synthesis in yeast caused by cycloheximide (Gross and Pogo, 1974). While many details of this model require experimental confirmation, it does explain our results, whereas the less complicated model of a rapidly turning over subunit of RNA polymerase does not.

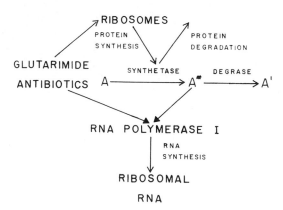

Fig. 5. Proposed model for the effect of cycloheximide and related glutarimide antibiotics on RNA synthesis in Achlya (redrawn from Timberlake and Griffin, 1974a). For explanation see text.

ACKNOWLEDGEMENT

This work has been supported by grants from the Brown-Hazen Fund of the Research Corporation, the SUNY Research Foundation and Sigma Xi.

REFERENCES

Adman, R., L. D. Schultz and B. D. Hall 1972. Transcription in yeast: separation and properties of multiple RNA polymerases. *Proc. Nat. Acad. Sci. U.S.A.* 69: 1702-1706.

Benecke, B. J., A. Ferencz, and K. H. Seifart 1974. DNA-dependent RNA polymerase species from rat liver tissue. These Proceedings.

Bottomley, W., H. J. Smith and L. Bogorad 1971. RNA polymerases of maize: partial purification and properties of the chloroplast enzyme. *Proc. Nat. Acad. Sci. U.S.A.* 68: 2412-2416.

Brewer, J. M. and R. B. Ashworth 1969. Disc electrophoresis. *J. Chem. Ed.* 46: 41-45.

Brogt, T. M. and R. J. Planta 1972. Characteristics of DNA-dependent RNA polymerase activity from isolated yeast nuclei. *F.E.B.S. Letters* 20: 47-52.

Cain, A. K. and E. W. Nester 1973. Ribonucleic acid polymerase in *Allomyces arbuscula*. *J. Bacteriol.* 115: 769-776.

Chambon, P., F. Gissinger, J. L. Mandel, Jr., C. Kedinger, M. Gniazdowski and M. Meihlac 1970. Purification and properties of calf thymus DNA-dependent RNA polymerase A and B. *Cold Spring Harbor Symposium on Quant. Biol.* 35: 693-707.

deKloet, S. R. 1966. Ribonucleic acid synthesis in yeast. The effect of cycloheximide on the synthesis of ribonucleic acid in *Saccharomyces carlsbergensis*. *Biochem. J.* 99: 566-581.

Dezélée, S., A. Sentenac and P. Fromageot 1972. Role of DNA-RNA hybrids in eukaryotes I. purification of yeast RNA polymerase B. *F.E.B.S. Letters* 21: 1-6.

Dezélée, S. and A. Sentenac 1973. Role of DNA-RNA hybrids in eukaryotes. Purification and properties of yeast RNA polymerase B. *Eur. J. Biochem.* 34: 41-52.

diMauro, E., C. P. Hollenberg and B. D. Hall 1972. Transcription in yeast: a factor that stimulates yeast RNA polymerases. *Proc. Nat. Acad. Sci. U.S.A.* 69: 2818-2822.

Farber, J. L. and R. Farmar 1973. Differential effects of cycloheximide on protein synthesis as a function of dose. *Biochem. Biophys. Res. Commun.* 51: 626-630.

Fiala, E. S. and F. F. Davis 1965. Preferential inhibition of synthesis and methylation of ribosomal RNA in *Neurospora crassa* by actidione. *Biochem. Biophys. Res. Commun.* 18: 115-118.

Flint, S. J., D. I. dePomerae, C. J. Chesterton, and P. H. Butterworth 1974. Template specificity of eucaryotic DNA-dependent RNA polymerases. Effect of DNA structure and integrity. *Eur. J. Biochem.* 42: 567-579.

Fukuhara, H. 1965. RNA synthesis of yeast in the presence of cycloheximide. *Biochem. Biophys. Res. Commun.* 18: 297-301.

Gong, C. S. and J. L. van Etten 1972. Changes in soluble ribonucleic acid polymerases associated with the germination of *Rhizopus stolonifer* spores. *Biochim. Biophys. Acta.* 272: 44-52.

Gornicki, S. Z., S. B. Vuturo, T. V. West, and R. F. Weaver 1974. Purification and properties of deoxyribonucleic acid-dependent ribonucleic acid polymerases from the slime mold *Physarum polycephalum*. *J. Biol. Chem.* 249: 1792-1798.

Gottlieb, D. and P. D. Shaw 1967. *Antibiotics*, Springer-Verlag, New York.

Gross, K. J. and A. O. Pogo 1974. Control mechanism of ribonucleic acid synthesis in eukaryotes. *J. Biol. Chem.* 249: 568-576.

Hildebrandt, A. and H. W. Sauer 1973. DNA-dependent RNA polymerases from *Physarum polycephalum*. *F.E.B.S. Letters* 35: 41-44.

Hollenberg, C. P. 1973. Ribosomal ribonucleic acid synthesis by isolated yeast ribonucleic acid polymerases. *Biochem.* 12: 5320-5325.

Horgen, P. A. 1971. In vitro ribonucleic acid synthesis in the zoospores of the aquatic fungus *Blastocladiella emersonii*. *J. Bacteriol.* 106: 281-282.

Horgen, P. A. and D. H. Griffin 1971a. Specific inhibitors of the three RNA polymerases from the aquatic fungus *Blastocladiella emersonii*. *Proc. Nat. Acad. Sci. U.S.A.* 68: 338-341.

Horgen, P.A. and D. H. Griffin 1971b. RNA polymerase III of *Blastocladiella emersonii* is mitochondrial. *Nature New Biol.* 234: 17-18.

Horgen, P.A. 1972. The DNA-dependent RNA synthesizing enzymes of the aquatic fungus *Blastocladiella emersonii*. Ph.D. Thesis. S.U.N.Y. College of Forestry, Syracuse, N.Y.

Horgen, P. A. and J. L. Key 1973. The DNA-directed RNA polymerases of soybean. *Biochim. Biophys. Acta* 294: 227-235.

Hossenlopp, P., P. Oudet, and P. Chambon 1974. Animal DNA-dependent RNA polymerases. Studies on the binding of mammalian RNA polymerases AI and B to Simian Virus 40 DNA. *Eur. J. Biochem.* 41: 397-411.

Jacob, S. T., E. Sajdel and H. N. Munro 1970. Different responses of soluble whole nuclear RNA polymerase and soluble nucleolar RNA polymerase to divalent cations and to inhibition by α-amanitin. *Biochem. Biophys. Res. Commun.* 38: 765-770.

Jacob, S. T. 1973. Mammalian RNA polymerases. *Progr. Nucl. Acid Res. Mol. Biol.* 13: 93-126.

Kedinger, C., M. Gniazdowski, J. L. Mandel, Jr., F. Gissinger and P. Chambon 1970. α-amanitin: a specific inhibitor of one of two DNA-dependent RNA polymerase activities from calf thymus. *Biochem. Biophys. Res. Commun.* 38: 165-171.

Küntzel, H. and K. P. Schäfer 1971. Mitochondrial RNA polymerase from *Neurospora crassa*. *Nature New Biol.* 231: 265-269.

Lindell, T. J., F. Weinberg, P. W. Morris, R. G. Roeder and W. J. Rutter 1970. Specific inhibition of nuclear RNA polymerase II by α-amanitin. *Science* 170: 447-449.

Meilhac, M. and P. Chambon 1973. Animal DNA-dependent RNA polymerases. Initiation sites on calf-thymus DNA. *Eur. J. Biochem.* 35: 454-463.

Mondal, H. R., K. Mandal and B. B. Biswas 1970. Factors and rifampicin influencing RNA polymerase isolated from chromatin of eukaryotic cell. *Biochem. Biophys. Res. Commun.* 40: 1194-1200.

Muramatsu, M., N. Shimada and T. Higashinakagawa 1970. Effect of cycloheximide on the nucleolar RNA synthesis in rat liver. *J. Mol. Biol.* 53: 91-106.

Murphy, S., M. N. and J. S. Lovett 1966. RNA and protein synthesis during zoospore differentiation in synchronized cultures of *Blastocladiella*. *Dev. Biol.* 14: 68-95.

Pong, S. S., and W. L. Loomis, Jr. 1973. Multiple nuclear ribonucleic acid polymerases during development of *Dictyostelium discoideum*. *J. Biol. Chem.* 248: 3933-3939.

Ponta, H., U. Ponta, and E. Wintersberger 1971. DNA-dependent RNA polymerases from yeast. Partial characterization of three nuclear enzyme activities. *F.E.B.S. Letters* 18: 204-208.

Ponta, H., U. Ponta, E. Wintersberger 1972. Purification and properties of DNA-dependent RNA polymerases from yeast. *Eur. J. Biochem.* 29:110-118.

Reid, B. D. and P. Parsons 1971. Partial purification of mitochondrial RNA polymerase from rat liver. *Proc. Nat. Acad. Sci. U.S.A.* 68: 2830-2834.

Roeder, R. G. and W. J. Rutter 1969. Multiple forms of DNA-dependent RNA polymerase in eukaryotic organisms. *Nature*

224: 234-237.

Roeder, R. G. and W. J. Rutter 1970a. Specific nucleolar and nucleoplasmic RNA polymerases. *Proc. Nat. Acad. Sci. U.S.A.* 65: 675-682.

Roeder, R. G. and W. J. Rutter 1970b. Multiple ribonucleic acid polymerases and ribonucleic acid synthesis during sea urchin development. *Biochem.* 9: 2543-2553.

Roeder, R. G., S. Chou, J. A. Jaehning, L. B. Schwartz, V. E. F. Sklar, and R. Weinmann 1974. Structure, function, and regulation of RNA polymerases in animal cells. These Proceedings, Vol. III.

Rudick, V. L. and R. A. Weisman 1973. DNA-dependent RNA polymerase from trophozoites and cysts of *Acanthamoeba castellanii*. *Biochim. Biophys. Acta* 299: 91-102.

Scragg, A. H. 1971. Mitochondrial DNA-directed RNA polymerase from *Saccharomyces cerevisae* mitochondria. *Biochem. Biophys. Res. Commun.* 45: 701-706.

Sebastian, J., M. M. Bhargava and H. O. Halvorson 1973. Nuclear deoxyribonucleic acid-dependent ribonucleic acid polymerases from *Saccharomyces cerevisiae*. *J. Bacteriol.* 114: 1-6.

Sebastian, J., F. Mian, and H. Halvorson 1973. Effect of the growth rate on the level of the DNA-dependent RNA polymerases in *Saccharomyces cerevisiae*. *F.E.B.S. Letters* 34: 159-162.

Seifart, K. H., P. P. Juhasz, and B. J. Benecke 1973. A protein factor from rat-liver tissue enhancing the transcription of native templates by homologous RNA polymerase B. *Eur. J. Biochem.* 33: 181-191.

Stein, H. and P. Hausen 1970. A factor from calf thymus stimulating DNA-dependent RNA polymerase isolated from this tissue. *Eur. J. Biochem.* 14: 270-277.

Strain, G. C., K. P. Mullinix and L. Bogorad 1971. RNA polymerase of maize: Nuclear RNA polymerases. *Proc. Nat. Acad. Sci. U.S.A.* 68: 2647-2651.

Tellez de Iñon, M. T., P. D. Leoni and H. N. Torres 1974. RNA polymerase activities in *Neurospora crassa*. *F.E.B.S. Letters* 39: 91-95.

Timberlake, W. E. 1972. Characterization of the DNA-dependent RNA polymerase I of *Achlya bisexualis*. M.S. Thesis. State Univ. College of Environmental Science and Forestry, Syracuse, N. Y.

Timberlake, W. E., G. Hagen, and D. H. Griffin 1972. Rat liver DNA-dependent RNA polymerase I is inhibited by cycloheximide. *Biochem. Biophys. Res. Commun.* 48: 823-827.

Timberlake, W. E., L. McDowell and D. H. Griffin 1972.

Cyclohexemide inhibition of the DNA-dependent RNA polymerase I of *Achlya bisexualis*. *Biochem. Biophys. Res. Commun.* 46: 942-947.

Timberlake, W. E. 1973. The direct inhibition of ribosomal RNA synthesis by cyclohexemide. Ph.D. Thesis. SUNY College of Environmental Science and Forestry. Syracuse, New York.

Timberlake, W. E. and D. H. Griffin 1974a. Differential effects of analogs of cyclohexemide on protein and RNA synthesis in *Achlya*. *Biochim. Biophys. Acta* 349: 39-46.

Timberlake, W. E. and D. H. Griffin 1974b. Differential effects of cyclohexemide and other inhibitors of protein synthesis on in vivo r RNA synthesis in *Achlya bisexualis*. *Biochim. Biophys. Acta.* In Press.

Tocchini-Valentini, G. P. and M. Crippa 1970. Ribosomal RNA synthesis and RNA polymerase. *Nature* 228: 993-995.

Tsai, M. J., G. Michaelis and R. S. Criddle 1971. DNA-dependent RNA polymerase from yeast mitochondria. *Proc. Nat. Acad. Sci. U.S.A.* 68: 473-477.

Widnell, C. C. 1965. Characterization of the product of the RNA polymerase of isolated rat-liver nuclei. *Biochem. J.* 95: 42P-43P.

Widnell, C. C. and J. R. Tata 1964. Evidence for two DNA-dependent RNA polymerase activities in isolated rat-liver nuclei. *Biochim. Biophys. Acta.* 87: 531-533.

Willems, M., M. Penman and S. Penman 1969. The regulation of RNA synthesis and processing in the nucleolus during inhibition of protein synthesis. *J. Cell. Biol.* 41: 177-187.

RNA POLYMERASE II OF *Escherichia coli*

LEE CHAO AND JOSEPH F. SPEYER
Genetics and Cell Biology Section
Biological Sciences Group
The University of Connecticut
Storrs, Connecticut 06268

ABSTRACT. A second form of the DNA-dependent RNA polymerase has been isolated and partially characterized. This new form, RNA polymerase II, like RNA polymerase I, contains the α and β subunits and the σ factor. The α subunits of polymerase II are slightly smaller than those of polymerase I. Another difference is that the holo or core enzyme of polymerase II has an extra polypeptide of about 50,000 daltons which is absent in polymerase I. RNA polymerase II (like I) is sensitive to rifampicin, but resistant when isolated from rifampicin resistant cells. The template and metal ion cofactor requirements of the two forms of this enzyme are different. RNA polymerase II uses Mn^{++} ion much better than Mg^{++} ion as a cofactor, and poly dAT is by far the best template for this enzyme. Polymerase II is almost inactive in transcribing natural DNA templates. However, it apparently binds to some of those templates because it has a repressor-like in vitro effect. It inhibits the transcription of T4 DNA, but not T7 DNA, by RNA polymerase I. This might reflect an in vivo regulatory role on transcription by RNA polymerase II. It occurs in late log-stationary phase cells, but apparently not in actively growing cells.

INTRODUCTION

This work on *E. coli* RNA polymerase was inspired by the work of Losick and Sonnenschein on the change from vegetative growth to sporulation in *B. subtilis*. The transition in *B. subtilis* growth pattern was attributed to changes in RNA synthesis resulting from structural alterations in the β subunit of the vegetative RNA polymerase (Losick and Sonnenschein, 1969; Losick et al, 1970). However, this structural alteration, the cleavage of the subunit and loss of σ factor in the vegetative *B. subtilis* RNA polymerase, is due to a protease that is associated with sporulation. These alterations occurred during the in vitro extraction of the enzyme (Linn et al, 1973; Leighton et al, 1973), and are thus an in vitro effect, rather than a cause, of sporulation. Recently, Holland and Whiteley (1973) have isolated a new form of RNA polymerase associated

89

with the late logarithmic, vegetative growth of *B. subtilis*. This new form contains an extra polypeptide of 60,000 daltons, which is present in addition to the other subunits of the normal vegetative enzyme and it does not seem to have resulted from cleavage of other subunits. The occurrence of the new RNA polymerase correlates with shifts in growth prior to sporulation.

The growth of *E. coli* cells also has various stages. The transition from logarithmic to stationary phase of growth is interesting. These two growth phases are different in many ways; for example, bacteriophages T4 and T7 both form plaques on logarithmically growing cells, yet only T7 but not T4 phage forms plaques on stationary cells. We explored the possibility that the RNA polymerase isolated from these two growth stages might be different in template specificity and in other ways. We found a new form of RNA polymerase in late logarithmic and stationary phase cells and have called this RNA polymerase II (Chao and Speyer, 1973). Both forms of the RNA polymerase, with more polymerase I than polymerase II, are present in stationary phase cells. In log phase cells there is apparently little or no polymerase II. Although RNA polymerases I and II share some subunits and properties, they differ in template specificity, metal ion requirements, and subunit structure. Polymerase II has an extra polypeptide and slightly smaller α subunits. In in vitro experiments polymerase II can inhibit polymerase I by acting like a repressor.

Recently there have also been other reports on the heterogeneity of *E. coli* RNA polymerase. Travers and Buckland (1973) observed some functional heterogeneity between RNA polymerases isolated from logarithmic and stationary phase cells. Other reports (Fukuda et al, 1974; Iwakura et al, 1974) have shown the existence of a new sigma factor in *E. coli* RNA polymerase.

PURIFICATION OF RNA POLYMERASE II

E. coli W3110, ton$^-$, was grown at 37°C with aeration in 16 liter M9 minimal medium (Champe and Benzer, 1962). Cell density was measured periodically at 550 nm until the growth was completely out of logarithmic phase. The absorbance of the culture at this point was between 5 and 7. Cells were then harvested using a continuous-flow refrigerated centrifuge. All subsequent operations were carried out at 4°C unless specified otherwise.

Harvested cells were washed once in Tris-EDTA buffer (0.01 M Tris-HCl, pH 7.7, 0.001 M EDTA) and resuspended in the same

buffer at a ratio of one gram wet cells per 4 ml Tris-EDTA
buffer. Lysozyme was added to the cell suspension at 0.2 mg
per ml and the mixture allowed to stand 30 min at 4°C. The
lysozyme treated cell suspension was frozen and thawed repeatedly (2-3X) to open the cells. The viscous lysate had a protein concentration of about 25 mg per ml. The lysate was
adjusted to 0.01 M magnesium acetate, 0.001 M dithiothreitol
and treated with pancreatic deoxyribonuclease (2 μg/ml)
for 1 hour to reduce the viscosity. Cell debris were removed
by 1 hour centrifugation at 27,000 x g. Ribosomes were partially removed by 90 min centrifugation at 100,000 x g. The

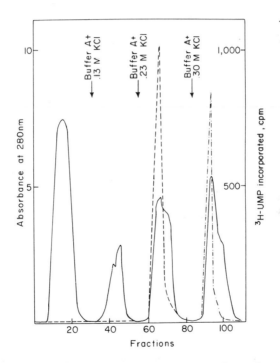

Fig. 1. DEAE-cellulose column chromatography of RNA polymerases. Polymerase I eluted at 0.23 M KCl was assayed according
tp Burgess (1969). Polymerase II, eluted at 0.3 M KCl, was
assayed with poly dAT as the template and MnCl$_2$ as cofactor.
————, absorbance at 280 nm; - - - -, polymerase I activity;
-•-•-, polymerase II activity.

91

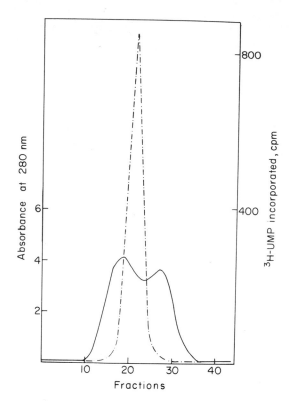

Fig. 2. Chromatography of polymerase II on Sepharose 6B column. ——, absorbance at 280 nm; –•–•–, enzyme activity.

proteins precipitated between 33% and 50% saturation with ammonium sulfate at pH 7.7 contain the RNA polymerase activity. The precipitate was dissolved in buffer A (Burgess, 1969) (0.01 M Tris-HCl, pH 7.7, 0.01 M magnesium acetate, 0.1 mM EDTA and 0.1 mM dithiothreitol, 5% glycerol) and adjusted to 12 mg/ml in protein concentration. The volume is about 60 ml.

The sample was chromatographed on a DEAE-cellulose column (2.5 x 20 cm) by step-wise elution with KCl in buffer A. Polymerase I and polymerase II were eluted from the column at 0.23 M KCl and 0.3 M KCl, respectively (Fig. 1). Fractions containing polymerase II activity were precipitated with 55% ammonium sulfate and dissolved in minimum amount of buffer B (0.01 M Tris-HCl, pH 7.7, 0.01 M magnesium acetate, 0.1 mM EDTA, 0.1 mM dithiothreitol, 5% glycerol, 0.3 M KCl) and 10% sucrose. This sample was chromatographed on a Sepharose 6B column

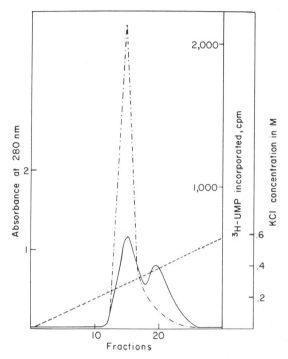

Fig. 3. DEAE-cellulose column chromatography of polymerase II
with a continuous KCl gradient. ———, absorbance 280 nm;
-•-•-, enzyme activity; - - - -, KCl concentration.

(3 x 38 cm) in buffer B (Fig. 2). Fractions containing polymerase II activity were diluted 3 fold with buffer A and
chromatographed on a small DEAE-cellulose column (1.6 x 12 cm)
with a continuous gradient of KCl in buffer A (Fig. 3). Polymerase II holoenzyme was further purified by chromatography
on Sephadex G-150 in buffer B (high salt), the activity is
largely in the excluded volume. This is followed by chromatography on Sepharose 6B in low salt. Detailed description
of these procedures will be published elsewhere. To obtain
polymerase II core enzyme, polymerase II was precipitated with
55% ammonium sulfate and dissolved in sufficient volume of

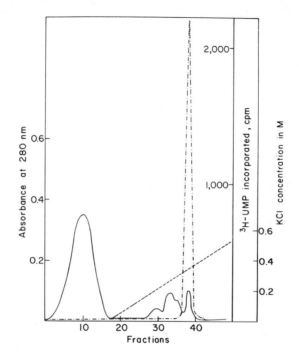

Fig. 4. Phosphocellulose column chromatography of polymerase II. ———, absorbance at 280 nm; —•—•—, enzyme activity; - - - -, KCl concentration.

buffer C (0.05 M Tris-HCl, pH 7.7, 0.1 mM EDTA, 0.1 mM dithio-threitol, 5% glycerol) to give a final salt concentration equivalent to 0.01 M KCl as measured by a conductivity bridge. This sample was chromatographed on a phosphocellulose column (1.6 x 8 cm) with a KCl gradient in buffer C (Fig. 4).

PROPERTIES OF POLYMERASE II

The holoenzyme of polymerase I and polymerase II are different in their template specificities (Table I). Polymerase II is relatively inactive with all natural templates tested. It is active in poly dAT directed poly rAU synthesis, but has reduced activity in poly dGC directed poly rGC synthesis. Polymerase II requires manganese as a cofactor (Chao and Speyer, 1973). In the case of polymerase I, both magnesium and manganese stimulate the enzyme activity although magnesium

TABLE I

TEMPLATE SPECIFICITY OF POLYMERASE I
AND POLYMERASE II

Templates	Incorporation of ^{3}H-UMP in 10 min at 37°C (pmole)	
	Polymerase I	Polymerase II
Poly dAT (2.5 µg)	636	612
Poly dGC (2.5 µg)	114	39
Calf thymus DNA (2 µg)	326	85
T4 DNA (3 µg)	510	46
T7 DNA (2.5 µg)	641	35
E. coli DNA (2.5 µg)	125	33

Polymerase I (3 µg) was assayed according to Burgess (1969).
Polymeras II (3 µg) was assayed similarly except that Mn^{++}
was used instead of Mg^{++}

is required for accurate transcription.

The inactivity of polymerase II is not due to a lack of
active sigma factor. We tested polymerase II for stimulation
of transcription of natural DNA templates by polymerase I core
enzyme. The results (Table II) show that polymerase II indeed
contains active sigma factor and can donate it to polymerase
I core enzyme.

We have shown previously that both logarithmic and stationary phase cells can support the productive infection by T7
while only logarithmic phase cells can support T4 growth.
This observation is correlated with the appearance of polymerase II in late logarithmic and stationary phase cells (unpublished results). Furthermore, polymerase II is able to inhibit the transcription of T4 and E. coli, but not T7 DNA by polymerase I (Chao and Speyer, 1973). This inhibition is observed
only if polymerase II is first incubated with limiting amount
of DNA before adding polymerase I. These results may be
interpreted as due to binding of polymerase II to certain promoters on the T4 DNA which are thus blocked for transcription
by polymerase I. The inhibition is about 50% using equal
amounts of polymerase II and polymerase I. This weak inhibition could be significant in terms of regulating RNA synthesis if the inhibition is specific for certain key promoters.

The difference between core enzymes of polymerase I and
polymerase II was examined by SDS polyacrylamide gel electrophoresis. RNA polymerase II purified by phosphocellulose
chromatography was precipitated with 10% trichloroacetic acid.
The precipitate was pelleted by centrifugation and washed
twice in acetone. The pellet was dissolved in electrophoresis

TABLE II
COMPLEMENTATION OF POLYMERASE I
CORE ENZYME BY POLYMERASE II

Polymerase II	Polymerase I	T4 DNA	T7 DNA	Incorporation of ^{3}H-UMP (pmole)
3 µg	---	3 µg		46
---	2.5 µg	3 µg		26
3 µg	2.5 µg	3 µg		240
3 µg	---		2.5 µg	35
---	2.5 µg		2.5 µg	44
3 µg	2.5 µg		2.5 µg	932

Enzyme activity was assayed at 37°C for 10 min according to
Burgess procedure (1969).

sample buffer (0.01 M sodium phosphate, pH 7.1, 1% sodium
dodecyl sulfate, 0.1% 2-mercaptoethanol and 10% sucrose) and
boiled for 2 min in a water bath. Polymerase I core enzyme
was purified (Burgess, 1969) and treated in the same manner.
SDS-polyacrylamide gel electrophoresis was carried out in
sodium phosphate buffer system at pH 7.1 (Shapiro et al, 1967).
The result (Fig. 5) shows that the core enzymes of polymerase
I and polymerase II are different. The α subunit of polymerase
II moves further toward the anode than the normal α subunit.
There is a new polypeptide of 50,000 daltons present in the
core enzyme of polymerase II (Fig. 5, arrow). This factor,
unlike the sigma, tau or omega factors, is not separable from
the polymerase II core enzyme by phosphocellulose chromatography. This polypeptide is absent in all polymerase I
preparations.

DISCUSSION

At present only the following is clear: There exists a new
form of RNA polymerase in stationary *E. coli*, polymerase II,
which is different in structure and activity from the standard
RNA polymerase. We know that the β subunit of both polymerases is coded by the same gene, and that the subunit composition of both is similar, with clear differences between them.
We do not know whether there is a precursor: product relationship between the two polymerases subunits. This is a
possibility since the α subunit of polymerase II is smaller
than that of polymerase I. We do not know the role of this
subunit or that of the extra polypeptide that characterizes
polymerase II.

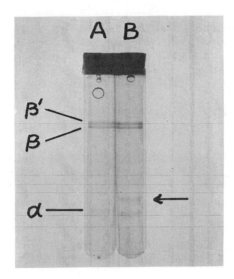

Fig. 5. SDS-polyacrylamide gel electrophoresis of polymerase
I and polymerase II core enzymes. The system, 5% gel and 0.1%
SDS in phosphate buffer, was according to Shapiro et al (1967).
Protein migration was toward the anode (bottom). Proteins
were stained with Coomassie brilliant blue R-250. Sample
size was 8 µg each. A. Polymerase I; B. polymerase II.

Perhaps the most interesting question is the function of
polymerase II. Since we do not know the answer what follows
is speculation. As an RNA polymerase, polymerase II is apparently almost inactive, except with poly dAT and Mn^{++} ion. We
may postulate a regulatory role, especially as polymerase II
seems associated with the termination of logarithmic growth.
Polymerase II might be a modified polymerase I designed to
transcribe those genes needed in stationary phase.

Alternatively, polymerase II may not be a transcriptase at
all, but function as a generalized in vivo repressor. This
would be a novel role for an RNA polymerase. It would mean
that the in vitro activity of polymerase II with poly dAT
is only a convenient assay and not indicative of transcription
in vivo. There is some evidence that suggests that such a
repressor-like role is not an unreasonable idea.

As reported here, polymerase II inhibits the in vitro RNA
synthesis of polymerase I. This occurs when the template,
T4 DNA, is limiting and the polymerases are in excess. This
resembles in vivo transcription more than the usual assay
conditions in which the template is in excess; with T4 DNA

in excess polymerase II stimulates transcription by core
polymerase I because it supplies the σ factor. In the cell
there is an excess of RNA polymerase molecules compared to
operons. This is evident from a comparison of the rates of
RNA synthesis with the rate of chain elongation, and many
RNA molecules can be seen simultaneously transcribing a gene
in electron microscope pictures (Miller et al, 1970). Template
competition leading to inhibition has been shown in vitro
between T7 and T3 RNA polymerases (Dunne and Bautz, 1971).
These two polymerases have quite different template specificities: each prefers its own T3 or T7 DNA template. The T3
enzyme will bind to T7 DNA without transcribing it. Dunne
and Bautz suggest that this non-productive binding is at least
partly responsible for the mutual exclusion of T3 and T7 phage
in coinfections.

Other evidence of competition for promoters by RNA polymerases is the behavior of *E. coli* that are merodiploid for
rifampicin resistance and sensitivity. The sensitive allele
is dominant (Austin and Scaife, 1970). This is explained **by the**
blocking of promoter sites by the rifampicin inactivated RNA
polymerase. These sites are then unavailable for transcription by the resistant enzyme. It has been shown that in
vitro rifampicin inactivated polymerase still binds to DNA
(Hinkel et al, 1972). In vitro the RNA synthesis by rifampicin
resistant polymerase is inhibited by the sensitive RNA polymerase with rifampicin, providing that the DNA template is
limiting (Ilyana et al, 1972). Those conditions are analogous
to the ones we have used for the in vitro "repression" of
polymerase I by polymerase II. This may account for the inhibition of RNA synthesis which occurs as growing cells enter
stationary phase.

ACKNOWLEDGMENT

This work is supported by a grant from the National Institutes of Health Number GM-15697.

REFERENCES

Austin, S. and J. Scaife 1970. A new method for selecting
    RNA polymerase mutants. *J. Mol. Biol.* 49: 263-267.
Burgess, R. R. 1969. A new method for the large scale purif-
    ication of *Escherichia coli* deoxyribonucleic acid-depen-
    dent ribonucleic acid polymerase. *J. Biol. Chem.* 244:
    6160-6167.
Champe, S. P. and S. Benzer 1962. Reversal of mutant pheno-
    type by 5-fluorouracil: An approach to nucleotide

sequences in messenger-RNA. *Proc. Natl. Acad. Sci. USA*
48: 532-546.

Chao, L. and J. F. Speyer 1973. A new form of RNA polymerase
isolated from *Escherichia coli*. *Biochem. Biophys. Res.
Commun.* 51: 399-405.

Dunne, J. J., F. A. Bautz, and **E. K. F.** Bautz 1971. Different
template specificity of phage T3 and T7 RNA polymerase.
Nature New Biol. 230: 94-96.

Fukuda, R., Y. Iwakura, and A. Ishihama 1974. Heterogeneity
of RNA polymerase in *Escherichia coli*. I. A new holoenzyme
containing a new sigma factor. *J. Mol. Biol.* 83: 353-367.

Hinkel, D. C., W. F. Mangel, and M. J. Chamberlin 1972.
Studies on the binding of *E. coli* RNA polymerase to DNA.
IV. The effect of rifampicin on binding and on RNA
chain initiation. *J. Mol. Biol.* 70: 209-220.

Holland, M. J. and H. R. Whiteley 1973. A new polypeptide
associated with RNA polymerase from *Bacillus subtilis*
during late stages of vegetative growth. *Biochem. Biophys. Res. Commun.* 55: 462-469.

Ilyana, T. S., M. I. Ovadis, S. Z. Mindlin, Z. M. Gorlenko,
and R. B. Khesin 1972. Interactions of RNA polymerase
mutations in haploid cells of *E. coli*. *Mol. Gen. Genetics*
110: 118-133.

Iwakura, Y., R. Fukuda, and A. Ishihama 1974. Heterogeneity
of RNA polymerase in *Escherichia coli*. II. Polyadenylate
. polyuridylate synthesis by holoenzyme II. *J. Mol.
Biol.* 83: 369-378.

Leighton, T. J., R. H. Doi, R. A. J. Warren, and R. A. Kelln
1973. Relationship of serine protease activity to RNA
polymerase modification and sporulation in *B. subtilis*.
J. Mol. Biol. 76: 103-122.

Linn, T. G., A. L. Greenleaf, R. G. Shorenstein, and R. Losick
1973. Loss of the sigma activity of RNA polymerase of
B. subtilis during sporulation. *Proc. Natl. Acad. Sci.
USA* 70: 1865-1869.

Losick, R. and A. L. Sonnenschein 1969. Changes in template
specificity of RNA polymerase during sporulation.
Nature 224: 35-37.

Losick, R., R. G. Shorenstein, and A. L. Sonnenschein 1970.
Structural alterations in RNA polymerase during sporulation. *Nature* 227: 910-913.

Miller, O. H., Jr., B. A. Hamkalo, and C. A. Thomas 1970.
Visualization of bacterial genes in action. *Science*
169: 392-395.

Shapiro, A. L., E. Viñuela, and J. V. Maizel, Jr. 1967.
Molecular weight estimation of polypeptide chains by
electrophoresis in SDS-polyacrylamide gels. *Biochem.*

Biophys. Res. Commun. 28: 815-820.

Travers, A. and R. Buckland 1973. Heterogeneity of *E. coli* RNA polymerase. *Nature New Biol.* 243: 257-260.

THE ROLE OF SUBUNIT INTERACTIONS IN THE GENESIS OF NON-BINOMIAL LACTATE DEHYDROGENASE ISOZYME DISTRIBUTIONS

JAMES B. SHAKLEE

Department of Biology, Yale University
New Haven, Connecticut 06520

ABSTRACT. Starch gel electrophoresis indicates that several species of teleost fishes in the family Clupeidae exhibit non-binomial lactate dehydrogenase isozyme distributions. Detailed quantitative cellulose acetate electrophoresis of extracts of several different tissues from the alewife reveals that both asymmetric heteropolymers--A_3B_1 and A_1B_3--are present in substantially reduced amounts in all tissues studied. In vitro molecular hybridization of the A_4 and B_4 isozymes of the alewife results in a binomial distribution of isozymes demonstrating that the A and B subunits are capable of random association in all possible tetrameric combinations. The isolated LDH isozymes of this fish exhibit little or no subunit exchange in several different buffer systems. However, both of the asymmetric heteropolymers are characterized by a marked differential instability under certain conditions. This instability of the A_3B_1 and A_1B_3 isozymes is directly influenced by pH and results in a rapid loss of the enzymatic activity of these heteropolymers by denaturation. It is suggested that the observed instability of the two isozymes is due to decreased inter-subunit interactions in these asymmetric heteropolymers. Minor changes in the primary structure of the A and/or B subunits, arising during evolution, are probably responsible for the diminished inter-subunit interactions observed.

INTRODUCTION

Recent investigations focusing on the nature of subunit interactions have provided many insights into the mechanisms of synthesis and assembly (Shapiro et al., 1966; Wolf et al., 1973; Shaeffer and Altenburg, 1974; Stevens, 1974; Whitt, 1970b; Rosenbaum and Carlson, 1969), turnover (Dice and Schimke, 1972; Lebherz et al., 1973), and regulation of function (Monod et al., 1965; Koshland and Neet, 1968; Antonini and Brunori, 1970; Frieden, 1971; Yang et al., 1974; Hamilton and Edelstein, 1974) of a large number of multi-subunit proteins. Such studies have served to emphasize the importance of epigenetic processes which operate at an intermediate level between the transcription and translation of the genetic message into polypeptides, and the actual expression of this information in the form of functional protein molecules.

The enzyme lactate dehydrogenase (E.C.1.1.1.27) is a tetrameric protein in all vertebrates studied (Markert and Ursprung,
1962; Cahn et al., 1962; Markert, 1963; 1964; 1965; Shaw and
Barto, 1963; Boyer et al., 1963; Pesce et al., 1967; Jaenicke
and Knof, 1968; Adams et al., 1973). Although at least three
different types of LDH subunits (A, B, and C) are known in
vertebrates, the C subunits are usually only present in one,
or at most, a few tissues (Zinkham et al., 1963; Goldberg,
1963; Markert and Faulhaber, 1965; Whitt, 1970; Sensabaugh and
Kaplan, 1972; Shaklee et al., 1973). As a result, the tissues
of many species contain only the A and B subunits which commonly associate to generate five isozymes of LDH--LDH-1 (B_4),
LDH-2 (A_1B_3), LDH-3 (A_2B_2), LDH-4 (A_3B_1), and LDH-5 (A_4). The
binomial nature of both in vivo and in vitro generated isozyme
distributions indicates that A and B subunit association is a
random phenomenon likely dependent on chance collisions of
these two subunits (Markert, 1963; Shaw and Barto, 1963;
Markert and Masui, 1969).

For some time it has been reported that one species of
teleost fish--the alewife--exhibits a non-binomial distribution
of these five A-B isozymes (Daugherty, 1966; Massaro and
Markert, 1968; Markert, 1968). These earlier studies suggested
that the two asymmetric heteropolymers--LDH-2 (A_1B_3) and LDH-4
(A_3B_1)--were present in much reduced quantities from those
expected for random subunit association. The existence of such
unusual isozyme distributions could not be readily explained,
and indeed, suggested the possible existence of a mechanism
differentially regulating isozyme levels. The present report
quantitatively documents these non-binomial LDH isozyme distributions and describes their biochemical basis.

MATERIALS AND METHODS

Alewives (*Alosa pseudoharengus*) (Clupeiformes, Clupeidae)
were obtained from local sources in Connecticut and were either
analyzed when fresh, or frozen at $-20^{\circ}C$ until they could be
processed. Lactate dehydrogenase activity was determined
spectrophotometrically at $25^{\circ}C$ by monitoring the oxidation of
NADH at 340 nm in a reaction mixture consisting of: 100mM
potassium phosphate buffer pH 7.0, 0.33mM lactate, and 0.14 mM
NADH. All enzyme activities are reported in International
Units.

Vertical starch gel electrophoresis was accomplished in
pH 8.6 EBT buffer (Boyer et al., 1963) using electrostarch
(Lot #98) with subsequent histochemical staining for LDH
activity. Quantitative cellulose acetate electrophoresis was
accomplished at 475 V for 50 minutes using a Beckman Microzone
apparatus. The membranes were then stained for LDH activity

under conditions in which the amount of dye deposited was linearly related to the amount of enzyme present (see Shaklee, 1972). Following staining and clearing, the membranes were scanned at 550nm using a Beckman Analytrol (Model RB) recording densitometer. The resulting optical density peaks were then quantitated and the relative distribution of enzymatic activity among the five isozymes recorded. From these observed isozyme distributions it was possible to determine the ratio of A to B subunits in each tissue extract and thereby calculate an "expected" distribution of isozymes based on the binomial expansion of the expression $(A + B)^4$.

Individual isozymes to be used in the in vitro stability experiments were isolated and purified both by preparative starch gel electrophoresis and by ion-exchange chromatography. Each isozyme preparation was then subjected to the following buffer conditions: 0.1 M citrate buffer pH 6.0; 0.1 M phosphate buffer pH 7.0 plus 10% ammonium sulfate; 0.1 M Tris-HCl buffer pH 8.0; and a "physiological" buffer pH 7.75 (containing: 84 mg/100ml sodium bicarbonate; 925 mg/100ml potassium chloride; 100 mg/100ml glucose; 2,000 mg/100ml bovine serum albumin; and 32 mg/100ml lactic acid).

In vitro molecular hybridization was accomplished using the NaCl freeze-thaw technique (Markert, 1963).

RESULTS

As illustrated in Fig. 1, the alewife, like virtually all other teleosts examined (Whitt et al., 1974), possesses three different kinds of lactate dehydrogenase subunits (A, B, and C). The specialized C subunit, which is restricted to eye and brain tissues in this species, will not be considered in the following discussion of A-B subunit interactions. Casual visual observation of the isozyme pattern in Fig. 1 suggests that, although all five expected A-B tetramers are present, the asymmetric heteropolymers (A_3B_1 and A_1B_3) are present in greatly reduced amounts.

A survey of clupeid fishes revealed that such non-binomial isozyme distributions were not restricted to the alewife but were characteristic of several other species as well. For example, the LDH pattern of the Atlantic thread herring (*Opisthonema oglinum*) shown in Fig. 2 is, if anything, more non-binomial than that of the alewife. In order to examine such naturally occurring non-binomial isozyme patterns in greater detail, the LDH isozymes of the alewife were further characterized as outlined below.

Using quantitative cellulose acetate electrophoretic techniques, the relative isozyme distributions characteristic of several alewife tissues were carefully measured. When these

JAMES B. SHAKLEE

LDH ISOZYMES OF THE ALEWIFE

(*Alosa pseudoharengus*)

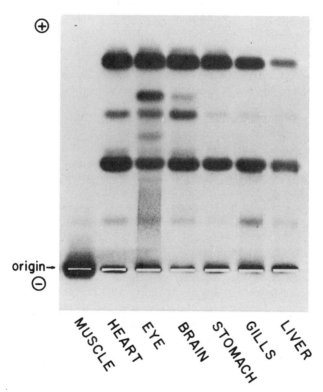

Fig. 1.   LDH isozymes of the alewife (*Alosa pseudoharengus*).
The A_4 homopolymer predominates in white skeletal muscle while
the B_4 homopolymer predominates in heart, eye, and brain.   Both
asymmetric heteropolymers (A_3B_1 and A_1B_3) are present in reduced amounts in all tissues.   Extracts of eye and brain also
possess the C_4 isozyme.

---

observed distributions were compared with "expected" distributions calculated assuming a binomial distribution of isozymes,
it was clear that, in all cases, the observed distributions
were non-binomial (Fig. 3).   Thus, in extracts of heart muscle,
stomach muscle, and brain from several different fish, the ob-

LDH ISOZYMES OF THE ATLANTIC THREAD HERRING
(*Opisthonema oglinum*)

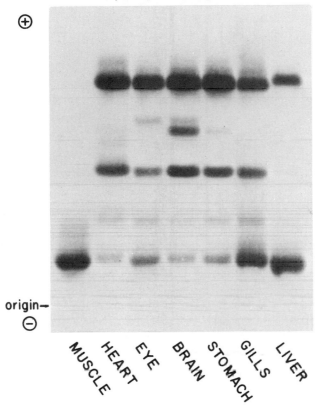

Fig. 2. LDH isozymes of the Atlantic thread herring (*Opisthonema oglinum*). The A_4 homotetramer predominates in white skeletal muscle and the B_4 homotetramer predominates in heart, eye, brain, and stomach. Only traces of the two asymmetric heteropolymers are visible in most tissues. The C_4 isozyme is present in both eye and brain.

---

served quantities of both asymmetric heteropolymers—A_3B_1(LDH-4) and especially A_1B_3(LDH-2)—were much less than expected.

Having defined the extent of the non-binomiality of alewife LDH isozymes, it was of particular importance to examine the nature of A-B subunit association in this fish. This was ac-

FIGURE 3

Relative Distribution of LDH Isozyme Activity in Alewife Tissues

TISSUE		LDH-1 (B_4)		LDH-2 (A_1B_3)		ISOZYME LDH-3 (A_2B_2)		LDH-4 (A_3B_1)		LDH-5 (A_4)	
Heart #1	Observed	35.2	(±1.0)	6.3	(±0.6)	56.1	(±1.4)	2.2	(±0.3)	0.6	(±0.2)
	Calculated	35.2		72.6		56.1		19.3		2.5	
Heart #6	Observed	41.3	(±0.8)	5.1	(±0.4)	52.4	(±1.2)	1.3	(±0.1)	0.8	(±0.2)
	Calculated	41.3		75.9		52.4		16.1		1.8	
Heart #8	Observed	43.7	(±1.3)	9.8	(±0.6)	45.7	(±1.4)	1.6	(±0.2)	0	
	Calculated	43.7		73.0		45.7		12.7		1.3	
Heart #11	Observed	46.6	(±1.1)	5.9	(±0.7)	46.4	(±1.2)	1.3	(±0.1)	0.6	(±0.3)
	Calculated	46.6		75.9		46.4		12.6		1.3	
Heart #13	Observed	53.9	(±0.8)	2.0	(±0.2)	43.6	(±0.6)	0.7	(±0.1)	0.2	(±0.1)
	Calculated	53.9		79.1		43.6		10.7		1.0	
Stomach Muscle #3	Observed	55.6	(±2.0)	4.1	(±0.8)	40.3	(±1.4)	0		0	
	Calculated	55.6		77.2		40.3		9.3		0.8	
Stomach Muscle #9	Observed	58.5	(±2.3)	1.7	(±0.2)	38.9	(±2.5)	1.1	(±0.8)	1.0	(±0.2)
	Calculated	58.5		71.7		38.9		6.7		0.5	
Brain #6	Observed	50.5	(±1.3)	16.1	(±1.0)	33.0	(±0.6)	1.1	(±0.2)	0	
	Calculated	50.5		66.7		33.0		7.3		0.6	
Brain #5	Observed	54.8	(±0.7)	8.1	(±0.8)	35.4	(±1.5)	1.3	(±0.4)	0.8	(±0.2)
	Calculated	54.8		64.2		35.4		5.5		0.4	

complished by subjecting a mixture of the purified A_4 and B_4 isozymes to in vitro freeze-thaw molecular hybridization. The results of this experiment are summarized in Fig. 4. The initial mixture contained only the A_4 and B_4 isozymes (in equal amounts). After hybridization all five expected isozymes were observed. Quantitation of the resulting isozyme distribution and comparison with that expected from the binomial expansion, with equal quantities of A and B subunits, unequivocally demonstrates that random subunit association has occurred. These experiments prove that the A and B subunits of alewife LDH are capable of associating at random to generate binomial isozyme distributions. It seems reasonable to extrapolate these in vitro results and conclude that alewife LDH isozymes initially are generated in binomial proportions in vivo.

The in vitro stability of the five alewife LDH isozymes was investigated by subjecting preparations of isolated isozymes to various ionic environments at $4^{\circ}C$ and measuring the rate of loss of enzymatic activity. These experiments, which are summarized in Fig. 5, clearly indicate that the five isozymes exhibit differential stabilities under several buffer conditions. Although all five A-B isozymes were characterized by a slow decay rate in both the citrate buffer and the phosphate buffer plus ammonium sulfate ($T_{\frac{1}{2}} > 60$ days), major differences in stability were observed in both the tris buffer and the "physiological" buffer. In each of these solutions, LDH-1, LDH-3, and LDH-5 each exhibited a slow loss of activity ($T_{\frac{1}{2}} > 60$ days) while LDH-4 (A_3B_1) and especially LDH-2 (A_1B_3) were characterized by marked instability, exhibiting approximate half-lives of 10 days and 2 days respectively. This experiment was repeated several times using different enzyme concentrations (0.2--30 µg protein/ml) for each isozyme preparation and similar values were obtained each time. These in vitro results are directly parallel to the observed in vivo non-binomial isozyme distributions characteristic of the tissues of this fish and provide an explanation for them: although the five LDH isozymes are originally formed in binomial proportions, the differential instability of the asymmetric heteropolymers

---

Fig. 3. Relative distribution of LDH isozyme activity in alewife tissues. The tabulated "observed" LDH distributions for each tissue sample represent averages based on from six to twelve independent determinations. The standard errors accompanying each average are presented parenthetically. The "observed" distributions measured by quantitative cellulose acetate electrophoresis are compared with "calculated" distributions obtained by the binomial expansion of $(A + B)^4$ (assuming random subunit association).

MOLECULAR HYBRIDIZATION OF LDH ISOZYMES OF THE ALEWIFE

(Alosa pseudoharengus)

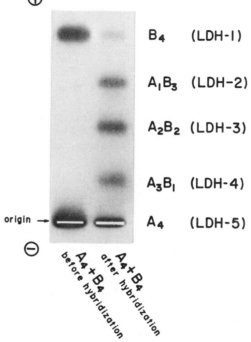

Fig. 4. Molecular hybridization of LDH isozymes of the alewife. A mixture containing equal amounts of the A_4 and B_4
homopolymers was subjected to in vitro freeze-thaw molecular
hybridization. The resulting distribution of five isozymes
was quantitated by cellulose acetate electrophoresis and the
"observed" distribution compared with a "calculated" distribution obtained from the binomial expansion of $(A + B)^4$.

PERCENTAGE DISTRIBUTION OF LDH ACTIVITY

		LDH·1	LDH·2	LDH·3	LDH·4	LDH·5
before hybridization	control mixture	50.0	—	—	—	50.0
after hybridization	observed	4.6	27.2	36.1	24.2	7.9
	calculated	6.3	25.0	37.5	25.0	6.3

(LDH-2 and LDH-4) results in the generation of a non-binomial
isozyme distribution.

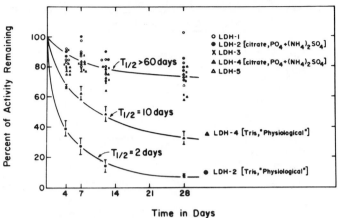

Fig. 5. Stability of alewife isozymes at 4°C. The stability of isolated LDH isozymes was determined in vitro by following the rate of loss of enzymatic activity under a variety of buffer conditions. The approximate half-lives (T½) of the three decay curves are presented to facilitate comparisons among isozymes.

In order to examine the fate of the A and B subunits present in the unstable heteropolymers, starch gel electrophoresis of the isozyme preparations from the stability experiments described above was accomplished. Fig. 6 shows the five isolated isozymes of alewife LDH after eleven days at 4°C in citrate buffer and in "physiological" buffer (cf. Fig. 5). The isozymes in citrate buffer after eleven days were indistinguishable from those present in all buffers on day zero so that this zymogram can also be considered as showing the day zero isozyme patterns for all buffers. The low levels of LDH-4 (A₃B₁) in citrate buffer seen in this figure were the same as those present at the beginning of the experiment and resulted from the inability to isolate larger quantities of this unstable isozyme and not from its decay in citrate buffer. Comparison of the relative quantities of LDH-2 present in citrate and in "physiological" buffer reveals a dramatic decline of this isozyme to nearly undetectable levels in the "physiological" buffer (cf. Fig. 5). It can be concluded that, under the conditions investigated, the alewife LDH-1, LDH-3, LDH-5, and probably LDH-4 retain their molecular identity and do not generate any new isozymes even over extended time periods. The LDH-2 solution, on the other hand, shows slight traces of LDH-1

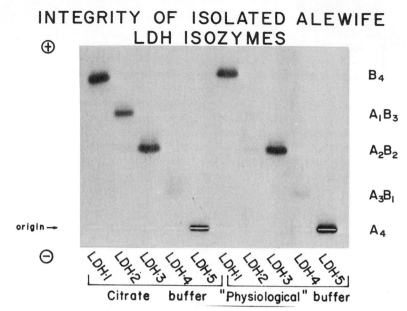

Fig. 6. Integrity of isolated alewife LDH isozymes. The isolated isozyme preparations used in the in vitro stability experiments were subjected to electrophoresis to determine their isozymic composition after 11 days. The five isozymes in citrate buffer were indistinguishable from the isozymes in all buffers on day zero.

and LDH-3 suggesting that a limited amount of recycling of subunits occurs as the LDH-2 decays. However, these newly generated LDH-1 and LDH-3 molecules are not in the ratio expected from the subunit composition of LDH-2 (A_1B_3) and, in addition, these "new" isozymes account for only a very small fraction of the initial LDH activity (cf. Fig. 5). Thus, although a slight amount of subunit reassociation does accompany the decay of LDH-2, nearly all disaggregated subunits are denatured and enzymatic activity is irreversibly lost.

In an attempt to elucidate the molecular basis for the observed instability of the asymmetric heteropolymers several physical and kinetic measurements of the A_4 and B_4 isozymes of alewife LDH were accomplished. However, although substrate dependent kinetics, molecular weight, amino acid composition, and peptide maps of these two homopolymers were investigated, no obvious basis for the instability could be detected (Shaklee, 1972).

The role of pH in the observed instability of LDH-2 in vitro was investigated by comparing the rates of disappearance of enzymatic activity from solutions of LDH-2 during storage in buffer systems of various pHs. The results are diagrammed in Fig. 7. It is clear from this experiment that pH plays an important role in the behavior of this asymmetric heteropolymer. Slightly acid pHs tend to stabilize the protein while, the more basic the pH, the greater the observed instability of the A_1B_3 isozyme. These pH effects correlate well with the stabilities discussed above (cf. Fig. 5) and no doubt provide at least a partial explanation for the observed behavior of LDH-2 under various buffer conditions.

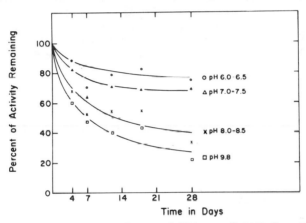

Fig. 7. Effect of pH on the stability of LDH-2. The rate of loss of enzymatic activity due to exposure to different pHs was investigated using both potassium phosphate and tris-HCl buffers.

DISCUSSION AND CONCLUSIONS

To the degree that the in vitro measured characteristics of the isolated isozymes reflect their in vivo behavior, it is now possible to describe the biochemical basis for the unusual lactate dehydrogenase isozyme patterns of several fishes. The three major dynamic processes responsible for the genesis of the non-binomial LDH isozyme patterns characteristic of the alewife are outlined in Fig. 8. Following their synthesis, the A and B subunits associate at random to generate a binomial distribution of all five possible tetrameric isozymes. The resulting tetrameric molecules of alewife LDH exhibit little or

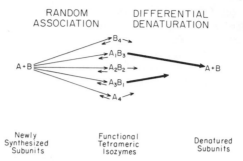

RANDOM ASSOCIATION   DIFFERENTIAL DENATURATION

Newly Synthesized Subunits   Functional Tetrameric Isozymes   Denatured Subunits

Fig. 8. Schematic representation of the interaction of alewife LDH subunits to generate the observed non-binomial isozyme distributions. The newly synthesized LDH subunits associate at random to generate all five tetrameric isozymes. Once the subunits are aggregated into tetramers there is little or no tendency for subunit exchange between isozymes. Both asymmetric heteropolymers--LDH-2 (A_1B_3) and LDH-4 (A_3B_1)--exhibit marked instabilities which lead to the genesis of non-binomial isozyme distributions.

---

no tendency to dissociate into free subunits capable of reassociation to generate other functional tetramers, thus, there is no subunit exchange between isozymes. However, the two asymmetric heteropolymers (A_3B_1 and A_1B_3) exhibit a marked instability which results in the rapid loss of enzymatic activity by these tetramers. Both the random subunit association and the virtual lack of subunit exchange between tetramers are characteristic of the LDHs of nearly all vertebrates. The dramatic differential instability of the two asymmetric heteropolymers is, on the other hand, unique to the LDHs of the alewife (and certain closely related fishes) and is no doubt the cause of the non-binomial isozyme distributions characteristic of this species.

The binomial nature of the isozyme distribution resulting from the in vitro molecular hybridization of alewife A and B subunits (cf. Fig. 4) clearly demonstrates that random subunit association is a property of the LDH subunits of the alewife (just as it is of the LDH subunits of higher vertebrates). Indeed, such random subunit association appears to be characteristic of many other vertebrate proteins including malate dehydrogenase (Bailey et al., 1970), pyruvate kinase (Cardenas and Dyson, 1973), and several hemoglobins (Bunn and McDonough, 1974).

For vertebrate LDHs, the equilibrium between monomers (and dimers) and tetramers strongly favors the tetrameric state.

Consequently, tetrameric LDH isozymes characteristically fail to exhibit measureable dissociation or subunit exchange under physiological conditions of ionic environment, pH, and enzyme concentration (Cohen and Mire, 1971; Bartholmes et al., 1973; but see, Fritz et al., 1971). Thus, it is not surprising that subunit recombination among vertebrate LDHs is only demonstrable after such molecular hybridization treatments as: salt mediated freeze-thaw (Markert, 1963; 1965; Chilson et al., 1965), high ionic strength (Levi and Kaplan, 1971), acid dissociation (Anderson and Weber, 1966), and extreme enzyme dilution (Markert and Massaro, 1968) (but see also Henderson et al., 1970 and Millar et al., 1971). In contrast to an earlier report (Rosenberg, 1971) the LDH isozymes of the alewife, like the LDH isozymes of other vertebrates (Markert, 1963; Rauch, 1969; Bartholmes et al., 1973), fail to exhibit a significant level of spontaneous subunit reassociation (cf. Fig. 6) even over an extended time period.

Although many kinetic, immunochemical, and physical properties of the heteropolymeric isozymes of vertebrate LDHs are intermediate between those of the two homopolymers (A_4 and B_4), the heteropolymers are more readily dissociated into subunits than is either homopolymer (Anderson and Weber, 1966). This observation led Anderson and Weber (1966) to conclude that inter-subunit interaction is considerably weaker in heteropolymers than it is in homopolymers. The marked instability of the two asymmetric heteropolymers (LDH-2 and LDH-4) observed in the present study is consistent with this observation and reveals that heteropolymers containing unequal numbers of A and B subunits can be especially unstable. The instability of these LDH isozymes appears similar to that noted for some asymmetric hemoglobin molecules (see, Bunn and McDonough, 1974).

Because the instability of the asymmetric heteropolymers is not noticeably modified by coenzyme (NADH), substrate (lactate), enzyme (LDH), or total protein concentration, it would appear to be an inherent property of the lactate dehydrogenase heteropolymers themselves. This is in contrast to the subunit exchange between glyceraldehyde-3-phosphate dehydrogenase isozymes which apparently occurs under physiological conditions only in the presence of adenine nucleotides (Lebherz et al., 1973).

The pH-dependence of the observed instability of the asymmetric heteropolymers of alewife LDH is similar to that reported for the dissociation of several hemoglobins in that changes in pH (toward one or both extremes) result in increased dissociation (Efremov et al., 1969; Hamilton and Edelstein, 1974; Bunn and McDonough, 1974). These observations, and those mentioned above concerning the tendency of LDH heteropolymers to dissociate more readily than homopolymers, suggest that the observed instability of alewife LDH-2 and LDH-4 results from the increased dissociation of the asymmetric heteropolymers followed by the secondary denaturation of the resulting free subunits. Thus, it is hypothesized that actual denaturation occurs mainly at the level of the disaggregated subunits and not the intact tetramers themselves. This hypothesis is also supported by the observation of small amounts of newly generated isozymes (resulting from subunit exchange) present in the LDH-2 preparations under certain conditions.

Although the physical analyses of the alewife A_4 and B_4 isozymes failed to reveal any obvious molecular basis for the observed instability of the asymmetric heteropolymers, they do allow certain tentative conclusions to be drawn. Thus, the instability of the A_3B_1 and A_1B_3 isozymes does not appear to result from any large alterations of the subunits involved. For this reason, it seems likely that this instability is the result of a minor change in the primary structure of one or both subunits which has led to a decrease in the inter-subunit binding affinity especially noticeable in asymmetric heteropolymers. That is, minor evolutionary changes in the amino acid sequences of the alewife LDH subunits have resulted in the inability of certain subunit combinations to form stable tetramers. Analogous changes have been described for maize alcohol dehydrogenase (Schwartz, 1971), and are well documented for several human hemoglobin variants which exhibit dramatic changes in inter-subunit binding affinity due to single amino acid substitutions (Efremov et al., 1969; de Jong et al., 1971; Smith et al., 1972; and Alberti et al., 1974). Similarly, the different stabilities of hybrid hemoglobins comprised of subunits from different species are directly correlated with the degree of sequence differences between the homologous and heterologous subunits (Jones and Steinhardt, 1974). A definitive test of this hypothesis regarding the unstable isozymes of alewife LDH will probably require the elucidation of the complete amino acid sequences and three-dimensional conformations of the molecules involved.

ACKNOWLEDGMENTS

The author is especially grateful to Professor Clement L. Markert for his interest and guidance throughout the course of this investigation. This investigation was supported by NSF grant GB 5440 X to C.L. Markert and an NSF graduate fellowship to the author. This research is a portion of a dissertation submitted to the faculty of the Department of Biology, Yale University, in partial fulfillment of the requirements for the degree of Doctor of Philosophy.

LITERATURE CITED

Adams, M.J., M. Buehner, K. Chandrasekhar, G.C. Ford, M.L.
    Hackert, A. Liljas, M.G. Rossmann, I.E. Smiley, W.S.
    Allison, J. Everse, N.O. Kaplan, and S.S. Taylor   1973.
    Structure-function relationships in lactate dehydrogenase.
    Proc. Natl. Acad. Sci. USA 70: 1968-1972.

Alberti, R., G.M. Mariuzzi, L. Artibani, E. Bruni, and L.
    Tentori   1974. A new hemoglobin variant: J-Rovigo alpha
    53 (E-2) alanine → aspartic acid. *Biochim. Biophys. Acta*
    342: 1-4.

Anderson, S., and G. Weber   1966. The reversible acid dis-
    sociation and hybridization of lactic dehydrogenase.
    Archiv. Biochem. Biophys. 116: 207-223.

Antonini, E., and M. Brunori. 1970. Hemoglobin. *Ann. Rev.
    Biochem.* 39: 977-1042.

Bailey, G.S., A.C. Wilson, J.E. Halver and C.L. Johnson
    1970. Multiple forms of supernatant malate dehydrogenase
    in salmonid fishes. Biochemical, immunological, and
    genetic studies. *J. Biol. Chem.* 245: 5927-5940.

Bartholmes, P., H. Durchschlag, and R. Jaenicke   1973.
    Molecular properties of lactic dehydrogenase under the
    conditions of the enzymatic test. Sedimentation analysis
    and gel filtration in the microgram and nanogram range.
    Eur. J. Biochem. 39: 101-108.

Blanco, A., and W.H. Zinkham   1963. Lactate dehydrogenases
    in human testes. *Science* 139: 601-602.

Boyer, S.H., D.C. Fainer, and E.J. Watson-Williams   1963.
    Lactate dehydrogenase variant from human blood: Evidence
    for molecular subunits. *Science* 141: 642-643.

Bunn, H.F., and M. McDonough   1974. Asymmetrical hemoglobin
    hybrids. An approach to the study of subunit inter-
    actions. *Biochem.* 13: 988-993.

Cahn, R.D., N.O. Kaplan, L. Levine, and E. Zwilling   1962.
    Nature and development of lactic dehydrogenases.
    Science 136: 962-969.

Cardenas, J.M., and R.D. Dyson   1973. Bovine pyruvate
    kinases. II. Purification of the liver isozyme and its
    hybridization with skeletal muscle pyruvate kinase.
    J. Biol. Chem. 248: 6938-6944.

Chilson, O.P., L.A. Costello, and N.O. Kaplan   1965. Studies
    on the mechanism of hybridization of lactic dehydrogenases
    in vitro. *Biochem.* 4: 271-281.

Cohen, R., and M. Mire   1971. Analytical-band centrifugation
    of an active enzyme substrate complex. 2. Determination
    of active units of various enzymes. *Eur. J. Biochem.*
    23: 276-281.

Daugherty, W.F.   1966.   Fish lactate dehydrogenase.   Ontogeny
and phylogeny. *Ph.D. Thesis*. Johns Hopkins.

De Jong, W.W.W., L.F. Bernini, and P.M. Khan   1971.   Haemo-
globin Rampa: α_{95} Pro → Ser. *Biochim. Biophys. Acta*
236: 197-200.

Dice, J.F., and R.T. Schimke   1972.   Turnover and exchange
of ribosomal proteins from rat liver. *J. Biol. Chem.*
247: 98-111.

Efremov, G.D., T.H.J. Huisman, L.L. Smith, J.B. Wilson, J.L.
Kitchens, R.N. Wrightstone, and H.R. Adams   1969.   Hemoglobin Richmond, a human hemoglobin which forms asymmetric
hybrids with other hemoglobins. *J. Biol. Chem.* 244:
6105-6116.

Frieden, C.   1971.   Protein-protein interaction and enzymatic
activity. *Ann. Rev. Biochem.* 40: 653-696.

Fritz, P.J., E.L. White, E.S. Vesell, and K.M. Pruitt.   1971.
New theory of the control of protein concentrations in
animal cells. *Nature New Biology* 230: 119-122.

Goldberg, E.   1963.   Lactic and malic dehydrogenases in human
spermatozoa. *Science* 139: 602-603.

Hamilton, M.N., and S.J. Edelstein   1974.   Cat hemoglobin.
pH dependence of cooperativity and ligand binding.
J. Biol. Chem. 249: 1323-1329.

Henderson, R.F., T.R. Henderson, and B.M. Woodfin   1970.
Effects of D_2O on the association-dissociation equilibrium
in subunit proteins. *J. Biol. Chem.* 245: 3733-3737.

Jaenicke, R., and S. Knof   1968.   Molecular weight and
quaternary structure of lactic dehydrogenase. 3. Comparative determination by sedimentation analysis, light
scattering, and osmosis. *Eur. J. Biochem.* 4: 157-163.

Jones, D. D. and J. Steinhardt 1974.   Comparison of the acid
denaturation of several hemoglobins which differ in amino
acid sequence. *Arch. Biochem. Biophys.* 161: 472-478.

Koshland, D. E., Jr., and K. E. Neet 1968.   The catalytic and
regulatory properties of enzymes. *Ann. Rev. Biochem.* 37:
359-410.

Lebherz, H. G., B. Savage, and E. Abacherli 1973.   Adenine
nucleotide-mediated subunit exchange between isoenzymes
of glyceraldehyde-3-phosphate dehydrogenase. *Nature New
Biol.* 245: 269-271.

Levi, A. S. and N. O. Kaplan 1971.   Physical and chemical
properties of reversibly inactivated lactate dehydrogenases. *J. Biol. Chem.* 246: 6409-6417.

Markert, C.L. 1962.   Isozymes in kidney development. In: *Hereditary, Developmental, and Immunologic Aspects of
Kidney disease*. Editor, J. Metcoff. Northwestern University Press, Evanston, pp. 54-64.

Markert, C. L. 1963. Lactate dehydrogenase isozymes: Disassociation and recombination of subunits. *Science* 140: 1329-1330.

Markert, C.L. 1964. Developmental Genetics. In: *The Harvey Lectures, Series 59*. Academic Press, Inc., New York. pp 187-218.

Markert, C.L. 1965. Mechanisms of cellular differentiation. In: (J.A. Moore, ed.) *Ideas in Modern Biology*. The Natural History Press, New York. pp 229-258.

Markert, C.L., and I. Faulhaber 1965. Lactate dehydrogenase isozyme patterns of fish. *J. Exp. Zool.* 159: 319-332.

Markert, C.L., and E.J. Massaro 1968. Lactate dehydrogenase isozymes: Dissociation and denaturation by dilution. *Science* 162: 695-697.

Markert, C.L., and Y. Masui 1969. Lactate dehydrogenase isozymes of the penguin (*Pygoscelis adeliae*). *J. Exp. Zool.* 172: 121-145.

Markert, C.L., and H. Ursprung 1962. The ontogeny of isozyme patterns of lactate dehydrogenase in the mouse. *Dev. Biol.* 5: 363-381.

Massaro, E. J. and C. L. Markert 1968. Isozyme patterns of salmonid fishes: Evidence for multiple cistrons for lactate dehydrogenase polypeptides. *J. Exp. Zool.* 168: 223-238.

Millar, D.B., M.R. Summers, and J.A. Niziolek 1971. Spontaneous *in vitro* hybridization of LDH homopolymers in the undenatured state. *Nature New Biology* 230: 117-119.

Monod, J., J. Wyman, and J.-P. Changeux 1965. On the nature of allosteric transitions: A plausible model. *J. Mol. Biol.* 12: 88-118.

Pesce, A., T.P. Fondy, F. Stolzenbach, F. Castillo, and N.O. Kaplan 1967. The comparative enzymology of lactic dehydrogenases. III. Properties of the H_4 and M_4 enzymes from a number of vertebrates. *J. Biol. Chem.* 242: 2151-2167

Rauch, N. 1969. The degradation of lactate dehydrogenase isozymes injected into frog eggs. *J. Exp. Zool.* 172: 363-368.

Rosenbaum, J.L., and K. Carlson 1969. Cilia regeneration in *Tetrahymena* and its inhibition by colchicine. *J. Cell Biol.* 40: 415-425.

Rosenberg, M. 1971. Epigenetic control of lactate dehydrogenase subunit assembly. *Nature New Biology* 230: 12-14.

Schwartz, D. 1971. Dimerization mutants of alcohol dehydrogenase of maize. *Proc. Natl. Acad. Sci. USA* 68: 145-146.

Sensabaugh, G.F., Jr., and N.O. Kaplan 1972. A lactate dehydrogenase specific to the liver of gadoid fish. *J. Biol. Chem.* 247: 585-593.

Shaeffer, J.R., and L.C. Altenburg 1974. Studies on the

mechanism of the effect of hydroxylamine on hemoglobin assembly in rabbit reticulocytes. *J. Biol. Chem.* 249: 2243-2248.

Shaklee, J. B. 1972. A genetic, biochemical, and evolutionary characterization of LDH isozyme structure in fishes. *Ph. D. Dissertation.* Yale University, New Haven, Conn.

Shaklee, J.B., K.L. Kepes, and G.S. Whitt 1973. Specialized lactate dehydrogenase isozymes: The molecular and genetic basis for the unique eye and liver LDHs of teleost fishes. *J. Exp. Zool.* 185: 217-240.

Shapiro, A.L., M.D. Scharff, J.V. Maizel, Jr., and J.W. Uhr 1966. Polyribosomal synthesis and assembly of the H and L chains of gamma globulin. *Proc. Natl. Acad. Sci. USA* 56: 216-221.

Shaw, C.R., and E. Barto 1963. Genetic evidence for the subunit structure of lactate dehydrogenase isozymes. *Proc. Natl. Acad. Sci. USA* 50: 211-214.

Smith, L.L., C.F. Plese, B.P. Barton, S. Charache, J.B. Wilson, and T.H.J. Huisman 1972. Subunit dissociation of the abnormal hemoglobins G Georgia ($\alpha_2{}^{95}$ Leu (G2) β_2) and Rampa ($\alpha_2{}^{95}$ Ser (G2) β_2). *J. Biol. Chem.* 247: 1433-1439.

Stevens, R.H. 1974. Translational control of antibody synthesis. Control of light chain production. *Eur. J. Biochem.* 42: 553-559.

Whitt, G.S. 1970a. Developmental genetics of the lactate dehydrogenase isozymes of fish. *J. Exp. Zool.* 175: 1-35.

Whitt, G.S. 1970b. Directed assembly of polypeptides of the isozymes of lactate dehydrogenase. *Archiv. Biochem. Biophys.* 138: 352-354.

Whitt, G.S., J.B. Shaklee, and C.L. Markert 1974. Evolution of the lactate dehydrogenase isozymes in fishes. *IV. Isozymes: Genetics and Evolution,* C.L. Markert, editor, Academic Press, New York.

Wolf, J.L., R. G. Mason, and G.R. Honig 1973. Regulation of hemoglobin β-chain synthesis in bone marrow erythroid cells by α chains. *Proc. Natl. Acad. Sci. USA* 70: 3405-3409.

Yang, Y.R., J.M. Syvanen, G.M. Nagel, and H.K. Schachman 1974. Aspartate transcarbamoylase molecules lacking one regulatory subunit. *Proc. Natl. Acad. Sci. USA* 71: 918-922.

SUBUNIT INTERACTION IN LACTATE DEHYDROGENASE

EMORY H. BRASWELL
Biological Sciences Group
University of Connecticut
Storrs, Connecticut

ABSTRACT. Ainslie and Cleland (1969) and Ainslie (1970) using pig and beef LDH have confirmed our earlier assertion made for chicken LDH (Rouslin and Braswell, 1968) that subunit interaction occurs for the substrate inhibition step of the reaction. That is, hybrids containing A subunits tend to display more substrate inhibition (more "B"-like nature) than would be expected for a tetramer of that subunit composition. Further, as the number of B subunits increases so does the effect, such that the A_1B_3 isozyme possesses substrate inhibition properties very similar to that of the B_4 isozyme. Their data also indicated that a number of the reaction steps other than substrate inhibition are also affected by this phenomenon, which they called the "oligomeric environment" effect. More general kinetic studies by Ainslie (1970) showed that some kinetic parameters of the hybrids possess an apparently higher "B"-like nature whereas other kinetic parameters evinced an apparently higher "A"-like nature, indicating the complex nature of the phenomenon. Ainslie (1970) also showed that the subunits of the tetramer hybrid possess closely similar values for their second substrate Michaelis constants. This implies that both B and A subunits mutually modify each other so that they now possess similar Michaelis constants for the second substrate which are closer to that typical of the B type than to that of the A type subunit.

INTRODUCTION

Lactate dehydrogenase (LDH) is a regulatory enzyme that catalyzes the reversible dehydrogenation of lactate, converting it to pyruvate. In this reaction, NAD^+ is used as the hydrogen receptor. Takenaka and Schwert (1956) found by equilibrium binding techniques that there are four sites per LDH molecule on which NAD^+ can be bound. Lactate and pyruvate are not bound to a measurable extent unless NAD^+ or NADH have first been bound. In other words, the reaction is an ordered one in which a ternary complex is formed which can then undergo the catalyzed reaction. The generally accepted reaction sequence is as follows (Takenaka and Schwert, 1956):

$$E + NAD^+ \rightleftharpoons E \cdot NAD^+$$

$$E \cdot NAD^+ + Lac \rightleftharpoons E \cdot NAD^+ \cdot Lac \rightleftharpoons E \cdot NADH \cdot Pyr \rightleftharpoons$$

$$E \cdot NADH + Pyr$$

$$E \cdot NADH \rightleftharpoons E + NADH$$

It was observed by a number of workers (see for example Hakala et al, 1956) that high concentrations of pyruvate (over ca. $5 \times 10^{-4}M$) inhibited the reverse reaction. Therefore, at a given pH, there is an optimum pyruvate concentration at which the backward reaction velocity is maximal. Nygard (1956) found that high concentrations ($5 \times 10^{-3}M$) of lactate inhibited the foreward reaction. The inhibition from the pyruvate side of the reaction is the result of the formation of an abortive ternary complex between pyruvate, LDH, and one of the products, i.e. NAD^+ (Gutfreund et al, 1968; Kaplan et al, 1968). This ternary complex has been isolated and used as the object of an X-ray crystallography study by Rossman's group at Purdue (see for example, Leberman, 1969). The formation of the abortive ternary complexes may be represented as follows:

$$E \cdot NAD^+ + Pyr \rightleftharpoons E \cdot NAD^+ \cdot Pyr$$

$$E \cdot NADH + Lac \rightleftharpoons E \cdot NADH \cdot Lac$$

Markert and Møller (1959) using zone electrophoretic techniques reported that a multiplicity of bands could be obtained from a variety of enzymes, including LDH, prepared from the same species or even from the same organ. These authors suggested that the term "isozyme" was a useful designation for the multiple molecular forms of an enzyme found in a single organism. Plagemann et al (1960) established that five isozymes of LDH were found in many mammalian species. Appella and Markert (1961) treated preparations of single isozymes with a number of reagents known to disrupt the structure of protein. Examination of the products revealed that the enzyme had been dissociated into four equal sized subunits. These subunits had no enzymatic activity. A mixture of the five isozymes which had been dissociated into subunits yielded two bands when subjected to electrophoresis on acrylamide gels. It therefore, became obvious that five tetrameric isozymes could be made by the random association of two different subunits

(Appella and Markert, 1961). Designating the two subunits as A and B, the five isozymes can be represented as A_4, A_3B_1, A_2B_2, A_1B_3, and B_4. The B_4 type is the predominant form present in the heart muscle, and moves most rapidly on gel electrophoresis in slightly alkaline buffers (pH = ca. 8), whereas A_4 predominates in extracts of most skeletal muscles and moves the slowest upon electrophoresis. The other LDH isozymes are found in varying quantities in nearly every tissue and move at intermediate speeds during electrophoresis depending on their subunit composition. The amount of each LDH isozyme present in a tissue is presumably controlled by the amount of A or B type subunits produced, the combination into the various tetramers being random. This was demonstrated by Markert (1963) who, upon freezing and thawing in 1 M NaCl equal quantities of A_4 and B_4 isozymes, was able to dissociate the parent enzymes into subunits and then recombine them into active isozymes. When this material was subjected to electrophoresis it was found that all five isozymes had been produced in amounts following the binomial distribution of 1:4:6:4:1 for the A_4 through B_4 isozymes respectively. This indicates that an A subunit has as much probability of reacting with another A subunit as it does with a B subunit and vice versa. Further, Markert (1964) was able to hybridize the subunits of beef heart LDH-B_4 with the LDH subunits from many types of animals. Earlier, Markert (1963a) showed that the A_4 and B_4 isozymes were quite different in amino acid composition, thereby indicating that the subunits are different.

Because there is a tissue specificity of the isozyme pattern, i.e. each tissue seems to produce a characteristic ratio of A to B subunits, there has been a considerable effort made to relate the function of the tissues to its LDH isozyme pattern. As a result of the rough correlation between the preponderance of B subunits and the constancy of the oxygen supply to the tissue, and because the heart type isozymes are inhibited by lower levels of lactate and pyruvate than the A type it has been felt that the ratio of A to B subunits helps to regulate the concentration of lactic acid to within the acceptable limits of the cell type in that tissue (Kaplan et al, 1968; Cahn et al, 1962; Dawson et al, 1964).

There have been some objections to this notion, however. Vesell and Pool (1966) and Vesell (1968) (see also Stambaugh, 1966) have pointed out that substrate levels high enough to inhibit the B_4 isozyme significantly do not occur in vivo. Vesell (1968) also states that at physiological temperatures this isozyme is inhibited much less than it is under normal assay conditions (25°C). Further, Vesell (1966) lists a number of tissues such as mature human erythrocytes, platelets,

and bovine lens fibre cells which pose contradictions to this
hypothesis; these tissues are anaerobic, but they contain
little of the A_4 isozyme. Wuntch et al (1970) claim that substrate inhibition is not detected when the LDH concentrations
are great enough to approach physiological values (10^{-7}-10^{-8}M).
But this was refuted by Everse et al (1970) who claim that at
the high enzyme concentrations at which the former work was
done, the reaction is complete before the ternary complex is
formed. By allowing NAD^+ to be present initially they were
able to demonstrate substrate inhibition at physiological
concentrations of LDH.

The Catalytic Interaction of Subunits

Whether or not substrate inhibition occurs in nature, it
is easily detected in vitro. As a result of the great difference in the concentrations of pyruvate needed to inhibit
the two parental isozymes (A_4 and B_4), a method of quantitatively determining the relative proportion of each of them in
a mixture of the two was developed. Further, through the
incorporation of several implicit assumptions, its use has
been extended to that of analyzing for the relative composition
of A and B subunits in a mixture of isozymes. For example,
in the method as developed by Plagemann et al., (1960a) the
steady state reaction velocity of the enzyme was measured
at two different concentrations of pyruvate. A linear relationship was observed to exist between the logarithm of
the ratio of activities "R" (i.e. activity at high pyruvate
to that at low pyruvate), exhibited by mixtures of A_4 and B_4
and the percentage of B_4 in the mixture. Because of the observed linearity it was assumed that this differential assay
procedure constituted a rapid means for the estimation of the
relative amounts of A and B subunits in the tetramer of a
mixture consisting of some or all of the isozymes. Subsequently other investigators adopted differential assay procedures
for the determination of subunit composition in a wide variety
of biological materials, using various pairs of assay conditions (e.g. see for example, Cahn et al, 1962 and Stambaugh
and Post, 1966a). Rouslin and Braswell (1968) showed however,
that implicit in the extension of the analytical method from
determining relative concentration of the two parental isozymes in mixtures of the parental isozymes, to one of determining the ratio of A to B subunits in a mixture of isozymes, is
the assumption that there exists in the tetrameric molecule
complete intersubunit catalytic independence. That is, an A
subunit has the same turnover number in the presence of B
subunits in the same tetramer as it does in the presence of

three A subunits, and vice versa. Using the turnover numbers of Kaplan and Cahn (1962), Rouslin and Braswell (1968) indicated that this assumption might be false, in that the "R" values observed for the isozyme hybrids change with the number of B type subunits present in the tetramer in a manner not in accordance with the assumption of intersubunit catalytic independence. This change is in such a direction that it implies that the A subunits become more like the B subunits (e.g. they are more easily inhibited by high substrate concentrations) as the number of B subunits increase in the tetramer. We speculated at that time (Rouslin and Braswell, 1968) that this alteration in A subunit catalytic behavior is probably the consequence of a B-induced conformational change in the A subunit, when A subunits become associated with B subunits in the tetramer.

Our paper, (Rouslin and Braswell, 1968) concluded with a request that careful turnover number determinations be made in order to make possible a study of subunit interaction. Subsequently, Ainslie and Cleland (1969) and Ainslie (1970) performed a thorough steady state kinetic study of beef and pig LDH isozymes and obtained more than enough data to permit a decision to be made on whether or not the subunits are catalytically independent. Some of the data of Ainslie (1970) will be used here in order to recapitulate more clearly the effect that Rouslin and Braswell (1968) reported. In Fig. 1 is shown the substrate inhibition curves for beef LDH isozymes in which is defined a high and a low substrate concentration which are the respective optima for the A_4 and B_4 activities. By dividing the activity (Vh) obtained at high concentration by that obtained at low substrate concentration (Vl), for each isozyme, one defines a ratio, $R = Vh/Vl$ which characterizes the substrate inhibition of that isozyme. In the example shown these ratios are 1.26/1.57 = 0.80 and 2.76/2.13 = 1.30 for the B_4 and A_4 isozymes respectively. If the subunits are independent of each other, i.e. there is no interaction between subunits, one would expect hybrids made from these parental types to have R values somewhere between these two extreme values. Various investigators used different methods to arrive at the *manner* in which the R value should vary with, for example, A subunit content. Kaplan and Cahn (1962) stated that R varied almost linearly with subunit composition, whereas Plagemann, Gregory and Wroblewski (1960) found a logarithmic relationship. In Fig. 2 is shown an illustrative example derived from Ainslie's (1970) data (including that shown in Fig. 1). The crosses indicate the R values that Kaplan's group assumed would result for mixtures and hybrids if the catalytic sites are independent of one another, i.e. the R values fall on a straight line connect-

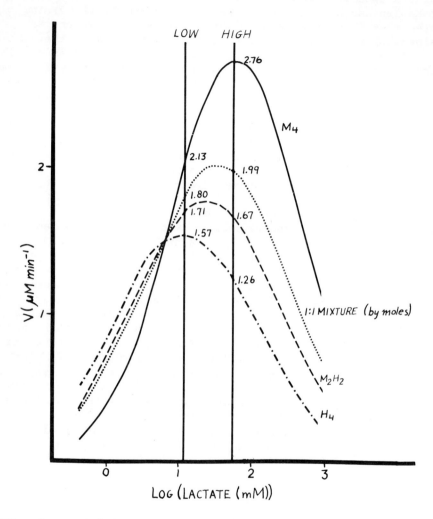

Fig. 1. Initial steady state velocity of reaction as a function of lactate concentration for certain beef LDH isozymes (from Ainslie, 1970). The 1:1 mixture refers to an equal molar mixture of A_4 and B_4 isozymes. The total enzyme concentration for each study was 0.158 nM. The numbers appearing on the figure refer to the velocities resulting at the two arbitrarily chosen substrate concentrations, i.e. high and low.

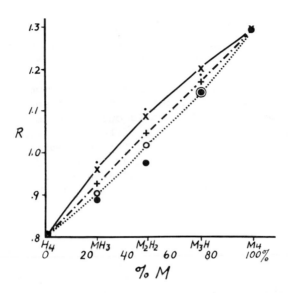

Fig. 2. Variation of R, the ratio of enzyme velocities at high and low substrate concentrations, with subunit type composition. These data were derived from substrate inhibition studies of Ainslie (1970) such as those shown in Fig. 1. Symbols defined as follows:

+ ; expected for mixtures and hybrids (assuming catalytic independence) if linear relation assumed (Kaplan and Cahn, 1962).

O ; expected for mixtures made up on activity basis (and also for hybrids by Kaplan and Cahn 1962).

X ; expected for mixtures made up on molar basis and hence hybrids if catalytic independence of subunits is assumed (Rouslin and Braswell, 1968).

• ; found for mixtures made up on molar basis (derived from Ainslie, 1970).

● ; found for hybrids (derived from Ainslie, 1970).

ing the extreme R values of the parental isozymes.

The various investigators reasoned that if the sites of the hybrid were catalytically independent, then the isozyme should display an R value similar to that which would be found for a mixture of the two parental isozymes made up so as to have the same % composition of each type of subunit as the hybrid in question. Thus far the reasoning is correct. The mistake these investigators made, was to prepare their mixtures on an *activity* basis rather than on a *molar* basis. For example, a mixture consisting of 50 enzyme units of A_4 and 50 enzyme units of B_4, each measured at their particular optimum second substrate concentration was assumed to be comparable with the A_2B_2 hybrid. Obviously this would only be true if the A_4 and B_4 isozymes had turnover numbers of the same value at their optimum substrate concentration. This has not been found for the admittedly rather limited number of species studied so far.

It was also shown by Rouslin and Braswell (1968) that a straight line relationship is not obtained with mixtures made up on an activity basis unless the value of R for one of the parental isozymes is close to one, i.e. no second substrate inhibition. For some species this is approximately true for the A_4 isozyme for the substrate concentrations chosen, which may account for the near linearity of some of the standard curves obtained from mixtures of parental isozymes found by earlier workers. In the example shown in Fig. 2, the point (open circle) that would have resulted from the 1:1 (by activity) mixture of the two parental isozymes is approximately 6% less than that expected from a linear relation. The calculation giving these values is performed as follows:

$$R = \frac{\text{Vh of Mixture}}{\text{Vl of Mixture}} = \frac{(\text{Activity of } A_4 + \text{Activity of } B_4) \text{ at}}{(\text{Activity of } A_4 + \text{Activity of } B_4) \text{ at}}$$

$$\frac{\text{high substrate}}{\text{low substrate}} = \frac{\text{Activity of } A_4 \text{ at optimum} + (\text{Activity}}{(\text{Activity of } A_4 \text{ at optimum}/R_{A_4}) + \text{Activ-}}$$

$$\frac{\text{of } B_4 \text{ at optimum} \times R_{B_4})}{\text{ity of } B_4 \text{ at optimum}}$$

The activities of the parental isozymes were taken from Fig. 1, remembering that in the formation of the mixtures the enzyme activities at their optimum substrate concentrations are used to create the ratios; for example: the 1:1 (A_4 and B_4) mixture is calculated as follows:

$$\frac{1 + (1 \times .8)}{(1/1.3) + 1} = 1.02$$

For activity ratios other than 1:1 (A_4 to B_4), one merely substitutes different values for the numeral one in the above illustration, e.g. for 3:1 (B_4 to A_4) the above would become

$$\frac{1 + (3 \times .8)}{(1/1.3) + 3} = .90$$

With subunit independence the R value of a 1:1 *molar* mixture of the parental isozymes should agree with that of the hybrid A_2B_2. Early workers usually plotted velocity vs. second substrate concentration as the fraction of maximum velocity. Fig. 1, on the other hand, is a plot of enzyme velocities measured with *equal molar* concentrations. Thus, the velocity axis is proportional to turnover number. Using these values, one can calculate the R value of a 1:1 mixture which should be equal to the A_2B_2 isozyme. The points shown as "X"s in Fig. 2 were calculated from "turnover numbers" (Tn) of the parental enzymes in Fig. 1 as follows:

$$R = \frac{(N_A \times TnA_4) + (N_B \times TnB_4) \text{ at high substrate}}{(N_A \times TnA_4) + (N_B \times TnB_4) \text{ at low substrate}}$$

where N represents molar concentration (or ratios) of the particular parental species. An example of this type of calculation for the 1:1 mixture on a molar basis is

$$R = \frac{(1 \times 2.76) + (1 \times 1.26)}{(1 \times 2.13) + (1 \times 1.57)} = 1.09$$

This value is about 8% above that based on an assumption of linearity and 14% above an incorrect mixture. The small points represent the values found for such mixtures (see Table 1). The agreement with calculation is good.

The solid circles represent the R values found from Ainslie's (1970) data for hybrids. For the A_2B_2 hybrid, R is 22% lower than predicted for catalytic independence but only 4% lower than the 1:1 activity mixture; those who used this type of standard curve assumed incorrectly that the subunits act independently in the A_2B_2 isozyme. In this figure it is apparent that as the number of B subunits increases, the tetramer exhibits a disproportionately large B-like nature. This suggests that in beef LDH isozymes, as in chicken LDH (Rouslin and Braswell, 1968), the A subunits display some B-like behavior in heteropolymers! This effect for A_2B_2 in Fig. 2 yields a hybrid R value 22% lower than predicted; thus it acts as though composed of 70% B subunits. Assuming that the 20% excess B-like character is contributed entirely by altered A subunits, one can see that these subunits are acting as though they

TABLE 1

Substrate Inhibition Studies of Beef LDH

pH 7.5 Phosphate Buffer

Varying Lactate Concentration (0.89 mM NAD$^+$)
(Ainslie, 1970)

Isozymes	R*	Optimum Substrate (mM)	$T_{max}(min^{-1})$ x 10^{-3}	K_S (mM)	K_{I_S} (mM)
A_4	1.30	58	5.4	8.0	450
A_3B_1	1.15	44	4.8	3.9	510
$1B_4:3A_4$	1.19	45	–	–	–
A_2B_2	0.98	22	3.4	1.9	290
$1A_4:1B_4$	1.11	38	3.8	–	–
A_1B_3	0.89	16	2.9	1.2	220
$3B_4:1A_4$	0.98	25	3.2	–	–
B_4	0.80	16	2.82	0.94	140

* Determined from Ainslie's data

possessed 40% B-like nature.

Since investigators are still using activities in order to describe the percent of A and B subunits in a mixture of hybrids (see for example Bishop et al, 1972), one then might ask, what errors would occur in a subunit assay if one used the "wrong" type (i.e. enzyme concentration by activity) of parental mixture to make up the standard curve? Actually most R values of hybrids determined by previous investigators agreed fairly well with those measured from the incorrect mixtures, as in fact is the case for the example shown in Fig. 2. This may seem peculiar but is easily explained. The turnover number of the A_4 isozyme has a higher value than that of the B_4 isozyme throughout most of the substrate inhibition region. Therefore, if one had, for example, a mixture containing equal activities of each parental isozyme, one would perforce, have a much higher molar concentration of the B_4 isozyme. When the value of R of this mixture is compared with that of the A_2B_2 isozyme, the difference will not be great because the A_2B_2 is also *displaying* an increased B character. In the latter case, however, it is due to the effect of the B subunit environment.

By examining their extensive data more generally, Cleland and Ainslie (1969) and Ainslie (1970) independently came to the conclusion that pig and beef LDH isozyme hybrid show considerable subunit interaction. They clearly show that A subunits take on some of the kinetic characteristics of the B subunit in hybrids, and have named the phenomenon the "oligomeric environment effect." In order to see how this effect evinces itself in a more general way one must re-examine Fig. 1. It should be noticed that the protein solutions were all at the same concentration which makes the Y axis proportional to turnover number. It then becomes evident that the kinetic behavior of the mixture is approximately intermediate between the extremes of the parental isozymes, whereas that of the hybrid is a good deal closer to the kinetic behavior of the B_4 type isozyme. The gradual transition from A type behavior to B type can be seen to consist of two types of change, first, a general decrease in the turnover numbers and second, a shift in the position of V_{max} to lower substrate concentrations.

Ainslie (1970) has similarly investigated each beef LDH hybrid isozyme and its comparable mixture of A_4 and B_4, the results of which can be seen in Table 1. The R values shown in the second column were calculated from their data and used to prepare Fig. 2. The values displayed in the last three columns were obtained by Ainslie by fitting kinetic data similar to that in Fig. 1 to the equation:

$$V = V_{max}S/ (K_S + S + S^2/K_{I_S})$$

where V is the reaction velocity, S the second substrate concentration, K_S is the Michaelis constant and K_{I_S} is the inhibition constant which is related to the dissociation constant of the abortive ternary complex. In this table one can see that for every parameter shown, i.e. R value, optimum second substrate concentration, maximum turnover number (V_{max}/enzyme concentration), second Michaelis constant, and second substrate inhibition constant, the value for the hybrid is lower than either that of the comparable mixture or that value which would result if one assumed a linear change between parental isozymes. These data also indicate that the A subunits develop increasing B-like nature as the number of A subunits in the tetramer decreases. Of course, only the last three columns represent basic kinetic parameters, since the R value and optimum substrate concentration are derivable from them (e.g. optimum substrate concentration $=\sqrt{K_S \cdot K_{I_S}}$). Ainslie found essentially similar results for the reverse reaction, and for the forward reaction at a higher pH. In addition, he found that the same generalization applied to both the forward and reverse reaction for the artificial hybrid consisting of beef heart muscle isozyme and pig skeletal muscle isozyme (BB_2PA_2).

There are so far then, three basic kinetic parameters that are affected by the oligomeric environment effect. They are: first, the inhibition constant, K_{Ilac} (or K_{Ipyr}) which is related to the dissociation constant of the abortive ternary complexes; second, the second Michaelis constant; and third, the maximum turnover number. Since the last two parameters are complicated functions of kinetic constants of steps of the reaction other than that of substrate inhibition, one might naturally ask: What additional steps of the reaction are affected by varying the subunit (oligomeric) environment? Ainslie (1970) attempted to answer this by means of a classical steady state kinetic analysis in which the concentrations of both substrates are varied methodically. The concentration of the second substrate, however, was not allowed to reach levels which would cause inhibition. Some of the results of this study can be seen in Table 2. In this table, are listed

TABLE 2
Steady State Kinetic Constants for Beef LDH at pH 9.0
(Ainslie 1970)

(All K's in μM)

Constant	B_4	A_2B_2	A_4	Shift
$T_{max} \times 10^{-3}$ (min^{-1})	4.6	5.0	16.8	B
K_{NAD^+}	17	20	20	A
K_{iNAD^+}	2.0×10^3	9.0×10^2	6.0×10^2	A
K_{lac}	3.0×10^2	1.2×10^3	5.0×10^3	B

the values of the maximum turnover number (T_{max}), the first Michaelis constant (K_{NAD^+}), the dissociation constant for the first complex (K_{iNAD^+}), the second Michaelis constant (K_{lac}) for the parental and the middle hybrid isozymes. In the last column is recorded the parental isozyme, to which the value of the particular kinetic constant for the A_2B_2 hybrid seems to have shifted. One can see that the values of T_{max} and K_{lac} (the second Michaelis constant) have shifted toward that of the B_4 isozyme (as found in the substrate inhibition studies), whereas the values of K_{NAD^+} and $K_{i_{NAD^+}}$ have shifted toward the A_4. Ainslie also performed the same type of study for the beef LDH reaction in the reverse direction and for pig LDH in both directions. Essentially they observed the same shifts for each parameter in these studies, as are shown in Table 2. Since these steady state kinetic parameters are complex functions of the velocity constants of the reaction (Takenaka and Schwert, 1956), it is therefore impossible at this time to resolve which of the individual steps of the reaction are being affected, other than that of the formation of the abortive ternary complex. The fact that some of the parameters display shifts toward the B behavior and others an apparently opposite shift (toward A behavior), indicates that the phenomenon is complicated.

There is a further question one might ask: Why should one assume that only the A subunits change? Perhaps, there is a general but unsymmetrical accomodation of both subunits toward each other's properties. This leads to the question, do all subunits in a hybrid appear identical? For an answer to this question we again turn to the work of Ainslie (1970). The kinetics of a mixture of two parental isozymes might be expected to obey a two-term Michaelis equation, resulting therefore, in curvilinear Lineweaver-Burk or Eadie-Hofstee plots. In a similar fashion, an enzyme with two kinds of independent sites each displaying a different value for the Michaelis constant for a given substrate, would also yield curved plots. Linear graphs result when there is one kind of site present. Ainslie (1970) investigated the kinetics of a number of hybrids and their comparable molar ratio mixtures and plotted the results in the form of Eadie-Hofstee graphs. Their choice of this type of plot over the Lineweaver-Burk type was based on the fact that the former was better suited for displaying curvature at the substrate concentration used. In Fig. 3 is depicted the data for the beef A_2B_2 and the 1:1 molar mixture of beef A_4 and B_4 isozymes in which the concentration of the second substrate (lactate) was varied. For substrate concentrations below that which causes substrate inhibition, two conclusions can immediately be reached. The first is that the hybrid is clearly different from the mixture; and second, the hybrid acts as though it possesses a single-valued second Michaelis constant indicating that its sites are quite similar with respect to this constant. The mixture, on the other hand, appears as expected, to be composed of enzyme sites displaying more than one value for K_s. All of Ainslie's data (pyruvate also varied) yielded similar results, some showing less difference between the hybrid and the mixture, others more. This study suggests that the second Michaelis constants of the A and B subunits mutually adjust to a more nearly equal value when the two sub-

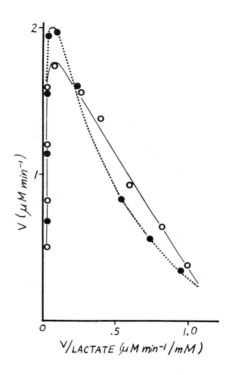

Fig. 3. Eadie-Hofstee plot of kinetic behavior of: mixture of .079 nM beef B_4 and .079 nM beef A_4 isozymes (solid points); and .152 nM beef A_2B_2 hybrid isozyme (open points); from Ainslie (1970).

units are in the same tetramer. The previously discussed studies showed that this value is more like that displayed by the B subunit in the B_4 tetramer.

One wonders if this mutual adjustment occurs for some other step of the reaction. Since the dissociation constant of the first substrate shows a shift toward A behavior for hybrids, perhaps a study of the binding of NADH and NAD+ to LDH would reveal if a hybrid displays more than one binding coefficient. R. Anderson and Gregario Weber (1965) studied the binding of NADH to the five beef LDH isozymes using fluorescence tech-

niques. They found that B_4 bound NADH with a single association constant independent of the number of substrate molecules
bound per tetramer. However, titration of the A_4 isozyme
with NADH showed marked changes in the binding constant after
the first substrate molecule was bound. An identical study
of the hybrids produced results like those found for the A_4
isozyme. They felt that their observations could be explained
in terms of molecular relaxation effects. A more recent investigation has been reported using similar methods by Stinson
and Holbrook (1973). They found that the binding of NADH to
ox muscle, pig heart, pig muscle, rabbit muscle, dogfish muscle, and lobster muscle LDH can in each case be expressed by
a *single* macroscopic dissociation constant, thus indicating the
independence of sites with respect to binding. The same was
found to be true for the binding of NAD^+ to the pig muscle and
heart LDH isozymes. Using calorimetric methods to investigate
the binding of NADH to pig muscle LDH, Hinz and Jaenicke (1973)
reported that they also found that their results could be
described by a theoretical binding curve calculated on the
assumption of four independent and *identical* binding sites.
Since neither of these groups have as yet reported on studies
with hybrids, the question of whether the *hybrids* display more
than one binding constant for the first step of the reaction
is still unanswered.

CONCLUSION

Upon reflection, it should not be surprising that the presence of another subunit, or the change in conformation of a
neighboring subunit should induce changes in the conformation
and hence in the chemical and physical properties of a particular subunit. The fact that active monomer LDH subunits have
not been found in solution indicates that the presence of four
subunits in a tetramer mutually modifies the conformations of
the inactive subunits enough to allow them to display enzymatic
activity.

For the future, it would seem necessary that Stinson and
Holbrook's (1973) work be extended to hybrids in order to
ascertain if A and B subunits in tetramers mutually adjust to
yield similar *first* substrate binding constants. In general,
further kinetic studies should be performed in order to reveal
which steps of the reaction are susceptible to the "oligomeric
effect," its direction and whether or not it is accomplished
through mutual adjustments of both types of subunit. In addition, the importance of supplementing the kinetic studies with
structural investigations must be emphasized. For example,
comparison of structure as determined by X-rays, or ORD-CD,

of the A_4, A_2B_2 and B_4 abortive ternary complexes should be
of great interest. In addition, ORD-CD should be used to compare the enzyme-coenzyme complexes for the same isozymes.
Such studies should reveal what kinds of structural changes
lie behind the kinetic changes observed.

The biological implications of the oligomeric effect are
not clear. Though it may be possible that, if it is necessary
for a tissue to transform from one containing predominantly
the A_4 isozyme to one containing largely the B_4 isozyme, the
"oligomeric environment" effect would enable the tissue to
possess a high degree of B-like capability, upon the production
of relatively small amount of B subunits. This would be gained
without the necessity of eliminating many of the A subunits.
This reasoning assumes that the subunits can in vivo, hybridize
in a rapid random fashion. Whether such a mechanism would be
an advantage or can actually occur in vivo, are questions
better left to the biologist.

REFERENCES

Ainslie, G. R. 1970. Kinetic Studies on Alcohol and Lactate
Dehydrogenases. Ph.D. Diss. University of Wisconsin.

Ainslie, G. R. and W. W. Cleland 1969. The Effect of Oligomeric
Environment on the Kinetics of Lactate Dehydrogenase
Subunits. *Fed. Proc. Abst., 53rd Meeting* 28: 468, Abstract
#1176.

Anderson, R. and G. Weber 1965. Multiplicity of Binding by
Lactate Dehydrogenases. *Biochem.* 4: 1948-1957.

Appella, E. and C. L. Markert 1961. Dissociation of Lactate
Dehydrogenase into Subunits with Guanidine Hydrochloride.
Biochem. Biophys. Res. Commun. 6: 171-176.

Bishop, M. J., J. Everse, and N. O Kaplan 1972. Identification
of Lactate Dehydrogenase Isozymes by Rapid Kinetics.
Proc. Natl. Acad. Sci. 69: 1761-1765.

Cahn, R. D., N. O. Kaplan, L. Levine and E. Zwilling 1962.
Nature and Development of Lactic Dehydrogenase. *Science*
136: 962-969.

Dawson, D. M., T. L. Goodfriend, N. O. Kaplan 1964. Lactic
Dehydrogenase: Functions of the Two Types. *Science*
143: 929-933.

Everse, J., R. L. Berger, and N. O. Kaplan 1970. Physiological
Concentrations of Lactate Dehydrogenases and Substrate
Inhibition. *Science* 168: 1236-1238.

Gutfreund, H., R. Cantwell, C. H. McMurray, R. S. Criddle and
G. Hathway 1968. The Kinetics of the Reversible Inhibition of Heart Lactate Dehydrogenase Through the Formation
of the Enzyme-Oxidized Nicotinamide-Adenine Dinucleotide-

Pyruvate Compound. *Biochem. J.* 106: 683-687.

Hakala, M. T., A. J. Glaid, and G. W. Schwert 1956. Lactic Dehydrogenase II. Variation of Kinetic and Equilibrium Constants with Temperature. *J. Biol. Chem.* 221: 191-209.

Hinz, H. J. and R. Jaenicke 1973. Calorimetric Investigation of Binding of NADH to Pig Muscle Lactate Dehydrogenase. *Biochem. and Biophys. Res. Commun.* 54: 1432-1436.

Kaplan, N. O. and R. D. Cahn 1962. Lactic Dehydrogenases and Muscular Dystrophy in the Chicken. *Proc. Natl. Acad. Sci.* 48: 2123-2130.

Kaplan, N. O., J. Everse, J. Admiraal 1968. Significance of Substrate Inhibition of Dehydrogenases. *Ann. N. Y. Acad. Sci.* 151: 400-412.

Leberman, R., I. E. Smiley, D. J. Hass, and M. G. Rossman 1969. Crystalline Ternary Complexes of Lactate Dehydrogenase. *J. Mol. Biol.* 46: 217-219.

Markert, C. L. 1963. Lactate Dehydrogenase Isozymes: Dissociation and Recombination of Subunits. *Science* 140: 1329-1330.

Markert, C. L. 1963a. Epigenetic Control of Specific Protein Synthesis in Differentiating Cells. in *Cytodifferentiation and Macromolecular Synthesis* (Locke, ed). Academic Press, New York, 165-184.

Markert, C. L. 1964. Hybrid Isozymes of Lactate Dehydrogenase. *Sixth Int. Congr. Biochem. Abstracts IV* 104: 320.

Markert, C. L. and F. Møller 1959. Multiple Forms of Enzymes: Tissue, Ontogenic and Species Specific Patterns. *Proc. Natl. Acad. Sci.* 45: 753-763.

Nygard, A. P. 1956. Reaction Velocities of DPN-linked Lactic Dehydrogenase. *Acta Chem. Scand.* 10: 408-412.

Plagemann, P. G. W., K. F. Gregory and F. Wroblewski 1960. The Electrophoretically Distinct Forms of Mammalian Dehydrogenase. *J. Biol. Chem.* 235: 2282-2287; 1960a. *loc. cit.* 2288-2293.

Rouslin, W. and E. Braswell 1968. Analysis of Factors Affecting LDH Subunit Composition Determinations. *J. Theoret. Biol.* 19: 169-182.

Stambaugh, R. and D. Post 1966. Substrate and Product Inhibition of Rabbit Muscle Lactic Dehydrogenase Heart (H_4) and Muscle (M_4) Isozymes. *J. Biol. Chem.* 214: 1462-1467.

Stambaugh, R. and D. Post 1966a. A Spectrophotometric Method for the Assay of Lactic Dehydrogenase Subunits. *Analyt. Biochem.* 15: 470-480.

Stinson, R. A., and J. J. Holbrook 1973. Equilibrium Binding of Nicotinamide Nucleotides to Lactate Dehydrogenases. *Biochem. J.* 131: 719-728.

135

Takenaka, Y. and G. W. Schwert 1956. Lactic Dehydrogenase III. Mechanism of the Reaction. *J. Biol. Chem.* 223: 157-170.

Vesell, E. S. 1966. pH Dependence of Lactate Dehydrogenase. Isozyme Inhibition by Substrate. *Nature* 210: 421-422.

Vesell, E. S. 1968. Introduction to the Conference on Multiple Molecular Forms of Enzymes. *Ann. N. Y. Acad. Sci.* 151: 5-13.

Vesell, E. S. and P. E. Pool 1966. Lactate and Pyruvate Concentration in Exercised Ischemis Canine Muscle: Relationship of Tissue Substrate Level to Lactate Dehydrogenase Isozyme Patterns. *Proc. Natl. Acad. Sci.* 55: 756-762.

Wuntch, T., R. F. Chen, and E. S. Vesell 1970. Lactate Dehydrogenase Isozymes: Kinetic Properties at High Enzyme Concentrations. *Science* 167: 63-65.

A STRUCTURAL COMPARISON OF PORCINE B_4 AND DOGFISH A_4 ISOZYMES OF LACTATE DEHYDROGENASE

W. EVENTOFF

M. L. HACKERT

S. J. STEINDEL

M. G. ROSSMANN

Department of Biological Sciences
Purdue University
West Lafayette, Indiana 47907

ABSTRACT. The LDH:NADH:oxamate complex of the B_4 isozyme of porcine heart lactate dehydrogenase crystallizes in the space group C2 with a=162.0 Å, b=60.7 Å, c=138.5 Å and β= 93.2°. The rotation function has been used to show that the P, Q, and R 2-fold axes found in dogfish A_4 lactate dehydrogenase are conserved in the B_4 isozyme. The tetramer is oriented in the unit cell with Q and R rotated 12.5° from the b and -c axes respectively and P along a*. The translation function has shown that the molecular center is positioned at (0.29, 0.5, 0.25).

A comparison of the structures of the pig B_4 LDH:NADH: oxamate complex and the dogfish A_4 LDH:NAD-pyruvate complex at 6 Å resolution indicates that they are very similar. The only difference occurs in the region around the essential histidine (from residue 190 to 199). The movement is discussed in terms of a change in the interaction of this region with the αH helix. The similarity of the two molecules is discussed in terms of the function of the conserved structural entities.

INTRODUCTION

The NAD dependent enzyme lactate dehydrogenase (E.C.: 1.1. 1.27, LDH) catalyzes the interconversion of lactate and pyruvate in the glycolytic pathway. The enzyme is a tetramer, having a molecular weight of 144,000 Daltons. The subunits are each capable of binding one molecule of coenzyme and substrate and show no cooperative effects (d'A Heck, 1969; Schwert et al, 1967).

Two subunit types are found in most tissues. The A type subunit (sometimes designated M) predominates in skeletal muscle while the B type (or H) predominates in heart tissue. The B and A type subunits differ in their immunological (Pesce et al, 1964), physical, and chemical properties (for a recent review see, Everse and Kaplan, 1973). Indeed, the interspecies similarity of one subunit type of LDH is greater than the intraspecies similarity between the two subunit types. Everse and Kaplan (1973) have suggested that the two types of LDH have different metabolic roles, and furthermore, that the B type, through the formation of an abortive ternary complex with oxidized coenzyme and substrate, is under a unique type of metabolic control.

Despite the differences between the two subunit types they do form hybrid tetramers in vivo and in vitro (Markert, 1963; Chilson et al, 1965), which would imply some degree of structional similarity. In order to determine the basis for the behavior of the two LDH subunit types a structural study of the B_4 isozyme from porcine heart was undertaken. In this paper a comparison of the structure of the B_4 isozyme with the known structure of the A_4 isozyme of dogfish muscle LDH will be presented.

EXPERIMENTAL

The structure of the A_4 isozyme of dogfish muscle LDH has been described elsewhere (Rossmann et al, 1971).

The B_4 isozyme of porcine heart LDH was purified by affinity chromatography (Eventoff et al, 1974). Crystals of the ternary complex LDH:NADH:oxamate were grown from sodium phosphate buffer (0.05M, pH = 7.8) which contained 8 mg/ml of LDH, 0.05M sodium oxamate, 1.6 mM NADH and 1.81 to 1.87M ammonium sulphate. The enzyme crystallizes in space group C2 with a = 162.0 Å, b = 60.7 Å, c = 138.5 Å, and β = 93.2°.

X-ray diffraction data were collected by the precession method. The films were processed using an Optronics Film Scanner interfaced with an IBM 7094 Computer (Ford, 1974). The final R factor for reflections appearing on more than one film after the film planes were scaled together was 10.5%.

RESULTS AND DISCUSSION

Rotation Function of Pig B_4 LDH

The rotation function (Rossmann and Blow, 1962) was used to determine the molecular symmetry elements and their orientation with respect to the crystallographic axes. Data from 10 to 6 Å resolution were used in the calculations. The original Patterson contained 2796 terms and was modified to remove the origin peak. It was compared with a large term Patterson which contained 346 terms. The radius of integration around the Patterson origin was limited to 50 Å. In all calculations the 27 nearest neighbors of the non-integral reciprocal lattice point were used for interpolation.

The 2/m Patterson symmetry causes the asymmetric unit in
polar coordinates which must be searched to lie within the
limits $0^\circ \leq \psi \leq 90^\circ$ and $0^\circ \leq \varphi \leq 180^\circ$ (where ψ is measured
from the b axis, φ is measured from the a* axis and K defines
the number of degrees of rotation).

The presence of 222 symmetry in dogfish A_4 LDH (Rossmann
et al, 1971), approximate 222 symmetry in GPD (Rossmann et al,
1972) and the approximate 222 symmetry found in pig A_4 LDH
(Hackert et al, 1973), caused the initial searches to be restricted to 2-fold axes.

The results of the rotation function are shown in Fig. 1.

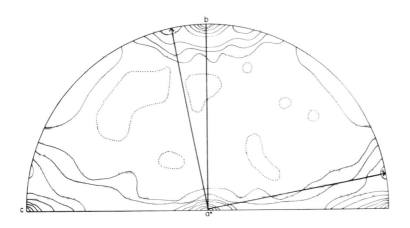

Fig. 1. Stereographic projection representing the results of
the rotation search for molecular 2-fold axis in pig B_4 LDH
within the asymmetric unit defined by $0^\circ \leq \psi \leq 90^\circ$ and $0^\circ \leq
\varphi \leq 180^\circ$. The arrows indicate the positions of the molecular
2-fold axes.

There are two peaks on the line $\varphi = 90^\circ$ beside the origin and
pseudo origin peaks. Searches for both 3-fold and 4-fold
axes were carried out to explore the background of the function, and were featureless in both cases. The line $\varphi = 90^\circ$
was searched in order to determine the exact position of the
molecular 2-fold axes. The results shown in Fig. 2 indicate
that two of the molecular 2-folds are rotated 12.5° from the
b and c crystallographic axes respectively. The third axis
represented by the peak at $\psi = 90^\circ$, $\varphi = 0^\circ$ is along the a*
direction of the crystallographic unit cell.

The three 2-fold axes of dogfish A_4 LDH have been labeled

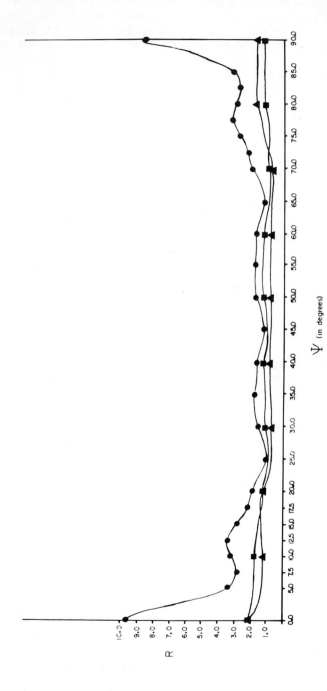

Fig. 2. Rotation function searches for 2-fold (●——●), 3-fold (▲——▲), and 4-fold (■——■) axes in pig B$_4$ LDH along the line $\varphi = 90°$.

P, Q, and R (Rossmann et al, 1973). Conservation of the Q
molecular axis in soluble malate dehydrogenase as well as the
conservation of the P, Q, and R molecular 2-fold axes in pig
A_4 LDH (Hackert et al, 1973) provide evidence for the conservation of both tertiary and quaternary structures among the
dehydrogenases. A comparison of the abortive ternary complex
LDH:NAD-pyruvate of dogfish A_4 LDH with the LDH:NADH:oxamate
complex of pig B_4 LDH was undertaken to see if the molecular
2-fold axes were conserved among the LDH isozymes. Since
preliminary comparisons of the dogfish A_4 LDH:NADH:oxamate
and LDH:NAD-pyruvate complexes show only minor differences,
the use of the pyruvate adduct is reasonable.

The known electron density of one tetramer of the dogfish
A_4 LDH:NAD-pyruvate adduct was isolated and placed in a cell
with lattice constants a = b = c = 160 Å in space group P222.
The molecular center was positioned at (0,0,0) and the 2-fold
P, Q, and R axes were positioned along a, -c, and b respectively. Structure factors were calculated to a resolution of
6 Å based on 5 Å phases sampled at approximately 2 Å intervals. The conditions for the rotation function were identical
to the rotation function of pig B_4 LDH described above except
that now the P222 Patterson, represented by 389 large terms,
was used for the comparison. The P, Q, and R axes were
aligned in turn with the molecular 2-fold axis of pig B_4
LDH:NADH:oxamate directed along a*, and K was searched from
0° to 90° in 2.5° intervals. The results are shown in Fig.
3. The large peak at 12.5° when P is aligned along the a*
direction of the pig B_4 cell is greater than twice the height
of any other peak. Thus the P axis of the dogfish A_4 isozyme
corresponds to the 2-fold axis along a* in the pig B_4 structure while the R and Q axes are rotated 12.5° from the b and
-c crystallographic axes respectively, in confirmation of the
results obtained by the comparison of the pig B_4 Patterson
with itself.

It has been shown previously (Rossmann et al, 1972) that
comparison rotation functions of identical structures give
significant peaks only when the axes are correctly superimposed. The good agreement found when the P axis of the
dogfish A_4 isozyme is aligned with one of the 2-fold axes of
the pig B_4 isozyme indicates that the structures of the two
LDH isozymes are similar.

Translation function of Pig B_4 LDH

Given that the structures of the dogfish A_4 isozyme and
the pig B_4 isozyme are similar, the known structure of the
A_4 isozyme can be used to determine the position of the molec-

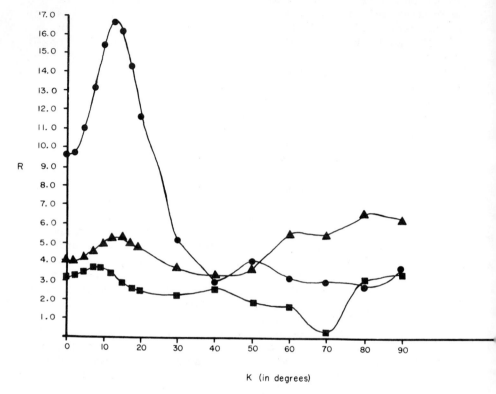

Fig. 3. Comparison rotation function of pig B_4 and dogfish A_4 LDH. The results of aligning the P, Q, and R axes of dogfish A_4 LDH with the molecular 2-fold of pig B_4 LDH along a* are represented by (●,●), (▲—▲), and (■—■) respectively.

ular center of the B_4 isozyme in its crystal cell (Tollin, 1966, 1969; Joynson et al, 1970; Rossmann et al, 1972; Crowther and Blow, 1967; Hackert et al, 1973).

Translation coefficients were calculated to a resolution of 6 Å using a sphere of 40 Å radius around the center of the dogfish A_4 LDH:NAD-pyruvate tetramer. The A_4 tetramer was positioned in the pig B_4 cell in the same manner as in the comparison rotation function. The origin of the pig B_4 cell was defined relative to the crystallographic symmetry axes as in the International Tables of X-Ray Crystallography (1962). The independence of the y coordinate in the C2 space group limits the search to two dimensions. The results of the translation function are shown in Fig. 4. The largest peak is found at x = 0.29, z = 0.25. In order to determine the

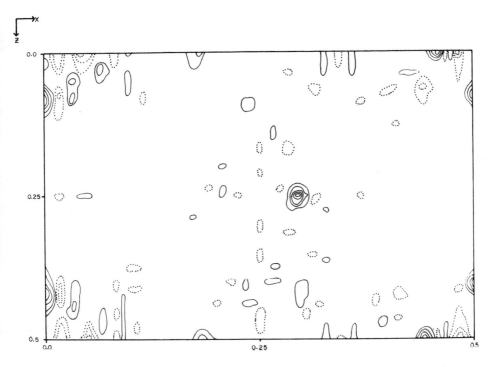

Fig. 4. Translation function search for the molecular center of pig B_4 LDH.

position of the molecular center accurately, R factors $R = \dfrac{\Sigma (\ |F_o| \ - \ |F_c|\ )}{\Sigma |F_o|}$ were calculated for positions of the molecular center around the largest peak found in the translation function. The R factors were calculated using both the previously described method and also by placing the properly oriented density of the dogfish A_4 tetramer in the pig B_4 cell and calculating the Fourier transform. The results obtained by both methods, shown in Figs. 5 and 6, indicate that the molecular center is positioned at x = 0.29, z = 0.25. R factor searches of the next largest features in the translation function gave only background values. The low R factors obtained when the dogfish A_4 tetramer is properly positioned again indicate that the structures of the two LDH isozymes are closely related.

Once the position of the molecular center was determined, phases to a resolution of 6 Å, based on a properly oriented and positioned tetramer of the dogfish A_4 LDH:NAD-pyruvate adduct, were calculated. An electron density map of the pig

143

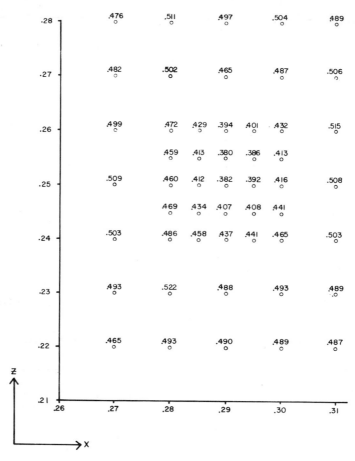

Fig. 5. R factors $\left(R = \dfrac{\Sigma \left(\left|F_O\right| - \left|F_C\right| \right)}{\Sigma \left|F_O\right|} \right)$ calculated from translation function coefficients for the area around the molecular center of pig B_4 LDH.

B_4 LDH:NADH:oxamate complex based on these phases and the observed amplitudes was calculated. The map was skewed so that the P, Q, and R axes were now directed along −x, z, and y respectively. The map was then averaged across the three 2-fold axes (Buehner et al, 1974).

   In order to compare the two structures the electron density

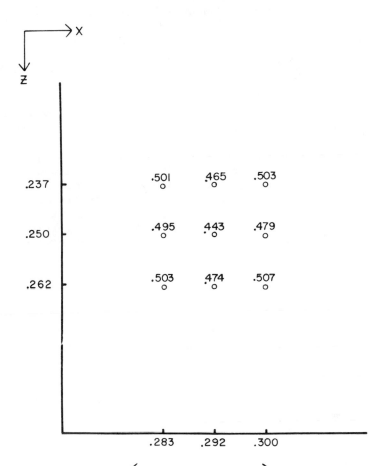

Fig. 6. R factors $\left(R = \dfrac{\Sigma (\; |F_O| \; - \; |F_C| \;)}{\Sigma \; |F_O|} \right)$ for the area around the molecular center of pig B_4 LDH calculated from the Fourier transform of the density of a tetramer of dogfish A_4 LDH.

---

of the dogfish A_4 molecule in the pig B_4 cell was regenerated from the calculated amplitudes and phases and skewed so that its orientation was identical to the pig B_4 map.

Comparison of the Structures of the B_4 and A_4 Isozymes of LDH at 6 Å Resolution

The positions of the α-carbons of the dogfish A_4 complex, after refinement using Diamond's model building program (Diamond, 1966), were plotted on the electron density map of the pig B_4 LDH:NADH:oxamate derivative. The fit of the atomic positions in the available density was then examined. Due to the low resolution of the present pig B_4 LDH:NADH:oxamate map, a more detailed comparison of the structures was not warranted.

The structures of the ternary complexes of the B_4 and A_4 LDH isozymes are very similar, in agreement with the results obtained with the rotation and translation functions. A schematic diagram of the structure of one subunit of the dogfish A_4 isozyme is shown in Fig. 7. Within the first half of the molecule there are no regions which show significant differences. The first 20 residues form the "arm" region of the subunit, which is involved in subunit contacts and essential for the formation of the tetramer. For example, in soluble malate dehydrogenase where the arm is not present only a dimeric structure is formed (Tsernoglou et al, 1972). The residues from 22 to 162 form the coenzyme binding site and "Loop" region. Conservation of the coenzyme binding site is not unexpected since this same structure has been found in many other proteins, all of which bind nucleotides (Rossmann et al, 1974). The conservation of both the primary structure (Taylor et al, 1973) and position of the loop region would imply a similar movement on formation of the ternary complex in the B_4 isozyme.

The only difference between the two structures is found in the second half of the molecule. The region from 190 to 199 (all numbers refer to the dogfish A_4 sequence) appears to move away from the molecular center. Histidine 195 has been implicated in the mechanism of both the dogfish A_4 and pig B_4 isozymes (Woenckhaus et al, 1969; Adams et al, 1973). The sequence of the peptide from the pig B_4 isozyme (191-203) has been determined by Woenckhaus (1969) and is identical to the peptide in the dogfish A_4 isozyme (Taylor et al, 1973) except for residue 194 which changes from glutamine in the A_4 sequence to glutamate in the B_4 sequence. A possible reason for the movement of the region from 190-199 may lie in the interaction of residue 194 with the αH helix. In the dogfish structure glutamine 194 interacts with serine 318. Substitution of glutamate for glutamine in the pig B_4 structure may favor interaction with residue 317 which is lysine in both the pig B_4 and dogfish A_4 isozymes. Indeed, inspection

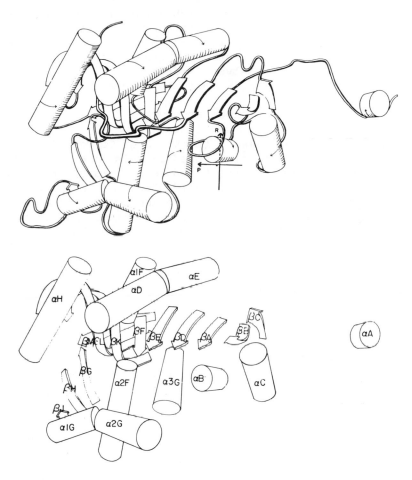

Fig. 7. A schematic diagram of one subunit of dogfish A_4 lactate dehydrogenase.

of the model of the high resolution dogfish structure indicates such an interaction is possible and would cause a movement of the chain in the correct direction. The effect of this distortion on the detailed arrangement of the active site cannot be determined at the present resolution.

The remainder of the second half contains secondary structures which are involved in subunit interactions. Conservation of these structures is not unreasonable in view of the lack of preferentiality found in the formation of mixed B and

A hybrids (Markert, 1963; Chilson et al, 1965).

The resolution of the pig B₄ LDH:NADH:oxamate structure is being extended to 2.5 Å. The high resolution structure should detail the differences in the structures of the two LDH isozymes. However, the above results suggest that the structures are very closely related and the basis of the differences may manifest themselves through changes in sequence and consequently interactions within the enzyme.

ACKNOWLEDGMENTS

The authors would like to thank Dr. G. C. Ford for writing many of the computer programs used in this work. The authors express appreciation to Mrs. S. Hurto for aid in the preparation of this manuscript. We would also like to thank C. L. M. Braun for preparing the illustrations used in the text.

This work was supported in part by National Institute of Health grant (GM 10704). W. E. would like to thank the National Institutes of Health for a post-doctoral biophysics training grant. S. J. S. expresses appreciation to the American Cancer Society for a post-doctoral fellowship (PF-870). M. L. H. would like to thank the National Institutes of Health for a post-doctoral fellowship (No.1 FO2 Am 52315-01).

REFERENCES

Adams, M. J., M. Buehner, K. Chandrasekhar, G. C. Ford, M. L. Hackert, A. Liljas, M. G. Rossmann, I. E. Smiley, W. S. Allison, J. Everse, N. O. Kaplan, and S. S. Taylor 1973. Structure-function relationships in lactate dehydrogenase. *Proc. Natl. Acad. Sci. U.S.A.* 70: 1968-1972.

Buehner, M., G. C. Ford, D. Moras, K. W. Olsen, M. G. Rossmann 1974. Structure determination of crystalline lobster D-glyceraldehyde-3-phosphate dehydrogenase. *J. Mol. Biol.* 82: 563-585.

Chilson, O. P., G. B. Kitto, N. O. Kaplan 1965. Factors affecting the reversible dissociation of dehydrogenases. *Proc. Natl. Acad. Sci. U.S.A.* 53: 1006-1014.

Crowther, R. A., D. M. Blow 1967. A method of positioning a known molecule in an unknown crystal structure. *Acta Cryst.* 23: 544-548.

Diamond, R. 1966. A mathematical model building program for proteins. *Acta Cryst.* 21: 253-266.

Eventoff, W., K. W. Olsen, M. L. Hackert 1974. The purification of porcine heart lactate dehydrogenase by affinity chromatography. *Biochem. Biophys. Acta,* in press.

Everse, J., N. O. Kaplan 1973. Lactate dehydrogenases: Structure and function. in *Advances in Enzymology* (Meister, ed), J. Wiley & Sons. 37: 61-133.

Ford, G. C. 1974. Intensity determination by profile-fitting applied to precession photographs. *J. Appl. Cryst.* Submitted for publication.

Hackert, M. L., G. C. Ford, M. G. Rossmann 1973. Molecular orientation and position of the pig M4 and H4 isoenzymes of lactate dehydrogenase in their crystal cells. *J. Mol. Biol.* 78: 665-673.

Heck, H d'A. 1969. Porcine heart lactate dehydrogenase. *J. Biol. Chem.* 244: 4375-4381.

International Tables for X-Ray Crystallography. 1962. Vol. I, Knoch Press, Birmingham, England.

Joynson, M. A., A. C. T. North, V. R. Sarma, R. E. Dickerson, L. K. Steinrauf 1970. Low-resolution studies on the relationship between the triclinic and tetragonal forms of lysozyme. *J. Mol. Biol.* 50: 137-142.

Markert, C. L. 1963. Lactate dehydrogenase isozymes: dissociation and recombination of subunits. *Science* 140: 1329-1330.

Pesce, A., R. H. McKay, F. Stolzenbach, R. D. Chan, N. O. Kaplan 1964. The comparative enzymology of lactic dehydrogenases. *J. Biol. Chem.* 239: 1753-1761.

Rossmann, M. G., M. J. Adams, M. Buehner, G. C. Ford, M. L. Hackert, P. J. Lentz, Jr., A. McPherson, Jr., R. W. Schevitz, I. E. Smiley 1971. Structural constraints of possible mechanisms of lactate dehydrogenase as shown by high resolution studies of the apoenzyme and a variety of enzyme complexes. *Cold Spring Harbor Symp. Quant. Biol.* 36: 179-191.

Rossmann, M. G., D. M. Blow 1962. The detection of sub-units within the crystallographic asymmetric unit. *Acta Cryst.* 15: 24-31.

Rossmann, M. G., M. J. Adams, M. Buehner, G. C. Ford, M. L. Hackert, A. Liljas, S. T. Rao, L. J. Banaszak, E. Hill, D. Tsernoglou, L. Webb 1973. Molecular symmetry axes and subunit interfaces in certain dehydrogenases. *J. Mol. Biol.* 76: 533-537.

Rossmann, M. G., G. C. Ford, H. C. Watson, L. J. Banaszak 1972. Molecular symmetry of glyceraldehyde-3-phosphate dehydrogenase. *J. Mol. Biol.* 64: 237-249.

Rossmann, M. G., D. Moras, K. W. Olsen 1974. Chemical and biological evolution of a nucleotide binding protein. *Nature.* Submitted for publication.

Schwert, G. W., B. R. Miller, R. J. Peanasky 1967. Lactic dehydrogenase. *J. Biol. Chem.* 242: 3245-3252.

Taylor, S. S., S. S. Oxby, W. S. Allison, N. O. Kaplan 1973. Amino acid sequence of dogfish M_4 lactate dehydrogenase. *Proc. Natl. Acad. Sci. U.S.A.* 70: 1790-1794.

Tollin, P. 1966. On the determination of molecular location. *Acta Cryst.* 21: 613-614.

Tollin, P. 1969. Determination of the orientation and position of the myoglobin molecule in the crystal of seal myoglobin. *J. Mol. Biol.* 45: 481-490.

Tsernoglou, D., E. Hill, L. J. Banaszak 1972. Cytoplasmic malate dehydrogenase - heavy atom derivatives and low resolution structure. *J. Mol. Biol.* 69: 75-87.

Woenckhaus, C., J. Berghäuser, G. Pfleiderer 1969. Markierung essentieller aminosaurereste der lactat-dehydrogenase aus schweineherz mit (carbonyl-^{14}C)3-(2-brom-acetyl)-pyridin. *Hoppe-Seyler Z. Physio. Chem.* 350: 473-483.

THE TWO-STAGE REVERSIBLE DENATURATION
OF LACTATE DEHYDROGENASE AT LOW pH

RICHARD B. VALLEE
Department of Biology
Yale University
New Haven, Connecticut   06520

ABSTRACT. Upon exposure to conditions of low pH, lactate dehydrogenase rapidly loses enzymatic activity, but this process can be completely reversed yielding 100 percent of the original activity if the enzyme is immediately returned to neutral conditions. As the time of exposure to low pH is increased the fraction of activity recovered rapidly declines to a value of 50 to 60 percent. Correlated with this behavior is a change in the kinetics of the recovery of activity. Recovery of activity has been shown to be a second-order process for enzyme exposed to low pH for brief periods of time. It is found to become first-order and considerably slower in the concentration range observed, after several minutes at low pH. The denatured enzyme is capable of hybridization, and this property disappears as renaturation progresses. Gel-filtration chromatography at low pH separates the protein into two fractions. The lower molecular-weight fraction is primarily monomeric, representing a single thermodynamic component, and is capable of recovering activity. The higher molecular-weight fraction is generated from the lower molecular-weight fraction, and is incapable of recovering activity. These results are interpreted to indicate that the enzyme exists sequentially in three denatured forms at low pH, the first two capable of being restored to the native state, and the third irreversibly denatured. The data are discussed in terms of a two-step dissociation and of a sequential dissociation and unfolding mechanism.

Lactate dehydrogenase (LDH) is a tetrameric protein of molecular weight about 140,000 daltons (Appella and Markert, 1961; Jaenicke and Knof, 1968; Adams et al, 1970). It appears to undergo reversible dissociation to subunits under a variety of conditions, as indicated by the technique of isozyme hybridization (Markert, 1963; see review by Jaenicke, 1970, and Everse and Kaplan, 1973). In general, dissociation is accompanied by irreversible denaturation of the protein, behavior which has complicated the interpretation of direct physicochemical data on the dissociated state. A number of reports have been published indicating that fully reversible dissociation of LDH may be induced by exposure of the enzyme to low pH (Anderson and Weber, 1966; Levitzki, 1972; Levitzki and Tenenbaum, 1974). Loss of catalytic activity at low pH may be completely reversed upon neutralization, and, if more than one isozyme is involved, the formation of hybrids results. Recovery of activity was found to follow second-order kinetics in the case of beef A_4 and B_4 LDH, further evidence that these isozymes become dissociated at low pH. However, for all isozymes examined, full recovery of the native condition may be obtained only if exposure of the enzyme to acidic conditions is of very short duration, exposure for longer than several minutes resulting in irreversible loss of activity. This suggests that the dissociated state is unstable and indicates that interpretation of physico-chemical data must still be approached with caution.

Direct characterization of the enzyme at low pH has shown the molecular weight to be decreased from that of the native tetramer (Deal et al, 1963; Jaenicke and Knof, 1968; Anderson and Weber, 1966; Millar et al, 1969). Millar and co-workers using enzyme prepared from beef heart presented ultracentrifugal evidence indicating the existence of a number of different molecular weight species at low pH in rapidly reversible equilibrium. The minimum molecular weight observed was about 18,000 daltons, about half the previously reported monomer molecular weight (Appella and Markert, 1961). The centrifugation data were found to be consistent with a scheme involving the reversible association of the species of 18,000-molecular weight to polymeric species of a size as large as the native tetramer. In contrast with the full recovery of activity obtained after brief exposure to low pH, enzyme used in this study was found to be incapable of recovering activity. This has suggested to us that two different denatured forms of the enzyme may have been under investigation in the different studies, a short-lived form capable of complete renaturation, and an irreversibly denatured form present after the long exposure to acidic conditions involved in the equilibrium studies. We report here results confirming this hypothesis and showing, in addition, that before becoming irreversibly denatured, the enzyme exists sequentially in two distinct forms that are both capable of being restored to the native state.

MATERIALS AND METHODS

The enzyme used in most experiments was the B_4 isozyme of lactate dehydrogenase (EC 1.1.1.27) purified from beef heart (Worthington, "H_4 isozyme"). The enzyme used was judged to be pure by vertical starch gel electrophoresis and acrylamide "disc" gel electrophoresis. Glut herring A_4 LDH (a gift of Dr. J. Shaklee) and pig A_4 LDH (Boehringer-Mannheim, "band 5," native) were also used.

All chemicals were reagent grade and solutions were prepared using glass-distilled water. Enzyme stored as an ammonium sulfate precipitate was prepared for use by extensive dialysis against several changes of 0.002 M sodium phosphate buffer, pH 7.4. Changes in pH were accomplished by dilution or by addition of concentrated (0.5 M) buffer at the desired value of pH. Enzyme was contained only in glassware treated with "Siliclad" (Clay-Adams), or polypropylene or Teflon vessels. Stirring of solutions of enzyme to initiate renaturation was kept to a minimum since it was found that recoverable activity can be reduced significantly by excessive stirring. Optimal conditions for the recovery of activity by acid-denatured enzyme were determined and include protein concentration less than 0.3 mg/ml, pH greater than 7.0, and buffer concentration less than 0.4 M.

Protein concentration was determined spectrophotometrically. Catalytic activity was determined at 25° using an assay solution consisting of 1.4×10^{-4} M NADH plus 5×10^{-4} M sodium pyruvate in 0.1 M sodium phosphate buffer, pH 7.0.

Vertical starch gel electrophoresis was performed according to the method of Smithies (1955). Staining for activity was as described by Massaro (1967).

Gel-filtration chromatography was performed at 4°. Sephadex G-100 and G-150 were products of Pharmacia. Native LDH, yeast alcohol dehydrogenase, bovine serum albumin, ovalbumin, sperm whale myoglobin, blue dextran 2000, and sodium chloride were used as standards. The molecular weight of the acid-denatured enzyme based on the assumption of a globular structure was estimated according to the method of Andrews (1964, 1965).

Equilibrium centrifugation was performed with the use of a Beckman-Spinco Model E analytical ultracentrifuge equipped with a helium-neon laser light source (Williams, 1972) capable of firing in synchrony with the rotor (Paul and Yphantis, 1973). Samples were loaded at 4°, and the rotor was prechilled. An external loading Rexolite centerpiece (Ansevin et al, 1970) and a top-loading charcoal-filled Epon centerpiece, both equipped with sapphire windows, were used. Blank runs using distilled water were done along with each experimental run.

Experiments were performed according to the meniscus depletion method of Yphantis (1964). The data were reduced with a computer program (Roark and Yphantis, 1969) which smoothes the

raw data and calculates a set of local average molecular
weights along the length of the solution column. A value of
0.75 mg/gm was used for the partial specific volume (Markert
and Appella, 1961). Samples were run at two or three concentrations. Plots of the local weight-average molecular weights
vs. protein concentration for different loading concentrations
will coincide for a homogeneous or reversibly associating
system, but will fail to do so for a non-interacting mixture
of protein species.

The sedimentation coefficient of renatured LDH was measured
at a single low concentration of protein with the use of the
Model E photoelectric scanner.

RESULTS

Renaturation as a Function of Time at Low pH

Upon exposure of beef B_4 LDH to pH 2.0 or 3.0 sodium phosphate buffer at 0°, catalytic activity is destroyed completely
in less than 15 seconds. If the inactivated enzyme is transferred into neutral sodium phosphate buffer at room temperature
activity reappears. The percent of the original activity that
may be recovered as a function of the duration of exposure of
the enzyme to pH 3.0 is shown in Fig. 1. It can be seen that
the activity recovered is initially quite high, but rapidly
declines and after approximately 10 minutes levels off at a
value of about 60 percent. Some slow decline in recovered
activity may be seen to occur over the remaining period of
time shown.

Preliminary kinetic measurements indicated that the rate of
recovery of activity as well as the extent is a function of
the duration of exposure to acidic conditions, the rate declining with increasing time of exposure to pH 2 or 3. The
decreased rate could be due either to a reduction in the fraction of enzyme capable of recovering activity, or to a change
in the kinetics of the renaturation process. To see which
of those possibilities might be correct, the following experiment was performed. Enzyme was exposed to pH 2.0 sodium
phosphate buffer for 20 minutes, sufficient time for the
rapid decrease in the extent and rate of recovery of activity
to be complete, and then neutralized at two different values
of concentration. The time-course of the recovery of activity
was determined and is shown in Fig. 2. Recovery of activity
can be seen to occur with a short induction period. The
over-all process is independent of concentration, indicating
it to be first-order with respect to the concentration of the
enzyme. A plot of the logarithm of the activity remaining to

154

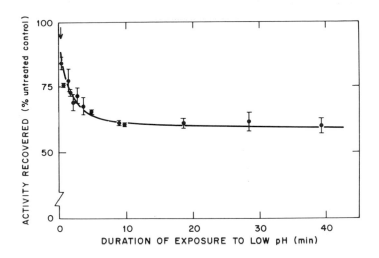

Fig. 1. Recovery of enzymatic activity as a function of the duration of exposure to low pH. Beef B_4 LDH was diluted 20-fold into 0.1 M sodium phosphate buffer, pH 3.0, containing 0.002 M dithiothreitol (DTT), at 0°C, at time zero, and then 20-fold to a concentration of 0.021 mg/ml in 0.2 M sodium phosphate buffer, pH 7.3, containing 0.002 M DTT, at 21°, at the times indicated. After four hours the samples were assayed for catalytic activity, which is plotted as the percent of the activity of an untreated control. Enzyme exposed to pH 3.0 loses activity completely at least as early as the time indicated by the arrow.

be recovered vs. time was linear following the induction period, the half-time for the linear phase being about 15 minutes.

The characteristics of this process are in contrast with the rapid (90-100 second half-time) second-order recovery of activity reported for enzyme exposed to low pH for brief periods of time (Anderson and Weber, 1966; Levitzki and Tenenbaum, 1974). Thus, as the duration of exposure to acidic conditions increases the kinetics as well as the extent of renaturation is seen to change. This behavior implies that beef B_4 LDH exists in two distinct denatured states at low pH. The first is a short-lived state from which complete recovery of activity is obtained and for which recovery of activity follows second-order kinetics. Over a period of 5-10 minutes this state gives rise to another from which only a fraction of the original activity is recovered, and for which

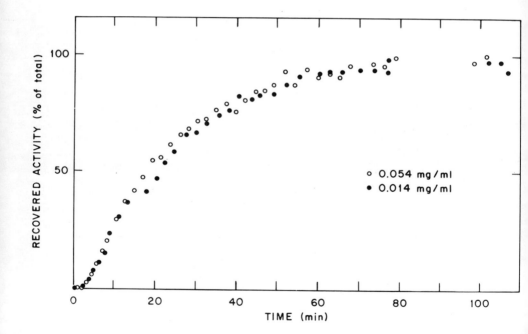

Fig. 2. Time-course of the recovery of enzymatic activity.
Beef B$_4$ LDH was incubated for 20 minutes in 0.1 M sodium phosphate buffer, pH 2.0, containing 0.002 M DTT, at 0°, and then
diluted into the renaturation buffer described in Fig. 1, at
20°. The final concentration of enzyme was 0.014 mg/ml (●),
and 0.055 mg/ml (○). Data are plotted as the percent of the
final activity recovered, 43 and 30 percent of the activity
of an untreated control, respectively (these values reflecting
a generally observed dependence of the extent of recovery of
activity on enzyme concentration).

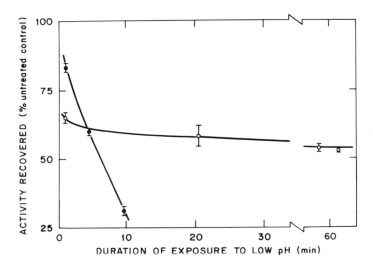

Fig. 3. Recovery of activity as a function of the duration of exposure of glut herring and pig A_4 LDH to low pH. Experimental procedure as described in Fig. 1. Glut herring A_4 LDH diluted to a concentration of 0.15 mg/ml in 0.1 M sodium phosphate buffer, pH 3.0, containing 0.002 M DTT, at 0°, and then to a concentration of 0.0075 mg/ml in the renaturation buffer described in Fig. 1, at 22° (○). Pig A_4 LDH diluted to a concentration of 0.35 mg/ml in 0.2 M glycine-H_3PO_4, pH 2.5, containing 0.001 M DTT and 0.001 M ethylenediamine tetra-acetate (EDTA), at 0°, and then to a concentration of 0.018 mg/ml in 0.2 M sodium phosphate buffer, pH 7.4, containing 0.001 M DTT and 0.01 M EDTA, at 25° (buffers after Levitzki. 1972) (●).

the recovery process follows the largely first-order kinetics described above. The second denatured state would appear to be rather stable, judging from the nearly constant recovery of activity obtained after about 10 minutes at low pH. The slow decline in recovery that is observed during this period would imply a further slow change in the state of the enzyme.

Recovery of activity as a function of the duration of exposure to low pH was also determined for glut herring and pig A_4 LDH (Fig. 3). Over a period of an hour at pH 3.0 the glut herring enzyme shows a nearly constant extent of recoverable activity, suggesting that the process of denaturation for this isozyme may be similar to that for the beef B_4 isozyme. The pig isozyme, on the other hand, shows a rapid decline in recoverable activity upon exposure to low pH, suggesting that the mechanism of denaturation in this case may be somewhat different. A comparison of the time-course of recovery of activity by beef B_4 and glut herring A_4 LDH after one hour at low pH (Fig. 4A) suggests that renaturation may follow a characteristic rate for each isozyme type.

Hybridization

The following experiment was performed to assess the ability of the beef B_4 and glut herring A_4 isozymes to hybridize during the course of the renaturation process. After incubation of both isozymes at pH 3.0 for an hour, one was diluted into neutral buffer, and the other then added after intervals of 1, 20, and 60 minutes. Neutralization was also carried out in the reverse order, and, in addition, the two isozymes were mixed just before neutralization. Recovery of activity was allowed to proceed for 18 hours after which time aliquots of each of the samples were subjected to vertical starch gel electrophoresis and stained for activity (Fig. 4B). Extensive hybridization is observed when the two enzymes are mixed before neutralization (sample 4). (A sixth hybrid band may be noted, the origin of which is not understood.) If the two isozymes are neutralized together within a period of a minute, hybridization is still extensive (samples 5 and 8), but is reduced when one of the isozymes is allowed to renature for an hour before addition of the second (samples 7 and 10). Bands corresponding to asymetric hybrids may also be noted in these latter samples.

Determination of the State of Association at Low pH

The apparent stability of the denatured state reached after 5 to 10 minutes at low pH (Fig. 1) suggested that characterization by equilibrium methods might be possible. Initial attempts to determine the molecular weight of the beef B_4 protein at low pH by equilibrium centrifugation indicated that it was composed of a mixture of non-interacting molecular weight species. To separate these components the protein was subjected to gel-filtration chromatography at pH 2.0. Chromatography on a 0.9 x 90 cm column packed with Sephadex G-100 showed the protein to consist of two major fractions, as can be seen in Fig. 5. The peak position of the trailing fraction corresponds to that expected for a globular protein of 65,000 daltons molecular weight. Aliquots from each of the collected samples were diluted into neutral buffer to test for the ability to recover activity. It can be seen (dotted

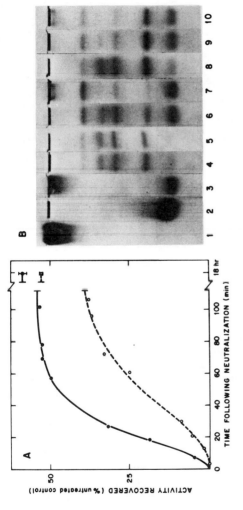

Fig. 4. A: Time-course of the recovery of activity by beef B_4 (●) and glut herring A_4 (○) LDH. Conditions for the denaturation and renaturation of both isozymes were as described in Fig. 3 for the glut herring isozyme, and the time of exposure to low pH was 1 hr.   B: Vertical starch gel electrophoresis of isozymes hybridized under conditions used in A.   (1) glut herring A_4; (2) beef B_4; (3) A_4 plus B_4 carried through the hybridization procedure at pH 7.3; (4) A_4 and B_4 denatured at pH 3.0 and mixed before neutralization; (5) - (7) denatured B_4 neutralized first, and denatured A_4 added after 1, 20, and 60 minutes; (8) - (10) denatured A_4 neutralized first, and denatured B_4 added after 1, 20, and 60 minutes.   The gel was stained for activity twice.

159

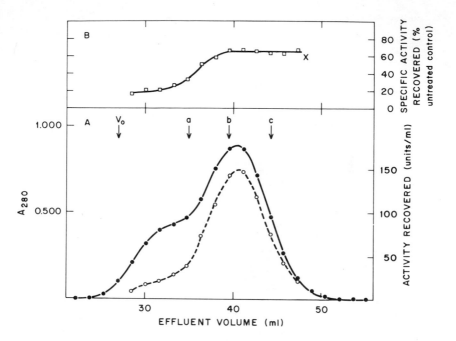

Fig. 5. Sephadex G-100 chromatography of beef B$_4$ LDH at pH
2.0. Conditions were 0.1 M sodium phosphate buffer, pH 2.0,
containing 0.002 M DTT, at 6°. Protein from each of the tubes
was diluted 10-fold into 0.2 M sodium phosphate buffer, pH 7.3,
containing 0.002 M DTT, at 16°, for renaturation. Samples
were assayed for catalytic activity after 24 hr. A small
correction for the dependence of recovered activity on protein
concentration was made, giving a somewhat increased peak height.
A; optical absorbance (●——●), and recovered activity, expressed as units of activity recovered per ml of column eluate
(O---O). Elution volumes of standards: V$_o$, blue dextran
2000; a, LDH; b, bovine serum albumin; c, ovalbumin. Standards were run at pH 7. B: "specific recovered activity,"
calculated from the recovered activity and protein concentration curves shown in A. Also shown is the percent of the
activity of an untreated control recovered by an unchromato-
graphed sample of enzyme, exposed for 2 hr to pH 2.0 and
then renatured (X).

line) that this property is associated with the trailing fraction. Chromatography of protein exposed to pH 2.0 for 25 hours
rather than for one hour as in Fig. 5 showed the relative pro-

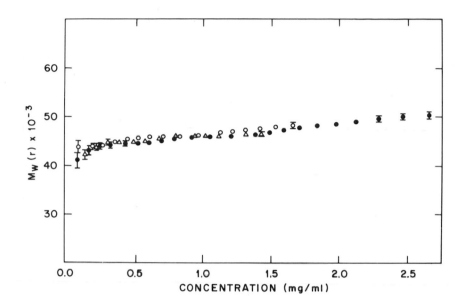

Fig. 6. Equilibrium sedimentation of protein from the peak tube of a chromatographic experiment at pH 2.0. The enzyme was in the buffer described in Fig. 5. Centrifugation was performed at 30,000 rpm, at a temperature of 5°. Concentrations loaded: 0.74 (●); 0.37 (○); 0.22 (△) mg/ml. Protein concentration in mg/ml was calculated from the fringe displacement using a value for the refractive increment at 633 nm calculated by the method of Perlmann and Longsworth (1948).

portions of the fractions to be reversed, suggesting that the leading fraction is the product of the aggregation of the material comprising the trailing fraction. Less aggregate formation was found to occur at pH 2.0 than at pH 3.0, and with decreasing concentrations of protein.

Samples from the trailing peak of a repeat of the experiment shown in Fig. 5 and from the leading peak of an experiment conducted at pH 3.0 were subjected to analysis by analytical ultracentrifugation. It can be seen in Fig. 6 that the molecular weight of the protein from the trailing fraction is close to that previously reported for the LDH monomer (Appella and Markert, 1961). The extrapolated molecular weight of about 42,000 daltons at infinite dilution is probably somewhat high due to the binding of buffer ions to the highly charged pro-

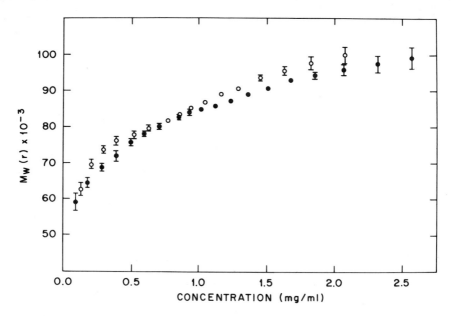

Fig. 7. Equilibrium sedimentation of the leading fraction of a chromatographic experiment at pH 3.0. Sample from the peak tube was used. The enzyme was in 0.1 M sodium phosphate buffer, pH 3.0, containing 0.002 M DTT, at 3°C. The speed was 26,000 rpm. Concentrations loaded: 0.62 (●); 0.31 (○) mg/ml.

tein (Williams et al, 1958). A weak tendency to associate to species of higher molecular weight is apparent, and this association is apparently reversible. The molecular weight of protein from the leading fraction (Fig. 7) is greater than that of the material from the trailing fraction at all values of concentration. Material from the leading fraction also appears to represent primarily a reversibly associating system, with a stronger tendency to associate. In this case a small amount of a non-interacting contaminant may also be present, as indicated by the imperfect overlap of the points from the two channels.

Characterization of the Renatured Protein

Enzyme that had been incubated at pH 3.0 for an hour and

then allowed to recover activity for 14 hours was concentrated and analyzed by gel-filtration chromatography on Sephadex G-150. The re-activated enzyme eluted at a position corresponding to that of the LDH tetramer and showed 89 percent of the native specific activity. It is not clear whether this represents a significant difference in the properties of the renatured enzyme, as has been reported for LDH renatured after exposure to concentrated lithium chloride (Levi and Kaplan, 1971). A second peak of inactive protein was also recovered and ran in the void volume.

The $s_{20,w}$ of the renatured enzyme was determined at a concnetration of 0.25 mg/ml to be 7.1 S, a value consistent with that previously reported for the native enzyme (Markert and Appella, 1961).

DISCUSSION AND CONCLUSIONS

The results presented here are consistent with the following scheme for the denaturation and renaturation of beef B_4 LDH:

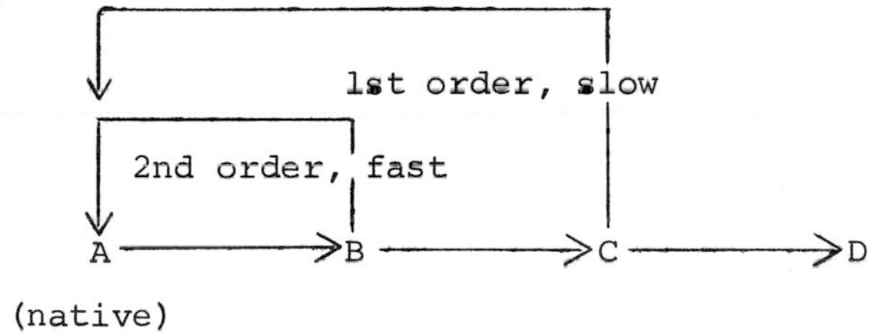

(native)

Upon exposure to low pH conditions, the native enzyme, A, is transformed to an inactive state, designated by B. This is the form studied by Anderson and Weber and by Levitzki and Tenenbaum and is characterized by complete recovery of activity following second-order kinetics. B undergoes further transformation to a state designated by C, from which recovery of activity is less complete (Fig. 1). Re-activation now occurs following an observable induction period after which it behaves as a first-order process (Fig. 2). In addition to these changes, C appears to be slowly transformed to a third denatured state, designated by D, which can no longer be restored to the native state. D is observed as the non-renaturable high molecular weight aggregate seen in Figs. 5 and 7 and its formation would account for the slow secondary decline in recovered activity observed after the first 5-10 minutes exposure to acidic pH (Fig. 1).

A suggested molecular mechanism for the denaturation process is shown below. The second-order kinetics of renaturation shown by B and the ability of enzyme in this form to hy-

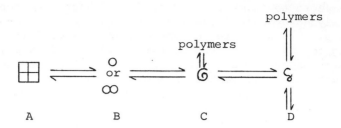

bridize are indications that B represents a dissociated state, but it is not clear whether the enzyme is fully dissociated to monomers as it is in the form of C. One possible denaturation mechanism would involve a two-step dissociation of the tetramer, B representing dimer which would further dissociate to monomer.

The capacity of enzyme in the form of C to hybridize throughout the duration of the re-activation process indicates that renaturation cannot proceed via a mechanism involving rapid reassociation of the subunits followed by a slow first-order activation step. The rate-limiting step in the recovery of activity may instead involve a change in the conformation of the subunit. The only direct evidence presented here indicating that C may represent a conformationally altered as well as a dissociated state lies in the fact that while the molecular weight as determined by equilibrium sedimentation corresponds to that of the monomer, the protein eluates at a position corresponding to that of a globular protein of molecular weight 65,000 on gel-filtration chromatography. One possible explanation for this behavior is that the protein is extensively unfolded at low pH and, therefore, occupies a larger effective volume than would a globular monomer. The fact that C recovers activity more slowly than B is consistent with the possibility that C might be extensively unfolded while B is not. Thus the conversion of A to B followed by the conversion of B to C may represent kinetically distinguishable dissociation and unfolding processes. This could reflect a general kind of behavior for LDH occurring under dissociating conditions. The transformation of the subunits to the forms represented by C or even D would account for the incomplete reversibility that is a general feature of dissociation and hybridization.

The structural change involved in the conversion of C to D is not fully understood. Enzyme in the form designated by D appears to be in rapidly reversible equilibrium with monomer (Fig. 7). The assumption of a difference in the structure of the monomer in state C as opposed to that in state D may be

required to reconcile this result with the slow conversion of
C to D. Enzyme was probably in the form designated by D in
the earlier study of the physical properties of beef heart
LDH at low pH conducted by Millar and co-workers (1969). In
addition to the fact that the enzyme was incapable of recovering activity, it also showed a strong reversible association
similar to that indicated by the results shown in Fig. 7. It
is interesting to note that despite the resemblance of the
centrifugation pattern seen in Fig. 7 to that expected for a
concentration dependent dissociation of the native LDH tetramer, D appears to be only distantly related structurally to
the native state. No evidence has been found in the present
study for species of molecular weight less than that of the
monomer as was found in the earlier study. The reason for
the earlier finding is not certain but could involve splitting
of the subunit under certain conditions (Fosmire and Timasheff,
1972) or the presence of a low molecular weight impurity in
some preparations (Huston et al, 1972).

FOOTNOTES

[1] This work represents part of a thesis submitted in partial
fullfillment of the requirements for the degree of Doctor of
Philosophy, Yale University. Present address: Laboratory of
Molecular Biology, University of Wisconsin, Madison, Wisconsin
53706.

ACKNOWLEDGMENTS

This work was supported by research grant HL 12901-05 of
the National Institutes of Health. R. V. was supported by
USPHS training grant HD-00032.
The advice and guidance of Dr. Robley C. Williams, Jr. are
gratefully acknowledged.

REFERENCES

Adams, M. J., G. C. Ford, R. Koekoek, P. J. Lentz, Jr., A.
McPherson, Jr., M. G. Rossmann, I. E. Smiley, R. W. Schevitz, and A. J. Wonacott 1970. Structure of lactate dehydrogenase at 2.8 Å resolution. *Nature* 227: 1098-1103.
Anderson, S. and G. Weber 1966. The reversible acid dissociation and hybridization of lactic dehydrogenase. *Arch.
Biochem. Biophys.* 116: 207-223.
Andrews, P. 1964. Estimation of the molecular weights of proteins by Sephadex gel-filtration. *Biochem. J.* 91:
222-233.

Andrews, P. 1965. The gel-filtration behavior of proteins related to their molecular weights over a wide range. *Biochem. J.* 96: 595-606.

Ansevin, A. T., D. E. Roark, and D. A. Yphantis 1970. Improved ultracentrifuge cells for high-speed sedimentation equilibrium studies with interference optics. *Anal. Biochem.* 34: 237-261.

Appella, E. and C. L. Markert 1961. Dissociation of lactate dehydrogenase into subunits with guanidine hydrochloride. *Biochem. Biophys. Res. Commun.* 6: 171-176.

Deal, W. C., W. J. Rutter, V. Massey, and K. E. Van Holde 1963. Reversible alteration of the structure of enzymes in acidic solution. *Biochem. Biophys. Res. Commun.* 10: 49-54

Everse, J. and N. O. Kaplan 1973. Lactate dehydrogenases: structure and function. *Adv. Enzymol.* 37: 61-133.

Fosmire, G. J., and S. N. Timasheff 1972. Molecular weight of beef heart lactate dehydrogenase. *Biochemistry* 11: 2455-2460.

Huston, J. S., W. W. Fish, K. G. Mann, and C. Tanford 1972. Studies on the subunit molecular weight of beef heart lactate dehydrogenase. *Biochemistry* 11: 1609-1612.

Jaenicke, R. 1970. Quaternary structure and conformation of lactic dehydrogenase and glyceraldehyde-3-phosphate dehydrogenase. In *Pyridine nucleotide-dependent dehydrogenases*. H. Sund, ed. Springer-Verlag, Berlin, pp. 71-90.

Jaenicke, R. and S. Knof 1968. Molecular weight and quaternary structure of lactic dehydrogenase. 3. Comparative determination by sedimentation analysis, light scattering and osmosis. *Eur. J. Biochem.* 4: 157-163.

Levi, A. S. and N. O. Kaplan 1971. Physical and chemical properties of reversibly inactivated lactate dehydrogenases. *J. Biol. Chem.* 246: 6409-6417.

Levitzki, A. 1972. The assembly pathway of lactic dehydrogenase isozymes from their unfolded subunits. *FEBS Lett.* 24: 301-304.

Levitzki, A. and H. Tenebaum 1974. Dimers as intermediates in the assembly of tetrameric proteins: a study on lactate dehydrogenase isozymes. *Israel J. Chem.*, in press.

Markert, C. L. 1963. Lactate dehydrogenase isozymes: dissociation and recombination of subunits. *Science* 140: 1329-1330.

Markert, C. L. and E. Appella 1961. Physicochemical nature of isozymes. *Ann. N. Y. Acad. Sc.* 94: 678-690.

Massaro, E. J. 1967. Induction of subunit reassociation among lactate dehydrogenase isozymes. *SABCO J.* 3: 51-62.

Millar, D. B., V. Frattali, and G. E. Willick 1969. The quaternary structure of lactate dehydrogenase. I. The subunit molecular weight and the reversible association at acid pH. *Biochemistry* 8: 2416-2421.

Paul, C. H. and D. A. Yphantis 1972. Pulsed laser interferometry (PLI) in the analytical ultracentrifuge: II. Clocked trigger Circuit. *Anal. Biochem.* 48: 605-612.

Perlmann, G. and L. Longsworth 1948. The specific refractive increment of some purified proteins. *J. Amer. Chem. Soc.* 70: 2719-2724.

Roark, D. E. and D. A. Yphantis 1969. Studies of self-associating systems by equilibrium ultracentrifugation. *Ann. N.Y. Acad. Sc.* 164: 245-278.

Smithies, O. 1955. Zone electrophoresis in starch gels: group variations in the serum proteins of normal human adults. *Biochem. J.* 61: 629-641.

Williams, J. W., K. E. Van Holde, R. L. Baldwin, and H. Fujita 1958. The theory of sedimentation analysis. *Chem. Rev.* 58: 715-806.

Williams, R. C., Jr. 1972. A laser light source for the analytical ultracentrifuge. *Anal. Biochem.* 48: 164-171.

Yphantis, D. A. 1964. Equilibrium ultracentrifugation of dilute solutions. *Biochemistry* 3: 297-317.

MECHANISM OF ISOZYME FORMATION AS RELATED TO SUBUNIT INTERACTIONS

DAVID B. MILLAR
Laboratory of Physical Biochemistry
Naval Medical Research Institite
National Naval Medical Center
Bethesda, Maryland 20014

ABSTRACT. In 50% ethylene glycol, lactate dehydrogenase exists as a reversible associating system which is best described as composed of monomers (MW 36,000), dimers (MW 72,000), and tetramers (MW 144,000). Thermodynamic and Cornish-Bowden-Koshland statistical analyses of the system show that the association of subunits follows and obligatory dimer pathway. The tetramer hence can be described as a dimer of dimers. The subunit interactions are isologous with one interdimer binding interface being weaker than the other by $\sim$1 Kcal. Hence, dissociation is topographically directed. Circular dichroic studies show that α helical content is preserved during dissociation but some β sheet structure is lost. It is concluded that dissociation preferentially occurs along the β sheet interface leading to dimers bonded along α helical interfaces. These in turn dissociate. Hybridization between LDH-A and LDH-B occurs in this solvent and is very temperature dependent. The kinetics of hybridization are second order and the rate of formation of A_2B_2 is proportional to weight fraction of dimer. Hybridization mechanisms involving a simultaneous four monomer collision or two tetramers colliding followed by subunit interchange are not compatible with the data. Thermal denaturation studies show the ease of denaturation to be monomer > dimer >tetramer. Equations are presented which allow estimates of monomer concentration and the dimerization equilibrium constant to be estimated from thermal denaturation data. Enzyme activity studies suggest that the monomer might be enzymatically active although at a very low level. A model is given for in vivo hybridization based upon these factors.

From the Bureau of Medicine and Surgery, Navy Department Research Subtask MR041.06.01.0005AOCK. The opinions and statements contained herein are the private ones of the writer and are not to be construed as official or reflecting the views of the Navy Department or of the Naval Service at Large.

INTRODUCTION

One of the reasons why the enzyme lactate dehydrogenase (LDH) is and has been so intensively studied is that in vivo its isozymic forms respond dramatically to varying environmental stresses and disease states. Understanding the basic mechanism by which hybridization takes place between A and B subunits of LDH would be of both practical and theoretical interest since it would allow investigators to make some testable speculations about the environment of the cell in which hybridization takes place. To the physical biochemist in vivo studies of hybridization have been, until recently, limited to either, freeze-thaw hybridization, guanidine or urea hybridization, or acid pH induced hybridization. In these denaturant solutions, either only monomer existed or the n-mer makeup of the solution was not known. This deffiency made interpretation of the data difficult. We would like to take this opportunity to present data we have collected employing 50% ethylene glycol as a gentle denaturant.

RESULTS

MOLECULAR WEIGHT STUDIES

Ethylene glycol has been found to split polymeric proteins to their constituent subunits (Yielding, 1971, and Contaxis, 1971). Not surprisingly, we found that the molecular weight of B_4LDH also decreased in 50% ethylene glycol, 0.15 M sodium phosphate, 0.005 M Cleland's reagent, pH 7.5 (Figure 1). The molecular weight of the intact tetramer is about 145,000. The purpose of the experiment shown in Figure 1 is to demonstrate that there is no time dependent nonspecific aggregation of LDH and therefore the n-mer composition of the solution is probably time invariant. To make an analysis of the kind of species present in an associating system it is necessary to make molecular weight measurements over a wide protein concentration range. Figure 2 presents such experiments done for LDH in 50% ethylene glycol, 0.15 M sodium phosphate, 0.005 M Cleland reagents, pH 7.5 (known hereafter as 50% E.G.). The data were obtained by means of Yphantis techniques (Yphantis, 1960 and Yphantis, 1964) utilizing both Schlieren and interference optics. The $\overline{V}$ of the native protein, 0.747 ml/g (Millar, 1962) was assumed not to change in the presence of ethylene glycol. This has been successfully suggested to be the case for two other proteins (Contaxis and Reithel, 1971 and Tanford et al., 1962). Since the data from different experiments are seen to be overlapping, we may conclude that

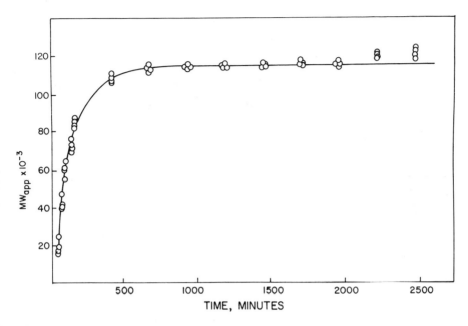

Fig. 1. Time required for equilibrium to be obtained in 50% ethylene glycol. Solution column 0.8 mm. The four symbols indicate the four channels observed.

LDH in this solvent system is a reversibly associating system. Prior to analysing the data we need to know the molecular weight at infinite dilution. This is generally the monomer molecular weight but does not necessarily have to be. Figure 3 shows the results of several high speed Yphantis meniscus depletion runs. A free hand line drawn through the data points would appear to intersect the y-intercept at a point equal to a molecular weight of 36,000. This intercept is equal to the accepted value of the monomer molecular weight. Such a result would lend support to the assumption concerning the constancy of the partial specific volume. Two reasonable possibilities must be laid to rest before this result can be accepted. The first is that these high speed miniscus de- pletion experiments are run at such low concentrations that inherent errors might possibly be magnified and prejudice the apparent intercept; secondly, lower molecular weight species contamination might be present which artifactually lowers the apparent molecular weight. Although acrylamide gel electro- phoresis indicated only one coomasie blue staining band present in our LDH-B preparations we decided to test the en- zyme with sodium dodecyl sulfate gel electrophoresis. We

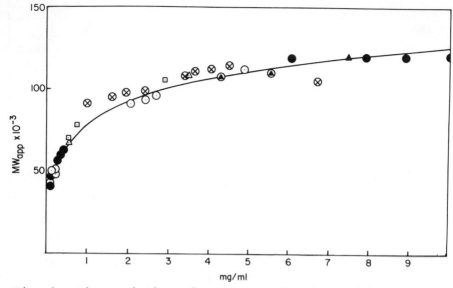

Fig. 2. The variation of apparent molecular weight of lactate dehydrogenase in 50% ethylene glycol with protein concentration at 20°C. Speeds of 6000 to 39,460 rev./min were used. The line connecting the points is theoretical. See text for details.

selected conditions designed to enhance the action of any endogenous proteases (Weber and Osborn, 1969). With each gel experiment a full set of molecular weight markers was included. The results of a number of such experiments are shown below:

1. 5 mg/ml enzyme in 0.15 M Na phosphate, 0.001 M Cleland's reagent, 1% sodium dodecyl sulfate, pH 7.5, incubated 5, 15, 30, 60, 90, and 120 min at 4]°C. Results: one band, MW = 34,500.

2. As above but with 20 mg/ml enzyme. Results: turbid solutions.

3. 15 mg/ml enzyme but in 0.05 M Na phosphate, 0.02 M Cleland's reagent, pH 7.5 and 1% sodium dodecyl sulfate. (The lowered phosphate and increased sulfhydryl reagent concentrations were used to avoid aggregation). Turbidity resulted at 41°C but after standing for 15 minutes at room temperature, turbidity was markedly reduced. Results: one band, MW = 38,000.

4. 10 mg/ml enzyme is 50% ethylene glycol dialyzed overnight at 20°C. This solution was brought to 1% sodium dodecyl

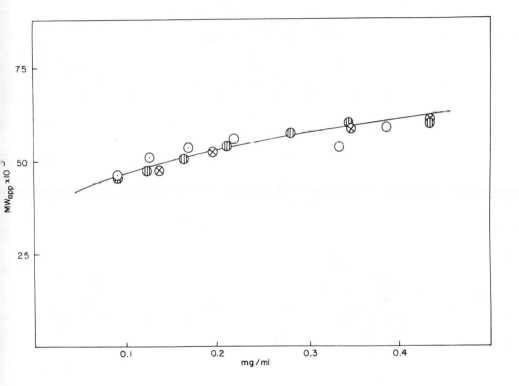

Fig. 3. Estimation of the subunit molecular weight of lactate dehydrogenase in 50% ethylene glycol by meniscus depletion experiments. Speeds of 35,600 to 39,460 were used. Temp 20°C.

sulfate and treated as shown:

A)	1 h at 20°C	MW = 40,000
B)	1 h at 41°C	MW = 40,000
C)	1 h at 30°C	MW = 41,000

To check that any protease which might be present in our

enzyme preparations was capable of functioning under the denaturing conditions employed we performed experiments in which various proteases were added to the enzyme solutions. The conditions and results are shown below:

5. 15 mg/ml enzyme + 0.3 mg/ml papain in 50% ethylene glycol was dialyzed overnight at 20°C, then brought to 1% sodium dodecyl sulfate, The solution was incubated 2 hr at 41°C. Results: one band MW = 38,000.

6. 20 mg/ml enzyme in 50% ethylene glycol was incubated 8 hr at 20°C, then dialyzed free of ethylene glycol at 4°C overnight. The following day, it was brought to 1% sodium dodecyl sulfate and incubated 1 hr at 41°C. Results: MW = 36,000.

7. 10 mg/ml enzyme in 1% sodium dodecyl sulfate, 0.15 M Na phosphate, 0.001 M Cleland's reagent pH 7.5. At protease levels of 0.3 mg/ml and 1 h at 41°C the following results were obtained.

Chymotrypsin: bands of 18,000, 20,000, 28,000, and 36,000 MW were seen.

Trypsin: bands of 17,800, 22,000, 29,000, and 35,700 MW were seen.

Pepsin: variable results obtained; occasionally bands of MW 18,000 were seen. These experiments show that 1) exogenous proteases can cleave lactate dehydrogenase to lower MW species under the conditions used and 2) no detectable protease contamination exists in our enzyme preparations.

Recently we reported that at acid pH and in high levels of guanidine the molecular weight of the lactate dehydrogenase subunit apparently decreased to 18,000 (Millar et al., 1969). We saw no evidence of 18,000 molecular weight species in the preparations studied here but only when exogenous protease was added. Since we do not find evidence of contaminating proteases in the present study it may be that the drop in molecular weight we previously observed was due to a hydrolysis catalyzed by an element in certain guanidine preparations. This was first suggested by Fosmire and Timasheff (1972) who have also observed this phenomenon with certain guanidine preparations. Huston, Fish, Mann, and Tanford (1972) who also observed this lowering of molecular weight suggested that it was due to a low molecular weight impurity. However, on the basis of the sodium dodecyl sulfate studies and the sedimentation data we conclude that the enzymes subunit molecular weight in 50% ethylene glycol is 36,000.

DETERMINATION OF THE CORRECT MODEL SYSTEM

To determine the model system which best fits the molecular

weight data we first must make certain assumptions. We have already assumed that the $\bar{V}$ of the native protein is unchanged in 50% E.G.*; secondly, we assume that the $\bar{V}$ of all n-mers is the same and that there is no pressure dependence of $\bar{V}$. We then determined the apparent weight fraction of monomer, fc_1, without *a priori* assumptions by use of the Steiner integral (Steiner, 1952) as modified by Adams and Williams (1964):

$$(1) \quad \text{Ln } fc_1 = \int_0^c \left(\frac{M_1}{M_{wapp}} - 1 \right) \frac{dc}{c} = \text{Ln } fc_1 + BM_1c$$

where M_1 is the monomer molecular weight, 36,000 g/mole, and B is the second virial coefficient.

The curve of M_{wapp} versus protein concentration goes above a value of 72,000 and does not recurve below this value. This result suggests two things: One, tetramers are probably present; and two, the value of the virial coefficient is not large. The likely models then are: monomer $\rightleftharpoons$ tetramer, monomer $\rightleftharpoons$ dimer $\rightleftharpoons$ trimer $\rightleftharpoons$ tetramer, and monomer $\rightleftharpoons$ dimer $\rightleftharpoons$ tetramer. We did not consider an isodesmic association since we did not have molecular weight data at high enough protein concentrations, but crystallographic data tend to rule out this possibility (Adams, et al., 1969).

Following Adam's (1967) guide lines we derived testable expressions for the above schemes. For the monomer $\rightleftharpoons$ tetramer model the final equation in which only one unknown is present is:

$$(2) \quad (4cM_1/M_{napp}) - c = 3\alpha \ (\exp(-BM_1c))$$

where M_{napp} is the apparent number average molecular weight and α is f_{app} x c. The means of determining cM_1/M_{napp} have been detailed elsewhere (Adams, 1967). All of the left hand side (L. H. S.) of Equation 2 is directly determinable from experimental data and in the right hand side (R.H.S.) only B is unknown. Computer programs were written in which a value of B was searched such that S $((\text{L.H.S.} - \text{R.H.S.})^2)$ versus L.H.S. was made and the result for the monomer $\rightleftharpoons$ tetramer model is shown in Figure 4A. S was 0.511 and B = 0.02. The large change in slope demonstrates that this model is not satisfactory.

The second model we tested is the monomer $\rightleftharpoons$ dimer $\rightleftharpoons$ tetramer. The testable equations are:

$$(3) \quad 8cM_1/M_{napp} - 6c = 3\alpha \ (\exp(-BM_1c)) + 4BM_1c^2 + 1/Y$$

---

*There is a small amount of lipid like material which is released by 50% E.G. treatment of the enzyme. The amount is too small to affect our calculations.

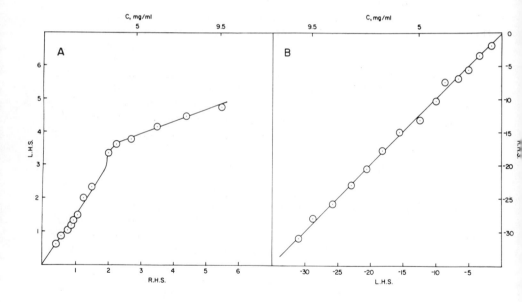

Fig. 4A. An L.H.S.-R.H.S. plot for a monomer-tetramer association model. See text.
Fig. 4B. An L.H.S.-R.H.S. plot for a monomer-dimer-tetramer
association model. See text.

where $1/Y = 1/((M_1/cM_{wapp}) - BM_1)$. The computer solutions
were $B = -.002$ and $S = 0.277$. For this minimum, Figure 4B
displays L.H.S. versus R.H.S. The plot is linear with slope
of 1.0 and intercept of zero. These conditions indicate the
aptness of model 2.

The third model monomer $\rightleftharpoons$ dimer $\rightleftharpoons$ trimer $\rightleftharpoons$ tetramer, for which the appropriate equation is:

(4) $24 \, cM_1/M_{napp} - 26c = 6\alpha(\exp(-BM_1 c)) + 12 \, BM_1 c^2 - \Psi - 9/Y$

Y is as defined above, and Ψ is equal to:

$$(5) \quad = \frac{d/dc(M_1 cM_{wapp})}{\dfrac{M_1}{cM_{wapp}} - BM_1} \quad 3$$

d/dc $(M_1 cM_{wapp})$ was evaluated via computer programming. A B of 0.0 gave a minimum of $S = 40$. This very large S and a curvilinear plot of L.H.S. versus R.H.S. confirmed the non-suitability of this model. The monomer $\rightleftharpoons$ dimer $\rightleftharpoons$ tetramer model was therefore judged best. The equilibrium constants for monomer to dimer, K_2, and that for dimer to tetramer, K_4, were determined by standard methods (Adams, 1967) and the line drawn through the data in Figure 2 was calculated with $K_2 = 3.91$, $K_4 = 18$, and $B = -.002$ (mg/ml scale). Figure 5 gives the n-mer distribution for this model.

VISCOSITY OF LDH IN 50% E.G.

Along with dissociation, 50% E.G. swells LDH. The intrinsic viscosity of native LDH is 0.039 dL/g while for LDH in 50% E.G. the value is 0.059 dL/g. No time dependence of viscosity was noted.

CIRCULAR DICHROIC SPECTRA OF LDH-B

Fig. 6 shows the CD spectra of control LDH-B and LDH-B in 50% E.G (.02 mg/ml.). The lines through the data were calculated by the method of Saxena and Wetlaufer (1971). There is no Cleland's reagent in either of these experiments due to its high UV absorbtion. The theory of Saxena and Wetlaufer (1971) gives for control LDH 25% α helix, 39% β sheet, and remainder 36%. For 50% E.G. we calculate 27% α helix, 31% β sheet, and 42% remainder.

EFFECT OF ETHYLENE GLYCOL ON ENZYMATIC ACTIVITY

We found that in 50% E.G. the K_m for NADH was 9.0×10^{-5}, that for pyruvate 4.7×10^{-4}. Control values were 4.2×10^{-5} and 4.5×10^{-4}. V_{max} showed nearly a 7 fold decrease in the presence of 50% E.G. Since some of the changes in V_{max} are probably diffusion controlled losses, and the remainder due to conformational effects, we decided to study the effect of ethylene glycol on the enzyme at 37°. Since activity dropped as glycol concentration increased, the phenomenological

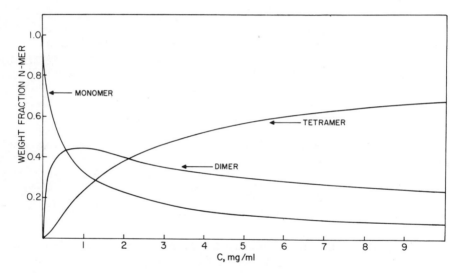

Fig. 5. The calculated distribution of monomer, dimer, and tetramer of lactate dehydrogenase in 50% ethylene glycol.

expression shown below was tested to see if it could fit the data:

$$(6) \quad \% \, A/A_O = K \, (\eta)^{-N}$$

where A_O is control activity, A is activity in the presence of glycol of viscosity η, K is a constant, and N is the order of interaction between glycol and enzyme.

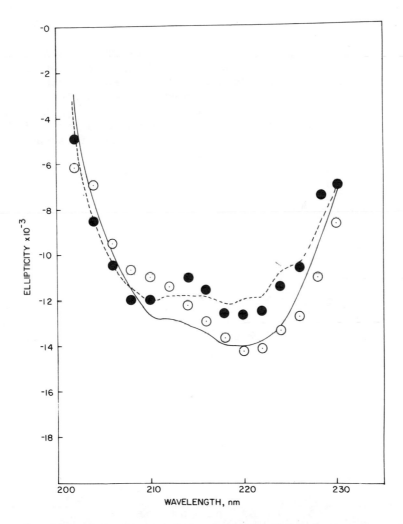

Fig. 6. The near ultraviolet circular dichroic spectra of lactate dehydrogenase in 50% ethylene glycol and in buffer. ●, 50% ethylene glycol; 0, control. The lines are calculated following Saxena-Wetlaufer.

In log form (6) becomes:

(7) $\log (\%A/A_O) = \log K + (-N \log \eta)$

if N does not change, the plot should be linear with slope -N. Figure 7 shows that a break occurs at a glycol concentration of about 35%. This suggests that a major conformational event occurs at this point. Here, N changes from 1.0 to about 4.

TIME DEPENDENCE OF LDH DENATURATION BY GLYCOL

When small aliquots of enzyme were added to 50% E.G. and tested for activity as a function of time in an assay mechanism not containing ethylene glycol, it was found that the activity declined in an unusual manner. A log plot of % activity versus time is shown in Figure 8. The plot is not rectilinear. Additional tests showed the loss not to be second order. We conclude that more than one first order process is occurring. This suggested that the n-mers might be denaturing at different rates. We tested this possibility by heating various concentrations of the enzyme 2 hr at 37° in 50% E.G. The results are shown in Figure 9. This figure shows that the amount of activity remaining is dependent upon

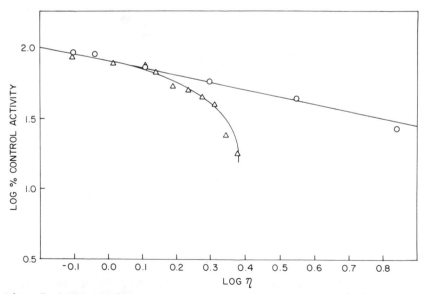

Fig. 7.   The influence of solution viscosity on activity of lactate dehydrogenase at 37°C. Δ, ethylene glycol; O, sucrose, η is viscosity in cpoise.

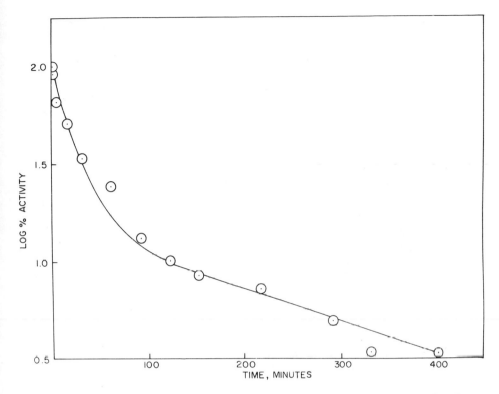

Fig. 8. Thermal denaturation of lactate dehydrogenase in 50% ethylene glycol. Temp 37°C.

---

protein concentration. Such a result is compatible with the notion that higher n-mers denature at slower rates than lower n-mers. A similar idea and supporting data have been previously advanced by Boross (1972) for D-glyceraldehyde-3-phosphate dehydrogenase. Note that the shape of this denaturation curve is similar to the tetramer weight fraction concentration curve. We suggest that the order of heat denaturability is monomer > dimer >tetramer. If we assume that each n-mer undergoes only a one step denaturation and that the denaturation of a monomer unit results in the same loss of catalytic efficiency then we can easily derive an expression which describes the denaturation of total enzyme, E. The expression is

$$(8) \quad E/E_o = c_{1,o}e^{-k_I t} + c_{2,o}e^{-k_2 t} + c_{4,o}e^{-k_4 t}$$

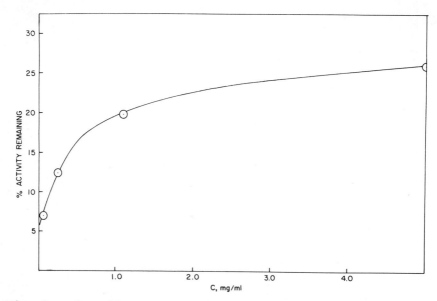

Fig. 9. The effect of enzyme concentration upon thermal denaturation at 37°C in 50% ethylene glycol.

---

where $c_{1,o}$, $c_{2,o}$, and $c_{4,o}$ refer to the percentages of the
respective n-mer present at 0 time and E_o is their sum. E
is enzyme present at time t. The k's are the respective first
order denaturation constants. If k_1, k_2, and k_4 have the
correct magnitudes then a plot of log % activity remaining
versus time will have at least two linear portions. Further
if $k_1 > k_2 > k_4$ and at a time long enough for the k_1 process
to be completed then the loss of activity at this point is
due only to the k_2 and k_4 processes. If the denaturation
data past the point of the completion of the k_1 process is
linear then extending this line back to t = 0 gives log
$(c_2 + c_4)_o/E_o$ and hence $c_{1,o}$ and from this the absolute concentration of monomer. A series of such denaturation experiments was made at different temperatures and fC_1 the weight
fraction monomer calculated. Figure 10 shows the results.
Included in these results is the value of fC_1 from equilibrium
data at 20°. The line drawn through the kinetic data is in
good agreement with the equilibrium data. These data show
that at 50°, the enzyme is virtually totally in the monomeric

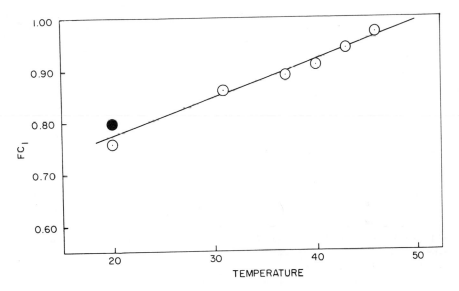

Fig. 10. Weight fraction, fc_1, determined from denaturation data versus temperature (0). O is fc_1 determined from ultracentrifuge data.

---

state. The rate constants were derived from the plots by computer fit. $k_2"$ (the sum of $k_2 + k_4$) was considerably less than k_1 as is seen in Table 1. An Arrhenius plot gave an energy of inactivation of the monomer of 73 Kcal. The $k_2"$ process was considerably less although the limited temperature range of observations prevented a good estimate.

We can evaluate k_2, the dimerization constant from the kinetic data. Since:

TABLE I
KINETICS AND ENERGETICS OF DENATURATION[A]

Temperature	K_1 (min^{-1})	K_2 (min^{-1})	ΔG_1 (KCal)	ΔG_2 (KCal)
20	1.04×10^{-4}	B	5.43	
25	4.03×10^{-4}	B	4.61	
30	1.56×10^{-3}	B	3.88	
33	3.06×10^{-3}	B	3.81	
37	5.01×10^{-2}	3.47×10^{-3}	1.84	+3.48
40	0.184	3.55×10^{-3}	1.05	+3.5
43	1.61	3.63×10^{-3}	0.32	+3.51
46	2.032	3.80×10^{-3}	-0.302	+3.52
50		C	-0.454	

[A]Concentration: 0.05 mg/ml
[B]Too slow to measure
[C]No evidence for K_2 process

(9) $\quad E_o = c_{1,o} + c_{2,o} + c_{4,o} = c_{1,o} + k_2 c_{1,o}^2 + k_4 c_{1,o}$

if E_o is small and $k_4 < 100\ k_2$ then a good approximation is:

(10) $\quad k_2 = \dfrac{E_o - c_{1,o}}{c_{1,o}^2}$

The equilibrium data on k_2 and k_4 and the kinetically demonstrated decrease of $c_2 + c_4$ versus temperature indicate that the constraint that $k_4 < 100\ k_2$ has been met. We calculated the k_2 values and converted them to a moles/liter basis (k_2'). Table 2 shows the results. At 20°, the kinetically determined value of k_2' 14.9 x 10^4 M^{-1} is in fair agreement with the equilibrium value 7.04 x 10^4 M^{-1}. As expected k_2 decreases with increasing temperature. The ΔH for the reaction is + 13.3 Kcal/mole. The entropy change is -22cal/mole suggesting that, along with dimerization, solvent reorganization along subunit surfaces may be occurring. The positive sign of ΔH suggests the dimerization reaction to be chiefly an electrostatic process in accord with other experimental data (Millar, 1962, and Jaenicke and Knof, 1968).

RATE CONTROL OF HYBRIDIZATION KINETICS
BY DIMER WEIGHT FRACTION

This laboratory had previously concluded (Millar et al., 1971) that hybridization between LDH-A$_4$ and LDH-B$_4$ which took place in the absence of denaturants or freeze-thaw, required the

TABLE 2
THERMODYNAMICS OF DIMER FORMATION[A]

Temperature	$K'_2 (M)$	ΔG
20	14.96×10^4	-7.0 KCal
37	4.72×10^4	-6.6 KCal
40	3.40×10^4	-6.5 KCal
43	2.45×10^4	-6.3 KCal
46	1.0×10^4	-5.8 KCal

$\Delta S = -22$ cal
$\Delta H = -13$ KCal
[A]Concentration: 0.05 mg/ml

enzymes to be in an activated state. We suggested that since coenzyme by itself or an abortive ternary complex could stop hybridization, these agents must prevent the transition to this complex. Since we have demonstrated in this report that appreciable conformational and destabilizing events occur in 50% E.G. We wondered whether one of the conformational events might place the enzyme in a configuration close to the activated state. An equimolar mixture of LDH-A and LDH-B was incubated at 30°. Rapid hybridization was observed to occur as shown by cellulose acetate electrophoresis and by enzyme activity stain (Millar et al., 1971). Analysis of the kinetics was complicated by the time dependent loss of activity discussed above, and also by the fact that we observed different rates of denaturation for LDH-A and LDH-B. We therefore decided to measure the initial velocity of hybridization at time zero utilizing A_2B_2 (band 3) as hybridization indicator. (We chose A_2B_2 as it is the first band to be observed and usually remains the dominant band throughout the experiment). In hybridization experiments many samples were analysed early during the incubation period and the apparent % A_2B_2 fitted to a curve of the third degree. Computer tests with a model system utilizing: (1) theoretical denaturation rates for all products and starting materials, (2) binomial distribution of the hybridization reaction, and (3) theoretical second order rate constants of formation indicated that even under adverse conditions the fitting procedure could get within 10% of the value of % A_2B_2 at 30 seconds (our arbitrarily chosen zero time). Keeping A/B LDH at 1, hybridization experiments were done at total protein concentrations between 0.8 mg/ml and 4 mg/ml. Figure 11 shows the results. The initial velocity-concentration profile resembles the

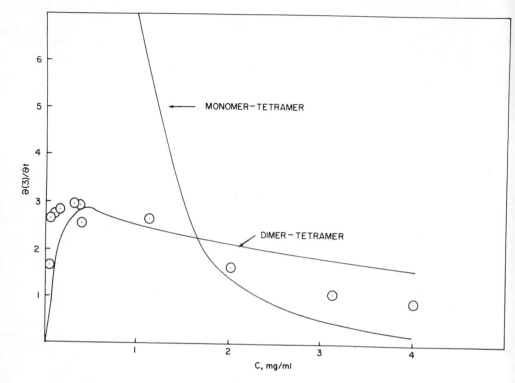

Fig. 11. Dependence of hybridization rate of A_4 and B_4 lactate dehydrogenase upon total enzyme concentration. 50% ethylene glycol at 30°C. The lines are calculated. See text for details.

dimer weight fraction curve. This suggested that hybridization kinetics were related to the weight fraction of dimer. If in 50% E.G. LDH-A is dissociated in the same way as LDH-B, then the formation of A_2B_2 can be described as: 2 dimers $\longrightarrow$ tetramer. Therefore, expressing band 3 (% A_2B_2 of totals bands) with D being dimer concentration expressed as weight fraction, the rate equation is:

$$(11) \quad d(3)/dt = k_{D-T}(D)^2$$

We utilized the 20° n-mer distribution (since we did not have the 30° distribution) and assumed that k_{D-T} was independent of concentration. k_{D-T} was solved for over the concentration range and was found to be 12.1 ± 1.8% (3)/min. The lower line in Figure 11 was calculated with this rate constant. Also shown in Figure 11 is a line calculated from the monomer tetramer case in which the equation is:

(12) $d(3)/dt = k_{M-T} (M)^4$

Here using either the kinetically derived value of A or the sedimentation equilibrium value of A at 20° gave a line similar to the upper line of Figure 11. For the latter case k_{M-T} has the value 46 ± 16. The M-T model gives a very bad fit to the data and we can reject it on these grounds. (This conclusion is in agreement with the rejection of an M-T model for the association of LDH-B as studied by sedimentation equilibrium). The shape of the hybridization curve also allows us to reject a mechanism involving two tetramers colliding and interchanging subunits (Jaenicke et al., 1971).

HYBRIDIZATION TEMPERATURE DEPENDENCE

LDH-A and LDH-B in 50% E.G. at a total concentration of 2 mg/ml (A/B= 1) were incubated at various temperatures, aliquots sequentially removed and then electrophoresed. Figure 12 shows the results expressed as band 3 formation as a function of time at different temperatures. Hybridization temperature dependence is quite pronounced and indeed hybridization is almost instantaneous at 37°. A_2B_2 was the dominant band at all temperatures. Jaenicke et al.,(71) in hybridization experiments conducted in slightly acid medium also have observed A_2B_2 dominance. The large activity losses prevented a precise estimate of the energy of activation but a value of 30-50 Kcal or greater would be compatible with the data.

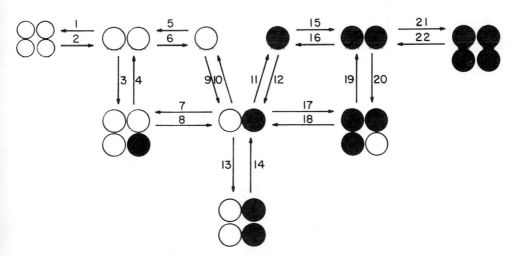

Fig. 12. A model for in vivo and in vitro hybridization between A_4 and B_4 lactate dehydrogenase.

OTHER ALCOHOLS AND HYBRIDIZATION

Methanol, ethanol, propanol, and butanol in 0.3 M sodium phosphate, pH 7.5 rapidly denatured the enzyme at 20° or 37°. No hybridization took place. However, molecular weight analysis showed considerable dissociation of the enzyme. If the phosphate concentration were reduced to 0.025 M, rapid hybridization occurred and the rate was proportional to alcohol concentration. At 15% W/V, denaturation was dominant.

DISCUSSION

Our molecular weight analysis and kinetic analysis of hybridization rates both indicate that the means of tetramer formation from monomer is via successive dimerization - that is, one may regard LDH (LDH-B at least) as a dimer of dimers. A scheme for hybridization involving such a concept is shown in Figure 12. It may be noted that such a scheme offers elaborate opportunity for environmental control by appropriate modification of one or more of the rate steps.

Dividing the free energy of association by the number of subunits involved ($\Delta G°/n$) gives a measure of the equivalence or non-equivalence of intersubunit binding edges (Klotz et al., 1970). The bonds joining two monomers to form a dimer differ about -1 Kcal from those for the association of two dimers to form a tetramer. Therefore there appears to be some difference in the nature of the binding areas on different edges of the subunit. Utilizing our association constants and the approach of Cornish-Bowden and Koshland (1971) and the methodology of Klotz et al., (1970) we can calculate whether intersubunit binding is Isologous (q site binds to q site, p site to p site, etc., etc.) or Heterologous (q site binds to p site). Making such calculations reveals that (assuming LDH-B is, like dog fish LDH, more planar than tetrahedral) Isologous intersubunit binding best describes LDH-B. In view of the fact that β sheet structure is partially lost in ethylene glycol, a binding domain is probably located there.

The existence of stable dimers (as compared to the interdimer bonds in a tetramer) leads to some interesting possibilities in predicting what hybridization products between an LDH-A and LDH-B from different sources. The products will clearly depend upon the relative stability of hybrid dimers to homologous dimers and the amount of monomer present.

We have seen that in a denaturing medium, the chances for successful hybridization depends on the relative rates of denaturation and hybridization. The conclusion that LDH must be in an activated state raises interesting constraints as to physiologically occurring hybridization. The enzyme must not be tightly membrane bound, the anti-hybridizing effects of coenzyme and substrate must be cancelled and activation must take place - all at (in humans) temperatures not much higher or lower than 37°. Such rigorous conditions lead one to speculate that naturally occurring accelerators of the hybridization reaction do exist.

REFERENCES

Adams, E. T., Jr. 1967. Analysis of self-associating systems by sedimentation equilibrium experiments. *Fractions* 3: 1-18.

Adams, E. T., Jr. and J. W. Williams 1964. Sedimentation equilibrium in reacting systems. II. Extensions of the theory to several types of association phenomena. *J. Am. Chem. Soc.* 86: 3454-61.

Adams, M. J., D. J. Haas, B. A. Jeffrey, A. McPherson, H. L. Mermall, M. G. Rossman, R. W. Schevitz, and A. J. Wonacott 1969. Low resolution study of crystalline L-lactate dehydrogenase. *J. Molec. Biol.* 41: 159-88.

Boross, M. V. L. 1972. Heat inactivation of D-glyceraldehyde-3-phosphate dehydrogenase apoenzyme. *Acta Biochim. Biophys. Acad. Sci. Hung.* 7: 105-14.

Contaxis, C. C. and F. J. Reithel 1971. Studies on protein multimers. II. A study of the mechanism of urease dissociation in 1,2-propanediol: comparative studies with ethylene glycol and glycerol. *J. Biol. Chem.* 246: 677-85.

Cornish-Bowden, A. J. and D. E. Koshland, Jr. 1971. The quaternary structure of proteins composed of identical subunits. *J. Biol. Chem.* 246: 3092-102.

Jaenicke, R. and S. Knof 1968. Molecular weight and quaternary structure of lactic dehydrogenase. 3. Comparative determination by sedimentation analysis, light scattering and osmosis. *Europ. J. Biochem.* 4: 157-63.

Jaenicke, R., R. Koberstein, and B. Teuscher 1971. The enzymatically active unit of lactic dehydrogenase. Molecular properties of lactic dehydrogenase at low-protein and high salt concentrations. *Eurp. J. Biochem.* 23: 150-9.

Klotz, I. M., N. R. Langerman, and D. W. Darnall 1970. Quaternary structure of proteins. *Ann. Rev. Biochem.* 39: 25-62.

Millar, D. B. 1962. Physico-chemical properties of lactic dehydrogenase. *J. Bio. Chem.* 237: 2135-9.

Millar, D. B., M. R. Summers, and J. A. Niziolek 1971. Spontaneous in vitro hybridization of LDH homopolymers in the undenatured state. *Nature* 230: 117-9.

Rossman, M. G., M. J. Adams, M. Buehner, G. C. Ford, M. L. Hackert, P. G. Lentz, Jr., A. M. Schevitz, and I. E. Smiley 1971. Structural constraints of possible mechaisms of lactate dehydrogenase as shown by high resolution studies of the apoenzyme and a variety of enzyme complexes. *Cold Spring Harbour Symposium Quan. Biol.* XXXVI: 179-91.

Saxena, V. P. and D. B. Wetlaufer 1971. A new basis for interpreting the circular dichroic spectra of proteins. *Proc. Nat. Acad. Sci. U.S.A.* 68: 969-72.

Steiner, R. F. 1952. Reversible association processes of globular proteins. I. Insulin. *Arch. Biochem. Biophys.* 39: 333-54.

Tanford, C., C. E. Buckley III, P. K. Doe, and E. P. Lively 1962. Effect of ethylene glycol on the conformation of γ-globulin and β-lactoglobuin. *J. Biol. Chem.* 237: 1168-71.

Weber, K. and M. Osborn 1969. The reliability of molecular weight determinations by dodecyl sulfate-polyacrylamide gel electrophoresis. *J. Biol. Chem.* 244: 4406-12.

Yielding, K. L. 1971. Modification by sucrose of the catalytic activity and physical properties of glutamic dehydrogenase. *Biochem. Biophys. Res. Commun.* 38: 546-51.

Yphantis, D. A. 1960. Rapid determination of molecular weights of peptides and proteins. *Ann. N.Y. Acad. Sci.* 88: 586-601.

Yphantis, D. A. 1964. Equilibrium ultracentrifugation of dilute solutions. *Biochemistry* 3: 297-317.

BIOCHEMICAL AND GENETIC STUDIES OF PEROXISOMAL MULTIPLE
ENZYME SYSTEMS: α-HYDROXYACID OXIDASE AND CATALASE

ROGER S. HOLMES and JOHN A. DULEY
Department of Biochemistry, La Trobe University
Bundoora, 3083, Victoria, Australia

ABSTRACT. L-α-hydroxyacid oxidase (HAOX) and catalase (Ct)
are differentially localized in the peroxisomes and cell
sap of mammalian liver and kidney cells and exist as
multiple forms. Rat liver and kidney HAOX isozymes were
isolated in a highly purified form. The liver isozyme
(HAOX-A$_4$) has a native MW of 179,000 and a subunit size of
43,000. The tetrameric subunit structure was further substantiated by genetic hybridization studies using BALB/C
and NZC inbred strains of mice with electrophoretically
distinct allelic isozymes HAOX-A$_4$ and A'$_4$ respectively.
The kidney isozyme has a native MW of 243,000 and a similar
subunit size (39,500) to HAOX-A$_4$ which suggests a hexameric
subunit structure.
    Genetic analyses using NZC and $C_s{}^b$ inbred strains of
mice which exhibited a homozygous phenotype for HAOX-A'$_4$/Ct^a
(Hao-1^b/Ce^a) and HAOX-A$_4$/Ct^b (Hao-1^a/Ce^b) respectively,
demonstrated genetic linkage on linkage group V of the
mouse for the loci encoding these enzymes. In addition,
the 5 major forms of catalase, which are differentially
localized in mouse erythrocytes, liver cell sap, and peroxisomes and in kidney, were shown to be encoded by a single
genetic locus. This result is consistent with the epigenetic basis of multiplicity reported for this enzyme.
HAOX-A and B polypeptides were shown to be encoded by
distinct genetic loci by this study (called HAOX-1 and
HAOX-2 respectively).
    Antibodies prepared in rabbits against rat HAOX-A$_4$ cross
reacted with both mouse HAOX-A and B isozymes which suggest
structural homology and a common evolutionary origin for
these enzymes.

INTRODUCTION

Peroxisomes (also called "microbodies") are single membrane
bound subcellular organelles which are widely distributed
among eukaryote organisms. In mammals, they occur predominantly within liver and kidney cells and contain enzymes which are
involved with the oxidation by molecular oxygen of a number of
substrates (see De Duve, 1969; 1973). Two of these peroxisomal
enzymes, L-α-hydroxyacid oxidase (HAOX; E.C. 1.1.3.1. OR
1.1.3.15) and catalase (Ct; E.C. 1.11.1.6) exhibit molecular

heterogeneity and are differentially distributed between the soluble fraction (cell sap) and the peroxisomes of these cells.

This communication reviews the extent and nature of this multiplicity and describes genetic, biochemical, and immuno-chemical studies which establish the genetic distinctness, molecular weights, subunit composition, and immunochemical homology of rodent HAOX isozymes and the epigenetic basis of multiplicity for mouse liver catalase. In addition, the genetic loci for mouse catalase and HAOX-A_4 are shown to be localized together on linkage group V of the mouse.

α-Hydroxyacid Oxidase (HAOX)

HAOX is a flavoprotein enzyme which catalyses the reaction:
$$R-CHOH-COOH + O_2 \rightarrow R-CO-COOH + H_2O_2$$
Previous studies have shown that the enzyme exists as two major forms in mammalian tissues and oxidizes a variety of α-hydroxyacids (Robinson et al., 1962; Allen and Beard, 1965; Domenach and Blanco, 1967; Duley and Holmes, 1974a). Rat and pig liver HAOX preferentially oxidize short chain aliphatic hydroxyacids and exhibit maximum activity with glycolate; they have been called glycolate oxidases (Kun et al., 1954; Nakano et al., 1968; Schuman and Massey, 1971). The rat kidney enzyme oxidizes long chain aliphatic L-hydroxyacids most ef-ficiently and also uses L-amino acids as substrates (Blanchard et al., 1944; Nakano and Danowski, 1966; Nakano et al., 1967). This enzyme has been called hydroxyacid oxidase or L-amino acid oxidase. Alternatively, pig kidney exhibits both forms of activity although the long chain form does not oxidize L-amino acids (Robinson et al., 1962). HAOX isozymes are designated in this communication as HAOX-A and B which refer to the short chain and long-chain α-hydroxyacid oxidases respectively.

Catalase

Catalase is a tetrameric haemenzyme which catalyses the reaction:
$$2H_2O_2 \rightarrow 2H_2O + O_2$$
Catalase multiplicity was first reported for the rat liver enzyme in 1954 by Price and Greenfield, and subsequently, other workers confirmed this phenomenon (Nishimura et al., 1964; 1966; Holmes and Masters, 1965; Higashi and Shibata, 1965). At this time, the enzyme was also shown to undergo artifactual changes in net surface charge following purifica-tion, storage, or exposure to oxidizing agents (Thorup et al., 1964; Heidrich, 1968; Morikofer-Zwez et al., 1970). As a re-sult, catalase was not readily accepted as a 'true' enzyme multiplicity system (Heidrich, 1968; Deisseroth and Dounce,

1970). More recently, however, by the use of procedures designed to obviate any artifactual modification of the enzyme, native catalase multiplicity has been established for some mammalian species (Holmes and Masters, 1969; 1970b; 1972).

In mouse, catalase exists as 5 major forms of activity which are differentially distributed among tissues and sub-cellular organelles (Holmes and Masters, 1970b; 1972; Feinstein and Peraino, 1970; Holmes, 1971; 1972; Kovama, 1971; Olivares et al., 1973). The most acidic form (CT-1) is localized in kidney and in the soluble fraction from liver peroxisomes. Mouse erythrocytes contain predominantly the slowest migrating form of the enzyme (Ct-5), while the liver cell sap exhibits four forms of activity (Ct 2-5). In contrast to many multiple enzyme systems, mouse catalase has an epigenetic basis of multiplicity (Holmes and Masters, 1970a; Holmes, 1972) which is associated with the removal of sialic acid residues from the molecule (Jones and Masters, 1974). The genetic studies reported in this paper confirm that a single genetic locus encodes the various forms of murine catalase.

MATERIALS AND METHODS

Animals. Three inbred strains of *Mus musculus* were used in these investigations. The original stocks of NZC and BALB/C strains were obtained from the Walter and Eliza Hall Institute, Melbourne, while those of the hypocatalasemic mouse strain C_s^b were kindly donated by Dr. R.N. Feinstein of the Argonne National Laboratory. An inbred strain of rat (*Rattus rattus*), DA/Ss Wehi, was also used in this study.

Matings between male BALB/C mice and female NZC mice were made and the F_1 offspring were backcrossed with NZC mice to examine segregation of the loci encoding HAOX-A and agouti phenotypes. Matings between C_s^b and NZC mice were also made and these F_1 offspring were backcrossed with C_s^b mice to examine segregation of the loci encoding HAOX-A and catalase.

Homogenate preparation and gel electrophoresis. Liver, kidney, and erythrocyte extracts were prepared and the supernatants subjected to zone electrophoresis on horizontal starch gels. The gels were then sliced, stained for HAOX or catalase activity, and photographed. Details of these methods have been previously described (Holmes and Masters, 1970a; Duley and Holmes, 1974a).

Sucrose-gradient centrifugation. Sucrose gradient fractionation of the large granule fraction from livers of female BALB/C mice previously injected with Triton WR 1339 (85 mg/100g)

(Leighton et al., 1968) was carried out as described in Holmes and Masters (1972). Cytochrome oxidase and α-hydroxy-acid oxidase assays were performed using the methods of Yonetani (1960) and Robinson et al. (1962), respectively.

Enzyme Purification and Immunochemical Studies. HAOX-A and B isozymes were purified from rat liver and kidney extracts, respectively. The method of purification will be described in detail elsewhere (Duley and Holmes, 1974b). Four mg of purified HAOX-A was injected into rabbits and the antisera obtained used in immunochemical titration experiments (see Markert and Holmes (1969) and Holmes and Scopes (1974) for details).

Native and Subunit MW Determinations. The native MW's of the purified rat HAOX-A and B isozymes were determined using a calibrated sephadex G-150 gel filtration column with the following proteins as standards: cytochrome c (13,000); ovalbumin (45,000); hexokinase (102,000); lactate dehydrogenase (140,000); catalase (200,000); α-globulin (205,000); cholinesterase (320,000); and thyroglobulin (670,000) (Andrews, 1965). The subunit size of the purified rat HAOX-A and B isozymes were obtained from SDS electrophoretic studies using the following subunits as standards: triose phosphate isomerase (27,000), glyceraldehyde phosphate dehydrogenase (36,000), phosphoglycerate kinase (45,000), bovine serum albumin (68,000), and phosphorylase b (100,000) (Weber and Osborne, 1969).

RESULTS AND DISCUSSION

Subcellular and Cellular Distribution of HAOX and Catalase

Figure 1 illustrates the sucrose gradient centrifugation of mouse liver large granules and the resolution of the 3 major organelles. Lysosomes (acid phosphatase) and the bulk of the mitochondria (cytochrome oxidase) are separated from the peroxisomes exhibiting urate oxidase, catalase and HAOX activities. This result confirms extensive studies by De Duve and co-workers (see De Duve, 1969; 1973) who have reported the properties and enzyme composition of peroxisomes from a variety of eukaryote organisms.

In addition to the peroxisomal activity, HAOX and catalase are also localized in the cell sap. In rat liver and kidney cells most of the HAOX activity is localized in the soluble fraction (Table 1). This contrasts with the distribution reported for this enzyme in pig kidney cells in which 2% or less of HAOX activity is found in the cell sap (Robinson et al.,

194

1962). A similar difference in enzyme activity distribution
between species has been reported for catalase (Holmes and
Masters, 1970b; 1972). Within rat and mouse liver, the bulk
of catalase activity was localized in the peroxisomes whereas
in other species (e.g. sheep, beef cattle, dog, cat) in excess of 80% of this activity occurred in the soluble fraction.

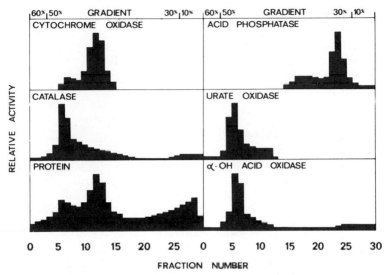

Fig. 1. Sucrose gradient fractionation of the large granules
from livers of female BALB/C mice injected with Triton
WR-1339. See Holmes and Masters, 1972 for details.

Figure 2. illustrates the electrophoretic resolution of
mouse soluble catalase activity by starch gel electrophoresis.
Five major forms of the enzyme may be distinguished which are
differentially distributed among the 3 cell types analyzed.
The most anodal migrating enzyme (Ct-1) is found in kidney
cells while the other 4 forms occur in the cell sap of liver
tissue. Erythrocytes contain 3 forms of activity of which 2
correspond in electrophoretic mobility to the slowest migrating
liver catalases observed. Figure 3 illustrates the comparative
electrophoretic properties of liver peroxisomal and cell sap
catalase activities. The enzyme extracted from peroxisomes
by hypotonic aqueous buffer solutions corresponds to Ct-1,
the soluble and peroxisomal enzyme of kidney cells. These
results confirm previous reports on the differential distribution of mouse catalase multiple forms between various
cellular and subcellular sources. (Holmes and Masters, 1970a
and b; 1972; Feinstein and Peraino, 1970; Holmes, 1971; 1972;
Koyama, 1971; Olivares et al., 1973).

TABLE 1

COMPARATIVE PROPERTIES OF α-OH ACID OXIDASE ISOZYMES

Property	HAOX-A	HAOX-B
Minimum MW (from FMN studies)	50,000[a]	49,300[b]
Minimum MW (from SDS gels)	43,000	39,500
Native MW (G-150 gel filtration)	179,000[c]	243,000
Postulated subunit structure	tetrameric[d]	hexameric
Crossreaction with anti HAOX-A	+++	++
Tissue distribution	liver (rat[b,e,f] mouse, pig[g]) kidney (pig[h])	kidney (rat[i], mouse, pig[h])
Subcellular distribution (ratios of peroxisomal/cell sap %)		
pig kidney[h]	66/2	63/<1
rat liver	24/74	-
rat kidney	-	16/83

Activity with various substrates (% of maximum rate)	rat[f]	pig[g]	rat[j]	pig[h]
glycolate	100	100	0	1
L-OH isocaproate	63	32	100	100
L-lactate	10	3	46	1
DL-OH isovalerate	2	6	36	23
LOH-β phenylactate	0	0	60	94
D-lactate	0	0	0	0

a - Dickinson, 1965; b - Nakano et al., 1968; c - 236,000 at low ionic strength; d - hexameric at low ionic strength; e - Kun et al., 1954; f - Ushijima, 1973; g - Schuman and Massey, 1971; h - Robinson et al., 1962; i - Nakano et al., 1967; j - Nakano and Danowski, 1966; rat kidney HAOX-B also oxidizes a variety of L-amino acids (Blanchard et al., 1944; Nakano et al., 1967).

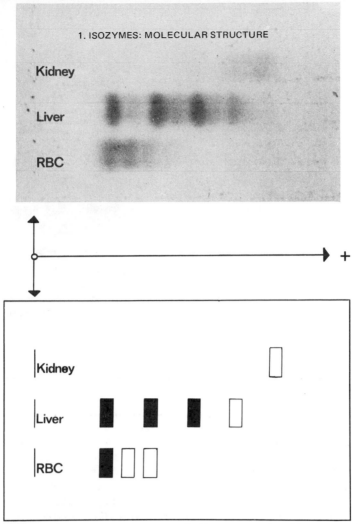

Fig. 2. Starch gel zymogram and diagrammatic illustration of mouse liver, kidney and erythrocyte soluble catalase. For details of electrophoretic method see Holmes and Masters, 1970a.

HAOX isozymes are also differentially distributed between rat and mouse liver and kidney cells (Figure 4). HAOX-A, which hydrolyses the short chain α-OH acids more effectively, is localized in liver and is electrophoretically distinct from the corresponding HAOX-B from kidney extracts. The latter isozyme may be visualized on the starch gel zymograms using longer chain α-hydroxyacids (e.g. L-α-hydroxy-β-phenyl lactate) whereas HAOX-A does not exhibit any observed activity with this substrate. Table 1 summarizes in more detail the distinct substrate specificities reported for these isozymes.

197

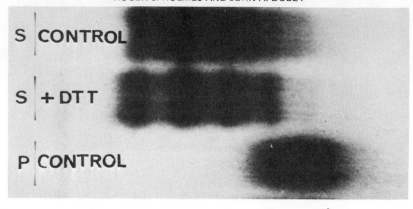

Fig. 3. Starch gel zymogram of peroxisomal and cell sap
mouse liver catalase. 5 control - liver cell sap fraction;
S + DTT - liver cell sap fraction with 5mM dithiothreotol;
P control - peroxisomal catalase. Note the reduction in
catalase multiplicity when the extract is prepared in the
presence of the reducing agent.

Mouse HAOX and Catalase Genetics

Duley and Holmes (1974a) have reported a polymorphism for
HAOX-A in inbred strains of the mouse. HAOX-A activity from
NZC mice migrated more slowly to the anode than the BALB/C
enzyme while the NZC x BALB/C hybrid liver isozyme exhibited
3 intermediate forms of activity in addition to the parental
forms (Figure 5). HAOX-B activity is electrophoretically invariant in these strains. Genetic studies demonstrated that
the A and A' phenotypes in BALB/C and NZC strains segregated
as though they were controlled by codominant alleles at a
single autosomal locus (Hao-1) which mapped closely (5.3%
recombination) to the agouti locus in linkage group V
(Table 2; Figure 6).

Using a hypocatalasemic mouse strain (C_s^b) (Feinstein
et al., 1966) and the NZC strain, further genetic studies
were performed to investigate the genetic basis of multiplicity
for catalase and the possible linkage of the loci encoding
HAOX-A and catalase. Soluble and peroxisomal C_s^b catalase
from liver and kidney are electrophoretically distinct from
multiple forms normally observed (Holmes, 1972; Figure 7).
The genetic mutation which occurred in C_s^b catalase has
altered the activities and electrophoretic mobilities of all
forms of the enzyme irrespective of the cellular or subcellular
source, which infers that the 5 major forms of catalase are
encoded by a single genetic locus. The genetic analyses con-

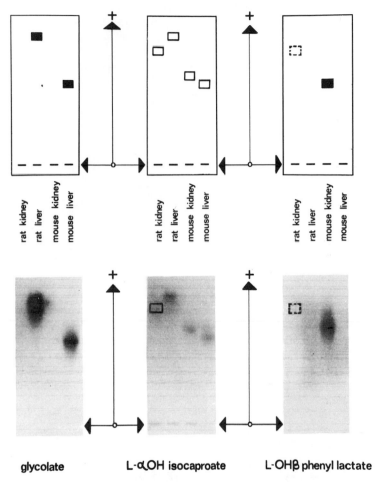

glycolate **L-αOH isocaproate** **L-OHβ phenyl lactate**

Fig. 4. Starch gel zymogram and diagrammatic illustration of mouse and rat L-α-hydroxyacid oxidase isozymes. Liver and kidney extracts were electrophoresed on the same gel and stained for activity using 3 separate substrates as described in Duley and Holmes, 1974a. Although the rat kidney HAOX activity is not apparent here, other experiments have confirmed its mobility and relative activity. Note the preference of the liver isozyme (HAOX-A) for short chain hydroxyacids and the kidney isozyme (HAOX-B) for long chain substrates.

firm this conclusion (Table 3). Segregation of the Ct^a and Ct^b phenotypes in F2 progeny of (NZC x C_s^b) x C_s^b matings revealed that liver, kidney, and erythrocyte catalase behaved identically. An epigenetic basis of multiplicity for mouse

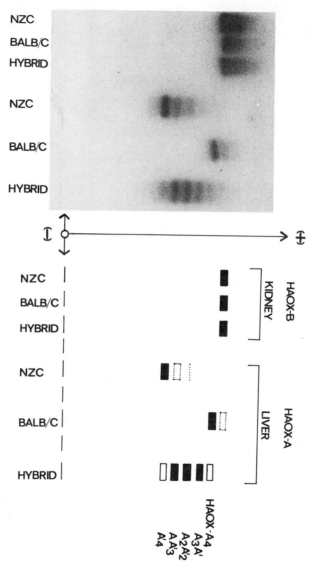

Fig. 5. Starch gel zymogram and diagrammatic illustration of α-hydroxyacid oxidase isozymes of mouse inbred strains. Modified from Duley and Holmes, 1974a. Proposed subunit structures for the allelic liver HAOX isozymes are given.

catalase is consistent with this observation. The segregation analyses of the $Hao-1^a/1^b$ alleles and the catalase (Ce^a/Ce^b)

TABLE 2

DISTRIBUTION OF HAOX-A AND A (AGOUTI) PHENOTYPES AMONG
BACKCROSS PROGENY FROM (NZC X BALB/C) F_1 AND NZC

	aa	Aa	Total
HAOX-A^1 HAOX-A^1	33	2	35
HAOX-A^1 HAOX-A	2	38	40
Total	35	40	**75**

Recombination frequency = 5.3%; Variation from 1:1:1:1 ratio
for unlinked loci was significant (χ^2 = 60.5, 3 d.f.) indicating that the 2 loci are linked.

(from Duley and Holmes, 1974).

)calization of HAOX-A and Catalase loci on linkage Group V of
the mouse

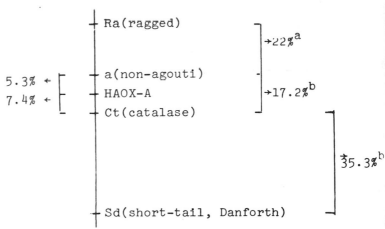

$\overline{V}$ (recombination frequencies given)

a - Carter and Phillips, 1954; b - Dickerman et al, 1968;
c - Duley & Holmes, 1974.

Fig. 6. Localization of L-α-hydroxyacid oxidase A (HAOX-A)
and catalase (Ct) loci on linkage group V of the mouse.
Other linkage data obtained from (a) Carter and Phillips,
1954; (b) Dickerman et al., 1968; and (c) Duley and Holmes,
1974a.

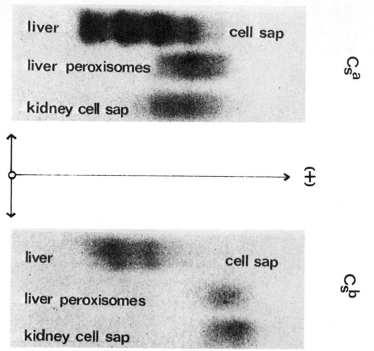

Fig. 7. Starch gel zymogram of catalase multiple forms from mouse inbred strains Cs^a (wild type) and Cs^b (hypocatalasemic). Note that the allelic variation has altered the electrophoretic mobilities and activities of all forms of the enzyme. See Holmes, 1972 for details.

TABLE 3

DISTRIBUTION OF HAOX-A AND CATALASE (CT) PHENOTYPES AMONG BACKCROSS PROGENY FROM (NZC X Cs^B) F_1 AND Cs^B

	$Ct^B Ct^B$	$Ct^B Ct^A$	Total
HAOX-A HAOX-A	30	4	34
HAOX-A^1 HAOX-A	1	33	34
Total	31	37	68

Recombination frequency = 7.4%; Variation from 1:1:1:1 ratio for unlinked loci was significant (χ^2 = 50.0, 3 d.f.) indicating that the 2 loci are linked.

alleles also demonstrated that the genetic loci for these
enzymes are localized together on linkage group V (Figure 6).
This result agrees with the studies of Dickerman et al.
(1968) who established the linkage of the catalase locus with
several linkage group V markers.

Subunit Composition and Immunochemical
Properties of HAOX Isozymes

Purified rat liver (A) and kidney (B) HAOX isozymes were
subjected to sephadex G-150 gel filtration chromatography and
SDS electrophoresis in order to determine their native
molecular weights and subunit composition (Figure 8) HAOX-A
has a molecular weight of 179,000 when chromatographed in
10mM Tris HCl buffer (pH 7.4) containing 50 mM KCl whereas in
the absence of KCl the MW was 236,000 (Table 1; Figure 8).
HAOX-B had a MW of 243,000 by gel filtration analysis when
chromatographed in the presence of KCl. The SDS polyacrylamide gel electrophoretic studies showed HAOX-A and B to contain subunits of similar sizes: 43,000 and 39,500 respectively
(Figure 9). This indicates that these isozymes are tetrameric
and hexameric respectively although HAOX-A appears to form a
hexameric structure under conditions of low ionic strength.
The tetrameric subunit structure of HAOX-A is further substantiated by the genetic variation studies (Figure 5). The
five isozyme patterns observed for the (NZC x BALB) and the
(NZC x C_s^b) hybrid mice are consistent with a tetrameric
structure: HAOX-A_4, A_3A'; $A_2A'_2$, AA'$_3$, and A'$_4$.

Previous reports on the MW and subunit composition of
HAOX-A and B reveal a wide divergence of results. Blanchard
and co-workers (1945) observed 2 molecular species of rat
HAOX-B with molecular weights of 138,000 and 552,000 which
contained 2 and 8 moles of FMN respectively. More recently,
Nakano et al. (1968a) determined the molecular weight of rat
kidney HAOX-B by gel filtration on sephadex G-200 and obtained a value of 310,000. The minimal MW of the enzyme based
upon FMN content was 49,300 which indicated that HAOX-B contains 6 moles of FMN per mole of enzyme and is hexameric
assuming 1 FMN per subunit. Although the molecular weight of
the native enzyme and the subunit are shown by our results to
be somewhat smaller, there is a correlation in the number of
subunits for the enzyme obtained by Nakano's group.

Pig liver HAOX-A has been reported by Dickinson (1965)
(referred to in Schuman and Massey, 1971) to have a molecular
weight of 100,000 and a minimal size (from FMN studies) of
50,000. Nakano and co-workers (1968b) determined the molecular
weight of rat HAOX-A by gel filtration analyses and reported

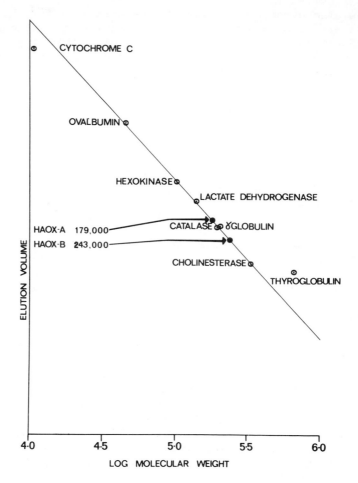

Fig. 8. Determination of the native molecular weights of HAOX-A and B isozymes using a calibrated sephadex G-150 chromatography column. See text for details.

a value of 300,000. It is relevant to note that these workers used a low ionic strength buffer in their gel filtration analyses which would account for their higher MW result in comparison to our finding. Ushijima and Nakano (1969) subsequently investigated the FMN content of this enzyme and obtained a result indicating 2 moles of FMN per mole of HAOX-A. This result of 150,000 subunit size is difficult to reconcile with the 43,000 obtained from SDS electrophoretic analyses. It is possible that HAOX-A may contain less than

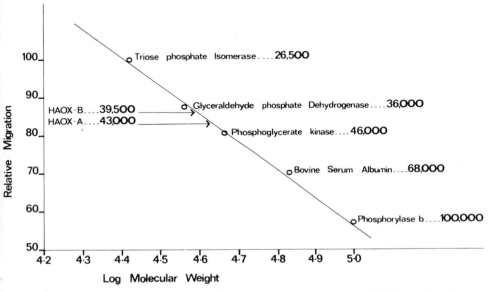

Fig. 9. Determination of the subunit sizes of HAOX-A and B isozymes using SDS electrophoresis and appropriate standards. See text for details.

one mole of FMN per subunit although this would contrast to the situation for HAOX-B. Alternatively, the enzyme may have progressively lost its prosthetic group FMN during purification in a similar way to that observed for spinach glycolate oxidase by Frigerio and Harbury (1958).

Figure 10 illustrates the immunochemical interactions of mouse HAOX-A and B with antibodies prepared in rabbits against purified rat HAOX-A. Both isozymes interacted with the antibodies and were removed from their appropriate zone on the gel zymogram following preincubation with the antisera. These results demonstrate that HAOX-A and B are immunochemically homologous even though they are encoded by distinct genetic loci. This suggests that a considerable degree of structural homology exists between these isozymes and provides firm evidence for their common evolutionary origin. Similar results have been reported for the vertebrate LDH isozyme system (Holmes and Markert, 1969; Whitt, 1970; Holmes and Scopes, 1974). There is one previous report on the immunochemical properties of HAOX. Nakano and co-workers (1967) have prepared antibodies against rat HAOX-B and used these to demonstrate immunochemical identity between the soluble and peroxisomal HAOX-B.

HAOX-B **HAOX-A**

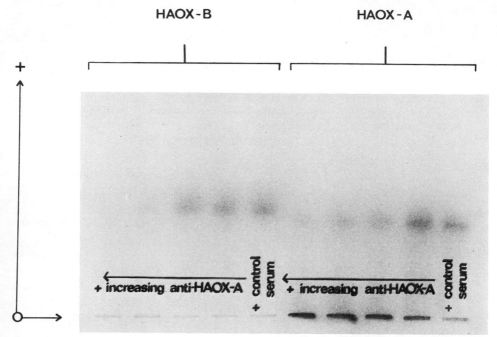

Fig. 10. Immunochemical precipitation of HAOX-A and B isozymes by antibodies prepared in rabbits against rat liver
HAOX-A_4.

Molecular Basis of Catalase Multiplicity

The genetic analyses reported earlier in this paper
demonstrate that the 5 multiple forms of mouse catalase are
encoded by a single genetic locus. This confirms previous
work which suggested an epigenetic basis of multiplicity for
this enzyme (Holmes and Masters, 1970a; Holmes, 1972).

Jones and Masters (1974) have recently investigated the
nature of this epigenetic modification and have obtained evidence that mouse liver catalase multiplicity is based upon a
process of progressive desialation. Incubation of the 4 forms
(Ct 2-5) of soluble liver catalase with sialic acid, CTP, and
microsomal extract (containing N-acetylneuraminicacid
transferase (O'Brien et al., 1966) resulted in the formation
of Ct-1. Alternatively, treatment with neuramidase shifted
the pattern toward the slowest migrating form, Ct-5. Since
catalase is a tetramer, the progressive attachment of sialic
acid residues to each subunit (A→A') may account for the
multiplicity (A_4, A_3A', $A_2A'_2$, AA'_3, and A'_4) (Figure 11).
Multiplicity of other enzymes such as human serum cholinesterase

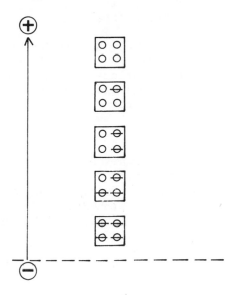

Fig. 11.   Diagram illustrating the proposed subunit structure of mouse liver catalase multiple forms.   O represents a sialated subunit; —Ө— represents a desialated subunit. Proposal based on the data of Jones and Masters, 1974.

(Harris et al., 1962) and acid phosphatase (Ostrowski et al., 1970) have also been explained in terms of the attachment of sialic acid.

These results indicate that a sialated form of catalase (Ct A'$_4$) is localized in the peroxisomes of mouse liver and kidney as well as in the kidney cell sap, whereas progressively desialated catalases exist in the liver cell sap, and the fully desialated enzyme predominates in erythrocytes.   Consequently, these multiple forms of catalase are not strictly "isozymes" under a recent definition of this term (Enzyme Nomenclature, 1972).

Additional multiplicity for mouse liver catalase is observed if the reducing reagent (5 mM dithiothreotol) is absent from the extraction medium (Figure 3).   The extra electrophoretic zones migrated intermediate to the 5 major forms of the enzyme.   These have presumably arisen from an irreversible oxidation of a free SH which would increase the net negative charge of the enzymes (Morikofer-Zwez et al., 1970) or from alternative oxidation states of SH groups contributing to catalase multiplicity (Heidrich, 1968).

ACKNOWLEDGMENTS

We are grateful to Dr. C.J. Masters for supplying a copy of his unpublished manuscript on murine liver catalase multiplicity and for his continued advice and encouragement of this work. We acknowledge also the invaluable assistance of Dr. R.K. Scopes in the determinations of the molecular weights of HAOX isozymes and Miss Terri Jenkins for technical assistance.

REFERENCES

Allen, J.M., and M.E. Beard. 1965. L-α-Hydroxyacid oxidase: localization in renal microbodies. *Science* 149: 1507-1509.

Andrews, P. 1965. The gel filtration behaviour of proteins related to their molecular weights over a wide range. *Biochem. J.* 96: 595-606.

Blanchard, M., D.E. Green, V. Nocito, and S. Ratner. 1945. Isolation of L-amino acid oxidase. *J. Biol. Chem.* 161: 583-598.

De Duve, C. 1969. The peroxisome: a new cytoplasmic organelle. *Proc. Roy. Soc. B.* 173: 71-83.

De Duve, C. 1973. Biochemical studies on the occurrence, biogenesis and life history of mammalian peroxisomes. *J. Histochem. Cytochem.* 21: 942-948.

Deisseroth, A., and A.L. Dounce. 1970. Catalase: physical and chemical properties, mechanism of catalysis, and physiological role. *Physiol. Rev.* 50: 319-375.

Dickerman, R.C., R.N. Feinstein, and D. Grahn. 1968. Position of the acatalasemia gene in linkage group V of the mouse. *J. Heredity* 59: 177-178.

Dickinson, F.M. 1965. *Doctoral dissertation*, University of Sheffield, England.

Domenech, C.E., and A. Blanco. 1967. Alpha-hydroxyacid oxidases in subcellular fractions from rat kidney. *Biochem. Biophys. Res. Commun.* 28: 209-214.

Duley, J.A., and R.S. Holmes. 1974a. α-hydroxyacid oxidase genetics in the mouse: evidence for two genetic loci and a tetrameric subunit structure for the liver isozyme. *Genetics* 76: 93-97.

Duley, J.A., and R.S. Holmes. 1974b. Purification and immunochemical properties of rat hydroxyacid oxidase isozymes (in preparation).

Enzyme Nomenclature. 1972. Recommendations of the Commission on Biochemical Nomenclature on the classification of enzymes. *Elsevier Scientific Publ. Co.*, Amsterdam.

Feinstein, R.N., and C. Peraino. 1970. Separation of soluble and particulate mouse liver catalase by isoelectric focusing. *Biochim. Biophys. Acta* 214: 230-232.

Feinstein, R.N., J.B. Howard, J.T. Braun, and J.E. Seaholm. 1966. Acatalasemic and hypocatalasemic mouse mutants. *Genetics* 53: 923-933.

Frigerio, N.A., and H.A. Harbury. 1958. Preparation and some properties of crystalline glycolic acid oxidase of spinach. *J. Biol. Chem.* 231: 135-157.

Harris, H., D.A. Hopkinson, and E.G. Robson. 1962. Two dimensional electrophoresis of pseudocholinesterase components in normal human serum. *Nature, Lond.* 196: 1296-8.

Heidrich, H.G. 1968. New aspects on the heterogeneity of beef liver catalase. *Z. Physiol. Chem.* 349: 873-880.

Higashi, T., and Y. Shibata. 1965. Studies on rat liver catalase. IV. Heterogeneity of mitochondrial and supernatant catalase. *J. Biochem., Tokyo* 58: 530-537.

Holmes, R.S. 1971. Ontogeny of mouse liver peroxisomes and catalase isozymes. *Nature New Biology* 232: 218-220.

Holmes, R.S. 1972. Catalase multiplicity in normal and acatalasemic mice. *FEBS letters* 24: 161-164.

Holmes, R.S., and C.J. Masters. 1965. Catalase heterogeneity. *Arch. Biochem. Biophys.* 109: 196-197.

Holmes, R.S., and C.J. Masters. 1969. On the tissue and subcellular distribution of multiple forms of catalase in the rat. *Biochim. Biophys. Acta* 191: 488-490.

Holmes, R.S., and C.L. Markert. 1969. Immunochemical homologies among subunits of trout lactate dehydrogenase isozymes. *Proc. Natl. Acad. Sci. U.S.* 64: 205-210.

Holmes, R.S., and C.J. Masters. 1970a. Epigenetic interconversions of the multiple forms of mouse liver catalase. *FEBS letters* 11: 45-48.

Holmes, R.S., and C.J. Masters. 1970b. On the latency, multiplicity and subcellular distribution of catalase activity in mammalian tissues. *Int. J. Biochem.* 1: 474-482.

Holmes, R.S., and C.J. Masters. 1972. Species specific features of the distribution and multiplicity of mammalian liver catalase. *Arch. Biochem. Biophys.* 148: 217-223.

Holmes, R.S., and R.K. Scopes. 1974. Immunochemical homologies among vertebrate lactate dehydrogenase isozymes. *Europ. J. Biochem.* 43: 167-180.

Jones, G.L., and C.J. Masters. 1974. On the nature of the multiplicity of murine liver catalase (unpublished results).

Koyama, S. 1971. Studies on the heterogeneity of catalase in rat liver. II. Catalase pI-Isozyme pattern and its intracellular distribution. *Tumour Res.* 6: 21-36.

Kun, E., J.M. Dechary, and H.C. Pitot. 1954. The oxidation of glycolic acid by a liver enzyme. *J. Biol. Chem.* 210: 269-280.

Markert, C.L., and R.S. Holmes. 1969. Lactate dehydrogenase isozymes of the flatfish, Pleuronectiformes: kinetic, molecular and immunochemical analysis. *J. Exp. Zool.* 171: 85-104.

Morikofer-Zwez, S., J.P. Von Wartburg, and H. Aebi. 1970. Heterogeneity of erythrocyte catalase: variability of the isoelectric point. *Experentia* 26: 945-947.

Nakano, M., and T.S. Danowski. 1966. Crystalline mammalian L-amino acid oxidase from rat kidney mitochondria. *J. Biol. Chem.* 241: 2075-2083.

Nakano, M., Y. Tsutsumi, and T.S. Danowski. 1967. Crystalline L-amino acid oxidase from the soluble fraction of rat kidney cells. *Biochim. Biophys. Acta* 139: 40-48.

Nakano, M., O. Tarutau, and T.S. Danowski. 1968a. Molecular weight of mammalian L-amino acid oxidase from rat kidney. *Biochim. Biophys. Acta* 168: 156-157.

Nakano, M., Y. Ushijima, M. Saga, Y. Tsutsumi, and H. Asami. 1968b. Aliphatic L-α-hydroxyacid oxidase from rat livers. Purification and properties. *Biochim. Biophys. Acta* 167: 9-20.

Nishimura, E.T., S.N. Carson, and T.Y. Kobara. 1964. Isozymes of human and rat catalases. *Arch. Biochem. Biophys.* 108: 452-459.

Nishimura, E.T., S.N. Carson, and G.W. Patton. 1966. Immunoelectrophoretic identification of catalase subcomponents in the homogenates of rat tissues. *Cancer Res.* 26: 92-96.

O'Brien, P.J., M. Canady, C.W. Hall, and E.A. Neufeld. 1966. Transfer of N-acetylneuramic acid to incomplete glycoproteins associated with microsomes. *Biochim. Biophys. Acta* 117: 331-341.

Olivares, J., V. Costa, and E. Montoya. 1973. Selective inhibitory effect of toxohormone on mice hepatic catalase forms. *FEBS letters* 36: 343-346.

Ostrowski, W., Z. Wasyl, M. Weber, M. Guminska, and E. Webter. 1970. The role of neuraminic acid in the heterogeneity of acid phosphomonoesterase from the human prostate gland. *Biochim. Biophys. Acta* 221: 297-306.

Price, V.E., and R.E. Greenfield. 1964. Liver catalase. II. Catalase fractions from normal and tumour bearing rats. *J. Biol. Chem.* 209: 363-376.

Robinson, J.C., L. Keay, R. Molinari, and I.W. Sizer. 1962. L-α-hydroxyacid oxidases of hog renal cortex. *J. Biol. Chem.* 237: 2001-2010.

Schuman, M., and V. Massey. 1971. Purification and characterization of glycolic acid oxidase from pig liver. *Biochim. Biophys. Acta* 227: 500-520.

Thorup, O.A., J.T. Carpenter, and P. Howard. 1964. Human erythrocyte catalase: demonstration of heterogeneity and relationship to erythrocyte ageing *in vivo*. *Brit. J. Haemat.* 10: 542-550.

Ushijima, Y. 1973. Identity of aliphatic L-α-hydroxyacid oxidase and glycolate oxidase from rat livers. *Arch. Biochem. Biophys.* 155: 361-367.

Ushijima, Y., and M. Nakano. 1969. Aliphatic L-α-hydroxyacid oxidase from rat liver. *Biochim. Biophys. Acta* 178: 429-433.

Weber, K., and M. Osborn. 1969. The reliability of molecular weight determinations by dodecyl sulphate-polyacrylamide gel electrophoresis. *J. Biol. Chem.* 244: 4406-4412.

Whitt, G.S. 1970. Developmental genetics of the lactate dehydrogenase isozymes of fish. *J. Exp. Zool.* 175: 1-36.

Yonetani, T. 1960. Studies on cytochrome oxidase. II. Steady state properties. *J. Biol. Chem.* 235: 3138-3143.

STUDIES ON THE ELECTROPHORETIC VARIANTS OF HUMAN HYPOXANTHINE-GUANINE PHOSPHORIBOSYLTRANSFERASE

WILLIAM J. ARNOLD
WILLIAM N. KELLEY
Division of Rheumatic and Genetic Diseases
Departments of Medicine and Biochemistry
Duke University Medical Center
Durham, North Carolina 27710

ABSTRACT. During the purification of hypoxanthine-guanine phosphoribosyltransferase (HGPRT) from normal human male erythrocytes, three distinct peaks of HGPRT activity (I-pI 5.65; II-pI 5.80; III-pI 6.01) have been reproducibly distinguished by preparative isoelectric focusing and subsequently purified to homogeneity. With these highly purified preparations of variants I, II, and III we have found that: (1) the three variants are interconvertible as determined by polyacrylamide gel electrophoresis; (2) the three variants are immunologically identical; (3) the three variants have similar substrate utilization and end-product inhibition; (4) the three variants have the same native molecular weight (68,000) and Stokes radius (36Å) and each is composed of two non-covalently bound subunits of equal molecular weight (34,000) and net charge; (5) the amino acid composition of variants II and III is nearly identical. We conclude that the electrophoretic variants of human erythrocyte HGPRT most likely result from a non-genetic post-transcriptional alteration of one or both subunits of the HGPRT enzyme molecule. Although the exact nature of the post-transcriptional alteration is not known, differential sialiation, differential binding of ribose-5-phosphate or ampholytes and association of the subunits into trimers, tetramers, etc. have been excluded.

Hypoxanthine-guanine phosphoribosyltransferase (HGPRT - E.C. 2.4.2.8.) has been extensively studied since the discovery that a virtually complete deficiency of this enzyme activity in humans (Seegmiller, Rosenbloom, and Kelley, 1967) is associated with a devastating neurological disorder, hyperuricemia, and hyperuricaciduria - the Lesch-Nyhan syndrome (Lesch and Nyhan, 1964). Initially characterized in brewer's yeast by Kornberg et al (1955), the kinetics and substrate specificity of the enzyme were thoroughly investigated by Krenitsky et al (1969) and Krenitsky and Papaioannae (1969) in a partially-purified preparation of normal human erythrocyte HGPRT. During the course of our studies on the purification and structural properties of normal human HGPRT (Arnold and Kelley, 1971), we reproducibly found three different peaks of HGPRT activity with preparative isoelectric focusing. It is the purpose of this paper to review the data we have accumulated to date on the nature of these electrophoretic variants of HGPRT and to summarize the observations of others.

METHODS

·The purification of normal human erythrocyte HGPRT to homogeneity is summarized in Table 1. A single peak of HGPRT activity was consistently observed after DEAE-cellulose chromatography (Fig. 1). However, three distinct peaks of HGPRT activity (designated I, II, III in order of proximity to the anode) were easily distinguished by preparative isoelectric focusing using pH 4-6 ampholytes in a 220 ml 0-40% sucrose gradient while the APRT activity coincidentally present on the same run clearly eluted as a single peak (Fig. 2).

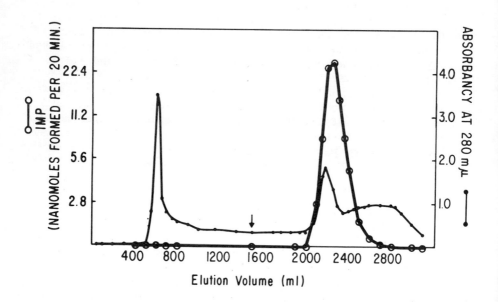

Fig. 1. DEAE-Cellulose Column Chromatography of HGPRT. Following a 1500ml wash with 35mM sodium phosphate buffer pH 7.0, the arrow indicates the beginning of the linear KCl gradient used to elute HGPRT activity. (From Arnold and Kelley, 1971).

TABLE 1

PURIFICATION OF HYPOXANTHINE-GUANINE PHOSPHORIBOSYLTRANSFERASE FROM HUMAN ERYTHROCYTES (MALE)

Step	Specific Activity Nmoles/mg protein/hr	Total Protein mg	Recovery % of initial activity	Fold-Purification X
1. Hemolysate	81	72,000	--	--
2. DEAE-eluate	9470	399	64	117
3. Heat-treatment	29206	85	42	361
4. Dialysis	36245	67	42	447
5. Isofocusing				
I			4.2 ⎫	
II			5.8 ⎬ 11.2	
III			1.2 ⎭	
6. Sephadex G-100				
I	229207	0.56	3.7 ⎫	8487
II	223883	1.44	5.3 ⎬ 10	8022
III	83000	0.47	1.0 ⎭	3141

(From Arnold and Kelley, 1971)

215

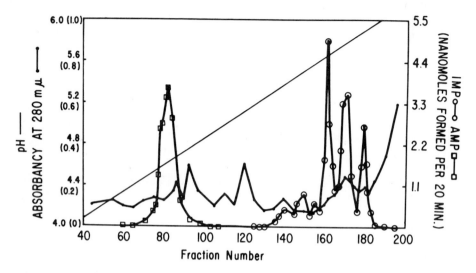

Fig. 2. Preparative Isoelectric Focusing of HGPRT. The anode is to the left. (From Arnold and Kelley, 1971).

Subsequent gel filtration of each peak on Sephadex G-100 afforded further purification and removed the ampholytes from the HGPRT samples. Homogeneity of peaks I and II was judged by the presence of a single band of protein after electrophoresis on three different cross-linkages of polyacrylamide gel (Arnold and Kelley, 1971) which corresponded to eluted HGPRT activity. A single minor contaminant was present in the preparation of peak III.

"Burst synthesis" of guanosine-5'-monophosphate (GMP) by impure and homogeneous preparations of HGPRT was assessed at 0° in the absence of magnesium by the method of Groth and Young (1971). Monospecific antisera to HGPRT was raised in rabbits using the homogeneous preparation of normal human HGPRT from erythrocytes (Arnold, Meade and Kelley, 1972).

RESULTS

Three electrophoretic variants of HGPRT (I-pI 5.65, II-pI 5.80, and III-pI 6.01) could be reproducibly distinguished by preparative isoelectric focusing after partial purification of the enzyme from human erythrocytes. Despite numerous attempts, we were unable to distinguish three peaks of HGPRT activity in crude hemolysate using polyacrylamide gel electrophoresis, agarose film electrophoresis (Arnold, Lamb, and Kelley, 1973) or DEAE–cellulose chromatography. The nature of the electrophoretic heterogeneity was investigated using highly purified preparations of the enzyme.

A. *Properties of Native Human HGPRT*

Interconversion of the electrophoretic variants of HGPRT was investigated with polyacrylamide gel electrophoresis. The three different homogeneous peaks of HGPRT activity could best be distinguished when run separately on 5% polyacrylamide gels using a Tris–glycine buffer system (Arnold and Kelley, 1971). However, as can be seen in Fig. 3, when the three

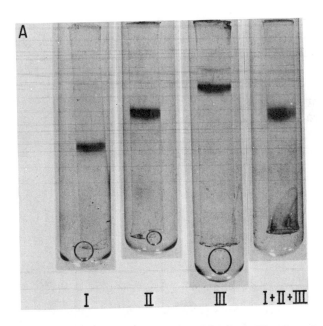

Fig. 3. 5% Analytical Polyacrylamide Gel Electrophoresis of the Three Electrophoretic Variants of HGPRT (I, II, III) and a Mixture of the Three Variants (I+II+III). The anode is at the bottom. (From Arnold and Kelley, 1971).

peaks are mixed together for one hour at 4° and then applied to a gel, only a single band is apparent, and this corresponds in mobility to peak II. Also, after two weeks of storage at -70°, the mobility of peak III changed to that of peak II. Additional evidence for this interconvertibility is shown in Table 2 which summarizes data gathered from seven separate but identical purifications of human HGPRT. As can be seen, despite the highly reproducible isoelectric points of each of the three peaks, the amount of each present varied considerably.

TABLE 2

COMPARISON OF ISOELECTRIC POINTS AND RECOVERY OF
PURIFIED ELECTROPHORETIC VARIANTS OF HUMAN HGPRT

Electrophoretic Variant	Per Cent of Total HGPRT Activity Recovered[+]		Isoelectric Point[+]
	(mean ± S.D.)	Range	(mean ± S.D.)
I	31 ± 11	15–41	5.65 ± 0.06
II	45 ± 4.5	38–52	5.8 ± 0.06
III	25 ± 12	11–39	6.01 ± 0.09

[+] Data calculated from 7 separate but identical purifications of HGPRT. (From Kelley and Arnold, 1973).

Each of the three peaks of HGPRT activity separated by isoelectric focusing as well as the HGPRT activity present in an impure preparation (STEP 2) elute in exactly the same position from a Sephadex G-100 column corresponding to a molecular weight of 68,000 and Stokes radius of $36\overset{\circ}{A}$ (Fig. 4).

The immunologic identity of each of the highly purified forms of the enzyme can be seen in Fig. 5. Once separated, the three forms of the enzyme could be distinguished by immunoelectrophoresis in 50mM Barbital buffer, pH 8.0 (Arnold and Kelley, 1971).

Other studies of the three native, intact peaks of HGPRT activity failed to reveal the presence of bound ribose-5-phosphate groups as judged by the absence of a "burst" of GMP synthesis before or after Sephadex G-100 chromatography. Also, the affinity of each variant for substrates (hypoxanthine and guanine) and sensitivity to end product (GMP) and substrate (PRPP) inhibition were nearly identical (Arnold and Kelley, 1971). Incubation of each of the three variants with neuraminidase failed to liberate detectable sialic acid and

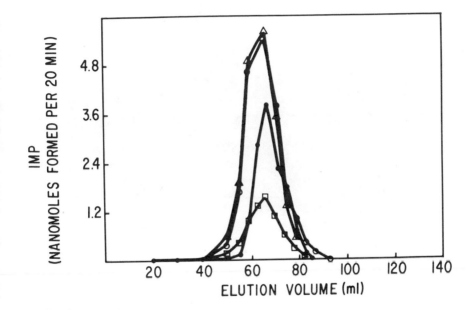

Fig. 4. Sephadex G-100 Column Chromatography of the Electro-
phoretic Variants of HGPRT and an Impure (STEP 2) HGPRT
Preparation. □———□ , I; △———△ , II; O———O , III; ●———● ,
STEP 2. V_0 - 40ml.

did not change the relative migration of each on polyacrylamide
gels. The three variants had similar stability to storage at
-70^0 despite the presence or absence of stabilizing agents
(Arnold and Kelley, 1971).

B. *Properties of Denatured Human HGPRT*

When each of the HGPRT variants is digested in 1% sodium
dodecyl sulfate (SDS) and 3% ß-mercaptoethanol and run separately on polyacrylamide gels, the molecular weight of the
single major protein band is approximately 34,000 for variants
I, II, and III (Fig. 6). The same result is obtained in the
absence of ß-mercaptoethanol (Kelley and Arnold, 1973). Also,
staining of the SDS gels for carbohydrate by the method of
Zacharius et al (1969) fails to reveal any carbohydrate bound

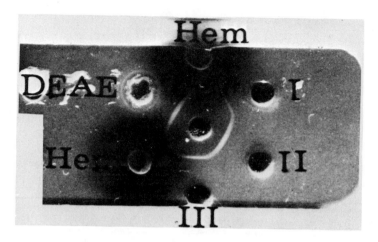

Fig. 5. Immunodiffusion of HGPRT. The center well contains monospecific anti-HGPRT rabbit serum. The peripheral wells contain: Hem - normal human hemolysate, I - electrophoretic variant I, II - electrophoretic variant II, III - electrophoretic variant III, and DEAE - STEP 2 partially-purified HGPRT. Antigen wells were filled twice; antisera well was filled once. The pattern was developed at 4°C for 36 hours. (From Arnold, Lamb, and Kelley, 1973).

to the HGPRT protein subunit. A mixture of the three peaks of HGPRT activity denatured in 8 M urea and 3% β-mercaptoethanol migrates as a single band on urea polyacrylamide gels (Arnold and Kelley, 1971). Finally, preliminary determinations of the amino acid composition of peaks II and III show a great degree of similarity (Table 3).

DISCUSSION

We, as well as Davies and Dean (1971) and Der Kaloustian et al (1973),have reported heterogeneity of HGPRT activity using isoelectric focusing. Both of the other groups have used analytical isoelectric focusing of crude hemolysate on polyacrylamide slabs and have distinguished three peaks of HGPRT activity. The isoelectric points of Der Kaloustian et

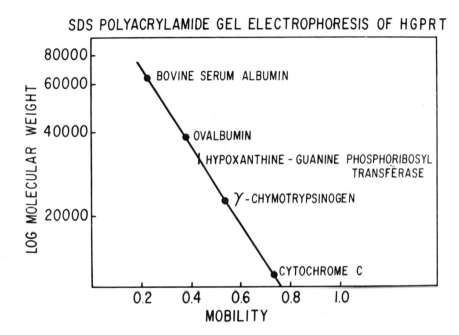

Fig. 6. Sodium Dodecyl Sulfate (SDS) Polyacrylamide Gel
Electrophoresis of the Electrophoretic Variants of HGPRT.
(From Arnold and Kelley, 1971).

al (1973) are nearly identical to those we have described
(pI 5.71, pI 5.83, and pI 6.00), while Davies and Dean (1971)
report isoelectric points somewhat higher (pI 6.0, pI 6.5,
pI 7.5). Additionally, other groups have also noted multiple
forms of HGPRT activity without using isoelectric focusing.
Bakay and Nyhan (1971) and Rubin et al (1971) have noted four
and two forms of HGPRT, respectively, utilizing different
techniques. Gutensohn and Guroff (1972) found three distinct
peaks of rat brain HGPRT during DEAE-cellulose chromatography.
Thus, the number of electrophoretic variants of HGPRT activity observed ranges from two to four, apparently depending on
the technique used.

Although multiple molecular forms of an enzyme may be
genetic in origin, it would appear that those of human HGPRT

TABLE 3

HYPOXANTHINE-GUANINE PHOSPHORIBOSYLTRANSFERASE: AMINO ACID
COMPOSITION OF ELECTROPHORETIC VARIANTS II AND III

Amino Acid	Number of Amino Acid Residues per Subunit	
	II*	III*
Lysine	11	13
Histidine	8	8
Arginine	18	15
Threonine	11	14
Serine	13	14
Glutamic Acid	30	29
Proline	18	19
Glycine	26	28
Alanine	22	21
Valine	20	18
Methionine	8	10
Isoleucine	8	8
Leucine	24	24
Tyrosine	7	8
Phenylalanine	12	12

(From Arnold, Lamb, and Kelley, 1973).

*The results shown represent the average of two 24-hour
digests of each variant.

are non-genetic or post-transcriptional in nature. Firstly,
since all available evidence suggests that the structural gene
coding for HGPRT is located on the X-chromosome and that a
single mutation results in the virtual absence of detectable
HGPRT activity in all tissues of the body (Lesch-Nyhan syndrome), it is unlikely that there are different tissue-
specific HGPRT enzyme proteins (Rosenbloom et al, 1967).
Secondly, since only single units of blood from individual
male donors were used for each purification, the electrophoretic variants cannot represent the product of allelic
genes. Thirdly, one of the hallmarks of a genetically-
distinct isozyme is the stability of its electrophoretic
behavior. However, the variants of human HGPRT are clearly
interchangeable as shown by the mixing experiments on polyacrylamide gels (suggesting an equilibrium favoring peak II),
changing electrophoretic behavior of peak III with storage at
-70^{0}, and the variable ratios of the amount of the three electrophoretic variants despite the reproducibility of the isoelectric points. Fourthly, in view of the nearly identical

substrate utilization and end-product inhibition of the three variants, it is unlikely that they are of physiologic significance to the cell.

Structural studies of the purified variants to date have not revealed any differences between I, II, and III in that all have a native molecular weight of 68,000 and are composed of two non-covalently bound subunits with identical molecular weight (34,000) and net charge. Additionally, immunologic techniques using monospecific anti-HGPRT serum demonstrate only a single preciptin line for impure HGPRT by both immunoelectrophoresis and immunodiffusion. Also, all three purified electrophoretic variants are immunologically identical. Perhaps the strongest evidence for the identity of the primary structure of each of the variants is the nearly identical amino acid composition of variants II and III.

The exact nature of the post-transcriptional alteration of the HGPRT molecule responsible for the observed electrophoretic variation has not been determined. The negative glycoprotein stain, absence of liberated sialic acid, and lack of changing electrophoretic behavior after neuraminidase digestion renders differential sialation as an unlikely explanation. Additionally, differential binding of ribose-5-phosphate is improbable in the absence of "burst" synthesis of GMP by any of the variants. Aggregation of the subunits into trimers, tetramers, etc. can be discounted since all three variants were found to have the same native molecular weight (68,000). Differential binding of ampholytes has been implicated in the production of electrophoretic variants; however, the reproducibility of the isoelectric points and the demonstration of heterogeneity by other techniques makes this unlikely. Other mechanisms of post-transcriptional modification of the primary structure of HGPRT cannot be excluded. For example, although interchain disulfide linkages appear to be lacking, variable sulfhydryl oxidation of intrachain disulfide bonds in the HGPRT subunits could be responsible for the observed electrophoretic heterogeneity. An example of electrophoretic heterogeneity on the basis of artifactual sulfhydryl oxidation has been reported in rabbit skeletal muscle phosphoglucose isomerase (Blackburn et al, 1972). Another possibility is non-enzymatic deamidation of glutamine or asparagine. This has been documented to produce electrophoretic heterogeneity of beef heart cytochrome C by Flatmark (1966). Differential phosphorylation, methylation or carbamylation of a single HGPRT enzyme protein are also possible mechanisms for the observed electrophoretic heterogeneity of HGPRT activity. Finally, Spencer and Gelehrter (1974) recently demonstrated variants of tyrosine aminotransferase which were different enzymes with overlapping substrate specificities. This study illustrates the necessity for a careful kinetic and structural assessment of "isozymes" while cautioning against observations on apparent isozymes drawn solely from electrophoretic techniques.

REFERENCES

Arnold, William J. and W. N. Kelley 1971. Human hypoxanthine-guanine phosphoribosyltransferase. Purification and sub-unit structure. *J. Biol. Chem.* 246: 7398-7404.

Arnold, William J., J. C. Meade, and W. N. Kelley 1972. Hypoxanthine-guanine phosphoribosyltransferase: characteristics of the mutant enzyme in erythrocytes from patients with the Lesch-Nyhan syndrome. *J. Clin. Invest.* 51: 1805-1812.

Arnold, William J., V. R. Lamb, and W. N. Kelley 1973. Human hypoxanthine-guanine phosphoribosyltransferase (HGPRT): purification and properties. Purine metabolism in man - enzymes and metabolic pathways.*Adv. Exp. Biol. Med.* (Sperling, DeVries, and Wyngaarden, eds) 41A: 5-14.

Bakay, Bohdan and W. L. Nyhan 1971. The separation of adenine and hypoxanthine-guanine phosphoribosyl transferase iso-enzymes by disc gel electrophoresis. *Biochem. Genet.* 5: 81-90.

Blackburn, M. N., J. M. Chirgwin, G. T. James, T. D. Kempe, T. F. Parsons, A. M. Register, K. D. Schnackerz, and E. A. Noltman 1972. Pseudoisoenzymes of rabbit muscle phosphoglucose isomerase. *J. Biol. Chem.* 247: 1170-1179.

Davies, Mark R. and Betty M. Dean 1971. The heterogeneity of erythrocyte IMP: pyrophosphate phosphoribosyl transferase and purine nucleoside phosphorylase by iso-electric focusing. *FEBS Letters.* 18: 283-286.

Der Kaloustian, Vazken M., Z. L. Awdeh, R. T. Hallal, and N. W. Wakid 1973. Analysis of human hypoxanthine-guanine phosphoribosyl transferase isozymes by iso-electric focusing in polyacrylamide gel. *Biochem. Genet.* 9: 91-95.

Flatmark, T. 1966. On the heterogeneity of beef heart cyto-chrome C. *Acta Chem. Scan.* 20: 1487-1496.

Groth, D. P. and L. G. Young 1971. On the formation of an intermediate in the adenine phosphoribosyltransferase reaction. *Biophys. Biochem. Res. Commun.* 43: 82-87.

Gutensohn, W. and G. Guroff 1972. Hypoxanthine-guanine phosphoribosyltransferase from rat brain (purification, kinetic properties, development and distribution). *J. Neurochem.* 19: 2139-2150.

Kelley, William N. and W. J. Arnold 1973. Human hypoxanthine-guanine phosphoribosyltransferase: studies on the normal and mutant forms of the enzyme. *Fed. Proc.* 32: 1656-1659.

Kornberg, A., I. Leeberman, and E. S. Simms 1955. Enzymatic synthesis of purine nucleotides. *J. Biol. Chem.* 215: 417-427.

Krenitsky, Thomas A. and R. Papaioannae 1969. Human hypoxanthine phosphoribosyltransferase. II. Kinetics and chemical modification. *J. Biol. Chem.* 244: 1271:1277.

Krenitsky, Thomas A., R. Papaioannae and G. B. Elion 1969. Human hypoxanthine phosphoribosyltransferase I. Purification, properties and specificity. *J. Biol. Chem.* 244: 1263-1270.

Lesch, Michael and W. L. Nyhan 1964. A familial disorder of uric acid metabolism and central nervous system function. *Amer. J. Med.* 36: 561-570.

Rosenbloom, Frederick M., W. N. Kelley, J. Miller, J. F. Henderson, and J. E. Seegmiller 1967. Inherited disorder of purine metabolism - correlation between central nervous system dysfunction and biochemical defects. *J.A.M.A.* 202: 175-177.

Rubin, Charles S., J. Dancis, L. C. Yip, R. C. Nowinski, and M. E. Balis 1971. Purification of IMP: pyrophosphate phosphoribosyltransferases, catalytically incompetent enzymes in Lesch-Nyhan disease. *Proc. Natl. Acad. Sci.* U.S.A. 68: 1461-1464.

Seegmiller, Jarvis E., F. M. Rosenbloom, and W. N. Kelley. Enzyme defect associated with a sex-linked human neurological disorder and excessive purine synthesis. *Science* 155: 1682-1684.

Spencer, Carolyn J. and Thomas D. Gelehrter 1973. Pseudoisoenzymes of hepatic tyrosine aminotransferase. *J. Biol. Chem.* 249: 577-583.

Zacharius, R. N., Z. E. Zell, J. H. Morrison, and J. J. Woodlock 1969. Glycoprotein staining following electrophoresis on acrylamide gels. *Anal. Biochem.* 30: 148-152.

HETEROGENEITY OF HUMAN ERYTHROCYTE CATALASE

H. AEBI, S. R. WYSS and B. SCHERZ
Medizinisch-chemisches Institut der Universität Bern
CH - 3000 B E R N E 9, Switzerland

ABSTRACT. 1. In human erythrocyte catalase three types of heterogeneity are observed: (A) Alternative molecular forms due to conformational changes (B) Partial dissociation of the active tetramer molecule into subunits (C) Genetic variants of the normal enzyme.
2. Catalase from human (and horse) erythrocytes can be separated into three fractions of equal specific activity, but differing with respect to electrophoretic mobility, the molarity of the buffer required for elution from DEAE-cellulose and regarding the IEP. Conversion of fraction A to B and C proceeds rapidly in presence of oxygen and heavy metal catalysts, but can be prevented by chelating agents (EDTA) and anaerobic conditions.
3. Preparations of normal catalase contain free sub-units (dimer and monomer) in small amounts (~4%); they can be separated from the active tetramer by exclusion chromatography on Sephadex G - 150. Different antigenic properties of tetramer, dimer, and monomer favor the assumption of "hidden" determinants.
4. Residual catalase activity in the blood of an individual homozygous for Swiss Type Acatalasemia (1-2% of normal level) consists of a catalase variant, which is unstable, but of normal specific activity. Since much more dimer than tetramer is found, it is assumed that in this variant the tendency of dissociation into subunits is greatly enhanced.

There are various mechanisms which may contribute to heterogeneity of catalase in mammalian tissues. The existence of true isozymes has been demonstrated for cytosol-catalase in liver of rabbit, rat and mouse, whereas in the liver of seven other species no such heterogeneity could be detected (Holmes & Masters, 1972). On the other hand, no isozymes occur in peroxisomal catalase. These findings indicate that in view of different results obtained for catalase in different cell fractions and species no extrapolations are feasible in this respect.

This report is restricted to human erythroctye catalase. This enzyme exhibits three types of isozymic heterogeneity: (1) Alteration of the *conformation* leading to the existence of alternative molecular forms (Thorup et al, 1964; Mörikofer et al, 1969); (2) Partial *dissociation* of the tetramer molecule, thereby producing dimer and

monomer particles (Aebi et al, 1974); (3) *Mutations,* which
are responsible for the transformation of the normal enzyme
to enzyme variants either of low stability or of low specific
activity (Takahara et al, 1967; Aebi et al, 1967).

Alternative molecular forms of erythrocyte catalase

This type of heterogeneity, first observed by Thorup et al
(1964) has been further investigated and extended in the
author's laboratory (Matsubara et al, 1967; Mörikofer et al,
1969). It has been shown that catalase from human and horse
erythrocytes can be separated into three fractions (A, B, and
C) of equal specific activity, but differing with respect
to electrophoretic mobility and the molarity of the buffer
required for elution from DEAE-cellulose (A < B < C). In
hemolysates, fractions B and C are rapidly formed out of fraction A in the presence of oxygen and heavy metal catalysts.
This conversion, however, can be prevented by chelating
agents, such as EDTA, in combination with anaerobic conditions. The mechanism responsible for the occurrence of alternative molecular forms of catalase was investigated by analyzing these fractions with respect to (1) their chromatographic and electrophoretic properties, (2) the number of
titratable SH-groups, and (3) the isoelectric point. The
methods used in these studies have been described previously
(Mörikofer et al, 1969; Aebi et al, 1972).

1. *Chromatography of Erythrocyte Catalase on DEAE-cellulose under Oxidative and Non-oxidative Conditions*

Three types of experiments were performed: (A) In the standard procedure according to Thorup et al (1964) the hemolysate was prepared with distilled water, freed from stroma by
filtration, dialysed against 0.003 M phosphate buffer pH 6.8
and applied to the column. In the case of the human enzyme
(Fig. 1, middle diagram) catalase activity is eluted in two
peaks, the first at 0.012 M consisting of fractions A and B,
and the second (fraction C) at 0.05 M phosphate pH 6.8, respectively.
(B) In another series of experiments the hemolysates were
prepared in an oxygen-free environment and in presence of
EDTA in order to prevent oxidation. All buffers were saturated with nitrogen, the DEAE-cellulose was washed with EDTA.
Fig. 1, top diagram, demonstrates that catalase from human
erythrocytes is eluted in one homogeneous peak by 0.012 M
phosphate.
(C) For the third type of experiment the hemolysate was dia-

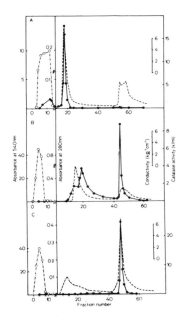

Fig. 1. Chromatography of human erythrocyte catalase on DEAE-cellulose under different conditions. The columns (1.6 x 18cm) were eluted stepwise with 3, 12, and 50 mM Na-K-phosphate buffer pH 6.8. Flow rate 50 ml/h; 10-ml fractions. All experiments were carried out at 4°. o-·-o, absorbance at 540 nm (hemoglobin); ●—●, catalase activity (k/ml); ----, absorbance at 280 nm; ····, conductivity ($k\Omega^{-1}cm^{-1}$). (Top) Chromatography under nitrogen and with EDTA: to 40 ml of hemolysate 40 ml of cold 100 mM EDTA (adjusted to pH 6.8 with NaOH) were added. The hemolysate was dialyzed for 45 hours against starting buffer and the final difference in conductivity corrected by dilution with distilled water. 98 ml of hemolysate, centrifuged at 23,000 x g (4.9 k/ml; 11.2 mg hemoglobin/ml) were chromatographed on a column of EDTA-washed cellulose. All steps were carried out under nitrogen. (Middle) Standard procedure according to Thorup et al (1964). The hemolysate was dialysed for 20 hours at 4° against 3 mM starting buffer. 15 ml (32 k/ml; 113 mg hemoglobin/ml) were applied to the column. (Bottom) The hemolysate was allowed to stand 70 hours at 4° and then dialysed against oxygen-saturated starting buffer. 16.5 ml (23.0 k/ml; 66 mg hemoglobin/ml) were applied to the column.

lysed for 60 hours at 4°C against oxygen-saturated starting buffer. The chromatographic pattern shown in Fig. 1, bottom diagram, also reveals one single peak, but here catalase is eluted only after the concentration of the buffer has been raised to 0.05 M phosphate. The overall yield in all chromatographic separation experiments was more than 80%, indicating that there were no gross changes in the specific activity. Experiments with horse erythrocyte catalase revealed analogous elution patterns.

The fractions obtained from human hemolysate by chromatography differ in electrophoretic mobility on starch gel at pH 8.0, as could be expected from the chromatographic behavior: catalase fraction A from the 0.012 M elution step (Fig. 1, top diagram) had a slower anodic electrophoretic mobility than catalase fraction C from the 0.05 M step (Fig. 1, bottom diagram). An analogous difference in electrophoretic mobility was observed when freshly prepared hemolysate was compared with hemolysate stored for several days at $+4^{o}$C, catalase in old hemolysates moving faster than the one in a fresh sample. A more detailed study of this "ageing" phenomenon observed first in 1962 (Blumberg et al, 1962) has disclosed that under aerobic conditions (i.e. in equilibrium with air) electrophoretic mobility of catalase in hemolysate steadily increases within a period of about 8 days. However, if iodoacetamide is added immediately after hemolysis (20 mM, which is estimated to be a 4-fold molar excess over the thiol groups in hemolysate; pH 8.7) no alteration of electrophoretic mobility is observed within 10 days. A similar effect is obtained by adding cystamine to the hemolysate (10-fold molar excess, 50 mM, pH 8.7). Consequently, alkylation of SH-groups or formation of a mixed disulphide seem to inhibit the alterations in the molecular structure of catalase responsible for this increase in electrophoretic mobility.

2. *Sulfhydryl Group Content of Purified Fractions A and C*

The above findings and the fact that carboxymethylation prevents the transition of fractions A to C suggest that the oxidation of sulfhydryl groups is involved in the conversion of fractions A to C. Therefore the sulfhydryl group content of pure fractions A and C was determined. The results of pCMB-titrations are summarized in Table 1.

Fraction A of horse erythrocyte catalase contains a total of about sixteen sulfhydryl groups. No further SH-groups can be detected after reduction with mercaptoethanol in 8 M urea, thus excluding the presence of a disulphide bridge in fraction A of horse erythrocyte catalase. In fraction A of human erythrocyte catalase thirteen to fourteen sulfhydryl groups per molecule are titratable after denaturation. Reduction with mercaptoethanol in 8 M urea yields two addi-

TABLE 1

SULFHYDRYL GROUP CONTENT AND ISOELECTRIC POINT OF DIFFERENT
MOLECULAR FORMS OF ERYTHROCYTE CATALASE

Species	Molecular form	IEP (pH)	Number of groups per molecule		
			Sulfhydryl	Disulphide	Irreversibly oxidized sulfhydryl
Human	Native enzyme	6.5	-	-	-
	Fraction A	5.95	13.8	1	0
	Intermediate fraction	5.8	11.8	1	~2
	Fraction C	5.2	3.5	1	~10
Horse	Native enzyme	7.3	-	-	-
	Fraction A	5.9–6.3	15.8	0	0
	Fraction C	5.7	5.6	0	~10

tional SH-groups, indicating the existence of a disulphide
bridge in fraction A of the human enzyme.

The conversion of fraction A of both species to fraction
C is characterized by a decrease of sulfhydryl groups titratable in 8 M urea. On the other hand, the same number of sulfhydryl groups is found in fraction C of the horse enzyme before and after reduction with mercaptoethanol in 8 M urea.
Since there is no formation of disulphide bridges during the
transition from A to C, the formation of higher oxidation products such as sulfinic or sulfonic acid has to be taken into
consideration.

3. *The Isoelectric Point (IEP) of Native Catalase and of Purified Fractions A and C*

The IEP of erythrocyte catalase varies considerably according to the species as well as the pretreatment of the enzyme
sample (Table 1). The result of an isoelectric focusing
experiment with freshly prepared hemolysate of human red
cells is shown in Fig. 2a. This catalase, considered to
represent the native form of the enzyme, was mainly focused
at pH 6.5. Fraction A of human erythrocyte catalase is chromatographically homogeneous. However, the isoelectric focusing experiment shown in Fig. 2b suggests the presence of several molecular forms of catalase within this fraction. The
main part of the activity is focused at pH 5.95. Only a
small amount of native catalase as found in hemolysates (IEP=
6.5) seems to be present. Similar results were obtained with
fraction A of the horse enzyme.

In the presence of oxygen, but in the absence of heavy
metal catalysts, fraction A of the human enzyme is converted
to a relatively stable intermediate, containing 2 irreversibly oxidized sulfhydryl-groups (Table 1). This intermediate
fraction is eluted from DEAE-cellulose as fraction C. It
has an IEP of 5.8 and apparently consists of only one molecular species (Fig. 2c). In a second oxidation step, taking
place in presence of oxygen and heavy metal catalysts, more
than 2 sulfhydryl groups may be irreversibly oxidized. This
results in a further lowering of the IEP of catalase fraction
C without affecting its activity or its chromatographic
properties. Fig. 2d shows the isoelectric focusing of a
fraction C which contained approximately 10 irreversibly
oxidized cystein residues. The main part of the active material is focused at pH 5.20.

In horse erythrocyte catalase, the transition from fraction A to fraction C takes place only in presence of oxygen
and heavy metal catalysts. A stable intermediate, as detected

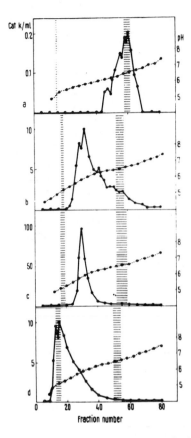

Fig. 2. Isoelectric focusing patterns of different molecular forms of human erythrocyte catalase. a) Fresh hemolysate (polarity of the electrodes inverted). b) fraction A (14 sulfhydryl groups); c) fraction C (2 sulfhydryl groups oxidized irreversibly); d) fraction C (10 sulfhydryl groups oxidized irreversibly). For experimental details see text. ●—● , catalase activity (k/ml); ●••••● , pH (measured at 0°C). ≡ , pH 6.45 - 6.55; ≡ , pH 5.10 - 5.20.

in human erythrocyte catalase, was not found. Fraction C, containing approximately 10 irreversibly oxidized cystein residues, has an IEP of 5.70 (Table 1).

The data presented indicate that the transition of fraction A of human as well as horse erythrocyte catalase to fraction C is paralled by an irreversible oxidation of sulfhydryl groups and a considerable shift of the IEP. These observations are compatible with the assumption that, during the

transition of fraction A to C, acid derivatives of cystein
are formed by oxidation of sulfhydryl groups present in fraction A.

The difference in IEP between the enzyme present in a
fresh hemolysate and the oxidized fraction C is larger than
the one between fractions A and C, and amounts to 1.3 pH-
units in the human enzyme (pH 6.5 - 5.2) and 1.6 in horse
erythrocyte catalase (pH 7.3 - 5.7). Thus, additional factors other than SH-oxidation affect the IEP of erythrocyte
catalase. Fraction A, although isolated under mild, non-
oxidizing conditions, has a lower IEP than catalase in a
fresh hemolysate and therefore does not correspond to the
native enzyme. Since fraction A contains no irreversibly
oxidized sulfhydryl groups, this part of the difference cannot be attributed to the formation of acid derivatives of
cystein. It is possibly due to conformational changes occuring during the purification of the enzyme.

The results are compatible with the assumption that only
one "native" molecular form of catalase exists in the intact
erythrocyte. In vivo this enzyme - like all other red-cell
constituents - is protected against oxidation by reduced
glutathione (H. Aebi & H. Suter, 1974). Hence, the heterogeneity of erythrocyte catalase observed after removal of the
enzyme from the cell and exposure to aerobic conditions, is
due to secondary reactions. These preparatory artifacts
which do not affect the specific activity of the enzyme, can
be demonstrated by taking the number of the SH-groups or the
IEP as an indicator. However, the causal relationship between these parameters is not yet established. Alterations
of this type may be of minor importance for most investigations of catalase. However, the variability of these parameters implies that such effects are watched carefully, if
the data are intended to be used for a differentiation
between the normal enzyme and enzyme variants.

Conformational alterations of similar nature have also
been observed with liver catalase (Heidrich, 1968), muscle
phosphoglucose isomerase (M. N. Blackburn et al, 1972) and
red-cell enzymes such as glucose-6-phosphate dehydrogenase
and aspartate aminotransferase, stressing the importance of
a reducing environment for maintenance of the structure of
intracellular enzymes (Walter et al, 1965).

Dissociation of erythrocyte catalase into subunits

1. *Detection of subunits in preparations of normal human
erythrocyte catalase*

Erythrocyte catalase is an enzyme of oligomer structure with a molecular weight of 240,000; it consists of 4 identical subunits corresponding to a particle size of approximately 60,000.

It has been known for some time that in stored hemolysate specimens, catalase activity decreases at a considerable rate. The relative loss of activity depends on the concentration and the degree of purification. In a concentrated hemolysate of normal human erythrocytes (60 mg Hb/ml), catalase is stable at +4°C for at least 24 hours. However, up to 80% of the original activity may be lost when kept in dilute solution (60 μg Hb/ml) (Hübl and Bretschneider, 1964; Aebi, 1970). On the other hand, activity measurements in erythrocyte populations fractioned as to density and age indicate that no significant decrease of catalase activity can be noticed throughout the life span of a normal human erythrocyte (Aebi & Cantz, 1966; Sass, 1963). Therefore, the question arises whether this phenomenon, which is of·practical relevance for the accuracy and the reproducibility of catalase activity measurements, might possibly be due to a slow, but steady dissociation of the active enzyme into inactive subunits. The possibility of oligomer dissociation of erythrocyte catalase has been investigated by means of exclusion chromatography and antibodies specific for total catalase and its subunits respectively (Shapira & Ben Yoseph, 1973).

When purifying erythrocyte catalase according to the procedure of Mörikofer et al (1969), the last step consists of a separation of the pure enzyme from inert material by chromatography on CM - cellulose. If such a freshly isolated sample of catalase is chromatographed on Sephadex G - 150, an elution pattern as shown in Fig. 3 is observed. There are three peaks, if absorbance at 280 nm is taken as an indicator. The protein eluted first exerts catalase activity and has a ratio A 405 / A 280 nm only slightly lower than the pure enzyme. This main peak is followed by two minor peaks exerting no detectable catalase activity and having a significantly lower ratio A 405 / A 280 nm (Table 2). Based on the respective elution volumes, a molecular weight of 240,000 was obtained for peak 1, 120,000 for peak 2 and 60,000 for peak 3. These figures indicate that peak 2 has half the molecular size of catalase, whereas peak 3 material is four times smaller than the active catalase molecule.

In a series of fractionation experiments performed with the same material under different conditions, notably the time of storage of the preparations, the relative amounts eluted as peak 2 and 3 varied considerably. On the other hand, in contrast to the yield, the elution volume as well as the ratio A 405 / A 280 nm showed only little variation

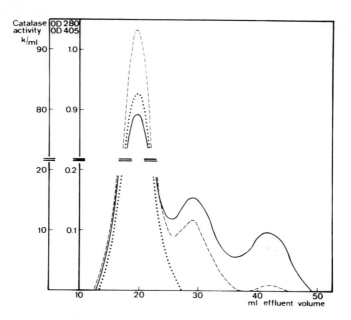

Fig. 3. Fractionation of normal human erythrocyte catalase
by exclusion chromatography on Sephadex G - 150 : Separation
of three molecular species. The catalase active fractions
obtained from chromatography on CM - cellulose were pooled
and concentrated by ultrafiltration. 0.5 ml of this solution, containing about 10 mg protein/ml was then used for
exclusion chromatography on Sephadex G - 150 (column 0.9 x
70 cm; flow rate 3.2 ml/h; 0.8 ml fractions; elution with PBS).
---- absorbance at 405 nm, ——— absorbance at 280 nm, ····
catalase activity (k/ml).

(Table 2). Tentatively, the peak 2 fraction may be considered
as an inactive catalase-dimer and peak 3 fraction as a monomer.
If catalase active material (= peak 1 fraction; see Fig. 3)
is rechromatographed on Sephadex G - 150 under the same experimental conditions, only one single peak is eluted. However,
if the sample is kept at $+4^{\circ}C$ for some time, the original
heterogeneous distribution pattern occurs. Although there is
no close proportionality between time lag and distribution
ratio, highest yields regarding peak 2 and 3 fractions are
seen after prolonged storage, e.g. 3 months at $+4^{\circ}C$. Repeated
freezing - thawing also brings about a redistribution of the
components, but not to a higher extent than prolonged storage.
The same phenomenon could be observed after isolating and
storing samples of peak 2 material (catalase-dimer), inasmuch

236

TABLE 2

PROPERTIES OF CATALASE FRACTIONS OBTAINED BY EXCLUSION CHROMATOGRAPHY
OF NORMAL HUMAN ERYTHROCYTE CATALASE ON SEPHADEX G – 150

Purified proteins used for calibration (Column size 0.9 x 70 cm):

Beef liver catalase,	mol. wt. 240,000,	effluent volume 19.7 ml
Lactate dehydrogenase,	mol. wt. 140,300,	effluent volume 25.5 ml
Malate dehydrogenase,	mol. wt. 70,000,	effluent volume 40.0 ml

Sample	Effluent volume (ml)		Ratio $\dfrac{A\ 405\ nm}{A\ 280\ nm}$		Catalase Activity
	x ± s	N	x ± s	N	
Fraction I (Tetramer)	19.7 ± 1.1	13	1.06 ± 0.10	18	+++
Fraction II (Dimer)	29.2 ± 1.8	9	0.75 ± 0.16	12	0
Fraction III (Monomer)	41.7 ± 1.2	10	0.34 ± 0.15	7	0

as increasing amounts of peak 3 material were obtained. These observations may be taken as evidence for a slow, but significant dissociation of the catalase tetramer molecule into subunits (dimer and monomer), at least in dilute solution under certain experimental conditions.

2. *Isolation of subunits from an unstable catalase variant*

Blood from individuals homozygous for acatalasemia contains only a small fraction of catalase active material. The level relative to that of normal blood, as well as the properties of this residual activity, vary from family to family. In Swiss Type Acatalasemia (propositus A.B.) 1-2% are found (Aebi & Suter, 1971). In order to study its oligomer-monomer distribution pattern, this enzyme variant was purified by the same isolation procedure as used for the normal enzyme. The pooled catalase active fractions eluted from the CM - cellulose column were subjected to exclusion chromatography on Sephadex G - 150. The result of this separation experiment is shown in Fig. 4. Contrary to the normal enzyme (see Fig. 3), the variant reveals a different elution pattern. The minor first peak, which has been identified as catalase - tetramer (elution volume = 19.7 ml; ratio A 405 / A 280 nm = 1.06; catalase equivalent to 52 µg), is followed by a large second peak. This material has no catalase activity and the ratio A 405 / A 280 nm is only 0.46. These data, as well as the elution volume of 29.2 ml, suggest that this represents the dimer-fraction of the unstable mutant enzyme. A small third peak has the same mobility as the monomer of normal catalase. A possible interconversion in vitro of the unstable tetramer into its subunits was investigated as follows: The catalase active peak fractions were pooled and concentrated. After one week of storage at +4°C, the material was rechromatographed in the same way. Upon analysis of the eluate fractions, no catalase activity could be detected. The occurrence of one single peak with an elution volume of 41.7 ml (see Fig. 4, legend B) indicates that the mutant enzyme had undergone an apparently complete dissociation into its monomer within a relatively short time.

3. *Comparison between the normal and the mutant enzyme*

The main differences between normal and mutant catalase are summarized in Table 3. Unlike normal human erythrocyte catalase, the mutant enzyme is very unstable. This is visualized by a pseudomosaicism in the red cell population as well as by a rapid decrease of catalase activity in reticulocytes,

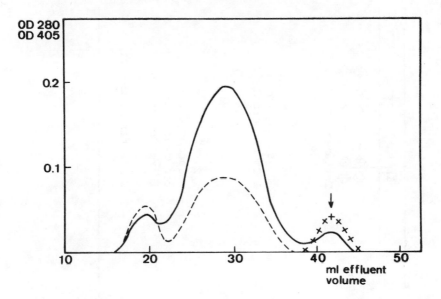

Fig. 4. Fractionation of variant human erythrocyte catalase
by exclusion chromatography on Sephadex G - 150 : Separation
of three molecular species. Blood obtained from an individual
homozygous for Swiss Type Acatalasemia (A.B.) A) The catalase
active fractions obtained from chromatography on CM - cellulose were pooled and concentrated by ultrafiltration. 0.5 ml
of this solution, containing 2.6 mg protein/ml were then chromatographed on Sephadex G - 150 (column 0.9 x 70 cm; flow rate
3.2 ml/h; elution with PBS). ---- absorbance at 405 nm,
_____ absorbance at 280 nm. B) The catalase active fractions
obtained from the above column (A) were again pooled and concentrated, stored at +4°C for one week and an aliquot was
rechromatographed under identical conditions : no absorbance
detectable at 405 nm, ++++ absorbance at 280 nm.

when incubated in vitro (Aebi & Cantz, 1966). Furthermore,
the variant enzyme is much more heat sensitive (Matsubara et
al, 1967). These differences are explained by the observation reported here, that dissociation into catalase inactive
subunits proceeds much faster in the variant than in the normal enzyme. In contrast to the differences mentioned above,
both enzyme species exert the same specific activity (Aebi &
Cantz, 1966) and there is also a complete antigenic identity
(Aebi et al, 1964; Aebi et al, 1974).

TABLE 3

FORMS OF HUMAN ERYTHROCYTE CATALASE

	Normal enzyme	Instable variant (A.B.)
molecular distribution pattern	Tetramer >> Dimer ~96% ~4%	Tetramer << Dimer ~2% ~98%
activity in red cells	2.2 mg catalase / g Hb (=100%)	20 μg / g Hb ~1-2% of normal
signs in vivo	t/2 > 120 d equal distribution in red cell population	t/2 ~ 24 h pseudomosaicism, residual activity mainly in reticulocytes
inactivation in vitro	50% after 10 min at 63°C	50% after 10 min at 55°C

4. *Immunochemical characterization of various catalase preparations with anti-catalase IgG and anti-catalase subunit IgG*

In double immunodiffusion tests, a strong enzymatically active precipitin line indicating antigenic identity is formed between purified normal catalase, purified variant catalase and anti-catalase IgG. However, this line does not appear with the acatalasemic hemolysate. These findings permit one to conclude that there is an apparently complete antigenic identity between normal and variant catalase; this is, however, detectable only when an appropriate antigen-antibody ratio is reached, which is accomplished by using solutions of the purified enzyme preparations.

An anti-catalase subunit IgG preparation was used for testing a variety of components isolated from normal and from acatalasemic blood. A typical experiment is shown in Fig. 5.

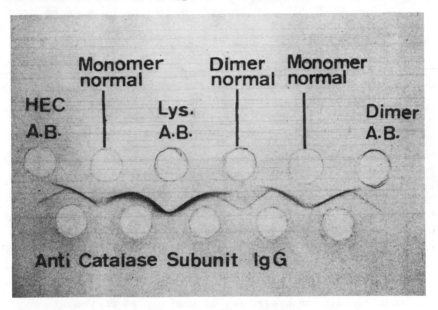

Fig. 5. The immunoprecipitin pattern of various catalase components isolated from normal and acatalasemic blood with anti-catalase subunit IgG. HEC A.B. = purified variant erythrocyte catalase preparation from A.B. Lys. A.B. = hemolysate from the blood of propositus A.B. Dimer normal, Monomer normal, and Dimer A.B. = preparations isolated by recycling exclusion chromatography on Sephadex G - 150 from normal human and variant human erythrocyte catalase respectively.

A precipitin line is formed with all the antigens tested. However, spurs formed at several antigen wells demonstrate a

distinct antigenic deficiency when using the anti-catalase
subunit IgG. Normal dimer, variant dimer (A.B.) as well as
variant catalase (HEC A.B.) are deficient in respect to normal monomer and acatalasemic hemolysate (Lys. A.B.). The
latter two components seem to possess identical antigenic
properties. In the hemolysate of acatalasemic blood, an
appreciable amount of catalase monomer is apparently present,
also some dimer, but only a small quantity of the tetramer.
On the other hand, the similarity of the precipitation pattern
of normal dimer, variant dimer (A.B.) and of variant catalase
(HEC A.B.) is due to the fact that purified variant catalase
consists mostly of dimer and of tetramer - the latter not being detectable with the anti-subunit antiserum used. Enzymatic activity was not found in any of the precipitin lines
developed with the anti-catalase subunit IgG. The availability of normal dimer and monomer preparations has enabled us
to investigate the antibody properties of the anti-catalase
subunit IgG (Fig. 5). At least two antibodies must be present
in this antiserum: one reacting with an antigenic determinant
common to dimer and monomer, and a second one specific for
monomer only (see Fig. 5, spur formation of the monomer over
the dimer). Both these determinants are accessible only
after the catalase molecule has been split, since this antiserum does not react with the intact tetramer molecule.
Therefore, both of them may be designated as "hidden" determinants.

CONCLUDING REMARKS

Both aspects of heterogeneity reported here reflect the instability of this enzyme. Recent progress made in analytical
enzymology discloses that secondary rearrangements of the
molecular structure leading to conformational changes, without affecting enzymatic activity, are more frequent than it
was originally believed. The tentative names given to such
multiple forms include conformers, pseudoisozymes, and metazymes. Thus, a more uniform and systematic classification of
such alterations is desirable. In this context it is not
unreasonable to include those steps or products which initiate inactivation or breakdown of the active enzyme. The
dissociation of oligomer enzymes, as this has been shown here
for red cell catalase, may proceed at such a rate that it may
ultimately lead to an apparent lack of enzyme.

There are various possible ways which may lead to acatalasemia. If this condition is due to a structural gene mutation, two different types of enzyme variants ("allozymes")
have to be considered: catalase mutants of (a) low specific
activity, or (b) of approximately normal specific activity,
but low stability. Obviously, the individual A.B., representing "Swiss-type" Acatalasemia, as well as the Argonne-strain
of acatalasemic mice (Aebi, Suter, and Feinstein, 1969),

belong to the latter group. This differentiation is of practical relevance inasmuch as it probably affects the level of
residual enzyme activity in tissues. In an unstable mutant
the relative level of enzyme activity in a given tissue
largely depends on its turnover; therefore the level relative
to normal is very low in blood, but almost unaltered in organs
such as liver. Before extrapolating these observations it
must be remembered that - in catalase as in hemoglobin - every
mutational event has to be considered as a separate genetic
entity, unless its identity can be proven at the molecular
level.

ACKNOWLEDGMENT

This investigation is part of project 3.8460.72 subsidized by
the Schweizerischer Nationalfonds zur Förderung der wissen-
schaftlichen Forschung. We wish to thank Dr. E. Shapira for
providing us with anti-catalase antisera. The helpful criti-
cism of Dr. F. Škvaril and Prof. S. Barandun, Institute for
clinical and experimental cancer research, University of
Bern, and the capable technical assistance of Miss Brigitta
Kissling are gratefully acknowledged.

REFERENCES

Aebi, H., M. Baggiolini, B. Dewald, E. Lauber, H. Suter,
    A. Micheli, and J. Frei 1964. Observations in two Swiss
    Families with Acatalasia II. *Enzymol. biol. clin.* 4:
    121-151.

Aebi, H. and M. Cantz 1966. Ueber die celluläre Verteilung
    der Katalase im Blut homozygoter und heterozygoter
    Defektträger (Akatalasie). *Humangenetik* 3: 50-63.

Aebi, H., E. Bossi, M. Cantz, S. Matsubara, and H. Suter 1967.
    Acatalas(em)ia in Switzerland in Hereditary Disorders of
    Erythrocyte Metabolism. *City of Hope Symposium Series*
    Vol. 1 41-65. Grune & Stratton, New York and London.

Aebi, H. 1970. Katalase. Buchbeitrag in "Methoden der enzymat-
    ischen Analyse", Ed. H. U. Bergmeyer, 2. Auflage, Band 1,
    S. 636-647.

Aebi, H., and H. Suter 1971. Acatalasemia. *Advances in Human*
    Genetics Vol. 2, Plenum Press, New York and London.

Aebi, H., S. Mörikofer-Zwez, and J. P. von Wartburg 1972.
    Alternative molecular forms of Erythrocyte Catalase.
    Structure and Function of Oxidation Reduction Enzymes,
    Pergamom Press, Oxford and New York, 345-351.

Aebi, H., H. Suter, and R. N. Feinstein 1969. Activity and
    stability of catalase in blood and tissues of normal and
    acatalasemic mice. *Biochem. Genet.* 2: 245-251.

Aebi, H., and H. Suter 1974. Protective function of reduced
    glutathione (G-SH) against the effect of prooxidative
    substances and of irradiation in the red cell. *Proceed-*

ings of the 16th Conference of the German Society of Biological Chemistry, Tübingen, 192-201. Georg Thieme Publishers, Stuttgart.

Aebi, H., S. R. Wyss, B. Scherz, and F. Skvaril 1974. Heterogeneity of Erythrocyte Catalase II; Isolation and characterization of normal and variant erythrocyte catalase and their subunits. / in press: *Eur. J. Biochem.* (1974).

Ben-Yoseph, Y. 1973. Antigenic Properties of human carbonic anhydrase and catalase. Thesis submitted to the Senate of the Hebrew University of Jerusalem, April 1973.

Blackburn, M. N., J. M. Chirgwin, G. T. James, T. D. Kempe, T. F. Parsons, A. M. Register, K. D. Schnacherz, and E. A. Noltmann 1972. Pseudoisoenzymes of Rabbit Muscle Phosphoglucose Isomerase. *J. Biol. Chem.* 247: 1170-1179.

Blumberg, A., H. R. Marti, F. Jeunet, and H. Aebi 1962. Katalase-und Hämoglobindifferenzierung bei Fällen von Akatalasie. *Schweiz. Med. Wochschr.* 92: 1324.

Heidrich, H. G. 1968. New aspects of heterogeneity of beef liver catalase. *Z. physiol. Chem.* 349: 873-880.

Holmes, R. S., and C. J. Masters 1972. Species specific features of the distribution and multiplicity of mammalian liver catalase. *Arch. Biochem. Biophys.* 148: 217-223.

Hübl, P., and R. Bretschneider 1964. Die Titanylsulfatmethode zur Bestimmung der Katalase in Blut, Serum und Harn. *Z. physiol. Chem.* 335: 146-155.

Matsubara, S., H. Suter, and H. Aebi 1967. Fractionation of erythrocyte catalase from normal, hypocatalatic and acatalatic humans. *Humangenetik* 4: 29-41.

Mörikofer-Zwez, S., M. Cantz, H. Kaufmann, J. P. von Wartburg, and H. Aebi 1969. Heterogeneity of Erythrocyte Catalase. Correlation between sulfhydryl group content, chromatographic and electrophoretic properties. *Eur. J. Biochem.* 11: 49-57.

Sass, M. D. 1963. Catalase activity in young redcells. *Nature* (London) 197: 503.

Shapira, E., Y. Ben-Yoseph, and H. Aebi 1973. Stabilizing effect of anti-normal erythrocyte catalase antibody on the labile catalase variant present in Swiss Type Acatalasemia. *Experientia* 29: 1402.

Takahara, S. 1967. Acatalasemia in Japan. in Hereditary Disorders of Erythrocyte Metabolism, *City of Hope Symposium Series Vol. 1*, 41-65. Grune & Stratton, New York and London.

Thorup, O. A., Jr., J. T. Carpenter, and P. Howard 1964. Human erythrocyte catalase: Demonstration of heterogeneity and relationship to erythrocyte ageing in vivo. *Brit. J. Haemat.* 10: 542.

Walter, H., F. W. Selby, and J. R. Francisco 1965. Altered electrophoretic mobility of some erythrocytic enzymes as a function of their age. *Nature* 208: 76-77.

AFFINITY CHROMATOGRAPHY RESOLUTION OF PLASMINOGEN ISOZYMES

FRANCIS J. CASTELLINO, JAMES M. SODETZ AND
GERALD E. SIEFRING, JR.
Department of Chemistry, University of Notre Dame
Notre Dame, Indiana   46556

ABSTRACT. Two major forms of rabbit plasminogen have been
isolated by affinity chromatography from rabbit plasma.
One form consists of 5 subforms with an isoelectric pI
range of 6.20-7.78. The other plasminogen form can also be
resolved into 5 subforms, but these possess an isoelectric
pI range of 6.95-8.74. All plasminogen forms and subforms
can be converted to plasmin by common plasminogen activators such as urokinase, streptokinase, and the streptokin-
ase-human plasmin(ogen) complex. Each major plasminogen
form, as well as all isolated subforms, possesses nearly
identical molecular weights of approximately 90,000 and is a
single chain molecule.   In addition, each major form contains identical amino- and carboxyl-terminal amino acids
and identical amino terminal amino acid sequences through
at least 12 residues. Carbohydrate analyses indicate differences in sialic acid, neutral sugars and hexosamines between the two major forms. Removal of the sialic acid from
each form abolishes the electrophoretic differences between
the forms and also leads to loss of resolution of some of
the component subforms. However, the affinity chromatography resolution of the two major forms is not affected by
removal of the sialic acid. Metabolic studies in intact
animals demonstrate that both major plasminogen forms possess dissimilar rates of turnover and synthesis and are not
interconverted.   Pre-steady state kinetic analyses, using
specific p-nitrophenolate esters, of the plasmins resulting
from the urokinase induced activation of each major rabbit
plasminogen form demonstrate that substrates bind more
tightly to plasmin form 1 than plasmin form 2. Acylation
rate constants for these substrates are very similar for
each plasmin form.

INTRODUCTION

Plasminogen is the inactive plasma protein precursor of
the enzyme plasmin. Plasmin functions in the fibrinolytic
system, digesting both fibrin and fibrinogen (Pizzo et al
'73); it activates the first component of complement (Ratnoff
and Naff '67) ,and it digests Hageman Factor, thereby divert-
ing the function of this factor from its role of initiation

of the intrinsic blood coagulation system to generation of kinins (Spragg et al '73). The activation of the proenzyme plasminogen to the enzyme plasmin is mediated by many agents, among which are the proteases,urokinase,and tissue activators and the otherwise inactive bacterial protein, streptokinase.

Although plasminogen was discovered many years ago, it is only recently that some idea of the extent of its heterogeneity in the circulation has come to light. At least three different types of heterogeneity of human plasminogen have been described: 1) two plasminogen forms with different NH_2-terminal residues which can be separated by DEAE-Sephadex chromatography (Wallén and Wiman '72); 2) two different forms of plasminogen from several species with identical NH_2-terminal residues, separated by affinity chromatography and polyacrylamide gel electrophoresis at pH 4.3, described by our laboratory (Brockway and Castellino '72); and 3) several plasminogen forms separated by isoelectric focusing from human (Summaria et al '72) and rabbit plasminogen (Sodetz, Brockway and Castellino '72). The first type of heterogeneity appears to be a result of proteolysis during the purification procedures (Wallén and Wiman '72) and this will not be further considered. In this report, we will attempt to unify some ideas on the nature of the other two types of heterogeneity that have been described.

MATERIALS AND METHODS

The preparation of both major forms of rabbit and human plasminogen and the methods of activation of rabbit plasminogen by streptokinase, urokinase or human plasminogen activator have been published by Brockway and Castellino ('72) and Sodetz, Brockway,and Castellino ('72). The electrophoresis and ultracentrifugation procedures were as described in these papers. Methods of carbohydrate analysis, amino terminal end group analysis and,amino terminal amino acid sequencing were as reported by Castellino et al ('73). Hexosamines were analyzed as described by Elson and Morgan ('33). Our methods for studying the metabolic turnover, biosynthesis,and interconversion characteristics of each major plasminogen form have been described in an earlier report (Siefring and Castellino '74). Pre-steady state kinetic analysis of each rabbit plasmin with the p-nitrophenolate esters were as previously reported for other plasmin species (Sodetz and Castellino '72). The substrates were synthesized as described by Glover et al ('73).

Sialic acid was removed from each rabbit plasminogen form by dissolving the plasminogen at 5 mg/ml in a buffer consist-

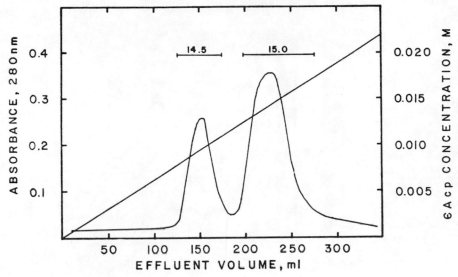

Fig. 1. Gradient elution of rabbit plasminogen from the Sepharose 4B-L-lysine affinity chromatography columns. This
graph shows the profile obtained after all other proteins
have been washed from the column. These peaks were identified as plasminogen by their esterolytic properties after
addition of urokinase.

ing of 0.1 M acetate, 0.1 M lysine, 0.002 M CaCl$_2$ at pH 5.6.
To 2.5 ml of this solution, 2.5 ml of *Vibrio cholera* neuraminidase (500 units/ml) in the same buffer was added and the
solution allowed to incubate for 9 hours at 37°. At the end
of this period diisopropyl fluorphosphate was added such that
the final concentration was 0.01 M. This solution was incubated for 2 hours at 4°. The solution was then dialyzed
against 0.1 M phosphate for 2 days and the plasminogen purified by affinity chromatography. Under these conditions, no
proteolytic cleavage is noted in the plasminogen.

RESULTS

It has previously been shown by Deutsch and Mertz ('70)
that plasminogen could be removed from human plasma by an
affinity chromatography method utilizing L-lysine coupled to
Sepharose 4B. We coupled L-lysine to this column and applied
a linear gradient of the fibrinolytic inhibitor, ε-amino
caproic acid, to selectively remove the plasminogen. The results of this chromatography are shown in Fig. 1. Clearly
two major forms of plasminogen are obtained by this treatment. Analytical polyacrylamide gel electrophoresis runs on

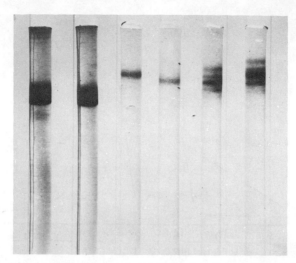

Fig. 2. Analytical polyacrylamide gels for each rabbit plasminogen form. Left to right: sodium dodecyl sulfate-mercaptoethanol gel of form 1 plasminogen; sodium dodecyl sulfate-mercaptoethanol gel of form 2 plasminogen; pH 4.3 gel of form 1 plasminogen; pH 4.3 gel of form 2 plasminogen; pH 9.5 gel of form 1 plasminogen; pH 9.5 gel of form 2 plasminogen. Form 1 and form 2 plasminogens are identified solely by their order of elution from the affinity chromatography columns.

TABLE I

ISOELECTRIC POINTS OF THE SUBFORMS
OF RABBIT PLASMINOGEN AT 22°

| Subform | pI | |
	Form 1	Form 2
1	6.20	6.95
2	6.56	7.18
3	6.85	7.89
4	7.24	8.42
5	7.78	8.74

each component are shown in Fig. 2. At pH 4.3, single bands are obtained for each component and these possess distinct electrophoretic mobilities. At pH 9.5, various subforms are resolved from each major band with staggered overlapping of the individual components. In sodium dodecyl sulfate, single bands with practically identical molecular weights are obtained. Isoelectric focusing profiles of each major form are shown in Fig. 3. The behavior of the subforms with this

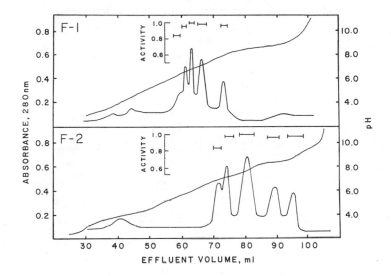

Fig. 3. Isoelectric focusing profiles of rabbit plasminogen.
Top: rabbit fraction 1 focused at 4° on a pH 3.0-10.0 gradient. The absorbance at 280 nm, the experimental pH gradient,
and the activities of each dialyzed and concentrated pool after conversion to plasmin are indicated on the graph. The
specific activities are relative to the central pool arbitrarily given a value of 1.0. The actual V_{max} of the central
pool was 13.4 µmoles Tos-arg-OMe cleaved min^{-1} mg^{-1} of
plasminogen originally taken for activation. Bottom: as top
except that form 2 was employed. The V_{max} of the central
pool was 13.7 µmoles Tos-arg-OMe cleaved min^{-1} mg^{-1} of
plasminogen originally activated.

technique closely parallels the results of the 9.5 gels in
that staggered overlapping of the subforms are evident. The
isoelectric points of the subforms are shown in Table I. A
wide range of isoelectric points is noted for the subforms
of each major form.

Table II shows some native and subunit molecular weight
data for each major form as well as for some selected subforms. In all cases single chain molecules of molecular
weights 88,000-94,000 are obtained. Quantitative amino and
carboxyl terminal end group analysis of each major plasminogen form is shown in Table III. Single amino acids were
identified at very high yields in each case. These results
would not only suggest that each major form possesses identical
end terminal amino acids but also suggest that each subform

TABLE II

MOLECULAR WEIGHT OF SELECTED PLASMINOGEN
FORMS AND SUBFORMS

Protein	Molecular weight
Form 1 - intact	89,000 - 93,000
Form 1 - pI 6.56 subform	90,000 - 94,000
Form 1 - pI 7.78 subform	90,000 - 94,000
Form 2 - intact	89,000 - 93,000
Form 2 - pI 7.18 subform	88,000 - 92,000
Form 2 - pI 8.42 subform	88,000 - 92,000

TABLE III

AMINO ACID END GROUP ANALYSIS
ON EACH RABBIT PLASMINOGEN FORM

Protein	Amino terminal (moles/mole)	Carboxyl terminal[a]
Plasminogen F-1	glutamic acid (0.95)	asparagine
Plasminogen F-2	glutamic acid (1.04)	asparagine

[a]determined for human plasminogen (Collen '73).

also possesses identical end groups. We have also determined
the amino terminal amino acid sequences of each major rabbit
plasminogen form. Both forms consist of identical sequences
through at least the first twelve amino acids. The sequence
for each form is NH_2-glu-pro-leu-asp-asp-tyr-val-asn-thr-gln-
gly-ala-. Furthermore, since only single amino acids were detected at each point during the sequencing, we feel that all
subforms must also possess this same sequence.

Carbohydrate analysis on the two major rabbit plasminogen
forms reveals some interesting differences. This data is
shown in Table IV. There appear to be greater amounts of
sialic acid, neutral sugars, and hexosamines in form 1 than in
form 2. With regard to the subforms, there are small differences in sialic acid but no differences in neutral or amino
sugars. We therefore were interested in determining what
effect removal of the sialic acid would have on some of the
distinct properties of the 2 major forms. We first determined the fate of a mixture of asialo-form 1 and 2 on affinity chromatography columns. This behavior is shown in Fig.
4. Almost identical profiles are noted for the asialo- and
sialo-plasminogens. We attempted to determine whether the
subform structure of each individual form of plasminogen was
altered by removal of the sialic acid and the results of
isoelectric focusing of asialo-form 1 rabbit plasminogen are
shown in Fig. 5. The isoelectric points are shifted almost

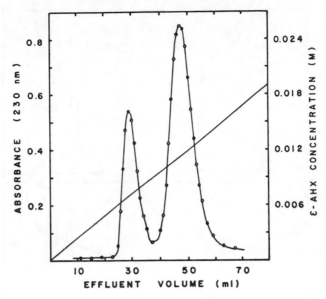

Fig. 4. Affinity chromatography of the asialo rabbit plasminogen forms. O, neuraminidase treated rabbit plasminogen
forms 1 and 2; ●, normal rabbit plasminogen forms 1 and 2
shown for comparison.

TABLE IV

CARBOHYDRATE ANALYSIS OF THE TWO MAJOR
FORMS OF RABBIT PLASMINOGEN

Moiety measured	Content in plasminogen	
	F-1	F-2
neutral sugar (%, w/w)	1.5-1.7	0.6-0.8
hexosamines (moles/mole)	6-8	3-4
sialic acids (moles/mole)	3.0-3.3	1.8-2.2

2 pH units higher but still at least 3 subforms are obtained.
It is significant to note from this experiment that removal
of the sialic acid appears to abolish the charge difference
between the two major rabbit plasminogen forms. However, as
can be seen from Fig. 4, removal of sialic acid does not effect the functional differences between the two forms.

We have performed some metabolic studies on the rates of
synthesis, degradation,and possible interconversion of the
two major plasminogen forms in the rat and rabbit (Siefring
and Castellino '74). The results of these studies are shown
in Table V. Rates of synthesis were obtained by injecting
tritiated L-leucine into rats, removing the blocd at times
when active synthesis was occurring,and isolating each form

251

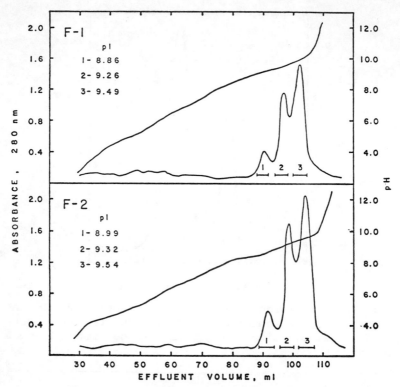

Fig. 5. Isoelectric focusing profiles of each asialo-rabbit plasminogen form. The details are as in Fig. 3. The insert gives the pI values, obtained at 4°, of each indicated subform.

of labeled plasminogen by affinity chromatography. The specific radioactivity of each form was then determined. It appears clear that the rate of biosynthesis of form 2 is twice as great as that of form 1. The half-lives of each plasminogen form in rats were determined by injecting either iodinated or sialic acid labeled plasminogen directly into the blood stream of the rat and removing aliquots at desired times to observe the rate of radioactive decay. The half-life of 10 hours for each form was practically the same but the small differences between the two forms are reproducible. We conclude that form 2 has a slightly shorter half-life in the circulation than form 1. The possible interconvertibility of the two forms in the animal was tested by injecting one labeled form of plasminogen into the rat and determining the extent of labeling in the other form at sufficient times after injection. It is clear that no significant interconver-

TABLE V

METABOLIC STUDIES ON THE TWO MAJOR FORMS OF
PLASMINOGEN IN THE RAT

Rates of synthesis		
Time after [^{3}H] L-leucine injection	Plasminogen form 1 (dpm/mg) ± 8%	Plasminogen form 2 (dpm/mg) ± 8%
1 hour	2450	4900
2 hours	3903	8072

Rates of degradation	
Plasminogen Form	Half life in circulation
1	12-14 hours
2	9-11 hours

Interconvertibility				
Form added (dpm/mg)	Quantity injected (mg)	Time (min)	dpm/mg recovered in F-1	F-2
F-1 (23,500)	0.3	90	2500	150
F-2 (35,100)	0.7	90	149	4007

sion of the forms occurs.

Finally, we have compared some kinetic parameters of the esterolytic activity of each form of rabbit plasmin. Although plasmin is a much larger protein than trypsin, its specificity and mechanism of action is similar to that of trypsin. The proposed mechanism of plasmin is summarized as:

$$E + S \underset{}{\overset{K_S}{\rightleftharpoons}} ES \xrightarrow{k_2} ES' \xrightarrow{k_3} E + P_2$$
$$+$$
$$P_1$$

where E is the enzyme; S, the ester substrate; ES, a complex of the enzyme and substrate; ES'; the acyl-enzyme covalent complex; P_1, the alcohol product and P_2, the acid product. The constant K_S refers to the binding constant of substrate to enzyme and the k_2 and k_3 are rate constants for acylation and deacylation, respectively. We have determined the pre-steady state constants K_S and k_2 for each rabbit plasmin with a series of specific p-nitrophenyl esters. These compounds formed relatively stable acyl intermediates with plasmin, so k_3 was extremely small. These results are shown in Table VI. For all substrates, the rate constants for acylation are not very different. However, significant differences are noted in some cases for K_S. It appears as though substrates bind more tightly to form 1 than form 2 plasmin and this may possess some significance in the function of plasmin.

TABLE VI

PRE-STEADY STATE CONSTANTS FOR EACH FORM OF RABBIT PLASMIN WITH SUBSTRATES OF THE

NO_2 class

X	Form 1 Plasmin		Form 2 Plasmin	
	k_2 (sec^{-1})	$K_S \times 10^4$ (M)	k_2 (sec^{-1})	$K_S \times 10^4$ (M)
$^+$N–CH$_2$– (pyridinium)	0.122	2.10	0.127	2.30
H$_2$N$^+$=C–NH– / NH$_2$ (amidinium)	0.125	0.052	0.125	0.113

$$\underset{C_2H_5}{\overset{CH_3}{>}}\overset{+}{S}-CH_2-$$

0.117   2.14   0.144   2.58

$$\underset{NH_2}{\overset{H_2N}{>}}\overset{+}{C}-S-CH_2-$$

0.364   0.270   0.333   0.426

DISCUSSION

As a result of the studies described in this manuscript, some conclusions can be drawn as to the nature of the multiple forms of the circulating plasminogen. We wish to emphasize the fact that the plasminogens which we are describing are purified in approximately 95-100% yield from the animal by a method which is mild and rapid. These observations would suggest that the structural multiplicity is not an artifact of the purification procedures. Furthermore, the multiplicity found in pooled plasma is also observed in plasma from individual animals.

The major structural differences in the two major plasminogen forms appear to reside in their extent of glycosylation. We do not mean to infer that there are no differences in the primary structures, but we have not found such differences. The major functional differences between these two plasminogen forms appear to be their differential affinities for antifibrinolytic amino acids. In fact, this is the basis of the separation of forms. Again, there may be other functional differences such as differential rates of activation (Sodetz, Brockway and Castellino '72), but these remain to be established. At this time, it would seem to be appropriate to classify the two major plasminogen forms as isozymes. We have not been able to demonstrate interconversion of the two forms either in vitro or in vivo. Furthermore, the metabolic studies indicate that each form is synthesized and degraded independently. We feel that it is extremely important to emphasize that the charge differences in the two forms are incidental to their classification as isozymes since we can abolish the charge differences in the forms by removal of the sialic acid without affecting their separation on affinity chromatography columns. With regard to the subforms within each major form, we feel that some of these may be a result of desialyation, since removal of the sialic acid causes a loss of resolution of 2 of the subforms in each major form. However, owing to the very high isoelectric points of the asialo-subforms, we have a practical problem with the resolving power of the commercial ampholines at these pH values.

Finally, we have observed some interesting and subtle differences in the kinetic properties of the plasmins which result from the urokinase-induced activation of each major rabbit plasminogen form. Thus, it appears that with some of the specific substrates listed in Table VI the K_s value of the nitrophenyl esters to plasmin form 1 is lower than plasmin form 2. However, the acylation rate constants (k_2) for each form are similar. This suggests that although the binding of

the substrate is tighter to plasmin form 1, the orientation
of the substrate on the enzyme for attack by the active site
serine residue is similar for each form.

ACKNOWLEDGEMENTS

This study was supported in part by grants HL-13423 and
HL-15747 as well as grants from the Indiana and American
Heart Associations.
    Figs. 1 and 3, as well as Table I were taken from Sodetz,
Brockway and Castellino ('72) and reprinted with the permis-
sion of the American Chemical Society.

REFERENCES

Brockway, W. J. and F. J. Castellino 1972. Measurement of
    the binding of antifibrinolytic amino acids to various
    plasminogens. *Arch. Biochem. Biophys.* 151: 194-199.
Castellino, F. J., G. E. Siefring, Jr., J. M. Sodetz, and
    R. K. Bretthauer 1973. Amino terminal amino acid se-
    quences and carbohydrate of the two major forms of rab-
    bit plasminogen. *Biochem. Biophys. Res. Commun.* 53:
    845-851.
Collen, D. 1973. De microheterogeneiteit van humaan plasmin-
    ogeen. Ph.D. thesis presented to Katholieke Universiteit
    Leuven.
Deutsch, D. G. and E. T. Mertz 1970. Plasminogen: purifica-
    tion from human plasma by affinity chromatography.
    Science 170: 1905-1906.
Elson, L. A. and W. T. J. Morgan 1933. A colorimetric method
    for the determination of glucosamine and chondrosamine.
    Biochem. J. 27: 1824-1828.
Glover, G., C.-C. Wang, and E. Shaw 1973. Aromatic esters
    which inhibit plasmin or thrombin by formation of rela-
    tively stable acyl enzymes. *J. Med. Chem.* 16: 262-266.
Pizzo, S. V., M. L. Schwartz, R. L. Hill, and P. A. McKee
    1973. The effect of plasmin on the subunit structure of
    human fibrin. *J. Biol. Chem.* 248: 4574-4583.
Ratnoff, O. D. and G. B. Naff 1967. The conversion of C'_{1s}
    to C'_{1s} esterase by plasmin and trypsin. *J. Exp.*
    Med. 125: 337-358.
Siefring, G. E., Jr. and F. J. Castellino 1974. Metabolic
    turnover studies on the two major forms of rat and rab-
    bit plasminogen. *J. Biol. Chem.* 249: 1434-1438.
Sodetz, J. M., W. J. Brockway,and F. J. Castellino 1972. The
    multiplicity of rabbit plasminogen. Physical character-
    ization. *Biochemistry* 11: 4451-4457.

Sodetz, J. M. and F. J. Castellino 1972. A comparison of the steady and presteady state kinetics on human and bovine plasmins. *Biochemistry* 11: 3167-3171.

Spragg, J., A. P. Kaplan, and K. F. Austen 1973. The use of isoelectric focusing to study components of the human plasma kinin-forming system. *Ann. N. Y. Acad. Sci.* 209: 372-385.

Summaria, L., L. Arzadon, P. Bernabe, and K. C. Robbins 1972. Studies on the isolation of the multiple molecular forms of human plasminogen and plasmin by isoelectric focusing methods. *J. Biol. Chem.* 247: 4691-4702.

Wallén, P. and Wiman, B. 1972. Characterization of human plasminogen. II. Separation and partial characterization of different molecular forms of human plasminogen. *Biochim. Biophys. Acta* 257: 122-134.

Warren, L. 1959. The thiobarbituric assay of sialic acids. *J. Biol. Chem.* 234: 1971-1975.

MULTIPLE MOLECULAR FORMS OF TRANSGLUTAMINASES IN HUMAN AND GUINEA PIG

SOO IL CHUNG
Laboratory of Biochemistry
National Institute of Dental Research
National Institutes of Health
Bethesda, Maryland 20014

ABSTRACT. The multiple molecular forms of transglutaminases in human and guinea pig may be classified into three distinct groups on the basis of physical, chemical, immunological, and enzymatic properties. These enzymes are: 1) the protransglutaminases (Factor XIII) of plasma, platelet, placenta, prostate gland, uterus, and liver, 2) the tissue transglutaminase of various organs and tissues, 3) the hair follicle transglutaminase. Each transglutaminase catalyzes a simple acyl-transfer reaction by a common catalytic mechanism (Folk, 1969). Tissue and plasma transglutaminases appear to have a common spatial arrangement within their glutamine side chain binding sites. However, each enzyme displays pronounced differences in its catalytic efficiency toward the various peptide substrates. Each of the transglutaminases plays a specified role in its respective physiological processes although the exact role of the tissue enzyme is not yet certain.

INTRODUCTION

Transglutaminase catalyzes a Ca^{2+}-dependent acyl-transfer reaction in which the γ-carboxamide groups of peptide-bound glutamine residues are the acyl donors. Primary amino groups in a variety of compounds may function as acyl acceptors with the subsequent formation of mono-substituted γ-amides of peptide-bound glutamic acid. In the presence of less than saturating levels of a suitable primary amine or in the absence of an amine, water can act as the acyl acceptor with formation of peptide-bound glutamic acid. These reactions may be denoted respectively as follows:

$$
\begin{array}{l}
\overset{\displaystyle NH_2}{\underset{\displaystyle |}{}} \\
-Glu- \ + \ R-NH_2 \ \rightleftharpoons \
\overset{\displaystyle NH-R}{\underset{\displaystyle |}{-Glu-}} \ + \ NH_3
\end{array}
$$

$$
\begin{array}{l}
\overset{\displaystyle NH_2}{\underset{\displaystyle |}{}} \\
-Glu- \ + \ H_2O \ \rightleftharpoons \ -Glu- \ + \ NH_3
\end{array}
$$

The name transglutaminase was assigned by Clark et al (1957) in order to distinguish this enzymatic activity from that of other enzymes that affect hydrolysis or transfer at the car- boxamide of free glutamine, i.e., glutaminase, γ-glutamyl transferase, glutamine dehydrogenase, etc.

Initial survey of transglutaminase activity as measured by [^{14}C]cadaverine incorporation into various tissue proteins (Clark et al, 1959) showed a wide distribution of enzyme activity among tissues and organs. Tyler and Lack (1964) re- ported that the amine-incorporating activities of various tissues are closely associated with fibrin clot stabilization. Later, with partially purified guinea pig liver transglutamin- ase, Lorand et al (1966) and Tyler and Laki (1966) were able to demonstrate the formation of insoluble fibrin clots.

The first indication of the multiple nature of transglu- taminase in tissues and organs was obtained by Tyler and Laki (1966) by the separation of two peaks of fibrin clot stabil- izing activity in the guinea pig liver extracts on DEAE-Cell- ulose.

In recent years considerable progress has been made in understanding the physico-chemical nature of both guinea pig liver transglutaminase and human plasma protransglutaminase (see review Folk and Chung, 1973). The very low levels of enzymes in human tissue and difficulty of obtaining sufficient amount of autopsy material limited the direct thorough compara- tive studies of the multiple nature of transglutaminases in same species. However, extensive studies involving a variety of different substrates have shown that human blood transglu- taminase and guinea pig red blood cell and liver transglutamin- ase possess almost identical kinetic properties (Chung et al, 1974). Particular emphasis will be placed on the human plate- let enzyme and guinea pig tissue transglutaminase. The char- acterization of these multiple forms of transglutaminases with respect to their catalytic, structural, immuno-chemical, and physiological properties and their mechanism of action will be the subject of this paper.

Distribution and Properties of Transglutaminases

Based upon the immuno assay, specific enzymatic assay, and ion-exchange and exclusion chromatography, three different forms of transglutaminase have been identified in both human and guinea pig (Chung, 1972).

The most widely distributed form, the tissue enzyme, was found to be present in almost all tissues we have examined. The second enzyme occurs as a zymogen (protransglutaminase) and is found in plasma, platelets, placenta, uterus, prostate gland and liver. This zymogen is also commonly known as coagulation Factor XIII. The third enzyme, hair follicle enzyme, was found only in hair follicles.

Two assays were used to determine the distribution of these enzymes; one measures the incorporation of $[^{14}C]$glycine ethyl ester at the γ-carboxamide group of benzyloxycarbonyl-α-L-glutaminylglycine; the other measures the incorporation of $[^{14}C]$putrescine into casein. Comparison of activities of purified liver, hair follicle and plasma transglutaminases as measured by these two assays is shown in Table 1. Liver transglu-

TABLE 1

COMPARISON OF TWO ASSAYS FOR TRANSGLUTAMINASE

Transglutaminase (TGase)	CBZ–Gingly GlyEE μmole/hr/μg	Casein Putrescine nmole/hr/mg
Liver TGase	8.94	341.0
Liver TGase+thrombin	8.90	340.2
Plasma proTGase	0	0
Plasma proTGase+thrombin	0	19.2
Hair follicle TGase	0	10.2
Hair follicle TGase+thrombin	0	10.2

taminase catalyzes amine incorporation into both dipeptide and protein substrates. The plasma and hair follicle enzymes showed no activity toward the dipeptide substrate but are active with casein substrate. Preincubation with thrombin does not alter the activity of either the liver or the hair follicle enzyme. The plasma enzyme is present as a zymogen and requires proteolytic activation (Buluk et al, 1966). Examination of the soluble fraction of perfused guinea pig organs and muscle by both assay procedures revealed the highest levels of transglutaminase activity in liver and spleen, and the lowest in muscle (Chung, 1972). With the use of ion-exchange chromatography on DE-52, the different forms of transglutaminases have been separated from the extracts of uterus, prostate gland, liver and hair follicle. Typical chromatograms are shown in Fig. 1. Two peaks of thrombin-activatable transglutaminases were obtained with erythrocyte-free plasma (the first peak coincided with the plasma zymogen and the second peak with platelet zymogen). Uterus extracts showed the presence of a transglutaminase indistinguishable from tissue enzyme. However, a significant amount of zymogen that was activated by thrombin was also found in this organ. Further, this zymogen is eluted in position identical with that of platelet zymogen.

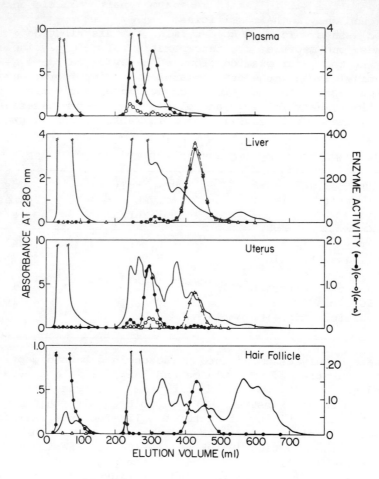

Fig. 1.   DEAE-Cellulose chromatography of transglutaminases
of guinea pig tissues.   Erythrocyte-free plasma, supernatants
of tissue homogenates were chromatographed on 2.5 x 25 cm
DE-52 column previously equilibrated with 0.01 M Tris-Acetate
buffer pH 7.4 containing 0.001 M EDTA.   After 150 ml of wash
with equilibrating buffer, 700 ml of 0 - 1 M NaCl linear gradient was applied.   Enzyme activities were expressed as n moles
of [^{14}C]putrescine incorporated into casein/hr/ml either in
the presence (●——●) or absence (o——o) of 1 unit of human
thrombin, and n moles of [^{14}C]glycine ethylester incorporated
into carbobenoxyglutaminylglycine/hr/ml (△——△).

Liver contains mostly tissue transglutaminase but a small
amount of zymogen that eluted at the same position with that
of platelet zymogen is also present. Examination of guinea
pig hair follicle extracts revealed the presence of two transglutaminases. One was found to be indistinguishable from
tissue enzyme. This enzyme is present in very limited quantity. The other, hair follicle transglutaminase, did not
catalyze amine incorporation into benzyloxycarbonyl-α-L-
glutaminylglycine and was not absorbed by DEAE-Cellulose.

Pretreatment of the soluble fractions of various organs
with antiserum to liver tissue transglutaminase caused almost
complete loss of enzyme activity as measured by both assays.
Double diffusion of partially purified tissue enzymes against
anti-serum to liver tissue transglutaminase showed, in each
case, a single precipitin band that fused with that given by
purified liver tissue transglutaminase (Chung, 1972). Guinea
pig plasma before or after thrombin treatment showed no reaction with anti-serum to tissue enzyme. Hair follicle enzyme
also did not react with this anti-serum. Uterus, prostate
gland, and liver zymogen (before or after treatment with
thrombin) again showed no reaction with anti-serum to tissue
transglutaminase.

These findings from the studies of guinea pig tissues may
be summarized as follows: A transglutaminase found in the
soluble fraction of all organs and muscle is indistinguishable
from the well-characterized liver tissue transglutaminase.
This enzyme is not found in plasma or platelets. The zymogen,
protransglutaminase, has been observed in plasma, platelets,
placenta, prostate gland, uterus, and liver. The third active
enzyme was present only in hair follicles and is distinctively
different from other enzymes and zymogens.

Examination of human organs showed, in general, low levels
of tissue transglutaminase as measured by specific assay with
dipeptide substrate. Overall distribution of the various
transglutaminases and protransglutaminases among the organs
of humans and guinea pigs appears to be the same. However,
the relative amounts of tissue enzyme present in these organs
are quite different. Among human organs, the lung and the
uterus were found to contain the highest levels of activity,
whereas low activity was observed in the liver and the spleen.

Human uterus, placenta, prostate gland, and liver contained
both an active transglutaminase and inactive zymogen; platelets
and plasma contain only a zymogen, and red blood cells contain
only a tissue transglutaminase. The elution pattern of human
tissue transglutaminase and plasma and platelet protransglutaminase appears to be similar to that of guinea pig enzymes
and zymogens, respectively. The order of elution with increas-

ing ionic strength is plasma zymogen, platelet zymogen, and the tissue transglutaminase.

Immuno-diffusion of purified protransglutaminase of uterus, placenta, prostate gland, platelets, plasma, and liver against anti-serum to platelet zymogen showed, in each case, a single precipitin band that fused with that of platelet zymogen (Fig. 2). Such fusion of the precipitin band indicates complete antigenic identity of the protransglutaminases of all these tissues. Similar immunogenic identity between platelet and placenta enzyme has been reported by Bohn (1971). However, a partial antigenic identity between the protransglutaminases of plasma and platelets was observed by Bohn (1970) and Israels et al (1973). Human red blood cell transglutaminase or liver tissue transglutaminase did not react with either anti-serum to plasma or platelet zymogen.

Apparent molecular weights of transglutaminases and pro-transglutaminases of various organs and tissues were estimated by exclusion chromatography on BioGel A.5m and 1.5m. With the exception of human hair follicle transglutaminase, which has not been examined, the enzymes and zymogens from the human show identical chromatographic properties with those from guinea pig tissues. The molecular weights of various trans-glutaminases estimated from the studies of sedimentation and diffusion, sedimentation equilibrium, and polyacrylamide gel electrophoresis in SDS, as well as by exclusion chromatography, are shown in Table 2.

The molecular weight of plasma protransglutaminase of 300,000 was estimated from both sedimentation equilibrium and exclusion chromatography and this value is in agreement with the value reported by Schwartz et al (1972) and Bohn et al (1972). On the basis of subunit molecular weights determined by polyacrylamide gel electrophoresis in SDS and sedimentation equilibrium in guanidine (Schwartz et al, 1972), it was concluded that the plasma zymogen is composed of catalytic dimer (a_2) and non-catalytic dimer (b_2) (Chung and Folk, 1974). In the presence of Ca^{2+} ion, the thrombin activated plasma transglutaminase dissociates into catalytic dimer (a_2') and non-catalytic dimer (b_2'). This dissociation is reversible by removal of Ca^{2+} ion. The isolated non-catalytic subunit B-chain dimer forms a complex with protransglutaminases or thrombin activated transglutaminase of platelets and placenta (Bohn et al, 1972; Chung and Folk, 1974). These molecular weight studies indicate that the catalytic subunit of plasma protransglutaminase is identical to that of the zymogens present in other tissues. The identity between the plasma catalytic subunit and the platelet zymogen was further supported by the identical amino acid sequence of the peptide cleaved

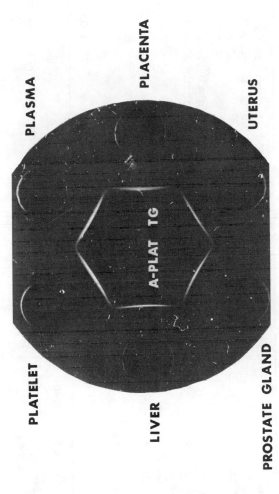

Fig. 2. Ouchterlony immunodiffusion pattern of protransglutaminases (from plasma, placenta, uterus, prostate gland and liver) against rabbit anti-serum to platelet protransglutaminase (A-PLAT TG). The level of transglutaminase activity in each of the wells was adjusted approximately equal to that of platelet enzyme.

TABLE 2

MOLECULAR WEIGHT AND SUBUNIT COMPOSITION OF TRANSGLUTAMINASES

Source	Molecular Weight			Subunit
	Exclusion Chromatography (Bio Gel A 6%)	Sedimentation equilibrium & diffusion	Polyacrylamide gel-electro-phoresis in SDS	
Plasma ProTGase	300,000 ± 20,000	296,700 ± 16,700	A chain 80,000 / B chain 86,000	2 A chains / 2 B chains
Plasma TGase* in Ca++ b	165,000 ± 15,000	160,000 ± 20,000	86,000	2 B chains
a	76,000 ± 4,000	160,000 ± 20,000	76,000	2 A' chains
Platelet ProTGase	76,000 ± 4,000	161,500 ± 19,300	A chain 80,000	2 A chains
Platelet TGase in Ca++	76,000 ± 4,000	161,500 ± 19,300	A chain 76,000	2 A' chains
Placenta, Uterus, Liver, Prostate Gland ProTGase	76,000 ± 4,000			
Liver Tissue TGase**	80,000 ± 4,000	76,900 ± 5,000	80,000	1 chain
Red Blood Cell TGase	80,000 ± 4,000			
Other Organs & Muscle Tissue TGase	80,000 ± 4,000			
Hair Follicle TGase	54,000 ± 2,000		27,000	2 chains

* Thrombin activated zymogen

** Sedimentation and Diffusion, and Polyacrylamide gel electrophoresis was carried out with only guinea pig enzyme.

from these zymogens by thrombin (Takagi and Doolittle, 1974).
Thus, these studies on human tissues have indicated that the
zymogens present in platelets, placenta, prostate gland,
uterus, and liver are identical to those of the catalytic subunit of plasma protransglutaminase. Only the plasma protransglutaminase contains an additional, non-catalytic subunit
of different composition. The tissue transglutaminases are
again present in all the tissues we have examined and appear
to be quite similar to that of guinea pig tissue transglutaminase in chromatographic as well as catalytic properties.

Catalytic Properties of Transglutaminases

Substrate Specificity. A survey of several glutamine-
containing peptides and derivatives as possible substrates for
transglutaminases showed that the B-chain of oxidized insulin
served as a substrate for all of these enzymes. This offered
a unique opportunity to determine, with each of the enzymes,
whether their catalytic action is directed only toward the
carboxamide group of glutamine. The B-chain was modified by
acetylation in order to mask the α-amino group, as well as the
single ϵ-amino group of lysine, either or both of which might
participate in the transfer reaction. The mechanism of the
transglutaminases was studied kinetically at pH 7.5 in the
presence of calcium ion, using the transfer of [^{14}C]methylamine
into the acetylated B-chain of oxidized insulin. Preliminary
to the kinetic studies, the single product of the transfer
reaction with each transglutaminase was identified as peptide
bound γ-glutaminc acid methylamide. Fig. 3 shows a diagram-
matical representation of a paper chromatogram, carried out
in methanol:pyridine:water (16: 0.6: 8), that served to identify this residue. Prior to chromatography, γ-glutamic acid
methylamide was liberated from enzyme-modified B-chain by
digestion with chymotrypsin C, carboxypeptidase A, pronase, and
leucine aminopeptidase. Identical chromatograms were obtained
with samples of acetylated B-chain into which [^{14}C]methylamine
was incorporated by the catalytic action of each of the transglutaminases. This is additional evidence that each of these
enzymes acts upon the carboxamide groups of glutamine residues
and not upon those of asparagine residues. The results of
kinetic studies with each of the transglutaminases are in
accord with a common mechanism for all of these enzymes. In
this mechanism, previously proposed for guinea pig liver tissue
enzyme (Folk, 1969; Chung et al, 1970), glutamine substrate
binds with enzyme in a binary complex, ammonia dissociates
with the formation of acyl-enzyme intermediate, and the acyl
group is transferred from the enzyme either to water (hydrol-

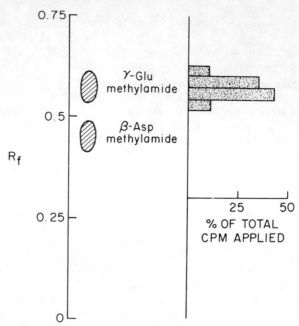

Fig. 3. Identification of γ-glutamic acid methylamide as the single radioactive amino acid formed by transglutaminase-catalyzed [^{14}C]methylamine incorporation into the acetylated B chain of oxidized insulin. Digests of [^{14}C]methylamine-labeled B chain were prepared as described in the text. The figure is a diagrammatic representation of a paper chromatogram showing the positions of markers and radioactivity in the digest. Chromatography in two other solvent systems showed similar results.

TABLE 3
KINETIC CONSTANT*

Transglutaminase	K_m for A$^+$ (at sat.B) mM	K_m for B† (at sat.A) mM	V_{max}
Plasma	2.08	0.62	142
Platelets	1.90	0.65	240
Hair follicle	0.49	1.04	38
Liver	0.082	0.28	250

*Maximum velocities are given in nmoles of [^{14}C]methylamine incorporated into acetylated B chain of oxidized insulin min^{-1} (mg^{-1} of each enzyme).

$^+$A = Acetylated B chain of oxidized insulin.   †B =[^{14}C]-methyl-amine.

ysis) or to a primary amine (transfer). The kinetic constants
(Table 3) obtained from these studies are significantly different for hair follicle, tissue, and plasma transglutaminases
(Chung and Folk, 1972), providing a further basis for distinction between members of this group of enzymes.

The identical Michaelis constants obtained for the plasma
and platelet enzymes with the acetylated B-chain of oxidized
insulin and methylamine suggest a close similarity in catalytic
properties of these enzymes. Also, the finding that the maximum velocity per milligram of enzyme obtained for plasma enzyme
was about one half the value for the platelet enzyme would
suggest that only one type of subunit of plasma protransglutaminase becomes catalytically active after thrombin activation.

Active Site Studies of Transglutaminases. The active sites
of transglutaminases were examined by selective alkylation of
the SH group of a single cysteine residue with labeled iodoacetamide at low pH in the presence of Ca^{2+}, and mapping of
the active-site by means of catalytic efficiency of glutamine
or aliphatic substrate analogues. The alkylation of the single
SH group results in the loss of all catalytic activity by the
enzymes (Folk and Cole, 1966b; Chung and Folk, 1974). The
active site amino acid sequence, *Gly-Gln-Cys-Try*, found in
guinea pig liver tissue transglutaminase (Folk and Cole, 1966a)
was shown also to be in the active site of human plasma protransglutaminase (Holbrook et al, 1973).

To extend these studies on the active site of transglutaminases, the glutamine binding sites of the enzymes was carried
out in Folk's laboratory by the use of substrate analogues,
including aliphatic amides and peptide derivatives containing
isomers of methylglutamine. The results of such studies are
shown in Table 4. Straight chain and γ-branched chain aliphatic amides served as substrates for tissue transglutaminase
and α- and β-branched chain amide substrate did not (Gross and
Folk, 1973a). Similar results were obtained with glutamine
derivatives; α-methylglutamine is a substrate for tissue transglutaminase, whereas β- and γ-methylglutamines are not (Gross
and Folk, 1973b). Plasma transglutaminase (thrombin activated
protransglutaminase) exhibits this same specificity toward
aliphatic amides, thus providing evidence for a common spatial
arrangement within the glutamine side chain binding sites of
transglutaminases (Gross et al, 1973).

The tissue and plasma transglutaminases, however, display
differences in requirements for amino acid residues surrounding the substrate glutamine in peptide substrates (Gross et al,
1973). The active site titration studies indicated a single

TABLE 4

ALIPHATIC AMIDE SUBSTRATES FOR TRANSGLUTAMINASES

Amide	Guinea Pig Tissue		Human Plasma	
	V_{max} (μmoles min^{-1})	K_m (M)	V_{max} (μmoles min^{-1})	K_m (M)
Foramide	0		0	
Acetamide	0.5	1.8	0.2	2.2
Propionamide	2.0	0.9	0.1	0.3
α-Methylpropionamide	0		0	
Butyramide	3.3	2.8	1.0	1.8
β-Methylbutyramide	0		0	
Valeramide	2.0	0.6	0.5	0.2
γ-Methylvaleramide	13	1.5	2.0	0.4
γ,γ-Dimethylvaleramide	+		+	
Caproamide	0.6	0.1	0.4	0.5
Heptamide	+		+	

site exists per molecule of tissue transglutaminase (one site
per 80,000 polypeptide chain (Folk and Cole, 1966b). However,
only one site per molecule of platelet transglutaminase, which
exists as dimer (one site per two 80,000 polypeptide chain)
was obtained suggesting that "half of the site" reactivity
exists in platelet enzyme (Chung and Folk, 1974).

Physiological Role of Transglutaminase. The physiological
role of tissue transglutaminase was the source of some early
speculation (Waelsch, 1962). To date, however, there is no
concrete evidence to support any special biological function
for this enzyme. In contrast, plasma, hair follicle, and
prostate gland transglutaminases appear to participate in
important biological reactions. After activation by thrombin,
the protransglutaminases of plasma, platelets (blood coagulation Factor XIII) can bring about the stabilization of fibrin
clots by catalyzing the formation of $\varepsilon(\gamma$-glutamyl)lysine
cross-links between peptide chains. This process of covalent
bond formation between polypeptide chains appears to play an
important role in hemostasis as well as wound healing (see
review, Finlayson, 1974). The enzyme isolated from hair
follicles of the guinea pig (Chung and Folk, 1972; Harding and
Rogers, 1972) is probably responsible for the formation of
the $\varepsilon(\gamma$-glutamyl)lysine cross-links found in the proteins of
hair (Harding and Rogers, 1971; Asquith et al, 1970). Recently,
Buxman and Wuepper (1974) and Goldsmith et al (1974) reported
the isolation of transglutaminase similar to hair follicle
enzyme, from cow snout epidermis and possible presence of
$\varepsilon(\gamma$-glutamyl)lysine in the proteins keratinized tissue.

The prostate gland transglutaminase (Vesiculase, has been
used to designate this enzyme activity) appears to play a
significant role in the coagulation of seminal plasma (Gotterer
et al, 1955; Notides and Williams-Ashman, 1967). The isolation
of $\varepsilon(\gamma$-glutamyl)lysine from the clotted protein of guinea pig
seminal vesicle secretions further supports the involvement
of prostate transglutaminases (Williams-Ashman et al, 1972).
Wing et al (1974) reported the isolation of transglutaminase
from the coagulating gland of guinea pig prostate gland, which
appears to be similar in the substrate specificity and molecular weight to that of tissue enzyme.

REFERENCES

Asquith, R. S., M. S. Otterburn, J. H. Buchanan, M. Cole,
    J. C. Flethcher, and K. L. Gardner 1970. The identifi-
    cation of $\varepsilon N(\gamma$-glutamyl)-L-lysine cross-link in native
    wool keratins. *Biochim. Biophys. Acta* 221: 342-348.
Bohn, H. 1970. Isolierung und Charakterisierung des fibrin-

stabilisierenden Faktors aus menschlichen Thrombozyten. *Thromb. Diath. Haemorrh.* 23: 455-468.

Bohn, H. 1971. Immunochemical studies on the fibrin stabilizing factors from human plasma and platelets. *Blut.* 22: 237-243.

Bohn, H., H. Haupt, and T. Krauz 1972. Die molekulare struktur der fibrinstabilisierenden faktoren des menschen. *Blut.* 25: 235-248.

Buluk, K., T. Januszko, and J. Olbromski 1961. Conversion of fibrin to desmofibrin. *Nature* 191: 1093-1094.

Buxman, M. M. and K. D. Wuepper 1974. Epidermal transglutaminase: Biochemical and immunochemical analysis. *Clin. Res.* 22: 575 Abs.

Chung, S. I., R. I Shrager, and J. E. Folk 1970. Mechanism of guinea pig liver transglutaminase. VII. Chemical and stereochemical aspects of substrate binding and catalysis. *J. Biol. Chem.* 245: 6424-6435.

Chung, S. I., J. S. Finlayson, and J. E. Folk 1971. Tissue transglutaminase and factor XIII. *Federation Proc.* 30: 1075 Abs.

Chung, S. I. 1972. Comparative studies on tissue transglutaminases and factor XIII. *Ann. N. Y. Acad. Sci.* 202: 240-255.

Chung, S. I. and J. E. Folk 1972a. Transglutaminase from hair follicle of guinea pig. *Proc. Natl. Acad. Sci.* 69: 303-307.

Chung, S. I., and J. E. Folk 1972b. Kinetic studies with transglutaminases, the human blood enzymes (activated coagulation factor XIII) and guinea pig hair follicle enzyme. *J. Biol. Chem.* 247: 2798-2807.

Chung, S. I., M. S. Lewis, and J. E. Folk 1974. Relationships of the catalytic properties of human plasma and platelet transglutaminases (activated blood coagulation factor XIII) to their subunit structures. *J. Biol. Chem.* 249: 940-950.

Chung, S. I., J. S. Finlayson, and J. E. Folk 1974. Manuscript in preparation.

Clark, D. D., A. Neidle, N. K. Sarkar, and H. Waelsh 1957. Metabolic activity of protein amide groups. *Arch. Biochem. Biophys.* 71: 277-279.

Clark, D. D., M. J. Mycek, A. Neidle, and H. Waelsh 1959. The incorporation of amines into proteins. *Arch. Biochem. Biophys.* 79: 338-354.

Finlayson, J. S. 1974. Cross-linking of fibrin. *Semin. Thromb. Hemos.* 1: 000-000 (in press).

Folk, J. E., and P. W. Cole 1966a. Identification of a functional cysteine essential for the activity of guinea pig liver transglutaminase. *J. Biol. Chem.* 241: 3238-3240.

Folk, J. E. and P. W. Cole 1966b. Mechanism of action of guinea pig liver transglutaminase. Purification and properties of the enzyme: identification of a functional cysteine essential for activity. *J. Biol. Chem.* 241: 5518-5525.

Folk, J. E. 1969. Mechanism of action of guinea pig liver transglutaminase. VI. Order of substrate addition. *J. Biol. Chem.* 244: 3707-3713.

Folk, J. E. and S. I. Chung 1973. Molecular and catalytic properties of transglutaminases. *Advances in Enzymology* 38: 109-191.

Goldsmith, L. A., H. P Baden, S. I. Roth, R. Coleman, L. Lee, and B. Fleming 1974. Vertebral epidermal transamidases and post-translational modifications of epidermal fibrous proteins. *Biochim. Biophys. Acta* (in press).

Gotterer, G. D., D. Ginsburg, T. Schulman, J. Banks, and H. G. Williams-Ashman 1955. Enzymic coagulation of semen. *Nature* 176: 1209-1211.

Gross, M. and J. E. Folk 1973a. Mapping of the active sites of transglutaminases. I. Activity of the guinea pig liver enzyme toward aliphatic amides. *J. Biol. Chem.* 248: 1301-1306.

Gross, M. and J. E. Folk 1973b. Mapping of the active sites of transglutaminases. II. Activity of the guinea pig liver enzyme toward methylglutamine peptide derivatives. *J. Biol. Chem.* 248: 6534-6540.

Gross, M., S. I. Chung, and J. E. Folk 1973. Transglutaminases: active site mapping. *Federation Proc.* 32: part I 437 Abs.

Israels, E. D., F. Paraskevas, and L. G. Israels 1973. Immunological studies of coagulation factor XIII. *J. Clin. Invest.* 52: 2398-2403.

Harding, H. W. J., and G. E. Rogers 1971. ε(γ-glutamyl)lysine cross-linkage in citrulline-containing protein fractions from hair. *Biochem.* 10: 624-630.

Harding, H. W. J., and G. E. Rogers 1972. Formation of the ε(γ-glutamyl)lysine cross-link in hair proteins. Investigation of transamidases in hair follicle. *Biochem.* 11: 2858-2863.

Holbrook, J. J., R. D. Cooke, I. B. Kingston 1973. The amino acid sequence around the reactive cysteine residue in human plasma factor XIII. *Biochem. J.* 135: 901-903.

Lorand, J., T. Urayama, and L. Lorand 1966. Transglutaminase as a blood clotting enzyme. *Biochem. Biophys. Res. Commun.* 23: 828-835.

Notides, A. C., and H. G. Williams-Ashman 1967. The basic protein responsible for the clotting of guinea pig semen.

Proc. Natl. Acad. Sci. 58: 1991–1995.

Schwartz, M. L., S. V. Pizzo, R. L. Hill, and P. A. Mckee 1972. Human factor XIII from plasma and platelets. *J. Biol. Chem.* 248: 1395–1407.

Takagi, T., and R. F. Doolittle 1974. Amino acid sequence studies on factor XIII and the peptide released during its activation by thrombin. *Biochem.* 13: 750–756.

Tyler, H. M., and C. H. Lack 1964. A tissue stabilizing factor and fibrinolytic inhibition. *Nature* 202: 1114–1115.

Tyler, H. M., and K. Laki 1966. Fibrin stabilizing enzymes from guinea pig liver. *Biochem. Biophys. Res. Commun.* 24: 506–512.

Williams-Ashman, H. G., A. C. Notides, S. S. Pabalan, and L. Lorand 1972. Transamidase reactions involved in the enzymic coagulation of semen: Isolation of γ-glutamyl-ε-lysine dipeptide from clotted secretion protein of guinea pig seminal vesicle. *Proc. Natl. Acad. Sci.* 69: 2322–2325.

Wing, D., C. R. Curtis, L. Lorand, and H. G. Williams-Ashman 1974. Isolation of a transglutaminase from the coagulating gland of the guinea pig prostate. *Federation Proc.* 33: part I 290 Abs.

THE ISOZYMES OF TWO HUMAN ADIPOSE TISSUE LIPOLYTIC ACTIVITIES

J. A. CORTNER and J. D. SCHNATZ
Departments of Pediatrics and Internal Medicine
State University of New York at Buffalo,
and The Children's Hospital, Buffalo, New York

ABSTRACT. Two lipolytic activities of human adipose tissue
have been shown to have multiple molecular forms. Alkaline
lipolytic activity (ALA, hydrolysis of tributyrin at 47°C,
pH 8.0) exists as three fractions separable by Sephadex gel
filtration: ALA-I, ALA-II and ALA-III. ALA-I is a lipid-
protein complex of ALA-II and has no mobility during starch
gel electrophoresis. ALA-II is a smaller molecule than
ALA-I and exists primarily as four electrophoretically separable bands of activity. ALA-III is considerably smaller
than ALA-II, is a subunit of ALA-II, and exists primarily
as a fifth band of electrophoretically separable activity.
A genetic variant of ALA in rabbit adipose tissue has been
characterized by slower migration during starch gel electrophoresis, a decreased tendency to dimerize, and reduced
enzymatic activity in contrast to the wild type. No genetic
variation has been observed in man, and the function of this
soluble esterase of human adipose tissue in unknown.
Neutral, hormone sensitive lipase (hydrolysis of [14]C-
triolein at 27°C, pH 6.8) is a large molecule, associated
with lipid, and separable from ALA-1 by DEAE Sephadex chromatography. Beta adrenergic stimulation with isoproteranol
increases lipase activity compared to control, an effect
which is not seen during beta adrenergic blockade by propranolol. Starch block and Cellogel electrophoresis separates
this lipase into two components, one of which migrates
toward the anode while the other remains in the region of
the origin at pH 8.6.

Two lipolytic activities, alkaline lipolytic activity (ALA)
and neutral lipolytic activity (NLA), have been studied in
human adipose tissue obtained at surgery and processed immediately[1] (Schnatz, 1964a; Cortner and Schnatz, 1967; and Schnatz
and Cortner, 1967). The substrate requirements suggest that
ALA is an esterase and NLA a true lipase as defined by Desnuelle (1961) and that neither represented activity of lipo-
protein lipase.

---

[1] 1 or 2 gm of adipose tissue was homogenized with 0.15 M KCl,
centrifuged, and the aqueous extract studied. Assay for ALA
involves incubation of aqueous tissue extract for 30 min at

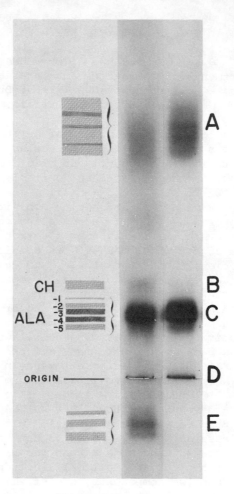

Fig. 1. Esterase zymogram of human adipose tissue. The diagram on the left represents a composite of the esterase zymograms that resulted from the staining of starch gels after the electrophoresis of human adipose tissue preparations. The anode is at the top of the figure. The origin represents the point of insertion of the sample. The capital letters A - E groups the bands of enzymatic activity which behaved similarly toward certain substrates and inhibitors. The center lane depicts the results using α-naphthyl acetate as substrate and the right lane shows the results using α-naphthyl butyrate as substrate and Fast Blue B as the coupling agent.

Esterase

Because ALA is an esterase, we investigated hydrolysis of
α-naphthol esters. Human adipose tissue preparations were
subjected to starch gel electrophoresis and subsequently
stained by histochemical tehcniques using naphthol esters and
Fast Blue B as the coupling agent. Fig. 1 depicts the results
using α-naphthyl acetate (pictured in the center) and α-naphthyl butyrate (on the right) (Cortner and Schnatz, 1967).
Five areas of staining were identified (labeled A - E). The
diagram on the left illustrates that these could be resolved
into as many as 17 bands of esterase activity. Starch block
electrophoresis studies showed the presence of these five
bands of activity in the same eluates proven to contain ALA
and direct assay of the starch gel after electrophoresis
showed that 86% of the ALA activity was recovered from Area C
and 9% from Area D (origin) (Cortner and Schnatz, 1967). Area
B or Ch was proven to be serum pseudocholinesterase. These
five bands, labeled ALA-1 through ALA-5, had similar substrate
and inhibitory specificity compatible with the isozyme concept; specifically they have increasing activity with α-naphthyl acetate, α-naphthyl propionate, α-naphthyl butyrate and
α-naphthyl valerate and decreased activity with α-naphthyl
myristate, caprylate and laurate as well as β-naphthyl stearate. They were inhibited by 10^{-2} M sodium fluoride and resistant to 10^{-4} M eserine, 10^{-5} M acetozolamide, 10^{-3} M EDTA
and 10^{-5} M calcium chloride, all of which are known inhibitory
and substrate characteristics of ALA. Fig. 2 (Cortner and
Schnatz, 1967) shows a closeup of the ALA isozymes obtained
from five different human adipose tissue preparations. Analysis of 200 human adipose tissue preparations showed that
bands 3 and 4 were always present and were the most intensely
staining. Bands 1 and 2 were present in 23 and 80% of the
specimens respectively, and band 5 occurred in 50% of the
specimens. These characteristic electrophoretic bands of ALA
have also been found in human liver, lung, kidney, heart, adrenal, spleen, thyroid, ovary, testis, brain and muscle (Cort-

---

pH 8.0, 47°, in an albumin-containing solution of tributyrin
(Schnatz, 1966). NLA was assayed by incubation of tissue extract for 30 min at pH 7.0, 37°, in an albumin-containing
emulsion of triolein (Schnatz, 1966). Aliquots of each assay
system were obtained before and after incubation, extracted
and titrated for free fatty acids (FFA) (Schnatz, 1964b).
Activities were expressed as µEq per ml of sample per hour.

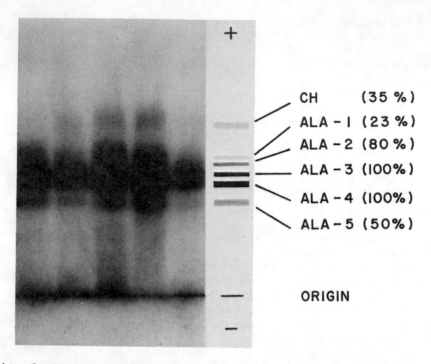

CH   (35 %)
ALA - 1 (23 %)
ALA - 2 (80 %)
ALA - 3 (100%)
ALA - 4 (100%)
ALA - 5 (50%)

ORIGIN

Fig. 2. Representative patterns seen on starch gel after electrophoresis of human adipose tissue. Fresh human adipose tissues obtained at surgery from 182 individuals were homogenized in 0.15 M KCl (300 mg/ml) and centrifuged at 30,000 x g
for 10 min. The aqueous extract was subjected to starch gel
electrophoresis and the lipolytic activity identified as described in the text. Pictured above is an area of a starch gel
which extends from the origin to approximately 8 cm toward
the anode. This area consistently has certain characteristic
bands which for reasons detailed in the text have been designated as Ch and ALA 1-5. A schematic representation of these
bands is seen to the right. The frequency with which these
bands occurred in 182 surgical specimens is noted in the
parentheses to the far right of the figure.

ner and Schnatz, 1967). Although their physiological function
is not known, the ability to hydrolyze β-naphthol stearate
suggests a possible role in lipolysis.

Fig. 3 (Schnatz and Cortner, 1967) shows the results of
G-200 Sephadex gel filtration of a human adipose tissue preparation. NLA is eluted in the area of the void volume, while

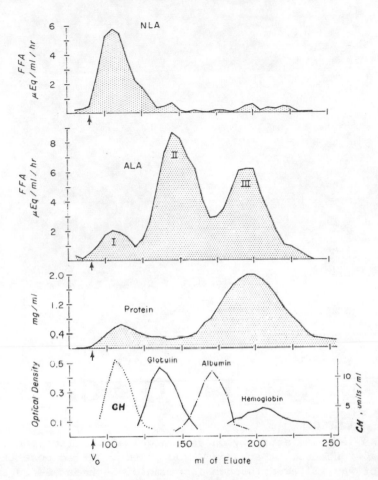

Fig. 3. Sephadex G-200 gel filtration of a human adipose tissue extract. Twenty-five ml of adipose tissue extract were filtered through Sephadex G-200 with 0.008 M phosphate-citrate buffer, pH 7.0, and collected in 5 ml fractions. Each fraction was analyzed for NLA, ALA and protein content. The elution of dextran blue (void volume, V_o), pseudocholinesterase (CH), γ-globulin, albumin, and hemoglobin was determined. The three fractions of ALA are designated by Roman numerals I, II and III.

ALA is eluted in three peaks labeled I, II and III. The protein markers in the bottom tier suggest the relative sizes of ALA-I, II and III. When representative samples from ALA-I, II and III were subjected to electrophoresis on starch gel and subsequently identified using α-naphthyl butyrate as substrate,

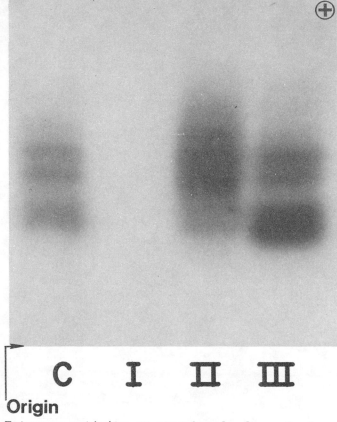

Origin

Fig. 4. Esterase activity on starch gel after electrophoresis of fractions from a Sephadex G-200 column. An adipose tissue extract was filtered through a column of Sephadex G-200 and analyzed as described. The values for NLA and ALA are similar to those depicted in Fig. 3. Starch gel electrophoresis was performed on single representative fractions for ALA-I, II and III and compared with an aliquot of the original adipose tissue extract (C) which was filtered through the column. After electrophoresis, the gel was sliced horizontally and stained for esterase activity.

---

results were obtained as depicted in Fig. 4 (Schnatz and Cortner, 1967). The control on the far left contains bands 5, 4, 3 and 2, band 1 not being visible in this reproduction. Eluates from the void volume area representing the peak of ALA-I and NLA activity show staining only at the origin. Samples from the peak of ALA-II, on the other hand, contain primarily bands 4, 3, 2 and 1 with a lesser amount of band 5, and repre-

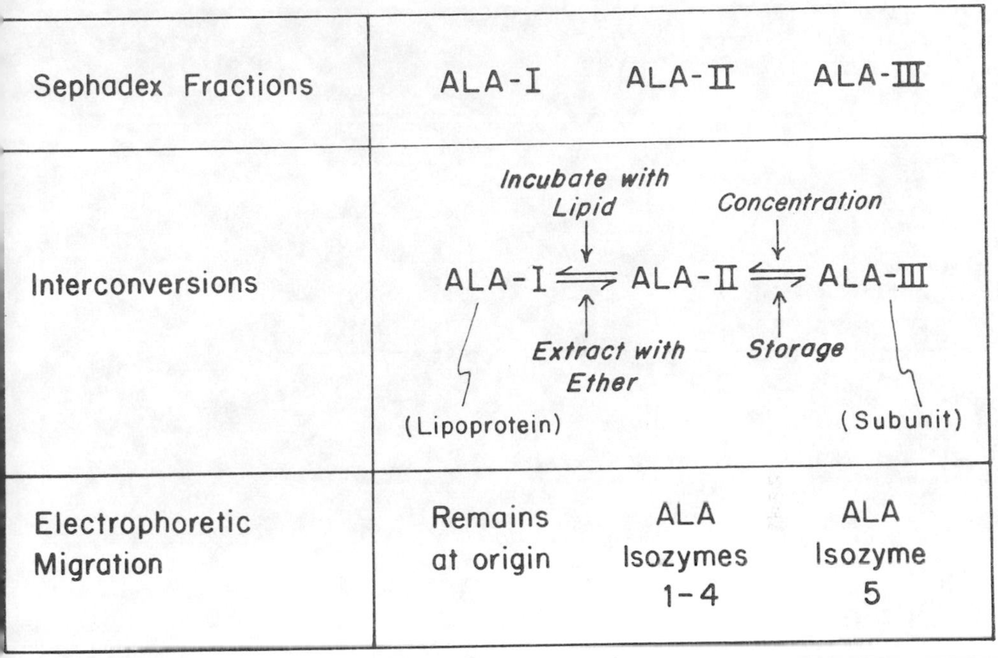

Sephadex Fractions	ALA-I	ALA-II	ALA-III
Interconversions		$\text{ALA-I} \rightleftarrows \text{ALA-II} \rightleftarrows \text{ALA-III}$ *Incubate with Lipid* ↓ / *Concentration* ↓ / *Extract with Ether* ↑ / *Storage* ↑ (Lipoprotein) (Subunit)	
Electrophoretic Migration	Remains at origin	ALA Isozymes 1-4	ALA Isozyme 5

Fig. 5. Tentative concept of human adipose tissue alkaline lipolytic activity.

---

sentative samples from the peak of ALA-III contain primarily band 5 and some 4 and 3.

Additional studies on these three peaks of enzymatic activity showed that the NLA, ALA-I fraction contained lipid. Extraction of this lipid with ether converted ALA-I to ALA-II, as defined by starch gel electrophoresis. Conversely, incubation of ALA-II with lipid (lipomul) produced ALA-I. These data indicate that ALA-I exists as a lipid-protein complex of the smaller molecule, ALA-II. The effects of time and temperature on the starch gel electrophoretic pattern suggested that band 5 was a subunit or a degradation product, and the association of ALA-III with increased amounts of band 5 suggested that ALA-III was similarly a degradation product. This hypothesis was confirmed in additional studies which showed that ALA-II produced ALA-III on standing. Conversely, ALA-III produced ALA-II with concentration. The fact that fractions from ALA-II always contain small amounts of band 5 and that fractions from ALA-III always contain small amounts of bands 4 and 3 suggests that a continuing equilibrium exists. Fig. 5 summarizes the information obtained from these studies (Schnatz and Cortner, 1967). ALA has been shown to exist as

281

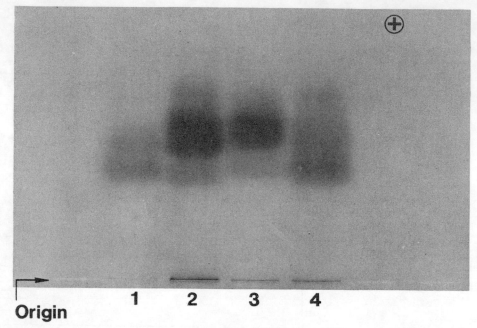

Origin

Fig. 6. ALA zymograms of four adult rabbits. Pictured above are the starch gel electrophoretic patterns of adipose tissue extracts from the two "fast" (Lanes 2 and 3) and two "slow" (Lanes 1 and 4) adult rabbits who were subsequently mated in the four possible combinations.

three separable fractions, ALA-I, II and III. ALA-I represents a small percentage of the total ALA, is eluted from Sephadex with NLA, exists as a lipid-protein complex of ALA-II, and does not move in starch gel electrophoresis. ALA-II is a smaller molecule than ALA-I and exists primarily as four electrophoretically separable bands of activity. ALA-III is considerably smaller than ALA-II, is a subunit of ALA-II, and exists primarily as a fifth band of electrophoretically separable activity.

ALA was shown to exist in the adipose tissue of all mammals tested with the exception of the steer, and the electrophoretic pattern observed in adipose tissue of white New Zealand rabbits appeared to be most similar to that of man (Rivello, Cortner and Schnatz, 1969). For that reason and since no electrophoretic variant was found in approximately 1,000 specimens of human adipose tissue, we attempted genetic studies in the rabbit. Two electrophoretic patterns were found (Fig. 6) (Schnatz and Cortner, 1968). The two "fast" migration patterns shown in the center of the illustration were very similar to those found in man; five bands of activity were seen with bands 4 and 3

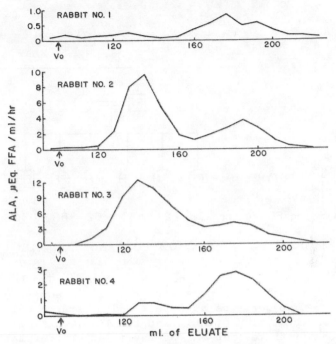

Fig. 7. Sephadex G-200 gel filtration. Adipose tissue extracts were prepared and filtered through Sephadex G-200. The elution patterns for extracts obtained from "fast" rabbits (Nos. 2 and 3) and "slow" rabbits (Nos. 1 and 4) are pictured above. V_o indicates the void volume for the column at the time that it was run.

having the most intense staining. The majority of the activities observed in lanes 1 and 4 did not migrate as rapidly and were called "slow." In addition, they were observed to have decreased activity. G-200 Sephadex gel filtration studies of these two phenotypic patterns are shown in Fig. 7 (Schnatz and Cortner, 1968). The "fast" phenotype contained predominantly ALA-II and a small amount of ALA-III, whereas the "slow" phenotype had reduced activity, the majority of which was present in the form of ALA-III. Fig. 8 (Cortner and Schnatz, 1970) shows a representative study in which known homozygous fast and heterozygous fast phenotypes had their Sephadex gel filtration pattern compared. The total ALA activity was less and the ratio of ALA-II to ALA-III less in the heterozygotes than in the homozygotes. Fig. 9 (Cortner and Schnatz, 1970) shows the difference in the electrophoretic pattern of the heterozygote on the left, the homozygote "slow" in the center, and the homozygote "fast" on the far right.

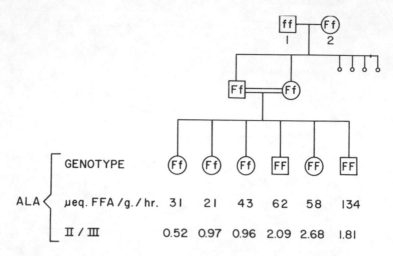

Fig. 8. Rabbits 1 and 2 of Fig. 3 were mated. Two heterozygous "fast" rabbits from this litter were subsequently mated four times. The genotype as determined by starch gel electrophoresis, quantitation of ALA, and the ALA-II/ALA-III ratio of Sephadex filtrates for one of these latter four matings are compared above.

In summary, the "slow" variant was characterized by slower migration during starch gel electrophoresis, a decreased tendency to dimerize, and a reduced enzymatic activity in contrast to the "fast" electrophoretic type. The heterozygous phenotype can be identified by the presence of a slow band identical to that of the homozygous slow phenotype. The inheritance was Mendelian. No gross abnormalties were noted among rabbits having the homozygous slow phenotype, and specific studies of metabolism have not been done.

Lipase

Partial separation of NLA from ALA was obtained using ammonium sulfate fractionation and Sephadex gel filtration, but complete separation of ALA-I and NLA was not obtained (Schnatz and Cortner, 1967). In addition, NLA could not be recovered following starch block electrophoresis although a small amount was recovered at the origin after starch gel electrophoresis. In vitro studies indicated that neither NLA nor ALA was increased by the action of adrenalin, whereas in vitro lipolysis and the hydrolysis of whole tissue homogenates were both increased in the same tissue (Schnatz and Downey,

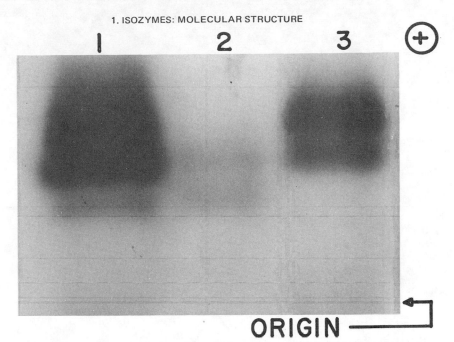

Fig. 9. Three electrophoretic phenotypes of ALA. Adipose tissue extracts from rabbits were subjected to starch gel electrophoresis and subsequent staining with α-naphthyl butyrate
and Fast Blue B. Pictured above are the starch gel electrophoretic patterns of adipose tissue extracts from a heterozygous fast parent (Ff, lane 1), a homozygous slow newborn
(ff, lane 2), and a homozygous fast newborn (FF, lane 3).

1968). In in vivo studies, no increase in either NLA or ALA
was seen after prolonged fasting or adrenalin infusion (Schnatz
and Cummiskey, 1969). Consequently, a sensitive assay$_2$for
lipase was developed using [^{14}C]-triolein as substrate^2. The

---

^{2}The assay system contained 5.5×10^4 DPM of glyceryl trio-
leate-1-^{14}C emulsified in 1% polyvinyl alcohol, 0.5 mg of bovine fraction V albumin (0.05%), and 50 μm (0.05 M) phosphate
buffer pH 6.8 in a total volume of 1 ml. On some occasions,
glyceryl trioleate-9,10-^{3}H was used as a substrate. An amount
of aqueous extract equivalent to 50 mg of adipose tissue was
used and the incubation carried out for two hours. After incubation at 27°C, during the linear phase of hydrolysis, radioactive oleic acid was extracted with a isopropanol-hexane mixture and isolated in 0.1 M KOH. A 1 ml aliquot of the KOH
phase was placed in a counting vial with 5 ml of Instagel,
counted, and DPM calculated with reference to external standard
technique. Release of radioactive FFA was determined in a

SEPHADEX GEL FILTRATION

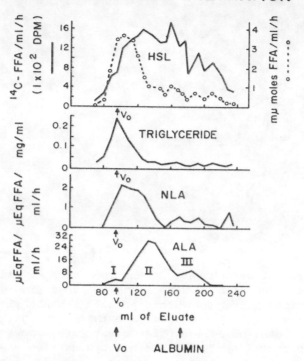

Fig. 10. Sephadex G-200 gel filtration of human adipose tissue extracts was carried out and the following determinations done: ALA, NLA, triglyceride and lipase (solid line). Calculations of free fatty acid (FFA) release in the lipase assays were made as indicated in the text (dotted line).

use of racioactive substrate, undiluted by non-radioactive substrate, proved to be the most sensitive assay, but when tissue triglyceride was present the results required correction for the dilutional effect of this triglyceride as demonstrated in Fig. 10 (Schnatz and Marotta, in preparation). This figure depicts the results of a G-200 Sephadex gel filtration of hu-

controlled system without tissue extracts and the increment above control reported as lipase activity. The results were expressed as DPM ^{14}C (or ^{3}H) FFA per gram of tissue per hour or, in the case of fractionation procedures, as radioactive FFA per ml of eluate per hour (Schnatz and Marotta, in preparation).

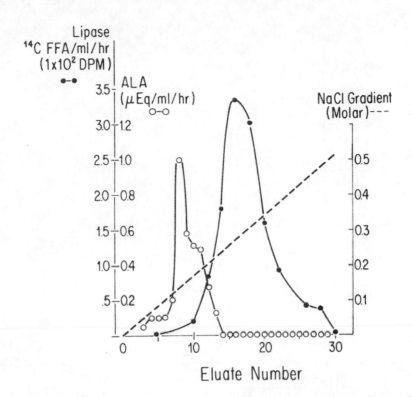

Fig. 11. DEAE Sephadex chromatography of the void volume eluate from a Sephadex G-200 gel filtration of an adipose tissue extract.

man adipose tissue as previously reported for NLA and ALA which are depicted in tiers 3 and 4. Tier 2 shows the trigly-ceride content of the eluates and the unbroken line of tier 1 demonstrates the raw data obtained from the lipase assay, un-corrected for the presence of triglyceride. The dotted line represents the corrected data, showing that lipase is eluted in the void volume in a manner similar to NLA. Additional studies which suggest that this is the same enzymatic protein as NLA include identical pH optima, substrate hydrolysis, in-hibition by NaF, fractionation by ammonium sulfate, and immobil-ity in starch gel electrophoresis.

Using this more sensitive assay system, increase in lipase compared to control was demonstrated after incubation of adi-pose tissue with isoproteranol, a known beta adrenergic stim-ulator. Beta adrenergic blockade with propranolol eliminated this effect. We therefore conclude that these studies are

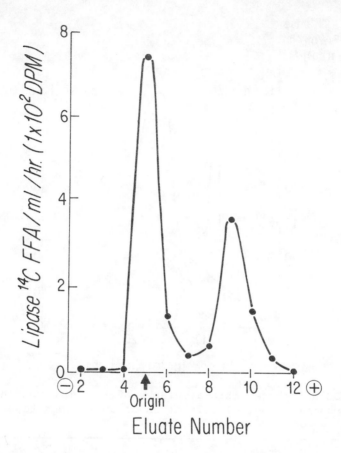

Fig. 12. Cellogel electrophoresis of an adipose tissue extract.

representative of human adipose tissue hormone sensitive lipase. As noted earlier, we had been unable to separate ALA-I,
a lipid-protein complex of ALA eluted in the void volume of
G-200 Sephadex, from NLA. However, using the radioactive lipase assay, we were able to demonstrate complete separation
from ALA-I after DEAE Sephadex chromatography (Fig. 11) (Cortner, Swoboda and Schnatz, in preparation).

This hormone-sensitive lipase activity of human adipose
tissue was separated into two components by starch block electrophoresis and by Cellogel electrophoresis (Fig. 12) (Cortner,
Swoboda and Schnatz, in preparation). One component migrated
towards the anode, whereas the other remained near the origin
or migrated slightly towards the cathode. These two electro-

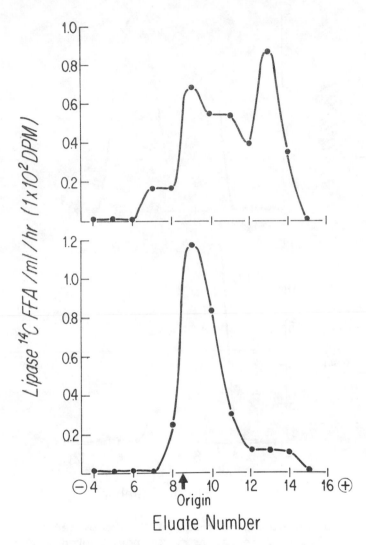

Fig. 13. Starch block electrophoresis of NLA recovered from
Sephadex G-200 gel filtration of an adipose tissue extract
(from Cortner, Swoboda and Schnatz, in preparation).
Upper tier: Electrophoresis of a void volume eluate.
Lower tier: Electrophoresis of an albumin-containing eluate.

phoretic components of lipase were present when either an adipose tissue extract or the void volume of a G-200 gel filtration were used. However, when the albumin-containing fractions

289

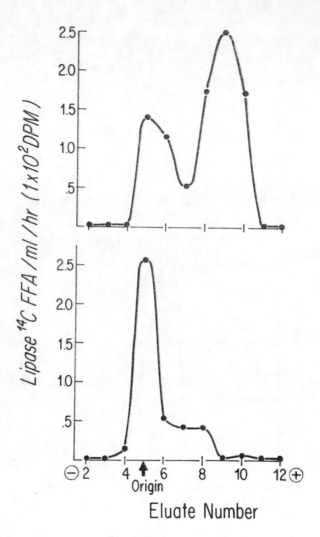

Fig. 14. Cellogel electrophoresis of NLA recovered from starch
block after electrophoresis (from Cortner, Swoboda and Schnatz,
in preparation). Upper tier: Cellogel electrophoresis of the
anodal component from starch block. Lower tier: Cellogel
electrophoresis of the cathodal component from starch block.

---

of a Sephadex G-200 gel filtration were used, only the cathodal
component was recovered (lower tier of Fig. 13). Finally,
when the anodal component from starch block electrophoresis
was subjected to Cellogel electrophoresis, both peaks were

observed (upper tier of Fig. 14). In comparison, Cellogel electrophoresis of the cathodal component from starch block revealed only the cathodal component (lower tier of Fig. 14).

We conclude that we are dealing with a human adipose tissue hormone sensitive lipase that has been separated into two components by starch block and Cellogel electrophoresis. By a combination of Sephdex G-200 gel filtration, DEAE Sephadex chromatography, and starch block electrophoresis, these two lipase components have been separated from all esterase activity. Although these two electrophoretic components originated in the void volume eluates from Sephadex G-200 gel filtration of human adipose tissue extracts, this does not necessarily mean that they are both large molecules. In fact, there is some evidence that the cathodal component is smaller than the anodal since starch block electrophoresis of G-200 Sephadex fractions which contain albumin produced only the cathodal component. In addition, these data suggest that the anodal component dissociates to form the cathodal since in no instance was the anodal component recovered without some cathodal component present, whereas the cathodal component could be recovered alone.

REFERENCES

Cortner, J. A. and J. D. Schnatz 1967. Electrophoretic behavior of alkaline lipolytic activity in human adipose tissue. *Biochim. Biophys. Acta* 139: 107-122.

Cortner, J. A. and J. D. Schnatz 1970. Alkaline lipolytic activity of rabbit adipose tissue: genotypes and their inheritance. *Biochem. Genet.* 4: 529-537.

Cortner, J. A., E. Swoboda, and J. D. Schnatz. In preparation. Characterization of human adipose tissue hormone sensitive lipase : separation into two components.

Desnuelle, P. 1961. Pancreatic lipase. *Adv. Enzymol.* 23: 129-161.

Rivello, R. C., J. A. Cortner, and J. D. Schnatz 1969. A comparative study of alkaline lipolytic activity in adipose tissue of various mammals. *Proc. Soc. Exper. Biol. & Med.* 130: 232-235.

Schnatz, J. D. 1964a. Lipolytic activity in human adipose. tissue. *Life Sci.* 3: 851-855.

Schnatz, J. D. 1964b. Automatic titration of free fatty acids. *J. Lipid Res.* 5: 483-486.

Schnatz, J. D. 1966. Neutral and alkaline lipolytic activities in human adipose tissue. *Biochim. Biophys. Acta* 116: 243-255.

Schnatz, J. D. and J. A. Cortner 1967. Separation and further characterization of human adipose tissue neutral and alkaline lipolytic activities. *J. Biol. Chem.* 242: 3850-3859.

Schnatz, J. D. and J. A. Cortner 1968. Genetic variation of alkaline lipolytic activity in rabbits. *Biochim. Biophys. Acta* 167: 367-372.

Schnatz, J. D. and T. C. Cummiskey 1969. Neutral and alkaline lipolytic activities in biopsies of human adipose tissue. *Life Sci.* 8: 1273-1279.

Schnatz, J. D. and D. Downey 1968. Human adipose tissue lipolytic activity and the in vitro effect of adrenaline and theophylline (abstract). *Diabetes* 17: 341.

Schnatz, J. D. and B. Marotta. In preparation. Hormone sensitive lipase in human adipose tissue.

NEW CHROMATOGRAPHIC APPROACHES TO THE SEPARATION OF HUMAN ALKALINE PHOSPHATASE ISOZYMES

GEORGE J. DOELLGAST and WILLIAM H. FISHMAN
Tufts Cancer Research Center and
the Department of Pathology (Oncology)
Tufts University School of Medicine
136 Harrison Avenue, Boston, Massachusetts   02111

ABSTRACT. Several new approaches to the chromatographic
separation of alkaline phosphatase isozymes are proposed.
Thus, salt-mediated hydrophobic chromatography is an alternative to salting-out in solution; detergent-mediated ion
exchange chromatography takes advantage of alterations in
DEAE-Sephadex elution profiles in the presence of detergent;
and agarose gel filtration can be used to isolate membrane-
bound alkaline phosphatases from neoplastic ascites fluid
and tissue homogenates. The system in which these approaches have been applied is in the isolation of human placental
alkaline phosphatase, but application to purification of
other alkaline phosphatases is indicated by an observed alteration in the chromatographic profile of liver alkaline
phosphatase on DEAE-Sephadex mediated by detergent and by
the separation of membrane-bound alkaline phosphatases from
cell homogenates of a variety of tissues on agarose gels.
The particular application of the agarose chromatography in
the search for new cancer-specific antigens and isozymes
is discussed.

Some of the difficulties that have hampered efforts to obtain highly purified alkaline phosphatase isozymes from human
tissues have been considered in a recent review by Fishman
('74). To summarize, the difficulty in obtaining large amounts
of human tissue, the heterogeneity of tissue specimens, phenotypic differences among alkaline phosphatase from different
individuals, and the membrane-bound character of alkaline
phosphatase, all hinder efforts to obtain homogeneous human
alkaline phosphatases from a single cell population.
  As a first step in preparation of human alkaline phosphatases from different tissues, we have explored several new approaches to the purification of human alkaline phosphatase by
column chromatography. The objective of these new approaches
has been two-fold: 1) To develop methods that would yield minimal losses during chromatography; 2) to ascertain whether the
membrane-bound character of human alkaline phosphatases could
be used to enhance, rather than prevent, chromatographic resolution of alkaline phosphatases.

As a first step in the development of these methods, we chose to study the chromatographic properties of human placental alkaline phosphatase and of Regan isozyme from ovarian cancer fluids. There were two reasons for this choice. First of all, placental tissue is the most readily obtainable human tissue, and therefore the limitation of availability is not restrictive in this case. Secondly, our laboratory has an interest in obtaining homogeneous human placental alkaline phosphatase and monospecific antiserum to it in order to develop sensitive radioimmunoassay and immunohistochemical procedures for placental-type (Regan) alkaline phosphatase.

The objective of developing methods that would yield minimal losses on chromatography was met in part by salt-mediated hydrophobic chromatography and by detergent-facilitated ion exchange chromatography.

The objective of taking advantage of the membrane-bound character of human alkaline phosphatase to enhance chromatographic resolution was reached through our experience with ovarian cancer fluid alkaline phosphatase excluded from Sepharose-4B.

In the following, we describe the experimental basis for the above statements and include a characterization of excluded alkaline phosphatase. The principles of the approach and the advantages offered for future work are discussed.

Materials and Methods

Human term placentae were prepared through step 3 of the large-scale procedure of Ghosh and Fishman ('68) and were stored at -20°C until used. The ammonium sulfate pellet prepared in this way was resuspended, dialyzed, and sonicated as described in a previous report (Doellgast and Fishman, '74).

Ovarian cancer ascites fluids were collected from patients on the Oncology services of Pondville, Lemuel Shattuck and University Hospitals. Normal human tissues were obtained from autopsies performed within 8 hours after the patients expired, and were frozen at -20°C until used. Fluids were centrifuged at 4,000 g for 30 minutes, and the supernatants were stored at -20°C until used.

D- and L-phenylalanine, phenyl phosphate (disodium salt), naphthol AS-MX phosphate (disodium salt) and adenosine-5' -monophosphate were purchased from Sigma. L-Homoarginine was a product of Nutritional Biochemical Company. Sepharose 4B and DEAE-Sephadex were products of Pharmacia. L-Phenylalanine and L-leucine Sepharoses were prepared as described previously (Doellgast and Fishman, '74). Ammonium sulfate and Tris base were products of Mann (Mann "ultra-pure" grade).

All other chemicals were of the highest grade commercially available.

Enzyme assays for alkaline phosphatase were performed at pH 9.8 using 18 mM phenyl phosphate as substrate by the manual method of Fishman and Ghosh ('67), or fluorometrically using 2 mM naphthol AS-MX phosphate as substrate, as described by Inglis et al ('73). Assay for total alkaline phosphatase from ovarian cancer ascites fluid was by the autoanalyzer method of Green et al ('71), and for placental-type (Regan) alkaline phosphatase was by the method of Anstiss et al ('71).

β-glucuronidase was by the method of Fishman ('67), 5' nucleotidase was by the method of Emmelot and Bos ('66) at pH 7.5 using adenosine-5'-monophosphate as a substrate. Na^+-K^+ dependent ATPase was determined by the method of Uesugi et al ('71). Leucine aminopeptidase was assayed by a modification of the fluorometric assay of Rockerbie and Rasmussen ('67).

Total DNA was determined by the method of Burton ('56). Phospholipid was extracted by the method of Folch et al ('57) and the phosphate content was determined by the method of Bartlett ('59). Protein was measured either by the absorbance at 280 mμ or by the method of Lowry et al ('51).

Disc-gel electrophoresis was performed as described in an accompanying report by Singer and Fishman (these Proceedings).

RESULTS

Salt-mediated Hydrophobic Chromatography

The chromatography of human placental alkaline phosphatase on L-phenylalanine Sepharose is seen in Fig. 1. 300 ml of solution containing enzyme was applied to the column in 1.25 M ammonium sulfate, 0.05 M Tris-acetate, pH 8.0. Elution with stepwise lower concentration of ammonium sulfate resulted in most of the protein being eluted in the 1.25 M and 1.00 M ammonium sulfate elutions, with most of the enzyme activity being contained in the 0.50 M and 0.25 M ammonium sulfate elutions. Table I shows the recovery of protein and activity in the pools of each of these successive elutions of enzyme. The 0.5 M ammonium sulfate and 0.25 M ammonium sulfate elutions had increases in specific activity of 3.8- and 4.9- fold, respectively, with recovery of 73% of the total activity in these two pools. 96% of the total placental alkaline phosphatase was recovered from this column in the various pools.

It should be emphasized that the results do not suggest a specific binding related to the phenylalanine binding site of the enzyme. The method is a general one for protein purification. The particular advantage for purification of placental

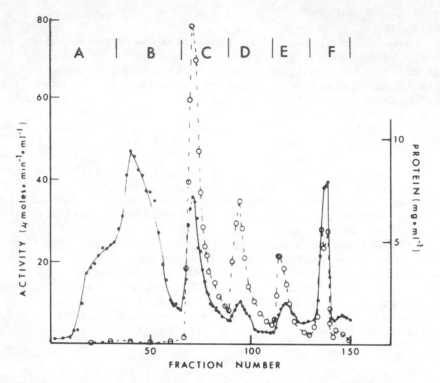

Fig. 1. Chromatography of human placental alkaline phosphatase (see Methods) on L-phenylalanine Sepharose. 300 ml of enzyme preparation containing 12.6 mg/ml of protein with a specific activity of 4.7 was applied to a 2.6 x 34 cm column of L-phenylalanine Sepharose in 0.05 M Tris acetate buffer, pH 8.0, containing 1.25 M ammonium sulfate (A). The column was then eluted successively with 300 ml each of 1.00 M ammonium sulfate, 0.04 Tris acetate, pH 8 (B), 0.5 M ammonium sulfate, 0.02 M Tris acetate, pH 8 (C), 0.25 M ammonium sulfate, 0.01 M Tris acetate, pH 8 (D), 0.25 M Tris acetate, pH 8.0 (E) and 0.25 M Tris base, pH 10.5 (F). Elution was at 4°C; 12 ml per fraction.

---

alkaline phosphatase is that it has a high absorption capacity (3.8 g of protein were applied to the column), and affords excellent recovery of enzyme activity with a considerable enhancement in specific activity.

Detergent-facilitated Ion Exchange Chromatography

Upon attempting chromatography of the dialyzed pool of the

TABLE I

Pool	Activity (μ moles/min)	Protein (mg)	Specific Activity (μ moles/min/mg)	% Recovery Acti-vity	Protein
Starting Material	17,670	3,780	4.68		
0.5M AmSO$_4$	9072	513	17.7	51.3	13.6
0.25M "	3783	165	22.9	21.4	4.4
0.25M NaCl	2737	316	8.6	15.5	8.4
0.25M Tris Base	1337	506	2.6	7.6	13.4
				95.8	39.8

Recovery of protein and activity in the various pools derived from the chromatography of human placental alkaline phosphatase on L-phenylalanine Sepharose. Enzyme activity was determined at pH 9.8 using phenyl-phosphate as substrate (see Methods), protein was determined by the method of Lowry et al ('51), using BSA as a standard.

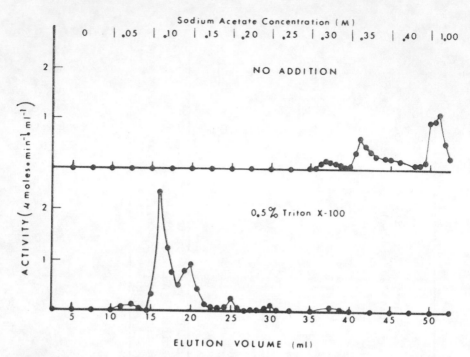

Fig. 2. Chromatography of human placental alkaline phosphatase
on DEAE-Sephadex A-50 in the absence (top) and in the presence
(bottom) of Triton X-100 detergent. 10-unit samples
(μ moles $\cdot$ min^{-1}) of placental alkaline phosphatase of the
0.50 M and 0.25 M ammonium sulfate elution from L-phe Sepharose
was applied to separate 0.9 x 11 cm columns of DEAE-Sephadex
A-50 which were equilibrated with 0.05 M Tris acetate buffer,
pH 8.0, and stepwise elutions of 5 ml each of the indicated
sodium acetate solutions in 0.05 M Tris-acetate, pH 8.0, were
used to elute the enzyme. Elution was performed at 4°C.

0.50 M and 0.25 M ammonium sulfate elutions of placental alkaline phosphatase on a DEAE-Sephadex ion-exchange chromatography
column, it was found that the enzyme did not elute as a well-
defined peak, but rather was strongly adsorbed to the ion-
exchange column, and was only eluted at very high salt concentrations.

To attempt to improve the resolution of the enzyme, Triton
X-100 detergent was included in the eluting buffer, and the
enzyme was now found to elute as a discrete peak in 0.10-0.15
M sodium acetate, 0.05 M Tris-acetate buffer, pH 8.0. Fig. 2
demonstrates the difference in the elution of placental alkaline phosphatase activity from DEAE-Sephadex A-50 in the absence and presence of Triton X-100 detergent. In the absence
of detergent, no enzyme activity is found to be eluted at a

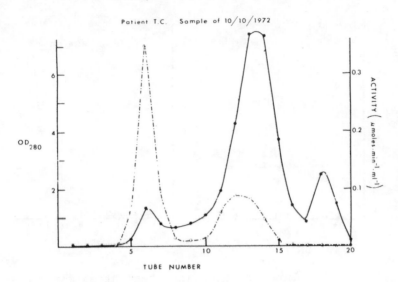

Fig. 3. Elution profiles of alkaline phosphatase and protein
from a sample of ovarian cancer ascites fluid chromatographed
on Sepharose 4B. 1 ml of fluid was chromatographed on a
0.9 x 27 cm column of Sepharose 4B equilibrated with 0.05 M
Tris acetate, pH 8.0, and 1.0 ml fractions were collected.
(-.-.-) Alkaline phosphatase activity (——) - protein.

sodium acetate concentration less than 0.30 M, while in the
presence of 0.5% Triton X-100 detergent, almost all of the
enzyme activity is eluted in 0.10 M and 0.15 M sodium acetate.
This formed the basis for the preparation of highly purified
human placental alkaline phosphatase (Doellgast and Fishman,
'74). The dialyzed pool of enzyme obtained from phenylalanine
Sepharose chromatography was subjected to three successive
ion-exchange steps on DEAE-Sephadex A-25: 1) a gradient of
0.00-0.15 M sodium acetate in 0.05 M Tris acetate, pH 8.0 containing 0.05% Triton X-100; 2) a gradient of 0.00-0.05% Triton
X-100 detergent in 0.10 M sodium acetate, 0.05 M Tris acetate,
pH 8.0; 3) a gradient of 0.10 M - 1.0 M sodium acetate in 0.05
Tris acetate buffer, pH 8.0, and subsequent washings with 1.0
M sodium acetate. The enzyme obtained was purified a further
10-fold relative to the pool of enzyme from the phenylalanine
Sepharose chromatography step with a 48% recovery, and the enzyme was homogeneous by immunodiffusion against antibody to
the crude placental homogenate, disc-gel electrophoresis, isoelectric focusing in polyacrylamide gels, and SDS-gel electrophoresis. Antisera prepared against this enzyme preparation
were highly specific for placental alkaline phosphatase but

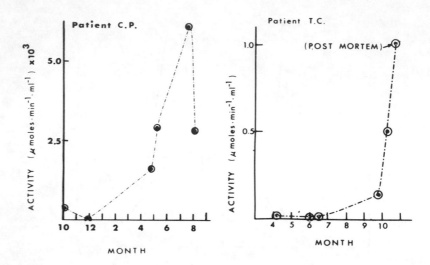

Fig. 4. The total amount of alkaline phosphatase which was excluded from Sepharose 4B in the ascites fluids of patients T.C. and C.P. during the course of their disease. The total excluded alkaline phosphatase was assayed by the column procedure outlined in Fig. 1. Patient C.P. - assay using naphthol AS-MX phosphate; patient T.C. - assay using phenyl phosphate.

did contain trace impurities that cross-reacted with normal human serum proteins. Immunoabsorption of this antiserum with serum yielded a final preparation of antiserum which was monospecific and is the basis for radioimmunoassay and immunohistochemical studies currently underway in our laboratory.

Chromatography of Ovarian Cancer Fluid Alkaline Phosphatase

Having developed a method for purification of human placental alkaline phosphatase that yielded good recovery of enzyme activity, we attempted to ascertain whether the Regan isozyme from ovarian cancer fluid could be purified using the same procedure. In this case, since the enzyme was presumably not bound to membrane, butanol extraction was not used as the first step in purification. When ascites fluid was chromatographed on L-phenylalanine Sepharose using the same protocol as above for the placental extract, however, it was found that most of the enzyme did not bind to the column, but in fact was eluted in the void volume of phenylalanine-Sepharose 4B. This indicated that most of the enzyme was present in a very large molecular weight aggregate.

300

Upon reviewing the literature, this observation appeared to coincide with the reports of Dunne et al ('67), Jennings et al ('70), Walker and Pollard ('71), and Shinkai and Akedo ('72) on the presence of a high-molecular weight form of alkaline phosphatase in the sera of patients with bone and liver cancer. Shinkai and Akedo ('72) demonstrated that the excluded alkaline phosphatase was immunologically similar to an isolated plasma membrane from liver, and was associated with leucine aminopeptidase, 5' nucleotidase, and Mg^{++}-dependent ATPase, all markers of the plasma membrane.

The amount of alkaline phosphatase in a form that is excluded from Sepharose 4B for a number of ovarian cancer fluids was assayed by the column procedure illustrated in Fig. 3. In this one case, 63% of the total alkaline phosphatase was excluded from Sepharose 4B. The excluded alkaline phosphatase in this case was inhibited 74% by 4 mM L-leucine and 71% by 5 mM L-phenylalanine, pH 9.8; the retained enzyme was inhibited 56% by L-leucine and 47% by L-phenylalanine. The excluded enzyme represents, therefore, principally the D-variant phenotype of Regan isozyme (Inglis et al, '73), which was present at very high concentrations in the fluid of this patient.

A summary of the alkaline phosphatase composition of a number of ovarian cancer fluids is given in Table II. In the case of patient T.C., the increase in the relative percentage of excluded alkaline phosphatase during the course of the disease was correlated with a considerable increase in the amount of placental-type (Regan) isozyme in these fluids. In the other three cases, there did not appear to be a correlation with the presence of the Regan isozyme, but all fluids did contain a significant amount of excluded alkaline phosphatase.

The progression of the total amount of excluded alkaline phosphatase for patients C.P. and T.C. during the course of the disease is shown in Fig. 4. At the terminal stages of the disease a considerable amount of excluded alkaline phosphatase is present in these fluids.

Since Shinkai and Akedo ('72) demonstrated the association of other plasma membrane markers (5' nucleotidase, ATPase, leucine aminopeptidase) with the alkaline phosphatase which was excluded from Sepharose 4B, we investigated the association of several plasma membrane markers with the excluded alkaline phosphatase. 25 ml ascites fluid from patient T.C. (sample of 10/10/1972) was centrifuged at 350,000 g_{max} for two hours, and the supernatant and resuspended pellet were assayed for alkaline phosphatase, 5'-nucleotidase, β-glucuronidase, Na^+K^+-dependent ATPase, DNA, phospholipid, and protein. As seen in Table III, most of the activity of

TABLE II

Patient	Date of Sample	Alkaline Phosphatases		
		Total (K-A units)	Placental (P.I. units)	%Ex-AP
T.C.	4/5/72	19.4	2.4	40.0
	5/31/72	13.1	5.1	37.1
	6/14/72	22.1	5.0	44.1
	9/22/72	49.8	81.0	48.9
	10/10/72	82.3	164.4	62.7
	10/19/72	139.2	304.0	67.6
C.P.	10/6/72	3.1	0.43	5.9
	11/27/72	1.7	0.77	0
	4/23/73	2.1	0.70	34.9
	5/7/73	4.8	1.03	45.4
	7/17/73	11.4	2.98	51.1
	8/3/73	3.4	1.84	45.6
K.S.	8/18/72	7.7	0.11	7.0
	10/24/72	16.2	0.06	30.7
I.B.	9/20/72	109.2	11.5	74.5

Analysis of the alkaline phosphatases of ascites fluids from patients with ovarian cancer. Ex-AP designates the alkaline phosphatase which is excluded from Sepharose 4B.

TABLE III

	Unit	Supernatant	Precipitate
Alkaline Phosphatase	μ moles phenol/min	9.51	1.31
5'Nucleotidase	μ moles phosphate/min	2.18	0.38
β-glucuronidase	μ moles phenolphthalein/min	0.119	0.110
Na^+K^+-ATPase	μ moles phosphate/min	ND*	ND*
DNA	μ g	ND*	ND*
Phospholipid Phosphate	μ moles phosphate	2.46	0.31
Protein	mg	910	50

*ND - not detectable

Total enzyme activities, DNA, phospholipid and protein of the precipitate and supernatant from centrifugation of ovarian cancer ascites fluid (patient T.C. 10/10/1972) at 350,000 g_{max} for 2 hours.

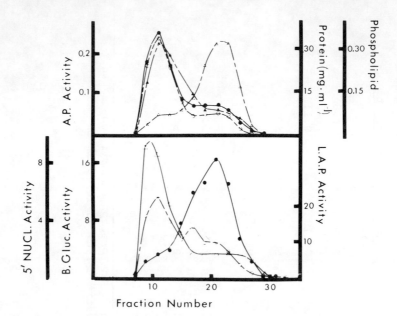

Fig. 5. The chromatographic profile of 2 ml of concentrated
ovarian cancer ascites fluid (see text) on Sepharose 4B. Column dimensions 1.6 x 21 cm, Buffer 10 mM Tris-HCl, pH 8.0-2.0
ml per fraction. Top-closed circles, alkaline phosphatase
activity, open circles-alkaline phosphatase activity of samples heated 5 minutes at 65°C; closed triangles-phospholipid
phosphate, open triangles - protein. Bottom-closed circles -
β-glucuronidase, open circles - leucine aminopeptidase; open
triangles - 5'-nucleotidase. Units: alkaline phosphatase,
μ moles phenol liberated per milliliter per minute; phospholipid - μ moles phospholipid phosphate per ml; 5' nucleotidase
- μ moles phosphate liberated per ml per hour; β-glucuronidase
- μg phenolphthalein per ml per hour; leucine aminopeptidase-
arbitrary units of fluorescence change per minute.

the plasma membrane marker enzymes and most of the phospholipid remained in the supernatant on high-speed centrifugation.
    After high-speed centrifugation, 20 ml of the supernatant
was lyophilized and redissolved in 3 ml of deionized water.
2 ml of this sample was then applied to a Sepharose 4B column
and the effluent fractions were assayed for enzymes, protein,
and phospholipid. The results for this chromatography are
shown in Fig. 5. Most of the plasma membrane markers, alkaline phosphatase, 5'nucleotidase, leucine aminopeptidase, and
phospholipid were excluded from the Sepharose 4B column, while
most of the non-plasma membrane associated β-glucuronidase and

Fig. 6

SCHEME FOR THE PREPARATION OF EXCLUDED
ALKALINE PHOSPHATASES FROM TISSUES

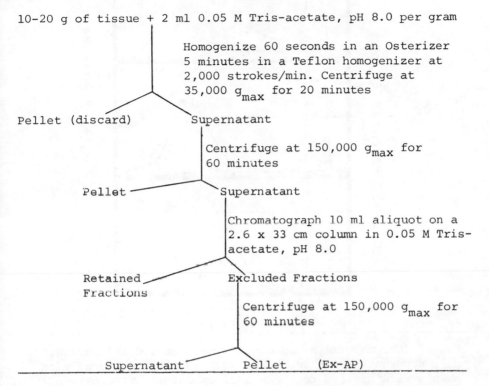

10-20 g of tissue + 2 ml 0.05 M Tris-acetate, pH 8.0 per gram

Homogenize 60 seconds in an Osterizer
5 minutes in a Teflon homogenizer at
2,000 strokes/min. Centrifuge at
35,000 g_{max} for 20 minutes

Pellet (discard) Supernatant

Centrifuge at 150,000 g_{max} for
60 minutes

Pellet Supernatant

Chromatograph 10 ml aliquot on a
2.6 x 33 cm column in 0.05 M Tris-
acetate, pH 8.0

Retained Excluded Fractions
Fractions

Centrifuge at 150,000 g_{max} for
60 minutes

Supernatant Pellet (Ex-AP)

protein were not. From these data, it appears that the excluded alkaline phosphatase did actually represent membrane-
associated enzyme.

The most surprising observation that resulted from this experiment was the finding that only about 15% of the total
membrane-bound alkaline phosphatase was able to pellet upon
high-speed centrifugation (350,000 g_{max} for 2 hours). After
separation of the excluded fractions, however, 90% of the excluded activity was obtained in a pellet of centrifugation at
150,000 g_{max} for 1 hour.

It therefore appeared that, in the presence of the components of the ascites fluid that are retained on the Sepharose
4B column, the membrane-associated alkaline phosphatase is unable to pellet in high-speed centrifugation. When separated
from the retained components, however, the excluded alkaline
phosphatase is able to pellet upon high-speed centrifugation.
There are two possible interpretations of this observation.

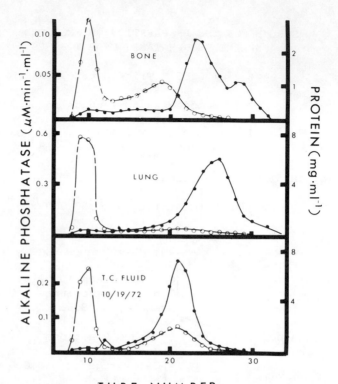

Fig. 7. Chromatography of bone, lung, and T.C. fluid 150,000 g_{max} supernatants on Sepharose 4B. 10 ml aliquots of each sample were applied to a 2.6 x 33 cm column of Sepharose 4B equilibrated with 0.05 M Tris acetate, pH 8.0, and the enzyme and protein eluted were assayed as described under Methods. 100 drops/tube were collected, using an LKB fraction collector; for buffer alone this represents 6.5 ml per tube, but variability in drop size occurred for the different tissue specimens. Temperature 4°C.

First, the association of alkaline phosphatase with membranes in ascites fluid is different in some respect from the association of alkaline phosphatase with membrane in cell homogenates from which membrane fractions are customarily prepared. Second, it is possible that what was observed to be the case with ascites fluid - that is, the inability of the membrane-bound alkaline phosphatase to pellet upon high-speed centrifugation is in actual fact a general phenomenon for any membrane fragment that is in the presence of human serum components.

To discriminate between these two possibilities, we prepared homogenates of a number of human tissues, centrifuged them at high-speed in the ultracentrifuge and chromatographed the supernatants from the high-speed centrifugation step on a Sepharose 4B column. The protocol for these experiments is given in Fig. 6. Two milliliters of buffer per gram (wet weight) of frozen tissues were homogenized in a Waring blender and in a Teflon homogenizer, centrifuged at 35,000 $\times$ g and then at 150,000 g, and the supernatant from the 150,000 $\times$ g step was chromatographed on a Sepharose 4B column. If the ascites fluid contained a unique form of membrane associated with alkaline phosphatase, then tissue extracts prepared by homogenization should not behave analogously to the ascites fluid.

Fig. 7 shows representative elution profiles for 150,000 g_{max} supernatants of bone, lung, and T.C. fluid chromatographed on Sepharose 4B. All of the samples chromatographed on this column contained a significant amount of alkaline phosphatase activity which was excluded from Sepharose 4B in the same fractions (9-11). Considerable variability was found in the relative recovery of alkaline phosphates in these fractions; as seen in Fig. 7, a much higher relative amount of bone alkaline phosphatase was retained on the column (57%) than was found for the lung alkaline phosphatase (24%) in these preparations. Most of the bone alkaline phosphatase which was retained was eluted in a position intermediate between the excluded fraction and the main peak of protein. The retained alkaline phosphatase of T.C. fluid, on the other hand, eluted with the main protein peak.

A summary of the preparation of excluded alkaline phosphatases is given in Table IV. As seen by a comparison of the specific activity in the 150,000 g_{max} supernatant, which was applied to the Sepharose 4B column, and the specific activity in the final preparation of excluded alkaline phosphatase, the increase in specific activity obtained from this procedure varied from a low of 8.3 for the placental excluded alkaline phosphatase to a high of 107-fold for the T.C. fluid excluded alkaline phosphatase. Recovery of activity in the excluded alkaline phosphatase, as seen most dramatically in the case of the placental sample, was proportional to the amount of total alkaline phosphatase which was excluded from the column. Thus, a 7.2% recovery of the placental enzyme in the excluded alkaline phosphatase obtained after centrifugation corresponded to the recovery of only 10% of the applied alkaline phosphatase in the excluded fractions. For the lung sample, which yielded 57% recovery of the applied activity in the Ex-AP, 76% of the enzyme recovered from the column was in the excluded fractions.

TABLE IV

Analysis of the preparation of excluded alkaline phosphatases from human tissues. Protein is expressed as mg total Lowry protein, activity as μmoles of phenol released per min, specific activity as μmoles phenol released per min per mg protein.

Tissue	35,000 g_{max} Supernatant			150,000 g_{max} Supernatant			Excluded Alkaline Phosphatase			
	Prot.	Act.	Sp. Act.	Prot.	Act.	Sp. Act.	Prot.	Act.	Sp. Act.	% Excl.*
Bone	88	8.1	0.09	77	4.0	0.05	1.1	0.9	0.84	43
Liver	188	6.0	0.03	164	3.5	0.02	5.6	2.0	0.35	69
Lung	241	22.3	0.09	228	11.0	0.05	2.4	6.3	2.63	76
Placenta	122	66.5	0.54	106	37.9	0.36	0.9	2.7	2.95	10
T.C.†	197	7.5	0.04	187	6.5	0.04	0.7	2.8	3.76	59

†Fluid obtained 10/19/72.

*This expression refers to the percentage of the total alkaline phosphatase which was recovered from the Sepharose 4B chromatography in the excluded fractions.

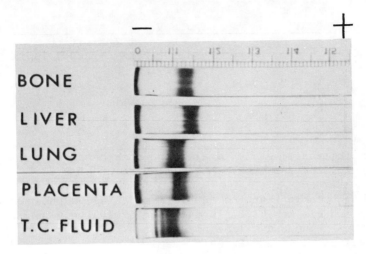

Fig. 8. Polyacrylamide disc gel electrophoresis, as described in Methods, of excluded alkaline phosphatases. 0.03 units of enzyme activity were applied per gel.

Characterization of Excluded Alkaline Phosphatase Preparations

The inhibition of all of the excluded alkaline phosphatase preparations at pH 9.8 by 8 mM L-homoarginine, an inhibitor of non-placental, non-intestinal isozyme, and by 5 mM L-phenylalanine, an inhibitor of placental and intestinal isozymes was measured. As seen in Table V, the bone, liver, and lung preparations were inhibited 80% by L-homoarginine and less than 20% by L-phenylalanine. The placental and T.C. fluid excluded alkaline phosphatases, on the other hand, were inhibited 80% by L-phenylalanine and less than 20% by L-homoarginine. The placental preparation was negligibly inhibited by L-homoarginine.

Disc-gel electrophoresis of the excluded alkaline phosphatases gave the results shown in Fig. 8. The bone, liver, and and lung preparations had single bands in distinct positions.

The placental sample showed one main band and a somewhat faster-moving second band. The T.C. fluid sample showed one main band and two somewhat slower-moving minor bands in this electrophoretic system. The main band of the T.C. fluid sample, which was a D-variant Regan isozyme, was somewhat slower-moving than the main band of the placental sample. With the exception of the T.C. fluid excluded alkaline phosphatase, all of the excluded alkaline phosphatases had some enzyme activity which was retained at the origin on disc-gel electrophoresis.

TABLE V

Tissue	Percent Inhibition	
	8 mM L-homoarginine	5 mM L-phenylalanine (D-L/D)
Bone	79.1	19.4
Liver	80.3	14.6
Lung	81.0	8.2
Placenta	2.7	77.9
T.C. Fluid	19.0	82.7

The inhibition of excluded alkaline phosphatases by 8 mM L-homoarginine and 5 mM L-phenylalanine at pH 9.80.

It is notable that the "B" form of alkaline phosphatase (Ghosh and Fishman, '68) is the form exclusively associated with the excluded alkaline phosphatase. No fast-moving enzyme was found in any Ex-AP's examined.

To test whether the excluded alkaline phosphatase serves as an effective antigen for preparation of antiserum that reacts specifically with the soluble alkaline phosphatase of that tissue, antisera were raised in male rabbits against placental and T.C. fluid excluded alkaline phosphatases. These antisera were found by Ouchterlony double diffusion to react only with the placental and Regan isozyme from T.C. fluid, and not with bone, liver, kidney, or intestinal alkaline phosphatases (data not shown).

Discussion

We have applied several new chromatographic approaches to the separation of human alkaline phosphatases. Chromatography of human placental alkaline phosphatase on phenylalanine-Sepharose is a general method for protein separation that is based upon principles similar to salting-out in solution and affords a considerable efficiency of separation of placental alkaline phosphatase (Doellgast and Fishman, '74).

What is apparently occurring in this system is an enhancement of interaction of the protein with the ligand covalently bound to the Sepharose column by addition of neutral salts. Some of the theoretical principles involved in this type of interaction have been considered by von Hippel and Schleich ('69) and by Hatefi and Hanstein ('69) and Hanstein et al ('71). By adding high concentrations of salts which enhance hydrophobic interactions (phosphate, sulfate) hydrophobic association of the protein and the phenyl group of phenylalanine can be enhanced. By using stepwise lower concentrations of sulfate, proteins are eluted due to a reduction in the salt-mediated association with the phenyl groups on the

310

column. By increasing the pH, a final elution of the protein which is strongly adsorbed to the column can be achieved.

This principle has been applied in our laboratory to the separation of human serum proteins and human placental extracts, preparation of rabbit immunoglobulins, and purification of horseradish peroxidase for use in immunohistochemical studies (unpublished work). It is therefore a generally applicable method for protein purification.

The general utility of salt effects for enhancement of hydrophobic interactions, whether specific or non-specific, has been considered (Doellgast, '73). It is likely that the effects which have been used for the purification of human placental alkaline phosphatase by salt-mediated hydrophobic chromatography will be of general utility in the study of hydrophobic associations of proteins with ligands or other proteins.

The effect of Triton X-100 detergent upon placental alkaline phosphatase chromatography on DEAE-Sephadex is that it permits a considerable purification of placental alkaline phosphatase by selectively altering the amount of salt and detergent in a series of steps, yielding a highly-purified final enzyme product with excellent recovery of enzyme activity. This approach in preliminary experiments has been found to be effective for the separation of human liver alkaline phosphatase.

The use of the protocol for preparation of excluded alkaline phosphatase from human tissues has several advantages for the initial preparation of human tissues for subsequent isolation of organ-specific-alkaline phosphatases. The combination of the pellet obtained by centrifugation at 150,000 x g and the excluded alkaline phosphatase represents most of the membrane-bound alkaline phosphatase obtainable from human tissues. If these then are used as the starting material for subsequent purification of alkaline phosphatase by butanol extraction and subsequent chromatography on salt-mediated hydrophobic chromatography and detergent-facilitated ion exchange chromatography, the starting material for the purification will be largely free of cytoplasmic and blood contaminants. This if of particular importance for tissue alkaline phosphatase purification since circulating serum alkaline phosphatases are frequently troublesome contaminants in any purification which is attempted using whole tissue homogenates as a starting material for purification. We have found the alkaline phosphatase in a butanol extract of 150,000 x g pellet + excluded alkaline phosphatase preparation of placenta has the same chromatographic characteristics on salt-mediated hydrophobic chromatography and detergent-facilitated ion-exchange chromatography as a butanol extract of whole placenta.

The utility of Sepharose 4B chromatography for separation

of the membrane-bound alkaline phosphatase in neoplastic fluids has an additional application in the search for new cancer-associated antigens. Since the Regan isozyme was found to be associated with membrane fragments isolated from one patient (T.C.), it is likely that other cancer-specific membrane-bound enzymes and antigens could be found in similar components isolated from cancer patients. The association of several carcinoplacental phenotypes with the Regan isozyme in cancer fluids as reported by Fishman (these Proceedings) argues for the feasibility of this approach.

ACKNOWLEDGMENTS

The authors appreciate the participation of Mr. Michael Mufson in the analysis of the ovarian cancer ascites fluid excluded alkaline phosphatases, and of Dr. Chi-Wei Lin, Mary Lou Orcutt, Sandra Kirley, and Normal Inglis in the analysis of association of other membrane markers with the excluded alkaline phosphatase of patient T.C.

This research was supported by grants in aid (CA-12924; CA-13332) of the National Cancer Institute, National Institutes of Health. W. H. Fishman is the recipient of Career Research Award K6-CA-18543 of the National Cancer Institute. The help of the University Cancer Committee (IN 23-P - American Cancer Society) is also acknowledged.

REFERENCES

Anstiss, C. L., S. Green, and W. H. Fishman 1971. An automated technique for segregating populations with a high incidence of Regan isoenzyme in serum. *Clin. Chim. Acta* 33: 279-286.

Bartlett, G. R. 1959. Phosphorus assay in column chromatography. *J. Biol. Chem.* 234: 466-468.

Burton, K. 1956. A study of the conditions and mechanism of the diphenylamine reaction for the colorimetric estimation of deoxyribonucleic acid. *Biochem. J.* 62: 315-323.

Doellgast, G. J. 1973. "Studies on the interaction of amino acid-Sepharoses and DNS-amino acids with α-IPM synthase from *Salmonella typhimurium;* salt effects." Ph.D. Thesis, Purdue University.

Doellgast, G. J. and W. H. Fishman 1974. Purification of human placental alkaline phosphatase: salt effects in affinity chromatography. *Biochem. J.*, in press.

Dunne, J., J. J. Fennelly, and K. McGeeney 1967. Separation of alkaline phosphatase enzymes in human serum using gel filtration (Sephadex G-200) techniques. *Cancer* 20: 71-76.

Emmelot, P. and C. J. Bos 1966. Studies on plasma membranes. III. Mg^{++}-ATPase, $(Na^+-K^+-Mg^{++})$-ATPase and 5'-nucleotidase activity of plasma membranes isolated from rat liver. *Biochim. Biophys. Acta* 120: 369-382.

Fishman, W. H. 1967. Determination of β-glucuronidase in "Method of biochemical analysis," ed. by D. Glick. 15: 77-145, Interscience Publishers, New York.

Fishman, W. H. and N. K. Ghosh 1967. Isoenzymes of human alkaline phosphatase. *Adv. in Clin. Chem.* 10: 256-370.

Fishman, W. H. 1974. Alkaline phosphatase isoenzymes. *Amer. J. of Med.*, in press.

Folch, J., M. Lees, and G. H. S. Stanley 1957. A simple method for the isolation and purification of total lipids from animal tissues. *J. Biol. Chem.* 226: 497-509.

Ghosh, N. K. and W. H. Fishman 1968. Purification and properties of molecular weight variants of human placental alkaline phosphatase. *Biochem J.* 108: 779-792.

Green, S., C. L. Anstiss, and W. H. Fishman 1971. Automated differential isoenzyme analysis. II. The fractionation of serum alkaline phosphatases into 'liver,' 'intestinal' and 'other' components. *Enzymologia* 41: 9-26.

Hanstein, N. G., K. A. Davis, and Y. Hatefi 1971. Water structure and the chaotropic properties of haloacetates. *Arch. Biochem. Biophys.* 147: 534-544.

Hatefi, Y. and W. G. Hanstein 1969. Solubilization of particulate proteins and monelectrolytes by chaotropic agents. *Proc. Natl. Acad Sci. USA* 62: 1129-1136.

Inglis, N. R., S. Kirley, L. L. Stolbach, and W. H. Fishman 1973. Phenotypes of the Regan isoenzyme and identity between the placental D-variant and the Nagao isoenzyme. *Cancer Research* 33: 1657-1661.

Jennings, R. C., D. Brocklehurst, and M. Hirst 1970. A comparative study of alkaline phosphatase enzymes using starch gel electrophoresis with special reference to high molecular weight enzymes. *Clin. Chem. Acta* 30: 509-517.

Lowry, O. H., N. J. Rosebrough, A. L. Farr, and R. J. Randall 1951. Protein measurement with the Folin phenol reagent. *J. Biol. Chem.* 193: 265-275.

Rockerbie, R. A. and K. L. Rasmussen 1967. An ultramicro method for the fluorometric determination of leucine aminopeptidase in serum. *Clin. Chem. Acta* 18: 183-185.

Shinkai, K. and H. Akedo 1972. A multienzyme complex in serum of hepatic cancer. *Cancer Research* 32: 2307-2313.

Uesugi, S., N. C. Dulak, J. F. Dixon, T. D. Hexum, J. L. Dahl, J. F. Perdue, and L. E. Hokin 1971. Studies on the char-

acterization of the sodium potassium transport adenosine
trip hosphatase. VI. Large scale partial purification
and properties of a Lubrol-solubilized bovine brain
enzyme. *J. Biol. Chem.* 246: 531-543.

Von Hippel, P. H. and J. Schleich 1969. "The Effects of
Neutral Salts on the Structure and Conformational Sta-
bility of Macromolecules in Solution." in *Structure and
Stability of Biological Molecules* (S. N. Timasheff and
G. D. Fasman, eds.) Marcel Dekker, New York. Chapter 6.

Walker, A. W. and A. C. Pollard 1971. Observations on serum
alkaline phosphatase electrophoretic patterns on polyacrylamide gel, with particular reference to the effects
of butanol extraction. *Clin. Chem. Acta* 34: 19-29.

ALKALINE PHOSPHATASE ISOZYMES IN THE ALIMENTARY CANAL OF THE SILKWORM

MASAHARU EGUCHI

Laboratory of Genetics, Kyoto Technical University
Matsugasaki, Kyoto, Japan

ABSTRACT. Alkaline phosphatases in the alimentary canal of two mutant and normal silkworms were studied. The phosphatase activity in the midgut tissue of A^-B^+ (fast moving band deficient) strain existed in the particulate fractions, whereas that of A^+B^- (slow moving band deficient) was mostly in the supernatant. By column chromatography in Sepharose 6B, phosphatase from A^-B^+ eluted faster than that from A^+B^-. The substrate specificity of digestive fluid phosphatase was similar to that of midgut phosphatase of A^+B^- strain, and the digestive fluid of A^+B^- strain was found to have only a trace of phosphatase activity.

By incubation of midgut extracts from A^-B^+ or normal strain with digestive fluid, C form phosphatase was produced. This enzyme was similar to the digestive fluid phosphatase in electrophoretic mobility, substrate specificity and elution position. The resistance of midgut alkaline phosphatase to digestion by digestive fluid was remarkably higher in A^-B^+ than in A^+B^-.

The antiserum to the C form was obtained by injection of purified phosphatase, and digestive fluid phosphatase was proved to be antigenically identical with C form. Results obtained suggest that the B isozyme in the midgut tissue was changed into digestive fluid phosphatase by the action of digestive fluid.

INTRODUCTION

Several studies of alkaline phosphatases in vertebrates, especially in mammals, have focused on aspects of development and physiology, intracellular localization, biochemistry, and clinical enzymology (Fishman, 1969). However, comparatively little is known about the genetics of alkaline phosphatases (Boyer, 1961; Robson and Harris, 1965; Beratis et al, 1970, 1971, 1972, etc.). In insects, genetic studies of phosphatases were carried out mainly in *Drosophila* (Beckman and Johnson, 1964; Johnson, 1966; Wallis and Fox, 1968) and *Musca domestica* (Ogita, 1968). In this case, however, physiological analysis seems difficult because the whole body was required as an enzyme source in most of the experiments. The silkworm, *Bombyx mori*, has the experimental advantage of large size and

315

many established strains for physiological and genetic studies.

The properties of two forms of alkaline phosphatase in the midgut of the silkworm were reported in earlier publications (Eguchi et al, 1972a, b) and a close relationship between isozymes in the midgut tissue and in the digestive fluid was suggested. The A form is considered to be free, while the B form is membrane associated, and these isozymes are different in optimum pH and Km.

It was shown by data from agar gel electrophoresis that phosphatase isozymes in the midgut are controlled by two separate genes on the same chromosome (Yoshitake et al, 1966). There are two mutants differing in the alkaline phosphatase isozymes of the midgut. One is the A^+B^- strain which has only a fast moving band corresponding with the A form, and the other is A^-B^+ having only a slow moving band (B form). These two mutants should be very useful for physiological and genetic studies of this enzyme. These mutants were originally discovered by agar gel electrophoresis; accordingly the evidence is qualitative and requires confirmation by quantitative methods.

Alkaline phosphatase in the digestive fluid of the silkworm is probably synthesized in midgut cells and transferred to the lumen, but so far as I know, nothing has been published on this problem. In the course of our experiments, we showed that a main enzyme component in the digestive fluid has an electrophoretic mobility similar to a new band produced from the B form isozyme by treatment with detergents or digestive fluid. These substances were added and incubated before electrophoresis. Eguchi et al (1972b) reported that the separation and purification of this isozyme (C form) was difficult. Later, we succeeded in the partial purification of the C form phosphatase. Midgut tissues of A^-B^+ strain and digestive fluid of A^+B^- strain with only a trace of alkaline phosphatase in the digestive fluid were utilized in this investigation.

The present paper describes the characterization and comparison of alkaline phosphatases from two mutants and from normal silkworms and deals with the separation and partial purification of the C form phosphatase produced by treatment of digestive fluids together with the immunological study of the C form and digestive fluid phosphatase.

In previous papers (Yoshitake, 1964; Yoshitake et al, 1966; Eguchi et al, 1972a, b) the designations F (fast) and S (slow) were used. In this paper the terms A and B are used, because these isozymes are controlled by non-allelic genes.

MATERIALS AND METHODS

The 5th instar larvae of *Bombyx mori*, 4 to 5 days after ecdysis, were employed. The silkworm strains used in these experiments were the A^+B^- mutant, Shinryukaku, which has only the A (fast migrating) band, and the A^-B^+ mutant, B7, which has only the B (slow migrating) band as seen on agar gel electrophoresis. The A^+B^+ (normal) strains, C105 and C124, were used for comparison.

A great part of the alimentary canal and most of the phosphatase activity is located in the midgut; therefore the midgut was used for the enzyme assay. The tissue was homogenized in a glass homogenizer with a Teflon pestle at 1,500 rpm at 0°C. A 10% homogenate in 0.25 M sucrose was centrifuged for 20 min at 900 g, and then for 20 min at 10,000 g, and for 90 min at 105,000 g. Digestive fluid was collected by applying a weak electric shock to the tissue in a test tube in the cold.

α-Naphthyl phosphate disodium salt (Nakarai) was used as substrate in the enzyme assay and on the zymograms; glucose-1-phosphate disodium salt (Sigma), glucose-6-phosphate disodium salt (Sigma), sodium-β-glycerol phosphate (Merck), and disodium p-nitrophenyl phosphate (Nakarai) were used to test substrate specificities.

In the enzyme assay, the Pi released from the substrate after incubation with enzyme at 30°C for 30 min was estimated by Takahashi's method (Takahashi, 1955). Phosphorus determination was carried out as follows: One ml of the filtrate from trichloroacetic acid precipitation was added to a mixture of of 1 ml 2% sodium molybdate, 1 ml 0.75 M H_2SO_4, and 4 ml isobutanol. After shaking and separating the butanol and water layers, 2 ml of butanol were mixed with 2 ml of 0.5% ascorbic acid, in $KHSO_4$, and 1 ml of ethanol. After 30 min incubation at 30°C, absorbance was measured at 720 nm in a Shimadzu-Bausch & Lomb Spectronic 20 colorimeter. Alkaline phosphatase activity was also measured by the rate of hydrolysis of p-nitrophenyl phosphate. The reaction mixture contained 0.5 ml, 5 mM p-nitrophenyl phosphate, 2 ml 0.1 M borate buffer (suitable pH for each isozyme), and 0.5 ml enzyme solution. After a 30 min incubation period at 30°C the reaction was stopped by adding 2 ml 0.1 M NaOH. The color formed was measured at 400 nm against appropriate blanks. Protein concentration was determined by the method of Lowry et al (1951). Detailed descriptions of electrophoresis, column chromatography, and enzyme assay were reported previously (Eguchi, 1968; Eguchi et al, 1972a).

The purification of the C form phosphatase was performed by the following procedure. The suspended precipitate of midgut extract after centrifugation at 105,000 g for 90 min was mixed with digestive fluid of the A^+B^- strain or with digestive fluid protease and was then incubated for 4 hr at 30°C. The mixture was then subjected to column chromatography in Sepharose 6B, and the C form was obtained as the main phosphatase peak (see Fig. 6). The eluates (fraction C) were concentrated in collodion bags under reduced pressure and loaded onto a DEAE-Cellulose column equilibrated with 0.01 M, pH 7 borate buffer. Elution was usually performed with a gradient of sodium chloride concentration (0.1 to 0.7 M). The eluate of a major peak in 0.2 M NaCl was concentrated and passed through a Sephadex G-100 column. In immunological studies, antiserum was produced in rabbits by intramuscular and subcutaneous injections of the phosphatase preparation mixed with Freund's adjuvant.

RESULTS

1) Comparison of midgut alkaline phosphatases of mutant and normal strains

Electrophoretic patterns on agar gel of midgut alkaline phosphatases of two mutants and of normal silkworms are illustrated in Fig. 1.

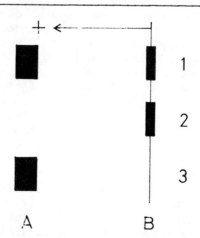

Fig. 1. Diagram showing zymograms of midgut alkaline phosphatases in three different strains of larvae. 1) A^+B^+ (normal), 2) A^-B^+, 3) strain A^+B^-.

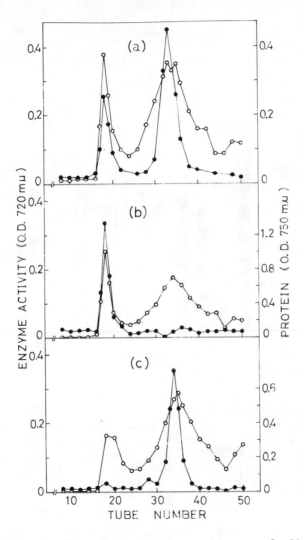

Fig. 2. Sepharose 6B column elution patterns of alkaline phosphatases from the midguts of three different strains. The 5 ml of supernatant from centrifugation at 10,000 g for 20 min were loaded onto a 2.64 x 36 cm column of Sepharose 6B equilibrated with 0.1 M, pH 9 borate buffer. Each fraction contains 4.5 ml of eluate. a) A^+B^+; b) A^-B^+; c) A^+B^- strain. Solid circles, enzyme activity; open circles, protein concentration.

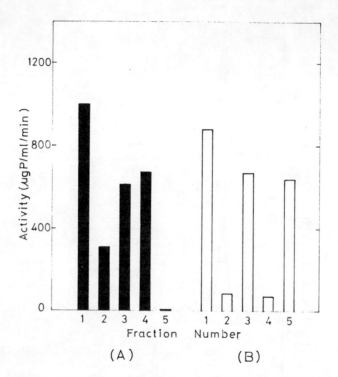

Fig. 3. Intracellular distribution of midgut alkaline phosphatases of two mutants. Midgut tissues were homogenized in 0.25
M sucrose solution. Fraction No. 1, supernatant of centrifugation at 900 g for 20 min; No. 2 and 3, precipitate (mitochondria) and supernatant at 10,000 g for 20 min; No. 4 and 5,
precipitate (microsome) and supernatant at 105,000 g for 90
min. A) A^-B^+, B) A^+B^- strain.

The diagram shows clearly the difference in zymograms from
these three different strains. Although two components of
this enzyme are found in the normal strain, only the A or B
band is detected in each mutant. In experiments using agar
gel electrophoresis, the B band did not migrate in the range
of pH 4.0 to 8.6 and no quantitative measurements were made
by this method.

Thus, column chromatography was used for the separation and
comparison of the isozymes. Fig. 2 shows the distribution of
alkaline phosphatases in the eluates from a Sepharose 6B column. The difference in elution pattern among the three strains
is apparent. Two phosphatase peaks are present in the normal

TABLE I

Substrate	Enzyme activity (μg P/min/mg protein)	
	A^+B^-	A^-B^+
α-Naphthyl phosphate	12.27	9.90
p-Nitrophenyl phosphate	13.69	13.18
β-Glycero phosphate	14.47	8.37
Glucose-1-phosphate	16.20	7.67
Glucose-6-phosphate	12.20	7.70

Hydrolysis of several substrates by midgut alkaline phosphatases from two mutants.

silkworm, while in the A^-B^+ strain one peak (tube no. 18) of alkaline phosphatase activity is observable, and the A^+B^- mutant has a slow eluting peak (tube no. 34). From results of electrophoresis on agar gel of these eluates, it was evident that the fast and slow elution positions correspond to the B and A bands respectively, as seen on agar gel. Each peak coincides with a peak in the elution pattern of the midgut extract from the normal strain. Two protein peaks are detected in each elution pattern of the three strains. In the A^-B^+ strain, the first peak of protein coinciding with the phosphatase activity is higher than the second peak, but the opposite result was obtained in A^+B^- strain.

The characterization of the two isozymes of midgut alkaline phosphatase in the normal strain has been reported (Eguchi et al, 1972a); therefore this report takes up the comparison between the two mutant strains. The intracellular distribution of alkaline phosphatases was investigated by differential centrifugation. The results are illustrated in Fig. 3. The enzyme activity in the mitochondrial fraction (fraction no. 2) is greater in A^-B^+ than in strain A^+B^-. Concerning the post-mitochondrial fraction (no. 4 and 5), more than 98% of the activity is in the microsomal fraction (no. 4) in the A^-B^+ strain; in the A^+B^- strain, most of the phosphatase occurs in the supernatant (no. 5).

The properties of the phosphatases in both strains were also compared, using the supernatant (A^+B^-) or the precipitate (A^-B^+) after centrifugation at 105,000 g for 90 min. The substrate specificities of the phosphatases in both strains are shown in Table I. With respect to strain A^+B^-, no striking difference in the hydrolysis of five substrates is seen, whereas in strain A^-B^+, the hydrolysis of p-Nitrophenyl phosphate is greater than for the other compounds.

TABLE II

Substrate	Enzyme activity (µg P/min/mg protein)		
	A^+B^-	A^-B^+	A^+B^+
α-Naphthyl phosphate	0.24	5.48	14.68
p-Nitrophenyl phosphate	0.27	5.68	13.88
β-Glycero phosphate	0.29	2.79	5.81
Glucose-1-phosphate	0.06	2.25	7.24
Glucose-6-phosphate	0.01	2.13	6.22

Hydrolysis of several substrates by alkaline phosphatases in the digestive fluid of mutant and normal larvae.

2) Comparison of digestive fluid alkaline phosphatase of mutant and normal strains

The hydrolyzing abilities for several substrates were compared among the three strains (Table II). The hydrolysis of five substrates in strain A^+B^- is remarkably weak; enzyme activity is highest in the normal strain. Nevertheless, the hydrolysis of the five substrates is similar in the three strains, that is, the hydrolysis of aromatic phosphomonoesters is higher than for aliphatic compounds. This enzymatic behavior is similar to that of the midgut alkaline phosphatase of the A^-B^+ mutant and to the B form of the enzyme of the normal strain.

The elution pattern of digestive fluid phosphatase from normal larvae is shown in Fig. 4. There is one peak of phosphatase activity, and the elution position is close to that of phosphatase C produced by incubating preparations with digestive fluid as shown in Fig. 6.

3) Relationship between midgut alkaline phosphatase and the digestive fluid enzyme

A. Effect of digestive fluid on midgut alkaline phosphatase activities:

Since the properties of alkaline phosphatases of the midgut and the digestive fluid have been described (Eguchi et al, 1972a, b), it seemed important to examine the relationships of the phosphatases found in the midgut and in the digestive fluid. Fig. 5 shows the changes in alkaline phosphatase activity of the midgut tissue caused by incubation with digestive fluid. In this experiment, digestive fluid of strain A^+B^- which has only a trace of phosphatase activity (see Table II) was used. As is apparent from this figure, the

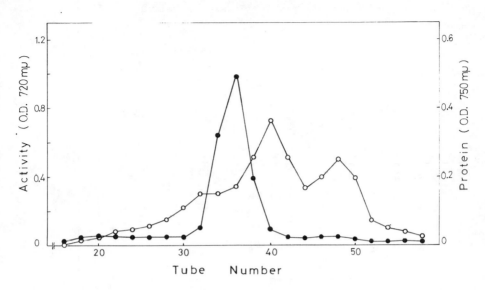

Fig. 4. Sepharose 6B column elution pattern of alkaline phosphatase in the digestive fluid of normal larvae. The 6 ml of digestive fluid were loaded onto a 2.64 x 36 cm column of Sepharose 6B equilibrated with 0.1 M, borate buffer, pH 9.0. Each fraction contains 4.5 ml of eluate. Solid circles, enzyme activity; open circles, protein concentration.

activity of midgut enzyme in strain A^+B^- decreases remarkably by incubation with digestive fluid. In the A^-B^+ strain, however, the phosphatase activity of the midgut is increased by addition of digestive fluid. Such a marked decrease in the phosphatase activity of the midgut from strain A^+B^- may be due to hydrolysis by protease in the digestive fluid.

B. Separation and properties of C form phosphatase:

As shown in our earlier study (Eguchi et al, 1972b), when the electrophoresis and enzyme staining were performed after incubation of extract from the midgut tissue with digestive fluid, a new alkaline phosphatase band (C) appeared between the A and B bands. The new band has an electrophoretic mobility similar to that of the enzyme from digestive fluid. The result suggests that there is a close relationship between the C isozyme and phosphatase of the digestive fluid. The C isozyme was therefore isolated by column chromatography, and its properties compared with those of the phosphatase from the

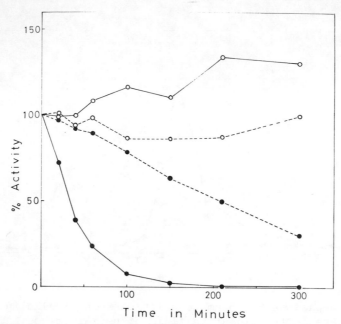

Fig. 5. Effect of digestive fluid on the midgut alkaline phosphatases of two mutants. The enzyme solution was obtained from midguts of each strain by differential centrifugation. Digestive fluid of strain A^+B^- with only a trace of phosphatase activity was utilized. The 4 ml of enzyme solution were mixed with 4 ml of digestive fluid and 4 ml of borate buffer, and incubated at 30°C. Phosphatase activities were measured at the indicated times. ●——● , A + digestive fluids; ●---● , A + distilled water; O——O , B + digestive fluid; O---O , B + distilled water.

digestive fluid. Fig. 6 shows the elution pattern of phosphatase C on a Sepharose 6B column. Midgut tissues of the A^-B^+ mutant were employed in this experiment, because the separation of the C and A isozymes by column chromatography was difficult when the normal (A^+B^+) strain was used. In this figure, minor and major peaks of phosphatase activity are discernible, and the former corresponds to the peak of the B form, and the latter peak is due to the C isozyme. These conclusions were confirmed electrophoretically. The new band (C) appeared on agar gel zymograms after incubation of the midgut extract with Triton X-100 (Eguchi et al, 1972b); hence column chromatography was performed after incubation of the midgut extract from strain A^-B^+ with Triton X-100. The separation of the C form, however, was unclear when compared with

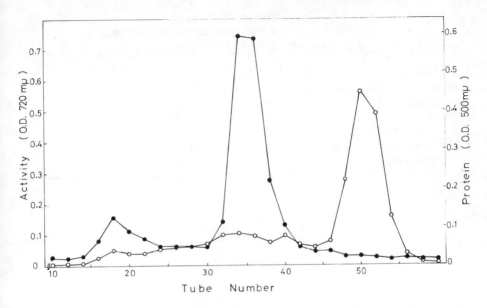

Fig. 6. Sepharose 6B column elution pattern of the C form of alkaline phosphatase produced by digestive fluid treatment. As a source of midgut alkaline phosphatase, 5 ml of supernatant from centrifugation at 10,000 g for 20 min of the midgut extract from strain A⁻B⁺ were used. The enzyme solution was incubated for 4 hrs at 30°C with 3 ml of digestive fluid of strain A⁺B⁻ with only a trace of alkaline phosphatase activity. The 5 ml of the mixture were loaded onto a 2.64 x 36 cm column equilibrated with 0.1 M borate buffer, pH 9. Each fraction contains 4.5 ml of eluate. Solid circles, enzyme activity; open circles, protein concentration.

the experiment using digestive fluid. The phosphatase activities of the C isozyme with five substrates are listed in Table III. The hydrolzying ability for aromatic phosphomonoesters is greater than for aliphatic compounds. The result is generally similar to that of digestive fluid phosphatase and to the B isozyme of the midgut tissue.

An immunological study was also carried out to elucidate the relationship between the C isozyme and the phosphatase from digestive fluid. Partially purified C isozyme was used as an antigen. Results of purification of the C form are outlined in Table IV. By incubation of the midgut extract with digestive fluid and by column chromatography using Sepharose 6B, DEAE-Cellulose, and Sephadex G-100, an enzyme preparation purified 345-fold was obtained. This phosphatase

TABLE III
HYDROLYSIS OF SEVERAL SUBSTRATES BY THE
C FORM OF ALKALINE PHOSPHATASE

Substrate	Enzyme activity (μg P/min/mg protein)
α-Naphthyl phosphate	74.94
p-Nitrophenyl phosphate	84.55
β-Glycero phosphate	44.20
Glucose-1-phosphate	44.47
Glucose-6-phosphate	39.80

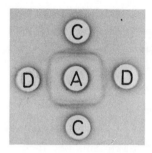

Fig. 7. Immunodiffusion with anti-C serum and purified enzymes.
A) anti-C serum, C) C isozyme of phosphatase, D) digestive
fluid phosphatase.

preparation was mixed with Freund's adjuvant and then injected
into rabbits. The agar diffusion method of Ouchterlony was
used for immunological analysis. As seen in Fig. 7, the antiserum to phosphatase C was placed in the central well. The
surrounding wells were filled with the purified C isozyme or
with the phosphatase from digestive fluid. The digestive
fluid phsophatase was purified 532-fold by ammonium sulfate
fractionation and column chromatography. After precipitation
bands developed and the agar plate was washed with sodium
chloride, phosphatase staining was performed with sodium
naphthyl phosphate and fast blue salt B at pH 11; accordingly
the dark bands in the plate represented alkaline phosphatase
activity of the immune precipitates.

Both purified phosphatases reacted to antiserum and immune
precipitates formed on the plate; the photograph shows the
converging arc of a homologous reaction, indicating that both
phosphatases are apparently antigenically identical.

TABLE IV

SUMMARY OF PURIFICATION PROCEDURE FOR THE C ISOZYME OF PHOSPHATASE

Step	Total activity (p-NPμg/min)	Total protein (mg)	Specific activity (p-NPμg/min/mg)	Ratio	Yield
10000 x g sup.	-----	-----	22.2	1	-----
M.G. + D.F.	1015	6.0	169.2	7.6	100
Sepharose 6B	6263	4.6	1362.0	61.4	617
DEAE-Cellulose	4062	1.0	3905.8	175.9	400
Sephadex G100	2314	0.3	7662.2	345.1	228

DISCUSSION

Normal silkworms and two mutant strains, A^+B^- and A^-B^+, are clearly useful for physiological and genetic studies of alkaline phosphatases. Experiments using differential centrifugation and column chromatography indicate that the post mitochondrial fraction of midgut extracts from strains A^+B^- and A^-B^+ provide the A and B forms of phosphatase, respectively.

From the phosphatase assay in the digestive fluid of normal and mutant strains, it was demonstrated that strain A^+B^- is deficient in digestive fluid phosphatase. This result supports the assumption that the source of digestive fluid phosphatase may be the B isozyme in the midgut tissue.

As shown in a previous paper (Eguchi et al, 1972b) and in the present report, the C isozyme appears after incubation of the midgut extract with detergents or digestive fluid; the origin of this enzyme is considered to be from the B form, since the new band is produced when strain A^-B^+ is used as the source of midgut phosphatase but is undetectable when strain A^+B^- is used. Moreover, the substrate specificity of the C form is similar to that of the B isozyme.

A close relationship between the C isozyme and digestive fluid phosphatase is suggested by the following facts: 1) The new band appearing on the phosphatase zymogram by incubation of midgut extracts with digestive fluid is similar to the phosphatase band of digestive fluid in electrophoretic mobility. 2) The C form has similar substrate specificity to that of the digestive fluid phosphatase. 3) The elution position of the C form on a Sepharose column corresponds to that of digestive fluid phosphatase.

The synthesis, intracellular transport, storage, and discharge of secretory proteins in and from pancreatic exocrine cell of the guinea pig has been thoroughly studied (Caro and Palade, 1964; Jamieson and Palade, 1967a, b). It was shown that the secretory proteins are synthesized on the rough surfaced endoplasmic reticulum, concentrated in the Golgi complex, and released into the glandular lumina. The relationship of alkaline phosphatases of serum and other organs has been investigated in mammals (Gould, 1944; Madsen and Tuba, 1952; Saini and Posen, 1969; Righetti and Kaplan, 1971, etc.). However, we have as yet very little information on the processing of secretion and absorption in insects.

From a consideration of the results described above, it is possible to conclude that the B form is synthesized in the midgut cells and at least part of it transferred to the lumen, changing into the C isozyme or digestive fluid alkaline phosphatase on the way. The effects of the digestive fluid on

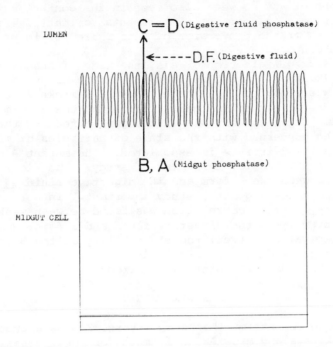

Fig. 8. Hypothetical mechanism for the production of digestive fluid phosphatase.

isozymes of alkaline phosphatase in the midgut tissue also support this assumption. As mentioned above, the resistance of the B form to digestive fluid is markedly greater than for the A isozyme of the midgut tissue; consequently, if the A form is discharged into the lumen, this phosphatase may be hydrolyzed by strong proteases in the digestive fluid, whereas the B form, at least in part, changes into the C form, with an increase in phosphatase activity.

The results of our immunological study show that the C isozyme and the digestive fluid phosphatase are antigenically identical. A hypothetical mechanism for the production of digestive fluid phosphatase from midgut phosphatase by the action of digestive fluid is shown in Fig. 8.

Digestive fluid protease will play an important role in the conversion of the B form to the digestive fluid phosphatase. In order to clarify this point, digestive fluid proteases were separated by column chromatography and then incubated with midgut phosphatase B. The production of the C form

329

phosphatase was confirmed by column chromatography, but some conversions of B to C were also seen in the control experiment after long incubation. Experiments to identify factors responsible for the conversion of B to C are now in progress in our laboratory.

Using agar gel electrophoresis, the mode of inheritance of the two alkaline phosphatase isozymes in the midgut of the silkworm was reported (Yoshitake, 1964; Yoshitake et al, 1966). However, the action of the gene controlling the B form is not clear. It is conceivable that a gene affecting the B form may be concerned with the state of particles or membranes with which the B isozyme is associated. The mutant A⁻B⁻ should be available for this kind of study. The close relationship between the B form and the digestive fluid alkaline phosphatase makes a genetic study important. In our crossing experiments, a clear correlation was found between weak phosphatase activity in the digestive fluid and a faint B band on the agar gel. However, the stainability of the B band after agar gel electrophoresis is in doubt. Therefore further investigation of this point is required.

ACKNOWLEDGMENTS

I wish to express my thanks to Messrs. Yasutaka Suzuki and Koichi Takizawa and Miss Akiko Iwamoto for their assistance in carrying out experiments.

REFERENCES

Beckman, L. and F. M. Johnson 1964. Variations in larval alkaline phosphatase controlled by Aph alleles in *Drosophila melanogaster*. *Genetics* 49: 829-835.

Beratis, N. G. and K. Hirschhorn 1970. Properties of placental alkaline phosphatase. I. Molecular size and electrical charge of the various electrophoretic components of the six common phenotypes. *Biochem. Genet.* 4: 689-705.

Beratis, N. G. and K. Hirschhorn 1972. Properties of placental alkaline phosphatase. III. Thermostability and urea inhibition of isolated components of the three common phenotypes. *Biochem. Genet.* 6: 1-8.

Beratis, N. G., W. Seegers, and K. Hirschhorn 1971. Properties of placental alkaline phosphatase. II. Interactions of fast- and slow-migrating components. *Biochem. Genet.* 5: 367-377.

Boyer, S. H. 1961. Alkaline phosphatase in human sera and placentae. *Science* 134: 1002-1004.

Caro, L. G. and G. E. Palade 1964. Protein synthesis, storage, and discharge in the pancreatic exocrine cell- an auto-radiographic study. *J. Cell Biol.* 20: 473-495.

Eguchi, M. 1968. Changes in zymograms of some dehydrogenases and hydrolyases in several organs of the silkworm, *Bombyx mori* L. *(Lepidoptera: Bombycidae)* during metamorphosis. *Appl. Ent. Zool.* (Tokyo) 3: 189-197.

Eguchi, M., M. Sawaki, and Y. Suzuki 1972a. Multiple forms of midgut alkaline phosphatase in the silkworm: Separation and comparison of two isozymes. *Insect Biochem.* 2: 167-174.

Eguchi, M., M. Sawaki, and Y. Suzuki 1972b. Multiple forms of midgut alkaline phosphatase in the silkworm: New band formation and the relationship between the midgut and digestive fluid enzymes. *Insect Biochem.* 2: 297-304.

Fishman, W. H. (ed.) 1969. The phosphohydrolases: Their biology, biochemistry, and clinical enzymology. *Ann. N.Y. Acad. Sc.* 166: 365-819.

Gould, B. S. 1944. Studies on the source of serum phosphatase, the nature of the increased serum phosphatase in rats after fat feeding. *Arch. Biochem. Biophys.* 4: 175-181.

Jamieson, J. D. and G. E. Palade 1967a. Intracellular transport of secretory proteins in the pancreatic exocrine cell. I. Role of the peripheral elements of the Golgi complex. *J. Cell Biol.* 34: 577-596.

Jamieson, J. D. and G. E. Palade 1967b. Intracellular transport of secretory proteins in the pancreatic exocrine cell. II. Transport to condensing vacuoles and zymogen granules. *J. Cell Biol.* 34: 597-615.

Johnson, F. M. 1966. *Drosophila melanogaster:* Inheritance of a deficiency of alkaline phosphatase in larvae. *Science* 152: 361-362.

Lowry, O. H., N. J. Rosebrough, A. L. Farr, and R. J. Randall 1951. Protein measurement with the Folin phenol reagent. *J. Biol. Chem.* 193: 265-275.

Madsen, N. B. and J. Tuba 1952. On the source of the alkaline phosphatase in rat serum. *J. Biol. Chem.* 195: 741-750.

Ogita, Z. 1968. Genetic control of isozymes. *Ann. N.Y. Acad. Sc.* 151: 243-262.

Righetti, A. B. and M. M. Kaplan 1971. The origin of the serum alkaline phosphatase in normal rats. *Biochim. Biophys. Acta* 230: 504-509.

Robson, E. B. and H. Harris 1965. Genetics of the alkaline phosphatase polymorphicsm of the human placenta. *Nature* 207: 1257-1259.

Saini, P. K. and S. Posen 1969. The origin of serum alkaline phosphatase in the rat. *Biochim. Biophys. Acta* 177: 42-49.

Takahasi, Y. 1955. Measurement of inorganic phosphate and
    creatine phosphate, and phosphoamidase and creatine
    phosphokinase activity in the sperm of the pig. *J. Jap.*
    Biochem. Soc. 26: 690-698 (in Japanese).
Wallis, B. B. and A. S. Fox 1968. Genetic and developmental
    relationships between two alkaline phosphatases in
    Drosophila melanogaster. *Biochem. Genet.* 2 : 141-158.
Yoshitake, N. 1964. Genetical studies on the alkaline
    phosphatase in the mid-gut of the silkworm, *Bombyx mori* L.
    J. Seric. Sci. Jap. 33: 28-33 (in Japanese).
Yoshitake, N., M. Eguchi, and M. Akiyama 1966. Genetic
    control of the alkaline phosphatase of the mid-gut in
    the silkworm, *Bombyx mori* L. *J. Seric. Sci. Jap.* 35: 1-7
    (in Japanese).

DIFFERENCES IN THE STRUCTURE, FUNCTION, AND FORMATION
OF TWO ISOZYMES OF *ESCHERICHIA COLI* ALKALINE PHOSPHATASE

MILTON J. SCHLESINGER, WILL BLOCH*, and PHILIP M. KELLEY
Department of Microbiology
Washington University School of Medicine
St. Louis, Missouri 63110

ABSTRACT. *E. coli* alkaline phosphatase isozyme 1 differs
from isozyme 3 by virtue of an extra arginine residue at
the amino-terminal position of the polypeptide chains of
isozyme 1. This difference is expressed functionally as
a two-fold increase in the steady state rate of phosphate-
ester hydrolysis at pH 5.5. Isozyme 1 is rapidly converted
to isozymes 2 and 3 in the bacterial cell, probably by a
protease. This conversion is inhibited by arginine and it
occurs after the alkaline phosphatase subunits have dimerized and attained a catalytically-active conformation.

INTRODUCTION

The isozymic character of *E. coli* alkaline phosphatase was
discovered during the initial biochemical studies on this
enzyme by Garen and Levinthal (1960). Subsequently, it was
clearly demonstrated that the isozymes were not the products
of different structural genes but represented some form of
epigenetic modification, and that the most important factors
in the production of isozymes for this protein were the growth
conditions of the bacteria (Signer, 1963). At the conference
on isozymes held in 1966, we reported on the physical, chemical, and enzymatic properties of two purified alkaline phosphatase isozymes (Schlesinger and Andersen, 1968). In that
study we offered preliminary data on a chemical difference
between two of the variants and on the conversion of one isozyme to another. In the intervening years we have been investigating the amino acid sequence of one of the isozymes and
have also studied in greater detail the catalytic mechanism of
the enzyme. From this work we have been able to firmly identify a chemical difference between two of the isozymes and to
detect a difference in their catalytic activities. Results
of these experiments have been detailed elsewhere (Bloch and
Schlesinger, 1974; Kelley et al, 1973). They are summarized
in this paper together with new information on the biosynthesis
of two isozymes.

* Present address: Department of Chemistry,
                   Reed College
                   Portland, Oregon 97202

EXPERIMENTAL PROCEDURES

Methods

Isolation, purification, and amino acid analyses of the two isozymes studied in this report have been described (Kelley et al, 1973). The stopped flow kinetic experiments have also been reported in detail (Bloch and Schlesinger, 1974).

Preparation and analysis of radioactive isozymes were carried out on 25 ml cultures of *Escherichia coli* K-10. ^{35}S-methionine (10 μl of 1 mc/ml; specific activity 40 Ci/mM; Amersham-Searle Lab.) was added and, at various times, cells were harvested by centrifugation at 4° in the presence of .1% sodium azide. Spheroplasts were prepared as previously described (Schlesinger, 1968) and 2 μl samples of the periplasmic fraction were applied to thin 10% starch gels. After electrophoresis in the Turner 310 apparatus (Schlesinger and Andersen, 1968), the gels were stained for alkaline phosphatase activity, then transferred to Whatmann 3 MM paper and dried under vacuo. Autoradiograms were prepared and the density of the film was recorded with a Gilford gel scanning attachment.

RESULTS

Structural Difference between isozymes 1 and 3

A report by Piggott et al (1972) indicated that alkaline phosphatase isozymes have different amino acids at the amino-terminal position of the protein. We have recently confirmed their finding--although the amino-terminal residues of one of our isozymes are different from those reported. Our data resulted from an investigation of the complete amino acid sequence of isozyme 3 (see Fig. 2 for numbering) *E. coli* alkaline phosphatase. This work has been carried out in collaboration with Dr. Ralph Bradshaw, Department of Biological Chemistry, Washington University School of Medicine,and is nearing completion. In determining the sequence of the large (Mol. Wt. = 43,000) polypeptide chain, we had first degraded the protein with cyanogen bromide and separated the fragments by gel filtration on Sephadex G-75 columns. We found that the amino-terminal cyanogen bromide fragment of isozyme 3 was a tetrapeptide (Table 1). We subsequently purified isozyme 1 by growing bacteria in a medium containing arginine because the addition of this amino acid leads to a selective accumulation of this isozyme (Piggot et al, 1972; see below). The amino-terminal cyanogen bromide fragment of isozyme 1 differed from that of isozyme 3 by virtue of an extra arginine which

TABLE 1

SEQUENCES OF THE AMINO-TERMINAL CYANOGEN BROMIDE
FRAGMENT FROM ISOZYMES 1 AND 3[a]

Isozyme 1	Arg-thr-pro-glu-hse
Isozyme 3	thr-pro-glu-hse

[a] Determined by Edman degradation (Kelley et al, 1973).

was at the amino-terminal position of the peptide (Table 1).
No significant differences were found in the amino acid compositions of the other cyanogen bromide fragments isolated from
the two isozymes. These fragments consist of a dipeptide, a
heptapeptide that is the carboxyl-terminal portion of the protein (Kelley et al, 1973) and polypeptides containing 37, 40,
45, 77, 81, and 128 amino acids. Single amino acid changes
can not be detected in the larger fragments. The previous
studies with isozymes 1 and 3 (Schlesinger and Andersen, 1968)
failed to show any significant differences between them in a
variety of physical properties such as UV, CD, and ORD spectra,
sedimentation coefficients, etc.

Analyses of the two isozymes for carbohydrates commonly
associated with proteins showed that neither contained significant amounts of N-acetylglucosamine, galactose, mannose,
fucose, or sialic acid (Kelley et al, 1973).

Functional differences between isozyme 1 and 3

The enzymatic activities of these two variants were reported
to be indistinguishable with regard to substrate specificity,
pH optima, temperature coefficients, and turnover numbers
measured at pH value above 7.4 (Schlesinger and Andersen,
1968). However, stopped flow kinetic measurements performed
at pH 5.5 have revealed a significant difference in the steady
state rates of the two isozymes (Table 2; Fig. 1). At this
lower pH, the hydrolysis of substrate proceeds by an initial
rapid transient burst followed by a much slower rate during
the steady state. From this and other kinetic data, it was
proposed that the burst rate measures a conformational change
in the protein during the initial turnover reaction of substrate with enzyme, and the steady state rate measures the
dephosphorylation of the enzyme phosphate complex (Halford et
al, 1969; Halford, 1971; Bloch and Schlesinger, 1974). At pH
values below 7.0, the dephosphorylation is rate-limiting in
the steady state but at pH values greater than 7.0, dephosphorylation is more rapid and the rate-limiting step becomes

TABLE 2

TRANSIENT KINETIC PARAMETERS OF ISOZYMES 1 AND 3 OF
E. COLI ALKALINE PHOSPHATASE

Substrate and Buffer	Parameter[a]	Isozyme 1	Isozyme 3	Ratio: Isozyme 1 / Isozyme 3
2,4 Dinitrophenyl Phosphate in pH 5.5 NaOAc	Total Burst Amplitude (ROH:E)[b]	1.38	1.58	0.87
	Transient Burst Rate Constant (sec^{-1})	22	21	1.1
	Steady-State Rate (moles ROH:mole active sites: sec)[c]	0.78	0.38	2.1
p-Nitrophenyl phosphate in pH 8.0 TrisCl	Steady-State Rate (moles ROH:mole active sites: sec)[c]	11.6	11.5	1.0

[a] Refer to Bloch and Schlesinger (1974) for the original operational definitions of these parameters.

[b] Values contain modest corrections, not used in Bloch and Schlesinger (1974), for the influence of the burst and steady-state rate constants on the burst amplitude. Correction procedure is that of Gutfreund (1965).

[c] Values normalized to the number of active sites, inferred from the Burst Amplitude, rather than to the number of enzyme moleucles, as was done in Bloch and Schlesinger (1974).

a conformational change in the protein.

It is the dephosphorylation rate that appears to be almost twice as fast for isozyme 1 relative to isozyme 3 (Table 2). Possibly the presence of the extra amino-terminal arginine facilitates removal of the phosphate from the catalytic site of isozyme 1.

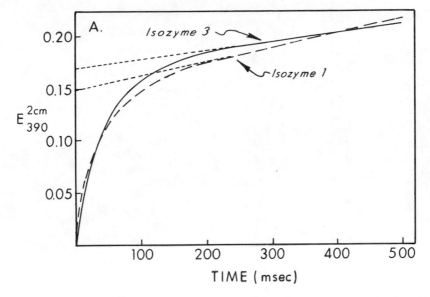

Fig. 1. Stopped-flow kinetic studies with isozymes 1 and 3 at pH 5.5. Consult Bloch and Schlesinger (1974) for experimental details.

Conversion of isozyme 1 to isozyme 3

Data from the previous report on these isozymes (Schlesinger and Andersen, 1968) suggested a precursor-product relationship between isozymes which occurred after maturation of the polypeptide chains into catalytically active dimers. We present additional data here to further support those conclusions. A short 1 min pulse of ^{35}S-methionine to an *E. coli* culture synthesizing alkaline phosphatase labels all three isozymes (Fig. 2A). The label in isozymes 1 and 2 decrease upon a subsequent chase with unlabeled amino acid (Fig. 2C). In one of the "chase" experiments we added inorganic phosphate to repress further synthesis of new alkaline phosphatase so we could follow conversion of the preexisting enzymes. Under these latter conditions, no new enzyme appears (Table 3) but the labeling pattern changes dramatically (Fig. 2B). The interconversion occurs quite rapidly as evidenced by the relatively equal amount of label in all three isozymes in a one minute label. The data presented in Fig. 2 show the conversion of radioactive isozyme 1 into isozymes 2 and 3. Identical patterns of conversion were observed if we assayed for enzymat-

337

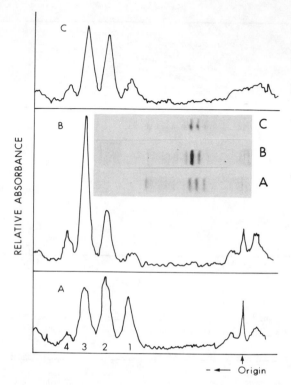

Fig. 2.   Densitometric tracings of ^{35}S-labeled isozymes.   Sample A is the pattern after a 1 min pulse of ^{35}S-methionine; Sample B is the pattern after a 1 min pulse of ^{35}S-methionine followed by addition of unlabeled methionine *and* inorganic phosphate; Sample C is the pattern after a 1 min pulse of ^{35}S-methionine followed by addition of unlabeled methionine.   Samples B and C were harvested 60 min after addition of label. Inset shows autoradiogram of starch gel electropherogram. Details are provided in Table 3 and experimental procedures.

---

ically active alkaline phosphate using a histochemical dye procedure (Schlesinger and Andersen, 1968); thus isozymes 2 and 3 are formed from isozyme 1 after the latter has assumed its catalytically active conformation.

If arginine is added to the culture five minutes prior to the 1 min labeling with ^{35}S-methionine, only isozyme 1 is detected in significant amounts and this form of the protein is not further converted to other isozymes when arginine is present (Fig. 3A and B).   We conclude that the enzyme involved in converting isozyme 1 to isozyme 3 is "product"-inhibited,

TABLE 3

ALKALINE PHOSPHATASE ACTIVITY
IN SAMPLES ANALYZED FOR ISOZYME CONTENT

Experiment	Time of Harvest after addition of ^{35}S-methionine	Enzyme[a] Activity (units/ml)	^{35}S-activity[b] cpm x 10^{-5}/ml
IA	1 min	3.4	2.7
IB	60 min (unlabeled met and Pi added 1 min after ^{35}S-met)	3.7	1.8
IC	60 min (unlabeled met added 1 min after ^{35}S-met)	6.4	1.8
IIA[c]	1 min	3.2	1.0
IIB	60 min (unlabeled met and Pi added 1 min after ^{35}S-met)	2.7	0.9

[a] Measured at pH 8.0 in $\underline{M}$ TrisCl with p-nitrophenyl phosphate as substrate; refer to Bloch and Schlesinger (1974) for details of the assay procedure. Activity was measured on the periplasmic fractions obtained from 10 ml of culture.

[b] Measured as hot trichloroacetic acid insoluble material in the periplasmic fraction. This value was 7 to 9% of the total ^{35}S-met , incorporated into bacterial cell protein in the 1 min label.

[c] In experiment II, 200 µg/ml of arginine was added 5 min before ^{35}S-methionine.

and this fact can account for accumulation of isozyme 1 in cultures containing arginine.

DISCUSSION

The data presented in this paper show that two of the major isozymes of *E. coli* alkaline phosphatase can be distinguished by (1) a difference in the amino acid at the amino-terminal portion of the polypeptide chain and (2) a difference in the rate of substrate hydrolysis at low pH values. The latter result suggests that a particular isozyme could provide a beneficial effect for bacteria able to grow at widely differing pH values. For example, an *E. coli* at pH 6.0 should grow

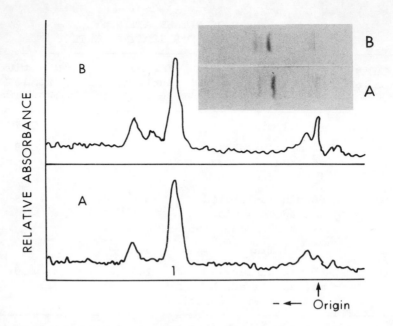

Fig. 3. Densitometric tracings of ^{35}S-labeled isozymes in presence of arginine. Sample A is the pattern after 1 min pulse of ^{35}S-methionine; Sample B is the pattern after 1 min pulse of ^{35}S-methionine followed by addition of unlabeled methionine *and* inorganic phosphate and harvested 60 min later. Details are in Table 3 and experimental procedures.

better with isozyme 1 than isozyme 3 when exposed to an environment deficient in inorganic phosphate.

In the work described here, we have further strengthened our earlier results that showed isozyme 1 is a precursor to isozymes 2 and 3. In addition, the data here suggest that this conversion proceeds via a step-wise proteolysis of the amino-terminal arginine from each of the identical subunits of the active dimer. The conversion may be accompanied by a significant conformational change; the amino-terminal threonine of isozyme 3 appears to be buried since it cannot be detected unless the protein is first denatured. This difference in conformation could account for the difference in steady-state rates of the low pH.

Under other growth conditions, a different array of alkaline phosphatase isozymes can be observed and as many as eight distinct isozymes have been found. We would hypothesize that these variants are the result of proteolytic activity occurring at the carboxyl-terminus of the molecule. We have, in fact, detected some variations in the composition of the 7-residue cyanogen bromide fragment that is the carboxyl-terminus of the protein and contains lysine at the carboxyl-terminal position of isozyme 3.

The protease activity responsible for converting isozyme 1 to 2 and 3 is inhibited by arginine and presumably functions in the periplasmic space where active alkaline phosphatase is localized. Attempts to find such a protease in periplasmic fractions of cultures have been unsuccessful. Partially-purified extracts of whole cells were able to convert the amino-terminal cyanogen bromide fragment of isozyme 1 to the analogous fragment of isozyme 3 by removing arginine but this activity failed to change native isozyme 1. We also tested the mutants described by Bukhari and Zipser (1973) as deficient in a degradation enzyme for their alkaline phosphatase isozyme pattern. The patterns from the mutants were identical to that of the parental wild type strains. Finally, we looked for membrane-bound proteases in bacteria making alkaline phosphatase isozyme 3 but failed to detect any activity that could convert isozyme 1 to 3.

LITERATURE CITED

Bloch, W. and M. J. Schlesinger 1974. Kinetics of substrate hydrolysis by molecular variants of *E. coli* alkaline phosphatase. *J. Biol. Chem.* 249: 1760-1768.

Bukhari, A. I. and D. Zipser 1973. Mutant of *Escherichia coli* with a defect in the degradation of nonsense fragments. *Nature New Biol.* 243: 238-241.

Garen, A. and C. Levinthal 1960. A fine-structure genetic and chemical study of the enzyme alkaline phosphatase of *E. coli*. I. Purification and characterization of alkaline phosphatase. *Biochem. Biophys. Acta* 38: 470-483.

Gutfreund, H. 1965. *An Introduction to the Study of Enzymes.* Blackwell, Oxford, 60-61.

Halford, S. E. 1971. *Escherichia coli* alkaline phosphatase. An analysis of transient kinetics. *Biochem. J.* 125: 319-327.

Halford, S. E., N. G. Bennett, D. R. Trentham, and H. Gutfreund 1969. A substrate-induced conformation change in the reaction of alkaline phosphatase from *Escherichia coli*. *Biochem. J.* 114: 243-251.

Kelley, P. M., P. A. Neumann, K. Shreifer, F. Cancedda, M. J. Schlesinger, and R. A. Bradshaw 1973. Amino acid sequence of *Escherichia coli* alkaline phosphatase. Amino- and carboxyl-terminal sequences and variations between two isozymes. *Biochemistry*. 12: 3499-3503.

Piggott, P. J., M. D. Sklar, and L. Gorini 1972. Ribosomal alterations controlling alkaline phosphatase isozymes in *Escherichia coli*. *J. Bact*. 110: 291-299.

Schlesinger, M. J. 1968.   Secretion of alkaline phosphatase subunits by spheroplasts of *Escherichia coli*. *J. Bact*. 96: 727-733.

Schlesinger, M. J. and L. Andersen 1968. Multiple molecular forms of the alkaline phosphatase of *Escherichia coli*. *Ann. N. Y. Acad. Sci*. 151: Art. 1, 159-170.

Signer, E. 1963. *Non-heritable factors in gene expression*. Ph.D. thesis, Mass. Inst. of Technology.

HORMONALLY INDUCED MODIFICATION OF HELA ALKALINE PHOSPHATASE WITH INCREASED CATALYTIC ACTIVITY

RODY P. COX
NIMAI K. GHOSH
Division of Human Genetics
Departments of Medicine and Pharmacology
New York University Medical Center
New York, N.Y. 10016

and

KIRSTAN BAZZELL
MARTIN J. GRIFFIN
Oklahoma Medical Research Foundation
Oklahoma City, Oklahoma 73100

ABSTRACT. Alkaline phosphatase (AP) activity of HeLa cells is increased 5 to 20 fold during growth in medium with cortisol. The kinetics of induction show a 12 to 20 hr lag period after adding hormone, followed by a rapid linear increase of enzyme activity which reaches a plateau after 60 to 90 hr. RNA and protein synthesis are required for "induction". However, physical and immunological methods show that the amount of AP protein in induced cells is not significantly increased, suggesting that an enhanced catalytic efficiency is responsible for the increase in enzyme activity.

Chemical and physical properties of the base-level and induced form of AP are similar including their molecular weights, isozyme patterns, zinc content and Michaelis-Menten constants (Km). The V_{max} and k_3 of the induced enzyme are much greater and the binding forces holding zinc at the active site are stronger than the base-level, suggesting an altered conformation at the catalytic site.

Comparison of the kinetics of enzyme synthesis to the kinetics of enzyme induction using radioactively labeled enzyme and specific precipitation by antiserum shows the AP is maximally labeled long before the increase of enzyme activity reaches a plateau. HeLa AP is a phosphoprotein and the induced form of the enzyme has approximately one-half of the phosphate residues associated with the base-level AP. Phosphorylation of enzyme apparently regulates AP activity by determining the nature of the binding of zinc at the active site and thereby controls its catalytic efficiency.

INTRODUCTION

The regulation of gene expression implies alteration of either the amount or kind of proteins synthesized by a cell. The spectrum and amount of protein synthesized determines the cell's character and metabolism. Proteins may have structural and functional (i.e. enzymatic) characteristics and the combination of subcellular localizations, time of appearance during development; and microheterogeneity determines the cell's epigenotype. Therefore, the control of gene expression is one of the central problems in biology with relevance to embryonic differentiation, heritable diseases, genetic diversity, and neoplasia.

Cellular, metabolic, and functional control involves regulation of enzyme activity and localization. Alterations in enzyme activity may occur either by altering the amount of enzyme protein or by altering the catalytic efficiency of the enzyme. Increase in the concentration of an enzyme may occur by accelerated synthesis or delayed enzyme degradation, or both. Cell culture provides a relatively simple model for investigating mechanisms of hormonal induction of enzyme activity. In hepatoma cells grown in culture, glucocorticoids have been shown to increase the synthesis and decrease the degradation of tyrosine amino transferase and tryptophan pyrrolase leading to the accumulation of enzyme protein within the cell (Granner, Hayashi, Thompson, and Tomkins, 1968; Johnson and Kenny, 1973). This accounts for the enhanced enzyme activity noted in these cell cultures. Evidence from certain studies suggests that these steroids exert their effect at the level of transcription by stimulating the synthesis of specific mRNA's and also possibly at the level of translation by increasing the rate of protein assembly from these RNA's. However, increased synthesis of enzyme is not the only mechanism whereby steroid hormones may increase the specific activity of an enzyme. Cortisol induction of alkaline phosphatase (AP) (E.C.:3.1.3.1) in HeLa$_{65}$ cells has been shown to be due to an enhanced catalytic efficiency of the enzyme rather than increases in the amount of enzyme protein content per cell (Griffin and Cox, 1966b; Cox, Elson, Tu, Griffin, 1971). The mechanism of this induction appears to have relevance to the control of several metalloenzymes. We will review the evidence for our current interpretation of the control of AP activity in HeLa cells.

Nature of HeLa Cell AP

AP synthesized by HeLa$_{65}$ cells is a membrane-associated fetal protein and is apparently the product of a single genetic locus which in HeLa cells is heterozygous for the F and S allele of placental AP (Elson and Cox, 1969). The highest specific activity of HeLa AP is found in the membranes of the smooth endoplasmic reticulum and plasma membranes, although all membrane fractions appear to have enzyme activity (Tu, Nordquist, and Griffin, 1972). Solubilization of HeLa AP requires homogenization with n-butanol or other biphasic agents. The enzyme differs in many of its chemical and physical properties from adult forms of AP (Cox and Griffin, 1967) but closely resembles the placental or fetal form of the enzyme (Ghosh, Rukenstein, Baltimore, and Cox, 1972). Fig. 1 shows that the electrophoretic pattern of HeLa cell AP

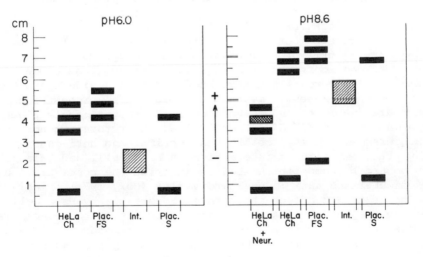

Fig. 1. Diagram of horizontal starch gel electrophoresis stained for AP activity (Kaplow, 1955). Electrophoresis was carried out at pH 6.0 and 8.6 for 5-6 hr at room temperature using the discontinuous buffer systems described by Robson and Harris (1965). HeLa Ch + Neur is HeLa AP treated with *Vibrio cholerae* neuraminidase 250 units/ml at 37°C for 18 hr.

---

has a triple band and resembles that of the placental isozyme of AP having the FS phenotype except that the HeLa enzyme migrates more slowly (Elson and Cox, 1969). In the placenta, this triple pattern is caused by the three possible combinations of a fast migrating monomer (F) and slow migrating monomer (S) uniting into a dimeric enzyme FF, FS, and SS. This triple banded pattern has not been observed with other human AP and constitutes evidence for a close similarity between the HeLa and placental enzymes. This relationship is strengthened by immunological studies. Immunodiffusion of the HeLa cell and placental AP using an antiserum against the HeLa cell enzyme shows a line of near identity at points of fusion (Elson and Cox, 1969). It is noteworthy that our antiserum against HeLa AP does not react with adult forms of the enzyme including human leukocyte and intestinal AP. These results suggest that the genome of HeLa cells is derepressed and these cells are producing a fetal form of AP. Moreover, the evidence supports the concept that HeLa AP is a product of a *single* genetic locus.

Relation of Hormone Structure to Induction of AP

Steroid hormones can be separated into optimal, suboptimal, and non-inducers on the basis of level of AP activity that they induce in HeLa$_{65}$ cells (Melnykovych, 1962; Cox, 1971). Optimal inducers (cortisol and prednisolone) have the 11β-, 17α-, and 21-trihydroxy configuration. Suboptimal inducers (aldosterone, corticosterone and 11β hydroxyprogesterone) lack the 17-hydroxyl but have an 11β- and 17α-side chain that contains a 20-ketone. Certain steroids (i.e. progesterone) that are unable to induce HeLa cell AP, competitively blocks the induction mediated by cortisol and therefore are anti-inducers. It seems probable that optimal and anti-inducer have the same site of interaction in HeLa cells and this is supported by competition studies in which the concentrations of inducer and anti-inducer are varied (Cox, 1971). Anti-inducers do not affect the uptake of cortisol by HeLa cells nor do they affect RNA or protein synthesis at the concentrations studied. This work suggests that steroids initiate an increased AP activity in HeLa$_{65}$ cells by interacting with a putative hormone receptor and the structure of the steroid determines the extent of induction.

AP Induction in Asynchronous Cell Cultures

The kinetics of induction of AP by cortisol or its analogues in asynchronous HeLa cell cultures are complex as shown in Fig. 2. There is a 12 to 20 hr lag phase after adding the steroid before the rapid increase in AP activity. The lag in induction cannot be explained by a delay in the uptake of cortisol since maximum accumulation of hormone is achieved within 30 min (Griffin and Ber, 1969; Cox, 1971). The dura-

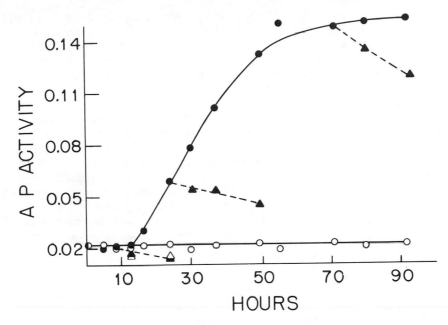

Fig. 2. Kinetics of AP induction in HeLa₆₅ cell suspension
cultures grown in medium with and without cortisol. AP activity is μ moles of p-nitrophenol released/30 min/mg of cell
protein. Control (O——O) cultures grown in Eagles Spinner
medium with 7% fetal calf serum. Induced (●——●) replicate
cultures grown in medium with 3 μg/ml hydrocortisone hemi-
succinate. Cycloheximide 1.0 μg/ml added to control (Δ---Δ)
and induced (▲---▲) cultures.

tion of the lag phase varies directly with the doubling time
of the culture and is followed by a linear increase of AP
activity reaching a level 5 to 20 fold greater than controls
after 50 to 90 hr of growth in medium containing the hormone.
This level of enzyme activity is maintained in the presence of
cortisol. When hormone is removed AP activity decreases over
the following 3 to 4 days due both to enzyme degradation and
dilution by cell multiplication (Griffin and Cox, 1966).

Requirement for RNA and Protein Synthesis

No gross changes in RNA or protein synthesis are observed
in suspension cultures of HeLa₆₅ cells grown in medium with
cortisol (Griffin and Cox, 1966a & b). However, the synthesis
of both of these macromolecules is required for hormone medi-

ated enhancement of AP activity (Griffin and Cox, 1966a & b; Cox, Elson, Tu, and Griffin, 1971). Actinomycin D (0.1 µg/ml) blocks the increase in AP activity when added as late as 8 hr after the hormone (Griffin and Cox, 1966a). However, Actinomycin D does not block induction during the most rapid phase of increase in AP activity, suggesting translation of pre-formed mRNA.

The role of RNA as a necessary intermediate for the induction of increased AP activity was further investigated by blocking protein synthesis in presence of the hormone and studying the accumulation of an intermediate (mRNA?) which can express itself after cycloheximide is removed by washing (Cox, Elson, Tu, and Griffin, 1971). The increase in AP activity occurs rapidly when the inhibition of protein synthesis is relieved without the usual 12 to 20 hr lag. The following observations further support the conclusion that hormone mediated RNA synthesis is required for an increase in AP activity. (1) The hormone must be present during the period of incubation with the inhibitor of protein synthesis since incubation with cycloheximide, alone produces no increase in AP specific activity. (2) Although the hormone must be present during the incubation with cycloheximide, after removal of the inhibitor enhanced AP activity occurs with or without further addition of hormone. (3) Addition of Actinomycin D after removal of cycloheximide from hormone-treated cultures does not prevent the increase in phosphatase activity, and indicates that during the reincubation period RNA synthesis is not required for induction. (4) If Actinomycin D is present with cycloheximide during the initial incubation with the hormone the increase of AP activity is completely inhibited.

The effect of inhibiting protein synthesis at various times during "induction" of AP by cortisol is shown in Fig. 2. Cycloheximide added as late as 7 hr after the addition of hormone completely inhibits the "induction". If protein synthesis is inhibited during the phase of rapid increase in enzyme activity, further increase is blocked and there is a gradual decrease in AP activity.

AP Induction in Synchronous Cultures

The lag or initial slow increase in AP activity might be the result of cells having to pass through a critical event in the cell cycle before becoming "induced". HeLa$_{65}$ cells were collected in mitosis by selective detachment of the rounded dividing cells from the monolayer (Griffin and Ber, 1969). When cortisol is added to synchronized cell cultures in mid G_1, AP induction occurs during the midportion of the

S phase of cell cycle, about 6 to 7 hr after adding hormone.
If however, the hormone is added to synchronous cultures in
late S phase or in G_2 then induction is delayed until the cells
have re-entered the S phase, about 18 to 20 hr after addition
of cortisol. The time of onset of induction, the rate of
increase of AP activity and the magnitude of induction were
similar under both conditions. The above findings cannot be
explained by alterations in cell permeability during the cell
cycle, since radioactive cortisol freely diffuses into cells
and achieves a concentration comparable to the levels in the
medium. Apparently the hormone interacts with the cell to
promote induction only when the cell traverses the S phase of
the cell cycle (Griffin and Ber, 1969; Cox, 1971).

Measurement of the AP content of induced HeLa$_{65}$ cells by immunoprecipitation

Quantitation of the amount of enzyme protein in control and
induced HeLa$_{65}$ cells was carried out by immunoprecipitation
with an antiserum against HeLa AP. The antiserum had the
characteristics described below (Cox, Elson, Tu, and Griffin,
1971). (1) Complete precipitation of HeLa AP was achieved
without neutralizing the enzyme activity in the antigen-
antibody complexes, allowing a high degree of accuracy in
quantitative precipitation reactions, since the enzyme activity removed from the supernatant can be recovered in the precipitate. (2) The antiserum was highly specific as shown by
a single broad band of precipitation on immunodiffusion with
lines of identity at points of fusion of base-level and induced AP. (3) Immunoelectrophoresis resolved three bands of
precipitate with both base-level and induced enzyme, two
major ones near the cathode with AP activity and a minor
anodal band without phosphatase activity that appears to be a
contaminating protein. When base-level and induced preparations labeled with radioactive leucine were used as antigens
in immunoelectrophoresis only 10 to 20% of the total radioactivity was recovered in the anodal contaminant band and the
remainder was recovered in the two other precipitin bands
with AP activity. These findings taken together indicate that
our antiserum is relatively specific.

Fig. 3 shows the titration of AP prepared from control and
prednisolone-treated cultures. The amount of enzyme activity
remaining in the supernatant is plotted and the bar graph
represents the AP activity of the antigen-antibody precipitate
when constant amounts of control and induced AP are reacted
with increasing concentrations of antiserum. With antibody
dilutions of 1 to 8 and less, the enzyme activity was nearly

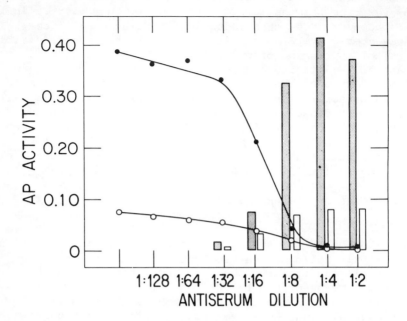

Fig. 3. Precipitation of base-level and induced AP preparations by increasing concentrations of antiserum. The enzyme
preparations were n-butanol extracts of HeLa$_{65}$ cells grown in
the presence or absence of prednisolone (1.5 μg/ml). The enzyme was further purified by gel filtration on Sephadex G-200.
The final purification was approximately 120-fold. Antigen
(0.2 ml containing about 22 μg protein) was incubated with
0.2 ml of diluted antiserum for 2 hr at 37°C. The reaction
mixture was then incubated at 4°C for 24 hr. The precipitate
was removed by centrifugation and the supernate decanted.
The precipitate was washed twice with ice-cold 0.9% NaCl and
then resuspended in 0.4 ml of 0.05 M Tris-HCL, pH 7.4. AP
activity of supernate-control, 0——0, and induced, ●——●.
AP activity of precipitate- control open bars ▯ and induced
solid bars ▮ .

quantitatively recovered in the antigen-antibody precipitate.
Both the base-level and induced preparations are completely
precipitated at the same antibody concentration (1 to 4)
despite a five fold difference in phosphatase activity. This
provides strong evidence that the enzyme protein content of
cells grown in the presence and absence of prednisolone are
the same despite the marked differences in AP activity. These

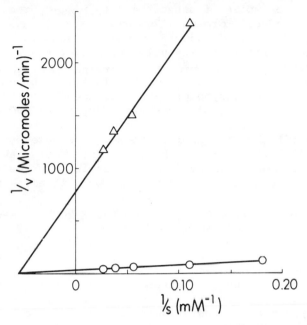

Fig. 4. Lineweaver-Burk plots for Km and V of base-level (Δ) and induced (O) HeLa_{65} AP. Both enzyme preparations were purified approximately 250 fold. The initial velocity (v) of the enzyme reaction is expressed in μ moles phenol released per min from phenylphosphate in 0.05 M carbonate-bicarbonate buffer (pH 10.7). s is the molar concentration of substrate.

findings further suggest that an increased catalytic efficiency of the induced enzyme molecules is responsible for the induction.

Comparison of kinetic properties of base-level and induced HeLa cell AP

An explanation for the increased catalytic activity of the induced form of HeLa AP was sought by comparing the kinetic properties of the base-level and induced enzymes (Ghosh, Rukenstein, Baltimore, and Cox, 1972). Fig. 4 shows the Lineweaver-Burk plots for base-level and induced HeLa cell AP both purified 250 fold. Michaelis constants (Km) of both the base-level and induced APs were 18 mM for the substrate phenylphosphate in 0.05 M carbonate-bicarbonate buffer at pH 10.7. However, the maximal velocities (V) for the base-level and

induced APs were markedly different. The base-level enzyme
had a V of 1.3 nmoles of substrate hydrolyzed per minute
whereas the induced form was 27.0 nmoles per minute. Since
changes in affinity of induced enzyme for substrate cannot
explain the increase in catalytic efficiency the first order
rate constant k_3 was measured to determine the rate of decomposition of enzyme-substrate complex (ES) into products.
Using the purified enzyme preparations and phenylphosphate as
substrate, the mean k_3 of base-level enzyme was 150.0×10^{-6}
and induced was 963.2×10^{-6} determined by the method of
Veibel and Lillelund as applied to β-glucosidase by Nath and
Rydon (1954). The increased k_3 value for induced AP suggests
that the O-P bond in induced ES complex is more easily cleaved
by nucleophilic attack by hydroxyl ions in alkaline medium.

Comparison of physical and chemical properties of base-level and induced AP

Immunological evidence clearly shows that the hormone-
mediated induction of increased AP activity in HeLa$_{65}$ cells
is *not* accompanied by an increase in the enzyme content of
induced cells but is due to an enhanced catalytic efficiency.
This is supported by studies on purification to apparent
homogeneity of AP from control and induced cells. Table 1
shows that the amount of enzyme protein in control and induced
cells is nearly the same and pI are also similar. The slight
increase in protein concentration and enzyme molecules in
induced cells may be the result of an increase in average cell
size mediated by the hormone (Cox and MacLeod, 1962). Calculation of the number of AP molecules per cell, based on a
molecular weight of 120,000 daltons, is also similar. Figs.
5a and b show the isoelectric focusing profiles of butanol
extracts of HeLa$_{65}$ control and induced cells respectively.
The enzyme activity peaks are symmetrical and are a small
percentage of the total applied protein. Enzyme preparations
purified this way or from preparative disc gel electrophoresis
gave a single protein band when analyzed by gel electrophoresis.

Gel filtration on Sephadex G-200 resolved both the base-
level and induced forms of HeLa AP into two similar peaks
(Cox, Elson, Tu, and Griffin, 1971). The first peak immediately after the void volume is probably enzyme tetramer and
the second peak is apparently dimer. The sedimentation
coefficients by sucrose gradient centrifugation of the first
AP peaks from the Sephadex G-200 column had an average value
of $S_{20W}^{0.725}$ 10.8×10^{-13} sec and the second was $S_{20W}^{0.725}$ $6.6 \times$

TABLE 1

PURIFICATION OF BASE-LEVEL AND INDUCED HELA AP BY ISOELECTRIC FOCUSING (IEF)

AP preparation	Protein conc. µg/10⁶ cells		pI^b	AP molecules (Millions) per cell[c]
	initial	final (IEF)[a]		
Base-level (4)	510	3.6 ± 0.4	4.2 ± 0.2	18
Induced (4)	750	4.4 ± 0.4	4.4 ± 0.2	24

[a] The final AP preparation was homogeneous as judged by a symmetrical peak on isoelectric focusing and a single protein band on disc gel electrophoresis. The fold purification could not be determined because AP is partially inactivated (70-80%) at the pI of 4.3.

[b] Average of three separate determinations.

[c] Calculations of enzyme molecules per cell are based on a molecular weight of 120,000 daltons.

353

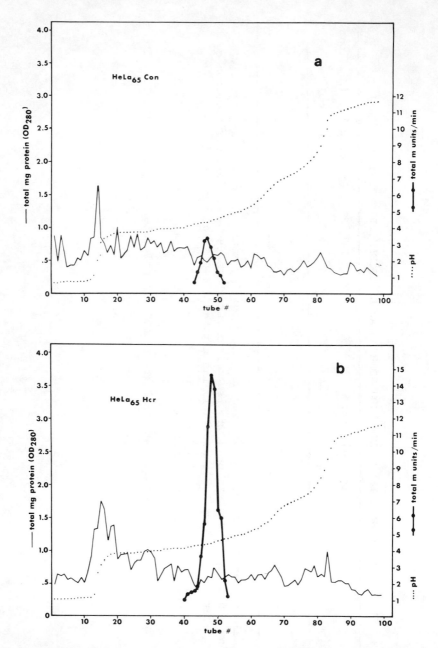

Fig. 5. Isoelectric focusing profile of (a) control (con) and (b) hydrocortisone induced (Hcr) HeLa$_{65}$ cell extracts prepared by homogenization with n-butanol. AP activity (●——●), protein concentration (——), and pH (····) are shown.

10^{-13} sec. The approximate molecular weights for these two forms of the enzyme are 240,000 ± 10,000 daltons and 120,000 ± 5,000 daltons respectively. The base-level and induced forms of the enzyme are apparently identical with respect to the above determinations and closely resemble placental AP.

The electrophoretic mobilities of AP prepared from control and induced cells have been studied by starch gel and disc gel electrophoresis (Cox and Griffin, 1967; Griffin and Bottomley, 1969; Ghosh, Rukenstein, Baltimore, and Cox, 1972). Both forms of the enzyme are resolved into identical fast and slow migrating bands with the faster band having greater activity. The mobilities of these bands in starch are 0.20 and 0.32 relative to albumin as a standard. A third minor band near the origin may be either enzyme aggregates or association of the enzyme with lipid or other macromolecules.

The identical electrophoretic mobilities and isozyme pattern of base-level and induced enzyme support the concept that they are the products of the same genetic locus. Moreover, when equal amounts of base-level and induced enzyme protein are applied to the gel, the induced form is much more enzymatically active when stained by histochemical methods showing there is no electrophoretically separable modulatory molecule present in either enzyme preparation.

Zinc content and binding forces of base-level and induced AP

AP is a metalloenzyme which requires zinc at the catalytic site for enzyme activity. The possibility that differences in catalytic efficiency might be the result of differences in the zinc content of base-level and induced AP was studied by further purifying the dimeric peaks from the Sephadex G-200 column by preparative gel electrophoresis. Zinc content was assayed by atomic adsorption spectroscopy. The base-level enzyme contained an average of 4.8 µg of zinc per mg of protein and the induced had 3.7 µg of zinc. The agreement between two different purifications was well within experimental error and showed the control and induced forms of AP contain similar amounts of zinc. These results suggest that differences in catalytic activity between these metalloproteins is not due to differences in the number of zinc atoms at the active site (Cox, Elson, Tu, and Griffin, 1971).

The binding of zinc to control and induced apoenzyme was studied by comparing the kinetics of enzyme inactivation by various concentrations of EDTA (Cox, Elson, Tu, and Griffin, 1971). The initial inactivation of control enzyme by EDTA is consistently much greater than observed with induced AP prep-

arations.   One min after adding 1 mM EDTA to control enzyme
approximately 40% of its activity was lost while induced enzyme lost only 15 to 20% under the same conditions.   After 1
hr incubation with EDTA the extent of inactivation of both
enzyme preparations was similar.   It should be noted that the
enzyme preparations were purified by column chromatography and
contained similar amounts of zinc and protein.   As time of
incubation with EDTA increases, the differences between inactivation of control and induced enzyme become less.   These
findings suggest that the zinc binding forces in the active
site of the control enzyme may be weaker than those for the
induced.   Formation of alternate bonds between apoenzyme and
zinc in the induced cells might lower the energy requirements of the enzyme substrate transition state and thereby
increase the catalytic efficiency (Cox, Elson, Tu, and
Griffin, 1971).   Vallee and Williams (1968) have described
alterations in the binding of metals to metalloenzymes that
change the catalytic activity of the enzyme by altering the
conformation of the catalytic site (so-called entatic effects).

Comparison of the kinetics of AP induction with the kinetics of AP synthesis

Several hypotheses can be advanced to explain the requirement for RNA and protein synthesis during induction of an
increased catalytic efficiency of AP without an increase in
the enzyme protein content.   One possibility is that the hormone in some way directly alters the conformation of AP during its synthesis.   If this hypothesis is correct one might
expect to see a direct correlation between the kinetics of
enzyme synthesis and the induction of increased enzyme activity.   Using radioactive labeling one can detect the accumulation of the induced form of the enzyme as it replaced the
base-level AP.   Alternatively the hormone may act indirectly
by interacting with modifiers which alter AP during its synthesis or by inducing the synthesis of modifiers which interact with AP to enhance its activity.   Under these latter cir-
cumstances there may be a dissociation between rates of AP
synthesis and the kinetics of enzyme induction.   Fig. 6 presents the results of an experiment designed to study these two
possibilities.   Immunological precipitation of radioactive AP
was used to compare the rates of increase of enzyme activity
with the kinetics of enzyme synthesis.   Fig. 6 (a) shows the
specific activity of the enzyme-antibody precipitate at various
times after adding prednisolone to the culture.   Maximum increase in enzyme activity was achieved by 96 hr.   Fig. 6 (b)
shows the radioactivity per mg of enzyme-antibody precipitate

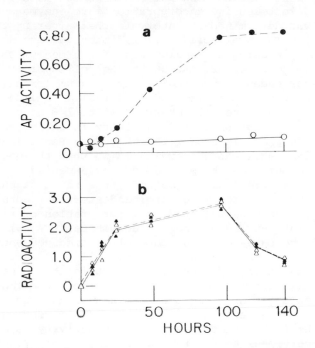

Fig. 6. Precipitation of radioactive AP by antiserum during induction of increased enzyme activity by prednisolone. (a) AP specific activity - μ mole of p-nitrophenol released/ min/mg of protein in enzyme-antibody precipitate. Control O——O and induced ●——●. (b) specific radioactivity (cts/min x 10^{-3}/mg of protein) in enzyme-antibody precipitate. Control Δ——Δ and induced ▲----▲.

HeLa cells were grown in monolayer culture in Weymouth's medium containing 10% calf serum. At zero time, 100 nc of [1-^{14}C] leucine/ml was added to all cultures and 1.5 μg prednisolone/ml was added to the steroid-treated cultures. Thereafter, cultures were harvested at the times indicated. After 48 hr, the medium was decanted and replaced with fresh medium containing radioactive leucine. At 96 hr, the radio-active medium was decanted from the remaining cultures and the cell monolayers were subcultured into nonradioactive medium. Prednisolone was readded to the hormone-treated cultures. AP preparations were extracted from cells and reacted with antiserum as described in the legend to Fig. 3.

during induction. The hormone and radioisotope were added at
time 0, and direct leucine incorporation into enzyme-antibody
precipitate was determined at intervals thereafter up to 96 hr
(lower panel). The availability of radioactive leucine was
maintained by replacing the medium with fresh medium containing leucine-1-^{14}C at 48 hr. At 96 hr the cells were subcultured and resuspended in non-radioactive medium (prednisolone
was readded to the hormone-treated cultures). Replicate cultures were then harvested at intervals up to 140 hr to estimate the rate of dilution of radioactivity in the enzyme as
detected by the precipitation with antibody. Reutilization
of incorporated label was not evaluated so that the decay
portion of the curve may be prolonged. As expected from previous results by more indirect methods, the rates of enzyme
synthesis and degradation in control and steroid-treated cultures are the same when measured by incorporation of radioactive label into enzyme-antibody precipitate (Fig. 6). The
half-life of the base-level and that of the induced enzyme
also appear to be the same, approximately 16 to 20 hr. An
unexpected observation of unusual interest is that a near
plateau in radioactivity in the enzyme-antibody complex is
achieved at about 30 hr, long before maximal induction of
alkaline phosphatase activity is reached. This finding suggests that the rate of increase in enzyme activity is *not*
directly proportional to the rate of synthesis of AP. The
discordance between the kinetics of AP synthesis and induction
of increased AP activity suggests that the modification of
the enzyme responsible for its enhanced catalytic efficiency
is complex.

Regulation of phosphorylation of HeLa AP as the mechanism of induction

Enzyme-mediated chemical modifications of other enzymes are
known to alter their substrate specificity or catalytic
activity. For example, deadenylation of *E coli* glutamine synthetase increases its catalytic activity (Brown, Segal, and
Stadtman, 1971) and phosphorylation or dephosphorylation of
glycogen phosphorylase alters its activity (Robison, Butcher,
and Sutherland, 1971). The hypothetical modification induced
by cortisol, as suggested by our experimental results, might
enhance the catalytic activity of HeLa cell AP by a chemical
modification of the enzyme. This possibility was studied by
measuring the phosphate content of highly purified base-level
and induced forms of AP, both of which were found to be phosphoprotein. As shown in Table 2, nine different AP preparations, two prepared by isoelectric focusing and seven purified

TABLE 2

PHOSPHATE CONTENT OF BASE-LEVEL AND INDUCED HELA$_{65}$ AP

Enzyme prepared by	Number of preparations	Fold purification[a]	Moles inorganic Pi/mole AP[b] base-level	induced
Isoelectric focusing	2	≈150	5.0 ± 0.2	2.0 ± 0.3
Preparative acrylamide gel electrophoresis	7	≈400	6.5 ± 0.2	3.3 ± 0.3

[a] Fold purification is based on protein determinations rather than AP activity since both isoelectric focusing and preparative acrylamide gel electrophoresis produce partial inactivation of enzyme activity. However, both preparations gave a single band on acrylamide gel electrophoresis.

[b] Phosphate was determined after pyrolysis of the enzyme preparation in 2.7 N perchloric acid in teflon lined tubes at 135°C for 15 hr. Inorganic phosphate was measured by reducing phosphomolybdate with ascorbic acid.

by preparative acrylamide gel electrophoresis were analyzed
for tightly bound inorganic phosphate. The base-level had
between 5 and 6.5 inorganic phosphate residues per enzyme
molecule while the induced had only between 2.7 and 3.3. The
"fold purification" shown in Table 2 is based on protein content of the enzyme preparations since enzyme activity was
partially lost during purification. The fold purification
calculations represent minimal values. The enzyme preparations used for phosphate determination were apparently homogeneous as shown by a symmetrical peak on isoelectric focusing and a single protein band on analytical acrylamide disc
gel electrophoresis. The difference in amount of tightly
bound inorganic phosphates is the most striking difference
between the physical properties of base-level and induced
forms of AP. For example, these enzyme preparations gave
similar amounts of hexosamine (8 ± 1), sialic acid (1 ± 0.5),
and free-sulfhydryl (4 ± 1) molecules per enzyme of 120,000
daltons. The simplest assumption is that the degree of phosphorylation of the enzyme regulates its AP activity presumably
by altering the binding of zinc at the catalytic site and
thereby enhancing its catalytic efficiency by an entatic
effect. It is noteworthy that in a recently completed study
the level of cAMP was not increased in HeLa cells during 72 hr
of growth in medium with cortisol, nor will this cyclic nucleotide induce an increase in AP activity (Griffin, Price,
Bazzell, Cox, and Ghosh, unpublished).

Several models for induction of increased AP activity in
HeLa cells can be proposed. The models must be compatible
with the following observations:

1) Three agents have been reported to induce an increase in
AP activity in HeLa cells. In addition to cortisol they are
sodium butyrate (Griffin, Price, Bazzell, Cox, and Ghosh,
1974) and increased medium osmolarity (Nitowsky, Herz,
and Geller, 1963). One possible common site of interaction of
these diverse inducers is the cell membrane.

2) HeLa AP is a membrane bound enzyme and only growing
cells in which new membrane synthesis occurs can be induced.

3) Several lines of evidence indicate that HeLa AP is a
stable enzyme and it takes several days of growth in the presence of inducers to achieve a maximum increase in enzyme
activity. Therefore, once the enzyme is integrated into membranes its intrinsic AP activity apparently cannot be changed.

4) Inhibition of protein synthesis during induction results
in a prompt cessation in the increase of AP activity, suggesting that only newly synthesized AP can be modulated by phosphorylation or dephosphorylation.

5) The requirements for RNA synthesis and for cell cycle

events as necessary prerequisites to induction suggest that
a mRNA required for mediating the increase in AP synthesis
may be synthesized only at a specific point of the cell cycle,
for example mid S. This might account for the observed requirement of cells to pass through DNA synthesis in the presence of hormone before manifesting the induced phenotype.

6) The completion of induction of increased AP activity
appears to be temporally independent of the synthesis of the
enzyme.

7) The catalytic activity of HeLa AP is apparently regulated by the degree of phosphorylation of enzyme protein and
this may affect the bond between protein ligand and zinc at
the active site. The induced form of HeLa AP is therefore
a metazyme as defined by Horecker (1974).

One possible model is that cortisol, and perhaps the other
inducers, alters the synthesis of membrane sites or membrane
subunits necessary for integration of AP into the membrane
mosaic. The altered membranes could favor incorporation of
AP with reduced phosphorylation. Alternatively, cortisol or
a cortisol-induced mediator may inhibit or repress a protein
kinase which phosphorylates AP or may stimulate a phosphoprotein phosphatase that hydrolyzes phosphate groups on the
enzyme after they are added. In either case there is
restricted time for phosphorylation or dephosphorylation,
suggesting that modification of AP preferentially occurs during enzyme synthesis or before its insertion into the membrane mosaic. Obviously, the above models are highly spec-
ulative and much more work is required to differentiate between them, or to develop alternate models to account for
the induction.

DISCUSSION

The production of increased enzyme activity by a mechanism
which increases the catalytic activity of an enzyme is also
found with APs derived from mammalian tissues other than HeLa
cells. Bottomley, Lovig, Holt, and Griffin (1969) have studied the marked alteration in human leukocyte AP activity that
occurs both in reactive granulocytosis and following administration of adrenal glucocorticoid hormones. They found that
normal leukocytes and reactive granulocytosis cells contain
similar amounts of AP protein despite marked differences in
the catalytic activity of the enzyme preparations. Etzler
and Moog (1966) have reported that during differentiation of
embryonic mouse intestine an enzymatically inactive protein
which immunologically reacts with a specific antibody against
intestinal AP is converted to the active enzyme. This activ-

ation, which results in a fifteen-fold increase in activity, normally occurs under the influence of endogenous corticoid hormones secreted by the developing adrenal but can also be made to occur prematurely by administering steroid hormones.

The regulation of the activity of a number of other metalloenzymes may involve mechanisms somewhat similar to those described for AP in HeLa cells. That is, the increase in enzyme activity mediated by inducers may cause modifications of the enzyme which alter the state of the metal at the active site, causing an enhanced catalytic activity without increasing the number of enzyme molecules. For example, the ten to forty-fold increase in aryl hydrocarbon hydroxylase activity induced by benz(a)anthracene in mice requires both RNA and protein synthesis; however, the amount of iron containing protein in the p450 complex is only modestly increased (Nebert and Bausserman, 1971). Current evidence suggests that the inducer interacts in some way with a genetic locus (Ah) causing the synthesis of induction-specific RNA and protein. The specific protein then interacts either in the cytoplasm or in the new membranes with newly synthesized heme or hemoprotein or with the aromatic hydrocarbon or both, such that the p450 enzyme complex is altered favoring a high-spin configuration of iron at the active site. (Nebert, Considine, and Kon, 1973).

Another interesting example of a well-studied increase in the catalytic efficiency of a metal-containing enzyme is the alcohol dehydrogenase in the scutellum of corn. This protein which in certain strains of corn exhibits either a high or a low catalytic activity. The level of enzyme protein is similar despite marked differences in alcohol dehydrogenase activity. Genetic experiments show that with one of the alleles at the structural gene locus for alcohol dehydrogenase, the difference in catalytic efficiency is determined by the product of a modifier locus which is located seventeen crossover units distant from the structural gene (Efron, 1970). Genetic variation in the catalytic efficiency of carbonic anhydrase (Tashian, 1969) and catalase (Ganschow and Schimke, 1969) also have been described. With these metalloenzymes the increase in catalytic efficiency is apparently the result of mutations in the structural gene which confers a higher specific activity.

The control of the activity of certain metalloenzymes appears to be mediated by protein modifiers synthesized in response to inducers. Thus, the catalytic activity may undergo marked changes without a change in the number of enzyme molecules. This situation contrasts with the regulation of a number of well-studied nonmetalloenzymes, where hormonal induction of an increase in enzyme activity is due to a proportionate increase in the enzyme protein content of cells.

ACKNOWLEDGMENT

Supported by research grants from the NIH GM 15508 and CA 10614.

REFERENCES

Bottomley, R. H., C. A. Lovig, R. Holt, and M. J. Griffin 1969. Comparison of alkaline phosphatase from human normal and leukemic leukocytes. *Cancer Res*. 29: 1866-1874.

Brown, M. S., A. Segal, and E. R. Stadtman 1971. Modulation of glutamine synthetase adenylation and deadenylylation is mediated by metabolic transformation of the P_{11}-regulatory protein. *Proc. Natl. Acad. Sci.* 68: 2949-2953.

Cox, R. P. 1971. Early events in hormonal induction of alkaline phosphatase in human cell cultures: Hormone-cell interactions and its dependence on DNA replication. *Ann. N.Y. Acad. Sci.* 179: 596-610.

Cox, R. P., N. A. Elson, S. H. Tu, and M. J. Griffin 1971. Hormonal induction of alkaline phosphatase activity by an increase in catalytic efficiency of the enzyme. *J. Mol. Biol.* 58: 197-215.

Cox, R. P., and M. J. Griffin 1967. Alkaline phosphatase I. Comparison of the physical and chemical properties of enzyme preparations from mammalian cell cultures, various animal tissues and *Escherichia coli*. *Arch. Biochem. Biophys.* 122: 552-562.

Cox, R. P., and C. M. MacLeod 1961. Alkaline phosphatase content and the effects of prednisolone on mammalian cell cultures. *J. Gen. Physiol.* 45: 439-485.

Efron, Y. 1970. Alcohol dehydrogenase in maize: Genetic control of enzyme activity. *Science*. 170: 751-753.

Elson, N. A., and R. P. Cox 1969. Production of fetal-like alkaline phosphatase by HeLa cells. *Biochem. Genet.* 3: 549-561.

Etzler, M. E., and F. Moog 1966. Inactive alkaline phosphatase in duodenum of nursling mouse: Immunologic evidence. *Science* 154: 1037-1038.

Ganschow, R. E., and R. T. Schimke 1969. Independent genetic control of the catalytic activity and the rate of degradation of catalase in mice. *J. Biol. Chem.* 244: 4649-4658.

Ghosh, N. K., A. Rukenstein, R. Baltimore, and R. P. Cox 1972.

Studies on hormonal induction of alkaline phosphatase in HeLa cell cultures. Kinetic, thermodynamic and electrophoretic properties of induced and base-level enzymes. *Biochim. Biophys. Acta* 286: 175-185.

Granner, D. K., S. Hayashi, E. B. Thompson, and G. M. Tomkins 1968. Stimulation of tyrosine aminotransferase synthesis by dexamethasone phosphate in cell culture. *J. Mol. Biol.* 35: 291-301.

Griffin, M. J. and R. Ber 1969. Cell cycle events in hydrocortisone regulation of alkaline phosphatase in HeLa S_3 cells. *J. Cell. Biol.* 40: 297-304.

Griffin, M. J., and R. H. Bottomley 1969. Regulation of alkaline phosphatase in HeLa clones of differing modal chromosome number. *Ann. N.Y. Acad. Sci.* 166: 417-432.

Griffin, M. J., and R. P. Cox 1966a. Studies on the mechanism of hormonal induction of alkaline phosphatase in human cell culture I. Effects of puromycin and Actinomycin D *J. Cell. Biol.* 29: 1-9.

Griffin, M. J., and R. P. Cox 1966b. Studies on the mechanism of hormonal induction of alkaline phosphatase in human cell cultures II. Rate of enzyme synthesis and properties of induced and base-level enzymes. *Proc. Natl. Acad. Sci.* 56: 946-953.

Griffin, M. J., G. Price, K. Bazzell, R. P. Cox, and N. K. Ghosh 1974. A study of adenosine 3':5'-cyclic monophosphate, sodium butyrate and cortisol as inducers of HeLa alkaline phosphatase. *Arch. Biochem. Biophys.* in press.

Horecker, B. L. 1974. Biochemistry of isozymes. *I. Isoyzmes: Molecular Structure* C. L. Markert, ed., Academic Press,N.

Johnson, R. W., and F. T. Kenney 1973. Regulation of tyrosine aminotransferase in rat liver XI. Studies on the relationship of enzyme stability to enzyme turnover in cultured hepatoma cells. *J. Biol. Chem.* 248: 4528-4531.

Kaplow, L. S. 1955. A histochemical procedure for localizing and evaluating leukocyte alkaline phosphatase activity in smears of blood and marrow. *Blood* 10: 1023-1029.

Melnykovych, G. 1962. Effects of corticosteroids in the formulation of alkaline phosphatase in HeLa cells. *Biochem. Biophys. Res. Commun.* 8: 81-86.

Nath, R. L., and H. N. Rydon 1954. The influence of structure on the hydrolysis of substituted phenyl-β-D-glucosides by emulsion. *Biochem. J.* 57: 1-10.

Nebert, D. W., and L. L. Bausserman 1971. Aryl hydrocarbon hydroxylase induction in cell culture as a function of gene expression. *Ann. N. Y. Acad. Sci.* 179: 561-579.

Nebert, D. W., N. Considine, and H. Kon 1973. Genetic differences in cytochrome P-450 during induction of monooxygenase activities. *Drug Metab. and Disposition.* 1: 231-238.

Nitowsky, H. M., F. Herz, and S. Geller 1963. Induction of alkaline phosphatase in dispersed cell cultures by changes in osmolarity. *Biochem. Biophys. Res. Commun.* 12: 293-299.

Robison, G. A., R. W. Butcher, and E. W. Sutherland 1971. in *Cyclic AMP* (Robison, Butcher, and Sutherland, eds). Academic Press, New York, 17-47.

Robson, E. B., and H. Harris 1965. Genetics of alkaline phosphatase polymorphism of human placenta. *Nature* 207: 1257-1259.

Tashian, R. E. 1969. Discussion in *CO2 Chemical, Biochemical, and Physiological Aspects* (Forester, Edsall, Otis, and Roughton, eds). NSAA, Sp. 188, Washington, D.C., 127-129.

Tu, S-H., R. E. Nordquist, and M. J. Griffin 1972. Membrane changes in HeLa cells grown with cortisol. *Biochim. Biophys. Acta* 290: 92-109.

Vallee, B. L., and R. J. P. Williams 1968. Metalloenzymes: The entatic nature of their active sites. *Proc. Natl. Acad. Sci. U.S.A.* 59: 498-505.

We appreciate permission from the publishers to reproduce Fig. 1 from *Biochem. Genet.* 3: 549, 1969; Fig. 2, 3, and 6 from *J. Mol. Biol.* 58: 197, 1971; and Fig. 4 from *Biochim. Biophys. Acta* 286: 175, 1972.

GENETIC AND NON-GENETIC VARIATION OF HUMAN ACID PHOSPHATASES

G. F. SENSABAUGH

Forensic Science Program
School of Criminology
University of California
Berkeley, Calif. 94720

ABSTRACT. Gel filtration studies on the acid phosphatases in human tissues show four distinct classes of activity that can be distinguished on the basis of molecular weight. Each of these classes can be further delineated and subdivided by electrophoresis and by response to several phosphatase inhibitors.

It has been found that a low molecular weight acid phosphatase formerly thought to be found only in red cells is in fact found in all tissues. This enzyme warrants further study, for it is genetically polymorphic and one expression of this polymorphism is large variation in the specific activity of the enzyme. It was found, in accord with other published reports, that the red cell enzyme has an affinity for flavin mononucleotide as a substrate. It was also found that the red cell enzyme was activated by adenine and adenine analogs, but only in the presence of little or no phosphate.

The existence of extensive variation among the acid phosphatases raised a number of questions regarding the evolution, the structural relationships, the catalytic mechanisms, and the biological functions of this group of enzymes.

Acid phosphatase activity is found in the cells and secretions of virtually all organisms, from *Escherischia coli* to *Homo sapiens* (Hollander, 1971). Work in many laboratories has demonstrated that the acid phosphatase activity in the tissues and body fluids of higher organisms is distributed among several distinct enzymes. The isozymes of acid phosphatase are differentially distributed between the cytoplasm and organelles within the cell. They are also differentially distributed between tissues. The differentiation of the several types of acid phosphatases has proven of practical value in clinical medicine (Latner and Skillen, 1968) and in forensic science (Adams and Wraxall, 1974).

In this report, some of the types of variation found in the acid phosphatases of human tissues will be illustrated.

This variation will be correlated with that which has been observed in other studies of animal acid phosphatases. In the course of this presentation, evidence will be presented that the low molecular weight acid phosphatase associated with red cells is present in significant proportion in other tissues as well.

MOLECULAR WEIGHT DISTRIBUTION OF ACID PHOSPHATASES IN HUMAN TISSUES

It has been shown that the acid phosphatases in human placenta (DiPietro and Zengerle, 1967), bovine liver (Heinrickson, 1969), and *Xenopus* tail (Filburn, 1973) can be partitioned by gel filtration into several fractions. Analysis by gel filtration of extracts of human tissues and body fluids -- liver, kidney, placenta, brain, prostate gland, red cells, and seminal plasma -- shows that four classes of acid phosphatase activity can be distinguished on the basis of molecular weight (Fig. 1).

A high molecular weight class elutes at or near the void volume of Sephadex G-100 and hence the acid phosphatases in this class are characterized by molecular weights of 200,000 daltons or more; it is present in all the tissues investigated but is absent in red cells and seminal plasma. This class may represent aggregates of lower molecular weight enzymes; further studies will be required to clarify this question.

The second class of acid phosphatase activity contains enzymes in the molecular weight range 100,000 - 130,000 daltons; this class of enzymes is represented in all tissues and in seminal plasma but not in red cells. Lysosomal acid phosphatase is known to have a molecular weight of about 100,000 and hence falls in this class (Brightwell and Tappel, 1968). Several distinct isozymes of acid phosphatase in this class can be separated by electrophoresis (see below); Swallow and Harris (1973) have found by electrophoresis three acid phosphatases of this class in human placenta with molecular weights of 120,000, 106,000, and 95,000. Thus the enzymes in this class are by no means uniform in their molecular properties although there does appear to be some uniformity in enzymatic properties (see below)

The third class of acid phosphatase activity is characterized by a molecular weight range of 30-60 thousand daltons. This class did not appear to be represented in all the tissues and fluids investigated. The enzymes in this class were activated by Mg^{++}. The extent of Mg^{++} activation varied considerably suggesting that there are several acid phosphatases in this molecular weight range.

A low molecular weight acid phosphatase is found in substantial proportion in all the tissues and fluids investigated except prostate and semen. The enzymes in this class have a molecular weight of 13,000-18,000 daltons. The principle acid

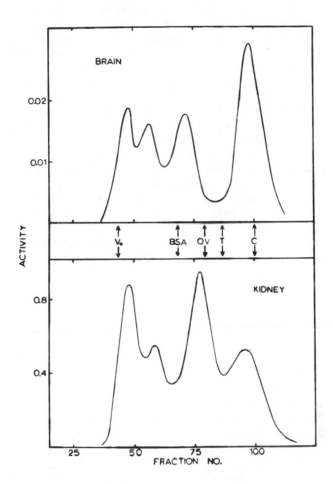

Figure 1. Gel filtration elution profile of acid phosphatase
activity in human brain and kidney. Tissues were homogenized
in 0.1 M potassium phosphate buffer pH 6.0 containing 0.25 M
sucrose and 0.1% Triton X-100. The homogenate was centrifuged
for 15 minutes at $30,000_{xg}$, 4°C, and 5 ml supernatant solution
was applied to a Sephadex G-100 column (3.5 x 55 cm) equilibrated with 0.1 M potassium phosphate buffer pH 6.0 containing
0.01% Triton X-100. The flow rate was about 20 ml/hr. The
column was calibrated with bovine serum albumin (BSA), ovalbumin (OV), trypsin (T) and cytochrome C (C); v_o signifies
the void volume as determined with blue dextran.

phosphatase in red cells is in this class; Fisher and Harris
(1971) have shown its molecular weight to be 14,800. Swallow,
Povey, and Harris (1973) have provided electrophoretic evidence

that the low molecular weight phosphatase found in human tissues is indeed the red cell enzyme; there is indirect evidence
on this question from other sources. Heinrickson (1969) and
Filburn (1973) have described phosphatases in this low molecular weight class in bovine liver and in *Xenopus* tadpole tail
respectively. These enzymes, unlike many other acid phosphatases, appear to have a restricted substrate specificity, preferring flavin mononucleotide (FMN) to any other natural substrate. This preference for FMN is also characteristic of
red cell acid phosphatase (Luffman and Harris, 1967). DiPietro
and Zengerle (1967) describe a low molecular weight acid phosphatase, "acid phosphatase III" which is characterized by an
activation of its p-nitrophenyl phosphatase activity by certain
purines such as 6-methyl adenine. An enzyme with this property
elutes as the low molecular weight phosphatase suggesting their
assignment of a 39,000 molecular weight may be in error. Further, all the low molecular weight acid phosphatases, including
the red cell enzyme, show similar activation by 6 methyl adenine
(see below), suggesting uniformity within the class and identity
of the low molecular weight tissue phosphatases with the red
cell acid phosphatase.

SPECIFIC INHIBITORS OF THE HUMAN ACID PHOSPHATASES

Characterization of the acid phosphatase activity in each
of the four molecular weight classes with respect to sensitivity to various inhibitors further delineates the four classes.
The acid phosphatase activity in the two high molecular weight
classes is nearly completely inhibited by both tartrate (10 mM)
and fluoride (10mM) but is unaffected by formaldehyde (1%).
In contrast, the low molecular weight red cell enzyme is rapidly and apparently irreversibly inhibited by formaldehyde but
not affected by either tartrate or fluoride. The Mg^{++} activated 30-60,000 dalton class shows no consistent pattern of inhibition with these three inhibitors.

These findings correlate very nicely with many earlier
studies on the tissue acid phosphatases which utilized substrate specificity and inhibitor sensitivity to discriminate
different forms of enzyme; these studies have shown a dichoto-
mous relationship between the acid phosphatases inhibited by
tartrate and fluoride and those inhibited by formaldehyde.
For example, Abul-Fadl and King (1949) used the tartrate-for-
maldehyde dichotomy to distinguish between red cell and seminal
acid phosphatase; the red cell enzyme was inhibited by formaldehyde and the prostatic enzyme by tartrate. They further
showed significant levels of both types of acid phosphatase in
other human tissues. Neil and Horner (1964) demonstrated that
the formaldehyde inhibited phosphatase activity in liver was
cytoplasmic and that this enzyme only weakly utilized β-glycer-

ophosphate as a substrate. The tartrate inhibited enzyme on the other hand tended to be associated with the microsomal fraction and had strong β-glycerophosphatase activity. It should be noted that red cell acid phosphatase is both cytoplasmic and has only slight β-glycerophosphatase activity (Luffman and Harris, 1967).

ELECTROPHORETIC VARIATION OF THE HUMAN ACID PHOSPHATASES

Analysis of the acid phosphatases in human tissues by electrophoresis adds several dimensions of variation that are independent of molecular weight and inhibitor sensitivities. Comparison of the four molecular weight classes of acid phosphatase shows that for the most part each class contains a distinct group of enzymes and that there is molecular heterogeneity within the classes. This is illustrated in Figure 2, which shows the electrophoretic patterns given by each of the four gel filtration fractions in kidney; two different substates were used to visualize the phosphatase activity on the gels. The pattern observed for whole kidney extract after staining with α-naphthyl phosphate (Fig. 2 left, track 1) is similar to that observed by Beckman and Beckman (1967); the bands correspond to their bands A, B and D in the order of decreasing electrophoretic mobility. The high molecular weight fraction (track 2) appears to contain only the A band whereas the 100-125 dalton fraction (track 3) contains predominantly the B and D bands with some A band contamination. Beckman and Beckman (1967) showed that the enzymes in the A, B, and D bands were observed in most of the tissues studied. In the prostate, however, the 100-125,000 dalton enzyme appears to be entirely A band. The relationship between these bands is under continuing study.

The 30-60,000 dalton fraction from kidney stained weakly with α-naphthyl phosphate in a position anodal to the A band; the location of this fraction is more easily seen on the gel stained with methyl umbelliferone phosphate (White and Butterworth, 1971) (Fig. 2 right). This molecular weight class is not expressed in a consistent fashion among the tissues studied; it appears that there are a number of enzymes with acid phosphatase activity in this molecular weight range and that these show variation in their tissue expression.

The electrophoretic pattern of the low molecular weight red cell acid phosphatase illustrates another kind of variation in electrophoretic mobility, that arising from a genetically determined polymorphism (Hopkinson, Spencer, and Harris, 1964). The low molecular weight red cell acid phosphatase phenotype is determined at a simple genetic locus at which there are three common alleles, P^a, P^b, and P^c. An added complexity of

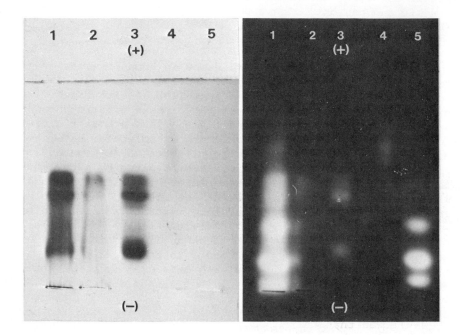

Figure 2. Electrophoretic characterization of the acid phosphatase fractions in kidney. The whole kidney extract was spotted in track 1; the four peak fractions obtained after gel filtration of the kidney extract (fig 1) were spotted in tracks 2 - 5 with the high molecular weight fraction in track 2 and the low molecular weight fraction in track 5. The sample positions in the gel on the left coincide with those in the gel on the right. The right gel was stained with α-naphthyl phosphate. The left gel was stained with methylumbelliferone phosphate; the fluorescent spots signify acid phosphatase activity. The samples were electrophoresed for 7 hours at 400 volts in the cold on a 12% starch gel buffered with 13 mM tris, 13 mM borate, pH 8.6; the electrode tanks contained the pH 8.6 buffer at 50 mM tris, 50 mM borate.

the system is that each allele appears to be responsible for two isozymes. The tissues used in this study were from a $P^a P^b$ heterozygous individual. This heterozygosity is illustrated in the figure (track 5, right) where there are four bands (the

two center bands have merged and appear as one in the photograph). The electrophoretic patterns of the low molecular
weight fraction in all tissues were identical to the pattern
observed with the red cell acid phosphatase; there was no
evidence of any low weight acid phosphatase distinct from the
red cell enzyme.

It is to be noted that the low molecular weight acid phosphatase stains strongly with the methylumbelliferone phosphate
but not at all with α-naphthyl phosphate (Fig. 2 right and left
respectively). This staining behavior is characteristic of red
cell acid phosphatase. Since most previous electrophoretic
studies of the tissue acid phosphatases have employed naphthyl
phosphate stains (cf. Lundin and Allison, 1966; Beckman and
Beckman, 1967), the presence of red cell acid phosphatase in
tissues has been overlooked until recently (Swallow, Povey,
and Harris, 1973).

Electrophoresis reveals another kind of variation among
the acid phosphatases; this is variation apparently arising
from post translational modification of the protein. This is
illustrated in Figure 3 where the semen acid phosphatase patterns of five individuals are shown; the seminal enzyme is
secreted by the prostate and falls in the 100-125,000 dalton
class. The multiple banding has been attributed to the attachment of different numbers of sialic acid residues to a single
enzyme protein (Smith and Whitby, 1968). It can also be seen
that there is a difference between the individuals in the general mobility of the group of acid phosphatase bands; it is
not known at present whether this difference is genetically
significant or is an artifact of the post-synthetic modification process.

FURTHER EVIDENCE FOR THE PRESENCE OF LOW MOLECULAR WEIGHT ACID PHOSPHATASE IN HUMAN TISSUES

Evidence for the presence of the low molecular weight red
cell acid phosphatase in tissue material is provided by gel
filtration, electrophoresis, purine activation, and inhibitor
sensitivity, as described above. However the presence of this
activity in tissue could be the result of the contamination
of the tissues with blood. To investigate this point, the
ratio of low molecular weight acid phosphatase activity to
hemoglobin concentration in four tissues was compared to the
ratio obtained from blood; the results are shown in Table 1.
It is clear that the ratio of enzyme activity to hemoglobin
is greater in the tissues than can be accounted for by contamination with blood. This provides convincing evidence that
the low molecular weight acid phosphatase thought to be specific
to red cells is in fact found in other cell types as well.

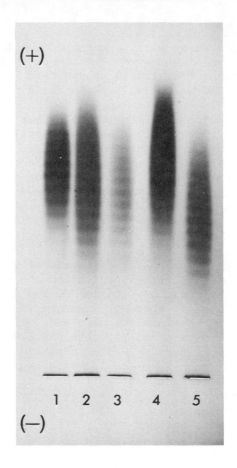

Figure 3. Electrophoretic distribution of acid phosphatase activity in the semen of 5 individuals. Semen samples were centrifuged to separate spermatozoa; the supernatant seminal plasma was diluted 1:100 with water and applied to the gels. Electrophoresis was on starch gels for 10 hours at 10 volts/ cm in the cold. The cathode tank buffer was 0.233 M tris, 0.086 M citrate, pH 5.9; in the anode tank, this buffer was diluted 1:5. The cathode tank buffer was diluted 3.5:100 in the gel.

STUDIES ON THE LOW MOLECULAR WEIGHT RED CELL ACID PHOSPHATASE

The presence of the "red cell" acid phosphatase in tissues is of biological interest in several respects. As noted above, the enzyme is genetically polymorphic. This genetical variation

TABLE I

LOW MOLECULAR WEIGHT ACID PHOSPHATASE IN TISSUES

	Activity in Tissue Extracts		Percent Low Molecular Weight AP	Low MW AP: Hb Ratio
Tissue	Hb (mg/ml)	AP (m units/ml)		(m units/mg)
Blood	61	152	100	2.5
Liver	4.4	282	38	24.2
Placenta	5.5	217	27	10.6
Kidney	12	630	25	13.1
Brain	12	156	40	5.2

Total acid phosphatase (AP) activity in tissue homogenates was determined in a standard assay using p-nitrophenyl phosphate as the substrate at pH 5.5. Hemoglobin (Hb) was determined as the cyanmet derivitive. The percent low molecular weight acid phosphatase activity in a homogenate was determined by gel filtration as in figure 1. Acid phosphatase activity is expressed in milliunits (m units); a unit of activity is defined as the turnover of one μmole substrate per minute at 25° C.

---

is expressed not only by differences in the electrophoretic mobility of the allelic products but also by differences in the levels of enzyme activity in red cells (Hopkinson, Spencer, and Harris, 1964). It remains to be determined whether the activity polymorphism is expressed in tissues as it is in erythrocytes; variations in the level of activity in tissues could be metabolically significant.

The exact physiological role of this enzyme is not known; to investigate this problem, we have initiated studies of the catalytic properties of the red cell acid phosphatase. For these studies, the low molecular weight acid phosphatase was partially purified by a two step procedure, involving chromatography on DEAE sephadex followed by gel filtration on sephadex G-75 or G-100.

The low molecular weight enzyme has been shown to be significantly active with only a few of the naturally occurring phosphoesters (Luffman and Harris, 1966; DiPietro and Zengerle, 1967; Heinrickson, 1969; Filburn, 1973); in particular, the enzyme appears to prefer flavin mononucleotide (FMN). If FMN were a natural substrate for the enzyme, the K_m and V_{max} for FMN should be at least comparable to the values determined for some of the more active synthetic substrates. This is found to be the case[2]; the K_m for FMN was determined to be about 0.06 mM as compared to 0.2 mM for p-nitrophenyl phosphate. Further studies have shown that the red cell enzyme will transfer phosphate from p-nitrophenyl phosphate to riboflavin to yield FMN. These rather preliminary findings suggest that the low molecular weight enzyme may have a role in riboflavin metabolism. However it should be noted that DiPietro (1968) has observed that this enzyme also acts preferentially on 17 β-estradiol 3-phosphate which looks nothing at all like FMN or ribloflavin. Despite the question about the natural substrate for this phosphatase, the restricted substrate specificity does suggest a more specific cellular function than that of a generalized phosphate scavenger.

A clue to the function of the low molecular acid phosphatase may be its activation by adenine analogs such as 6-methyl adenine (Di Pietro and Zengerle, 1967). This activation phenomenon is dependent upon the concentration of phosphate (Fig 4).

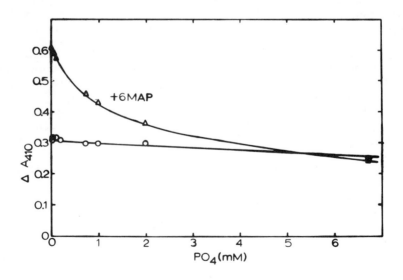

Figure 4. Effect of phosphate concentration on the activation of red cell acid phosphatase by 6-methyl adenine. Enzyme activity in the presence of 6-methyl adenine (Δ) is compared to activity in the absence of this purine (0). Enzyme activity was assayed by measuring the p-nitrophenol released in 10 minutes in an assay mixture containing initially 1 mM p-nitrophenyl phosphate and 1 mM 6-methyl adenine in 0.1 M sodium

Fig. 4 (con't)   acetate buffer pH 5.5; the absorbance of p-nitrophenol was read at 410nm after the addition of base.

In the presence of no or low phosphate concentration, high levels of activation are observed; as the concentration of phosphate increases, the activation effect diminishes and above 15 mM phosphate, a weak inhibition is observed in the presence of "activator." The physiological significance of an inter-action between the red cell phosphatase, purines, and phosphate is not immediately clear.

QUESTIONS RAISED BY THE VARIATION IN ACID PHOSPHATASES

It has been shown that there is extensive variation in the expression of acid phosphatase activity in tissues. At least four classes of acid phosphatase can be defined on the basis of molecular weight differences; each of these classes can be further subdivided into molecularly distinct acid phosphatase enzymes by electrophoresis. The existence of this extensive variation raises questions regarding the structural and evolu-tionary relationships between the multiple forms and questions regarding the biological function of the acid phosphatases.

Two basic questions can be asked about the structural rela-tionships between the acid phosphatases: a) are the isozymes within a molecular weight class structurally and evolutionarily related? and b) is there any structural or evolutionary rela-tionship between the enzymes in different molecular weight classes? With respect to the first question, some information is available in the literature but there is by no means a com-plete story for the acid phosphatases in any molecular weight class. The isozymes in the 100-130,000 dalton class are postu-lated to be dimers and to form hybrids (Swallow and Harris, 1972); Beckman and Beckman (1967) indicate that these isozymes are immunologically cross-reactive. Both of these findings suggest great structural similarities between these isozymes; this point requires further biochemical and immunological stu-dies for clarification. The low molecular weight red cell acid phosphatase is relatively well defined with respect to its ge-netic status, i.e., it is known to be the product of a single genetic locus, but the relationship between the two isozymes apparently determined by a single allele is obscure. Very little is known about any of the other tissue acid phosphatases.

There is no information bearing on the relationships between the different classes of phosphatases. It is not unreasonable to assume that the acid phosphatases have a common catalytic mechanism despite the considerable differences in their other molecular properties and in their substrate specificities. However this is probably not a sufficient condition upon which to posit a common evolutionary origin for all the acid phosphatases.

The question of relationships between classes of acid phosphatases raises the question of possible relationships between the acid phosphatases and alkaline phosphatases. In terms of catalytic mechanism, it would appear that the two may be different. First, alkaline phosphatase is known to act via a phosphoenzyme intermediate where the site of phosphorylation is a serine residue (Fernley, 1971). However phosphoserine is stable at acid pH and hence is an unlikely catalytic intermediate for the acid phosphatases. There is recent evidence suggesting the site of phosphorylation in acid phosphatase may be a histidine residue (Igarashi, Takahashi, and Tsuyama, 1970). Another potential difference in mechanism between the acid and alkaline phosphatases relates to the charge on the phosphate substrate; the hydrolysis of phosphate mono-anions appears to proceed by a somewhat different mechanism than the hydrolysis of phosphate di-anions (Kirby and Varvoglis, 1967).

A well defined biological function for the tissue acid phosphatases has not been indicated by prior studies of substrate specificities. For the most part, the acid phosphatases have rather broad specificities although distinct differences have been noted (DiPietro and Zengerle, 1967; Filburn, 1973); these differences are suggestive of differing roles in the overall cellular metabolism. As has been noted, an exception to this broad specificity is found with the low molecular weight acid phosphatase (Luffman and Harris, 1966; DiPietro and Zengerle, 1967; Heinrickson, 1969; Filburn, 1973). One aspect of phosphatase activity that may shed light on phosphatase function is the phosphotransferase activity of many of the acid phosphatases (Hollander, 1969). It is possible that some of the phosphatases may catalyze transphosphorylations between specific donor-acceptor pairs: such a phosphotransferase-phosphatase enzyme from *E. coli* has been described by Brunngraber and Chargaff (1973).

At the present time the outlines of a general pattern of variation among the acid phosphatases are becoming apparent. This pattern seems common to higher organisms; at least a similar pattern has been observed (or can be extrapolated from observations) in organisms from *Xenopus* to man. But much more needs to be known about this interesting class of enzymes to fill in the detail. Further biochemical studies on the different isozymes of the acid phosphatases are required.

ACKNOWLEDGEMENT

Published contribution 163 of the Forensic Science Program, School of Criminology, University of California, Berkeley. The assistance of E. T. Blake and D. Lofgren is gratefully acknowledged. This work was supported in part by a grant from the Committee on Research, University of California, Berkeley.

REFERENCES

Abul-Fadl, M. A. M. and E. J. King 1949. Properties of Acid phosphatases of Erythrocytes and of the human prostate gland. *Biochem. J.* 45: 51-60.

Adams, E. G. and B. G. Wraxall 1974. Phosphatases in Body fluids: the differentiation of semen and vaginal secretion· *Forensic Science* 3: 57-62.

Beckman, L. and G. Beckman 1967. Individual and Organ Specific variations of human Acid phosphatase. *Biochem. Genet.* 1: 145-153.

Bottini, E. 1973. Favism: Current problems and investigations. *J. Med. Genetics.* 10: 154-157.

Brightwell, R. and A. L. Tappel 1968. Lysosomal Acid pyrophosphatase and acid phosphatase. *Arch. Biochem. Biophys.* 124: 333-343.

Brunngraber, E. F. and E. Chargaff 1973. A nucleotide phosphotransferase from *Escherichia coli.* Purification and properties. *Biochem.*]2: 3005-3012.

DiPietro, D. L. 1968. 17 β Estradiol 3 phosphate Phosphohydrolase Activity of Human Placental Acid phosphatase III. *J. Biol. Chem.* 243: 1303-1305.

DiPietro, D. L. and F. S. Zengerle 1967. Separation and properties of three acid phosphatases from human placenta. *J. Biol. Chem.* 242: 3391-3396.

Fernley, H. N. 1971. Mammalian Alkaline Phosphatases, in *The Enzymes* 3rd ed. (P. D. Boyer, Ed.). Vol. IV pp. 417-447.

Filburn, C. R. 1973. Acid Phosphatase Isozymes of *Xenopus laevis* Tadpole tails. I. Separation and Partial Characterization. *Arch. Biochem. Biophys.* 159: 683-693.

Fisher, R. A. and H. Harris 1971. Further studies on the molecular size of red cell acid phosphatase. *Ann. Hum. Genet.* 34:449-453.

Heinrickson, R. L. 1969. Purification and characterization of a low molecular weight acid phosphatase from bovine liver. *J. Biol. Chem.* 244: 299-307.

Hollander, V. P. 1971. Acid Phosphatases in *The Enzymes* 3rd ed. (P. D. Boyer, Ed.) Vol. IV pp. 449-498.

Hopkinson, D. A., N. Spencer, and H. Harris 1964. Genetical Studies on Human Red Cell Acid Phosphatase. *Amer. J. Hum. Genet.* 16: 141-154.

Igarashi, M., H. Takahashi, and N. Tsuyama 1970. Acid phosphatase from rat liver: studies on the active center. *Biochem. Biophys. Acta.* 220: 85-92.

Kirby, A. J. and A. G. Varvoglis 1967. The reactivity of

phosphate esters. Monoester hydrolysis. *J. Amer. Chem. Soc.* 89: 415–423.

Latner, A. L. and A. W. Skillen 1968. *Isoenzymes in Biology and Medicine.* Academic Press (London and New York). p. 161.

Luffman, J. E. and H. Harris 1967. A comparison of some properties of human red cell acid phosphatase in different phenotypes. *Ann. Hum. Genet.* 30: 387–401.

Lundin, L. G. and A. C. Allison 1966. Acid phosphatase from different organs and animal forms compared by starch gel electrophoresis. *Acta. Chem. Scand.* 20: 2579–2592.

Neil, M. W. and M. W. Horner 1964. Studies on acid hydrolases in adult and foetal tissues. *Biochem. J.* 92: 217–224.

Oesper, P. 1950. Sources of the high energy content in energy rich phosphates. *Arch. Biochem.* 27: 255–270.

Smith, J. K. and L. G. Whitby 1968. The heterogeneity of prostatic acid phosphatase. *Biochem. Biophys. Acta.* 151: 607–618.

Swallow, D. and H. Harris 1972. A new variant of the placental acid phosphatases: its implication regarding their subunit structures and genetical determination. *Ann. Hum. Genet.* 36: 141–152.

Swallow, D. M., S. Povey and H. Harris 1973. Activity of the "red cell" acid phosphatase locus in other tissues. *Ann. Hum. Genet.* 37: 31–38.

White, I. N. H. and P. J. Butterworth 1971. Isoenzymes of human Erythrocyte acid phosphatase. *Biochem. Biophys. Acta.* 229: 193–201.

PHOSPHOGLUCOMUTASE ISOZYMES CAUSED
BY SULFHYDRYL OXIDATION

DAVID M. DAWSON and JERRY M. GREENE
Department of Neurology
Children's Hospital Medical Center and
Harvard Medical School
Boston, Massachusetts 02115

ABSTRACT. Phosphoglucomutase (PGM) exists in a series of
different molecular forms which can be resolved by gel electrophoresis, by chromatography, or by isoelectric focusing.
This series of isozymes is distinct from genetic variation
in the enzyme, and from expression of several PGM alleles,
both of which are also known to occur.

We present evidence that the isozymes of PGM differ from
each other by a single unit of charge, and that a closely
comparable set of isozymes can be produced by sulfhydryl
substitution with mercuribenzoate or by sulfhydryl oxidation. Succinylated PGM appears to differ by 2 units of
charge.

PGM isozymes have been observed in a wide variety of
extracts from mammals, birds, and amphibians, and in crude
extracts as well as purified enzymes.

Phosphoglucomutase (PGM) from mammalian sources is subject
to heterogeneity of several types. Genetic heterogeneity is
reflected in several different enzymes (PGM_1, PGM_2, and so
forth) which share many properties but may differ in molecular
weight (McAlpine, Hopkinson, and Harris, 1970). Mutations of
these enzymes are also known and PGM can also exist as phosphorylated and dephosphorylated derivatives.

Enzyme preparations which are homogeneous with respect to
the above criteria are subject to another source of heterogeneity with which this paper is concerned. This fourth type
of heterogeneity causes an enzyme preparation to appear on
starch-gel electrophoresis as a series of isozymic forms.
We now believe, based on evidence presented here, that the
appearance of these isozymic forms is caused by progressive
oxidation of the free sulfhydryl groups of PGM. According to
this hypothesis the major enzyme form has six reduced -SH
groups, and the more anodic forms have 1, 2 or more sulfur
acid groups. Adjacent electrophoretic forms would differ by
one in the number of negatively charged residues.

A typical electrophoretic separation of the multiple forms
of phosphoglucomutase activity is shown in Fig. 1. All the
active enzyme forms moved toward the positive pole at pH 7.0
and were equally spaced. In the figure, a maximum of four

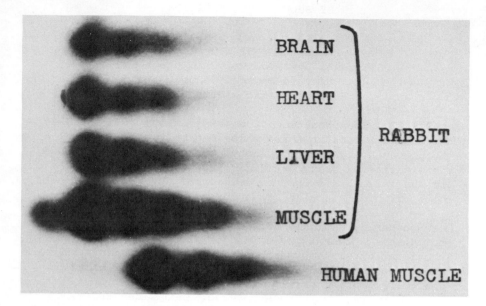

Fig. 1. Starch gel electrophoresis, stained for PGM (Dawson and Mitchell, 1969). Anode is to the right, origin near left hand margin.

spots are shown, but with heavier staining, as many as six or seven spots could be demonstrated, with progressively less activity in those migrating farthest. When genetic mutations occur in PGM$_1$ all the spots migrate in concert, as shown in Fig. 2. There was some variation in the intensity of the different spots in extracts from different tissues, but almost always the slowest migrating enzyme was the largest in amount. For convenience, we have called these isozymes A, B, C, D, etc.

A number of experiments were tried under different conditions in an effort to purify the individual isozymes, using DEAE-cellulose (Dawson and Mitchell, 1969). Enzyme A was the first to appear on gradient elution, while the faster migrating enzymes, B–D, were retained. Although peaks of enzyme activity were often obtained, particularly with the human enzyme, which was better retained, these peaks did not correspond to the distinct electrophoretic entities. It seemed possible that there was conversion of one isozyme into another after separation on the column, but a more likely explanation was that the initial separation itself was incomplete. We found it more convenient to prepare separate isozymes by isoelectric focusing as shown in Fig. 3. Purified commercial rabbit PGM was diluted and used

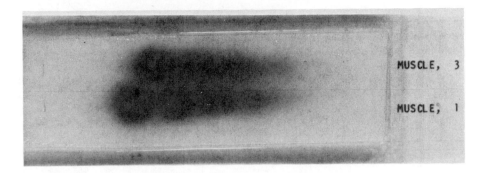

Fig. 2. Starch gel electrophoresis of human muscle extracts obtained at autopsy from two individuals. The difference in migration of PGM reflects the genetic polymorphism described by Spencer et al (1964).

as the light solution throughout the gradient, which was a shallow pH 6-8 gradient. A broad poorly defined activity peak was observed, even with shallower gradients. By judicious choice of fractions it was possible to choose fractions which were largely or wholly composed of a single isozyme (Fig. 4).

Refocusing an isolated isozyme gave sharp peaks which were stable on storage in the cold. They were not altered by freezing, thawing, or treatment with 0.1 molar mercaptoethanol. The isolated isozyme A was subjected to a number of modifications after purification by electrofocusing and removal of sucrose and ampholytes by dialysis.

Addition of mercuribenzoate (HMB) to isozyme A or to unfractionated PGM causes a progressive appearance of more anodally migrating enzyme (Fig. 5). The intervals between spots matches the intervals found with native enzyme. With additions of larger amounts of HMB the enzyme is less active, and hence cannot be visualized well. However, it is worth noting that we never observed more than seven isozymes (corresponding to the various states of substitution on six sulfhydryl groups). The addition of HMB was freely reversible, and on dialysis or addition of mercaptoethanol the original migration pattern was restored.

Additions of hydrogen peroxide produced alterations of isozyme pattern quite similar to those produced by HMB addition (Fig. 6). A number of possibilities exist for the site of action of hydrogen peroxide. That the site of action was sulfhydryl groups is suggested by a protective effect from HMB.

(•)
ACTIVITY (% of maximum)
100
80
60
40
20
0
(---)
9.0
8.5
8.0
7.5
7.0
6.8
6.0
5.5
5.0
4.5
pH
10
15
20
25
30
35
40
45
50
55
60
FRACTION NUMBER

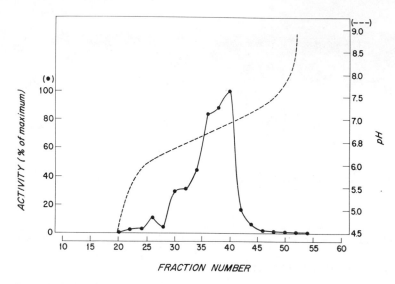

Fig. 3. Iso-electric focusing of rabbit muscle PGM (Boehringer-Mannheim Co.) using the 110 ml column of LKB-Productor Co. The PGM was incorporated into the sucrose gradient throughout.

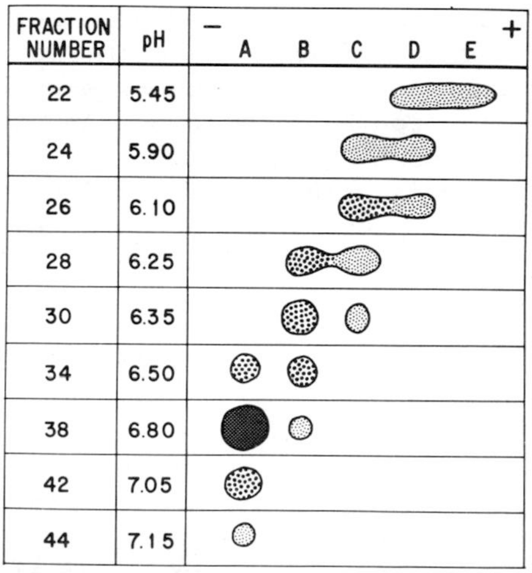

FRACTION NUMBER
pH
−
A B C D E
+
22 5.45
24 5.90
26 6.10
28 6.25
30 6.35
34 6.50
38 6.80
42 7.05
44 7.15

Fig. 4. Gel electrophoresis of separate fractions obtained from isoelectric focusing runs illustrated in Fig. 3.

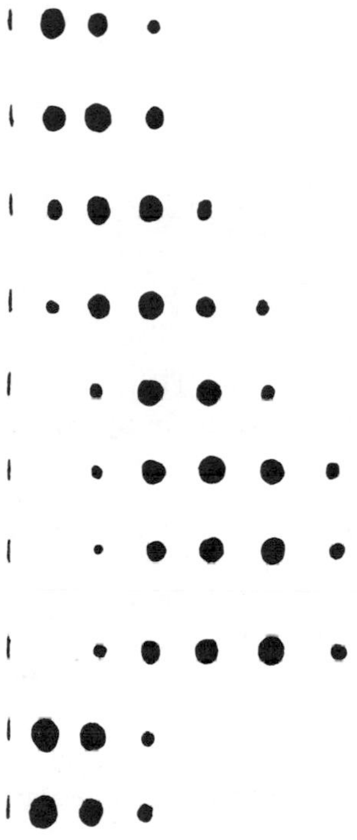

	HMB bound, moles/mole	Activity, percent
	0.39	100%
	0.73	90%
	1.1	94%
	1.5	85%
	2.2	61%
	2.5	68%
	3.2	90%
	3.4	68%
	untreated	--
	add mercaptoethanol	

Fig. 5. Gel electrophoresis patterns of hydroxymercuribenzoate (HMB) substituted PGM. The amount of HMB bound was assayed by change in absorption at 255 mu. The enzyme source was rabbit muscle PGM (Boehringer-Mannheim).

Enzyme that was pretreated with HMB, then treated with peroxide, and then dialysed against 0.1 M mercaptoethanol to remove the HMB went through this sequence essentially unaltered, except for a decrease in activity. Enzyme that was treated with peroxide and then with mercaptoethanol gave the same pattern as that treated with peroxide alone. Thiol titration of enzyme before and after peroxidation showed a decline in the number of mercuribenzoate titratable -SH from an average of 5.5 for unfractionated enzyme before peroxidation to 4.0 per PGM molecule, or a 27% decrease in titratable -SH after incubation for 1 hr at 20°C with 0.05% hydrogen peroxide. We interpret

385

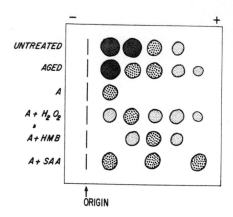

Fig. 6. Gel electrophoresis patterns of rabbit muscle PGM.
The first slot is of untreated unfractionated enzyme; after
months of storage the aged pattern would sometimes change.
Isozyme A was prepared by electrofocusing. It was oxidized by
addition of 0.05% H_2O_2 for 30 min, or HMB was added at a ratio
of 2 moles of mercurial per mole of enzyme. These patterns
are to be compared with that of succinylated isozyme A, in
which the spacing between spots is twice as large.

these results as indicating that the peroxide effect is on PGM
sulfhydryl groups, and is not reversible with mercaptoethanol.
The HMB addition reaction, on the other hand, is freely reversible (Greene and Dawson, 1973).

Isozyme A was also altered by succinylation. We dissolved
succinic anhydride in dioxane and reacted dilute PGM with small
amounts of the anhydride. The migration pattern of succinylated enzyme is compared with that of the sulfhydryl-altered
enzyme in Fig. 6. With the succinylated enzyme we observed
that the spacing between spots was twice as great.

Addition of HMB to a sulfhydryl group on PGM should result
in a change in the net charge on the enzyme molecule of -1
(that is, from $-SH$, uncharged at pH 7.0, to $-SHgC_6H_4COO-$).
Isozyme A, when treated with HMB, shows a stepwise shift in
its electrophoretic pattern to more anodal forms with the same
interval between adjacent spots as that observed with unfractionated enzyme. This would indicate that the unfractionated
isozymes differ by the same amount of charge as the HMB treated
enzyme--that is, by a single unit charge.

The succinylation experiments seem to confirm this deduc-
tion. Succinic acid anhydride reacts with free amino groups
such as the NH_3^+ of a lysine residue to give a peptide bond

and freeing a carboxyl group. The expected net charge change from the addition of one succinic acid anhydride molecule is -2 (that is, from $-NH_3^+$ to $-NHCOCH_2CH_2COO-$). The spacing of the succinylated form A is twice that of the unfractionated or HMB treated enzyme. Succinylation of unfractionated enzyme can produce a multiple of electrophoretic spots. In Fig. 7, 12 spots may be counted rather than the maximum of seven seen with alteration of the sulfhydryl groups.

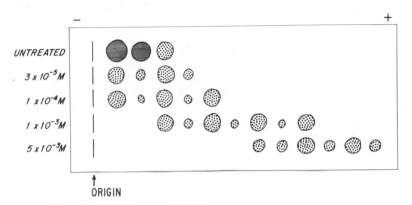

Fig. 7. Succinylation of rabbit muscle PGM, by adding the indicated concentrations of succinic anhydride in dioxane to rabbit muscle PGM, followed directly by gel electrophoresis.

Comparable patterns have been obtained with amphibian, reptile, and crustacean tissues (Dawson and Jaeger, 1970). We have not analyzed these in any other way than by electrophoresis, but we suggest that the phenomenon observed with rabbit and human PGM is widely present in the animal kingdom.

It now seems likely to us that the naturally occurring PGM isozymes, as distinct from different PGM enzymes, result from partial oxidation of sulfhydryl groups to sulfur acid groups. We do not know if such partially oxidized enzymes exist in vivo, but a sample of muscle taken from a freshly killed rabbit and homogenized immediately in deionized water and 0.1 M mercaptoethanol contained the additional isozymes just as do other tissue extracts. Hopkinson and Harris (1969) found a comparable phenomenon with aged red cell adenosine deaminase. In that case, additional electrophoretic bands could be produced by oxidized glutathione and other sulfhydryl reagents; however, the changes could be reversed by mercaptoethanol, which is not the case with the PGM isozymes.

Noltmann's group reported (Blackburn et al, 1972) that phosphoglucose isomerase was also subject to a similar type of multiplicity caused by sulfhydryl group oxidation, and detected by CM-Sephadex chromatography. In this symposium Noltmann summarizes evidence that rabbit muscle phosphoglucose isomerase is subject to -SH oxidation to produce distinguishable multiple forms. Our results resemble his except that we have been, thus far, unable to reduce the oxidized enzyme.

To be detectable, of course, such a change must not lead to any substantial change in enzyme activity and must possess a degree of stability, both of which criteria are met in the case of PGM.

REFERENCES

Blackburn, M. N. et al 1972. Pseudoisoenzymes of rabbit muscle phosphoglucose isomerase. *J. Biol. Chem.* 247: 1170-1179.

Dawson, D. M., and S. Jaeger 1970. Heterogeneity of phosphoglucomutase. *Biochem. Genet.* 4: 1-9.

Dawson, D. M., and A. Mitchell 1961. The isoenzymes of phosphoglucomutase. *Biochem.* 8: 609-614.

Greene, J. M., and D. M. Dawson 1973. Altered electrophoretic mobility of oxidized phosphoglucomutase. *Ann. Hum. Genet.* 36: 355-361.

Hopkinson, D. A., and H. Harris 1961. The investigation of reactive sulfhydryl in enzymes and their variants by starch gel electrophoresis. *Ann. Hum. Genet.* 33: 81-87.

McAlpine, P. J., D. A. Hopkinson, and H. Harris 1970. Molecular size estimates of the human phosphoglucomutase isozymes by gel filtration chromatography. *Ann. Hum. Genet.* 34: 117-186.

Noltmann, E. A. These proceedings.

Spencer, N. A., D. A. Hopkinson, and H. Harris 1964. Phosphoglucomutase polymorphism in man. *Nature* 204: 742-744.

CARBOXYLESTERASE ISOZYMES IN THE RABBIT LIVER: PHYSICOCHEMICAL PROPERTIES*

D. J. ECOBICHON
Department of Pharmacology
Faculty of Medicine
Dalhousie University
Halifax, N. S., Canada

ABSTRACT. The hepatic esterases of the albino New Zealand rabbit were identified, on the basis of substrate specificity, inhibitor sensitivity and subcellular fractionation studies as nonspecific carboxylesterases of microsomal origin. Gel filtration and molecular weight estimation on Sephadex G-200 yielded two distinct molecular forms having a mol. wt. of 125,000 and 60,000 respectively, suggesting a monomeric-dimeric relationship. While such reducing agents as mercaptoethanol and iodoacetamide were ineffective, the higher molecular weight form could be dissociated into units of 60,000 mol. wt. by a combination of acidic pH and high salt concentration. Electrostatic forces appeared to be involved in the aggregation phenomenon. Electrophoretically different multiple forms of 60,000 mol. wt. suggested the existence of isozymic carboxylesterases.

As part of a study to characterize the tissue esterases of different mammalian species, the liver and kidney of the albino New Zealand rabbits were studied. Exceedingly high carboxylesterase (E.C. 3.1.1.1) activity was detected in rabbit liver, activity much higher than that in any other species studied (Ecobichon, 1972b). This observation prompted a detailed study employing such techniques as starch electrophoresis, differential centrifugation and Sephadex gel filtration for separation and molecular weight determinations, as well as substrate specificity and inhibitor sensitivity studies in an attempt to characterize the hepatic esterases of the rabbit.

---

*The following abbreviations have been used: α-NA, alpha-naphthyl acetate; α-NP, alpha-naphthyl propionate; α-NB, alpha-naphthyl butyrate; α-NV, alpha-naphthyl valerate; p-NPA, p-nitrophenyl acetate; p-NPP, p-nitrophenyl propionate; p-NPB, p-nitrophenyl butyrate; DFP, diiosopropylfluorophosphate; DDVP, Dichlorvos or 0,0-dimethyl-2, 2-dichlorovinyl phosphate; EDTA, ethylenediaminetetraacetic acid, disodium salt.

MATERIALS AND METHODS

Chemicals. The substrates used in this study (α-naphthyl and p-nitrophenyl esters) as well as the pure globular proteins (horse heart cytochrome c, bovine pancreatic chymotrypsinogen A, ovalbumin, bovine serum albumin and rabbit heart lactate dehydrogenase) were purchased from Sigma Chemical Co., St. Louis, Mo. Iodoacetamide was purchased from Nutritional Biochemical Corporation, and 2-mercaptoethanol was purchased from Eastman Distillation Products, Rochester, N.Y. Diisopropylfluorophosphate (DFP) was purchased from Mann Research Laboratories, Inc., New York, N.Y. while the Dichlorvos (dimethyl dichlorovinyl phosphate, DDVP) was a gift of the Shell Oil Company of Canada.

Tissue Preparation. Male and female albino New Zealand rabbits weighing approximately 3-4 kg were used in all experiments. The animals were killed by cervical dislocation and samples of fresh liver and kidney were removed, chilled in ice-cold 0.9% saline, weighed, minced and washed thoroughly with cold 0.9% saline to remove residual blood in the tissue. The minced tissues were homogenized at 0°C with a Potter-Elvehjem glass homogenizer and motor-driven Teflon pestle, using sufficient cold distilled water to produce a final homogenate concentration of 20% (w/v). The homogenates were centrifuged at 0°C for 60 min at 11,000 x g, saving the supernatant fraction for enzyme studies.

For the subcellular fractionation, samples of fresh liver were minced and thoroughly washed with 0.25M sucrose containing 10^{-3}M EDTA (ethylenediamine tetraacetic acid, disodium salt). The tissue was then homogenized in sufficient 0.25M sucrose solution to make a 20% homogenate. The various subcellular fractions were prepared by differential centrifugation using the technique of Booth and Boyland (1958) as described by Schwark and Ecobichon (1968). The particulate fractions, nuclei-debris (N), mitochondria (Mt), and microsomes (Mc) were isolated successively by sedimentation at 600 g for 60 min, 20,000 g for 10 min and 90,000 g for 30 min respectively. At each fractionation step, the pellet was separated from the supernatant, washed twice by resuspension in a known volume of 0.25M sucrose and resedimented. The washings were combined with each supernatant fraction (stored at 0°C) and then processed for the isolation of the next subcellular fraction. The final supernatant obtained following isolation of the microsomes was called the cell sap (CS) fraction. Washed mitochondria were disrupted by resuspending the pellet in 1.0% Triton X-100 for 10 min prior to enzymatic analysis. The nuclei-debris and microsome fractions were resuspended in known volumes of 0.25M sucrose.

Electrophoresis and Staining. Tissue homogenates centrifuged at 11,000 g for 60 min were separated by vertical zone electrophoresis in starch gel prepared with 16% starch in 0.006M phosphate buffer, pH 7.4. Electrophoresis was carried out for 16-18 hr at room temperature using a constant current of 30mA (Ecobichon and Kalow, 1961, 1962). After electrophoresis, the gels were suitably sliced and stained with substrates to localize the esterase activity according to methods described previously (Ecobichon and Kalow, 1962).

Gel Chromatography. Gel filtration of the cell sap fraction of 20% liver homogenates was carried out on a 2.5 x 35 cm column of Sephadex G-200 equilibrated with 0.025M phosphate buffer containing 0.02% sodium azide as described previously (Ecobichon, 1969, 1972a). The flow rate (10 ml/hr) was controlled by a metering pump. Aliquots of 0.5 ml, containing 10-15 mg protein per ml, were applied to the column through a two-way valve using the upward flow technique. Fractions of 5.0 ml were collected and assayed for protein and esterase activity. Molecular weight estimates of the hepatic esterases were determined on the same column following calibration with purified lactate dehydrogenase bovine serum albumin, ovalbumin, chymotrypsinogen A and cytochrome c and preparation of suitable calibration curves for the pure globular proteins (Ecobichon, 1969, 1972a).

Enzyme Assays. Esterase activity was determined quantitatively using the spectrophotometric techniques described by Ecobichon (1970), the substrates being the acetate, propionate and butyrate esters of p-nitrophenol or α-naphthol. Initial rates of p-nitrophenol or α-naphthol liberation were determined at 322 nm and 345 nm respectively at 25°C and pH 7.4 and were corrected for spontaneous hydrolysis of the substrate.

Glucose-6-phosphatase was determined by the method of de Duve et al (1955), involving incubation of the enzyme with glucose-6-phosphate for 30 min at 37°C and determining the inorganic phosphate released by the method of Chen et al (1956).

The protein concentration in tissue preparations was determined colorimetrically by the Lowry method as described by Hartree (1972).

RESULTS

Fig. 1 shows representative starch gel esterase patterns of rabbit hepatic and renal extracts (11,000 g supernatants) prepared from three animals, the enzymatic activity being localized by the substrate α-NA. The esterases were localized in two distinct zones, one rapidly-migrating zone composed of

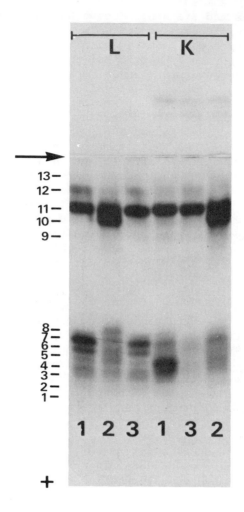

Fig. 1. Esterases of the liver (L) and kidney (K) of 11,000 g supernatant extracts from three adult albino New Zealand rabbits following electrophoresis in starch gel at pH 7.4. The esterase activity was localized with α-NA. The arrow indicates the point of sample application. The direction of migration was toward the bottom of the photograph.

6-8 distinct bands on a slowly-migrating zone composed of 1-2 faint and 2 intensely-stained bands. While the renal bands were not as distinct as those in liver, it was observed that the esterase patterns were qualitatively similar with the exception of two bands observed in the renal preparations which migrated toward the cathode. Quantitatively, the hepatic activity was approximately 10-fold greater than that detected in the kidney, necessitating a 10-fold dilution of the hepatic preparations before applying them to the gel. No significant differences were observed in esterase patterns of male and female tissues. It was noted, however, that when an electrophoretic difference was observed in hepatic esterases, the kidney esterases of that animal showed the same change (Fig. 1, rabbit 2, liver and kidney bands 10 and 11).

Using a series of esters of α-naphthol to stain the gels it was observed that α-NB was hydrolyzed much more rapidly than were either α-NA, α-NP, or α-NV. This observation was confirmed spectrophotometrically for hepatic esterases using a series of p-nitrophenyl esters, p-NPB and p-NPA being hydrolyzed at rates of 1.39 and 0.60 μmoles/min/ml of preparation respectively. The specificity of the rabbit hepatic esterases toward longer-chain esters was confirmed by studies of apparent K_m values, the range of values for p-NPB (0.6-0.9×10^{-4}M) being lower than that for p-NPA (2.0-4.0×10^{-4}M).

Inhibition studies were carried out using 1:10 dilutions of the hepatic 11,000 g supernatants and two organophosphorus esters (DFP and DDVP), the residual activity after a 5 min incubation with the inhibitor being determined with α-NA. The results (Fig. 2) showed that the esterases were extremely sensitive to inhibition, low I_{50} values being observed with both agents. Approximately 5-10% of the hepatic activity was resistant to inhibition by organophosphates at concentrations as high as 10^{-4}M.

Fig. 3 shows the results of a typical experiment to determine the intracellular distribution pattern of carboxylesterase in rabbit liver, the data being presented as a percent of the activity of the whole homogenate and also as proposed by de Duve et al (1955). The major portion of the hepatic carboxylesterase activity was found localized in the microsomal (Mc) fraction, though some activity was found in the cytoplasmic (CS) fraction and, despite rigorous washing, considerable activity was associated with the nuclei-debris fraction (N).

Fig. 4 shows the elution patterns of the proteins and esterases of a 0.5 ml aliquot of the cell sap fraction from a 20% homogenate of rabbit liver fractionated on a 2.5 x 35 cm column of Sephadex G-200. Alpha-naphthyl acetate was the substrate used to detect esterolytic activity in the 5.0 ml fractions collected. Three peaks of esterase activity were detected. Peak I was eluted with the void volume (Vo = 75 ml, determined with Blue Dextran 2000) while peaks II and III

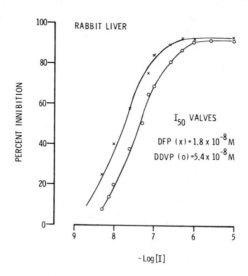

Fig. 2. Inhibition of hepatic esterases in a 10-fold dilution of a 11,000 g supernatant by DFP and DDVP. Aliquots (0.1 ml) of preparation were incubated with different concentrations of the inhibitor for 5 min before the residual activity was determined with 10^{-3} M α-NA. Each point is the mean of duplicate determinations.

had elution volumes of 105 and 135 ml respectively. Peak fractions were pooled, concentrated by dialysis against Carbowax 4000 (Ecobichon and Kalow, 1961) and were subjected to starch gel electrophoresis. A schematic diagram of the electrophoretic patterns observed is shown in Fig. 5. Diagram A shows the electrophoretic pattern of the hepatic cell sap fraction, the pattern being similar to that observed in Fig. 1, though the slowly-migrating faint bands were not visible. Peak I, eluted with the void volume, contained only the slowly-migrating bands (bands 9-11). Peak II contained the rapidly-migrating as well as the same activity in the slow-moving zone (band 9). Peak III contained only the rapidly-migrating bands of activity (bands 1-8).

To estimate the molecular weights of the rabbit hepatic esterases, pure globular proteins of known molecular weight were used to calibrate the 2.5 x 35 cm Sephadex G-200 column used in Fig. 3. Fig. 6 is a representative plot of the ratio of Ve/Vo against the molecular weights of the pure proteins. Peak I was too large in size to be estimated by this tech-

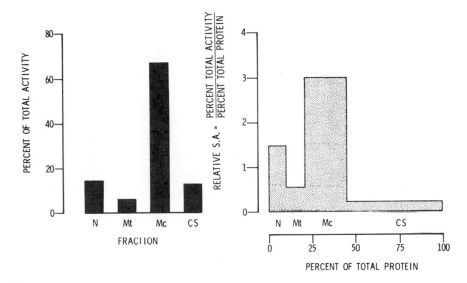

Fig. 3. A representative experiment showing the subcellular distribution pattern of esterases in the nuclei-debris (N), mitochondrial (Mt), microsomal (Mc) and cell sap (CS) fractions of liver from an adult rabbit. The activities are presented as a percent of the total esterase activity of the whole homogenate (left half) and also in the manner proposed by de Duve et al (1955). See text for details.

nique. Peak II and peak III were estimated to have molecular weights of 125,000 and 60,000 respectively, suggesting a monomer-dimer relationship.

If such a dimerization of rabbit hepatic carboxylesterases occurs, it should be possible to dissociate the dimeric form into active monomeric units. This was accomplished in the author's laboratory for human and equine liver carboxylesterases, using the technique reported by Barker and Jencks (1969) for the dissociation of porcine microsomal carboxylesterases. This technique involved incubating the enzyme preparation with 0.01M, sodium acetate (pH of 4.5) for varying lengths of time, adding sodium chloride sufficient to yield a concentration of 0.5-1.0M rapidly adjusting the pH to 7.5 and applying the preparation to a 2.5 x 35 cm Sephadex G-200 column equilibrated with 0.025M phosphate buffer pH 7.0 containing 0.5M sodium chloride. The results of a typical experiment are shown in Fig. 7. Fig. 7A presents the elution pattern of untreated

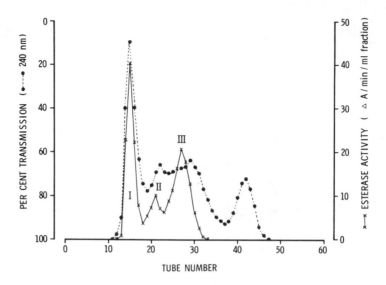

Fig. 4. The protein (●) and esterase (x) elution pattern of a
0.5 ml aliquot (12.8 mg protein per ml) of a 90,000 g ctyo-
plasmic fraction of rabbit liver following gel filtration on
a 2.5 x 35 cm column of Sephadex G-200 equilibrated with 0.025M
phosphate buffer pH 7.0. See text for details.

rabbit cell sap (cytoplasmic) carboxylesterases showing the
three peaks of activity normally detected. Fig. 7B shows the
elution pattern obtained following prolonged centrifugation
(90,000 g for 120 min) of the cell sap fraction, demonstrating
that the high molecular weight form will sediment. Fig. 7C
shows the results of incubating the cytoplasmic carboxylesterases with 1.0M sodium chloride at 25°C for 60 min prior to
applying an aliquot to the column. A slight shift in enzymatic
activity from peak II into peak III, the 60,000 molecular
weight form, was observed. Fig. 7D shows the results of incubating the enzyme mixture in 0.01M sodium acetate (pH 4.5)
for 40 min (solid dots) and 120 min (open dots), adding solid
sodium chloride to produce a 0.5M concentration and adjusting
the pH to 7.5 and separating a 0.5 ml aliquot by gel filtration. There was a progressive loss of activity in peak II,
the activity being found in peak III. With a longer incubation, even more activity was lost from peak II, suggesting
that the dimeric form can be dissociated into active monomeric
forms having a molecular weight of approximately 60,000.
Treatment of the hepatic cytoplasmic esterases with 0.05M 2-

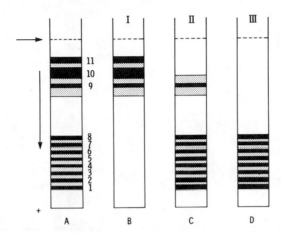

Fig. 5. A schematic diagram of the starch gel electrophoretic patterns of hepatic carboxylesterases in the active fractions obtained following gel filtration on Sephadex G-200 as in Fig. 4. "A" represents the pattern of the starting cytoplasmic preparation. "B-D" represent the esterase patterns obtained for eluted peaks I, II, and III following collection, concentration by dialysis and electrophoretic separation on starch gels. The enzymatic activity was localized in the gel by α-NA.

mercaptoethanol in the presence of 5.0M urea followed by neutralization with 0.06M iodoacetamide had no effect on the gel filtration patterns (Chow and Ecobichon, 1973).

DISCUSSION

On the basis of the electrophoretic separation (Fig. 1) and staining with esters of α-naphthol, it would appear that the hepatic and renal esterases of the rabbit possess very similar physical properties, though quantitatively the activity in the liver was some 10-fold higher than that in the kidney (Ecobichon, 1972b). From the results of substrate specificity, inhibition sensitivity (Fig. 2) and subcellular distribution (Fig. 3) studies, it would appear that the major proportion of the hepatic esterases of the rabbit are of microsomal origin and belong to the highly heterogeneous carboxylesterase group of enzymes (Augustinsson, 1961). The resistance of a small quantity of hepatic esterase activity to high concentrations of organophosphorus esters suggested the presence

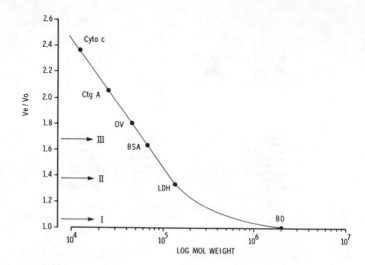

Fig. 6. Relationship between Ve/Vo and log. molecular weight for purified globular proteins separated on a 2.5 x 35 cm column of Sephadex G-200. The Ve/Vo ratio of the three hepatic esterase peaks are indicated by the arrows. The purified standard proteins are abbreviated as follows: Cyto c, horse heart cytochrome c.; Ctg A, bovine pancreatic chymotrypsinogen A; OV, ovalbumin; BSA, bovine serum albumin; LDH, rabbit muscle lactate dehydrogenase; BD, Blue Dextran 2000.

of arylesterase activity, a fact which has been subsequently confirmed in a separate study, appreciable hydrolysis of para-oxon (diethyl p-nitrophenyl phosphate) being observed with liver homogenates (Whitehouse, 1973).

The molecular weight estimates determined for the cytoplasmic hepatic esterases of the rabbit were comparable to those obtained for hepatic carboxylesterases of other mammalian species in this laboratory using the same technique (Ecobichon, 1972a, Chow and Ecobichon, 1973). The molecular weight estimates of 60,000 and 125,000 suggested a monomeric-dimeric relationship, though electrophoretic separation of these two forms revealed a complex mixture of bands, possibly isozymic, monomeric forms and at least one, slowly-migrating, higher molecular weight aggregate (Fig. 5C). The reduction of the rabbit hepatic carboxylesterases with 0.05M 2-mercaptoethanol in the presence of 5.0M urea and subsequent treatment with 0.06M iodo-acetamide had no effect on the Sephadex G-200 elution patterns,

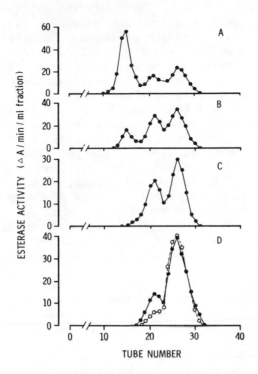

Fig. 7. The elution patterns of rabbit hepatic cytoplasmic esterases from a 2.0 x 35 cm column of Sephadex G-200 equilibrated with 0.025M phosphate buffer pH 7.0. Aliquots of 0.5 ml (approximately 7.0 mg protein) were applied to the column. Additional conditions were: A, untreated cytoplasmic fraction; B, prolonged centrifugation at 90,000 g for 120 min; C, enzymes incubated with 1.0M NaCl at 25°C for 60 min; D, enzymes incubated with 0.01M sodium acetate at pH 4.5 for 40 min (•) and 120 min (•). Sodium chloride was added to give a 0.5M concentration, pH was adjusted to 7.5 and an aliquot was applied immediately to the column equilibrated with buffer containing 0.5M NaCl. Esterase activity was measured spectrophotometrically using α-NA.

---

indicating that sulfhydryl groups did not appear to be involved in the aggregation. Several studies have demonstrated that aggregated forms of acetylcholinesterase (Hargreaves et al, 1963, Grafius and Miller, 1965, 1967, Miller et al, 1973) and hepatic carboxylesterases (Ecobichon, 1972a, Chow and Ecobichon, 1973, Miller et al, 1973, Benöhr and Krisch, 1967, Franz and Krisch, 1968, Kingsbury and Masters, 1970, Heymann

et al, 1971, Arndt et al, 1973, Wynne and Shalitin, 1973)
can be dissociated into active low molecular weight forms by
increasing the ionic strength of the medium while polymerization can be enhanced by reducing the ionic strength. Using
the technique described by Barker and Jencks (1969) the successful dissociation of high molecular weight forms of equine
and human hepatic carboxylesterases into active, lower molecular weight forms was achieved (Ecobichon, 1972a). Similar
results were obtained for the rabbit hepatic carboxylesterases,
suggesting that electrostatic forces appear to play an important role. The combination of acidic pH, altering the charge
on essential groups, and sodium chloride, subsequently interacting with these highly ionized groups, appears to be effective in dissociating the protein. There probably exists an
equilibrium between the dimeric (and higher forms) and the
monomeric forms, dependent upon protein concentration, pH and
ionic strength of the medium. This equilibrium appears to be
disrupted toward the formation of enzymatically-active monomers
by acidic pH and high ionic strength. The model proposed by
Barker and Jencks (1969) to explain the influence of such
treatment on porcine hepatic carboxylesterases can be applied
to the hepatic carboxylesterases of other mammalian species
since this phenomenon appears to be characteristic of this
type of enzyme.

Isozymes commonly occur in the same organ or tissue and have
the same substrate specificity and kinetic behavior but differ
in electrophoretic mobility. It would appear that the elec-
trophoretically-distinct, rapidly-migrating, cytoplasmic carboxylesterases of rabbit liver, isolated in peak III (Fig. 4)
and having a molecular weight of 60,000, should be classified
as isozymes. It is also apparent that these highly heterogeneous forms can aggregate in dimeric or higher molecular
weight forms, yielding electrophoretically-slow, broad bands
which are still heterogeneous but which cannot be resolved
further because of the molecular size which modifies the net
negative charge on the protein, influencing the rates of
migration.

REFERENCES

Arndt, R., E. Heymann, W. Junge, K. Krisch, and H. Hollandt
     1973. Purification and molecular properties of an un-
     specific carboxylesterase (E_1) from rat liver microsomes.
     Eur. J. Biochem. 36: 120-128.
Augustinsson, K. -B. 1961. Multiple forms of esterase in ver-
     tebrate blood plasma. *Ann. N.Y. Acad. Sc*. 94: 844-860.
Barker, D. L. and W. P. Jencks 1969. Pig liver esterase.

Physical properties. *Biochemistry* 8: 3879-3889.

Benöhr, H. C. and K. Krisch 1967. Carboxylesterase aus rinderlebermikrosomen. II. Dissoziation und assoziation, molekulargewicht, reaktion mit E600. *Hoppe-Seyler's Z. Physiol. Chem.* 348: 1115-1119.

Booth, J. and E. Boyland 1958. Metabolism of polycyclic compounds 13. Enzymic hydroxylation of naphthalene by rat-liver microsomes. *Biochem. J.* 70: 681-688.

Chen, P. S., T. Y. Toribara, and H. Warner 1956. Microdetermination of phosphorus. *Anal. Chem.* 28: 1256-1758.

Chow, A. Y. K. and D. J. Ecobichon 1973. Characterization of the esterases of guinea pig liver and kidney. *Biochem. Pharmacol.* 22: 689-701.

deDuve, C., B. C. Pressman, R. Gianetto, R. Wattiaux, and F. Appelmans 1955. Tissue fractionation studies 6. Intracellular distribution patterns of enzymes in rat-liver tissue. *Biochem J.* 60: 604-617.

Ecobichon, D. J. 1965. Multiple forms of human liver esterases. *Can. J. Biochem.* 43: 595-602.

Ecobichon, D. J. 1969. Bovine hepatic carboxylesterases: chromatographic fractionation, gel filtration and molecular weight estimation. *Can. J. Biochem.* 47: 799-805.

Ecobichon, D. J. 1970. Characterization of the esterases of canine serum. *Can. J. Biochem.* 48: 1359-1367.

Ecobichon, D. J. 1972a. Cytoplasmic carboxlesterases of human and domestic animal liver: aggregation, dissociation and molecular weight estimation. *Can. J. Biochem.* 50: 9-15.

Ecobichon, D. J. 1972b. Relative amounts of hepatic and renal carboxylesterases in mammalian species. *Res. Commun. Chem. Pathol. Pharmacol.* 3: 629-636.

Ecobichon, D. J. and W. Kalow 1961. Some properties of the soluble esterases of human liver. *Can. J. Biochem. Physiol.* 39: 1329-1332.

Ecobichon, D. J. and W. Kalow 1962. Properties and classification of the soluble esterases of human liver. *Biochem. Pharmacol.* 11: 573-583.

Franz, W. and K. Krisch 1968. Carboxylesterase aus schweine-nierenmikrosomen, I. Isolierung, ergenschaften und substratspezifität. *Hoppe-Seyler's Z. Physiol. Chem.* 349: 575-587.

Grafius, M. and D. B. Millar 1965. Reversible aggregation of acetylcholinesterase. *Biochim. Biophys. Acta* 110: 540-547.

Grafius, M. and D. B. Millar 1967. Reversible aggregation of acetylcholinesterase II. Interdependence of pH and ionic strength. *Biochemistry* 6: 1034-1046.

Hargreaves, A. B., A. G. Wanderley, F. Hargreaves, and H. Goncalves 1963. A simplified and improved method of preparation of acetylcholinesterase of the eel's electric organ. *Biochim. Biophys. Acta* 67: 641-646.

Hartree, E. F. 1972. Determination of protein. A modification of the Lowry method that gives a linear photometric response. *Anal. Biochem.* 48: 422-427.

Heymann, E., W. Junge, and K. Krisch 1971. Subunit weight and N-terminal groups of liver and kidney carboxylesterase (E.C. 3.1.1.1). *FEBS Lett.* 12: 189-193.

Kingsbury, N. and C. J. Masters 1970. Molecular weight interrelationships in the vertebrate esterases. *Biochim. Biophys. Acta* 200: 58-69.

Millar, D. B., M. A. Grafius, D. A. Palmer, and G. Millar 1973. Enzymatically active half-monomers of acetylcholinesterase. *Eur. J. Biochem.* 37: 425-433.

Schwark, W. -S. and D. J. Ecobichon 1968. Subcellular localization and drug-induced changes of rat liver and kidney esterases. *Can. J. Physiol. Pharmacol.* 46: 207-212.

Whitehouse, L. W. 1973. Doctoral Thesis, "Biotransformation of parathion by mammalian hepatic tissue." Dalhousie University.

Wynne, D. and Y. Shalitin 1973. Beef liver esterase. I. Isoelectric point and molecular weight. *Arch. Biochem. Biophys.* 154: 199-203.

STRUCTURAL CLASSES OF JACKBEAN UREASE VARIANTS AND THEIR RELATION TO A STRUCTURAL CLASSIFICATION OF ISOZYMES

WILLIAM N. FISHBEIN, K. NAGARAJAN,
and WARREN SCURZI
Biochemistry Branch
Armed Forces Institute of Pathology
Washington, D. C. 20306

ABSTRACT. A dozen molecular forms of purified jackbean urease have been characterized. Preparations of the enzyme at low ionic strength in the presence of thiols yield pure α urease, the traditional form of the enzyme with molecular weight 480,000 and Svedberg constant 18.3. All of the other forms can be generated from, and converted back to, this form. Each of the forms can also be separated in solution, or eluted from gels, and re-electrophoresed after several days with no change in mobility. The α form undergoes slow air oxidation to form a linear arithmetic, disulfide-bonded polymer series, which can be reconverted to α by addition of thiols. The polymerization rate increases with protein concentration, temperature, ionic strength, and other factors which lower the surface charge, and/or increase ionization of thiol groups. The α form can also be dissociated by non-aqueous solvents into two equal and active half-units, which reassociate to α at the isoelectric point, or upon increasing ionic strength. Application of such solvents to the polymer series generates hemipolymers, by removing a half-unit from each of the n-mers. Finally three pairs of separable isomers are demonstrated that exhibit differences in covalent bonding sites. Since on isoelectric focusing all soluble forms of the enzyme had the same isoelectric pH, and since the electrophoretic mobility and sedimentation rate differences of these isomer pairs cannot be explained by a charge difference, they are probably separated because of a difference in shape.

Urease has been little studied since its classic crystallization (Sumner, 1926) and molecular weight determination (Sumner, et al., 1938). The great size and catalytic efficiency of the enzyme remain noteworthy, but the simplistic views of a unique active form and its absolute substrate specificity have given way in the past decade to a greater appreciation of the enzyme's complexity, due largely to the development of a sensitive stain for ureolytic activity

(Fishbein, 1969a) and its application in acrylamide gel electrophoresis. Three additional substrates have been identified:
hydroxyurea (Fishbein, et al., 1965), dihydroxyurea (Fishbein,
1969b), and semicarbazide (Gazzola, et al., 1973), although
all are far less efficient than the natural substrate, urea.
The enzyme can be readily crystallized by the procedures of
Sumner (1926) and Dounce (1941) from jackbean meal, and purified to constant specific activity of 165 ± 15 Sumner units/mg.
However, the redissolved crystals displayed a dozen catalytically active forms on acrylamide gel electrophoresis at pH 8.5.
Jackbean meal extracts showed a similar array of forms plus a
number of enzymatically inactive protein bands which were
absent in the purified preparations (Fishbein, 1969c).

The forms to be discussed here are shown schematically in
Figure 1. Our nomenclature for the urease isozymes was based
on the six forms shown in black, which occurred regularly and
displayed a geometric mobility sequence, that is, the mobility
ratio of any two successive forms was constant at a given gel
strength. The fastest member of this series, which was also
the dominant form in almost all preparations, was denoted α
urease, and the succeeding members were coded by successive
letters of the Greek alphabet. The remaining forms were then
coded by their mobility relative to the next slower member of
the geometric series.

The geometricity displayed by the Greek letter series,
indicated that the members were related by a simple continuous
function, so that a relationship established for the first
three members could be extrapolated to the remaining three.
That relationship was shown to involve the molecular weight by
the use of simultaneous electrophoresis at two gel strengths,
together with a number of protein molecular weight standards
(Fishbein, et al., 1970). It was found that α urease was the
traditional form of the enzyme studied by Sumner, with molecular weight 480,000, and that β was its dimer and γ its
trimer. The remaining forms, then, were the tetramer, pentamer, and hexamer, the full series covering a molecular weight
range of 0.5-3.0 million daltons. The geometricity also provided information as to the polymer architecture, since
according to the Ogston (1958) model theory of gel electrophoresis, this pattern would obtain for linear, but not for
compact polymers (Scurzi and Fishbein, 1973).

By maintaining mercaptoéthanol and a low ionic strength in
the isolation procedures, pure α urease could be prepared.
Upon removing the mercaptoethanol by dialysis or column chromatography, and adding salt, or lowering the pH toward the isoelectric point, the polymers appeared over the subsequent hours,
days, or weeks, depending on the temperature, protein

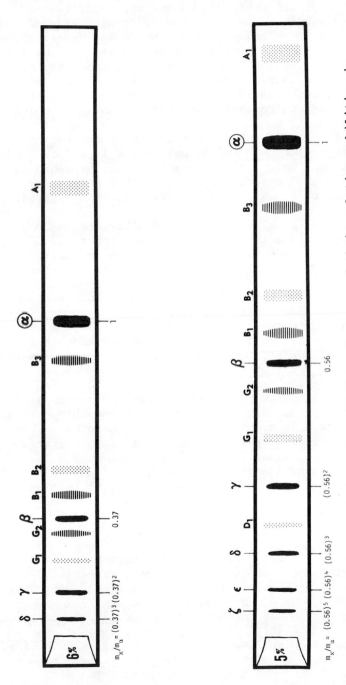

Fig. 1. Representation of the urease isozymes in terms of their relative mobilities in simultaneously run 5% and 6% acrylamide gels (5% cross-linking) in 0.05 M Tris-acetate buffer, pH 8.5. The *black* bands are a polymer series with geometric progression of mobilities; the *stippled* bands are a hemipolymer series; and the *striped* bands are a sitoisozyme series. See the text for discussion.

concentration, and pH (Fishbein and Nagarajan, 1972a). The
kinetics of polymerization, as evaluated by densitometry of
individual n-mers in acrylamide gels (Fishbein and Nagarajan,
1971), were similar to the theoretic Flory (1936) distribution
for simple linear bifunctional condensation. On addition of
mercaptoethanol, at levels as low as 50 μM, the polymers were
reconverted to α urease, indicating that oxidation with inter-
molecular disulfide bond formation was the mechanism of polymerization. Increasing ionic strength, and approaching the
isoelectric pH of 5.1 promotes polymerization by decreasing
surface charge repulsion. An additional factor is the rate of
sulfhydryl oxidation, which will increase as the thiol groups
ionize, and this would account for the fact that polymerization is faster at pH 9 than at pH 7.4 (Fishbein and Nagarajan,
1972a).

Polymerization was so slow that individual n-mers could be
eluted from homogenized gel zones and re-electrophoresed
several days later without alteration in forms or mobility,
thus eliminating any possibility of electrophoretic artifacts.
Notably slow reaction times are also characteristic of the
reaction of urease with its specific hydroxamate inhibitors
(Fishbein and Daly, 1970; Blakeley, et al., 1969).

Sedimentation constants were obtained for the first five
members of the geometric series, and these gave ratios that
agreed with the interpretation of integral arithmetic polymers.
The geometric arrangement of the polymers could then be
approached independently from the hydrodynamic data, using an
equation developed by Kirkwood (1954). This is based on a
model consisting of equal-sized spheres in any possible arrangement, so long as each sphere touches at least one other. From
the sedimentation coefficient of the dimer, its center-to-
center distance could be calculated, as shown in Figure 2, and
turned out to be almost identical to the Stokes diameter of α
urease calculated from its own sedimentation coefficient or
from its own diffusion coefficient (Fishbein, et al,, 1970,
1973). Urease thus provided an excellent fit to the Kirkwood
model, and α urease could be reliably considered to be essentially spherical. For successive polymers, additional spheres
could be added to the model, and geometries satisfying the
n-mer sedimentation coefficients could be determined (Fig. 2)
and compared to a strictly linear model. The trimer, tetramer,
and pentamer models were all quasilinear, and easily accommodated by non-rigid linear condensations, i.e., necklace-like
arrangements.

The monomer of the disulfide-bound polymer series, α urease,
could itself be dissociated into two equal and fully active
half-units, which turned out to be our A_1 band. The disso-

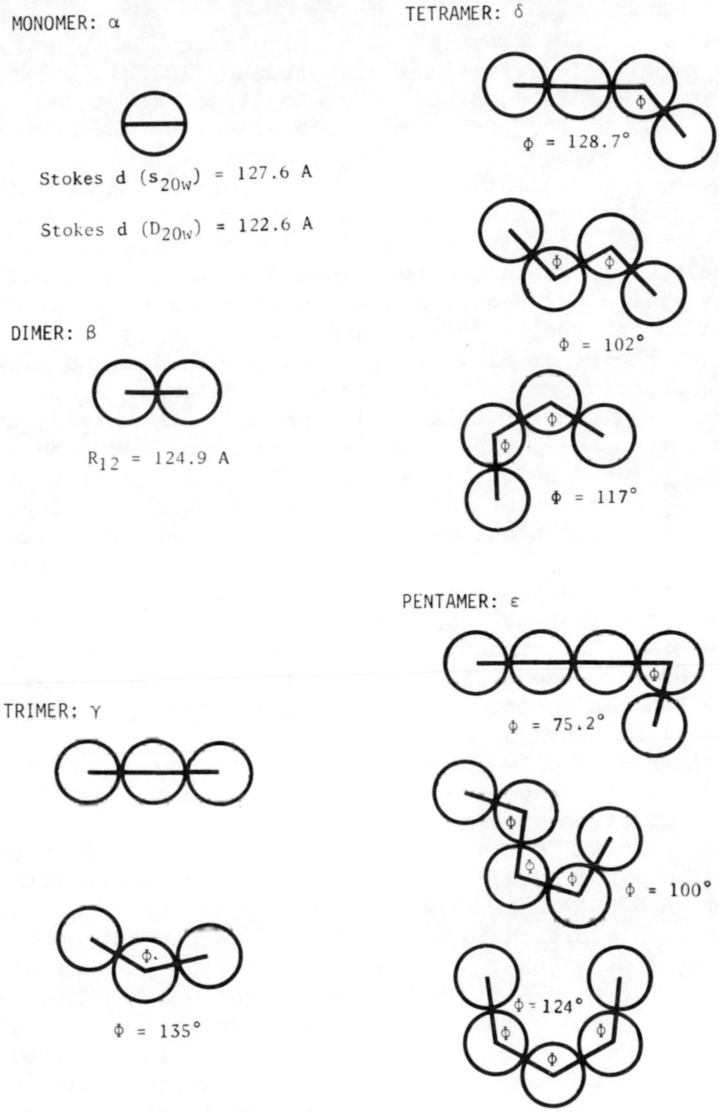

Fig. 2. Coplanar arrangements satisfying Kirkwood equation calculations from the sedimentation coefficients of the urease polymers. The close agreement between the center-to-center distance of β (R_{12}) and the Stokes diameter of α validates the Kirkwood model for urease. For data and formulae see Fishbein, et al., (1970, 1973).

ciation could be accomplished by mild acid (Tanis and Naylor, 1968) or alkali (Fishbein and Nagarajan, 1972b), by 4M citrate buffer (Lynn, 1970), or by ethylene glycol (Blattler, et al., 1967). All of the procedures were effective whether or not thiols were present, indicating that the A_1 units were joined solely by noncovalent bonds. The A_1 form could be maintained in low ionic strength aqueous solutions of pH 7-9, but otherwise reverted rapidly to α (Nagarajan and Fishbein, 1972).

Urease could therefore be prepared in three distinct quaternary states for negative stain electron microscopy with sodium phosphotungstate: A_1 urease, α urease, and high polymers (Fishbein, et al., 1972). The major electron microscopic features may be summarized as follows: A_1 urease appears as a regular triangular pyramid with three equal faces and a larger base. The base measures about 100 Å on a side, and the pyramid height is about 50 Å. The α urease molecule is composed of two of these units joined base to base, with a 10-15 Å cleavage plane between the bases, providing an overall hexahedron with dimensions about 100 x 110 Å. The forms are thus compact and symmetric in the anhydrous state, with α urease approximating a sphere, and A_1 urease approximating a hemisphere. The urease polymers appear as necklace-like arrays of α urease molecules joined along their cleavage planes at 120° angles to form straight-chain, loosely coiled, and rosette-like aggregates.

Considering the two distinct types of bonding structure involved in the urease polymers, it may be surmised that additional electrophoretic bands could be explained as hemipolymers, in which a single A_1 unit was missing from the polymer chain. Upon treating the urease polymers with 50% glycol in 0.05 M NaCl, the G_1 and B_2 forms could be regularly produced in addition to the A_1 form. Upon electrophoresis at two gel strengths, these were identified as hemipolymers, i.e., the G_1 form had a molecular weight halfway between that of β and γ, and the B_2 form had a molecular weight halfway between α and β (Fishbein, et al., 1973). The individual forms were fully separated by sucrose density-gradient sedimentation, and all had equal specific activities.

The forms remaining to be elucidated were the G_2, B_1, and B_3 bands. These could be generated, along with all of the other urease isozymes, by treatment of the polymer series with 60% glycol at pH 8. Electrophoresis at two gel strengths now indicated that three pairs of isomers were involved: B_3 had the same molecular weight as α, B_1 had the same molecular weight as B_2, and G_2 had the same molecular weight as β. Since isoelectric focusing (Fishbein and Nagarajan, 1972a) had previously shown that all soluble forms of urease had the same

pI = 5.15 ± 0.1, the only basis for electrophoretic separation
of these isomeric pairs would seem to be difference in shape.
This view was also supported by the two-dimensional separation
of all the urease isozymes by sucrose density-gradient
sedimentation followed by electrophoresis, as shown in
Figure 3. B_3 is about 5% slower than α in sedimentation rate
and in electrophoretic mobility; the same is true of B_1 relative to B_2, and of G_2 relative to β. A lower surface charge
would explain a lower electrophoretic mobility- but should
produce a greater (if not equal) sedimentation rate. However,
the more linear of two isomeric molecules with the same charge
would sediment *and* electrophorese more slowly, and we conclude
that the separation effected between these pairs is due
essentially to a difference in shape.

The structural relationship between these isomeric pairs was
established by comparing the effect on each, of treatment with
thiol and with 90% glycol. The experimental details will be
presented in a forthcoming publication; the results are summarized in Table 1, together with the sedimentation coefficients and molecular weights of all forms. Glycol treatment
dissociates α into A_1 units, but leaves B_3 unaffected, whereas
thiol reduction has the reverse effect, leaving α unaffected,
but producing A_1 from B_3. The conclusion is thus that B_3 consists of two A_1 units which are disulfide bound, in contrast
to α which contains two A_1 units which are noncovalently
bound. These forms therefore differ in the nature of their
bonding sites, constituting sitoisozymes in our initial designation of possible structures to explain the urease variants
(Fishbein, 1969c). The same applies to the B_1 and B_2 forms
(see Figure 4). The first is unaffected by glycol, but yields
A_1 on thiol reduction; the second yields B_3 and A_1 on glycol
treatment, and α and A_1 on thiol reduction. Inferentially,
the same conclusion would apply also to the G_2 and β forms,
although their separation was insufficient to permit a deci-
sive demonstration.

The structural relationships between the urease variants may
now be summarized by the simple two-dimensional model shown in
Figure 4, which explains the experimental observations, and
accounts for the difference in shape between the sitoisozyme
pairs. The model involves the formation of asymmetrical polymers from a symmetrical starting unit, since only one of the
two A_1 units in each α undergoes disulfide bond formation,
and each participating A_1 unit can react bifunctionally. If
disulfide bond formation causes rotation of the α units in the
third dimension, this would put the opposite A_1 units out-of-
plane with one another, and explain their lack of participation
in the polymerization process. The model indicates that each

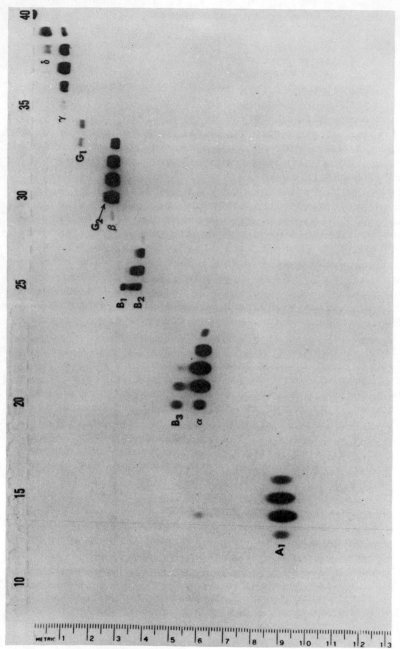

Fig. 3. Two-dimensional separation of the urease isozymes by sucrose density-gradient sedimentation and electrophoresis. Sedimentation rates increase *rightward*, and electrophoretic mobilities *downward* . The 6% gels were stained for ureolytic activity. See text for discussion.

411

TABLE 1

Some Physical and Chemical Characteristics of the Urease Isozymes

Form (in order of decreasing μ)	Product in 90% Glycol	Product of Thiol Reduction	10^{-6} x MW		10^{13} x Sedimentation coefficient		
			Mean ± SD	(No.)	Mean ± SD	(No.)	Concentration Dependence
A_1	0	0	0.24 ±.01	(5)	11.49 ±.34	(22)	0
α	A_1	0	0.48 ±.01	(5)	18.34 ±.29	(17)	0
B_3	0	A_1	0.49 ±.02	(5)	17.3	(1)	0
B_2	B_3,A_1	$α,A_1$	0.73 ±.02	(5)	23.28 ±.63	(6)	0
B_1	0	A_1	0.75 ±.09	(3)	21.7	(1)	
β	B_3,A_1	α	0.96 ±.03	(5)	27.09 ±.49	(11)	0
G_2	[B_1,A_1]		[0.96]		26.6	(1)	
G_1	[B_1,A_1]		1.24 ±.04	(5)	29.77 ±.59	(3)	
γ	[B_1,A_1]	α	1.45 ±.04	(5)	32.93 ±.45	(8)	0
δ		α	[1.94]		39.8	(1)	
ε		α	[2.22]		44.4	(1)	

Information in brackets is inferential, or otherwise non-definitive. A zero means the system was tested and there was no effect. Molecular weights are from two-gel electrophoresis. Sedimentation coefficients are corrected to water at 20^0, and include 11 values obtained by sucrose density-gradient procedures. The D_1 form is not listed because no information is available except its electrophoretic mobility, which suggests it is a hemipolymer analogous to B_2 and G_1.

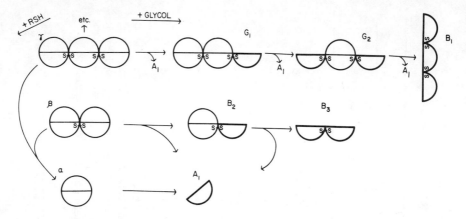

Fig. 4. Schematic representation of the urease isozymes. Non-covalent bonding between the two hemispherical A_1 units in each α is implicit. In the three sitoisozyme pairs, B_3 -α, B_1 - B_2, and G_2 - β, the former member is, in each case, more linear than the latter. See text for discussion.

---

urease n-mer should yield a unique endstage, on glycol treatment, consisting of disulfide-bound A_1 units, that would represent a sitoisozyme of some other form.

It may be surmised by this point that we do not agree with the IUB recommendation that the term isozyme be restricted to genetic variants. Not only are variants often erroneously considered genetic, but genetic forms can be subsequently modified, leading to redundant terminology. We believe the term "isozyme" should be used in the broadest operational sense to designate forms, separable *from the same solution,* of a protein from the same species, catalyzing the same stoichiometric reactions. The underlined qualification is necessary to eliminate artifacts due to the solution environment, which might contain ligands, proteases, or redox agents. Differential mobility of forms from separate sources might therefore disappear when the solutions are mixed. It should also be clear from the behavior of the urease polymers, that separability does not exclude the possibility of slow interconversion of the forms. The other class of artifacts which must be excluded, involve the alteration of forms during separation procedures, which, in the case of gel electrophoresis, is due primarily to oxidants, alkaline pH, or high ionic strengths. Therefore, the possibility of disulfide bond formation (and subsequent hydrolysis), and of oxidative or hydrolytic removal of amino groups must be evaluated by re-running

eluted forms, by orthogonal electrophoresis, or other procedures that can verify the stability of individual variants.

A list of the possible types of isozymes is shown in
Table 2, classified according to the level of chemical structure involved. At the primary structural level are included
not only variations in sequence and chain length, but also
any other molecules covalently bonded to the main peptide
chain, since hydrolysis would yield separate components, which
may be identified chemically. The amino acid pairs aspartic
acid - asparagine; and glutamic acid - glutamine are included
under end-group-side-chain variants to emphasize that, while
these pairs are coded by separate triplets differing by only
a single nucleotide, they are also interconvertible by (de)-
amidation; hence, even a sequence analysis would not be sufficient to guarantee a genetic origin for such variants.

Chiroptical and absorption spectroscopy in the far ultraviolet would seem to be the best approach to evaluate differences in secondary structure, and we believe that partial
denaturation, without full inactivation, is a possibility to
account for differences encountered here. At the level of
tertiary structure must be considered not only alternate conformational states, but also covalent bridges between amino
acids in the primary chain, since hydrolysis in this case
would yield no extra components, despite the differences in
bonding sites. Although the disulfide bond is the classic
example of this, more obscure possibilities are included in
Table 2. All of these are isomeric forms, that is, macromolecular isomers, neglecting slight alterations due to ligands
or water groups. The same types of isomers may also be
encountered at the quaternary level, and indeed are exemplified by the isomeric pairs of urease. It must be emphasized that an alteration in bonding structure will not, of
itself, explain the separability of the forms, which must
depend instead upon a coincident alteration in surface charge,
or, as in the case of urease, upon differences in molecular
shape. The urease variants, are not conformers, nevertheless,
since they exhibit distinguishable differences in covalent
bonding.

The possibility of epimeric isozymes should be considered
whenever the basal active assembly consists of six or more
subunits, since multiple active assemblies might exist,
differing by a single subunit, and hence by only about 10% in
molecular weight. Horse apoferritin, for example, exists in
the native state as an assembly of some 20-24 identical subunits (Crichton and Bryce, 1970). If multiple assemblies of

TABLE 2

CLASSIFICATION OF ISOZYMES ACCORDING TO THE LEVEL
OF CHEMICAL STRUCTURE INVOLVED

PRIMARY (Covalent with separable components for each bond)
1. Sequence Variants
   a. Protomeric (or unique oligomers)
   b. Oligomeric hybrids
2. End-group-Side-chain Variants
   a. Asp$\leftrightarrow$Asn; Glu$\leftrightarrow$Gln
   b. (de)Amidation, acylation, PO_4, SO_4
   c. Lipid, sugar, nucleotide conjugates
   d. Covalent prosthetic groups
3. Chain-length Variants
   a. Genetic
   b. Hydrolytic
      (1) Uncatalyzed
      (2) Proteolytic
      (3) Autocatalytic
SECONDARY (Non-covalent & isomeric)
4. Helicomers
   a. Ligand, protein interactions
   b. Partial denaturation
TERTIARY
5. Isomers
   a. Covalent bridge, internal
      (1) Disulfide
      (2) Amide, (thio) ester/lactone
   b. Conformeric (non-covalent)
      (1) Ligand-induced
      (2) Metastable states (true)
QUATERNARY
6. Polymers
   a. Covalent
   b. Non-covalent
7. Epimers
   a. Covalent
   b. Non-covalent
8. Isomers
   a. Covalent bridge, external (as 5a)
   b. Conformeric (non-covalent) (as 5b)

this protein do exist, they would differ by only 4% or 5% in
molecular weight, which would normally be regarded as experimental error. Since α urease may contain as many as 16
identical subunits (Contaxis and Reithel, 1972), we considered

this possibility early in our work (Fishbein, 1969c), and Blattler and Reithel, (1970) published data suggesting intermediate forms of urease differing by 60,000 in molecular weight. However, we consider that evidence unconvincing, since the size estimates were made at a single gel strength, and believe their intermediate forms are not epimers, but isomeric isozymes.

ACKNOWLEDGEMENT

This investigation was supported in part by United States Public Health Service Research Grant AM-10960 from National Institute of Arthritis, Metabolic and Digestive Diseases, Department of Health, Education, and Welfare, under the auspices of Universities Associated for Research and Education, in Pathology, Inc. The opinions or assertions contained herein are the private views of the authors and are not to be construed as official or as reflecting the views of the Department of the Army or the Department of Defense.

REFERENCES

Blakeley, R. L., J. A. Hinds, H. E. Kunze, E. C. Webb, and B. Zerner 1969. Jackbean urease (EC 3.5.1.5). Demonstration of a carbamoyl-transfer reaction and inhibition by hydroxamic acids. *Biochemistry* 8: 1991-2000.

Blattler, D. P. and F. J. Reithel 1970. Urease oligomers: a polyacrylamide gel electrophoresis study. *Enzymologia,* 39: 193-199.

Blattler, D. P., C. C. Contaxis, and F. J. Reithel 1967. Dissociation of urease by glycol and glycerol. *Nature,* 216: 274-275.

Contaxis, C. C. and F. J. Reithel 1972. A study of the structural subunits of urease obtained during controlled dissociation. *Can J. Biochem.,* 50: 461-473.

Crichton, R. R. and C. F. A. Bryce 1970. Molecular weight estimation of apoferritin subunits. *FEBS Letters,* 6: 121-124.

Dounce, A. L. 1941. An improved method for recrystallizing urease. *J. Biol. Chem.,* 140: 307-308.

Fishbein, W. N. 1969a. A sensitive and non-inhibitory catalytic gel stain for urease. *Fifth International Symposium on Chromatography and Electrohporesis,* Ann Arbor-Humphrey Science Press, Ann Arbor, p. 238-241.

Fishbein, W. N. 1969b. Urease catalysis III. Stoichiometry, kinetics, and inhibitory properties of a third substrate: Dihydroxyurea. *J. Biol. Chem.,* 244: 1188-1193.

Fishbein, W. N. 1969c. The structural basis for the catalytic complexity of urease: interacting and interconvertible molecular species (with a note on isozyme classes), *Ann. N. Y. Acad. Sci.*, 147: 857-881.

Fishbein, W. N. and J. E. Daly 1970. Urease inhibitors for hepatic coma II. Comparative efficacy of four lower hydroxamate homologs in vitro and in vivo. *Proc. Soc. Exp. Biol. Med.*, 134: 1083-1090.

Fishbein, W. N. and K. Nagarajan 1971a. Urease catalysis and structure VII. Factors involved in urease polymerization and its kinetic pattern. *Arch. Biochem. Biophys.*, 144: 700-714.

Fishbein, W. N. and K. Nagarajan 1972. Urease catalysis and structure VIII. Ionic strength dependence of urease polymerization and polymer pI. *Arch. Biochem. Biophys.*, 151: 370-377.

Fishbein, W. N. and K. Nagarajan 1972b. Production of the urease half-unit at alkaline pH. *Proc. Soc. Exp. Biol. Med.*, 140: 505-506.

Fishbein, W. N., K. Nagarajan, and W. Scurzi.1970. Urease catalysis and structure. VI. Correlation of sedimentation coefficients and electrophoretic mobilities for the polymeric urease isozymes. *J. Biol. Chem.*, 245: 5985-5992.

Fishbein, W. N., K. Nagarajan, and W. Scurzi 1973. Urease catalysis and structure IX. The half-unit and hemipolymers of jackbean urease. *J. Biol. Chem.*, 248: 7870-7877.

Fishbein, W. N., T. S. Winter, and J. D. Davidson 1965. Urease catalysis I. Stoichiometry, specificity, and kinetics of a second substrate: hydroxyurea. *J. Biol. Chem.*, 240: 2402-2406.

Fishbein, W. N., J. L. Griffin, W. F. Engler, and G. F. Bahr 1972. Electron microscopy of negatively stained urease at different levels of quaternary structure. *Fed. Proc.*, 31: 641 abs.

Flory, P. J. 1936. Molecular size distribution in linear condensation polymers. *J. Amer. Chem. Soc.*, 58: 1877-1885.

Gazzola, C., R. L. Blakeley, and B. Zerner 1973. On the substrate specificity of jackbean urease. *Can. J. Biochem.*, 51: 1325-1330.

Kirkwood, J. G. 1954. The general theory of irreversible processes in solutions of macromolecules. *J. Polymer Sci.*, 12: 1-14

Lynn, K. R. 1970. Urease: the 12S form. *Can. J. Biochem.*, 48: 631-632.

Nagarajan, K., and W. N. Fishbein 1972. Evaluation of the stability and association rate of the urease half-unit by gel electrophoresis. *Fed. Proc.*, 31: 889 Abs.

Ogston, A. G. 1958. The spaces in a uniform random suspension of fibers, *Trans. Faraday Soc.*, 54: 1754-1757.

Scurzi, W., and W. N. Fishbein 1973. The geometric mobility sequence in polyacrylamide gel electrophoresis of polymeric proteins: its significance for polymer geometry and electrophoretic theory. *Trans. N. Y. Acad. Sci.*, 35: 396-416.

Sumner, J. B. 1926. The isolation and crystallization of the enzyme urease. *J. Biol. Chem.*, 69: 435-441.

Sumner, J. B., N. Gralén, and I.-B. Eriksson-Quensel 1938. The molecular weight of urease. *J. Biol. Chem.*, 125: 37-44.

Tanis, R. J. and A. W. Naylor 1968. Physical and chemical studies of a low-molecular-weight form of urease. *Biochem. J.*, 108: 771-777.

RAT LIVER TYROSINE AMINOTRANSFERASE: PROBLEMS OF
INTERPRETING CHANGES IN MULTIPLE FORM PATTERNS

RONALD W. JOHNSON AND ALBERT GROSSMAN
Department of Pharmacology
New York University School of Medicine
New York, New York 10016

ABSTRACT. Following cortisol administration, three peaks of
rat liver tyrosine aminotransferase activity can be resolved
by CM-Sephadex chromatography. The proportion of enzyme
activity in the three forms is dependent on the method of
cell disruption and the pH and ionic strength of the buffers
used. Homogenization with glass-teflon tissue homogenizers
in 0.01-0.05 M potassium phosphate buffer (pH 6.5) yields
predominatly form I of the enzyme. This form of the enzyme
is the last to be eluted from the column. Increasing ionic
strength and/or elevating pH causes most of the enzyme to
appear as form III, the first major peak to be eluted from
the column. Further results indicated that freeze-thawing
or sonication cause conversion of form III to form I. By
pressing liver slices between glass slides and analyzing
the exudate, almost all of the tyrosine aminotransferase
activity was in form III. Presumably, this is the predominant form of the enzyme in vivo. The manner in which other
forms arise in vitro is unresolved.

The addition of 10 mM cyanate to a fresh liver homogenate
has been shown previously to convert tyrosine aminotransferase form I to forms II and III. The response to cyanate
disappears rapidly when the homogenate is preincubated at
4°. The amount of form III produced in response to cyanate
is half as great after 15 min. This suggests there is a
labile component, possibly the enzyme itself, in the cyanate-
dependent conversion.

Many unknown facets concerning the origin of these various enzyme forms remain. At present, one must proceed cau-
tiously when trying to interpret the significance of multiple
forms, and when attempting to compare patterns obtained from
different laboratories.

INTRODUCTION

The discovery by Lin and Knox (1957), that glucocorticoids
could induce rat liver tyrosine aminotransferase, initiated a
series of investigations whose object was to gain insight
into the mechanisms of mammalian enzyme regulation. Among the
findings from these studies are: (a) the half-life of tyrosine aminotransferase under basal conditions is about 2-6 hours.
(Kenney, 1967; Auricchio et al., 1969; Boctor and Grossman, 1970;

and Hershko and Tomkins, 1971); (b) following cortisol administration there is increased synthesis of enzyme as well as a
reduction in enzyme turnover (Kenney, 1962; Granner et al.,
1968; Boctor and Grossman, 1970; Levitan and Webb, 1970); and
(c) with declining concentration of hormone the rate of enzyme
synthesis returns to the basal rate and induced enzyme is rapidly inactivated (Levitan and Webb, 1970; and Grossman and Boctor,
1972). With regard to the latter observation, it still remains
to be determined how induced tyrosine aminotransferase is inactivated and how its components are returned to the protein
precursor pool.

A number of reports have indicated that tyrosine aminotransferase exists in multiple isozymic forms and that the proportion of each form can be affected by various hormones and
changes in diet (Holt and Oliver, 1969; Sadleir et al., 1970;
Blake and Broner, 1970; Iwasaki and Pitot, 1971; Gelehrter et
al., 1972; Iwasaki, 1973; Mertvetsov, 1973; and Johnson et al.,
1973). It has been inferred that these forms are inter-con-
vertible and that one form may be the substrate for the tyrosine aminotransferase inactivation system (Johnson et al.,
1973). Although this is an interesting hypothesis, the physiological role of these multiple enzyme forms in unclear. In
fact, it is uncertain, at present, whether multiple enzyme
forms exist within the cell. The observations presented here
indicate that the distribution of multiple forms of tyrosine
aminotransferase can be altered by several in vitro procedures.
It is appropriate, at this time, to stress that caution should
be observed in interpreting the significance of multiple forms
of rat liver tyrosine aminotransferase, and perhaps, of multiple
forms of other enzymes as well.

RESULTS

EFFECTS OF pH AND IONIC STRENGTH OF THE HOMOGENIZING BUFFER ON DISTRIBUTION OF MULTIPLE FORMS OF TYROSINE AMINOTRANSFERASE

Homogenates of rat liver (from adrenalectomized males) prepared
3-4 hours after administration of hydrocortisone phosphate (5
mg IP/100 g body weight) can be shown to contain at least three
reproducibly separable forms of tyrosine aminotransferase following CM-Sephadex column chromatography. The distribution of
enzyme activity in each peak, however, can be altered readily
simply by varying the composition of the homogenizing medium.
The data in panels A, B, and C of Fig. 1 demonstrate a shift
from enzyme form I to form III when rat liver samples are broken
with an electrically driven glass-teflon tissue homogenizer in
increasingly concentrated potassium phosphate buffers having
the same pH (6.5). It should be noted that the ordinate scale
of panel A is one-fourth that of the other panels. In this

experiment there was approximately a 4-fold difference in
total soluble tyrosine aminotransferase activity between liver

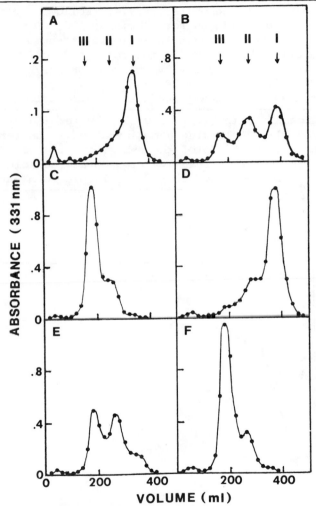

Fig. 1 The effects of homogenizing liver samples in various
buffers on tyrosine aminotransferase multiple form patterns
as analyzed by CM-Sephadex chromatography. Six 1 g samples
of liver, obtained from a hydrocortone phosphate treated animal, were homogenized separately at 0°in glass-teflon tissue
grinders in the following media: panel A, 4 ml 0.01 M potassium phosphate buffer, pH 6.5; panel B, 4 ml 0.05 M potassium phosphate buffer, pH 6.5; panel C, 4 ml 0.25 M potassium phosphate buffer, pH 6.5; panel D, after homogenizing
in 0.01 M potassium phosphate buffer, pH 6.5, as in panel A,

(Fig. 1 Con't.) 1.22 ml of 1 M potassium phsophate (pH 6.5) was added to make the final potassium phosphate concentration 0.2 M; panel E, 4 ml of 0.03 M potassium phosphate, pH 7.4; panel F, 0.05 M potassium phosphate, pH 7.4. Following homogenization, each was centrifuged for 35 min at 105,000 x g at 4°. One ml aliquots were brought to an ionic strength of 0.05 with ice cold distilled water and chromatographed on CM-Sephadex columns. Total enzyme activity charged (µmoles/hr/ml): A, 11.8; B, 55.1; C 74.3; D, 69.4; E 64.3; F, 88.1. Tyrosine aminotransferase was assayed by the procedure of Diamondstone 1966.

---

samples homogenized in 0.01 and 0.05 M buffer (compare panels A and B, Fig. 1). Following homogenization in buffers of low ionic strength, about 80% of the enzyme remains bound to particulate structures but is readily released as the solute concentration is increased (Fig. 2). The multiple form pattern of

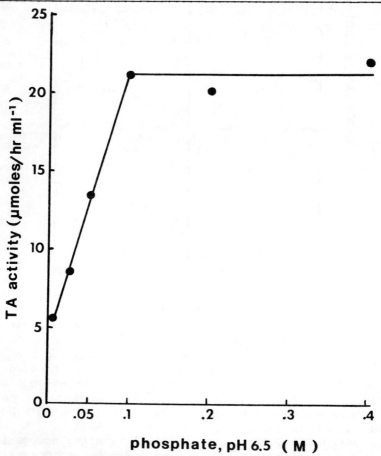

Fig. 2. Ionic strength-dependent solubilization of rat liver

422

(Fig. 2 Cont'd) tyrosine aminotransferase. Six g of liver from an animal treated with hydrocortone phosphate were homogenized in a glass-teflon tissue grinder with 24 ml ice cold 0.01 M potassium phosphate buffer, pH 6.5. To 1.2 ml aliquots of this homogenate was added varying amounts of ice cold 1 M potassium phosphate (pH 6.5) to give molar concentrations of 0.01, 0.025, 0.05, 0.1, 0.2 and 0.4 when each was brought to a final volume of 2.0 ml. Each sample was centrifuged at 4° for 35 min at 105,000 x g and an aliquot of the supernatant assayed for tyrosine aminotransferase activity.

released enzyme is dependent on the ionic strength of the medium in which the liver cells are initially broken. When liver homogenates were prepared initially in 0.01 M phosphate buffer pH 6.5, (as in panel A. Fig. 1) and then brought to 0.2 M, pH 6.5, an increase in total tyrosine aminotransferase activity was observed as described above, but the enzyme was predominantly in form I (panel D, Fig. 1). Homogenates prepared directly in 0.25 M phosphate buffer, pH 6.5, yielded enzyme mainly in form III (panel C, Fig. 1). Similar changes were observed in homogenates prepared in 0.2 M sucrose or KCl (pH ranged between 6.5-6.7) indicating that increased phosphate was not essential for the changes observed. Form III predominated not only when homogenates were prepared in solutions of high solute content, but also when pH was increased and ionic strength kept within narrow limits (compare panels B, E, and F, Fig. 1).

EFFECT OF FREEZING AND THAWING AND SONICATION

Liver homogenates prepared in 0.05 M phosphate buffer (pH 7.4) containing potassium chloride (0.07 M) were shown to contain tyrosine aminotransferase predominantly as form III (panel A, Fig. 3). Quick freezing of the liver, either before or after homogenizing in the same buffer, caused almost all the enzyme to shift to form I (panels B and C, Fig. 3).

Liver slices disrupted by short burst of sonic oscillation in buffers of varying ionic strengths produce essentially the same multiple form patterns as those observed after homogenization in glass-teflon tissue grinders (see panels A, B, and C of Fig. 1). However, if the initial sonicate, which was prepared in 0.05 M phosphate buffer (pH 6.5), was centrifuged and the supernatant exposed again to three 15 sec sonic bursts, a shift in the multiple form pattern of tyrosine aminotransferase was observed (Fig. 4). No such shift was seen when these procedures were carried out at pH 7.4 instead of 6.5.

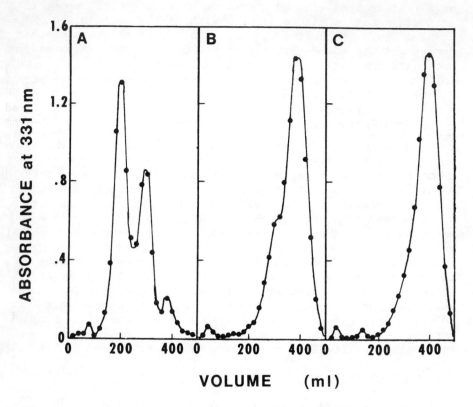

Fig. 3. Effect of freezing and thawing on tyrosine aminotransferase multiple forms as analyzed by CM-Sephadex chromatography.
Two grams of liver from a hydrocortone phosphate induced rat
was homogenized in 8 ml of 0.05 M potassium phsophate buffer
(pH 7.6) containing 0.07 M potassium chloride (final pH 7.4;
final ionic strength, about 0.15). Half the homogenate was
frozen in liquid nitrogen and then thawed at 20°. Both samples
were centrifuged for 35 min at 105,000 x g at 4° and 1 ml of
each supernatant chromatographed on CM-Sephadex columns. Panel A, control (187 μmoles/hr/ml); panel B, frozen-thawed sample
(176 μmoles/hr/ml). Another 1 g sample of the same liver was
frozen directly in liquid nitrogen, thawed at room temperature;
and then macerated with a cold porcelain mortar and pestle.
The resultant paste was extracted with 4 ml of the above buffer,
centrifuged, and 1 ml of the supernatant (167 μmoles/hr/ml)
chromatographed (panel C). In all cases the ionic strength
was adjusted to 0.05 prior to charging the sample on the column.

ENZYME PATTERN IN INTACT CELL

By the very nature of the information sought, it is difficult to say with certainty what forms of the enzyme, as presently described, actually occur in the hepatocyte. To gain

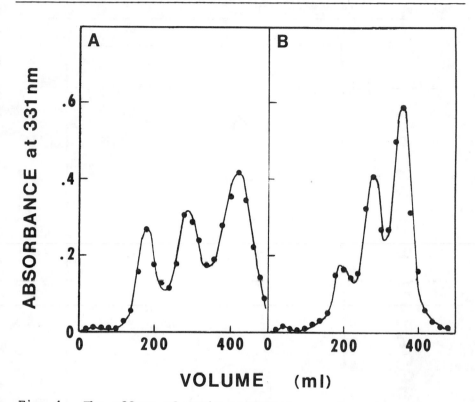

Fig. 4. The effect of sonic oscillation on tyrosine aminotransferase multiple forms as analyzed by CM-Sephadex chromatography. One g of rat liver, obtained from an animal that
received hydrocortone phosphate 3 hrs prior to sacrifice, was
minced with scissors and then sonicated by three 15 sec bursts
(65 Watts) in 4 ml of 0.05 M potassium phosphate buffer (pH
6.5). (Sonication was carried out with a Bronson Sonifier,
Model 185C, Heat Systems, Inc., Melville, L.I., N.Y.) The
tube in which the liver sonicate was prepared was kept surround-
ed by ice during this procedure. After centrifuging this material for 35 min at 105,000 x g at 4°, a 1 ml aliquot (94.9 μmoles/
hr/ml) was chromatographed directly (panel A) and a second 1 ml
aliquot (98.6 μmoles/hr/ml) was sonicated again, as above, and
then applied to a CM-Sephadex column (panel B).

425

some insight into this question liver slices were crushed between glass slides and the exudate transferred to centrifuge
tubes. After centrifuging for 35 min at 105,000 x g the small
supernatant volume was chromatographed in the usual way. The data
in Fig. 5 indicate that the enzyme obtained by this procedure
was almost entirely form III.

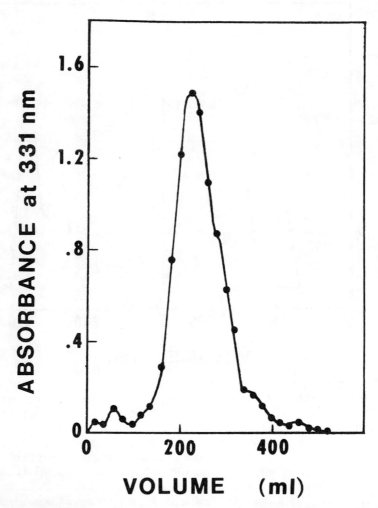

Fig. 5. CM-Sephadex chromatography of rat liver exudate.
Slices of rat liver obtained from a hydrocortone phosphate induced animal were crushed between glass slides. As much of
the broken tissue mass as possible was transferred by Pasteur
pipet to a centrifuge tube and spun for 35 min at 105,000 x g.
The small supernatant volume was carefully removed and added

Fig. 5 (Cont'd) to 4 ml ice cold distilled water. This entire volume was charged onto a CM-Sephadex column, eluted, and
analyzed for tyrosine aminotransferase activity.

In other experiments, biopsy samples of the liver were taken
from pentobarbital anesthetized rats prior to, and 3 and 9 hours
after, glucocorticoid administration and treated in a manner
similar to that described above. At each time the enzyme patten
resembled the one shown in Fig. 5 indicating predominantly one
form of the enzyme (form III).

INTRINSIC FACTORS INFLUENCING THE DISTRIBUTION
OF TYROSINE AMINOTRANSFERASE MULTIPLE FORMS

Johnson et al. 1973 observed that potassium cyanate caused
predominance of form III under conditions in which form I was
the expected major enzyme species. As shown in panel A of Fig.
6, homogenization of the liver in 0.01 M phosphate (pH 6.5),
followed immediately by an increase in phosphate concentration
to 0.05 M, yielded predominantly form I of tyrosine aminotransferase. If such a homogenate was brought quickly to 10 mM
with respect to cyanate, form III of tyrosine aminotransferase predominated (panel B, Fig. 6). This effect of cyanate
was specific since it could not be duplicated by addition of
equivalent amounts of other salts. The reason for having
carried out this experiment with haste, even at 0-4°, was that
the capacity of cyanate to convert form I to form III was time
dependent. Panels C-F (Fig. 6) demonstrate the progressively
diminished effect of cynate over a 15 min period at 0°. The
fraction of total tyrosine aminotransferase converted to form
III, expressed as a percent, declined at a first order rate,
with a half-time of about 15 min (Fig. 7). Sometimes, the
addition of cyanate caused an increase in tyrosine aminotransferase activity (compare panels A and B of Fig. 7). Other
experiments indicated this affect was not related to the capacity to convert form I of tyrosine aminotransferase to form III.

DISCUSSION

The data presented demonstrate that multiple form patterns
of tyrosine aminotransferase can easily be altered by varying
the medium in which the liver is homogenized and by procedures
commonly used to disrupt cells such as sonication and freeze-
thawing. As clearly shown in Fig. 1, elevating ionic strength
or pH of the homogenizing medium yields predominantly form III
of tyrosine aminotransferase, while lowering these parameters
favors form I. Of the buffer concentrations tested, 0.25 M
potassium phosphate (final concentration in homogenate about
0.2 M) is probably closest in ionic strength to that found within the cell. This would suggest, then, that form III should

427

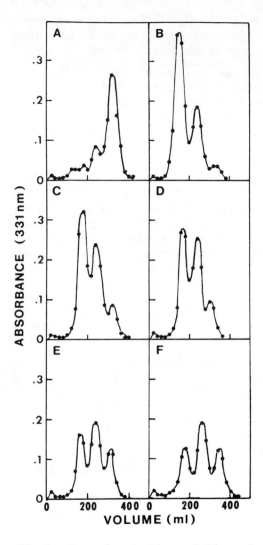

Fig. 6. The effect of preincubation of liver homogenate at 0-4° on the KOCN-dependent interconversion of tyrosine amino-transferase multiple forms as analyzed by CM-Sephadex chromatography. Six g of liver from a hydrocortone phosphate induced rat were homogenized with 24 ml of cold 0.01 M potassium phosphate (pH 6.5) A control sample (panel A) was prepared by adding 0.8 ml of 0.05 M potassium phosphate to 1.2 ml of the homogenate. At various times after preparation of the homogenate 1.2 ml aliquots were combined with 0.8 ml of 0.025 M KOCN in 0.05 M potassium phosphate (pH 6.5) Panel B; zero time; C, 2 min; D, 5 min; E, 10 min; and F, 15 min.

predominate in vivo. Breaking liver cells between glass slides and analyzing the exudate indicated that most of the enzyme was in this form (Fig.5). Essentially the same pattern as that shown in Fig. 5 was observed before and after glucocorti- coid admininstration. This observation is at variance with a

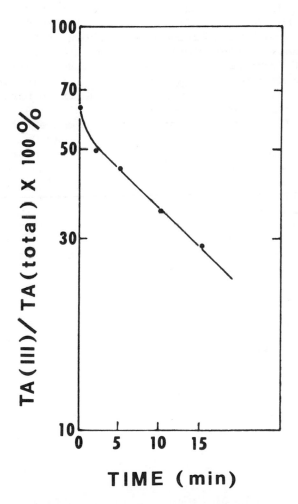

Fig. 7 The effect of preincubation of a liver homogenate at 0-4° on the KOCN-dependent conversion of tyrosine aminotransfer- ase forms I and II to form III. These data were calculated from those in figure 6.

number of reports that indicate significant variation in multiple forms of tyrosine aminotransferase resulting from changes in diet or hormone administration. It is possible that some of the conditions presently reported to alter multiple form patterns were responsible, at least in part, for the changes in distribution of tyrosine aminotransferase previously reported. Our data suggest that forms I and II are at lower levels within the liver. The manner in which they appear following cell rupture remains unresolved.

Once formed, forms I and II can be converted to a more rapidly moving protein similar to form III. This can be accomplished by the addition of cyanate as shown in Fig. 6. It is not clear whether carbamylation with cyanate results in re-formation of form III or formation of another multiple form having the same choromatographic mobility. It is clear, however, that a labile factor present in the fresh homogenate is required in order for cyanate to be effective. As shown in Figs. 6 and 7, the capacity of cyanate to convert tyrosine aminotransferase forms I and II to form III at 0° disappeared with a half-life of 15 min. During this time there was usually no loss of enzyme activity. Clearly, the labile component is the enzyme itself or some unknown cofactor. It should be mentioned that sulfhydryl reagents have no effect on the distribution of multiple forms of tyrosine aminotransferase.

It is apparent that problems exist in interpreting multiple form patterns of tyrosine aminotransferase. We wish to stress that detailed conditions of tissue preparation are necessary to allow evaluation and comparison of data from different laboratories.

REFERENCES

Auricchio, F.,D. Martin and G. Tomkins, 1969. Control of degradation and synthesis of induced tyrosine aminotransferase studied in hepatoma cells in culture, *Nature*, 224, 806-808.

Blake, R. L. and J. Broner, 1970. Deficiency of glucocorticoid inducible isoenzyme of liver tyrosine aminotransferase in the obese C57BL/6J-*ob* Mutant Mouse, *Biochem. Biophys. Res Commun*, 41, 1443-1451.

Boctor, A. and A. Grossman, 1970. Alteration of tyrosine aminotransferase turnover in rat liver following gluocorticoid administration, *J. Biol. Chem.*, 245, 6337-6345.

Diamondstone, T. I., 1966. Assay of tyrosine aminotransferase activity by conversion of p-hydroxyphenylpyruvate to p-hydroxybenzaldehyde, *Anal. Biochem.*, 16, 395-401.

Gelehrter, T. D., J. R. Emanuel and C. J. Spencer, 1972. In duction of tyrosine aminotransferase by dexamethasone, insulin, and serum, *J. Biol. Chem.*, 247, 6197-6203.

Granner, D. K., S. Hayashi, E. B. Thompson and G. Tomkins 1968, Stimulation of tyrosine aminotransferase synthesis by dexamethasone phosphate in cell culture, *J. Mol. Biol.*, 35, 291-301.

Grossman, A. and A. Boctor 1972. Evidence for reversible inactivation of induced tyrosine aminotransferase in rat liver in vivo, *Proc. Nat. Acad. Sci. USA*, 69, 1161-1164.

Hershko, A. and G. Tomkins, 1971. Studies on the degradation of tyrosine aminotransferase in hepatoma cells in culture, *J. Biol. Chem.*, 246, 710-714.

Holt, P. G. and I. T. Oliver 1969. Multiple forms of tyrosine aminotransferase in rat liver and their hormonal induction in the neonate, *Fed. Eur. Biochem. Soc.*, 5, 89-91.

Iwasaki, Y. , C. Lamar, K. Danenberg and H. C. Pitot, 1973. Studies on the induction and repression of enzymes in rat liver, *Eur. J. Biochem.*, 34, 347-357.

Iwasaki, Y. and H. C. Pitot, 1971. The regulation of 4 forms of tyrosine aminotransferase in adult and developing rat liver, *Life Sciences.*, 10, 1071-1079.

Johnson, R. W., L. E. Roberson and F. T. Kenney 1973. Regulation of tyrosine aminotransferase in rat liver, *J. Biol. Chem.*, 248, 4521-4527.

Kenney, F. T. 1967., Turnover of rat liver tyrosine transaminase: Stabilization after inhibition of protein synthesis, *Science*, 156, 525-527.

Kenney, F. T. 1962. Induction of tyrosine-α-ketoglutarate transminase in rat liver. Immunochemical analysis, *J. Biol. Chem.*, 237, 1610-1614.

Kenney, F. T. 1962, Induction of tyrosine-α-ketoglutarate transaminase in rat liver evidence for an increase in the rate

431

of enzyme synthesis, *J. Biol. Chem.*, 237, 3495-3498.

Levitan, I. B. and T. E. Webb, 1970. Hydrocortisone-mediated changes in the concentration of tyrosine aminotransferase in rat liver: An immunochemical study, *J. Mol. Biol*, 48, 339-348.

Lin, E. C. C. and W. E. Knox, 1957. Adaptation of the rat liver tyrosine-α-ketoglutarate transaminase, *Biochim. Biophys. acta*, 26, 85-88.

Mertvetsov, N. P., V. N. Chesnokov and R. I. Salganik, 1973. Specific changes in the activity of tyrosine aminotransferase isoenzymes in rat liver after cortisol treatment and partial hepatectomy, *Biochim. Biophys. Acta*, 315, 61-65.

Sadleir, J. W., P. G. Holt and I. T. Oliver, 1960. Fractionation of rat liver tyrosine aminotransferase during the course of purification. Further evidence for multiple forms of the enzyme, *Fed. Eur. Biochem. Soc.*, 6, 46-48.

MOLECULAR FORMS OF HEPATIC TYROSINE AMINOTRANSFERASE ACTIVITY

THOMAS D. GELEHRTER and CAROLYN J. SPENCER
Department of Human Genetics
Yale University School of Medicine
New Haven, Connecticut 06510

ABSTRACT. Dexamethasone, insulin, and serum induce an additive increase in the activity of tyrosine aminotransferase
in hepatoma cells in tissue culture. By the criteria of
immunotitration, heat stability, and electrophoresis on polyacrylamide gels, we have shown previously that all three
effectors increase the cellular concentration of the same
protein. During the course of these studies, however, we
observed an apparent isozyme of tyrosine aminotransferase
whose activity is not enhanced by these three inducers.
   This minor, more cathodal form is also a heat-stable,
cytoplasmic enzyme which differs from the major anodal form
in net charge but not molecular weight. Unlike the anodal
form, however, this enzyme can utilize oxaloacetate in place
of α-ketoglutarate in the transamination of tyrosine, and
can transaminate aspartate as well as tyrosine. A similar,
heat-stable, cathodal form with these same biochemical properties has also been found in cytoplasmic extracts of adult
and fetal rat liver. Incubation of extracts of hepatoma
cells, adult, and fetal rat liver with antiserum to purified
rat liver soluble tyrosine aminotransferase inactivates
only the anodal enzyme; whereas incubation with antiserum
to purified pig heart soluble aspartate aminotransferase
inactivates only the cathodal form. Thus the latter activity presumably represents cytosol aspartate aminotransferase
which, because of its broad substrate specificity for aromatic
amino acids, was detected as a "pseudoisozyme" of tyrosine
aminotransferase. These observations stress the need for
caution in the interpretation of apparent multiple forms
of enzymes.

Dexamethasone and other adrenal steroid hormones induce the
synthesis of tyrosine aminotransferase (L-tyrosine: 2-oxoglu-
tarate aminotransferase, EC 2.6.1.5) in HTC cells, an established line of rat hepatoma cells in tissue culture (Gelehrter,
'73). The addition of dialyzed bovine serum (Gelehrter and
Tomkins, '69) or insulin (Gelehrter and Tomkins, '70) to HTC
cells previously induced in a chemically-defined, serum-free
medium results in a rapid further doubling of tyrosine aminotransferase activity. Although both effectors are similar in
their actions, the simultaneous addition of maiximally effect-

ive concentrations of insulin and serum causes an additive stimulation of tyrosine aminotransferase activity (Gelehrter and Tomkins, '70). These data suggested that either insulin and serum affect different aspects of the control of tyrosine aminotransferase, or, alternatively, affect the synthesis of different isozymic forms of tyrosine aminotransferase.

In order to distinguish between these alternatives, extracts of HTC cells induced with dexamethasone alone, dexamethasone plus insulin, and dexamethasone plus serum were analyzed by titration with antibody prepared to purified rat liver tyrosine aminotransferase, by heat stability, and by electrophoresis on polyacrylamide gels. By all three criteria the tyrosine aminotransferase induced by dexamethasone, insulin, and serum appears to be identical, suggesting that these three humoral factors all enhance the synthesis of the same protein, possibly by affecting different aspects of its control (Gelehrter et al, '72).

During the course of these studies, however, we observed apparent heterogeneity of tyrosine aminotransferase activity in HTC cells which was not previously recognized. We demonstrated electrophoretically a second enzyme capable of transaminating tyrosine, which is not affected by dexamethasone, insulin, or serum. Preliminary experiments suggested that this enzyme is capable of utilizing oxaloacetate in place of α-ketoglutarate in the transamination of tyrosine, and can actively transaminate aspartate (Gelehrter et al, '72). These two properties differentiate this enzyme from the soluble tyrosine aminotransferase described in rat liver, since purified rat liver tyrosine aminotransferase manifests an absolute requirement for α-ketoglutarate and does not transaminate aspartate (Kenney, '59). Aspartate aminotransferase (L-aspartate: 2-oxoglutarate aminotransferase, EC 2.6.1.1), on the other hand, can utilize oxaloacetate, but there is disagreement in the literature as to whether this enzyme can (Shrawder and Martinez-Carrion, '72) or cannot (Miller and Litwack, '71) transaminate tyrosine. Therefore, although this electrophoretic variant might represent aspartate aminotransferase, it might also represent a previously undescribed tyrosine aminotransferase isozyme, or possibly the expression in the neoplastic hepatoma cell of a fetal transaminase not expressed in adult liver (Criss, '71).

In order to distinguish between these possibilities and to characterize the basis for the apparent heterogeneity of tyrosine aminotransferase activity in HTC cells, we have utilized the technique of polyacrylamide gel electrophoresis coupled with histochemical staining to analyze the heat stability, amino acid and ketoacid specificity, and antigenic nature of

434

the tyrosine and aspartate aminotransferase activities of cytoplasmic extracts of HTC cells, fetal rat liver, and adult liver. We have found an enzyme in both fetal and adult rat liver cytosol identical in all these respects to the one found in HTC cells, and have demonstrated immunologically that this enzyme appears to be aspartate aminotransferase. Since transaminases may exhibit rather broad substrate specificities (Shrawder and Martinez-Carrion, '72; Miller and Litwack, '71; Jacoby and LaDu, '62) and since aspartate aminotransferase is present in hepatocytes in large amounts, our sensitive histochemical assay detected this activity as an "isozyme" of tyrosine aminotransferase. This finding may also explain several other reported examples of what appeared to be multiple molecular forms of tyrosine aminotransferase (Blake, '70; Holt and Oliver, '69; Miller et al, '72).

METHODS

Preparation of cell extracts

Cell culture, enzyme induction and preparation of cell extracts in the HTC cells were performed as described previously (Gelehrter et al, '72). Liver homogenates from adult male Sprague-Dawley rats, or fetal rats (one or two days before term) were prepared as described previously (Spencer and Gelehrter, '74). All extracts were routinely heated at 65° for 6 min to achieve a 3-4 fold purification of tyrosine aminotransferase. After heat treatment the preparations were centrifuged at 48,000 x g for 20 min. These supernatants were used for all enzyme assays and electrophoretic experiments.

Enzyme assays

Tyrosine aminotransferase was assayed by a modification (Spencer and Gelehrter, '74) of the method described by Diamondstone ('66). One milliunit (mU) of tyrosine aminotransferase catalyzes the formation of one nmole p-hydroxyphenylpyruvate per min at 37°. Aspartate aminotransferase was assayed by the malate dehydrogenase-linked assay of Sizer and Jenkins ('62). One milliunit (mU) of aspartate aminotransferase catalyzes the conversion of one nmole of oxaloacetate per min at 25°. Protein was measured by the method of Lowry et al ('51).

Polyacrylamide gel electrophoresis

Sample preparation and electrophoresis were carried out as

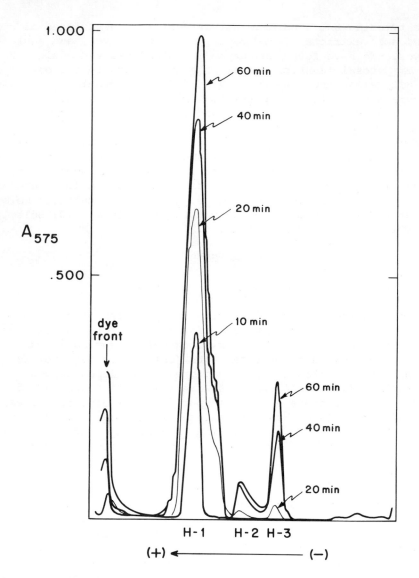

Fig. 1. Polyacrylamide gel electrophoresis of HTC cell tyrosine aminotransferase activity; effect of increasing staining time.

200 μl portions of a heated cytoplasmic extract of dexamethasone-induced HTC cells, containing 28 mU of enzyme activity, were subjected to electrophoresis for 90 min on 6% gels. Histochemical assay of tyrosine aminotransferase was performed at 60° for the times indicated. The stained gels were scanned

Fig. 1 legend continued
at 575 nm and the scans superimposed for comparison. The
stained bands have been designated H-1, H-2, and H-3 on the
basis of their mobility toward the anode. In this and subsequent figures, the origin of the gel is at the extreme right
of the figure. The direction of migration, from cathode toward
anode, is indicated by the arrow under the figure (Fig. 1A
from Spencer and Gelehrter, '74).

---

described previously (Gelehrter et al, '72) except that α-ketoglutarate was omitted from the cathode chamber, resulting in
a more rapid electrophoretic separation. Electrophoresis was
routinely performed at 3 mA per gel for 90 min at room temperature, using 0.5 x 7.5 cm, 6% acrylamide running gels. Histochemical staining of tyrosine aminotransferase was performed
as reported previously (Gelehrter et al, '72). Aspartate
aminotransferase was assayed histochemically by a modification
of the method of DeLorenzo and Ruddle ('70).

Inactivation of aminotransferase activities by specific
    antisera

Antiserum prepared in sheep to highly purified soluble rat
liver tyrosine aminotransferase was provided by Dr. David W.
Martin, Jr. (Gelehrter et al, '72), and antiserum prepared in
rabbits to purified soluble pig heart aspartate aminotransferase was kindly given to us by Dr. Marino Martinez-Carrion
(Shrawder and Martinez-Carrion, '72). Enzyme preparations
were mixed with the antibody and incubated at 37° for two
hours. The samples were then kept at 4° overnight and
centrifuged at 17,000 x g for 20 min at 4°. Supernatants were
assayed for enzymatic activity and aliquots applied to
acrylamide gels.

RESULTS

Heterogeneity of tyrosine aminotransferase activity in HTC
    cells

The apparent heterogeneity of tyrosine aminotransferase
activity in HTC cells is shown in Fig. 1. In this experiment,
an extract of dexamethasone-induced HTC cells was subjected
to electrophoresis, followed by histochemical staining at 60°
for tyrosine aminotransferase activity. After 10 min, a single band of activity, designated H-1, is seen near the anode.
After 20 min of staining, this band has become more prominent

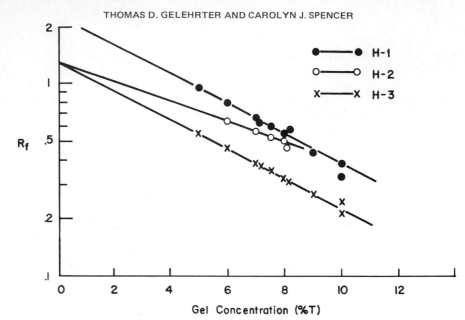

Fig. 2. Polyacrylamide gel electrophoresis of HTC tyrosine aminotransferase at different gel concentrations.

Heated cytoplasmic extracts of dexamethasone-induced HTC cells were analyzed by electrophoresis on polyacrylamide gels ranging in concentration from 5 to 10% acrylamide. Staining of tyrosine aminotransferase was performed at 60° for 65 to 85 min. The relative mobility (R_F) of the three stained bands was measured as described by Rodbard and Crambach ('71). H-1, H-2, and H-3 refer to the same bands described in Fig. 1 (Fig. 7 from Gelehrter et al, '72).

and a more cathodal band, designated H-3 is now apparent. With increasing time of staining, both bands become more intense. A band of intermediate mobility, H-2, is occasionally seen, and can be appreciated in this figure.

The increase in histochemically-assayable tyrosine amino-transferase activity with time of staining, expressed as the area under the peaks can yield a quantitative measure of enzyme activity under carefully defined conditions. However, if the gels are stained for a longer period such as 60 to 120 min, the assay becomes more sensitive in that smaller amounts of tyrosine aminotransferase activity can be detected, but quantitation is lost. In the experiments described below, the longer staining periods have usually been employed in order to define the qualitative characteristics of the minor form of tyrosine aminotransferase activity.

If HTC cell extracts are subjected to electrophoresis and stained for aspartate aminotransferase, a single band of histochemical activity is demonstrated in a position identical to H-3. Furthermore, if such extracts are analyzed by electrophoresis at different gel concentrations, the aspartate aminotransferase and tyrosine aminotransferase activities are not separated suggesting that a single protein is responsible for both enzyme activities.

In order to examine further the properties of the two tyrosine transaminating activities, the enzymes were electrophoretically eluted from the gel, using the technique described by Kirkman and Hanna ('68). When the material eluted from band H-1 was subjected to re-electrophoresis it yielded a single band of histochemical activity in the position of H-1. The material eluted from H-3 yielded on re-electrophoresis a single band of tyrosine aminotransferase activity in the expected position of H-3, and in addition exhibited aspartate aminotransferase activity in the same position. Thus there was no evidence of interconversion of forms H-1 and H-3, indicating that these forms were stable in vitro.

In order to determine whether the three bands of tyrosine aminotransferase activity represented isozymes or aggregates, the electrophoretic mobilities of these bands from an extract of dexamethasone-induced HTC cells were determined at several gel concentrations. A plot of log R_F against gel concentration ("Ferguson Plot," Fig. 2) showed that bands H-1 and H-3 have parallel slopes, i.e. essentially identical retardation coefficients (K_R). These data suggest that H-1 and H-3 differ in charge but not in molecular weight. Fig. 2 also shows that bands H-2 and H-3 have the same Y_O (R_F when % T = 0). These data suggest that forms H-2 and H-3 have the same net charge (valence) but differ in size (molecular weight). The simplest explanation for this finding might be that band H-2 represents an enzymatically active oligomeric form of band H-3.

Amino acid and ketoacid specificities of tyrosine aminotransferase activities in HTC cells

Since cytosol tyrosine and aspartate aminotransferases reportedly differ in their amino acid and ketoacid specificities, we have examined these properties of H-1 and H-3 from HTC cell extracts. Fig. 3 compares the histochemical assay of tyrosine aminotransferase using oxaloacetate in place of α-ketoglutarate as the ketoacid. Both H-1 and H-3 activities are demonstrable in the presence of α-ketoglutarate; however, only H-3 is able to transaminate tyrosine using oxaloacetate as the amino group acceptor, confirming that H-1 requires α-ketoglutarate. In the absence of either ketoacid there is no histochemical tyrosine

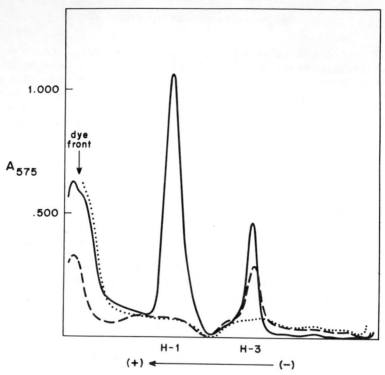

Fig. 3. Ketoacid specificity of the tyrosine aminotransferase activities in HTC cell cytoplasm.

A heated, cytoplasmic extract of dexamethasone-induced HTC cells was dialyzed against buffer lacking α-ketoglutarate. 15.4 mU of tyrosine aminotransferase was applied to each gel and subjected to electrophoresis for 96 min. All buffers used during electrophoresis were free of α-ketoglutarate. (————) stained at 60° for 120 min with 20 mM α-ketoglutarate; (-----) stained at 60° for 65 min with 20 mM oxaloacetate in place of α-ketoglutarate; (-----) stained at 60° for 120 min without either oxaloacetate or α-ketoglutarate (Fig. 2 from Spencer and Gelehrter, '74).

aminotransferase activity.

Since H-3 is capable of transaminating aspartate we have studied the effect of L-aspartate on the transamination of tyrosine. In the presence of aspartate, the transamination of tyrosine by H-1 is not inhibited and in fact may actually be enhanced somewhat; in contrast, the transamination of tyrosine by H-3 is markedly inhibited by aspartate in the staining mixture. At 8 mM L-aspartate, a concentration half that of the substrate monoiodotyrosine, tyrosine aminotransferase activity in H-3 is inhibited by approximately 50%; at 32 mM aspartate, the activity in H-3 is inhibited by more than 80%.

(Spencer and Gelehrter, '74).

Differentiation of cytoplasmic and mitochondrial aminotransferases

It has been demonstrated recently that rat liver mitochondrial tyrosine aminotransferase activity and mitochondrial aspartate aminotransferase activity represent the same enzyme (Shrawder and Martinez-Carrion, '72; Miller and Litwack, '71). It seems very unlikely, however, that H-3 represents this mitochondrial transaminase. The mitochondrial enzyme is quite labile to heating at 65°; in contrast, the two forms of tyrosine aminotransferase activity demonstrated in Fig. 1 are clearly heat stable. These extracts have been subjected to heating at 65° for 6 min prior to electrophoresis, and the histochemical staining itself is performed at 60°. Furthermore, the cell extracts were prepared under conditions in which mitochondria should be sedimented by centrifugation. In order to confirm this latter point, mitochondria were prepared from HTC cells and a mitochondrial extract subjected to electrophoresis. On histochemical staining at 37°, a single rather broad band of tyrosine aminotransferase activity was observed near the cathode, not in the position of any of the previously described tyrosine aminotransferase activities. After heating the mitochondrial preparation for 6 min at 65° no histochemical tyrosine aminotransferase activity was demonstrable (Spencer and Gelehrter, '74).

Therefore, it appears that H-3, like H-1, represents a heat-stable, cytoplasmic transaminase, which, unlike H-1, is capable of transaminating aspartate as well as tyrosine, and of utilizing oxaloacetate in place of α-ketoglutarate in the transamination of tyrosine.

Heterogeneity of tyrosine aminotransferase activity in adult and fetal rat liver

In order to determine whether this heterogeneity of tyrosine aminotransferase activity might represent the expression in the malignant hepatoma cell of a fetal enzyme not expressed in adult liver, we have performed similar studies on liver from adult and fetal rats. Cytoplasmic extracts of adult rat liver were found to have two major heat-stable forms of tyrosine aminotransferase activity as shown in Figs. 4 and 5. The more anodal band, designated AL-1, corresponded to the HTC band H-1, and the more cathodal band, designated AL-3, corresponded to H-3 in electrophoretic mobility. In addition there was a minor band, migrating more slowly than AL-3, which

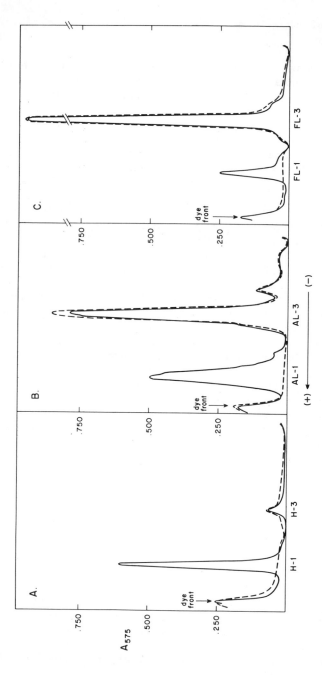

Fig. 4. Inactivation of the anodal tyrosine aminotransferase activity in cytoplasmic extracts of HTC cells, adult and fetal rat liver by antiserum to purified rat liver tyrosine aminotransferase.

Heated cytoplasmic extracts of adult and fetal rat liver and dexamethasone-induced HTC cells were prepared and analyzed by electrophoresis on polyacrylamide gels. Histochemical assay was performed at 60° for 60 min in A, and 50 min in B and C.

A. HTC cells. (——) no antibody treatment. 10.4 mU tyrosine aminotransferase activity

Fig. 4 legend, cont.

applied to gel; (-----) approximately 12 mU of enzyme activity was incubated with 30 mU of antiserum as described in "Methods," and the supernatant containing 0.4 mU activity applied to the gel.

    B. Adult rat liver. (———) no antibody treatment. 11.4 mU tyrosine aminotransferase applied to the gel; (-----) approximately 15 mU enzyme activity incubated with 37 mU antiserum, and the supernatant containing 1.6 mU activity applied to the gel.

    C. Fetal rat liver. (———) no antibody treatment. 2.1 mU tyrosine aminotransferase activity applied to the gel; (-----) approximately 5 mU activity incubated with 37 mU anitserum, and the supernatant containing no biochemically-assayable activity applied to the gel (Fig. 5 from Spencer and Gelehrter, '74).

---

did not correspond to any activity seen in the HTC cells. The relative activity of AL-3 to AL-1 appears to be considerably greater than the relative activities of H-3 to H-1.

    Tyrosine aminotransferase is present in very low levels in fetal rat liver, the amount rising rapidly at birth (Greengard, '69). In only one out of five preparations of fetal rat liver (each prepared from the pooled livers of a single litter) was tyrosine aminotransferase activity demonstrable by biochemical assay. This preparation was used for the experiments shown in Figs. 4 and 5. Nevertheless, all five cytoplasmic extracts of fetal liver showed a major band of heat-stable tyrosine aminotransferase activity which migrated in the same position as H-3 and is designated as FL-3. In the single preparation which showed tyrosine aminotransferase activity biochemically, there was also a band of histochemical activity, designated FL-1, which migrated in the same position as H-1 in HTC cells.

    Analysis of the electrophoretic mobilities of the two forms of tyrosine aminotransferase activity in HTC cells, adult, and fetal liver at different gel concentrations indicates that H-1, AL-1 and FL-1 are identical in their electrophoretic behavior as are H-3, AL-3, and FL-3. Furthermore, the plot of log R_F against gel concentration for the anodal and cathodal bands from each source yields parallel slopes suggesting that they differ in net charge but not in molecular weight. Histochemical assay of aspartate aminotransferase in extracts of adult and fetal liver demonstrated a single band of activity migrating in the position of AL-3 and FL-3, respectively. Furthermore, as is the case in HTC cells, this

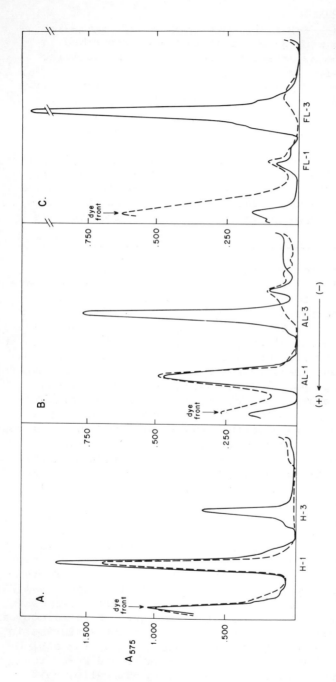

Fig. 5. Inactivation of the cathodal tyrosine aminotransferase activity in cytoplasmic extracts of HTC cells, adult and fetal rat liver by antiserum to purified pig heart soluble aspartate aminotransferase.

Preparation of cell extracts and incubation with antiserum is described in "Methods." Histochemical assay was performed at 60° for 96 min in A, and 55 min in B and C. Note the different absorbance scale in A compared with B and C.

A. HTC cells (———) no antibody treatment. 25.3 mU tyrosine aminotransferase activity and 16.7 mU aspartate aminotransferase activity applied to the gel; (-----) approximately 16.1 mU

444

Fig. 5 legend, cont.

aspartate aminotransferase activity incubated with 17.7 mU antibody, and the supernatant containing 20.5 mU tyrosine aminotransferase and negligible aspartate aminotransferase activity applied to the gel.

B. Adult rat liver. (————) no antibody treatment. 20.5 mU tyrosine and 198.8 mU aspartate transaminase activity applied to the gel; (-----) approximately 115.1 mU aspartate aminotransferase activity incubated with 120.8 mU antibody, and the supernatant containing 5.9 mU tyrosine and negligible aspartate aminotransferase activity applied to the gel.

C. Fetal rat liver. (————) no antibody treatment. 4.3 mU tyrosine and 198.4 mU aspartate aminotransferase activity applied to the gel; (-----) approximately 115.9 mU aspartate transaminase incubated with 131.2 mU antibody, and the supernatant containing 2 mU tyrosine and no aspartate aminotransferase activity applied to the gel (Fig. 6 from Spencer and Gelehrter, '74).

------

cathodal form was capable of using oxaloacetate in place of α-ketoglutarate in the transamination of tyrosine, and the histochemical activity of tyrosine aminotransferase in this band was inhibited by the addition of aspartate to the staining mixture (Spencer and Gelehrter, '74).

In summary, both adult and fetal liver as well as rat hepatoma cells in tissue culture contain two forms of heat-stable cytosol tyrosine aminotransferase activity. The more anodal form appears to be specific for α-ketoglutarate and does not transaminate aspartate, whereas the more cathodal form is capable of utilizing oxaloacetate and of transaminating aspartate. The presence of this form in both fetal and adult rat liver clearly eliminates the possibility that this form represents the expression by the malignant hepatoma cell of a fetal gene not expressed in adult liver.

Immunologic inactivation of the cytoplasmic tyrosine aminotransferase activities in HTC cells, adult and fetal liver

Because of the controversy in the literature regarding the ability of cytosol aspartate aminotransferase to transaminate tyrosine, it was unclear whether or not the cathodal enzyme was aspartate aminotransferase. In order to answer this question, we have examined the antigenic nature of the two forms of tyrosine aminotransferase activity. In the experiment shown in FIg. 4, extracts of HTC cells, adult, and fetal rat liver have been incubated with antibody to purified rat liver

445

TAT. Following inactivation the extracts were subjected to electrophoresis and histochemical staining as described in Methods. Incubation with anti-tyrosine aminotransferase antiserum in amounts sufficient to inactivate the enzyme activity in each extract resulted in complete disappearance of histochemical activity in the anodal band; whereas staining of the cathodal band was unaffected. In contrast, when these extracts were incubated with antiserum to purified pig heart cytosol aspartate aminotransferase and then subjected to electrophoresis, histochemical tyrosine aminotransferase activity was completely abolished in the cathodal band but was unaffected in the anodal band (Fig. 5). As expected, histochemical aspartate aminotransferase was simultaneously eliminated. These experiments demonstrate that biochemically and immunologically the more anodal histochemical activity is tyrosine aminotransferase whereas the more cathodal form represents aspartate aminotransferase.

DISCUSSION

Ideally, isozymes should be characterized by both genetic and biochemical techniques as naturally occurring, non-interconvertible, independent protein species arising from genetically determined differences in structure (IUPAC-IUB, '71). Frequently, however, isozymes are postulated simply on the basis of the histochemical or biochemical demonstration of multiple forms of enzyme activity separated by electrophoretic or other techniques.

Two major problems arise from such less rigorous characterization of isozymes. The first is the problem of pseudo-isozymes which represent different physiochemical states of the same enzyme arising from modification of a common primary sequence. This situation can arise from in vitro artifacts such as sulfhydryl oxidation during enzyme preparation or storage; an apparent example of this problem are the pseudo-isozymes of rabbit skeletal muscle phosphoglucose isomerase reported by Noltman and co-workers ('72). A similar situation could arise from in vivo, post-translational modification of an enzyme by phosphorylation, methylation, carbamylation, addition of carbohydrate groups, aggregation, allosteric modification, etc. This kind of modification could be modulated by hormones, and might account for the apparent isozymes of hepatic tyrosine aminotransferase reported by Iwasaki et al ('73), Holt and Oliver ('69), and Johnson et al ('73).

It is unlikely that our observations can be explained in this way. When HTC cell extracts are prepared and subjected

446

to electrophoresis in the presence or absence of the reducing agent dithiothreitol, the results are identical. Furthermore, when the material eluted from H-1 and H-3 is subjected to re-electrophoresis we observe histochemical activity of the appropriate kind in a single band with the expected mobility. Finally, the marked differences in H-1 and H-3 catalytic activity, relative substrate specificity, and antigenic behavior make it unlikely that they could represent inter-changeable forms of a single protein.

A second class of pseudoisozymes can arise when two or more electrophoretically distinguishable forms of an enzyme activity are indeed genetically different, but are different enzymes with overlapping substrate specificities. We believe that this represents the most likely explanation for our findings. It is also quite possible that this explanation accounts as well for the electrophoretically-demonstrated apparent iso-zymes of tyrosine aminotransferase in mouse liver (Blake, '70), and in preparations of rat liver purified by affinity chrom-atography (Miller et al, '72).

The anodal transaminase we have demonstrated in extracts of HTC cells (H-1), adult rat liver (AL-1), and in one of five pools of fetal rat liver (FL-1) clearly appears to be tyrosine aminotransferase (L-tyrosine: 2-oxoglutarate amino-transferase EC 2.6.1.5). It is a cytoplasmic, heat-stable enzyme with an absolute requirement for α-ketoglutarate as ketoacid, and transaminates tyrosine but not aspartate. It accounts for nearly all (>95%) of the soluble tyrosine trans-aminating activity in HTC cells and is completely inactivated by anti-serum to purified rat liver tyrosine aminotransferase. The concentration of this enzyme in HTC cells is increased by incubation of the cells with dexamethasone, insulin, and serum (Gelehrter et al, '72).

The more cathodal tyrosine aminotransferase activity in HTC cells (H-3), adult rat liver (AL-3), and fetal liver (FL-3), on the other hand, most probably represents aspartate aminotransferase (L-aspartate: 2-oxoglutarate aminotransferase EC 2.6.1.1). It is also a cytoplasmic, heat-stable enzyme with the same apparent molecular weight as the anodal enzyme described above, but differs in net charge. Unlike the anodal enzyme, it can utilize oxaloacetate in place of α-ketoglutarate, and can transaminate aspartate as well as tyrosine. It is inactivated by incubation with antiserum prepared against pig heart soluble aspartate aminotransferase but not by antiserum to tyrosine aminotransferase, and its activity in HTC cells is apparently not affected by corticosteroids, insulin, or serum.

As noted earlier, neoplastic cells may show differences

447

in isozyme patterns from their normal counterparts (Criss, '71; Farron et al, '72). We have ruled out the possibility that the cathodal form of tyrosine aminotransferase (H-3) in HTC cells might represent the expression by the malignant hepatocyte of a gene expressed in fetal liver but repressed in adult liver, since this enzyme is found in both adult and fetal liver. Furthermore, the observation that there is less aspartate aminotransferase activity in HTC cells than in normal liver is consistent with studies on other hepatomas (Sheid, '65).

In conclusion, we have demonstrated that the apparent heterogeneity of tyrosine aminotransferase activity in HTC cells represents a pseudoisozyme resulting from the rather promiscuous substrate specificity of soluble aspartate transaminase. These studies again stress the need for caution in characterizing isozymes by histochemical and electrophoretic techniques alone.

REFERENCES

Blackburn, M. N., J. M. Chirgwin, G. T. James, T. D. Kempe, T. F. Parsons, A. M. Register, K. D. Schnackerz, and E. A. Noltman 1972. Pseudoisoenzymes of rabbit muscle phosphoglucose isomerase. *J. Biol. Chem.* 247: 1170-1179.

Blake, R. L. 1970. Control of liver tyrosine aminotransferase expression: enzyme regulatory studies on inbred strains and mutant mice. *Biochem. Genet.* 4: 215-235.

Criss, W. E. 1971. A review of isozymes in cancer. *Cancer Res.* 31: 1523-1542.

DeLorenzo, R. J. and F. H. Ruddle 1970. Glutamate oxalate transaminase (GOT) genetics in *Mus musculus*. *Biochem. Genet.* 4: 259-273.

Diamondstone, T. I. 1966. Assay of tyrosine transaminase activity by conversion of p-hydroxyphenylpyruvate to p-hydroxybenzaldehyde. *Anal. Biochim.* 16: 395-401.

Farron, F., H. H. T. Hsu, and W. E. Knox 1972. Fetal-type isoenzymes in hepatic and non-hepatic rat tumors. *Cancer Res.* 32: 302-308.

Gelehrter, T. D. 1973. Mechanisms of hormonal induction of enzymes. *Metabolism* 22: 85-100.

Gelehrter, T. D., J. R. Emanuel, and C. J. Spencer 1972. Induction of tyrosine aminotransferase by dexamethasone, insulin, and serum: characterization of the induced enzyme. *J. Biol. Chem.* 247: 6197-6203.

Gelehrter, T. D. and G. M. Tomkins 1969. Control of tyrosine aminotransferase synthesis in tissue culture by a factor in serum. *Proc. Natl. Acad. Sci. USA* 64: 723-730.

Gelehrter, T. D. and G. M. Tomkins 1970. Posttranscriptional control of tyrosine aminotransferase synthesis by insulin. *Proc. Natl. Acad. Sci. USA* 66: 390-397.

Greengard, O. 1969. Enzymic differentiation in mammalian liver. *Science* 163: 891-895.

Holt, P. G. and I. T. Oliver 1969. Multiple forms of tyrosine aminotransferase in rat liver and their hormonal induction in the neonate. *FEBS Lett.* 5: 89-91.

IUPAC-IUB Commission on biochemical nomenclature 1971. The nomenclature of multiple forms of enzymes: recommendations 1971. *J. Biol. Chem.* 246: 6127-6128.

Iwasaki, Y., C. Lamar, K. Danenberg, and H. C. Pitot 1973. Studies on the induction and repression of enzymes in rat liver. *Eur. J. Biochem.* 34: 347-357.

Jacoby, G. A. and B. N. LaDu 1962. Non-specificity of tyrosine transaminase: an explanation for the simultaneous induction of tyrosine, phenylalanine, and tryptophan transaminase activities in rat liver. *Biochem. Biophys. Res. Commun.* 8: 352-356.

Johnson, R. W., L. E. Roberson, and F. T. Kenney 1973. Regulation of tyrosine aminotransferase in rat liver. X. Characterization and interconversion of the multiple enzyme forms. *J. Biol. Chem.* 248: 4521-4527.

Kenney, F. T. 1959. Properties of partially purified tyrosine α-ketoglutarate transaminase from rat liver. *J. Biol. Chem.* 234: 2707-2712.

Kirkman, H. N. and J. E. Hanna 1968. Isozymes of human red cell glucose-6-phosphate dehydrogenase. *Ann. N.Y. Acad. Sc.* 151: 133-148.

Lowry, O. H., N. J. Rosebrough, A. L. Farr, and R. J. Randall 1951. Protein measurement with Folin phenol reagent. *J. Biol. Chem.* 193: 265-275.

Miller, J. E. and G. Litwack 1971. Purification, properties and identity of liver mitochondrial tyrosine aminotransferase. *J. Biol. Chem.* 246: 3234-3240.

Miller, J. V., Jr., P. Cuatrecasas, and E. B. Thompson 1972. Purification of tyrosine aminotransferase by affinity chromatography. *Biochim. Biophys. Acta* 276: 407-415.

Rodbard, D. and A. Crambach 1971. Estimation of molecular radius, free mobility, and valence using polyacrylamide gel electrophoresis. *Anal. Biochem.* 40: 95-134.

Sheid, B., H. P. Morris, and J. S. Roth 1965. Distribution and activity of aspartate aminotransferase in some rapidly proliferating tissues. *J. Biol. Chem.* 240: 3016-3022.

Shrawder, E. and M. Martinez-Carrion 1972. Evidence of phenylalanine transaminase activity in the isoenzymes of aspartate transaminase. *J. Biol. Chem.* 247: 2486-2492.

Sizer, I. W. and W. T. Jenkins 1962. Glutamic aspartic
    transaminase from pig ventricles. *Methods Enzymol.*
    5: 677-684.
Spencer, C. J. and T. D. Gelehrter 1974. Pseudoisozymes of
    hepatic tyrosine aminotransferase. *J. Biol. Chem.*
    249: 577-583.

THE DIFFERENT ORIGINS OF MULTIPLE MOLECULAR
FORMS OF PHOSPHOGLUCOSE ISOMERASE

ERNST A. NOLTMANN
Department of Biochemistry
University of California
Riverside, California 92502

ABSTRACT. Crystalline phosphoglucose isomerase from rabbit muscle can be resolved into several different protein components by column chromatography on carboxymethyl Sephadex. These multiple forms are indistinguishable with respect to over-all amino acid composition except for a different accessibility of their sulfhydryl groups to p-mercuribenzoate. They can be interconverted both by treatment with dithiothreitol and by exposure to oxidative conditions, indicating that they represent different states of the same protein rather than genetically dissimilar protein species. In contrast, the isozymes of phosphoglucose isomerase from brewers' yeast exhibit significant differences in their primary structures as reflected in considerable variations in their tryptic peptide maps.

INTRODUCTION

The question of when it is appropriate to define multiple forms of an enzyme as isozymes (Markert and Møller, 1959) and when to use other terminologies has been with us since the first observation that the same enzymatic activity may be associated with different, physically separable proteins within in the same organism or even within the same tissue (e.g. Markert and Whitt, 1968). Phosphoglucose isomerase[1] [D-glucose-6-phosphate ketol-isomerase, EC 5.3.1.9] has had its share of this controversy. Yoshida and Carter (1969) postu-

---

[1]Phosphoglucose isomerase is the enzyme catalyzing the reversible interconversion between glucose 6-phosphate and fructose 6-phosphate. It is ambiguous and incorrect to designate this enzyme by the now obsolete name "phosphohexose isomerase" which was introduced (Bodansky, 1953) when the reaction catalyzed by the second phosphohexose isomerase, *viz.* phosphomannose isomerase, was not yet fully understood. After Slein's explicit description (Slein, 1955) of two enzymes under the heading "phosphohexoisomerases" (plural!) and clarification of the phosphomannose isomerase reaction (Slein, 1955; Noltmann and Bruns, 1958), use of the name phosphohexose isomerase for phosphoglucose isomerase is misleading and should be discontinued (see also Noltmann, 1972).

lated a number of years ago that the crystalline rabbit muscle enzyme can be resolved chromatographically into isozymes of different primary structures which they suggested to originate from different genes in the rabbit. We have not been able to reproduce their chromatographic resolutions but have found instead that rabbit muscle phosphoglucose isomerase may appear as multiple chromatographic species dependent upon how much care is taken to prevent oxidation of sulfhydryl groups during the work-up procedures. We have called these multiple forms pseudo-isozymes to indicate their nongenetic nature and to emphasize that they represent merely different states of the same enzyme protein (Noltmann, 1972; Blackburn et al., 1972). Various adjectives, some more cumbersome than others, have also been used to describe this category of multiple enzyme forms (e.g., artifactual, nongenetic, epigenetic, post-transcriptional, post-translational). "Pseudo-Isozyme" appears to us to fulfill the need for a short term to distinguish these multiple forms from genetically determined isozymes (Enzyme Nomenclature, 1972).

On the other hand, there are also several examples of multiple forms of phosphoglucose isomerase which appear to be genetically determined isozymes. Some of these have been discussed at this Conference (Gracy, 1974). In various mammalian species, for example, the existence of isozymes seems to be often associated with genetic aberrations. In fact, phosphoglucose isomerase variants have become an important object of study in human genetics and are now considered to be the third most common cause of enzyme deficient nonspherocytic hemolytic disease (Blume et al., 1972; Tilley et al., 1974).

Because of the significance which isozymes of different primary structures have for elucidating the molecular basis of genetic diseases, it is imperative to ascertain, with use of the tools of protein chemistry, in each case, whether such molecular differences do indeed exist. The simple observation that activity staining produces multiple bands after electrophoresis on agar gel is not sufficient evidence to postulate genetically determined isozymes!

This paper will be concerned with a discussion of the nongenetic basis for the multiple forms of rabbit muscle phosphoglucose isomerase as opposed to the isozymes of phosphoglucose isomerase from brewers' yeast which are characterized by different primary structures.

THE NONGENETIC MULTIPLE FORMS OF PHOSPHOGLUCOSE ISOMERASE FROM RABBIT MUSCLE

Phosphoglucose isomerase was isolated and crystallized from rabbit skeletal muscle ten years ago (Noltmann, 1964) and has since been studied in some detail in our laboratory with respect to both its physical and chemical properties (Pon et al., 1970; Blackburn and Noltmann, 1972) and its mechanism of action (for review, see Noltmann, 1972).

Although detailed and convincing evidence had been obtained that the 5-times crystallized enzyme with which most of our studies were made was a homogeneous protein, ion exchange column chromatography of crystalline or highly purified phosphoglucose isomerase yielded several chromatographic species when very shallow sodium phosphate gradients were employed for elution (Fig. 1). These species, however, did not remain single peaks on rechromatography. More significantly, when the chromatography was performed in the presence of dithiothreitol, the major chromatographic band, on rechromatography, did remain a single chromatographic species eluting in the same location from the column as the first time and did not produce any of the other peaks (Fig. 2).

We were convinced of the"pseudo"-isozymic nature of these multiple enzyme species when preparations made from rabbit skeletal muscle with dithiothreitol present throughout the procedure yielded consistently only one chromatographic peak and when, in addition, we could show that exposure to either reductive or oxidative conditions resulted in interconversions among the various species. This is shown in Fig. 3 in which rechromatography of peak G-1 (from Fig. 3A) in the absence of dithiothreitol (Fig. 3B) practically repeats the original pattern. In the presence of dithiothreitol Fig. 3C, however, a large portion of the same peak is now converted to the more reduced G-II species. Finally, when the so-called G-II peak was chormatographed in the presence of dithiothreitol it remained one species without conversion into any of the others (Fig. 3D).

Amino acid analyses yielded minimal differences between the chromatographic forms (Blackburn et al., 1972), definitely not of a magnitude that would justify proposing different primary structures.

Because of the pronounced effect of dithiothreitol on the distribution pattern of the chromatographic fractions as well as the finding that treatment with dithiothreitol of oxygen could interconvert the chromatographic species, it was felt that it might be the varying extent of oxidation of sulfhydryl groups per se which was responsible for the different

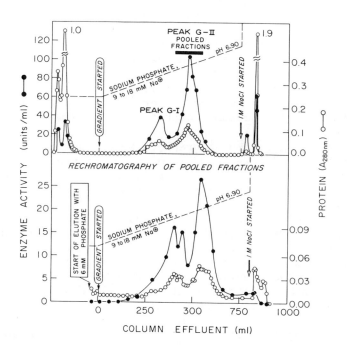

Fig. 1. Chromatography of twice-crystallized rabbit muscle phosphoglucose isomerase on carboxymethyl Sephadex with a sodium phosphate gradient in the _absence_ of dithiothreitol. From Blackburn et al., 1972.

chromatographic behavior. This was confirmed by numerous sulfhydryl analyses a summary of which is given in Table 1. It is conceivable that the different elution patterns are not directly caused by an obvious charge difference but that the overall net charge of the proteins is indirectly affected because the primary difference in the extent of sulfhydryl oxidation produces a secondary conformational charge which is then responsible for the difference in chromatographic mobility.

Our experience with rabbit muscle phosphoglucose isomerase (Blackburn et al., 1972) and other cases reported elsewhere

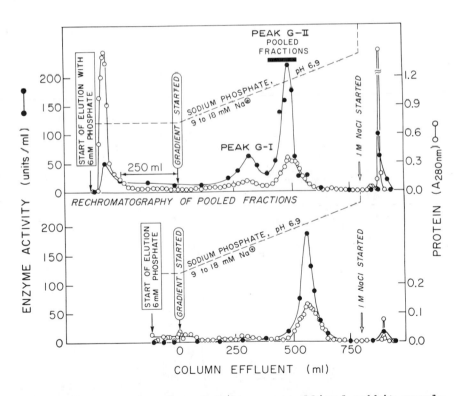

Fig. 2. Chromatography of twice-crystallized rabbit muscle phosphoglucose isomerase in the _presence_ of dithiothreitol. All conditions were as described for Fig. 1, except that the enzyme was dialyzed and chromatographed in the presence of 1 mM dithiothreitol. From Blackburn et al., 1972.

Figure 3.

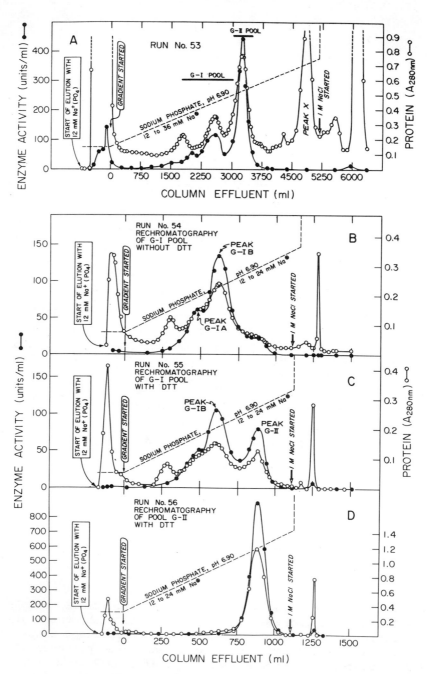

Fig. 3. (cont.) Chromatography of partially purified rabbit muscle phosphoglucose isomerase on carboxymethyl Sephadex. A, partially purified enzyme was dialyzed against gradient starting buffer, applied to a carboxymethyl Sephadex column, and eluted with a linear sodium phosphate gradient, pH 6.90; B, rechromatography of the combined G-I fractions without dithiothreitol; C, rechromatography in the presence of 1 mM dithiothreitol of the combined G-L fractions kept in 10 mM dithiothreitol for 6 days; D, rechromatography in the presence of 1 mM dithiothreitol of the G-II fraction (pooled as shown in A) kept in 1 mM dithiothreitol for 12 days. From Blackburn et al., 1972.

(e.g. Broder and Srere, 1963; Meizel and Markert, 1967; Mörikofer-Zwez et al., 1969; Payne et al., 1972; Jörnvall, 1973) and at this Conference are a strong indication that such non-genetically derived multiple enzyme forms may not be un-common. We believe, for example, that they are also the ex-planation for the multiple electrophoretic patterns frequently found on "ageing" of tissues or physiological fluids.

THE ISOZYMES OF YEAST PHOSPHOGLUCOSE ISOMERASE

In contrast to the interconvertible molecular forms of phosphoglucose isomerase from rabbit muscle, yeast phospho-glucose isomerase isozymes have been found from the beginning to be stable entities. We reported the resolution of crystal-line yeast phosphoglucose isomerase into three isozymes in 1967 (Nakagawa and Noltmann, 1967). These have since been subjected to detailed physical and chemical studies to de-termine the extent of differences and similarities among them.

Comparison of chromatographic patterns of yeast extract in the presence or absence of dithiothreitol indicated that the thiol had no effect on the distribution of the yeast phospho-glucose isomerase isozymes. The yeast isozymes are therefore not interconvertible, at least not under the conditions under which the rabbit muscle multiple forms are interconverted.

All of the physical studies led to the general picture that in terms of physical properties, other than overall net charge, remarkable similarity prevails between the three yeast phos-phoglucose isomerase isozymes. Fig. 4 shows the results of high speed equilibrium sedimentation ultracentrifugation ex-periments. The $\log J$ versus r^2 plots are linear, indicating homogeneity with respect to size. The molecular weights of all three isozymes determined by this method are close to 120,000 (Kempe et al., 1974b).

Measurement of electrophoretic mobilities of the isozymes

TABLE I

CYSTEINE AND TOTAL HALF-CYSTINE CONTENTS OF THE DIFFERENT MOLECULAR FORMS OF RABBIT MUSCLE PHOSPHOGLUCOSE ISOMERASE*

Method of analysis	Number of preparations analyzed	Total number of analyses	Chromatographic species					
			Peak G-IA		Peak G-IB		Peak II	
			Residues per molecule (mol. wt. 132,000)					
			without SDS	with SDS	without SDS	with SDS	without SDS	with SDS
Cysteine by p-mercuri-benzoate assay	10	80	3.8	9.8	4.5	11.5	5.5	12.9
Cysteic acid after performic acid oxidation	7	26	12.8		12.9		12.2	

*Condensed from Blackburn et al., 1972.

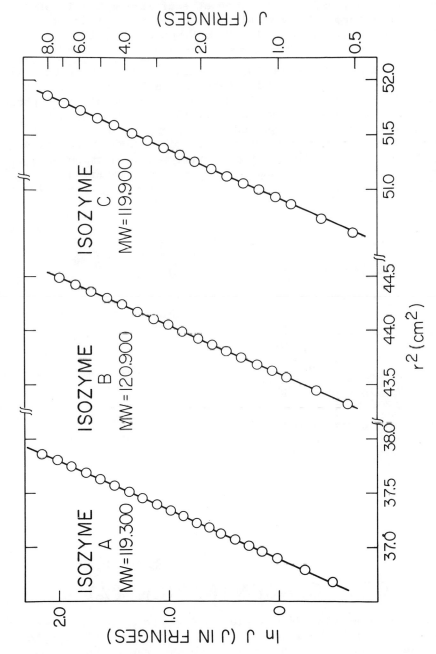

Fig. 4. Multiple same equilibrium sedimentation ultracen-

Fig. 4 (cont.) trifugation of native yeast phosphoglucose isomerase isozymes. Rotor speed at equilibrium was 17,980 rpm; temperature 20°; initial protein concentrations were approximately 0.5 mg per ml. From Kempe et al., 1974b.

---

as a function of pH by cellulose acetate strip electrophoresis yielded apparent isoelectric points at pH 5.4, 5.2, and 5.0, respectively, for Isozymes A, B, and C of brewers' yeast phosphoglucose isomerase (Kempe et al., 1974b). The yeast isozymes have also been subjected to polyacrylamide gel electrophoresis according to the method of Hedrick and Smith (1968) which involves determining the mobility as a function of gel concentrations. Fig. 5 shows that each isozyme migrates as a distinct, single band and that the mixture is perfectly

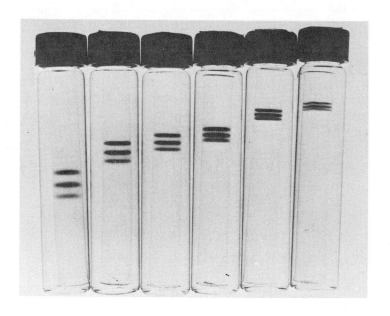

5   6   7   8   9   10
GEL   CONCENTRATION (%)

Fig. 5. Polyacrylamide gel electrophoresis of a mixture of native yeast phosphoglucose isomerase isozymes. Electrophoresis was conducted in the pH 7.3 Tris-asparagine buffer system of Hedrick and Smith (1968). From Kempe et al., 1974b.

resolved at all gel concentrations. Fig. 6 represents a plot of these mobilities as a function of gel concentrations. A family of parallel lines is obtained that is characteristic for protein species of the same molecular weight which differ only in charge. For proteins differing in size, the lines

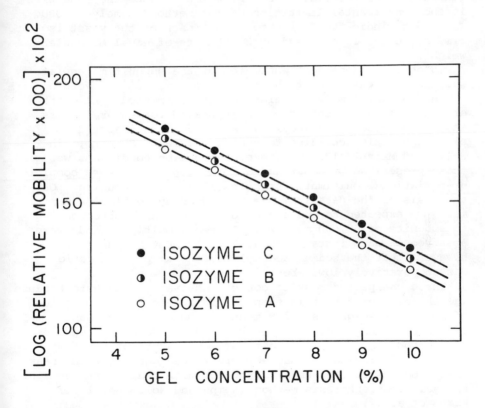

Fig. 6. Electrophoretic mobilities of native yeast phospho-glucose isomerase isozymes as a function of polyacrylamide gel concentration. Relative mobilities of each isozyme were determined from experiments analogous to those shown in Fig. 5. Each line represents a least squares fit of the points. From Kempe et al., 1974b.

would converge at low gel concentrations. It is appropriate, therefore, to classify the yeast phosphoglucose isomerase isozymes as "charge isomers."

This technique was also used in a quantitative fashion to obtain values for the molecular weights by comparing the mobilities as a function of gel concentration with those of several proteins of known molecular weights. Apparent molecular weights for all three yeast phosphoglucose isomerase isozymes were found by this technique to be between 120,000 and 125,000.

Subunit molecular weights of the yeast isozymes were determined by equilibrium sedimentation in the presence of 6 M guanidine hydrochloride. They were found to be the same within the experimental limitation of the method, namely 60,000. These data indicated furthermore that each of the yeast isozyme molecules is composed of 2 subunits of equal molecular weight.

To obtain independent estimates of the subunit molecular weights, use was made of the sodium dodecyl sulfate version of the polyacrylamide gel electrophoresis method. Aliquots of the individual isozymes were incubated for several hours at 37° in 1% sodium dodecyl sulfate and then subjected to electrophoresis according to the procedure of Shapiro, Viñuela, and Maizel (1967). The preincubation conditions were those prescribed by Weber and Osborn (1969) to assure complete dissociation of multimeric proteins into their subunits. On the basis of the data from the guanidine hydrochloride experiments we expected to see one band in each case corresponding to subunits of 60,000 mol. wt. Instead, multiple bands were observed in each instance (Fig. 7). It must be emphasized, however, that the sodium dodecyl sulfate to protein ratio was comparatively low, about 2.5 to 1.

These unexpected results could conceivably originate because (a) the denatured protein samples did indeed contain species of molecular weights smaller than 60,000 or (b) the bands represented charge differences between the molecular species. All methods employed to investigate the nature of the faster moving bands indicated that we were indeed dealing with breakdown products of the subunits and not with charge artifacts of the gel electrophoresis method (Kempe and Noltmann, 1971; Kempe et al., 1974a).

On the basis of these results we then reasoned that if the smaller units resulted from further dissociation of the 60,000-dalton monomer, more rigorous dissociation conditions should enhance the appearance of the smaller fragments at the expense of the larger ones. When isozyme samples were incubated at higher sodium dodecyl sulfate to protein ratios,

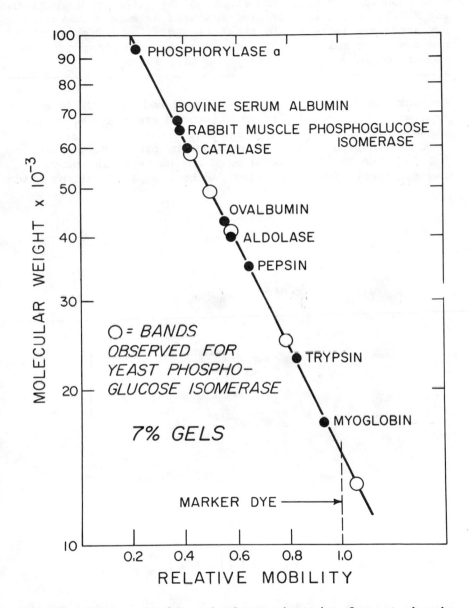

Fig. 7. Polyacrylamide gel electrophoresis of yeast phosphoglucose isomerase isozymes in the presence of sodium dodecyl
sulfate. Yeast phosphoglucose isomerase isozymes or marker
proteins were incubated with 1% sodium dodecyl sulfate in
0.01 M phosphate (Na⁺), pH 7.0, containing 1% β-mercapto-

Fig. 7(cont.) ethanol. The final weight ratio of SDS/protein
was 2.5:1. Mobilities of proteins used as molecular weight
standards are indicated by the filled circles. Open circles
represent the multiple bands observed for the phosphoglucose
isomerase isozymes. From Kempe et al., 1974a.

---

however, the fast moving bands disappeared (Fig. 8) in agreement with the concept that the isozymes are composed of two
subunits of 60,000 molecular weight.

Contamination of our preparations by proteolytic activity
had not been seriously considered in the beginning since
rigorous tests for homogeneity had been applied, all of which

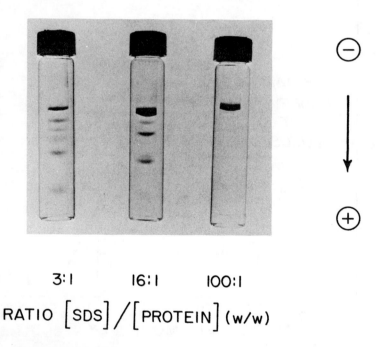

3:1   16:1   100:1

RATIO $\left[\text{SDS}\right]/\left[\text{PROTEIN}\right]$ (w/w)

Fig. 8. Effect of sodium dodecyl sulfate concentration on
the number of bands produced from yeast phosphoglucose isomerase during polyacrylamide gel electrophoresis. Samples of Isozyme B were incubated in 0.01 M phosphate (Na^+), pH 7.0, 1%
β-mercaptoethanol, at different weight ratios of SDS/protein
(as indicated). From Kempe et al., 1974a.

had indicated that the isozymes were homogeneous. However, at about that time Cassman and Schachman (1971) reported proteolytic activity as contamination in supposedly pure preparations of beef glutamate dehydrogenase.

We then tested our isozyme preparations by incubating them with denatured casein. Aliquots were removed from the incubation mixture at various time intervals and pipetted into trichloroacetic acid. After centrifugation, the absorbance at 280 nm in the supernatant fluid was determined as a measure of proteolytic activity. It was found that the absorbance increased steadily with time while a control of casein alone showed no such increase. Also, a second control in which the isozyme sample was incubated without casein did not show any indication of proteolytic degradation (Kempe et al., 1974a).

Thus, the multiple-size sub-sub-units were simply the result of trace contamination of our enzyme preparation by proteases. It is important to stress that these proteases did not act on the native isozyme protein. Only when phosphoglucose isomerase was denatured by low concentrations of sodium dodecyl sulfate did it become an active substrate for the protease. However, the protease functioned and caused a further breakdown of the dissociated phosphoglucose isomerase only as long as the sodium dodecyl sulfate concentration was not high enough to also denature the protease. The latter occurs when either the sodium dodecyl sulfate concentration or the temperature, or both, are raised, each of which will result in destroying the proteolytic activity and stop any further breakdown of the 60,000-dalton monomers of any of the isozymes. We have since eliminated the problem of proteolytic contamination by improving on our chromatographic purification procedures and have adopted, as routine procedure, preincubation conditions for sodium dodecyl sulfate gel electrophoresis that will avoid the aforementioned problems. Nevertheless, our experience with phosphoglucose isomerase illustrates that trace contamination by proteases may be more common than is usually assumed. Our findings would also appear to make it desirable to reinvestigate other proteins for which unexplainable multiple bands have been observed after sodium dodecyl sulfate polyacrylamide gel electrophoresis.

Our final data regarding the chemical characterization of the yeast phosphoglucose isomerase isozymes we believe show unequivocally that they have distinctly different primary structures. Comparison of the amino acid compositions yielded only small differences although we consider them to be significant since they were based on several hundred individual analyses (Kempe et al., 1974b). However, tryptic peptide maps of the three isozymes allowed us to pinpoint the characteristic

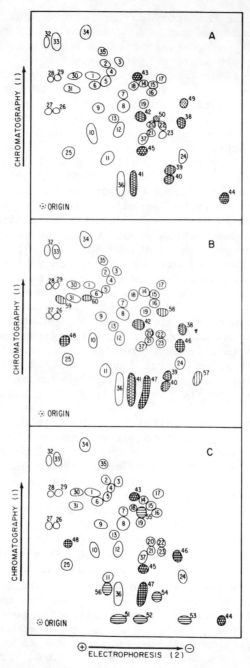

Fig. 9. Tryptic peptide maps of guanidine-hydrochloride-de-
natured phosphoglucose isomerase isozymes from yeast. Upper

Fig. 9 (cont.)   map, Isozyme A; center map, Isozyme B; lower
map, Isozyme C.   Ascending chromatography in butanol:acetic
acid:water (4:1:5 by volume).   Electrophoresis at pH 3.5 in
pyridine:acetic:water (1:10:289 by volume). Peptides were
visualized by fluorescamine spray.   Open spots refer to peptides present in all three isozymes, dotted spots to peptides
present in Isozyme A only, vertically lined spots to peptides
present in Isozyme B only, and horizontally lined spots to
peptides present in Isozyme C only.   Spots with dual coding
refer to peptides present in two of the three isozymes.   From
Kempe et al., 1974a.

CONCLUSIONS

The following conclusions may be derived from the study of
multiple forms of phosphoglucose isomerase.   It should be
emphasized that some of our observations are of general significance and applicable to other enzyme systems, phosphoglucose
isomerase merely being the case studied in detail in our laboratory.

(1)   The general phenomenon of enzymes existing as multiple
forms which have the same catalytic specificity is more subject to artifacts and erroneous interpretation than perhaps
any other area of biochemical investigation.

(2)   Multiple chromatographic species of an enzyme may not
necessarily be distinct, genetically specified protein species
with different primary structures, but they may simply represent different oxidative states of the same basic protein
molecule or, alternatively, reflect various extents of
deamidation or other epigenetic modifications.   Caution should
therefore be exercised in relying on multiple-band activity
staining patterns as the sole criterion for postulating genetically determined isozyme identities.

(3)   As amply demonstrated for phosphoglucose isomerase of
various origins, each set of multiple forms existing in or
originating from a specific source may be different and no
conclusions may be drawn as to the nature of such multiple
forms from one tissue and applied to another.

(4)   Electrophoretic techniques, long being the preferred
tool of isozymologists, have the inherent potential for producing artifacts such as the one described for the widely used
sodium dodecyl sulfate gel electrophoresis method.

ACKNOWLEDGEMENTS

The experimental work described in this paper was supported in part by research grants from the U.S. Public Health Service (AM 07203) and the National Science Foundation (GB 2236) and by Cancer Research Funds of the University of California. The author wishes to acknowledge the contributions of his collaborators who have participated in various stages of these investigations: M. N. Blackburn, J. M. Chirgwin, D. M. Gee, G. M. Hathaway, G. T. James, T. D. Kempe, Y. Nakagawa, T. F. Parsons, A. M. Register, and K. D. Schnackerz.

REFERENCES

Blackburn, Michael N., J. M. Chirgwin, G. T. James, T. D. Kempe, T. F. Parsons, A. M. Register, K. D. Schnackerz, and E. A. Noltmann 1972. Pseudoisoenzymes of rabbit muscle phosphoglucose isomerase. *J. Bio. Chem.* 247: 1170–1179.

Blackburn, Michael N. and E. A. Noltmann 1972. Physical studies on the subunit structure of rabbit muscle phosphoglucose isomerase. *J. Biol. Chem.* 247: 5668–5674.

Blume, K. G., W. Hryniuk, D. Powars, F. Trinidad, C. West, and E. Beutler 1972. Characterization of two new variants of glucose-phosphate isomerase deficiency with hereditary nonspherocytic hemolytic anemia. *J. Lab. Clin. Med.* 79: 942–949.

Bodansky, Oscar 1953. Serum phosphohexose isomerase. *J. Biol. Chem.* 202: 829–840.

Broder, I. and P. A. Srere 1963. Starch-gel electrophoresis of citrate-condensing enzyme from pig heart. *Biochim. Biophys. Acta* 67: 626–632.

Cassman, M. and H. K. Schachman 1971. Sedimentation equilibrium studies on glutamic dehydrogenase. *Biochemistry* 10: 1015–1024.

Enzyme Nomenclature 1972. Recommendations of the International Union of Pure and Applied Chemistry and the International Union of Biochemistry. Elsevier Publishing Company, Amsterdam, pp. 23–25.

Gracy, Robert W. 1974. Nature of the multiple forms of glucose-phosphate and triosephosphate isomerases. Proceedings of this Conference.

Hedrick, Jerry L. and A. J. Smith 1968. Size and charge isomer separation and estimation of molecular weights of proteins by disc gel electrophoresis. *Arch. Biochem. Biophys.* 126: 155–164.

Jörnvall, Hans 1973. Differences in thiol groups and multiple forms of rat liver alcohol dehydrogenase. *Biochem. Biophys. Res. Commun.* 53: 1096–1101.

Kempe, Thomas D., D. M. Gee, G. M. Hathaway, and E. A. Nolt-
    mann 1974a. Subunit and peptide composition of yeast
    phosphoglucose isomerase isoenzymes. *J. Biol. Chem.* 249:
    4625-4633.
Kempe, Thomas D., Y. Nakagawa, and E. A. Noltmann 1974b.
    Physical and chemical properties of yeast phosphoglucose
    isomerase isoenzymes. *J. Biol. Chem.* 249: 4617-4624.
Kempe, Thomas D. and E. A. Noltmann 1971. Unexpected artifacts
    in the determination of the subunit molecular weights of
    yeast phosphoglucose isomerase isoenzymes by the SDS poly-
    acrylamide gel electrophoresis method. *Fed. Proc.* 30:
    1122.
Markert, C. L. and F. Møller 1959. Multiple forms of enzymes:
    tissue, ontogenetic, and species specific patterns. *Proc.*
    Nat. Acad. Sci. 45: 753-763.
Markert, C. L. and G. S. Whitt 1968. Molecular varieties of
    isozymes. *Experientia* 24:977-991.
Meizel, S. and C. L. Markert. 1967. Malate dehydrogenase iso-
    zymes of the marine snail, *Ilyanassa obsoleta*. *Arch.*
    Biochem. Biophys. 122: 753-765.
Mörikofer-Zwez, S., M. Cantz, H. Kaufmann, J. P. von Wartburg,
    and H. Aebi. 1969. Heterogeneity of erythrocyte catalase:
    correlations between sulfhydrol group content, chromato-
    graphic and electrophoretic properties. *Eur. J. Biochem.*
    11: 49-57.
Nakagawa, Yasushi and E. A. Noltmann 1967. Multiple forms of
    yeast phosphoglucose isomerase I. Resolution of the
    crystalline enzyme into three isoenzymes. *J. Biol. Chem.*
    242: 4782-4788.
Noltmann, Ernst A. 1964. Isolation of crystalline phospho-
    glucose isomerase from rabbit muscle. *J. Biol. Chem.*
    239: 1545-1550.
Noltmann, Ernst A. 1972. Aldose-ketose isomerases. *The En-*
    zymes 6: 271-354.
Noltmann, E. and F. H. Bruns 1958. Phosphomannose-isomerase
    II. Anreicherung des Enzyms aus Hefe und Abtrennung
    von Phosphoglucose-isomerase durch Säulenchromatographie
    an Hydroxylapatit. *Biochem. Z.* 330: 514-520.
Payne, D. Michael, D. W. Porter, and R. W. Gracy 1972. Evi-
    dence against the occurrence of tissue-specific variants
    and isoenzymes of phosphoglucose isomerase. *Arch. Bio-*
    chem. Biophys. 151: 122-127.
Pon, Ning G., K. D. Schnackerz, M. N. Blackburn, G. C. Chatter-
    jee, and E. A. Noltmann 1970. Molecular weight and amino
    acid composition of five-times-crystallized phosphoglucose
    isomerase from rabbit skeletal muscle. *Biochemistry* 9:
    1506-1514.

Shapiro, Arnold L., E. Viñuela, and J. V. Maizel Jr. 1967. Molecular weight estimation of polypeptide chains by electrophoresis in SDS-polyacrylamide gels. *Biochem. Biophys. Res. Commun.* 28: 815-820.

Slein, Milton W. 1955. Phosphohexoisomerases from muscle. *Methods Enzymol.*, editors, S. P. Colowick and N. O. Kaplan, Academic Press, N. Y. 1: 299-306.

Tilley, Bill E., R. W. Gracy, and S. G. Welch. 1974. A point mutation increasing the stability of human phosphoglucose isomerase. *J. Biol. Chem.* 249: 4571-4579.

Weber, Klaus and M. Osborn 1969. The reliability of molecular weight determinations by dodecyl sulfate-polyacrylamide gel electrophoresis. *J. Biol. Chem.* 244: 4406-4412.

Yoshida, Akira and N. D. Carter. 1969. Nature of rabbit phosphoglucose isomerase isozymes. *Biochim. Biophys. Acta* 194: 151-160.

NATURE OF THE MULTIPLE FORMS OF GLUCOSEPHOSPHATE AND TRIOSEPHOSPHATE ISOMERASES

ROBERT W. GRACY[1]
Department of Chemistry
North Texas State University
Denton, Texas 76203

ABSTRACT. Human glucosephosphate isomerase, in the presence of reducing agents, exists as a single component in a variety of electrophoretic systems. However, the enzyme is easily oxidized resulting in multiple electrophoretic forms with catalytic activity. An allelic isozyme of human glucosephosphate isomerase ("Singh variant") has been isolated and shown to be the result of a single amino acid substitution which does not affect the catalytic properties of the enzyme, but markedly alters its stability. A survey has shown the absence of true isozymes of glucosephosphate isomerase in most species with a notable exception being the presence of distinct liver and muscle isozymes in bony fish. The liver and muscle enzymes have been isolated from the freshwater catfish and their chemical, physical, and catalytic properties compared.

Three isozymes exist for human triosephosphate isomerase and are the result of an AA, AB, and BB distribution of dimers. Amino acid analyses and peptide fingerprints have indicated that the two types of subunits are very similar but contain several differences in their primary structures. The three isozymes have similar catalytic properties and are found in all human tissues. All vertebrates examined showed three isozymes, while microorganisms and most invertebrates exhibited only a single form of the enzyme.

INTRODUCTION

Triosephosphate isomerase (EC 5.3.1.1) and glucosephosphate isomerase (EC 5.3.1.9) catalyze reversible aldose-ketose isomerizations and exhibit the highest catalytic rates of all enzymatic steps of glycolysis. The isomerases have generally been considered as bifunctional enzymes, participating in both glycolysis and gluconeogenesis and unlikely points for metabolic regulation. The high catalytic rates of these enzymes and their in vivo levels have led to the concept that these enzymes are in great excess in vivo and not likely to become metabolically rate limiting. However, Schneider et al (1965) and Baughan et al (1968) have described genetically transmitted isomerase deficiencies in which homozygous patients exhibited 5-20% normal levels of these enzymes. These studies indicated that a decrease in the levels of the isomerases by 5-10 fold can result in a severe metabolic block, even though the anticipated levels of these enzymes (based on in vitro assays) would suggest that they still should not be rate limiting. Both isomerases have been disputed by various groups to be composed of iso- zymes (see Noltmann, 1972).

HUMAN GLUCOSEPHOSPHATE ISOMERASE

Previous studies on human glucosephosphate isomerase have been limited due to the lack of a suitable method for the isolation of the enzyme. Purification procedures were lengthy, resulting in poor recoveries, and in some cases resulted in only partial purification or the appearance of multiple forms of the enzyme (Tsuboi et al, 1958; Carter and Yoshida, 1968; Tsuboi et al, 1971). All previous attempts to obtain the enzyme in a crystalline form have been unsuc- cessful.

We have recently developed a very simple, rapid procedure for the isolation of human glucosephosphate isomerase by specific elution of the enzyme from cellulose phosphate with substrate (Gracy and Tilley, 1974). The pure enzyme can thus be obtained with an overall yield of 75-85% in two steps requiring only one to two days. The crystalline enzyme iso- lated by this procedure was shown to be homogeneous by ana- lytical ultracentrifugation and appeared as a single compo- nent upon polyacrylamide or starch gel electrophoresis or isoelectric focusing (Tilley et al, 1974). However, after storage of the crystalline suspension in the absence of reducing agents, multiple electrophoretic forms with cata- lytic activity were observed (Fig. 1). These new electro- phoretic forms could be converted back to the single native form by treatment with reducing agents such as dithiothreitol. Thus, the multiple forms of human glucosephosphate isomerase previously observed (Tsuboi et al, 1971) seem to be the result of oxidative artifacts similar to those described by Blackburn et al (1971) for the rabbit muscle enzyme. Ex- tracts from human brain, liver, heart, kidney, spleen, and erythrocytes when subjected to starch gel or polyacrylamide electrophoresis in the presence of 2-mercaptoethanol also yielded single components with identical electrophoretic mobilities. In addition, coelectrofocusing of various human tissue extracts (Payne et al, 1972) resulted in a single component with a pI of 9.3. Thus, it would appear that in

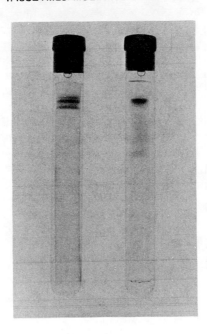

Fig. 1. Polyacrylamide gel electrophoresis of crystalline human glucosephosphate isomerase. Glucosephosphate isomerase from human erythrocytes (specific activity 850 μmoles fructose 6-phosphate isomerized per min per mg at 30°) was stored for two weeks in the presence (right gel) or absence (left gel) of 10 mM 2-mercaptoethanol. Standard 7.5% alkaline gels were run at pH 8.9 for 4 hours, 2 ma per gel, and stained with 1% w/v amido black. (Reprinted with permission from Tilley et al, 1974.)

man, neither true isozymes nor tissue specific variants of the enzyme exist.

ALLOZYMES OF HUMAN GLUCOSEPHOSPHATE ISOMERASE

At least ten different allozymes (allelic isozymes) of human glucosephosphate isomerase have been described in a variety of studies (see Tilley et al, 1974). Some of the allozymes exhibit reduced catalytic activity, and the homozygous patients suffer a chronic nonspherocytic hemolytic anemia. Glucosephosphate isomerase deficiency is now believed to be the third most common cause of such enzyme-

TABLE I

ELECTROPHORETIC PROPERTIES OF GLUCOSEPHOSPHATE AND TRIOSEPHOSPHATE
ISOMERASES OF VARIOUS SPECIES

Species[a]	Glucosephosphate Isomerase		Triosephosphate Isomerase	
	Number of Isozymes	Electrophoretic[b] Mobility mm hr^{-1}	Number of Isozymes	Electrophoretic[c] Mobility mm hr^{-1}
Human				
Erythrocytes	1	25	3	2.8, 2.2, 1.6
Skeletal muscle	1	25	3	2.8, 2.2, 1.6
Brain	1	25	3	2.8, 2.2, 1.6
Liver	1	25	3	2.8, 2.2, 1.6
Cardiac muscle	1	25	3	2.8, 2.2, 1.6
Spleen	1	25	3	2.8, 2.2, 1.6
Rhesus	1	50	3	2.8, 2.2, 1.6
Bovine	1	21	3	1.6, 1.4, 1.2
Porcine	1	29	3	1.6, 1.2, 1.0
Dove	1	17	3	2.8, 2.2, 1.5
Turtle	1	24	3	2.0, 1.5, 1.0
Frog	1	24	3	2.0, 1.5, 1.0
Catfish				
Muscle	1 (+1M)	14(33)	3	1.5, +0.8, −.8
Liver	1 (+1M)	33(14)	3	1.5, +0.8, −.8
Kidney	1 (+1M)	33(14)	3	1.5, +0.8, −.8
Brain	1 (+2M)	33(21, 14)	3	1.5, +0.8, −.8
Heart	1 (+2M)	33(21, 14)	3	1.5, +0.8, −.8
Crab	1	30	2	6.8, 6.0
Lobster	−−	−−	2	6.8, 6.0
Shrimp	1	24	2	6.5, 6.0

Beetle	1	44	1	11.3
Cricket	--	--	1	12.0
Grasshopper	--	--	1	14.6
Clam	1	27	1	0.9
Snail	--	--	1	2.3
Squid	2	20,35	1	2.5
Ascaris suum	--	--	1(+1M)	0.9
Sea Anemonae	4	20,23,26,30		3.8, (2.0)
Euglena gracilis				
Light-grown	1	34	2	0.66, 0.78
Dark-grown	1	34	1	0.66
Escherichia coli	--	--	1	12.5
Bacillus subtilis	--	--	1	15.0
Pseudomonas aeruginosa	1	34	1	10.0
Staphylococcus aureus	--	--	1	14.0

[a] In cases where no specific tissue is listed, samples were from skeletal muscle. For insects the entire organism was extracted. All mobilities are given as migration toward the anode. Minor bands are indicated with "M", with their mobilities given in parentheses.

[b] Electrophoresis on cellulose acetate plates under the conditions described in Fig. 2. Glucosephosphate isomerase was located by an activity stain (Payne, Porter, and Gracy, 1972).

[c] Electrophoresis was on 13% w/v starch gels under conditions described in Fig. 4. Triosephosphateisomerase activity was located by the activity stain of Kaplan et al, 1968.

deficient anemias (Blume and Beutler,1972). Other allozymes have been observed with normal activity levels, and one allozyme (designated "Singh-variant" by Welch, 1971) has been observed with elevated glucosephosphate isomerase levels in erythrocytes.

We have isolated glucosephosphate isomerase in crystalline form from an individual heterozygous for the Singh allozyme and resolved the variant homodimer from the normal homodimer and the heterodimer by isoelectric focusing (Tilley et al. 1974). Detailed structural studies at the subnanomole level permitted isolation of a single aberrant peptide, and revealed that the modification is the result of a single point mutation, causing an acidic → neutral amino acid substitution. The mutation does not affect the catalytic turnover rate or the binding of substrates or inhibitors, but results in a protein which is more stable than the wild type enzyme. Thus, the increased levels of glucosephosphate isomerase activity observed in human erythrocytes with the Singh allozyme seem to result from the greater stability and longer half life of the variant enzyme in these cells.

GLUCOSEPHOSPHATE ISOMERASE IN OTHER SPECIES

We have also examined a number of different species for the presence of isozymes of glucosephosphate isomerase. Table I shows that when caution is taken to maintain the enzyme in a reduced form, essentially all species examined exhibit only a single electrophoretic form of the enzyme.[2] However, in fish distinctly different isozymes were observed in various tissues.

Fig. 2 shows zymograms of glucosephosphate isomerase from extracts of various tissues of the freshwater catfish. The enzyme from muscle exhibited a much lower anodal migration than the enzyme from liver or kidney. Evidence of hybrid forms were also seen in heart and brain tissues. These observations of multiple electrophoretic forms of glucosephosphate isomerase in fish are consistent with the electrophoretic studies of Avise and Kitto (1973) who surveyed over 50 species of bony fishes and found widespread occurrence of two forms of the enzyme.

We have isolated glucosephosphate isomerase from catfish liver and muscle and shown each isozyme to be homogeneous by analytical ultracentrifugation and polyacrylamide electrophoresis. Table II summarizes several of the properties of the two isozymes. Although both the liver and muscle isozymes are of identical molecular weights and composed of two subunits of 65,000 daltons each, they differ in their electro-

CELLULOSE ACETATE ELECTROPHORESIS OF CATFISH PGI IN TISSUE EXTRACTS

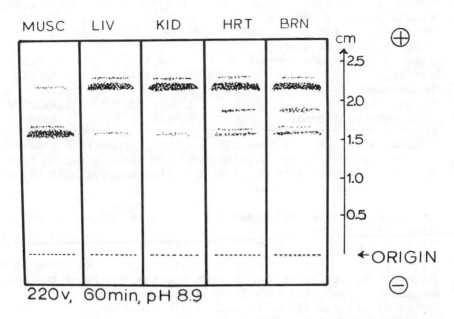

Fig. 2. Glucosephosphate isomerases from catfish tissues. Cell free tissue extracts (25% w/v) were prepared in 10mM triethanolamine buffer, pH 8.5, containing 1 mM EDTA and 10 mM 2-mercaptoethanol. Samples were placed on 2.5 x 10 cm cellulose acetate plates and subjected to electrophoresis at pH 8.9, 220 V, 2 ma/plate, for 60 min. The glucosephosphate isomerase was located by a specific activity stain (Payne, Porter, and Gracy, 1972). Abbreviations are Musc (skeletal muscle), Liv (liver), Kid (kidney), Hrt (heart), and Brn (brain).

phoretic and chromatographic properties. The two isozymes also differ with respect to their sensitivity toward a variety of denaturants, and preliminary data suggest significant differences in their amino acid compositions and peptide fingerprints. On the other hand, the catalytic properties thus far examined of the two isozymes show only slight differences. Carter and Dando (1973) have reported similar differences in liver and muscle glucosephosphate isomerase isolated from *Conger conger L.*

TABLE II
COMPARISON OF LIVER AND MUSCLE
GLUCOSEPHOSPHATE ISOMERASE

Property	Liver	Muscle
Molecular weight	130,000	131,000
Subunit molecular weight	65,000	65,000
Isoelectric point	6.2	7.0
Ionic strength to elute from DEAE-Sephadex (pH 9.2)	0.11	0.04
Stability[a]	stable	labile
Km (Fructose 6-P)[b]	9.4×10^{-5}M	9.8×10^{-5}M
Km (Glucose 6-P)	3.6×10^{-4}M	1.9×10^{-4}M
Ki (6-phosphogluconate)	5.5×10^{-5}M	3.8×10^{-5}M

[a]Stability experiments were carried out at identical conditions of protein concentration, pH, and ionic strength in a variety of denaturants, including sodium dodecyl sulfate, guanidinium chloride, and urea.

[b]All kinetic values are given for pH 8.2 at 30° and at an ionic strength of 0.10.

These distinct differences suggest that the two forms of glucosephosphate isomerase in fish are the result of two independent cistrons, and this interpretation is strongly supported by the studies of Avise and Kitto (1973), which revealed individual fish which were heterozygous for each of the two isozymes. They have further postulated that the multiplicity of glucosephosphate isomerase in teleosts is a result of the high degree of polyploidy exhibited by these species. Thus, it would appear that while true isozymes or tissue-specific variants do not exist in most species, a number of species of fish do possess true isozymes with tissue-specific localization. Further comparisons of the structural and catalytic differences in these isozymes are required in order to ascertain the physiological significance of the muscle and liver type glucosephosphate isomerases.

HUMAN TRIOSEPHOSPHATE ISOMERASE

Crystalline human triosephosphate isomerase was first isolated in our laboratory from erythrocytes (Rozacky et al, 1971) and resolved into three components with catalytic activity by isoelectric focusing (Fig. 3). Each of the three components refocused as a single species (Sawyer et al, 1972), and the three forms were observed in a variety of other

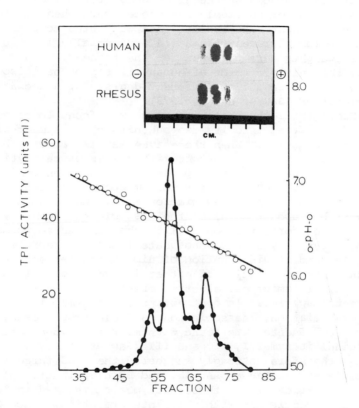

Fig. 3. Electrofocusing and electrophoresis of human erythrocyte triosephosphate isomerase. Crystalline triosephosphate isomerase from human erythrocytes (specific activity
10,500 μmoles glyceraldehyde 3-phosphate isomerized per min
per mg at 30°) was electrofocused in 1% ampholines (pH 5-7).
Fractions of 1.0 ml each were collected and assayed for enzyme activity (•) and pH (o). In the upper part of the
figure is shown a zymogram from human erythrocyte lysates
separated by starch gel electrophoresis (13% w/v gels, pH
9.3, running time 20 hours, 10 V/cm). The cathode is to the
left and anode to the right. The gels were stained for triosephosphate isomerase activity by an activity stain (Snapka
et al, 1974).

electrophoretic systems (Snapka et al, 1974). The electrophoretic and electrofocusing profiles were not altered by
the addition of reducing agents such as 2-mercaptoethanol or
protease inhibitors. Identical profiles were also obtained
from electrofocusing or electrophoresis of fresh hemolysates
(Fig. 3). Thus, it did not appear that the three forms were
artifacts produced by the electrophoretic methods, oxidation,
proteolysis, or the isolation procedure. When each of the
three components was isolated and subjected to sedimentation
equilibrium ultracentrifugation, essentially identical molecular weights were obtained for the three species suggesting
that the multiple forms of human triosephosphate isomerase
were not the result of different degrees of association or
aggregation (Sawyer et al, 1972).

Human triosephosphate isomerase was found to be very readily dissociated into its two subunits by treatment with guanidinium chloride. When the enzyme was titrated with this
denaturant, there was a concomitant loss of catalytic activity and a change in the apparent sedimentation coefficient
from 4.1S to 1.8S at a titration midpoint of 0.75 M guanidinium chloride. This dissociation was found to be completely
reversible upon removal of the dissociating agent by dialysis. Therefore, the possibility that the three forms of
human erythrocyte triosephosphate isomerase represented an
AA, AB, and BB distribution of dimers was tested. Dissociation and reassociation of either isozyme I or III followed
by polyacrylamide gel electrophoresis resulted in only the
parental species. In contrast, when isozyme II was subjected
to dissociation, reassociation, and electrophoresis, all
three components were found with migration identical to the
original isozymes I, II, and III (Sawyer et al, 1972). These
data, therefore, strongly suggested that the three forms
indeed represented an AA, AB, and BB distribution of dimers.
The properties of the three isozymes summarized in Table III
are also consistent with this interpretation, since in all
cases values for isozyme II (the AB heterodimer) were found
to be intermediate between those of the two homodimers.

Examination of the basic catalytic properties of the three
forms revealed small but reproducible differences in the Km
values for the two substrates. Although these differences
are small and difficult to assess in terms of their physiological significance, it should be noted that the in vivo
levels for glyceraldehyde 3-phosphate and dihydroxyacetone
phosphate in the human erythrocyte are 0.006 mM and 0.017 mM,
respectively, (Yoshida, 1973). Therefore, slight differences
in these catalytic constants could be physiologically important.

TABLE III

COMPARISON OF HUMAN ERYTHROCYTE
TRIOSEPHOSPHATE ISOMERASE ISOZYMES

Property	Isozyme I	Isozyme III	Isozyme III
Molecular Weight	56,000	56,000	56,000
Subunit Molecular Weight	28,000	28,000	28,000
Isoelectric pH	6.7	6.4	6.1
Distribution	5-10%	70-75%	20-25%
Km (G3P)	11.2µM	14.5µM	18.4µM
Km (DHAP)	82.5µM	45.2µM	38.2µM
pH optimum	7.8	7.8	7.8
(Residues mole^{-1})			
Histidine	14.8	10.1	7.9
Glycine	42.7	45.7	50.2
Cysteine	9.4	10.1	9.7
Valine	37.5	32.7	30.1
Leucine	41.5	45.0	48.3

The amino acid compositions of isozymes I, II, and III were determined and are partially shown in Table III. The contents of most amino acids were essentially identical in the three forms of the enzyme, but significant differences were found in histidine, glycine, serine, valine, and leucine. The composition of isozyme II, in every case, was intermediate between that of components I and III, and the higher histidine content of isozyme I is consistent with its more basic isoelectric value. It should also be noted that the cysteine content of the three isozymes was essentially identical, further arguing against different states of oxidation as a possible basis for the multiplicity.

When the three isozymes were S-carboxymethylated, digested with trypsin, and subjected to peptide fingerprinting, the patterns shown in Fig. 4 were obtained (Sawyer et al, 1972). Although a high degree of homology was observed for most peptides, at least four peptides were found with significantly altered mobilities. These structural differences were substantiated from the fingerprints of isozyme II (the AB heterodimer) which exhibited essentially a composite of isozymes I and III. These different peptides do not correspond to the carboxyl or amino terminal peptides, or the SH-containing peptides. Furthermore, a single active-site hexapeptide was found after labeling a mixture of the isozymes with the substrate analogue, chloroacetolphosphate and digesting with pepsin (Hartman and Gracy, 1973). These data suggest distinct, yet not extensive, differences in the A and B type sub-

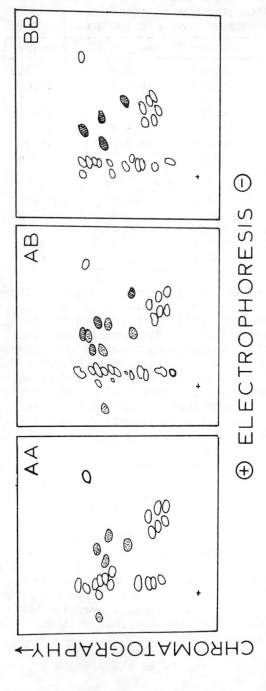

Fig. 4. Tryptic peptide fingerprints from human triosephosphate isomerase isozymes. The three isolated isozymes were reduced and S-carboxymethylated with 14C-iodoacetic acid prior to tryptic digestion. Thin layer peptide mapping was carried out as described by Sawyer et al, 1972. The unique peptides are shaded. (Reprinted with permission from Sawyer et al, 1972.)

units of human erythrocyte triosephosphate isomerase. Studies of Krietsch et al (1971) have shown similar differences in the isozymes from rabbit muscle. These data thus argue against conformational isomers, and seem to be most amenable with a two cistron - two subunit hypothesis.

When extracts of human heart, liver, skeletal muscle, kidney, brain, spleen, lung, and blood were subjected to starch gel electrophoresis and stained with an activity stain, or electrofocused in narrow range (pH 5-7) ampholines, three forms of triosephosphate isomerase activity were found in each case (Snapka et al, 1974). The electrophoretic migration, isoelectric points, and relative distribution of the three isozymes were identical for each of the eight tissues. In all human tissues isozyme I (AA) accounted for only 5-10% of the total activity with 70-75% as isozyme II and 20-25% as isozyme III. The basis for this somewhat unusual distribution is not understood; however, when isozyme II was dissociated and reassociated in vitro (Sawyer et al, 1972) the amount of isozyme I was essentially the same as isozyme III.

TRIOSEPHOSPHATE ISOMERASE DEFICIENCY DISEASE

Schneider et al (1965) showed that individuals homozygous for triosephosphate isomerase deficiency exhibited only 5-10% of the normal activity in their red cells. If the mutation were in a cistron coding for the B subunit, such that isozyme III (BB) and the hybrid isozyme II (AB) were both totally inactive leaving only isozyme I (AA) fully active, this could account for the 5-10% activity levels in these patients. Kaplan et al (1968) have conducted starch gel electrophoresis on hemolysates from a triosephosphate isomerase-deficient patient. The zymograms revealed only activity corresponding to the most basic isozyme (isozyme I), and the virtual absence of any activity corresponding to components II and III. These data further support a genetic basis of the isozymes of human triosephosphate isomerase and also suggest that the mutation in triosephosphate isomerase deficiency may be in the cistron coding for the B subunit. Furthermore, since it appears that isozyme I accounts for only 5-10% of the total triosephosphate isomerase activity in most other human tissues (Snapka et al, 1974), this would account for the generalized tissue defect and overall severity of the disease.

TRIOSEPHOSPHATE ISOMERASE FROM OTHER SPECIES

Table I also summarizes the results of a survey of a variety of species for isozymes of triosephosphate isomerase. In general, all vertebrates exhibit three isozymes. In micro-organisms and most invertebrates, only a single form of tri-osephosphate isomerase was observed, but in a few higher in-vertebrates two widely separated isozymes were found, suggest-ing the absence of heterodimer formation. Whether or not this is the result of tissue-specific compartmentalization of the two isozymes has not been determined. Therefore, these data further support a genetic basis for the multiple forms and suggest that gene duplication may have occurred early in ver-tebrate evolution.

The only species examined which showed identical electro-phoretic mobilities with the human was the Rhesus monkey. Although the three isozymes from the two species exhibited identical electrophoretic mobilities and coelectrofocused, the relative distribution of the three isozymes differed markedly. In the monkey the sequence of appearance and the relative in-tensity of isozymes were I > II > III, while in the human it was II > III > I. Further comparative studies of these two species may reveal the basis for the unusual distribution of the triosephosphate isomerase isozymes.

An interesting situation was found to exist in the uni-cellular green protist, *Euglena gracilis*. When the organism was grown in the dark on an organic carbon source, a single form of triosephosphate isomerase with an isoelectric pH of 4.8 was found. However, when *Euglena* was grown photoauto-trophically, a second, more acetic form (pI = 4.4) of the en-zyme was also observed (Mo et al, 1973). Furthermore, sub-cellular fractionation revealed that the pI 4.8 enzyme was of cytoplasmic origin, while the pI 4.4 enzyme was localized in the chloroplast. Growth of the organism under mixotrophic conditions in the presence of chloramphenicol to inhibit chloroplastic 70S ribosomal synthesis blocked the formation of the chloroplastic isomerase, but did not affect synthesis of the cytoplasmic enzyme. These studies also revealed that the synthesis of the two isozymes of *Euglena* triosephosphate isomerase paralleled the synthesis of the cytoplasmic (Class II) and chloroplastic (Class I) aldolases.

SUMMARY

In the case of glucosephosphate isomerase, true isozymes do not exist in man, but several allelic isozymes (allozymes) have been described with different catalytic and stability properties. The newly developed methods for isolation of the enzyme and methods for structural studies at the microgram level (Tilley et al, 1974) have permitted comparative struc-tural studies on these allozymes. In most other species, isozymes are also absent with the notable exception of the bony fish, where two distinct isozymes exist in muscle and liver tissues.

Triosephosphate isomerase from human and most vertebrate species exists as three isozymes, representing an AA, AB, and BB distribution of dimers. Structural studies on the human isozymes suggest that the proteins are very similar but contain several differences in their primary structure. In the case of *Euglena gracilis*, chloroplastic and cytoplasmic isozymes exist and their synthesis is regulated by growth conditions of the organism. Although the bulk of data favor a two-cistron basis for the isozymes, the overall structural similarity of the isozymes does not completely rule out a post transcriptional basis for the multiplicity.

ACKNOWLEDGMENTS

This work was supported in part by grants from the National Institute of Arthritis, Metabolic and Digestive Diseases, National Institutes of Health (AM 14638), the Robert A. Welch Foundation (B-502), and NTSU Faculty Research Funds.

FOOTNOTES

1. Recipient of National Institutes of Health Career Development Award (KO4 AM70198).

2. True isozymes of glucosephosphate isomerase have been observed and studied in detail from yeast and are discussed elsewhere in this series (Noltmann, 1974).

REFERENCES

Avise, J. C., and G. B. Kitto 1973. Phosphoglucose Isomerase Gene Duplication in the Bony Fishes: An Evolutionary History. *Biochem. Genet.* 8: 113-132.

Baughan, M. A., N. N. Valentine, D. E. Paglia, P. O. Ways, E. R. Simon, and Q. B. DeMarsh 1968. Hereditary Hemolytic Anemia Associated with Phosphoglucose Isomerase Deficiency. *Blood.* 32: 236-259.

Blackburn, M. N., J. M. Chirgwin, G. T. James, T. D. Kempe, T. F. Parsons, A. M. Register, K. D. Schnackerz, and E. A. Noltmann 1971. Pseudoisozymes of Rabbit Muscle Phosphoglucose Isomerase. *J. Biol. Chem.* 247: 1170-1179.

Blume, K. G., and E. Beutler 1972. Detection of Glucose-Phosphate Isomerase Deficiency by a Screening Procedure. *Blood.* 39: 685-687.

Carter, N. D., and P. Dando 1973. Phosphoglucose Isomerase

in Teleostean Fish. *Biochem. Soc. Trans.* 1: 17-18.

Carter, N. D., and A. Yoshida 1968. Purification and Characterization of Human Phosphoglucose Isomerase. *Biochim. Biophys. Acta* 181: 12-19.

Gracy, R. W., and B. E. Tilley 1974. Phosphoglucose Isomerase from Human Erythrocytes and Cardiac Tissue. *Methods in Enzymol.* 41: Part B (In press).

Hartman, F. C., and R. W. Gracy 1973. An Active-Site Peptide from Human Triose Phosphate Isomerase. *Biochem. Biophys. Res. Commun.* 52: 388-393.

Kaplan, J. C., L. Teeple, N. Moore, and E. Beutler 1968. Electrophoretic Abnormality in Triosephosphate Isomerase Deficiency. *Biochem. Biophys. Res. Commun.* 31: 768-773.

Krietsch, W. K. G., P. G. Pentchev, and H. Klingenburg 1971. The Isolation and Characterization of the Isoenzymes of Rabbit Muscle Triosephosphate Isomerase. *Eur. J. Biochem.* 23: 77-85.

Mo, Y., B. G. Harris, and R. W. Gracy 1973. Triosephosphate Isomerases and Aldolases from Light- and Dark-Grown *Euglena gracilis. Arch. Biochem. Biophys.* 157: 580-587.

Noltmann, E. A. 1972. Aldose Ketose Isomerases. *The Enzymes* VI: 271-354.

Noltmann, E. A. 1974. The different origins of multiple molecular forms of prosphoglucose isomerase. *I Isozymes: Molecular Structure.* C. L. Markert, Ed. Academic Press, N. Y.

Payne, D. M., D. W. Porter, and R. W. Gracy 1972. Evidence Against the Occurrence of Tissue-Specific Variants and Isozymes of Phosphoglucose Isomerase. *Arch. Biochem. Biophys.* 151: 122-127.

Rozacky, E. E., T. H. Sawyer, R. A. Barton, and R. W. Gracy 1971. Studies of Human Triosephosphate Isomerase I. Isolation and Properties of the Enzyme from Erythrocytes. *Arch. Biochem. Biophys.* 146: 312-320.

Sawyer, T. H., B. E. Tilley, and R. W. Gracy 1972. Studies on Human Triosephosphate Isomerase II. Nature of the Electrophoretic Multiplicity in Erythrocytes. *J. Biol. Chem.* 247: 6499-6505.

Schneider, A. S., W. N. Valentine, M. Hattori, and H. L. Heins 1965. Hereditary Hemolytic Anemia with Triosephosphate Isomerase Deficiency. *N. Engl. J. Med.* 272: 229-235.

Snapka, R. M., T. H. Sawyer, R. A. Barton, and R. W. Gracy 1974. Comparison of the Electrophoretic Properties of Triosephosphate Isomerases of Various Tissues and Species. *Comp. Biochem. Physiol.* 46B (In press).

Tilley, B. E., R. W. Gracy, and S. G. Welch 1974. A Point Mutation Increasing the Stability of Human Phosphoglucose Isomerase. *J. Biol. Chem.* 249: (In press).

Tsuboi, K. K., J. Estrada, and P. B. Hudson 1958. Enzymes of the Erythrocyte IV. Phosphoglucose Isomerase, Purification and Properties. *J. Biol. Chem.* 231: 19-29.

Tsuboi, K. K., K. Fukanaga, and C. H. Chervenka 1971. Phosphoglucose Isomerase from Human Erythrocytes. *J. Biol. Chem.* 246: 7586-7594.

Welch, S. G. 1971. Qualitative and Quantitative Variants of Human Phosphoglucose Isomerase. *Hum. Heredity* 21: 467-477.

Yoshida, A. 1973. Hemolytic Anemia and Glucose 6-Phosphate Dehydrogenase Deficiency. *Science* 179: 532-537.

THE USE OF THIOL REAGENTS IN THE
ANALYSIS OF ISOZYME PATTERNS

D. A. HOPKINSON
MRC Human Biochemical Genetics Unit
Galton Laboratory
University College London, England

For the past ten years our laboratory has been engaged in
a study of the extent and character of genetically determined
enzyme variation in human populations. Our principal analytical approach has been to use the method of zone electrophoresis in starch gels together with sensitive specific enzyme
staining techniques in a search for inherited electrophoretic
variations in the isozyme patterns exhibited by human red
cell lysates and tissue extracts. To date we have examined
the isozymes determined by more than forty separate structural
gene loci and information on the frequency of the genetic
variants which have been identified in this project will be
discussed at this conference (Harris, H., these proceedings).
The present paper is concerned with another aspect of this
project and it discusses the significance of the reactive
sulfhydryl groups of isozymes in the investigation of genetically determined electrophoretic variation.

During the course of our work we have been at pains to distinguish clearly between truly genetic individual variation
in isozyme patterns and spurious variations which might arise
during the period when a blood or tissue sample is being sent
or brought to the laboratory or stored for some reason prior
to analysis. It was from a consideration of this problem,
when we were faced with unwanted and somewhat aggravating in
vitro storage changes in isozyme patterns, that the importance
of the sulfhydryl groups of isozymes emerged.

Sulfhydryl modified isozymes

We noticed that the isozymes of several enzymes (Table I),
about a sixth of all the loci that we have tested, exhibited
a progressive change in electrophoretic mobility when cell or
tissue extracts were kept for a few days in the refrigerator.
The speed at which this change occurred varied from one enzyme
to another and also from one sample to another but a characteristic general feature of the change was an apparent discrete increase in the anodal electrophoretic mobilities of
the isozymes, usually without noticeable loss in the total
enzyme activity. Characteristically all the isozymes seen in
the fresh material were affected by the change so that even-

TABLE I
SOME HUMAN ENZYMES WHICH READILY
EXHIBIT SULFHYDRYL MODIFIED
ISOZYMES ON STORAGE IN VITRO

Enzyme	Locus	Reference
1. Red cell acid phosphatase	ACP_1	Fisher & Harris (1969)
2. Peptidase C	PEP-C	Lewis & Harris (1967)
3. Peptidase D	PEP-D	Lewis & Harris (1967, 1969)
4. Adenosine deaminase	ADA	Hopkinson & Harris (1969)
5. Phosphogluco-mutase	PGM_3	Fisher & Harris (1972)
6. Inosine triphos-phatase	ITP	Hopkinson (unpublished data)

tually the stored samples exhibited patterns in which the isozymes showed the same relative staining intensities and relative electrophoretic mobilities as the fresh samples but their
over-all anodal electrophoretic mobilities were greater. For
most of the enzymes the change in electrophoretic mobility
appeared to be a one step reaction but for two enzymes, peptidase D and inosine triphosphatase- subsequently shown to be
dimeric proteins, the change appeared to involve two steps
forward in anodal electrophoretic mobility.

It was found that if the sulfhydryl reducing agents mercaptoethanol, dithiothreitol or reduced glutathione were added to
the older samples then the isozyme patterns usually reverted
to those of the fresh samples; on the other hand preincubation of a fresh sample with oxidized glutathione led to the
production of a pattern similar to that seen in stored material, also the effect of oxidized glutathione could be reversed
by the subsequent addition of excess mercaptoethanol. It was
known that oxidized glutathione accumulates in stored red cell
lysates and in crude tissue extracts contaminated with red
cells and it was therefore concluded from these experiments
that the type of in vitro storage change described above probably involved a reaction between the free sulfhydryl groups
of cysteine residues on the isozymes and the oxidized glutathione to form a mixed disulfide:

$$Enz\text{-}SH + G\text{-}S\text{-}S\text{-}G \rightleftharpoons Enz\text{-}S\text{-}S\text{-}G + G\text{-}SH$$

The formation of such a product, linking half a molecule of oxidized glutathione to an isozyme, would increase the net negative charge by one unit for each sulfhydryl (-SH) group participating in the reaction (see Table II) and thus the isozyme would acquire a greater anodal electrophoretic mobility.

Thiol reagents

This hypothesis was tested by examining the effects of a number of thiol reagents, in addition to oxidized glutathione, on the electrophoretic mobilities of the isozymes of those enzymes (Table I) which, on the basis of their storage characteristics, were believed to contain one or more reactive -SH groups. The test system which was used is summarized in Table II.

The thiol reagents fall into three categories: disulfides which, like oxidized glutathione, react reversibly with sulfhydryls to form mixed disulfides; alkylating agents which react to form stable addition compounds; and organic mercurials which react reversibly with -SH groups to form mercaptides. In each case the nature of the net charge on the group added to the isozyme, via a reaction with a sulfhydryl group, will be expected to influence the electrophoretic mobility of the isozyme. Thus the monocarboxylic thiol reagents like iodoacetate and p-chloromercuribenzoate will add one extra negative charge per -SH group and the dicarboxylic reagents like maleate will add two extra negative charges and the modified isozymes will exhibit a one or two step increase in anodal electrophoretic mobility respectively. Reactions with the basic thiol reagents, such as bromoethylammonium bromide, ethylenimine and cystamine, however will be expected to have the opposite effect on electrophoretic mobility since a positive group will be acquired by the isozymes at each reactive -SH group. Neutral reagents such as n-ethylmaleimide and iodoacetamide in contrast will not be expected to change the electrophoretic properties of the isozymes but, since these particular reagents form stable products with sulfhydryls, they will be expected to block subsequent reactions with charged reagents such as iodoacetic or oxidized glutathione and also should abolish the normal in vitro storage effects on the isozyme mobilities.

Some results obtained from experiments with the isozymes of human red cell adenosine deaminase (ADA) provide a suitable model example to illustrate this general method of investigating the reactive sulfhydryl groups of isozymes.

When freshly prepared haemolysates are examined by starch gel electrophoresis, three different adenosine deaminase (ADA)

TABLE II

The effects of nine different thiol reagents on the electrophoretic mobility of enzymes containing a free reactive sulfhydryl group (Enz-SH) and the reactions postulated to explain
these effects.

Thiol reagents		
Type	Structure	Reaction with Enz-SH
Disulfide -glutathione	$\overset{+}{\text{glu}}-\overset{-}{\text{cys}}-\overset{-}{\text{gly}}$ $\mid$ S $\mid$ S $\mid$ $\overset{+}{\text{glu}}-\overset{-}{\text{cys}}-\overset{-}{\text{gly}}$	reversible
-cystamine	$S-CH_2.CH_2.NH_3^+$ $\mid$ $S-CH_2.CH_2.NH_3^+$	reversible
Alkylation -iodoacetate	$I.CH_2.COO^-$	non-reversible
-iodoacetamide	$I.CH_2.CONH_2$	non-reversible
-N-ethylmaleimide	$CH-CO$ $\Vert \quad\quad N.C_2.H_5$ $CH-CO$	non-reversible
-maleate	$CH.COO^-$ $\Vert$ $CH.COO^-$	(reversible)

Product formed with Enz-SH			
Structure	Charged group(s) acquired	Net charge difference from untreated enzyme	Observed anodal electrophoretic mobility - relative to untreated enzyme
$\overset{+}{\text{glu}}\text{-}\overset{-}{\text{cys}}\text{-}\overset{-}{\text{gly}}$ $\mid$ S $\mid$ S $\mid$ (Enz)	2 acidic 1 basic	-1	fast
$\overset{\mid}{\text{S}}\text{-CH}_2\text{.CH}_2\text{.NH}_3^+$ $\overset{\mid}{\text{S}}\text{-(Enz)}$	1 basic	+1	slow
$(\text{Enz})-\text{S}-\text{CH}_2\text{.COO}^-$	1 acidic	-1	fast
$(\text{Enz})-\text{S}-\text{CH}_2\text{.CONH}_2$	none	0	unchanged
$(\text{Enz})-\text{S}-\text{CH}-\text{CO}$ $\mid \qquad \diagdown^{\text{NC}_2\text{H}_5}$ $\text{CH}_2-\text{CO}\diagup$	none	0	unchanged
$(\text{Enz})-\text{S}-\text{CH.COO}^-$ $\mid$ $\text{CH}_2\text{.COO}^-$	2 acidic	-2	very fast

TABLE II, continued

Thiol reagents		
Type	Structure	Reaction with Enz-SH
-2-bromoethyl-ammonium bromide	$Br.CH_2.CH_2.NH_3^+$ (Br^-)	non-reversible
-ethylenimine		non-reversible
Mercurial -p-chloro-mercuri-benzoate		reversible

	Product formed with Enz-SH		
Structure	Charged group(s) acquired	Net charge difference from untreated enzyme	Observed anodal electrophoretic mobility – relative to untreated enzyme
$(Enz)-S-CH_2.CH_2.NH_3^+$ (Br^-)	1 basic	+1	slow
$(Enz)-S-CH_2.CH_2.NH_3^+$	1 basic	+1	slow
$(Enz)-S-Hg\langle\rangle COO^-$	1 acidic	−1	fast

Each reagent may be tested by incubating 1 volume fresh haemolysate with 1 volume 10mM thiol reagent at pH 6.0 for 30 minutes at 37°C, and the isozymes are then examined by starch gel electrophoresis. The reversibility of the reactions may be tested either by dialysing the treated sample against 10mM mercaptoethanol (pH 6.0) or, in the case of iodoacetamide and N-ethylmaleimide, by dialysing the treated sample against 10mM iodoacetate (pH 6.0).

495

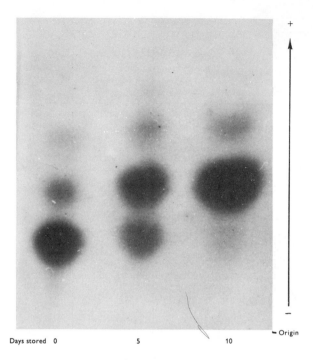

Days stored   0   5   10   ⤙ Origin

Fig. 1. Photograph of starch gel showing the electrophoretic pattern of the adenosine deaminase (ADA) isozymes observed in three haemolysates which were prepared from the same ADA 1 individual and then stored for 0, 5 and 10 days at +4°C.

isozyme patterns can be recognized in different individuals, and family studies show that these three types represent different genotypic combinations of two common alleles ($\underline{ADA}^1$ and $\underline{ADA}^2$) at an autosomal locus (Spencer, Hopkinson, and Harris, 1968). However, if the red cell lysates are stored for a few days in a refrigerator and then re-examined, the isozyme patterns are found to be altered, The main changes are a diminution in the relative staining intensities of the slower isozymes in all three types of pattern, an increase in the intensities of the faster isozymes and the appearance of extra isozymes anodal to the usual components. These changes are progressive with the time of storage and are similar in all three phenotypes. Fig. 1 shows a typical example of the storage change in an ADA 1 red cell lysate.

    The ADA isozymes were found to react with each of the different thiol reagents given in Table II (Hopkinson and Harris, 1969). In each case the results obtained were exactly as

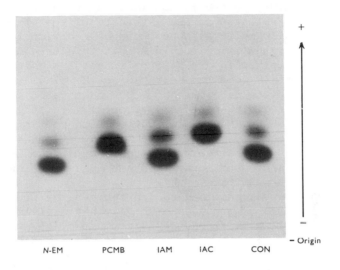

N-EM   PCMB   IAM   IAC   CON
+
− Origin

Fig. 2. Photograph of starch gel showing the electrophoretic pattern of adenosine deaminase (ADA) isozymes observed after treatment of the same ADA 1 haemolysate with four different thiol reagents: iodoacetate (IAC), iodoacetamide (IAM), p-chloromercuribenzoate (PCMB) and n-ethyl maleimide (N.EM). CON = control sample, untreated haemolysate.

predicted on the hypothesis that each ADA isozyme contains a single free reactive sulfhydryl group: with each reagent the reaction product was found to exhibit a characteristic electrophoretic mobility that was either the same as the untreated enzyme when the added moiety was neutral, or faster moving towards the anode when the added moiety was acidic, or slower than the untreated enzyme when the added moiety was basic. Fig. 2 shows a representative photograph of the results obtained with four of the thiol reagents tested. The stabilities of the ADA-thiol reagent complex were also found to be as pre= dicted: the effects of the disulfide, cystamine, and the organic mercurial, PCMB, were found to be reversible by dialysis against mercaptoethanol or one of the uncharged alkylating agents; whereas the effects of the alkylating agents were, except for the rather weak reagent maleate, found to be non-reversible.

These experiments were carried out at relatively low (c. 10 mM) concentrations of thiol reagents, at neutral pH and for relatively short periods of incubation. Under these

conditions most of the reagents produced no apparent change
in the ADA activity and so it may be argued that this particular free sulfhydryl group is not located at the active site
and is possibly situated on the external surface of the enzyme.
Incubation with 10 mM PCMB however did on occasion lead to
loss of activity and it is possible that this reagent either
reacted with other sulfhydryl groups in the ADA molecule which
were not accessible to the disulfides or the alkylating agents
under the conditions used or that PCMB inactivated ADA by
reactions not involving sulfhydryl groups. It is also important to note that prolonged incubation with most of the alkylating agents resulted in loss of ADA activity and electrophoretic changes in the isozyme patterns but these changes
were variable, non-specific and did not appear to involve -SH
group interactions.

No differences have been detected between the isozymes determined by the two common adenosine deaminase alleles ADA^1
and ADA^2 or between the isozymes determined by several different
rare ADA alleles (e.g. ADA^4 and ADA^5) in their reactions with
the various thiol reagents. Thus there is no evidence so far
for mutations of the ADA locus which have led to a change in
the sulfhydryl group reactivity of the red cell ADA isozymes.
However, such allelic variants of ADA may occur and in principal should be detectable by the thiol reagent technique
(see below).

Variation between enzyme loci

Results similar to those described for adenosine deaminase
have been obtained for each of the other enzymes (Table I)
where the occurrence of a reactive cysteine residue was suspected from in vitro storage changes of the isozyme patterns.
In each case it was possible to predict precisely the changes
in the electrophoretic mobilities of the isozymes from the
known chemistry of the different reagents and from the subunit
structures of the isozymes. Thus this method of testing for
reactive sulfhydryls has become a useful general tool in isozyme analysis which we have applied to a wide variety of enzymes in addition to those given in Table I. Of course, it
is not universally applicable since in certain cases (e.g.,
glyceraldehyde-3-phosphate dehydrogenase) the sulfhydryl groups
are essential for normal enzyme activity, but for most of the
enzymes that we have tested inactivation has not been a
serious problem.

It is interesting to note that reactive -SH groups detected
by this method are not characteristic of any one particular
class of enzymes and there appear to be quite striking differ-

498

ences between the products of loci which determine apparently functionally similar enzymes. For example, out of the several loci which determine the human peptidases, only the PEP-C and PEP-D isozymes exhibit a freely reactive -SH group (Lewis and Harris, 1967, 1969) except for the isozymes determined by a rare variant allele at the PEP-A locus (see below). Also among the human acid phosphatases only the ACP_1 isozymes and not the ACP_2 or ACP_3 isozymes exhibit reactive -SH groups by this method of testing (Swallow and Harris, 1972). In the case of human phosphoglucomutase (PGM) however the isozymes determined by alleles at all three loci (PGM_1, PGM_2, and PGM_3) show evidence of at least one reactive -SH per isozyme (Fisher and Harris, 1972). But the PGM_3 isozymes were found to be relatively much more reactive with the thiol reagents than the PGM_1 and PGM_2 isozymes. This agrees with common experience that typing for the PGM_3 polymorphism is difficult in placental tissue extracts unless mercaptoethanol is added to the tissue extracts, due to the formation of sulfhydryl modified isozymes, whereas PGM_1 and PGM_2 phenotyping is much easier since these isozymes are fairly stable in vitro.

Investigation of subunit structure

One rather unusual application of the thiol reagents which has arisen out of experiments with a wide range of different enzymes is the possibility of determining the subunit structure of an enzyme which contains a reactive -SH group.

Two electrophoretically different but stable forms of the enzyme under scrutiny are prepared by reacting one portion of the enzyme with an acidic alkylating agent, such as iodoacetate, and another portion of the enzyme with a neutral reagent, such as iodoacetamide. A 50 : 50 mixture of the two chemically altered forms of the enzyme is then subjected to one of the now standard in vitro "hybridization" procedures, such as urea treatment or freezing and thawing in strong salt solution, which leads to dissociation and subsequent recombination of any subunits. The mixture is then analyzed electrophoretically for the presence of new "hybrid" isozymes. The principles are illustrated in Fig. 3 which is a diagram of the results recently obtained by Fisher et al (1973) on experiments with human red cell inorganic pyrophosphatase. In this case one new hybrid isozyme was seen in the treated mixture which did not occur in the control samples thus suggesting that this particular pyrophosphatase is a dimeric enzyme. Subsequent data obtained from Chinese hamster-human somatic cell hybrids confirm this conclusion (Fisher, R. A. - personal communication, 1974).

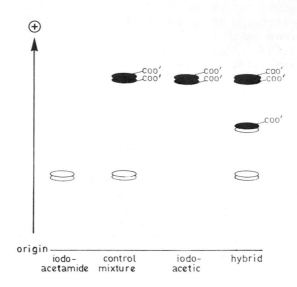

Fig. 3. Diagram of a starch gel showing the erythrocyte inorganic pyrophosphatase isozymes in a partially purified enzyme
preparation treated with iodoacetamide and iodoacetate to
manufacture two electrophoretically different species of the
enzyme. The 'hybrid' sample is a 50:50 mixture of these two
preparations which has been subjected to treatment with a
mixture of urea and sodium chloride--a new hybrid isozyme
was generated, which was not seen in the control mixture.

Sulfhydryl mutants

An application of the thiol reagents which is of particular
genetical interest is in the detection and characterization
of variants due to mutations which have resulted in the
substitution of a cysteine residue with a reactive -SH group
for some other residue in an enzyme protein or have led to the
replacement of a reactive cysteine residue. Such variants
may be individually rare but collectively may be not too infre-
quent. We have identified two probable examples of such mutants, one is a rare variant of human phosphohexose isomerase,
the PHI 5-1 phenotype (Detter et al, 1968; Hopkinson, 1970)
and the other is a rare variant of human peptidase A, the
PEP-A 5-1 phenotype (Lewis, Corney and Harris 1968; Sinha and
Hopkinson, 1969). Other possible examples are the human glu-
cose-6-phosphate dehydrogenase variant Gd$^{\text{Tel Hashomer}}$ (Kirkman,

Ramot and Lee, 1969), the phosphogluconate dehydrogenase var-
iant $\underline{PGD}^{Natal}$ and the phosphoglucomutase variant $\underline{PGM}_2^{10}$ (Blake,
Omoto and Kirk - personal communication, 1974). There are
also variants of this type affecting mouse and human haemoglobins (Bonaventura and Riggs, 1967).

The PHI 5-1 and PEP-A 5-1 variants provide an interesting
contrast since in the former it is suspected that the reactive
cysteine residue has replaced another neutral amino acid residue, whereas in the latter the reactive cysteine appears to
have replaced a positively charged basic amino acid residue,
arginine.

a) PHI 5-1 (genotype $\underline{PHI}^1$ $\underline{PHI}^5$)

The PHI 5-1 phenotype of a fresh red cell lysate is indistinguishable from the common PHI 1 phenotype. However if the
haemolysate is stored for about a week in the refrigerator and
then re-examined, a characteristic triple banded isozyme pattern is observed: one isozyme shows the same electrophoretic
mobility as the main isozyme of the PHI 1 phenotype; one has
a relatively much greater anodal electrophoretic mobility and
a third, rather prominent isozyme has intermediate electrophoretic mobility. This triple banded "stored" pattern can
however be reversed to the single band of the fresh lysate by
adding mercaptoethanol. The PHI 5-1 variant is unique among
other known electrophoretic variants of PHI in these respects.

The most likely interpretation is that the polypeptide determined by the variant $\underline{PHI}^5$ allele differs from the polypeptide determined by the common $\underline{PHI}^1$ allele by the substitution
of a cysteine residue which can react with oxidized glutathione
to form a mixed disulfide. This interpretation was tested by
examining the effects of the several different thiol reagents
on the electrophoretic properties of the isozymes of the PHI
5-1 lysate.

The results (Fig. 4) were exactly as predicted: the neutral
reagents iodoacetamide and n-ethylmaleimide produced a single
banded isozyme pattern identical with the fresh control but
they abolished the invitro storage change; the acidic reagent
iodoacetate produced a triple banded isozyme pattern identical
to that observed on storage or after incubation with oxidized
glutathione; the decarboxylic acid maleate produced an extended triple banded pattern in which the middle band had the same
electrophoretic mobility as the most anodal zone seen with
iodoacetate; treatment with the basic disulfide cystamine led
to the formation of a slow (i.e. less anodal) triple banded
isozyme pattern. Also as expected, the effects of the disulfides were found to be reversible by dialysis against mercap-

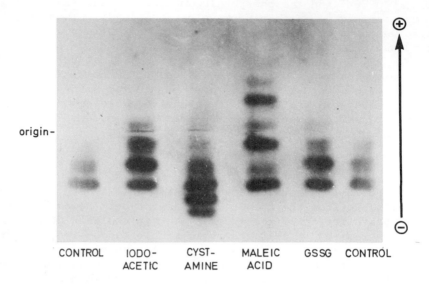

origin-

CONTROL   IODO-   CYST-   MALEIC   GSSG CONTROL
           ACETIC   AMINE   ACID

Fig. 4. Photograph of a starch gel showing phosphohexose isomerase (PHI) isozymes in a PHI 5-1 haemolysate treated with
various thiol reagents (10mM) at pH 6.0. The control is a
freshly prepared PHI 5-1 haemolysate.

---

toethanol but the effects of the alkylating agents were not
reversible. There was not enough material to test the effects
of PCMB.

These results fully support the idea that the variant polypeptide subunit α^5 differs from the usual subunit α^1 by the
possession of a single reactive cysteine residue which has
replaced another neutral amino acid residue as illustrated in
Fig. 5. It is suggested that the single main isozyme seen in
the fresh PHI 5-1 haemolysate consists of three different mol-
ecular species: a simple dimer $\alpha^1 \alpha^1$, a simple dimer $\alpha^5 \alpha^5$ and
a hybrid dimer $\alpha^1 \alpha_5$, each with the same electrophoretic mobility. However the $\alpha^5 \alpha^5$ dimer has two reactive sulfhydryl
groups, the hybrid isozyme $\alpha^1 \alpha^5$ has one and these groups can
interact with acidic, basic and neutral thiol reagents in the
manner illustrated in Fig. 5.

b) PEP A 5-1 (genotype $\underline{PEP-A^1}$ $\underline{PEP-A^5}$)

A freshly prepared lysate of PEP-A 5-1 red cells exhibits
a triple banded isozyme pattern which is electrophoretically

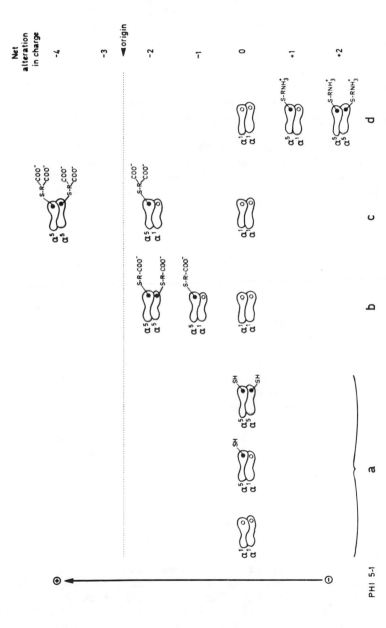

Fig. 5. Diagram illustrating the postulated subunits in the four different kinds of electrophoretic patterns that occur after treatment of a PHI 5-1 haemolysate with various thiol reagents due to reaction with the free -SH groups of the α^5 subunit. The net alteration in charge is taken as relative to the isozymes of the fresh untreated haemolysate and these have the same mobilities as those in the (a) pattern of this figure.

Fig. 5 legend continued

Pattern	Reagents
(a)	Iodoacetamide, N-ethyl maleimide
(b)	Iodoacetate, oxidized glutathione
(c)	Maleate
(d)	Cystamine

very similar to the PEP-A 2-1 variant. However if the haemolysate is kept for a few days and re-examined a more anodally extended triple banded isozyme pattern is observed in which the intermediate zone of the triplet now has the same electrophoretic mobility as the most anodal zone seen in the fresh haemolysate. The same change can be brought about by the addition of oxidized glutathione to fresh lysates and the reverse effect can be accomplished by adding mercaptoethanol to old lysates. The isozymes of PEP A 1 and other PEP A variants in contrast are not affected in this way either by storage or by mercaptoethanol and oxidized glutathione.

Thus it seems likely that the α^5 polypeptide determined by the PEP-A^5 allele differs from the α^1 polypeptide determined by the common PEP-A^1 allele by the presence of a reactive cysteine residue, the sulfhydryl of which can undergo an exchange reaction with oxidized glutathione, thus acquiring an extra negative charge and greater anodal electrophoretic mobility. Furthermore this cysteine residue must have replaced a basic amino acid residue since the variant isozymes in the PEP-A 5-1 phenotype have greater anodal electrophoretic mobilities than the usual PEP-A 1 isozyme.

This hypothesis was investigated by studying the electrophoretic properties of the PEP-A 5-1 isozymes after treatment with various thiol reagents. The results are shown in Fig. 6. The acidic reagents PCMB and iodoacetate were found to have the same effect on the PEP-A 5-1 isozymes as oxidized glutathione; maleate, as expected for a dicarboxylic reagent, produced a more widely spaced triplet isozyme pattern; whereas the basic reagent cystamine altered the electrophoretic properties of the variant isozymes so that they had the same mobility as the PEP-A 1 isozyme. The neutral reagents iodoacetamide and n-ethylmaleimide had the effect of fixing the PEP-A 5-1 isozymes in the same triplet pattern as that shown by the fresh lysate.

A diagrammatic representation of these findings is given in Fig. 7. They are in conplete agreement with the suggestion that the mutant PEP-A^5 allele codes for a variant polypeptide which contains a cysteine residue, with a reactive sulfhydryl group, in place of a particular basic amino acid residue.

In theory this cysteine residue could have replaced either

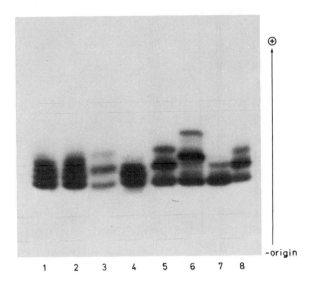

Fig. 6. Photograph of starch gel (10mM tris-maleate pH 7.5)
showing peptidase A isozymes in a PEP-A 5-1 haemolysate treated
with various thiol reagents at pH 6.0.

Sample: 1. Fresh untreated haemolysate*
           2. Mercaptoethanol*
           3. Oxidized glutathione
           4. Iodoacetamide
           5. Iodoacetate
           6. Maleate
           7. Cystamine
           8. PCMB

*Extra zone seen anodal to the three major zones is probably
$\alpha^1\alpha^5$ modified by the maleate present in the gel buffer.
This zone is not seen on phosphate gels.

a particular lysine or a particular arginine residue. However,
if the mutation has arisen from a single base change then the
substitution could not have involved lysine, according to the
genetic code (Crick, 1967). It must have resulted from a
point mutation in an arginine codon, as follows:

CGU (arg) → UGU (cys) or CGC (arg) → UGC (cys).

Further examination of the genetic code shows that cysteine
residues can only be acquired through single base changes as
a result of point mutations in these two codons for arginine
or in the codons for a fairly limited number of neutral amino
acid residues (serine, glycine, phenylalanine, tryptophan or

505

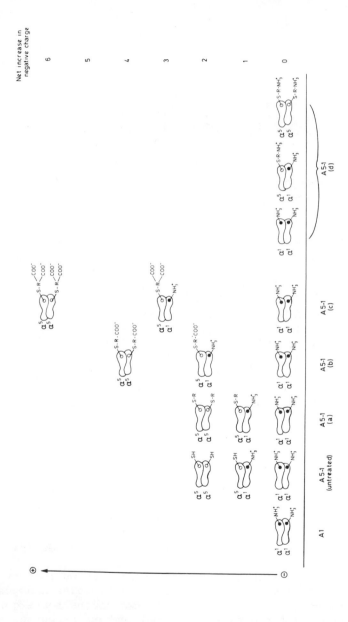

Figure 7. Diagram illustrating the postulated subunits in a freshly prepared PEP-A 5-1 haemolysate (untreated) and in the four different kinds of electrophoretic patterns that occur after treatment with various thiol reagents due to reactions with the free -SH group of the α^5 subunit. The net alteration in charge is taken as relative to the PEP-A 1 isozyme $\alpha^1\alpha^1$.

Fig. 7 legend continued

Pattern	Reagents
(a)	Iodoacetamide, N-ethylmaleimide
(b)	Iodoacetate, oxidized glutathione, PCMB
(c)	Maleate
(d)	Cystamine

tyrosine). The former type of change, a neutral for a basic residue (like the PEP-A^5 example) or vice versa, would be expected to produce a variant protein which is electrophoretically different, and thus easily detectable. For the latter group, however, (and PHI^5 is such an example) no electrophoretic differences would be expected to occur if the samples were fresh and such variants might therefore escape recognition. The use of stored samples would probably lead to their detection, as was the case with PHI^5, but this might be inconvenient and difficult to standardize as a routine screening procedure. The systematic use of specific thiol reagents and subsequent electrophoresis appears to offer a more definite and reliable means of detecting these variants.

REFERENCES

Bonaventura, J. and A. Riggs 1967. Polymerization of hemoglobins of mouse and man: structural basis. *Science* 158: 800-802.

Crick, F. H. C. 1967. The genetic code. *Proc. Roy. Soc.* (B) 167: 331-347.

Detter, J. C., P. O. Ways, E. R. Giblett, M. A. Baughan, D. A. Hopkinson, S. Povey, and H. Harris 1968. Inherited variations in human phosphohexose isomerase. *Ann. Hum. Genet.*, London 31: 329-338.

Fisher, R. A. and H. Harris 1969. Studies on the purification and properties of the genetic variants of red cell acid phosphohydrolase in man. *Ann. N.Y. Acad. Sci.* 166: 380-391.

Fisher, R. A. and H. Harris 1972. "Secondary" isozymes derived from the three PGM loci. *Ann. Hum. Genet.*, London 36: 69-77.

Fisher, R. A., B. M. Turner, H. Harris, and H. L. Dorkin 1973. An investigation of some of the structural properties of erythrocyte inorganic pyrophosphatase. *Ann. Hum. Genet.*, London 37: 341-353.

Hopkinson, D. A. 1970. The investigation of reactive sulphydryls in enzymes and their variants by starch gel electrophoresis: studies on the human phosphohexose isomerase variant PHI 5-1. *Ann. Hum. Genet.*, London 34: 79-84.

Hopkinson, D. A. and H. Harris 1969. The investigation of

reactive sulphydryls in enzymes and their variants by starch gel electrophoresis. Studies on red cell deaminase. *Ann. Hum. Genet.*, London 33: 81-87.

Kirkman, H. N., B. Ramot, and J. T. Lee 1969. Altered aggregational properties in genetic variant of human glucose 6-phosphate dehydrogenase. *Biochem. Genet.* 3: 137-150.

Lewis, W. H. P., G. Corney, and H. Harris 1968. Pep A 5-1 and Pep A 6-1: two new variants of peptidase A with features of special interest. *Ann. Hum. Genet.*, London 32: 35-42.

Lewis, W. H. P. and H. Harris 1967. Human red cell peptidases. *Nature*, London 215: 351-355.

Lewis, W. H. P. and H. Harris 1969. Peptidase D (prolidase) variants in man. *Ann. Hum. Genet.*, London 32: 317-322.

Sinha, K. P. and D. A. Hopkinson 1969. The investigation of reactive sulphydryls in enzymes and their variants by starch gel electrophoresis. *Ann. Hum. Genet.*, London 33: 139-147.

Spencer, N., D. A. Hopkinson, and H. Harris 1968. Adenosine deaminase polymorphism in man. *Ann. Hum. Genet.*, London 32: 9-14.

Swallow, D. and H. Harris 1972. A new variant of the placental acid phosphatases: its implications regarding their subunit structures and genetical determination. *Ann. Hum. Genet.*, London 36: 141-152.

SULFHYDRYL OXIDATION AND MULTIPLE FORMS
OF HUMAN ERYTHROCYTE PYRUVATE KINASE

JOHN A. BADWEY
E. W. WESTHEAD
Biochemistry Department
University of Massachusetts
Amherst, Massachusetts 01002

ABSTRACT. Human erythrocyte pyruvate kinase (ATP:
pyruvate-phosphotransferase, E.C. 2.7.1.40) has been
partially purified from outdated erythrocytes and some of
its kinetic and physical properties have been studied.
Two different kinetic forms of this enzyme are obtainable depending upon the presence or absence of EDTA-magnesium complex during isolation or storage. Enzyme purified in the absence of the EDTA-magnesium complex exhibited
grossly hyperbolic behavior with respect to the substrate
phosphoenolpyruvate (PEP) and was insensitive to activation by the effector fructose-1,6-diphosphate (FDP).
Samples exposed to EDTA-Mg showed sigmoid kinetics with
PEP and sensitivity to FDP.
Upon storage, both of these forms were rapidly oxidized
to forms which displayed greatly reduced activity and very
high Km (PEP) values. Dissociation of the enzyme to lower
molecular weight species was also observed in solutions
not containing EDTA. These oxidized forms were highly
susceptible to activation by FDP. The positive effectors
of this enzyme were found to prevent or retard this oxidation, probably by inducing a conformational change in which
the sulfhydryl groups become less accessible to the solvent.
We propose as a hypothesis that reversible oxidation of
the red blood cell enzyme serves as a control mechanism;
under oxidative stress the enzyme is inhibited to spare
NADH for use in the hemoglobin reductase reaction.

INTRODUCTION

Comparative studies of pyruvate kinases from different
tissues have shown well documented differences in their
immunological, electrophoretic and kinetic properties.
These differences have provided the basis for various classification schemes for these isozymes. Kinetically, two types
of pyruvate kinases are generally referred to, the L and M
forms. The L, or liver, form is an allosteric enzyme, showing sigmoid saturation by the substrate PEP and activation

509

by fructose diphosphate, whereas the M, or muscle, form shows hyperbolic saturation by PEP and is insensitive to the activator.

Classification of the enzyme from the human erythrocyte, however, presents difficulties. Kinetic data for PEP has been reported which can be interpreted in terms of cooperative (Koler and Vanbellinghen, 1967) and non-cooperative (Campos et al, 1965) behavior and both sensitivity (Koler and Vanbellinghen, 1967) and insensitivity (Jaffé, 1968) to FDP have been reported. The present study was undertaken to resolve some of these ambiguities.

METHODS AND MATERIALS

Outdated red blood cells stored in acid-citrate-dextrose solution were gifts from the Cooley Dickinson Hospital, Northampton, Massachusetts.

Deionized water was used for all experiments. Phosphoenolpyruvate (PEP)-tricyclohexylamine salt, adenosine-5'-diphosphate (ADP)-disodium salt, D-fructose-1,6-diphosphate (FDP)-tetracyclohexylammonium salt, reduced nicotinamide adenine dinucleotide (β-NADH)-disodium salt, tris(hydroxylmethyl)aminomethane (Tris), potassium chloride (enzyme grade) and all enzymes used were products of Sigma Chemical Company, Mo. Dithiothreitol (DTT) and N-2-hydroxyethylpiperazine-N-2-ethanesulfonic acid (Hepes) were obtained from Calbiochem, Calif. Ultrapure ammonium sulfate (enzyme grade) was purchased from the Schwartz/Mann Corp., N.Y. All other inorganic chemicals utilized were of reagent grade.

Assay Procedure

Pyruvate kinase (PK.) activities were measured by a modification of the coupled assay procedure proposed by Bücher and Pfleiderer (1955). The reaction mixture of 0.7 ml contained in addition to the PK. to be assayed: 3.7 mM ADP, 4.3 mM PEP, 0.31 mM NADH, 6.9 mM $MgCl_2 \cdot 6H_2O$, 570 mM KCl, 127 mM Tris-Cl, pH 7.25 and 25 units of lactate dehydrogenase. Variations from this are detailed in the corresponding figure legends. The disappearance of NADH was followed at 340 nm on a Gilford 240 spectrophotometer equipped with a Honeywell 19 recorder. Temperatures were maintained at 32°C. A unit of activity is defined as the amount of enzyme which causes formation of 1 μmole of NAD^+ per minute, under these conditions. Specific activity is units/mg of protein. Protein concentrations were estimated from the absorbancy ratios at 260 and 280 nm by means of the equation proposed by Layne (1957) or by the procedure of Lowry (1951).

Partial Purification of Pyruvate Kinase

The partial purification of the enzyme which was employed was basically similar to that described by Chern, Rittenberg, and Black (1972) up to and including their second ammonium sulfate fractionation (Step 2). The product from this step was taken up in a minimum volume of 20 mM Hepes buffer, pH 7.25, containing 150 mM KCl, 7 mM $MgCl_2 \cdot 6H_2O$, 2 mM EDTA and 1 mM DTT (Buffer C) and stored at $4^{\circ}C$ under a swab soaked with toluene to prevent microbial growth. The specific activities of these preparations varied from 0.8 to 1.8 units/mg and yields of 30 to 60% were generally achieved. Since controls showed these samples to be free of phosphatases and other substances which could affect the assay system, this degree of purification was adequate for the present investigation. Prior to use in any of the subsequent experiments, the enzyme was desalted on Sephadex G-25 equilibrated with Buffer C.

Molecular Weight Studies

Molecular weights were determined by descending gel chromatography at $4^{\circ}C$ on a Sephadex G-150 column (2.5 x 45 cm) equilibrated with 10 mM Hepes buffer, pH 7.5, containing 150 mM KCl, 5 mM Mg^{++} and 0.1 mM DTT. The columns were eluted at a flow rate of approximately 20 ml/hr and fractions of 3.0 mls were collected. The following reference proteins were used: human catalase, yeast alcohol dehydrogenase, human hemoglobin and cytochrome C.

RESULTS

Variable Kinetic Forms

During kinetic studies with this enzyme, it was noted that different saturation curves could be obtained for PEP depending upon the presence or absence of EDTA (2 mM) in the buffers used to store and de-salt the enzyme. Fresh enzyme prepared in the absence of EDTA exhibits hyperbolic saturation with PEP. Curves obtained in this fashion exhibit half-saturation constants of approximately 0.7 mM PEP and are insensitive to activation by FDP. They are almost indistinguishable from the curve shown in Fig. 1 for the sample in the presence of FDP.

Freshly isolated samples exposed to EDTA showed sigmoid kinetics (Fig. 1), with half-saturation constants of approximately 2.5 mM. These curves were susceptible to modulation

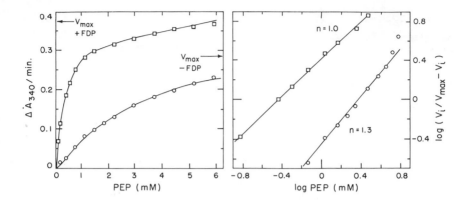

Fig. 1. PEP-Saturation curves for enzyme purified in the presence of 2 mM EDTA: (o) no FDP; (□) 2.0 mM FDP. The concentrations of ADP and EDTA were 4.4 mM and 29 μM, respectively, and all other conditions were the same as those described in the text. The reaction mixtures were allowed to incubate for 3 min at 32°C before the initiation of the reaction by the addition of PEP.

by FDP, which converted them to normal rectangular hyperbolas (n = 1.0) with greatly reduced Michaelis constants (0.4 mM).

Patterns similar to Fig. 1 have been reported for pyruvate kinases from adipose tissue (Pogson, 1968) and rat liver cells in culture (Walker and Potter, 1973). They may result from a direct binding of EDTA to the enzyme at anionic sites such as the ATP loci, or by displacing tightly-bound FDP molecules which are not normally removed by the desalting procedure. A third possibility is that EDTA mediates these interconversions by its ability to chelate metal ions. It is significant to note in this respect that Leonard (1972) has recently shown that both cooperative substrate saturation and activation effects of this enzyme are dependent upon the divalent cations present. EDTA in combination with excess Mg^{++} could remove a tightly bound metal such as Mn^{++} or Zn^{++} to give an enzyme with altered properties. Studies to differentiate among these possibilities were not performed.

Upon storage at this stage of purification the enzyme prepared in either fashion was found to be labile and undergo pronounced changes in the saturation curves for the substrate PEP. One of the most unusual curves generated in this fashion is shown in Fig. 2. This "curve" appears to be linear for almost all of the concentrations of PEP examined and does not

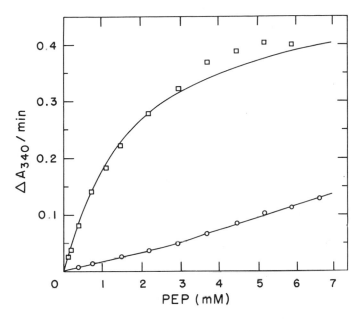

Fig. 2. PEP-Saturation curve for the aged enzyme: (o) no FDP; (□) 2.0 mM FDP. The concentration of ADP was 4.4 mM and all other conditions were the same as those described in the text and in the legend to Fig. 1.

reach either K_m or V_{max} values in the substrate range employed. The addition of FDP to this sample converts the curve to a normal Michaelis-Menten form with elevated activity at all concentrations studied.

In contrast with the changes in the PEP saturation curves described above, ADP saturation curves were essentially invarient, with a Km of 0.4 mM. No substantial difference is found in this curve whether the enzyme is prepared in the presence or absence of EDTA-Mg, whether the enzyme is aged to 90% loss of activity, or whether or not FDP is present.

Studies were undertaken to define conditions which would stabilize this enzyme. The results are shown in Fig. 3. It can be seen that those reactants and effectors which activate the enzyme stabilize it, whereas the others do not. This relationship became more significant when it was found that the 10 day old inactive enzyme could be reactivated to between 70-95% of its original activity by incubation with dithio-threitol (Fig. 4). These results suggest that the inactivation and kinetic changes that occur during aging result from oxidation of sulfhydryl groups and that FDP prevented this

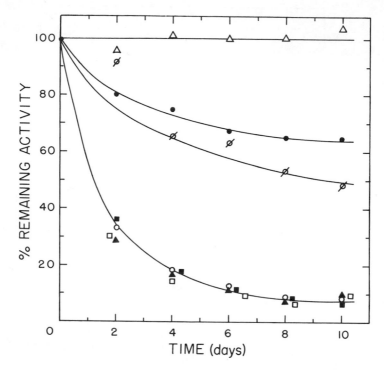

Fig. 3. Effect of substrates and modulators on stability;
(o) control; 20 mM Hepes buffer, pH 7.3, 50 mM KCl, 7 mM Mg^{++},
1 mM DTT and 2 mM EDTA. Additions to this buffer are as fol-
lows: ($\triangle$) 3 mM FDP, ($\bullet$) 3 mM glucose-6-phosphate; (ϕ) 3 mM
PEP; ($\blacksquare$) 3 mM pyruvate; ($\blacktriangle$) 3 mM ADP and ($\square$) 3 mM ATP.
The samples were stored at 4°C.

oxidation by inducing a conformational change in which these
groups became less accessible.
    To test this hypothesis, the rates of inactivation of the
enzyme with the irreversible sulfhydryl-inhibitor, N-ethylmaleimide (NEM), were studied in the presence and absence of
FDP (Fig. 5). After 3 hrs, the enzyme incubated in the absence of FDP had lost approximately 80% of its activity,
whereas that incubated with FDP showed only a 20% loss. The
NEM-inactivated enzyme is not activated by FDP. In contrast,
when the air-oxidized enzyme is assayed in the presence of
FDP, the enzyme returns at once to a fully active form if the
enzyme has not lost more than about 90% of its FDP-free activity.

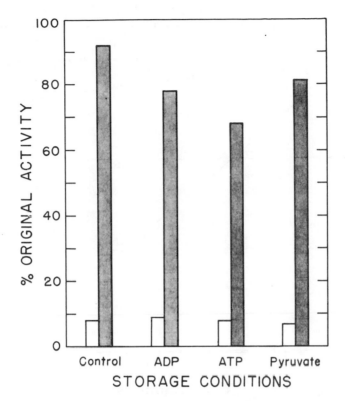

Fig. 4.   Reactivation by DTT.   The 10-day old samples of Fig. 3 were incubated for 16 hrs with 50 mM DTT at 4°; (□) control, no DTT; (■) plus DTT.

Variable Molecular Weight

Studies were also undertaken to determine the effects of aging on the molecular weight.   Previous reports have shown molecular weights of 205,000 (Staal et al, 1971) and 150,000 (Koler et al, 1964).   We found that fresh enzyme purified either in the presence or absence of EDTA gave a single, sharp band on Sephadex G-150 chromatography with no evidence of multiple forms.   The molecular weights in both cases were found to be 266,000 ± 9,000.   Upon aging, 2 major peaks of activity (Fig. 6) were observed.   Their molecular weights were calculated to be 210,000 ± 10,000; and 175,000 ± 10,000; a small peak of activity under hemoglobin was also usually observed, corresponding with the expected elution position of a monomer.   These data confirm the results reported by

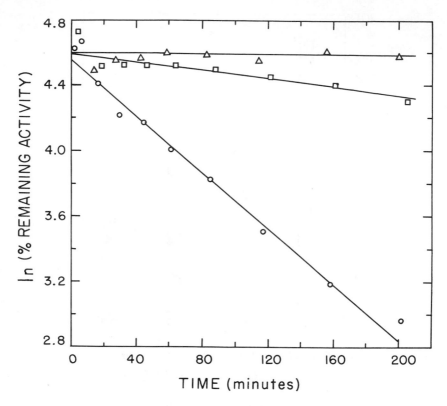

Fig. 5. Inactivation by N-Ethyl Maleimide (NEM). Inactivations were carried out in the presence of 20 mM Hepes buffer, pH 7.25, containing 150 mM KCl, 7 mM Mg^{++}, 2 mM EDTA and 1 mM NEM at 23°C: ($\triangle$) control, no NEM nor FDP; ($\square$) 1 mM NEM and 3 mM FDP; (o) 1 mM NEM and no FDP.

Ibsen, Schiller, and Haas (1971) that the enzyme is pleomorphic and *suggest* that these alterations may be the result of sulf-hydryl oxidation. We have found that enzyme stored in the presence of FDP retains the native molecular weight indefinitely. Studies to ascertain whether oxidation is indeed the primary event leading to dissociation are currently in progress. Our results to date show that sulfhydryl oxidation alone does not lead to immediate dissociation of the enzyme and further experiments are planned to determine the other factors involved in dissociation.

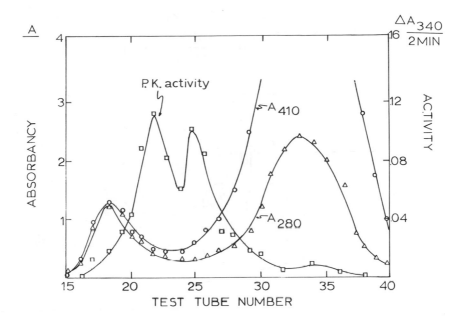

Fig. 6. Sephadex G-150 profile for aged (4-day-old) pyruvate kinase. For details, see Methods.

DISCUSSION

The results reported herein are summarized in the model shown in Fig. 7. Enzyme purified in the presence of EDTA would represent the inactive, high Km (PEP) form of a conformational pair which shows cooperative saturation by PEP. This form is converted to the active, low Km (PEP) form by the addition of FDP. Enzyme purified in the absence of EDTA probably represents this low Km form. The differences between these curves (i.e. Km (PEP) 0.4 vs. 0.7 mM) may reflect either mixtures of these two forms or be the result of a slight oxidation of this enzyme catalyzed by metal contaminants during purification in the absence of EDTA. The position of the sulfhydryl group(s) in the uppermost forms of the model were assigned on the basis of their reactivity towards NEM (Fig. 5). Their positions in the very-high K_m forms were assigned arbitrarily.

On storage, both of these conformational states undergo oxidation and are converted to forms which display greatly reduced activity and very high half saturation constants of PEP. These oxidized forms are highly sensitive to activation

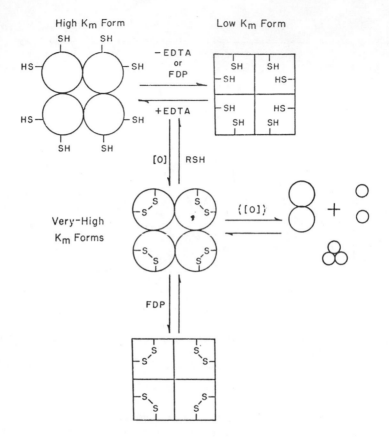

Fig. 7. Proposed model for redox control of pyruvate kinase. For details, see Discussion.

by FDP and may dissociate to yield the forms seen during Sephadex G-150 chromatography (Fig. 6). Oxidation may there-fore have been responsible for the interconvertible forms noted by other investigators (Ibsen et al, 1971).

The kinetic patterns of the air-oxidized enzyme reported here are very similar to those reported by Staal and co-work-ers (1973) obtained after treatment of this enzyme with oxi-dized glutathione. These investigators noted that their inhibited form could be reactivated and the initial kinetic patterns restored by treatment with mercaptoethanol.

These results parallel the effects we have reported on the restoration of activity of aged (air oxidized) enzyme by dithiothreitol. It will be interesting to learn whether the glutathione-induced inhibition is the result of the formation of a disulfide linkage between two of the enzyme's cysteine residues or is the result of an enzyme-glutathione disulfide.

In summary, the data we have reported shows that the saturation pattern with the substrate PEP depends on the isolation conditions. Aging of the enzyme causes a shift to a form with very low affinity for PEP. This aged enzyme can be reactivated by dithiothreitol or restored to maximum activity *without* covalent change by FDP. FDP is also able to protect the enzyme against oxidation and against the sulfhydryl reagent N-ethyl maleimide.

We propose that the redox model shown in Fig. 7 may have physiological significance. Under conditions in which erythrocytes are subjected to oxidative stress, the Fe^{+2} of hemoglobin is converted to the functionally inactive Fe^{+3} state. It is subsequently reduced back to the Fe^{+2} state by the action of methemoglobin reductases which require pyridine nucleotides as the electron donor. These enzymes are relatively specific for NADH. Oxidative inhibition of the pyruvate kinase reaction would retard the production of pyruvate. This self-imposed metabolic block would enable NADH normally used in the lactate dehydrogenase reaction to be utilized by the reductases. It is also possible that this reducing power could be channelled to the reduction of glutathione and ascorbate to restore high concentrations of intracellular reducing power.

After termination of the stress and restoration of normal redox-levels of the red-cell, pyruvate kinase could be re-reduced, possibly by glutathione.

A similar oxidation-reduction pattern has recently been demonstrated by Van-Berkel et al (1973) for the L-type pyruvate kinase from rat liver. These investigators have suggested that such a system of redox-control may play an important role in the balance between gluconeogenesis and glycolysis in this tissue.

The FDP level would serve as a fine control mechanism in this scheme by both its ability to protect the enzyme from oxidation and its ability to activate it even after oxidation. Specificity is found here in that the enzyme inactivated by either air oxidation or oxidized glutathione is activated by FDP, whereas the enzyme inactivated by NEM is not.

The mediator of pyruvate kinase oxidation and reduction could be glutathione and there could well be a role for a sulfhydryl transferase enzyme in this reaction. Another possibility is that superoxide radical may be the oxidizing agent. Misra and Fridovich (1972) have recently reported that the autooxidation of hemoglobin results in the production of superoxide radicals, which could serve as the direct oxidant of pyruvate kinase as a control mechanism. We are currently devising experiments to test this hypothesis.

We should like to note that Dr. K. H. Ibsen showed evidence at this Conference that FDP can bind so tightly to at least one site on the enzyme that it is not entirely removed during electrofocusing. Dr. Janet Cardenas reported in personal communication that she had made similar observations.

ACKNOWLEDGMENTS

Supported by U.S. Public Health Service grant GM 14945. This work represents part of the Ph.D. thesis of John A. Badwey, which will be available from University Microfilms, Ann Arbor, Michigan.

LITERATURE CITED

Bücher, T. and G. Pfleiderer 1955. Pyruvate Kinase from Muscle. in *Methods in Enzymology* (Colowick and Kaplan, eds) Vol. I. N.Y., Academic Press, 435-440.

Campos, J. O., R. D. Koler, and R. H. Bigley 1965. Kinetic Differences between Human Red Cell and Leukocyte Pyruvate Kinase. *Nature* 208: 194-195.

Chern, C. J., M. B. Rittenberg, and J. A. Black 1972. Purification of Human Erythrocyte Pyruvate Kinase. *J. Biol. Chem.* 247: 7173-7180.

Ibsen, K. H., K. W. Schiller, and T. A. Haas 1971. Interconvertible Kinetic and Physical Forms of Human Erythrocyte Pyruvate Kinase. *J. Biol. Chem.* 246: 1233-1240.

Jaffé, E. R. in *Hereditary disorders of erythrocyte metabolism* (Beutler, ed). N.Y., Grune and Stratton, I, 262.

Koler, R. D., R. H. Bigley, R. T. Jones, D. A. Rigas, P. Vanbellinghen, and P. Thompson 1964. Pyruvate Kinase: Molecular Differences Between Human Red Cell and Leukocyte Enzyme. *Cold Spring Harbour Sym. Quant. Biol., 29, 213* (1964).

Koler, R. D. and P. Vanbellinghen 1968. The Mechanism of Precursor Modulation of Human Pyruvate Kinase I by Fructose Diphosphate. *Advances Enzym. Regulation* 6: 127-140.

Layne, E. 1957. Spectrophotometric and Turbidimetric Methods for Measuring Proteins in *Methods in Enzymology* (Colowick and Kaplan, eds) Vol. III. N.Y. Academic Press, 447-454.

Leonard, H. A. 1972. Human Pyruvate Kinases. Role of the Divalent Cation in the Catalytic Mechanism of the Red Cell Enzyme. *Biochem.* 11: 4407-4414.

Lowry, O. H., N. J. Rosebrough, A. L. Farr, and R. J. Randall

1951. Protein Measurement with the Folin Phenol Reagent. *J. Biol. Chem.* 193: 265–275.

Misra, H. P. and I. Fridovich 1972. The Generation of Superoxide Radical During the Autooxidation of Hemoglobin.
J. Biol. Chem. 247: 6960–6962.

Pogson, C. I. 1968. Two Interconvertible Forms of Pyruvate
Kinase in Adipose Tissue. *Biochem. Biophys. Res. Commun.*
30: 297–302.

Pogson, C. I. 1968. Adipose-Tissue Pyruvate Kinase; Properties
and Interconversions of Two Active Forms. *Biochem. J.*
110: 67–77.

Van Berkel, Th. J. C., J. F. Koster, and W. C. Hülsmann 1973.
Two Interconvertible Forms of L-Type Pyruvate Kinase
from Rat Liver. *Biochim. Biophys. Acta* 293: 118–124.

Van Berkel, Th. J. C., J. F. Koster, J. K. Kruty, and G. E. J.
Staal. On the Molecular Basis of Pyruvate Kinase Defi-
ciency: I. Primary Defect or Consequence of Increased
Glutathione Disulfide Concentration. *Biochim. Biophys.*
Acta 321: 496–502.

Walker, P. R. and V. R. Potter 1973. Allosteric Properties of
a Pyruvate Kinase Isoenzyme from Rat Liver Cells in Culture. *J. Biol. Chem.* 248: 4610–4616.

BOVINE AND CHICKEN PYRUVATE KINASE ISOZYMES
INTRASPECIES AND INTERSPECIES HYBRIDS

J. M. CARDENAS, R. D. DYSON, and J. J. STRANDHOLM
Department of Biochemistry and Biophysics
Oregon State University
Corvallis, Oregon 97331

ABSTRACT. Bovine tissues, like those of other mammals, contain at least three non-interconvertible pyruvate kinase isozymes, designated type K or K_4, type L or L_4, and type M or M_4. The K- and L-type isozymes have sigmoidal kinetics with P-enolpyruvate and are activated by fructose 1,6-diphosphate, while type M pyruvate kinase has hyperbolic kinetics even in the absence of fructose diphosphate.

Hybrid species, produced in vitro from bovine L_4 and M_4 and designated L_3M, L_2M_2, and LM_3, were separated from each other by isoelectric focusing and characterized kinetically at varying concentrations of P-enolpyruvate, ADP, and ATP. LM_3, like M_4, has hyperbolic kinetics in the absence of fructose diphosphate, with neither product inhibition by ATP nor substrate inhibition by ADP. L_2M_2 has hyperbolic kinetics with P-enolpyruvate, is activated by fructose diphosphate, and is slightly inhibited by ATP or high concentrations of ADP. L_3M and L_4 have sigmoidal kinetics with P-enolpyruvate and are strongly inhibited by ATP and by high concentrations of ADP.

The activities of the hybrids are markedly different from the activities of equivalent L_4-M_4 mixtures, suggesting that the kinetic properties of a subunit are modified by the nature of its neighbors. That is, in the L_3M hybrid, the type M subunit behaves more like its type L neighbors, while in the LM_3 hybrid, the type L subunit behaves more like its type M neighbors.

Zymograms performed on chicken tissues revealed the presence of only two pyruvate kinase isozymes, apparently corresponding to the mammalian types K and M. Considerable homology between bovine and chicken pyruvate kinases was demonstrated by immunochemical cross-reactivity and by in vitro hybridization between chicken type M and bovine type L.

INTRODUCTION

Three distinct, non-interconvertible isozymes of pyruvate kinase have been isolated from rat tissues (Imamura and Tanaka, 1972; Imamura et al, 1972). The type M isozyme is apparently limited to striated muscle and brain, the type L isozyme to liver, kidney, intestines, and possibly erythrocytes. The third isozyme, designated here as type K (for kidney), is the only isozyme of the embryo and early fetus, and is found in all adult tissues examined except striated muscle and erythrocytes.

Because of the easy availability of large quantities of starting material, we chose to isolate and study the equivalent isozymes from bovine rather than rat tissues. We have reported that although significant kinetic and structural differences exist between the bovine type M and type L pyruvate kinases, each is comprised of four similar or identical subunits. The type M and type L subunit types are similar enough to allow an apparently random reassociation in vitro, and thus the formation of hybrid isozymes (Cardenas et al, 1973; Cardenas and Dyson, 1973). These hybrids were isolated and characterized in order to determine how the presence of nonhomologous subunits affects the kinetic behavior of subunits from the normally allosteric (type L) and non-allosteric (type M) parental isozymes (Dyson and Cardenas, 1973). The results of that study are summarized here, along with the presentation of new kinetic data on the hybrids.

In addition, we report here preliminary studies on pyruvate kinase isozymes of chicken tissues, where we find two, rather than three, native isozymes. Included are the results of an interspecies hybridization between the bovine type L and chicken type M isozymes that suggest the presence of considerable homology between mammalian and avian pyruvate kinases.

METHODS

Extracts for the zymograms were obtained by homogenizing tissues in 20 mM tris-HCl, 4 mM $MgCl_2$, 1 mM EDTA, 1 mM fructose di-P, 10 mM β-mercaptoethanol, at pH 7.0, then centrifuging at 10,000 x g for 20 min. Erythrocytes were obtained from heparinized blood by centrifuging it for 5 min at 3,000 x g, then aspirating off the plasma and buffy coat. The erythrocytes were washed three times with 1% NaCl, then lysed by adding an equal volume of 1 mM EDTA, 1 mM β-mercaptoethanol to the packed cells. Cell ghosts were removed by centrifugation at 10,000 x g. All extracts were dialyzed 1-2 hr against electrophoresis buffer (20 mM tris-HCl, 0.5 M sucrose, 4 mM $MgCl_2$, 0.5 mM EDTA, 1 mM fructose di-P, 10 mM β-mercaptoethanol, pH 7.5 at 4°C for bovine extracts; 20 mM tris-HCl, 0.5 mM $MgCl_2$, 1 mM fructose diphosphate, 10 mM β-mercaptoethanol, pH 8.0 at 4°C for chicken extracts). Electrophoresis was performed on cellulose acetate strips at 250 volts for 6 hr with bovine tissues or up to 16 hr with chicken tissues. Bands of

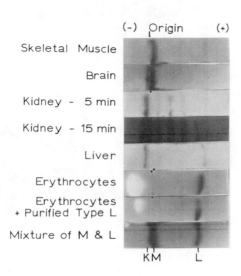

Skeletal Muscle

Brain

Kidney - 5 min

Kidney - 15 min

Liver

Erythrocytes

Erythrocytes
+ Purified Type L

Mixture of M & L

(-) Origin (+)

Fig. 1. Tissue distribution of pyruvate kinase isozymes in bovine tissues. The patterns shown here were obtained by electrophoresis on cellulose acetate followed by the specific
activity detection procedure described in the Methods section.
Shown here are the patterns obtained for kidney photographed
after allowing the activity assay to develop 5 min for better
resolution of the heavy pyruvate kinase bands, and 15 min for
more sensitive detection of the lighter bands.

pyruvate kinase activity were visualized and photographed by
the method of Susor and Rutter (1971), as modified by Cardenas
and Dyson (1973). During electrophoresis at pH 7.5, the type
L isozyme migrated toward the positive electrode, while types
K and M migrate toward the negative electrode, type K having
the greater cathodic mobility.

Bovine muscle and liver isozymes were purified and hybridized in vitro by methods previously described (Cardenas et al,
1973; Cardenas and Dyson, 1973). Separation of the hybrids
was accomplished by isoelectric focusing (Dyson and Cardenas,
1973). Purification of chicken skeletal muscle pyruvate kinase was performed using a procedure similar to that for the
bovine enzyme, but omitting the heat step.

Enzyme assays were performed at 25° by coupling the pyruvate
kinase reaction to that of lactate dehydrogenase and observing
the decrease in absorbance at 340 nm. The assay medium con-

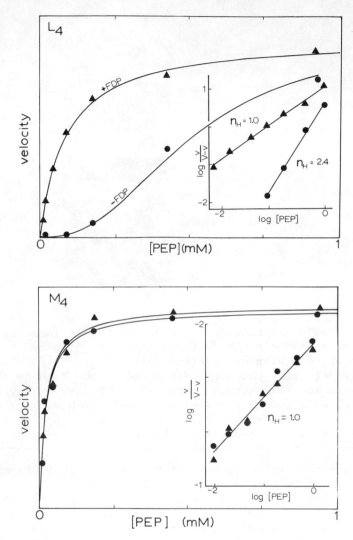

Fig. 2. Velocity profile of pyruvate kinase type L (L_4) and of type M (M_4) as a function of P-enolpyruvate (PEP) concentration. The data were obtained in the absence (●) or presence (▲) of 1 mM fructose diphosphate.

---

tained 0.05 M imidazole-HCl, 0.1 M KCl, 0.01 M $MgCl_2$, 0.16 mM NADH, and 20 units of lactate dehydrogenase at pH 7.5 plus specified amounts of ADP and P-enolpyruvate.

Antiserum was obtained from rabbits, as described earlier (Cardenas et al, 1973).

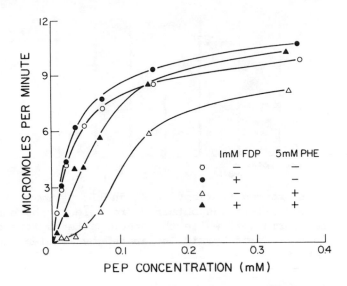

Fig. 3. Velocity profile of type M pyruvate kinase as modified by the presence of 5 mM phenylalanine and/or 1 mM fructose diphosphate.

Kidney isozymes were separated from each other by chromatography on a 2.5 x 40 cm column of CM-Sephadex (C-50) in 0.03 M potassium phosphate, 10 mM β-mercaptoethanol, pH 6.0, using a 0-0.3 M KCl linear gradient of one liter volume. The isozymes were further purified by chromatography on a 1.5 x 10 cm column of DEAE cellulose in 10 mM potassium phosphate, 0.5 M sucrose, 2 mM dithiothreitol, pH 7.0, eluting the isozymes with a 0-0.25 M KCl linear gradient of 300 ml volume.

RESULTS AND DISCUSSION

Bovine Pyruvate Kinases: The distribution of pyruvate kinase isozymes in bovine tissues, shown in Fig. 1, was quite similar to the isozyme distribution in rat tissues. Type K, or K_4, occurs in all early embryonic tissue, in liver, uterus, and brain, as well as in kidney. Type L, or L_4, occurs in liver, kidney, and erythrocytes, while the type M isozyme is found in heart, brain, and skeletal muscle. Adult liver homogenates appear to contain two distinct isozymes, K_4 and L_4, with no evidence for intermediate electrophoretic bands. On the other hand, adult kidney extracts contain, in addition to types K and L pyruvate kinases, three species of intermed-

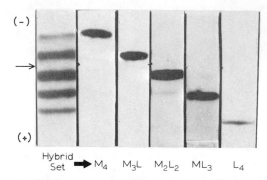

Fig. 4. Hybrids formed from bovine M_4 and L_4 pyruvate kinases. The pattern on the left was obtained from electrophoresis of the complete hybrid set, while the other patterns were obtained from electrophoresis of hybrids separated by isoelectric focusing. Arrow indicates origin.

iate electrophoretic mobility.

As had been reported for other mammals, bovine type L pyruvate kinase has sigmoidal kinetics with P-enolpyruvate, but the kinetic curve becomes hyperbolic when fructose 1, 6-diphosphate is added (see Fig. 2). Under the same assay conditions, type M pyruvate kinase has hyperbolic kinetics (Fig. 2) with little or no alteration by fructose diphosphate. However, type M pyruvate kinase does have sigmoidal kinetics when 5 mM phenylalanine is added to the assay medium, as shown in Fig. 3. The phenylalanine-induced sigmoidicity is lessened, and the enzyme appears to be activated, by the addition of 1 mM fructose diphosphate. These results agree with the data of Kayne and Price (1973) for rabbit skeletal muscle pyruvate kinase.

We have purified pyruvate kinase types L and M and have compared their kinetic and physical properties; these data are shown in Table I. Both M_4 and L_4 appear to be homogeneous tetramers with molecular weights of 230,000 and 215,000, respectively; however, considerable differences exist in their isoelectric points, amino acid analyses, and kinetic properties, and they are immunologically distinct. While the specific activity of bovine type M pyruvate kinase is 225 units (micromoles P-enolpyruvate utilized per min) per mg protein, the usual maximum specific activity attained for purified bovine type L pyruvate kinase is only 75 units per mg. We note, however, that due to the relative instability of the liver isozyme, its true specific activity could be higher than the measured

TABLE I
COMPARISON OF THE PROPERTIES OF
BOVINE SKELETAL MUSCLE AND LIVER ISOZYMES

	Skeletal muscle (M_4)	Liver (L_4)
Kinetic		
K_m (ADP)	0.35	0.18
Maximum Activity	225 μMoles/min/mg	75 μMoles/min/mg
Hill coefficient	1.0	2.5 (1.0 with FDP)
K_m or $S_{0.5}$ for PEP	0.03 mM	0.5 (0.09 with FDP)
Physical		
Electrophoretic Mobility at pH 7.5	toward (−)	toward (+)
Molecular weight	230,000	215,000
$S_{20,w}$	9.95	9.3
Subunits	4(57,000)	4(52,000)
Isoelectric Point	8.9	5.1
PEP Binding Sites	4	4

value.

Formation of Hybrid Isozymes: When electrophoresis is performed on a mixture of M_4 and L_4, the two isozymes migrate
as discrete bands, independent of each other (see Fig. 1).
However, when the M_4 and L_4 isozymes are denatured together
in guanidine hydrochloride and then renatured by dilution into
buffer containing dithiothreitol or β-mercaptoethanol, electrophoretic analysis of the products reveals the pattern
shown in Fig. 4. In addition to recovery of the two parental
isozymes, three pyruvate kinase species having electrophoretic
mobilities evenly spaced between L_4 and M_4 are formed. We
have designated these hybrids as L_3M, L_2M_2, and LM_3 according
to their subunit composition (Cardenas and Dyson, 1973).

The hybrids were separated from one another by isoelectric
focusing, the results of which are seen in Fig. 4. The separated species showed no tendency to interconvert when stored
in buffer, frozen in a sucrose solution, precipitated with
ammonium sulfate, or incubated with substrates. However,

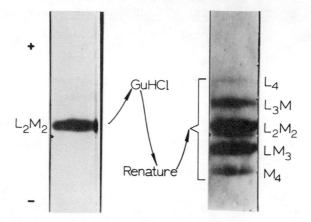

Fig. 5. Reassociation of the subunits from L_2M_2. The isolated hybrid which produced the pattern shown at the left was denatured in guanidine HCl and then renatured by dilution into buffer. After renaturation, the pattern at the right was obtained.

---

denaturation of L_2M_2 in guanidine hydrochloride, followed by renaturation as before, resulted in regeneration of the full five-membered set as shown in Fig. 5. These results are consistent with random redistribution of two kinds of subunits among tetrameric molecules.

To test the possibility that the intermediate forms might be species of varied degrees of aggregation, the sedimentation velocity of the L_2M_2 hybrid in a sucrose gradient was compared to that of the native M_4 and L_4. L_2M_2 sedimented at a rate midway between the parental forms, indicating that it, like like M_4 and L_4, is a tetramer. Further evidence supporting our designation of the intermediate electrophoretic species as hybrids was obtained through immunological studies: antibodies produced against M_4 cross-reacted with LM_3, L_2M_2, and L_3M, but not with L_4.

Kinetic Properties of the Hybrids: A kinetic analysis has been performed for each of the hybrids with P-enolpyruvate as the varied substrate. Fig. 6a contains computer summaries of their velocity profiles in the absence of the activator, fructose diphosphate, while Table II summarizes the kinetic parameters obtained both in the absence and in the presence of fructose diphosphate. The velocity profiles for all the hybrids

TABLE II
KINETIC PARAMETERS OF PYRUVATE KINASE HYBRIDS

Species	K_m for ADP[1]	Data with P-enolpyruvate[2] (−Fructose Diphosphate)		Data with P-enolpyruvate[2] (+Fructose Diphosphate)	
		$S_{0.5}$[3]	n_H[4]	K_m	n_H[4]
L_4	0.14 (mM)	0.52 (mM)	2.4	0.087 (mM)	1.0
L_3M	0.16	0.45	1.5	0.076	1.0
L_2M_2	0.37	0.13	1.1	0.060	1.0
LM_3	0.39	0.034	1.1	0.035	1.0
M_4	0.36	0.023	1.1	0.023	1.0

[1]Determined at 0.25 mM P-enolpyruvate

[2]Determined at 2 mM ADP

[3]$S_{0.5}$ is the substrate concentration required for half maximal velocity

[4]n_H is the Hill coefficient

531

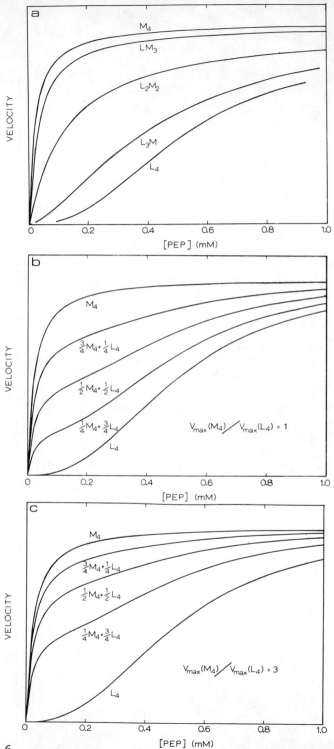

Fig. 6.

Fig. 6. Comparison of kinetic curves of the hybrids with equivalent mixtures of L_4 and M_4. The curves of part a are computer summaries of the actual hybrid kinetics, while parts b and c are computer-simulated mixtures representing relative ratios for V_{max} of 1 and 3, respectively.

---

become hyperbolic in the presence of fructose diphosphate, but both M_4 and LM_3 also have hyperbolic kinetics in the absence of fructose diphosphate, which does not affect their activities. In contrast, L_2M_2 has a hyperbolic velocity profile and a Hill coefficient of 1.0 in the absence of fructose diphosphate, but is nevertheless activated by fructose diphosphate through a reduction in the K_m for P-enolpyruvate. L_3M behaves more like L_4, with a sigmoidal velocity profile in the absence of fructose diphosphate and hyperbolic kinetics with it. The P-enolpyruvate concentrations producing half maximal activity range from 0.023 mM for M_4 to 0.52 mM for L_4 in the absence of fructose diphosphate (Table II). With fructose diphosphate present, we find a fairly uniform monotonic distribution of K_m values ranging from 0.023 mM for M_4 to 0.087 mM for L_4, with a Hill coefficient of about 1.0 in each case.

That the kinetic parameters of the hybrids fall within the limits set by the parental L_4 and M_4 led us to wonder whether the kinetics of the hybrids correspond to the kinetics of an equivalent ratio of L_4 and M_4 molecules. In order to answer this question, we have used a computer to simulate the kinetics of M_4-L_4 mixtures, verifying these simulations by comparing them with the kinetics obtained with actual mixtures of M_4 and L_4. In this analysis, one must consider that some uncertainty exists in the relative specific activities of M_4 and L_4, so that the theoretical contributions of these two subunit types to the overall activity of the hybrids is unclear. For this reason, we have simulated mixtures assuming equal values of V_{max} for M_4 and L_4, and mixtures with a V_{max} for M_4 that is three times that of L_4, the ratio actually obtained when comparing preparations of the two isozymes. The results of these comparisons are shown in Fig. 6. Neither assumption provides a simulated mixture of L_4 and M_4 corresponding to the actual kinetics obtained with the hybrids, suggesting that the properties of the subunits are affected by the parental origin of other subunits within the molecule.

Since type L pyruvate kinase is strongly inhibited by ATP while the type M isozyme is only slightly inhibited, the response of the hybrids to ATP should be an additional way of studying the subunit interactions. The results of kinetic

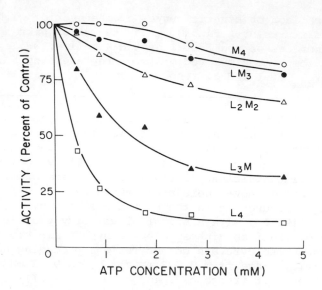

Fig. 7. The effect of ATP on the velocity profiles of the hybrids at 0.25 mM P-enolpyruvate and 1.0 mM ADP.

studies carried out at 0.25 mM P-enolpyruvate, 1.0 mM ADP, and varying concentrations of ATP at pH 7.5 are shown in Fig. 7. Like M_4, LM_3 is inhibited to only a slight degree by ATP, and L_2M_2 only a little more so. However, L_3M is inhibited nearly as strongly as L_4. Thus, the degree of inhibition does not appear to vary in direct proportion to the relative numbers of type L or type M subunits present in a hybrid.

High concentrations of ADP have been found by us and by others (Kutzbach et al, 1973; Ibsen and Trippet, 1973) to cause substrate inhibition with type L pyruvate kinase. Shown in Fig. 8 is the effect of ADP on the hybrids, assayed in the presence of 0.25 mM P-enolpyruvate. While M_4 and LM_3 show little or no substrate inhibition by ADP, a significant degree of inhibition is seen with L_2M_2 at the higher ADP concentrations. Both L_3M and L_4 show a very dramatic degree of substrate inhibition. The mechanism of inhibition by ADP is unknown, except that P-enolpyruvate seems to reverse it: type L pyruvate kinase appears to be inhibited by substrate only when the ADP concentration exceeds the concentration of P-enolpyruvate.

Table II shows the values for the K_m of ADP, calculated from velocities at noninhibitory concentrations of ADP. Rather than a uniform distribution of K_m's, we find relatively similar values for M_4, LM_3, and L_2M_2, and values for L_3M and L_4 that

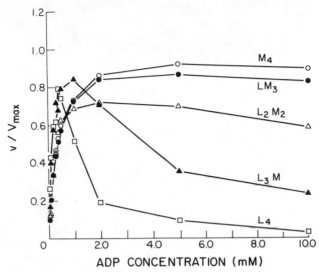

Fig. 8. The effect of ADP concentration on the velocity profiles of the hybrids at 0.25 mM P-enolpyruvate.

---

are much lower.

Thus, in all cases, the L subunit in the LM_3 hybrid seems
to behave more like a type M subunit than one would expect
and, conversely, the M subunit in L_3M appears to take on many
of the properties of its type L neighbors. In contrast, L_2M_2
is decidedly different from either of the two native isozymes.
These results could be interpreted in terms of either a sequential or concerted model for conformational transitions.
Application of the two-state model described by Monod et al
(1965) to this system is diagrammed schematically in Fig. 9.
(In considering the hypothesized equilibrium of the M_4 isozyme,
recall that sigmoidal kinetics with this species was demonstrated with phenylalanine in Fig. 3, suggesting that phenylalanine shifts the enzyme toward a less active conformation.)
We must emphasize, however, that the two-state model is presented only because it is the simplest way to conceptualize
the behavior of the hybrids; further analysis of the data may
require invoking more complex models.

<u>Naturally</u> Occurring <u>Hybrids</u>: Although the preceding studies
provide some insight into the nature of the L and M isozymes,
the hybrids themselves apparently do not form in vivo, probably because type L and type M pyruvate kinases are never
found within the same cell. However, as noted earlier and
documented in Fig. 1, kidney tissues contain multiple pyruvate
kinases that behave electrophoretically as if they were an

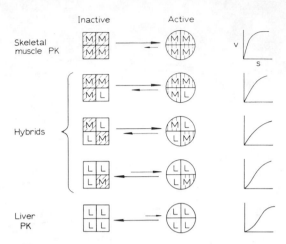

Fig. 9. A schematic representation for the kinetics of type M and type L pyruvate kinases and their hybrids, interpreted according to the two-state model of Monod et al (1965). It is assumed that any of the species can exist in either an active or less active configuration, with type M subunits favoring the active state and type L subunits favoring the less active state.

L-K hybrid set. To test this thesis, the five kidney species were separated from each other by ion exchange chromatography. The separated species did not interconvert in dilute buffer, but when the middle band was denatured in guanidine hydrochloride and renatured by dilution into Tris-HCl buffer, a new hybrid set was generated. This result, shown in Fig. 10, indicates that the intermediate forms found in kidney are, indeed, naturally occurring hybrids, consisting of KL_3, K_2L_2, and K_3L, formed from native K_4 and L_4.

We note, on the other hand, that zymograms of adult liver do not reveal hybrid forms, consistent with the studies of Crisp and Pogson (1972) and Van Berkel et al (1972), who found that the L and K isozymes of rodent liver are distributed to different cell types. It appears, therefore, that the extraction and zymogram techniques used in this study do not produce hybrids in vitro. Thus, the appearance of hybrids in a zymogram can be used as an indication of whether or not two isozymes found within the same tissue are synthesized within the same cells and at the same time, an argument presented also by Garnett et al (1974).

Evolution of the Pyruvate Kinases: Imamura and Tanaka (1972)

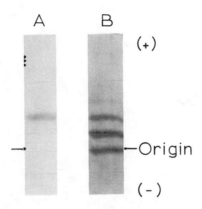

Fig. 10. Denaturation and renaturation of K_2L_2 isolated from kidney tissue. Recombination of subunits from this species (shown on the left) resulted in the formation of a set of pyruvate kinase species (shown on the right) with the same electrophoretic mobilities as was found in kidney extracts.

proposed that the type K mammalian isozyme represents the primordial pyruvate kinase from which the other two identified isozymes evolved. This thesis is supported by the observation that the type L and type M isozymes appear only late in fetal life, by the similar mass and quaternary structure of the three isozymes, and by the ability of any of the three isozymes to hybridize with any of the others. (The K-L and L-M hybrid sets have already been discussed. The third hybridization, between K and M, was reported in rat tissues by Susor and Rutter in 1971).

If L and M are, indeed, more highly evolved forms, then non-mammalian species may very well lack one or the other of these isozymes. We note, for example, that yeasts, which are among the simplest eucaryotes, apparently contain only one isozyme (Hunsley and Suelter, 1969). Although pyruvate kinases have been studied in diverse organisms, there have been few direct comparisons with mammalian forms. Accordingly, we have examined pyruvate kinases from chicken tissues, where we were able to identify only two distinct and non-interconvertible isozymes, as shown in Fig. 11. As in bovine tissues, a type M isozyme is found in skeletal muscle, heart, and brain. Type K pyruvate kinase occurs by itself in early embryonic tissue, just as it does in mammals, but only type K pyruvate kinase is found in adult kidney, liver, and erythro-

(-)   Origin (+)

Skeletal muscle

Brain

7-day Embryo

Kidney

Erythrocytes

Liver

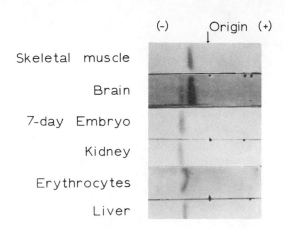

Fig. 11. Zymograms of pyruvate kinase isozymes from chicken tissues. Tissue extracts and electrophoresis (16 hr) were performed as described in the text. Note that the L isozyme is not seen here, nor was it seen after 1, 2, or 4 hours of electrophoresis.

cytes of chickens. In contrast, bovine liver and kidney contain both types K and L pyruvate kinases, and bovine erythrocytes apparently contain the type L isozyme only. Thus, the zymograms indicate that chickens do not possess the type L isozyme.

Kinetic studies performed on centrifuged, dialyzed homogenates of fetal chicken, and on similar homogenates of adult chicken liver, indicate that the pyruvate kinases of both tissues have sigmoidal kinetics with respect to P-enolpyruvate and a Hill coefficient in each case of about 1.7 in the absence of fructose diphosphate. The addition of fructose diphosphate converts the velocity profile from sigmoidal to hyperbolic, with a concomitant lowering of the Hill coefficient to 1.0. These similarities support the electrophoretic indication that fetal and adult liver tissues of the chicken have the same pyruvate kinase isozyme content - i.e., type L is not present in adult tissues.

Chicken skeletal muscle pyruvate kinase has been purified to homogeneity, and some of its properties are compared with the bovine and rabbit skeletal muscle isozymes in Table III. The chicken enzyme has hyperbolic kinetics with respect to P-enolpyruvate, and it appears to be a tetramer with a molecular weight slightly lower than that of the corresponding

TABLE III

COMPARISON OF SKELETAL MUSCLE PYRUVATE KINASES

	Chicken	Bovine	Rabbit
K_m for ADP	0.385mM	0.4	0.36[1]
K_m for PEP	0.036mM	0.04	0.048[1]
Molecular Weight	211,000	230,000	237,000[2]
Subunit Molecular Weight	51,000	57,000	57,100[3]
Partial Specific Volume	0.740 ml/g	0.740 ml/g	0.740 ml/g[2]
Sedimentation Coefficient	9.5	9.94	10.04[2]
Isoelectric Point	8.77	8.9	

[1] from McQuate, J. T. and M. F. Utter (1959) *J. Biol. Chem.* 234: 2151-2157.

[2] from Warner, R. C. (1958) *Arch. Biochem. Biophys.* 78: 494-496.

[3] from Steinmetz, M. A. and W. C. Deal, Jr. (1966) *Biochemistry* 5: 1399-1405.

mammalian skeletal muscle isozyme. Antobodies produced in rabbits against bovine skeletal muscle pyruvate kinase cross-react with the chicken skeletal muscle enzyme, but amino acid analyses and a slightly lower isoelectric point suggest that the chicken and bovine skeletal muscle isozymes have at least minor structural differences.

In view of the structural and kinetic similarities between the chicken and bovine type M isozymes, we sought to establish further homologies through interspecies hybridization of bovine liver pyruvate kinase with the skeletal muscle isozyme from chicken. The resulting electrophoretic pattern (Fig. 12) shows that pyruvate kinases from the two vertebrate classes hybridize with essentially random redistribution of subunits.

Thus, in addition to the similarity of the three bovine isozymes of pyruvate kinase, there appears to be a significant degree of mammalian-avian homology. It is possible that specific evolutionary changes through gene duplication have occurred in the original type K isozyme, thus providing mammals with the specialized L and M isozymes that are better fitted to metabolic requirements of the tissues in which they function. For instance, type M pyruvate kinase is well adapted to tissues that depend heavily on the Embden-Meyerhof pathway but have little need for control at the level of pyruvate

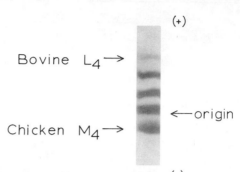

Fig. 12. Electrophoretic pattern obtained when bovine liver and chicken skeletal muscle pyruvate kinases are denatured and renatured together, using procedures described by Cardenas and Dyson (1973).

---

kinase, while the type L isozyme is better suited to gluconeogenic tissues such as liver and kidney, where control of pyruvate kinase is particularly important. Type K pyruvate kinase has many properties intermediate between those of types L and M, enabling it to meet the metabolic requirements of many different cell types, including smooth muscle, fetal, and regenerating liver. The type K pyruvate kinase of chickens may have greater potential for kinetic control than does the corresponding mammalian isozyme, according to preliminary data. One is tempted to think of it as a compromise version, able to meet the needs of those tissues where type K is found in mammals, in addition to meeting the needs of tissues where mammals employ the more specialized type L pyruvate kinase.

ACKNOWLEDGMENTS

The excellent technical assistance of Joan Miller and the financial support through National Institutes of Health research grants A.M. 15645 and G.M. 15715 are gratefully acknowledged. Figs. 2, 4, 5, 6 and 9 are reprinted from Dyson and Cardenas (1973) by permission of the American Society of Biological Chemists.

REFERENCES

Cardenas, J. M., R. D. Dyson, and J. J. Strandholm 1973. Bovine pyruvate kinases. I. Purification and characterization of the skeletal muscle isozyme. *J. Biol. Chem.*

248: 6931-6937.

Cardenas, J. M. and R. D. Dyson 1973. Bovine pyruvate kinases. II. Purification of the liver isozyme and its hybridization with skeletal muscle pyruvate kinase. *J. Biol. Chem.* 248: 6938-6944.

Crisp, D. M. and C. I. Pogson 1972. Glycolytic and gluconeogenic enzyme activities in parenchymal and non-parenchymal cells from mouse liver. *Biochem. J.* 126: 1009-1023.

Dyson, R. D. and J. M. Cardenas 1973. Bovine pyruvate kinases. III. Hybrids of the liver and skeletal muscle isozymes. *J. Biol. Chem.* 248: 8482-8488.

Garnett, M. E., R. D. Dyson, and F. N. Dost 1974. Pyruvate kinase isozyme changes in parenchymal cells of regenerating rat liver. *J. Biol. Chem.* 249: 5222-5226.

Hunsley, J. R. and C. H. Suelter 1969. Yeast pyruvate kinase. I. Purification and some chemical properties. *J. Biol. Chem.* 244: 4815-4818.

Ibsen, K. H. and P. Trippet 1973. A comparison of kinetic parameters obtained with three major non-interconvertible isozymes of rat pyruvate kinase. *Arch. Biochem. Biophys.* 156: 730-744.

Imamura, K. and T. Tanaka 1972. Multimolecular forms of pyruvate kinase from rat and other mammalian tissues. I. Electrophoretic studies. *J. Biochem.* 71: 1043-1051.

Imamura, K., K. Taniuchi, and T. Tanaka 1972. Multimolecular forms of pyruvate kinase. II. Purification of M_2-type pyruvate kinase from Yoshida Ascites hepatoma 130 cells and comparative studies on the enzymological and immunological properties of the three types of pyruvate kinases L, M_1, and M_2. *J. Biochem.* 72: 1001-1015.

Kayne, F. J. and N. C. Price 1973. Significance of phenylalanine control of muscle pyruvate kinase. Abstracts, Ninth Intl. Congress of Biochemistry, p. 61.

Kutzbach, C., H. Bischofberger, B. Hess, and H. Zimmermann-Telschow 1973. Pyruvate kinase from pig liver. *H. S. Z. Physiol. Chem.* 354: 1473-1489.

Monod, J., J. Wyman, and J. P. Changeux 1965. On the nature of allosteric transitions: A plausible model. *J. Mol. Biol.* 12 : 88-118.

Susor, W. A. and W. J. Rutter 1971. A method for the detection of pyruvate kinase, aldolase, and other pyridine nucleotide linked enzyme activities after electrophoresis. *Anal. Biochem.* 43: 147-155.

Van Berkel, J. C., J. F. Koster, and W. C. Hülsmann 1972. Distribution of L- and M-type pyruvate kinase between parenchymal and Kupffer cells of rat liver. *Biochim. Biophys. Acta* 276: 425-429.

PROPERTIES OF RAT PYRUVATE KINASE ISOZYMES

KENNETH H. IBSEN, PATRICIA TRIPPET, AND JOANN BASABE

Department of Biological Chemistry
California College of Medicine
University of California
Irvine, California 92664

ABSTRACT. Data describing the relationship among six variant forms of the K-isozyme and at least three forms of the L-isozyme of the rat are reviewed. A few physical and kinetic parameters describing these variant forms as well as those of the M-isozyme are presented. It is shown that the pH 6.4 variant of the K-isozyme can exist in a higher and lower molecular weight form depending upon the presence of magnesium ion. Evidence is presented supporting the idea that the red cell enzyme is an L-isozyme variant rather than a fourth isozyme. The kinetic properties of what are believed to be the major in vivo forms of the three isozymes are briefly reviewed. Evidence is presented suggesting that K^+ saturation follows a sequential pattern.

INTRODUCTION

There appear to be three basic rat pyruvate kinase isozymes, namely: the $K(M_2)$-isozyme found to predominate in the fetus, in many adult organs, and in dedifferentiated tissues; the M-isozyme of adult muscle and brain; and the L-isozyme of adult liver (Pogson 1968a, Jiménez de Asúa et al., 1971, Farron et al., 1972, Walker et al., 1972, Imamura and Tanaka 1972, Imamura et al., 1972, Ibsen and Trippet 1972, Ibsen and Krueger 1973). Isoelectrofocusing has proven to be a convenient method of separating these isozymes in semi-preparative quantities. Using fresh extracts, the K-isozyme was found to have pI values of 6.4 or 6.8, the L-isozyme of 5.4 or 5.7, and the M-isozyme of 7.5. Such focusing studies have shown liver extracts to contain about 90% of the total pyruvate kinase as L-isozyme and 10% as K-isozyme while kidney cortex extracts contain about 80% K-isozyme and 20% L-type (Ibsen and Trippet 1972). These data agree with those obtained using other techniques.

Each of the three isozymes are metastable in that they can be changed to other pI forms (Ibsen and Trippet 1972). Data will be presented to support the contention that the two major forms of the K and L isozymes found in fresh extracts are R, T conformation sets. Moreover, physical and kinetic properties of the major suspected in vivo variants of all three isozymes will be compared while the properties of the other "pseudozymic" variants will be described in lesser detail.

pI VARIANTS OF THE RAT ISOZYMES

L-Isozyme: Most of the pyruvate kinase activity in fresh liver
extracts from fed rats appears at about pH 5.4 after electrofocusing (Fig. 1B). Fasting the animals or incubating extracts
above pH 7.0 (with or without added FDPase) and then isoelectrofocusing causes most of the activity to appear at about pH
5.7 (Fig. 1B, Ibsen and Trippet, 1972). Moreover, the pH 5.7
peak yields sigmoidal kinetics with P-enolpyruvate and is activated by fructose-1, 6-diphosphate (FDP) while pH 5.4 enzyme
is not (Table 1). These data suggest that the 5.4 enzyme is
the R form and the 5.7 enzyme the T form of a conformational
pair. This conclusion is further confirmed by isolating the
pH 5.7 peak and incubating it in the presence or absence of
FDP, thereby converting it to the pH 5.4 form (Fig. 2).

Enzyme with pI values of 6.1 - 6.2 and 7.5 were reported
to be derived from pH 5.4 or 5.7 L-isozyme (Ibsen and Trippet
1972). In no subsequent preparation was pH 7.5 enzyme again
obtained. However, enzymes with pI values ranging from 6.0 -
6.4 have been obtained routinely from pH 5.4 L-isozyme. These
high pI forms also are driven to lower pI forms using FDP.
Fig. 3 shows data obtained using a partially purified preparation electrofocused with and without FDP. The activity curve
and pI values obtained without FDP appear similar to that reported by Sussor et al., (1973) who also used purified enzyme.

Thus, there are at least three forms of the L-isozyme,
derived or native, which may be isolated by isoelectric focusing procedures. These results are similar to data obtained by
Hess and Kutzbach (1971) for the pig liver enzyme. They found
that during purification a low pI form was converted to an intermediate pI form and then to a high pI form. FDP converted the
highest pI form to the lowest pI form which bound labeled FDP.
They postulated the intermediate pI form would be half saturated with FDP. Fig. 2 shows that incubation of pH 5.7 enzyme
with ^{14}C-FDP yields labeled pH 5.4 enzyme. There is radioactivity in the area of the pH 5.7 enzyme peak, but there is
not complete correspondence. Thus, from these data it is not
clear whether the pH 5.7 enzyme also binds FDP. However, if
the pH 6.0 - 6.4 enzyme were the true T form of enzyme (i.e.,
the form with no bound FDP), one would expect to find some evidence for L-type pH 6.0 - 6.4 enzyme in isoelectrofocusing
studies of incubated extracts. This does not occur (Fig. 1B),
even when exogenous FDPase is added (Ibsen and Trippet 1972).
It also is difficult to explain why this high pI liver isozyme
yields variable pI values when focused in the absence of FDP.
It seems probable that these higher pI forms are derived forms:
i.e., the purified preparations may be partially denatured.

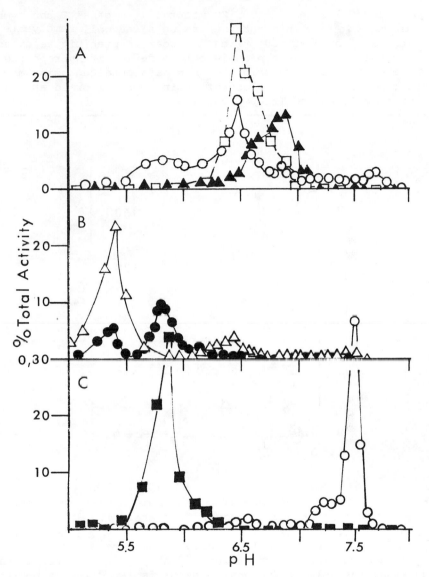

Figure 1. Typical Isoelectrofocusing Patterns. In panel A the symbols indicate: ○ = data obtained using freshly prepared extracts of kidney cortex enzyme; ▲ = pattern obtained when the three peak tubes, pH 6.3 to 6.5, were re-focused after incubation with 5×10^{-5}M FDP; □ = pattern obtained after electrofocusing epididymal fat pad extracts. In panel B the symbols indicate: △ = data obtained using a freshly prepared liver extract; and ● = data obtained using a liver extract incuba-

bated at pH 7.2 and 32° for 20 minutes. In panel C the symbols
indicate: ■ = data obtained using a red cell hemolysate
which was frozen at -60° twice in order to precipitate excess
hemoglobin; cells were made leucocyte deficient by repeated
differential centrifugation; ◯ = data obtained using a
heart muscle extract. Electrofocusing was performed as described by Ibsen and Trippet 1972.

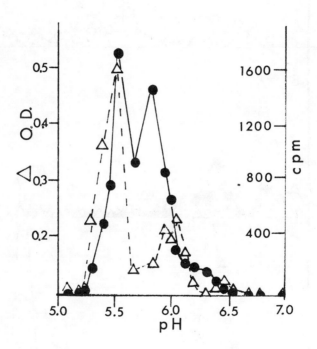

Figure 2. Pattern of Enzymatic and Radioactivity Obtained by
Electrofocusing pH 5.7 L-Isozyme Incubated with ^{14}C-FDP. The
symbols indicate: ● = enzymatic activity; △ = cpm per
0.1 ml aliquot added to 10 ml Bray's solution. In this experiment a liver extract was incubated at pH 7.2, 37° for 15 minutes and then electrofocused. Enzyme isolated in the pH range
of 5.5 to 6.3 (peaking at 5.7) then was incubated with 7.5 X
10^{-5}M ^{14}C-FDP (10 µC/µmol) for 20 minutes at 37°. It was then
re-focused in a 2% pH 5-8 ampholine solution containing 1 µM
[14]C-FDP.

Since pH 6.0 - 6.4 L-enzymes have pI values similar to the
K-isozyme, could the K and L isozymes be interconvertible?
The answer would appear to be no since, as discussed below,

incubation of pH 6.4 K-isozyme with FDP drives it to higher pI
forms, whereas incubation of liver pH 6.4 form with FDP drives
it to lower pI values (Fig. 3).
K-Isozyme: Kinetic studies on undialyzed enzyme showed the pH
6.8 form of the K-isozyme not to be activated by FDP and to give
hyperbolic rate plots with P-enolpyruvate, while the pH 6.4
form was activated and gave sigmoidal plots (Table 1). These
data suggest the pH 6.8 form to be the R conformer and the pH
6.4 form the T conformer. Moreover, incubation of 6.4 enzyme
with FDP drove it to pH 6.8 (Fig. 1A). Such refocusing studies
with the pH 6.4 or 6.8 K-isozymes were made somewhat confusing
by the concomitant production of pH 5.1, 5.4, 7.4 and/or 7.7
enzyme (Table 1, Fig. 3). The significance of the lower pI
forms may be compromised by the fact they are found in association with a precipitate (Ibsen and Trippet 1972). Variable
but usually small amounts of these variants of the K-isozyme
can be obtained using fresh homogenates (Fig. 1A). All four
derived forms can be distinguished from the pH 6.4 - 6.8
enzymes by their high K_m values for P-enolpyruvate (Table 1)·

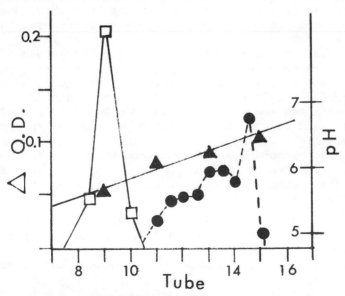

Figure 3. Isoelectrophoretic Pattern Obtained from a Highly
Purified Preparation of L-Isozyme in the Presence and Absence
of FDP. The symbols indicate: ☐ = liver enzyme purified
by $(NH_4)_2SO_4$ precipitation (26-45% satn. cut); elution from a
Val substituted Sepharose column with a 1 to 0.005 M PO_4 buffer
pH 6.8; and elution from a Bio Gel P-200 column. Overall increase in specific activity was 67-fold. ● = data obtained
by incubating all the tubes indicated above in the presence of

Fig. 3. (con't). 7.5 X 10^{-5}M FDP and then refocusing; ▲ = pH value.

Since the forms are derivable from pH 6.4 enzyme, and in some cases may be converted back to the 6.4 or 6.8 enzyme (Ibsen and Trippet 1972), it is clear that they are not different isozymes, but represent metastable forms of the same isozyme. Studies of whole homogenates suggest the predominant enzyme present has properties in common with pH 6.4 K-type isozyme (Ibsen and Trippet 1972), lending further credence to the concept it represents the major in vivo form of the enzyme.

In an attempt to better define the variant forms of the K-isozyme, the pH 6.4 form of the K-isozyme was incubated at pH 7.0 - 7.2 with 7.5 X 10^{-5}M ^{14}C FDP. In the three experiments attempted, most of the radioactivity migrated to the anode and trailed throughout the column, but a significant peak of radioactivity remained at about pH 6.2 and is not associated with a peak of pyruvate kinase activity (Fig. 4). This is at about the same pH as the anomolous peak of radioactivity obtained with the liver enzyme. More pH 7.7 enzyme than pH 6.8 enzyme was generated in these three experiments, but there does seem to be a second peak of radioactivity associated with the pH 6.8 enzyme, as would be expected if it were the R form.

The tubes around the pH 6.2 (pH 5.7 to 6.3) peak of radioactivity were combined, concentrated and passed through Bio Gel P-6. The activity divided into two fractions, high and low molecular weight. The high molecular weight fraction again was concentrated and passed through Bio Gel P-100. Radioactivity was found in association with a peak having an apparent molecular weight of about 25,000. Since the enzyme was not purified one cannot necessarily relate this peak to the pyruvate kinase. However, this may be the same substance responsible for the secondary peak of radioactivity found with the L-enzyme (Fig. 2).

M-Isozyme: Heart muscle extracts yielded a sharp peak at pH 7.5 and a smaller shoulder at pH 7.2. Incubation appeared to increase the pI value of the pH 7.5 peak slightly. Otherwise, no variant forms of the muscle isozyme were found (Ibsen and Trippet 1972).

Red Cell Enzyme: Some studies have suggested that the red cell enzyme represents a fourth isozyme, either an L_3K or L_2K_2 hybrid (Imamura and Tanaka 1972, Sussor and Rutter 1971, Whittell et al., 1973), while others have suggested it is identical to the liver L-isozyme (Tanaka et al., 1967, Ibsen and Krueger 1973). Fig. 1 shows red cell enzyme to have the same pI value as the T form of the liver L-isozyme. As discussed subsequently, this conclusion is substantiated by kinetic data.

TABLE 1. PROPERTIES OF VARIOUS pI FORMS OF THE THREE MAJOR ISOZYMES

PROPERTY	K-ISOZYME						L ISOZYME			M-ISOZYME	
mean pI value	~5.1	~5.4	6.41	6.77	7.44	7.67	5.46	5.73	6.0-6.4	7.48	7.55
Major Forms Obtained from fresh extracts	-	-	+	+	-	-	+	+	-	+	-
Derived Forms	+	+	-	-	+	+	-	+	+	-	+
K_m for P-enolpyruvate without FDP (x 10^4 M) (a)	6.5	~3.3	1.5(d)	0.63	16.0	12.0	0.69	1.6(d)	-	0.52	0.68
K_m for P-enolpyruvate with 5 x 10^{-5} M FDP (x 10^4 M) (a)	6.5	~3.3	0.60	0.65	16.0	12.0	0.67	0.71	-	0.52	0.58
Apparent Mol. Wt. (x 10^{-3}) with exposure to Mg^{++} (b)	115	-	199	210	-	-	190	171	-	-	-
without exposure to Mg^{++}	-	-	114	-	104	106	-	-	-	228	209
Electrophoresis (c)	C	-	C	C	C	C	A	A	-	O	-

(a) Undialyzed enzyme assayed at pH 7.0, 32° as described by Ibsen and Trippet 1972.

(b) Lower pH forms were diluted in assay solution prior to sedimenting in order to reduce their density.

(c) On cellulose acetate, pH 7.5 in the presence of FDP. A= migrated toward anode, C = migrated toward cathode, O = remained at origin as described by Ibsen and Krueger 1973.

(d) Rate plots were sigmoidal.

pH 7.2 Enzyme: Of the dozen or so pI forms of rat pyruvate kinase, all but one, a pH 7.2 variant, can be related to the K, L or M-isozyme. This variant was found as a minor component (5-10% total activity) in heart extracts (Fig. 1C) and occasionally as a trace component in kidney cortex extracts. It is not found in skeletal muscle extracts nor is it derived from the major M-isozyme by incubation and re-focusing. Since heart extracts also contain small amounts of apparent K-isozymes (Fig. 1C), it is conceivable that the pH 7.2 enzyme is a hybrid. Its pI value and its physical properties (Ibsen and Trippet 1972) are compatible with it being an M_3K hybrid.

PROPERTIES OF THE MAJOR FORMS OF THE RAT K, L, AND M-ISOZYMES

Physical: It has been demonstrated that the three apparent isozymes differ with respect to solubility (Pogson 1968a, Ibsen and Trippet 1972), and stability (Taylor et al., 1969, Ibsen and Trippet 1973). The sedimentation data summarized in Table 1 suggest that the predominant forms of the isozymes have molecular weights in the 200,000 range, compatible with them being

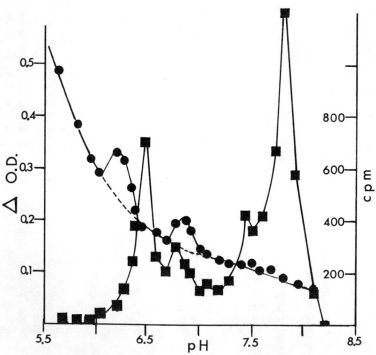

Figure 4. Pattern of Enzymatic and Radioactivity Obtained by

Fig. 4 (con't). reelectrofocusing pH 6.4 K-Isozyme Incubated with ^{14}C-FDP. The symbols indicate: ● = counts per minutes per 0.1 ml aliquot; ■ = enzyme activity. Conditions as in Fig. 2 except enzyme was incubated 15 minutes prior to FDP addition. The sample incubated with ^{14}C-FDP had a pH range of 6.26 to 6.63 and showed peak activity at pH 6.36.

---

tetramers. However, Pogson (1968b) obtained S values for the adipocyte enzyme more compatible with a dimeric structure. We also find the pH 6.4 form of the K-isozyme from adipocytes to have molecular weights of about 114,000 if sedimented immediately after isoelectrofocusing. Further investigation suggests the pH 6.4 K-isozymes in general exist in a dimer-tetramer equilibrium in which Mg^{++} causes a shift toward the tetrameric form. At 4° C the Mg^{++} -induced shift from apparent dimer to tetramer is slow but the rate of change is stimulated by increased temperature or the presence of FDP. This latter phenomenon may account for the marked increase in molecular weight induced by FDP (Pogson 1968b, Ibsen et al., 1971).

The variant forms of the K-isozymes migrate toward the cathode when electrophoresed at pH 7.5, even if their pI values are less than 7.0 (Table 1, Ibsen and Krueger 1973). This suggests they are converted to the pH 7.7 form. Pogson 1968b has shown EDTA treatement induces the K-isozyme to run toward the anode, suggesting Mg^{++} loss converts it to a low pI form.

Kinetic: Ibsen and Trippet (1973) and (1974) compared many of the kinetic properties of the three isozymes. The more salient differences include the following. As isolated the M-isozyme markedly favors the R conformer, while the L-isozyme most strongly favors the T conformer. At pH 7.4 and in the absence of FDP, an allosteric activator, the K_m values for P-enolpyruvate are: L-form, 0.7 to 1.7 mM; K-form, 0.2 mM; and M-form 0.03 mM, while in the presence of 5×10^{-5} M FDP the K_m values for the L and K-isozymes are 0.03 and 0.07 mM, respectively. The K_A values for FDP are approximately 8×10^{-6} M and 8×10^{-7} M for the K and L-isozymes, respectively. All three isozymes appear to have two different allosteric sites for amino acids (Ibsen and Trippet, 1973 and 1974, Carbonell et al., 1973), but the amino acids have different effects. Non-polar short side chain amino acids inhibit both the L and K-isozymes, except Ser which activates the latter enzyme and inhibits the former (5 mM inhibits 37%, under the conditions employed). The K and M-isozymes are more sensitive to inhibition by non-polar amino acids with bulky side chains than is the L-isozyme. K_i values for inhibition of the K, M and L-isozymes by Phe are: 0.15, 1.4, and 2.8 mM, respectively, while 5 mM Trp inhibits the three isozymes approximately 90, 40, and 0%, respectively. ADP does not act as a homotropic activator of any of the isozymes, but under the conditions used, ADP has the least affinity for the L-isozyme; the K_m values are 0.25 mM for the K and M-isozymes and 1.2 mM for the L-isozyme.

The kinetic properties of the pH 5.7 red cell enzyme which have been studied are similar to those of the L-isozyme, namely: K_m value for P-enolpyruvate, 1.5 mM without FDP, 0.08 mM with FDP; K_i for Ala, 0.15 mM; K_i for Phe 2.5 mM; no apparent sensitivity to Trp, but 35% inhibited by 5 mM Ser. These data indicate the red cell enzyme to be L-type. However, it is conceivable that an L_3K hybrid would show minimal K character since Dyson and Cardenas (1973) have demonstrated that the single monomer of LM hybrids has less influence on kinetic characteristics than would be expected statistically.

All three isozymes also require a mono- and a divalent cation for catalytic activity. NH_4^+ can substitute for the presumed natural monovalent activator K^+. Indeed, it is the more effective activator for both the K and L-isozymes: in the former case because of a lower K_m value and in the latter case because of a higher V_m value (Ibsen and Trippet 1973). The kinetic constants obtained with respect to K^+ and NH_4^+ were derived by drawing the best straight line through points obtained using the Hill equation. However, this approach appears to be an oversimplification. Thus, very careful K^+ saturation studies were conducted. The results summarized in Fig. 5 clearly show discontinuities in the plots. The human red cell enzyme, or the rat K or L-isozymes all yield a break in the plots occurring at the point marked II. At this point, cooperative effects are enhanced. Another unambiguous disjunction occurs at the points marked III. At this point, inhibitory effects of K^+ were interrupted by a second transitory phase of activation. Conceivably, there also are disjunctions at the points marked IL and IIM.

The increase in slope at the points labeled II appear to be analogous to those reported by others (Teipel and Koshland 1969; Winicov and Pizer 1974), which have been interpreted as being caused by a sequential saturation of subunits. Other explanations are possible, but if this mechanism did operate it should be possible to change the quantitative response of the one subunit to saturation of the second by addition of known effectors. The data of Fig. 6 show this to be the case. Activation of the K-isozyme with either FDP or P-enolpyruvate reduces the magnitude of the discontinuity at point II. The data thus become more analogous to those observed using the M-isozyme. Reduction of the P-enolpyruvate level increases the level of K^+ required to promote the apparent sequential subunit interaction. Addition of the inhibitor L-alanine apparently prevents the saturation of the first subunit from activating the second (Fig. 6). The subunits now show a negative cooperative relationship

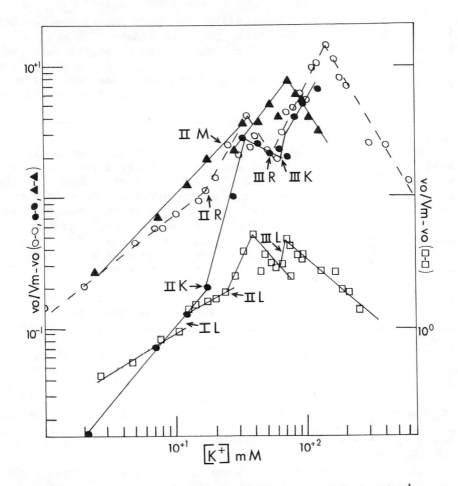

Figure 5. Influence of Increasing Concentrations of K⁺ on Rates Obtained With Human Red Cell of the Rat K-, L-, and M-Isozymes of Pyruvate Kinase. The ◯ - represent data obtained with human red cell enzyme (partially purified as described by Ibsen et al. 1968); the ⬤ , with the pH 6.4 kidney cortex K-isozyme; the ☐ , with pH 5.7 liver L-isozyme; and the △ with pH 7.5 heart muscle M-isozyme. The latter three preparations were exhaustively dialyzed after electrofocusing. Kinetic and physical studies showed no indication of cross-contamination among three preparations. The P-enolpyruvate concentrations were: 1.65 mM, 0.4 mM, 1.2 mM and 0.4 mM for the red cell, K-, L- and M-isozymes, respectively. The red cell enzyme was incubated at pH 7.0 at 32° and 0.11 mM ADP (Ibsen et al., 1968). The reactions were started by adding cold enzyme. The other assays were done at 37°, pH 7.5 and 0.43 mM ADP (Ibsen

Fig. 5 (con't) and Trippet 1973). In these latter experiments assay solution was preincubated for approximately 10 minutes, enzyme was added and allowed to equilibrate an additional five minutes and then the reaction was started by ADP addition. The possibility of discontinuities occurring at places where a more concentrated solution of K^+ was added was avoided by overlapping points.

(Levitzki and Koshland 1969).

The disjunctions at III appear to be unique. Obviously, inhibition begins before the enzymes are fully saturated at the activator site by K^+. Possibly inhibition begins when the first three monomers are approaching full saturation. Once full saturation is achieved a conformation change could then take place, which could increase the affinity of the final monomer for K^+ and enhance its activity. If such interaction between the monomeric units is discernible, it might be predicted that a discontinuity would be observed when the first monomer became saturated. This could account for the possible discontinuity marked LI (Fig. 5).

A sequential reaction mechanism was previously suggested because of the unique saturation kinetics of the M-isozyme by FDP and the time dependent effects of certain ligands (Ibsen and Trippet 1973).

Table 2 shows all three isozymes to be somewhat inhibited by nucleotides, the M-isozyme to the least extent. Both the L and K-isozymes were found to be strongly inhibited by Mg-ATP. As indicated, these effects were quite variable, particularly with respect to the K-isozyme. This variability also is reflected in the literature. For instance, Carbonell et al.(1973) use lack of sensitivity of K-isozyme to ATP as a criterion to distinguish between the K and L-isozymes.

It is postulated that this variability results because ATP can inhibit by three mechanisms: chelation of Mg^{++}, competition at the catalytic site, and allosterically. In the experiments reported herein, Mg^{++} effects were minimized by adding the amount calculated to chelate, using the data of Rose (1968). However, preliminary studies indicate that the % maximal inhibition decreases and the apparent K_i value increases as preparations age. Since K_i values were calculated by the method of Jensen and Nester (1966) (which should correct the K_i value for the fraction of desensitized enzyme), this observation is compatible with the concept that the observed increase in apparent K_i is caused by a greater proportion of inhibition not due to allosteric effects. That is, the percentage maximal inhibition observed with increasingly desensitized preparations reflects a greater proportion of competitive effect at the active site, true maximal inhibition never having been achieved.

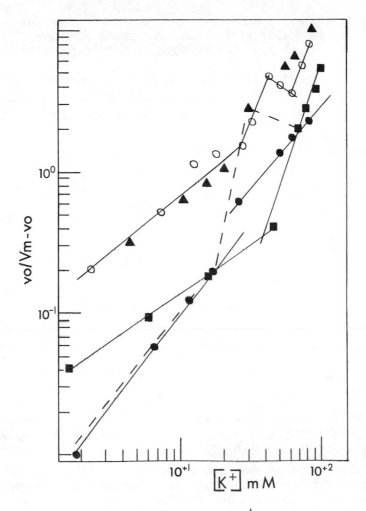

Figure 6. Changes in the Pattern of K⁺ Activation of the K-Isozymes Caused by L-Alanine, FDP or P-Enolpyruvate. The ● indicate data obtained at 0.4 mM P-enolpyruvate plus 0.08 mM L-alanine (a concentration causing about 40% inhibition at 65 mM K⁺). The ○ were obtained using 0.4 mM P-enolpyruvate plus 5×10^{-5}M FDP. The ▲ represent data obtained at 3.2 mM P-enolpyruvate and the ■ at 0.2 mM P-enolpyruvate. The broken line represents data obtained with 0.4 mM P-enolpyruvate and no additions. This curve was transposed from Fig. 6.

Adenosine appears to activate the K and L-isozymes when inhibited by Mg-ATP but does not release inhibition caused by amino acids nor does it alter cooperative effects of P-

TABLE 2. RELATIVE SENSITIVITY OF THE T FORMS OF THE
PRINCIPLE VARIANTS OF THE THREE PYRUVATE
KINASE ISOZYMES TO POTENTIAL INHIBITORS [a]

Compound	% inhibition at a 5 mM concentration for the		
	K-isozyme	L-isozyme	M-isozyme
Mg-ATP	68 (K_i=0.3-2.2 mM)	90 (K_i=0.2-0.6 mM)	15
Mg-GTP	68	73	15
Mg-UTP	31	6	10
Mg-CTP	20	None (Activates)	10
Mg-AMP	60	None (Activates)	7
Adenosine	None (Activates)	None (Activates)	None
Mg-2,3 diP-glycerate	75	60	33
Mg-P-Creatine	18 (45) [c]	24 (40) [d]	None (15) [c]

[a] All values were obtained at pH 7.5 and 37°. Enzyme was pre-incubated with assay mix and the reaction started by adding ADP (Ibsen and Trippet 1973).

[b] Except for the data obtained for the K_i value, all comparisons of data involving nucleotides or nucleosides were obtained using the same preparation.

[c] % inhibition at the optimal pH of 6.2.

[d] % inhibition at the optimal pH of 7.0.

enolpyruvate. These data suggest adenosine acts by competing for an allosteric ATP site. AMP seems to have a similar effect on the L-isozyme but inhibits the K-isozyme. In general, the K-isozyme is susceptible to inhibition by a greater spectrum of nucleotides (Table 2).

The three isozymes also are inhibited by Mg-P-creatine. As was true of Mg-ATP, the M-isozyme is least susceptible to P-creatine. This would not be expected physiologically and appears to differ from the situation obtained with rabbit M-isozyme.

Conceivably the rat M-isozyme is desensitized extremely easily with respect to these effectors. P-creatine inhibition is optimal at pH values below 7.0, (Table 2) as was true for rabbit enzyme (Kemp 1973). In contrast, H^+ acts to activate the isozymes when initially inhibited by amino acids or Mg-ATP at pH 7.5.

Table 2 also shows the three isozymes to be susceptible to Mg-2, 3-di-P-glycerate inhibition. Such inhibition previously was demonstrated for the human red cell enzyme by Ponce et al., (1971).

ACKNOWLEDGEMENT

The work described was supported by U.S. Public Health Service Grant No. CA-07883.

REFERENCES

Anderson, J. W., and L. Stowring 1973. Regulation of gut pyruvate kinase by amino acids. *Fed. Proc.* 32: 2512.

Carbonell, Juan, J. E. Felíu, R. Marco and A. Sols 1973. Pyruvate kinase. Classes of regulatory isoenzymes in mammalian tissues. *Eur. J. Biochem.* 37: 148-156.

Costa, Lidia, L. Jiménez de Asúa, E. Rozengurt, E. G. Bade, and H. Carminatti 1972. Allosteric properties of pyruvate kinase from rat kidney cortex. *Biochim. Biophys. Acta* 289: 128-136.

Dyson, Robert D., and J. M. Cardenas 1973. Bovine pyruvate kinases III: Hybrids of the liver and skeletal muscle isozymes. *J. Biol. Chem.* 248: 8482-8488.

Farron, Francoise, H. H. T. Hsu, and E. Knox 1972. Fetal-type isoenzymes in hepatic and nonhepatic rat tumors. *Cancer Res.* 32: 302-308.

Hess, Benno, and C. Kutzbach 1971. Identification of two types of liver pyruvate kinase. Hoppe-Seyler's Z. *Physiol. Chem.* 352: 453-458.

Ibsen, Kenneth H., and E. Krueger 1973. Distribution of pyruvate kinase isozymes among rat organs. *Arch. Biochem. Biophys.* 157: 509-513.

Ibsen, Kenneth H., K. W. Schiller, and T. A. Haas 1971. Interconvertible kinetic and physical forms of human erythrocyte pyruvate kinase. *J. Biol. Chem.* 246: 1233-1240.

Ibsen, Kenneth H., K. W. Schiller, and E. A. Venn-Watson 1968. Stabilization, partial purification, and effects of activating cations, ADP, and phosphoenolpyruvate on the reaction rates of an erythrocyte pyruvate kinase. *Arch. Biochem. Biophys.* 128: 583-590.

Ibsen, Kenneth H., and P. Trippet 1972. Interconvertible and noninterconvertible forms of rat pyruvate kinase. *Biochem.* 11: 4442-4450.

Ibsen, Kenneth H., and P. Trippet 1973. A comparison of kinetic parameters obtained with three major noninterconvertible isozymes of rat pyruvate kinase. *Arch. Biochem. Biophys.* 156: 730-744.

Ibsen, Kenneth H., and P. Trippet 1974. Effects of amino acids on the kinetic properties of three noninterconvertible rat pyruvate kinases. *Arch. Biochem. Biophys.* 163: 570-580.

Imamura, Kiichi, and T. Tanaka 1972. Multimolecular forms of pyruvate kinase from rat and other mammalian tissues. I. Electrophoretic studies. *J. Biochem.* (Tokyo). 71: 1043-1051.

Imamura, Kiichi, K. Taniuchi, and T. Tanaka 1972. Multimolecular forms of pyruvate kinase II. Purification of M_2-type pyruvate kinase from Yoshida Ascites Hepatoma 130 cells and comparative studies on the enzymological and immunological properties of the three types of pyruvate kinases, L, M, and M_2. *J. Biochem.* (Tokyo). 72: 1001-1015.

Jensen, Roy A., and E. W. Nester 1966. Regulatory enzymes of aromatic amino acids biosynthesis in *Bacillus subtilis*. I. The enzymology of feedback inhibition of 3-deoxy-D-arabino-heptulosonate-7-phosphate synthetase. *J. Biol. Chem.* 241: 3373-3380.

Jiménez de Asúa, Luis, E. Rozengurt and H. Carminatti 1971. Two different forms of pyruvate kinase in rat kidney cortex. *FEBS Letters* 14: 22-24.

Kemp, Robert G. 1973. Inhibition of muscle pyruvate kinase by creatine phosphate. *J. Biol. Chem.* 248: 3963-3967.

Levitzki, Alexander, and D. E. Koshland, Jr. 1969. Negative cooperativity in regulatory enzymes. *Proc. Nat'l. Acad. Sci.* (USA) 62: 1121-1128.

Pogson, C. I. 1968(a). Adipose-tissue pyruvate kinase. Properties and interconversion of two active forms. *Biochem. J.* 110: 67-77.

Pogson, C. I. 1968(b). Two interconvertible forms of pyruvate kinase in adipose tissue. *Biochem. Biophys. Res. Commun.* 30: 297-302.

Ponce, J., S. Roth, and D. R. Harkness 1971. Kinetic studies on the inhibition of glycolytic kinase of human erythrocytes by 2,3-diphosphoglyceric acid. *Biochim. Biophys. Acta.* 250: 63-74.

Rose, Irwin A. 1968. The state of magnesium in cells as estimated from adenylate kinase equilibrium. *Proc. Nat'l. Acad. Sci.* (USA) 61: 1079-1086.

Sussor, Walter A., M. Kochman, and W. J. Rutter 1973. Structure determinations of FDP aldolases and the fine resolution of some glycolytic enzymes by isoelectric focusing. *Ann. N.Y. Acad. Sci.* 209: 328-344.

Sussor, Walter A., and W. J. Rutter 1971. Method for the detection of pyruvate kinase, aldolase, and other pyridine nucleotide linked enzyme activities after electrophoresis. *Anal. Biochem.* 3: 147-155.

Tanaka, Takehiko, Y. Harano, F. Sue, and H. Morimura 1967. Crystallization, characterization and metabolic regulation of two types of pyruvate kinase isolated from rat tissues. *J. Biochem.* (Tokyo) 62: 71-91.

Taylor, C. B., H. P. Morris and G. Weber 1969. A comparison of the properties of pyruvate kinase from Hepatoma 3924-A, normal liver and muscle. *Life Sciences* 8 part II, 635-644.

Teipel, John, and D. E. Koshland, Jr. 1969. The significance of intermediary plateau regions in enzyme saturation curves. *Biochem.* 11: 4656-4663.

Van Berkel, Th. J. C., J. F. Koster, and W. C. Hülsman 1973. Some kinetic properties of the allosteric M-type pyruvate kinase from rat liver; influence of pH and the nature of amino acid inhibition. *Biochim. Biophys. Acta* 321: 171-180.

Walker, P. Roy, R. J. Bonney, J. E. Becker and V. R. Potter 1972. Pyruvate kinase, hexokinase and aldolase isozymes in rat liver cells in culture. *In vitro.* 8: 107-114.

Whittell, Neil M., D. O. K. Ng, K. Prabhakararao, and R. S. Holmes 1973. A comparative electrophoretic analysis of mammalian pyruvate kinase isozymes. *Comp. Biochem. Physiol.* 46B: 71-80.

Winicov, Ilga, and L. I. Pizer 1974. The mechanism of end product inhibition of serine biosynthesis. IV: Subunit structure of phosphoglycerate dehydrogenase and steady state kinetic studies of phosphoglycerate oxidation. *J. Biol. Chem.* 249: 1348-1355.

MULTIPLE MOLECULAR FORMS OF *NEUROSPORA* MITOCHONDRIAL MALATE DEHYDROGENASE

KATHY B. BENVENISTE and KENNETH D. MUNKRES
Laboratory of Molecular Biology
University of Wisconsin
Madison, Wisconsin 53704

ABSTRACT. Mitochondrial malate dehydrogenase (L-malate: NAD oxidoreductase, E.C. 1.1.1.37) of *Neurospora crassa* was purified 600 fold. Both the purified and unpurified enzymes exhibited 16 isoelectric points, 11 electrophoretic mobilities, and 16 molecular weights. Proportions and properties of the isozymes were not affected by purification.

The calculated molecular weight for the monomer of the weight series of *Neurospora crassa* isozymes is 13,000. This is equivalent in size to either the α or β subunit of malate dehydrogenase. All multiples of the monomer from 2 through 15 were active. The isozymes' weights, number, and proportion were functions of the ionic and pH environment. Both an increase in isoelectric point with increase in molecular weight and the occurrence of all possible integral weights (both odd and even multiples of the monomer), led us to conclude that the aggregation forms asymmetrical structures.

INTRODUCTION

In 1965, Whitehead proposed an empirical rule to describe the spectrum of molecular weights of dehydrogenases from a variety of organisms. Dehydrogenases might be composed of aggregated monomers, and the molecular weight (M) of the enzyme can be expressed by the formula:
$$M = nM_m,$$
where $\underline{n}$ is an integer equal to the number of monomers in the enzyme and M_m is the weight of that monomer.

The molecular weight of malate dehydrogenase (MDH) from many organisms has been determined (for review see Benveniste, 1971). There are discrepancies in the values reported by several laboratories and in some cases within a single laboratory.

We have purified mitochondrial MDH (m-MDH) from *Neurospora crassa* and examined its molecular weight under a variety of conditions. Multiple molecular weight isozymes are observed and can be described by Whitehead's formula when M_m is assigned a value of 13,000 daltons. These data suggest that the discrepancies in reported molecular weights for MDH's can be attributed to a difference in the state of aggregation.

Three properties of the isozymes will be discussed: molecular weight, isoelectric point, and electrophoretic mobility. According to these properties, three subclasses are defined, respectively, as M-isozyme, I-isozyme, and E-isozyme.

PURITY OF THE ENZYME

The purification scheme used was based on the method of Munkres and Richards (1965a). The m-MDH obtained by them has been analyzed by several methods and judged to be highly (90%) purified. The purity of this preparation is further confirmed by electrofocusing and electrophoresis in polyacrylamide gels which indicated that all protein bands had malate dehydrogenase activity (Fig. 3 and Benveniste, 1971).

Previous studies indicate that m-MDH associated specifically with a fraction of mitochondrial protein called fraction P ("structural protein") (Munkres and Woodward, 1966; Benveniste and Munkres, 1970; Swank and Munkres, 1971). Since the small oligopeptides which comprise 60% by weight or fraction P_O (Swank et al., 1971) would probably not be observed by the conventional electrophoresis technique, purified enzyme was analyzed by the high resolution SDS-polyacrylamide gel electrophoresis system specifically designed to determine molecular weights below 10,000 (Swank and Munkres, 1971). No oligopeptides were found. Therefore the multiple isozymes were not caused by an association with fraction P_O.

The acetone extraction of the mycelia should have effectively removed lipids from the preparation. Thus, it seemed unlikely that the heterogeneity in molecular weight and isoelectric point was due to contamination by phospholipids.

I-ISOZYMES IN PREPARATION OF DIFFERENT PURITY

The isoelectric points of the I-isozymes in different states of purity were determined by electrofocusing in polyacrylamide gels. When gels containing crude or purified preparations were stained for enzymatic activity, 11 isozymes with isoelectric points differing in the two preparations by less than 0.05 units, the limit of routine resolution, were observed (Fig. 2.). Five additional I-isozymes whose isoelectric points differ by as much as 0.5 units may be in common to both preparations. These differences may not have been significant, however, since these I-isozymes were at the end of the gels where measurements were less accurate due to the steep pH gradient. The presence of at least the same 16 I-isozymes in the crude and purified enzymes preparations lead us to conclude that they were not created by the purification procedure.

Twenty five mg of purified enzyme were electrofocused in a column stabilized by a sucrose gradient. Seven I-isozymes whose isoelectric points ranged from 3 to 7 were found (Fig.1)

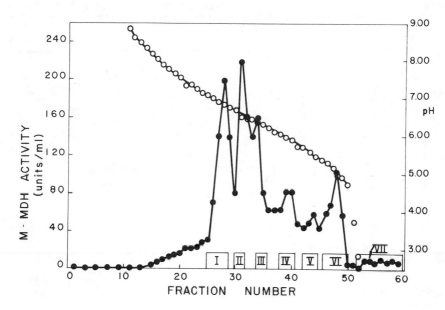

Fig. 1. Isoelectric Components of Purified *Neurospora* Mitochondrial Malate Dehydrogenase in Preparatory Scale. Mitochondrial MDH was purified 600-fold from mycelial acetone powder as described (Benveniste, 1971). A sample containing 4.25×10^4 enzymatic units and 24.6 mg protein was mixed with sucrose and ampholyte (pH 5-8) solution and used to form the sucrose gradient in 110 ml LKB column. Focusing was at 300 v. for 24 hours at 15°. Fractions of 2 ml were collected. Enzyme activity (●--●) and pH (O--O) are indicated. Roman numerals indicate pooled fractions which were dialyzed against Na phosphate buffer (pH 7.0) in preparation for the experiments described in Figures 5 and 6.

Additional I-isozymes with isoelectric points greater than 7 were low in activity and were not further examined. When larger quantities of enzymes were electrofocused, a total of 15 I-isozymes were found.

The isoelectric points of the isozymes, determined in either the preparatory column with sucrose or the analytical columns with polyacrylamide gel, were similar; hence the sucrose did not alter the isoelectric points. In comparable experiments with human hemoglobin, Drysdale et al. (1971) reached the same conclusion.

Four basic I-isozymes observed in the gel were not observed in the sucrose; however, that discrepancy was related to the difference in the capacity of the two systems to separate the proteins and to the fact that the pH range of the gel was from 3 to 10, while that of the sucrose column was from 5 to 8.

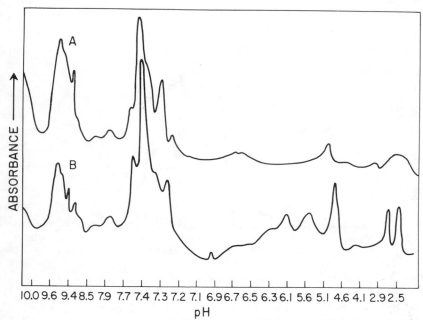

Fig. 2. Isoelectric Components of Crude and Purified *Neurospora* Mitochondrial Malate Dehydrogenase in Analytical Scale. A sample containing (A) 2.5 units of activity and 1 mg protein of step 1 enzyme and (B) 2.8 units of activity and 0.46 µg protein of step 5 enzyme was mixed and polymerized with polyacrylamide and ampholyte (pH 3-10). Electrofocusing was at 12° for 3 hours beginning with 2 mA/gel. Gels were stained for MDH activity by the tetrazolium method (Benveniste 1971). Absorbance: 540 nm.

ELECTROPHORESIS IN POLACRYLAMIDE GEL

Previously, when either fresh cytosol (Munkres et al., 1970) or purified enzyme (Munkres, 1968) was analyzed by electrophoresis in polyacrylamide gel, 3 to 5 isozymes were observed. Improved methods have permitted the detection of additional isozymes (Benveniste, 1971).

The E-isozyme patterns were identical at all stages of purification. The patterns were highly consistent and are illustrated in Figure 3. At least 11 E-isozymes were observed in both purified and crude preparations. The number, relative proportions and electrophoretic mobilities of the isozymes were constant throughout purification. Thus, the conclusion from the electrofocusing analyses were confirmed. The isozymes were not modified by the purification procedure. Furthermore, these results indicated that the majority of the I-isozymes were not artifacts arising from an interaction between protein and the ampholytes, such as has been observed with other proteins (Kaplan and Foster, 1971; Frater, 1970).

The relative specific activities of either the I- or the E-isozymes could not be determined quantitatively because neither the protein nor the enzyme stain was quantitative.

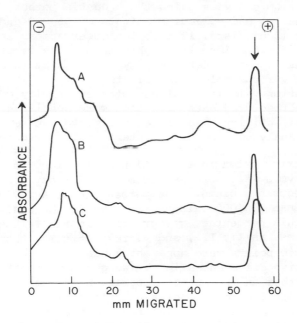

Fig. 3. Electrophoretic Components of Mitochondrial MDH in Analytical Scale. A sample containing (A) 2.5 units of activity and 1 mg protein of step 1 enzyme; (B) 300 units of activity and 100 µg protein of step 5 enzyme; or (C) 2.8 units of activity and 0.46 µg protein of step 5 enzyme was layered on top of the stacking gel. Electrophoresis was at 12° for 1 hour at 2.5 mA/gel. Gels were stained: A and C with tetrazolium dye for MDH activity, or B with Coomassie Blue for protein. The arrow indicates a bromphenol blue marker. Absorbance: A and C - 540 nm; B - 550 nm.

565

These experiments also indicated that neither the number nor the activity of the individual I-isozymes was dependent upon the protein concentration; although gel B (Fig. 3) contained over 200 times more protein than gel C, the patterns were similar.

MULTIPLE MOLECULAR WEIGHT SPECIES

The possibility of molecular weight heterogeneity has been previously indicated by Munkres and Richards (1965b). In two out of five experiments using the Archibald method of sedimentation equilibrium analysis a 160,000 dalton species in addition to the main 54,000 species was observed; the heavier component comprised as much as 30% of the total protein. In our sedimentation equilibrium analyses using the Yphantis method, the major and minor species had molecular weights of 65,000 and 52,000, respectively (Benveniste, 1971). The observations in previous experiments of two molecular weights in a given preparation substantiate our proposal that the enzyme can exist in an equilibrium between various molecular weights.

Sedimentation velocity analyses of the activity of purified enzyme in sucrose gradients, a method more sensitive to very low levels of activity than either sedimentation velocity or equilibrium analyses, indicated heterogeneity of molecular weight.

Upon centrifugation in a solution of low ionic strength several M- isozymes were obtained from each I-isozyme isolated from an electrofocusing column. The percentage of activity equal to or greater than 200,000 daltons is shown in the shaded area (Fig. 4). Since this percentage was greatest in the I-isozyme fraction I and less in II, III, and V, respectively, those with higher isoelectric points were apparently heavier. The absence of more definitive differences in the weight distributions may be caused by either an incomplete separation of the I-isozymes in the electrofocusing column, or an equilibrium among them. Such an equilibrium may be either inherent to the I-isozymes, or induced by dialyzing them against the buffer at pH 7 prior to centrifugation.

If the correlation between molecular weight and isoelectric point were absolute, acidic I-isozymes VI and VII should contain low percentages of heavy M-isozymes. This was not observed. However, the acidic environment of these I-isozymes was sufficient to dissociate them to their monomers (Munkres, 1965a, b). When the monomers reaggregated during dialysis against pH 7 buffer, a molecular weight series may have formed which is similar to that observed in the I-isozyme whose isoelectric point was 7.0. The similarity in the molecular weight distribution of the I-isozymes, I, VI, and VII supports this hypothesis.

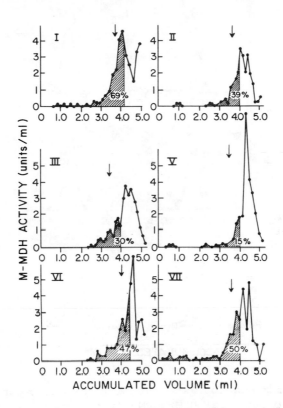

Fig. 4   Sedimentation Patterns of Isoelectric Components from
Purified Mitochondrial MDH at Low Ionic Strength.   Linear sucrose density gradients (5-20%) were formed with distilled water
as the solvent.   Catalase (50 µg) and isoelectric components of
m-MDH (2-5 units of activity; see Figure 1) were centrifuged at
178,000 g for 4 hours at 4° in a Spinco L2-65B centrifuge.
Fractions were collected from the bottom of the gradient and
assayed for MDH activity and catalase (at 410 nm).   Activities
and volumes are normalized.   The arrow indicates a molecular
weight of 250,000 and the shaded area indicates a molecular
weight of 200,000 or greater.

Only one M-isozyme, with a molecular weight of about 90,000,
was obtained from each of the I-isozymes after centrifugation
in a medium of moderate ionic strength (Fig. 5).   An additional 5-fold increase in the ionic strenght of the medium led to
a further decrease in the number of the M-isozymes in each of
the I-isozymes fractions.   Thus, the equilibrium concentrations

567

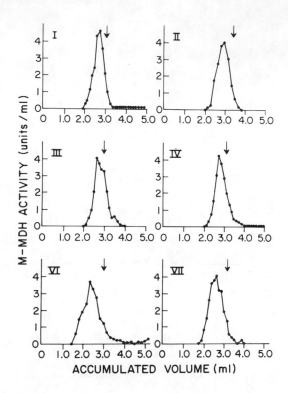

Fig. 5. Sedimentation Patterns of Isoelectric Components from Purified Mitochondrial MDH at Moderate Ionic Strength. Linear sucrose density gradients (5-20%) were formed with solution containing 50 mM phosphate buffer, EDTA and dithiothreitol (1 mM) at pH 7.0. Hemoglobin (50 µg) and isoelectric components of m-MDH (2-5 units of activity; see Figure 1) were centrifuged at 178,000g for 16 hours at 4°. Fractions were collected from the bottom of the gradient and assayed for MDH activity and hemoglobin (at 410 nm). Activities and volumes are normalized. The arrow indicates a molecular weight of 65,000.

of heavy and light isozymes were a direct function of the ionic strength of the medium during centrifugation.

MODE OF AGGREGATION

In a solution of low ionic strength, both crude and purified malate dehydrogenase existed as multiple M-isozymes. Collectively, the data from all of the sedimentation analyses fit an integral weight series based upon a monomer of 13,000. The weight (X) of the monomer was determined theoretically by trying various values of X to minimize the expression:

$$\left| \Sigma \ (Z - MW_{obs \ i}) \right|$$

where $Z = nX$, n was an integer, and the value of n for each observed isozyme was chosen such that $Z - MW_{obs \ i}$ was minimized. In these calculations, only weights up to 115,000 were used. Heavier ones were omitted because of 10% deviation in their values (the limit of resolution by sucrose gradient analysis) was greater than the weight of the monomer. Moreover, cases in which the value of $Z - MW_{obs \ i}$ was equal to X/2 were excluded. This term had a marked minimum when X had a value of 13,000.

Although an active monomer was never observed in these studies, Munkres (1965a, b) demonstrated that upon exposure to either acid or urea, purified malate dehydrogenase dissociates into inactive subunits weighing 13,500 daltons. These data strongly support the proposed molecular weight series and the theoretically calculated monomer weight.

With the exception of the monomer, active M-isozymes having molecular weights which were integral multiples (i.e. 2 through 15) of the monomer have been observed. These observations did not agree precisely with Whitehead's dehyrogenase hypothesis (1965), which proposes that n may be only 1,2,3,4,6,8,12, or 20. Additional values of n would be observed, however, with an enzyme such as glutamate dehydrogenase. The subunits of that enzyme aggregate into trimers which can stack to form long linear molecules (Eisenberg, 1970; Krause et al., 1970). Thus n might be any multiple of 3. Similarly, an even value of n can be observed if an enzyme's subunit forms dimers which aggregate in a specific way. Therefore, the prime numbers of n (5,7,11, and 13) for *Neurospora* malate dehydrogenase were irregular. These irregular n values indicated that the aggregation was not symmetrical. Rather the aggregate was asymmetrical in either one or two planes as in a filament or a sheet (G. Borisy, personal communication).

In a low ionic environment, both crude and purified enzyme existed as multiple weight and isoelectric forms. Since malate dehydrogenase of *N. crassa* is dissociated by treatment with acid (Munkres, 1965a) to monomers of 13,500 daltons (Munkres, 1965b), the isozyme at the anode in the electrofocusing column probably represented these monomers.

A linear correlation between the molecular weights and isoelectric points was obtained by assuming that the most acidic I-isozyme represented the monomer as suggested above. Each successively more basic I-isozyme represented a M-isozyme with an additional subunit. This correlation suggested that acidic groups became non-titratable as the degree of aggregation increased. Such an asymmetrical masking of acidic groups is consistent with the hypothesis of asymmetrical aggregation just discussed.

OTHER MULTIMERIC MALATE DEHYDROGENASE ENZYMES

Murphey et al. (1967a, b) proposed a weight of 60,000 and 120,000 for eucaryote and procaryote MDH respectively. This proposal is based upon data from gel filtratration under one set of conditions. However, our results indicate that molecular weight is highly dependent upon both ionic environment and the method used to determine the weight.

Subsequent to the proposal by Murphey et al.(1967a, b), multiple molecular weight forms of MDH have been documented in several organisms, both procaryotic and eucaryotic: *B. subtilis*, porcine heart, bean leaves, corn roots, and *Lemma minor* (Table V) O'Sullivan and Wedding (1972) have recently reported the presence of multiple molecular weight forms of malate dehydrogenase in cotton leaves. They also observe an interconversion of the various M-isozymes. The interconversion readily occurs in tris-glycine buffer but not tris-HCl buffer which suggests that the aggregation may be dependent upon the presence of chloride ion. However, in their studies no attempt is made to test the effect of ionic concentration or specific anions upon the interconversion.

In *Lemma minor*, both the number and the relative proportions of M-isozymes are a function of the ionic environment (Jeffries et al., 1969). Under appropriate conditions, a preparation yields a series of isozymes whose weights are multiples of 17,000. Low calcium ion concentrations favor the occurrence of heavier isozymes while high calcium ion concentrations favor that of the lighter ones. In addition, the lighter ones require higher calcium ion concentrations for maximum activation than do the heavier. These weight changes occur in vivo also and allow *Lemma* to adapt physiologically to various calcium concentrations in the environment.

THE ROLE OF ISOZYMES IN THE ASSEMBLY OF MITOCHONDRIA

A mitochondrial matrix enzyme is defined as one which is readily dissolved from mitochondria by osmotic lysis (Schnaitman et al., 1967), detergent lysis (ibid.), or sonic lysis of mitochondria (Sottocosa, 1967). However, these methods do not dissolve matrix enzymes from the insoluble mitochondrial residue (Cassady and Wagner, 1969; 1971; McKay et al., 1969). Such matrix enzymes may be associated with the mitochondrial membrane or some other large insoluble structure. This hypothesis is supported by the following observations: (1) electron microscopy reveals that the mitochondrial matrix is an organized network or skelton physically associated with the inner membrane (Hackenbrock, 1968; Pihl and Bahr, 1971; Wakabayashi, 1971); (2) mitochondrial malate dehydrogenase from *Neurospora* specifically associates with a protein fraction presumably derived from either the insoluble membrane of the matrix skeleton, called fraction P_0 (Munkres and Woodward, 1966, 1967; Benveniste and Munkres, 1970; Swant et al., 1971); (3) the association of this dehydrogenase and fraction P_0 appears to be physiologically significant (Munkres and Woodward 1966, 1967); (4) association of this dehydrogenase with the membrane may be a function of ionic concentration in, and the energetic state of, mitochondria (Rendon and Waksman, 1971); and (5) several of the tricarboxylic acid cycle enzymes of bovine mitochondria can be isolated together as a complex (D.E. Green, personal communication). Hence, the multiple molecular forms of mitochondrial malate dehydrogenase of *N. crassa* and other organisms could reflect an assembly process which leads to a highly organized matrix. The high concentration of malate dehydrogenase in mitochondria, about 2% of the weight of total mitochondrial protein or 20 μM where a molecular weight of 54,000 is assumed (Munkres, 1965a), may reflect such an assembly process.

Like *N. crassa* mitochondrial malate dehydrogenase, bovine glutamate dehydrogenase occurs in high concentrations in the mitochondrial matrix (Sund, 1964). This enzyme also forms aggregates (Eisenberg, 1970; Krause et al., 1970). In contrast, the TPN-glutamate dehydrogenase, an extramitochondrial enzyme in *N. crassa* (Flavel, personal communication), does not aggregate to form high molecular weight aggregates (Fincham, 1966). These circumstantial observations may indicate that the aggregation of an enzyme is important both to the underlying mechanism of its incorporation into mitochondria and to the assembly of a mitochondrial matrix complex.

In conclusion, we submit a model in which the in vivo assembly of the mitochondrial matrix depends upon the isozyme aggregation phenomenon. Mitochondrial malate dehydrogenase is synthesized by cytoribosomes (Howell et al., 1971; Benveniste and Munkres, 1970) and is incorporated subsequently into mitochondria. Perhaps only the monomer or the smaller isozymes are transported across the membrane into the matrix. Transport would be favored if the intramitochondrial environment shifted the equilibrium among the isozymes towards the larger aggregates. This would simultaneously lower the concentration of the smaller isozymes and trap the larger isozymes within the mitochondia. The progressive increase in isoelectric point which accompanies aggregation would augment aggregation if in the matrix the larger, basic isozymes associated with either an acidic macromolecule or an electronegative region of the inner membrane.

ACKNOWLEDGEMENTS
This research was supported by grants from the National Science of Health (GM-15751) and an NIH predoctoral fellowship (GM-36592) to K. B.

REFERENCES

Benveniste, K. B. P. 1971. Isoenzymes of malate dehydrogenase in *Neurospora crassa*. Ph.D. dissertation. University of Wisconsin, Madison.

Benveniste, K. and K. D. Munkres 1970. Cytoplasmic and mitochondrial malate dehydrogenases of *Neurospora*. Regulatory and enzymatic properties. *Biochim. Biophys. Acta,* 220: 161-177.

Benveniste, K. B. P. and K. D. Munkres 1973. Effects of ionic concentration and pH upon the state of aggregation of *Neurospora* mitochondrial malate dehydrogenase. *Biochem. Biophys. Res. Commun.* 50: 711-717.

Cassady, W. E. and R. P. Wagner 1969. Localization of isoleucine-valine biosynthetic enzymes in *Neurospora* mitochondria. *J. Cell. Biol.* 43: 18a-19a.

Cassady, W. E. and R. P. Wagner 1971. Separation of mitochondrial membranes of *Neurospora crassa*. I. Localization of L-kynurenine-3-hydroxylase. *J. Cell. Biol.* 49: 536-541.

Drysdale, J. W., P. Righetti and H. R. Bunn 1971. The separation of human and animal hemoglobins by isoelectric focusing in polyacrylamide gel. *Biochim. Biophys. Acta,* 229: 42-50.

Eisenberg, H. 1970. Structure and association of glutamate dehydrogenase solutions. *In Pyridine Nucleotide-Dependent Dehydrogenase,* Sund, H. (ed.). Springer-Verlag, New York, 293.

Fincham, J. R. S. 1966. *Genetic Complementation.* New York, W. A. Benjamin, Inc.

Frater, R. 1970. Artifacts in isoelectric focusing. *J. Chromatog.* 50: 469-474.

Hackenbrock, C. R. 1968. Chemical and physical fixation of isolated mitochondria in low-energy-and high-energy states. *Proc. Natl. Acad. Sci. U.S.A.* 61: 598-605.

Howell, N., C. A. Zuiches and K. D. Munkres 1971. Mitochondrial biogenesis in *Neurospora crassa*. I. An ultrastructural and biochemical investigation of the effects of anaerobiosis and chloramphenicol inhibition. *J. Cell Biol.* 5: 721-736.

Jefferies,R. L., D. Laycock, G. R. Stewart and A. P. Sims 1969. The properties of mechanisms involved in the uptake and utilization of calcium and potassium by plants in a relation to an understanding of plant distribution. *In Ecological Aspects of the Mineral Nutrition of Plants.* Symp. British Ecol. Soc., Rorison, I. H. (ed.). Oxford, Blackwell Scientific Pub., 281.

Kaplan, L. J. and J. F. Foster 1971. Isoelectric focusing behavior of bovine plasma albumin mercaptoalbumin, and β-lactoglobulins A and B. *Biochemistry* 10: 630-636.

Kraus, J., K. Markau, M. Minssen and H. Sund 1970. Quaternary structure and enzymatic properties of beef liver glutamate dehydrogenase. *In Puridine Nucleotide-Dependent Dehydrogenases*, H. Sund (ed.). Springer-Verlag, New York, 279.

McKay, R., R. Durgan, G. S. Getz and M. Rabinowitz 1969. Intramitochondrial localization of β-aminolaevulate synthetase and ferrochelatase in rat liver. *Biochem. J.* 114: 455-461.

Munkres, K. D. 1965a. Structure of *Neurospora* malate dehydrogenase. I. Reconstitution from acid and urea. *Biochemistry* 4: 2180-2185.

Munkres, K. D. 1965b. Structure of *Neurospora* malate dehydrogenase. II. Isolation and partial characterization of polypeptide subunits. *Biochemistry* 4: 2186-2196.

Munkres, K. D. 1968. Genetic and epigenetic forms of malate dehydrogenase in *Neurospora*. *Ann. N. Y. Acad. Sci.* 151: 294-306.

Munkres, K. D. 1970. Allosteric and multifunctional properties of *Neurospora* mitochondrial malate dehydrogenase. *Biochim. Biophys. Acta.* 220: 149-160.

Munkres, K. D., K. Benveniste, J. Gorski and C. A. Zuiches, 1970. Genetically induced subcellular mislocation of *Neurospora* mitochondrial malate dehydrogenase. *Proc. Natl. Acad. Sci. U.S.* 67: 263-270.

Munkres, K. D. and D. O. Woodward, 1967. Interaction of *Neurospora* mitochondrial structural protein with other proteins and coenzyme nucleotides. *Biochim. Biophys. Acta.* 143-150.

Munkres, K. D. and D. O. Woodward 1966. On the genetics of enzyme locational specificity. *Proc. Natl. Acad. Sci. U.S.A.* 55: 1217-1224.

Munkres, K. D. and F. M. Richards 1965a. Genetic alteration of *Neurospora* malate dehydrogenase. *Arch. Biochem. Biophys.* 109: 457-465.

Munkres, K. D. and F. M. Richards 1965b. The purification and properties of *Neurospora* malate dehydrogenase. *Arch. Biochem. Biophys.* 109: 466-479.

Murphey, W. H., C. Barnaby, F. J. Lin and N. O. Kaplan 1967a. Malate dehydrogenases. II. Purification and properties of *Bacillus subtilis, Bacillus stearothermophilus*, and *Escherichia coli* malate dehydrogenases. *J. Biol. Chem.* 242: 1548-1559.

Murphey, W. J., G. B. Kitto, J. Everse and N. O. Kaplan 1967b. Malate dehydrogenase. I. A survey of molecular size measured by gel filtration. *Biochemistry* 6: 603-609.

Pihl. E. and G. F. Bahr, 1971. Matrix structure of critical-point dried mitochondria. *Exptl. Cell. Res.* 63: 391-403.

Rendon, A. and A. Waksman, 1971. Intramitochondrial release and binding of mitochondrial aspartate aminotransferase and malate dehydrogenase in the presence and absence of monovalent and bivalent cations. *Biochem. Biophys. Res. Commun.* 42: 1214-1220.

Schnaitman, C. A., V. G. Erwin and J. W. Greenwalt 1967. The submitochondrial location of monoamine oxidase. An enzymatic marker for the outer membrane of rat liver mitochondria. *J. Cell. Biol.* 32: 719-735

Sottocase, G. L., B. Kuylenstierna, L. Ernster, J. Bergstrand, 1967. An electron-transport system associated with the outer membrane of liver mitochondria. *J. Cell Biol.* 32: 415-438.

Sund, H. 1964. *In Mechanismen Enzymatischer Reaktionen.* (14. Colloquim der Gesellschaft für Physiologische Chemie, 1963. Springer-Verlag, Berlin, 318.

Swank, R. T. and K. D. Munkres 1971. Molecular weight analysis of oligopeptides by electrophoresis in polyacrylamide gel, with sodium dodecyl sulfate. *Anal. Biochem.* 39: 462.

Swank, R. T., G. I. Shier and K. D. Munkres 1971. In vivo synthesis, molecular weights, and proportions of mitochandrial proteins in *Neurospora crassa. Biochemistry* 10: 3931-3939.

Wakabayashi. T. 1971. Structured nonmembranous systems in mitochondrial spaces. *Fed. Proc.* 30: 1226 abs.

Whitehead, E. P. 1965. A theory of the quaternary structure of dehydrogenases, dehydrogenating complexes, and other proteins. *J. Theor. Biol.* 8: 276-306.

STRUCTURAL RELATIONSHIP OF HUMAN ERYTHROCYTE
CARBONIC ANHYDRASE ISOZYMES B AND C

B. NOTSTRAND, I. VAARA and K. K. KANNAN
Department of Molecular Biology
The Wallenberg Laboratory
Uppsala University
Uppsala, Sweden

ABSTRACT. The three-dimensional structures of human carbonic anhydrase isozymes B and C have been determined to 2 Å resolution. The structural relationship of these isozymes has been demonstrated and postulated to extend to carbonic anhydrase from other species. The human carbonic anhydrases show closely similar secondary and tertiary structures and side chain interactions. Inhibitors bind in a similar fashion to the active sites of the two isozymes. However, there are some differences in their interaction with side chains, principally due to the non-homologous residues located in the active site cavities. Those differences probably account for the differences in the catalytic and inhibitor binding properties.

INTRODUCTION

Two carbonic anhydrase (E. C. 4.1.1.2, $CO_2 + H_2O \rightleftharpoons HCO_3^- + H^+$) isozymes with different catalytic activity were isolated from human erythrocytes by Nyman (1961). The low-activity enzyme was designated the B form (HCAB) and the high activity enzyme the C form (HCAC) (Rickli et al., 1964). Carbonic anhydrases occur in mammals, bacteria, and plants. These enzymes have been investigated by different physicochemical techniques to understand, among other things, their catalytic mechanism. Among others Maren (1967) and Lindskog et al., (1971) have reviewed the available literature on the carbonic anhydrases.

Carbonic anhydrases have a molecular weight of about 30,000 and contain one zinc atom per molecule. Carbonic anhydrases catalyze the hydration of CO_2 and the dehydration of carbonic acid or bicarbonate ion (Lindskog et al., 1971), the reversible hydration of aldehydes (Pocker and Meany, 1967) and the hydrolysis of various esters like p- and o-nitrophenyl acetate (Tashian et al., 1964; Pocker and Stone, 1967 and Verpoorte et al., 1967). Carbonic anhydrases are inhibited by monovalent anions and by aromatic or heterocyclic sulfonamides.

Human carbonic anhydrase B and C, however, differ in their specific activity, amino acid composition, and immunological specificity. The two enzymes show a considerable difference

in their esterase activity. HCAC catalyses the hydrolysis of
p-nitrophenyl acetate considerably faster than HCAB. However,
HCAB is a slightly better catalyst of o-nitrophenyl acetate.
The isozymes also show differences in certain other properties
like stability towards different denaturants (Edsall, 1968).
The amino acid sequence of the human carbonic anhydrases have
been reported from different laboratories (Andersson et al.,
1972; Henderson et al., 1973; Marriq et al., 1973). The considerable sequence homology of HCAB and HCAC indicate that
the two forms have developed from a common ancestor (Nyman
et al., 1968).

Proteins having similar catalytic functions, like the
serine proteases (e.g. *The Enzymes*, vol. III, 1971), have
been shown to have very similar tertiary structure. It would
thus be natural to expect that the carbonic anhydrases would
have similar three-dimensional structures. The molecular
structures of the human carbonic anhydrases B and C have been
determined in our laboratory. Preliminary results of the
high resolution structure investigations of HCAC have been
reported (Kannan et al., 1971; Liljas et al., 1972) and those
of HCAB by Kannan et al., (1973). This article presents a
comparative study of the three-dimensional structures of the
two human isozymes.

Experimental methods: Carbonic anhydrases B and C were purified from human erythrocytes by mild separation methods
(Armstrong et al., 1966). The purified isozymes were crystallized from 2.3 M ammonium sulfate in 0.05 M Tris-HCl, pH 8.5
(Strandberg et al., 1962; Kannan et al., 1972).

The preparation of the heavy atom derivatives for the two
enzymes have been reported elsewhere (Tilander et al., 1965;
Liljas et al., 1972; and Kannan et al., 1972 and 1973).

The unit cell dimensions and space groups are given below
for the two isozymes (Kannan et al., 1972; Liljas et al.,1972)
HCAB: a= 81.5, b = 73.6, c = 37.1 $\overset{o}{A}$

space group P 2_1 2_1 2_1 and one molecule per
asymmetric unit.
HCAC: a = 42.7, b = 41.7, c = 73.0 $\overset{o}{A}$, β = 104.6^{o}

space group P 2_1 and one molecule per asymmetric
unit.

The X-ray diffraction data were collected photographically
on precession cameras using filtered Cu K_α radiation and the
intensities were measured by automatic microdensitometers designed and constructed by Mr. V. Klimecki, Chemistry Department,
Uppsala University, in collaboration with our laboratory.

The "Best Fourier" (Blow and Crick, 1959) were calculated
for the respective isozymes to about 2 $\overset{o}{A}$ resolution and inter-

preted in an optical comparator (Richards, 1968) using Kendrew-Watson skeletal model parts.

HCAC interpretation: The electron density map of HCAC was interpreted (Kannan et al., 1971; Liljas et al., 1972) using the known sequence pieces of the two isozymes on the assumption that the single cysteinyl residues were homologously located   (Henderson et al., 1971).   The chemical sequences of the two isozymes are now completed (Andersson et al., 1972; Henderson et al., 1973).   A comparison of the two sequences, aligned to achieve maximum homology is given in Table 1, where they are numbered according to the B enzyme sequence.   The cysteinyl residues in the two enzymes do not occur in the same position as had been assumed, but 6 residues apart.   This necessitates a revision of the structure in a few places which in fact give a much better fit of the chemical sequence to the electron density.   Thus the loop above the active site assumed to consist of residues 203-209, which was quite invisible in the electron density map (Liljas et al., 1972), no longer exists.   One of the two consecutive prolyl residues, Pro 202, could be built only in cis conformation.

The electron density map in the region immediately before the well defined residue Met 241 is discontinous and had been interpreted as solvent molecules so as to account for a total number of 260 amino acids.   (Henderson et al., 1971).   However, with the help of the completed chemical sequence of HCAC (Henderson et al., 1973) residues 230-239 could be fitted in this region.   Many of these are hydrophilic residues.   The fit of the model to the electron density map was poor for the residues 29-31 when Pro 30 was built in trans conformation and the side chain of Ser 29 had no electron density.   However, if Pro 30 was built in the cis conformation, not only was a very good fit obtained for that residue but also the side chain of Ser 29 could be fitted to density.

Apart from these minor changes, the rest of the tertiary structure of HCAC is almost identical to that reported earlier (Kannan et al., 1971; Liljas et al., 1972).   The side chains have been built-in according to the sequence information (Henderson et al., 1973) and compared with the electron density maps.   A more complete description of this work will be published elsewhere.

HCAB interpretation: The electron density map of the B form was interpreted with the complete chemical sequence of HCAB. The fit between the electron density and the sequences was very good and in general encountered no special problems.   The best fit was obtained using the sequence reported by Andersson

TABLE I

THE PRIMARY STRUCTURES OF HCAB AND HCAC

The sequences are numbered to match exactly homologous regions.

1	5	10	15	20	25	30

HCAB: Ac-Ala-Ser-Pro-Asp-Trp-Gly-Tyr-Asp-Asp-Lys-Asn-Gly-Pro-Glu-Gln-Trp-Ser-Lys-Leu-Tyr-Pro-Ile-Ala-Asn-Gly-Asn-Asn-Gln-Ser-Pro-
HCAC: Ac-Ser-His-His-Trp-Gly-Tyr-Gly-Lys-His-Asp-Gly-Pro-Glu-His-Trp-His-Lys-Asp-Phe-Pro-Ile-Ala-Lys-Gly-Glu-Arg-Gln-Ser-Pro-

35	40	45	50	55	60

Val-Asp-Ile-Lys-Thr-Ser-Glu-Thr-Lys-His-Asp-Thr-Ser-Leu-Lys-Pro-Ile-Ser-Val-Ser-Tyr-Asn-Pro-Ala-Thr-Ala-Lys-Glu-Ile-Ile-
Val-Asp-Ile-Asp-Thr-His-Thr-Ala-Lys-Tyr-Asp-Pro-Ser-Leu-Lys-Pro-Leu-Ser-Val-Ser-Tyr-Asp-Gln-Ala-Thr-Ser-Leu-Arg-Ile-Leu-

65	70	75	80	85	90

Asn-Val-Gly-His-Ser-Phe-His-Val-Asn-Phe-Glu-Asp-Asn-Asp-Ser-Arg-Gly-Val-Leu-Lys-Gly-Gly-Pro-Phe-Ser-Asp-Ser-Tyr-Arg-Leu-
Asx-Asx-Gly-His-Ala-Phe-Asn-Val-Glu-Phe-Asp-Asp-Ser-Glx-Asx-Lys-Ala-Val-Leu-Lys-Gly-Gly-Pro-Leu-Asp-Gly-Thr-Tyr-Arg-Leu-

95	100	105	110	115	120

Phe-Gln-Phe-His-Phe-His-Trp-Gly-Ser-Thr-Asn-Glu-His-Gly-Ser-Glu-His-Thr-Val-Asp-Gly-Val-Lys-Tyr-Ser-Ala-Glu-Leu-His-Val-
Ile-Gln-Phe-His-Phe-His-Trp-Gly-Ser-Leu-Asp-Gly-Gln-Gly-Ser-Glu-His-Thr-Val-Asp-Lys-Lys-Lys-Tyr-Ala-Ala-Glu-Leu-His-Leu-

125	130	135	140	145	150

Ala-His-Trp-Asn-Ser-Ala-Lys-Tyr-Ser-Ser-Leu-Ala-Ala-Ser-Leu-Ala-Asp-Gly-Leu-Ala-Val-Ile-Gly-Val-Leu-Met-Lys-Val-
Val-His-Trp-Asn-Thr------Lys-Tyr-Gly-Asp-Phe-Gly-Lys-Ala-Val-Gln-Gln-Pro-Asp-Gly-Leu-Ala-Val-Leu-Gly-Ile-Phe-Leu-Lys-Val-

155	160	165	170	175	180

Gly-Glu-Ala-Asn-Pro-Lys-Leu-Gln-Lys-Val-Leu-Asp-Ala-Leu-Gln-Ala-Ile-Lys-Thr-Lys-Gly-Lys-Arg-Ala-Pro-Phe-Thr-Asn-Phe-Asp-
Gly-Ser-Ala-Lys-Pro-Gly-Leu-Gln-Lys-Val-Val-Asp-Val-Leu-Asp-Ser-Ile-Lys-Thr-Lys-Gly-Lys-Ser-Ala-Asp-Phe-Thr-Asn-Phe-Asp-

185	190	195	200	205	210

Pro-Ser-Thr-Leu-Leu-Pro-Ser-Ser-Leu-Asp-Phe-Trp-Thr-Tyr-Pro-Gly-Ser-Leu-Thr-His-Pro-Pro-Leu-Tyr-Glu-Ser-Val-Thr-Trp-Ile-
Pro-Arg-Gly-Leu-Leu-Pro-Glu-Ser-Leu-Asp-Tyr-Trp-Thr-Tyr-Pro-Gly-Ser-Leu-Thr-Thr-Pro-Pro-Leu-Leu-Glu-Cys-Val-Thr-Trp-Ile-

215	220	225	230	235	240

Ile-Cys-Lys-Glu-Ser-Ile-Ser-Val-Ser-Ser-Glu-Gln-Leu-Ala-Gln-Phe-Arg-Ser-Leu-Leu-Ser-Asn-Val-Glu-Gly-Asp-Asn-Ala-Val-Pro-
Val-Leu-Lys-Glu-Pro-Ile-Ser-Val-Ser-Ser-Glu-Gln-Val-Leu-Lys-Phe-Arg-Lys-Leu-Asn-Phe-Asn-Gly-Glu-Gly-Glu-Pro-Glu-Glu-Leu-

245	250	255	260

Met-Gln-His-Asn-Asn-Arg-Pro-Thr-Gln-Pro-Leu-Lys-Gly-Arg-Thr-Val-Arg-Ala-Ser-Phe
Met-Val-Asp-Asn-Trp-Arg-Pro-Ala-Gln-Pro-Leu-Lys-Asn-Arg-Gln-Ile-Lys-Ala-Ser-Phe-Lys

et al (1972). The homologous prolyl residues 30 and 202 discussed above for HCAC had cis conformation also in HCAB. The
loop region, 230-239 was very well defined in the HCAB electron density map. Residues 70 to 74 have discontinuous density. The availability of the complete chemical sequence was
very useful in the model building, especially in some surface
regions of the molecule. A fuller account of the B enzyme
structure will be published elsewhere.

Tertiary structure of HCAB and HCAC: It had been expected on
the basis of the highly homologous primary structures and
similar physicochemical properties that the carbonic anhydrase
isozymes would also have a similar tertiary structure (Kannan
et al., 1971; Lindskog et al., 1971). Indeed a visual comparison of the completed crystal structures showed great similarity in the tertiary structures. To facilitate a proper
comparison, the coordinates of all the non-hydrogen atoms of
the two enzymes were measured and compared by the method of
Rao and Rossmann (1973). In this method one of the structures
is rotated and translated into the other so as to minimize the
sum of the squared distances between the atoms of the two
structures that are assumed to be equivalent. The root mean
square deviation of these distances gives a measure of the
similarity between the structures.

The Cα coordinates of HCAB and HCAC were used in the computer program, kindly provided by Drs. Rao and Rossmann, to
determine the orientation of the B molecule with respect to
the C molecule. Residues 1-4 and 125 of the B enzyme and
residues 2-4 and 261 of the C enzyme were left out of this
calculation. Distances between corresponding α-carbon atoms
were calculated and are listed in Table 2. The root mean
squared deviation in the distances for 255 equivalent Cα atoms
used in these calculations was 1.7 Å. The largest deviation
in these distances was about 5 Å.

The stereo-drawings of the α-carbon atoms of HCAB and HCAC
are shown in Fig. 1 a and 1 b. A simplified drawing of the
tertiary structure of carbonic anhydrase is shown in Fig. 2.

Description of the HCAB and HCAC molecule: The ellipsoidal
molecule is about 41 x 42 x 55 Å. The carbonic anhydrase
isozymes contain about 20% helix as compared to the low percentage (about 1% - 10%) predicted from ORD and CD studies
(Timasheff, 1970). Most of the helices are less than two
turns and have irregular hydrogen bond formation. There are
a number of reverse bends that occur on the surface of the
molecule. However, the most predominant feature of the carbonic anhydrase molecule is a large twisted β-structure formed

TABLE 2

THE DISTANCES (A) BETWEEN CORRESPONDING α-CARBONS IN HCAB AND HCAC

Cα	Distances	Cα	Distances	Cα	Distances	Cα	Distances	Cα	Distances
		55	1.987	106	0.822	157	1.818	208	0.875
5	1.383	56	1.525	107	0.348	158	1.700	209	0.716
6	1.151	57	1.239	108	1.036	159	1.332	210	1.308
7	1.736	58	0.539	109	0.759	160	1.081	211	1.210
8	2.939	59	0.982	110	1.059	161	1.673	212	0.746
9	3.922	60	2.626	111	0.833	162	1.318	213	1.446
10	1.444	61	0.422	112	0.613	163	0.644	214	1.229
11	0.934	62	0.993	113	1.562	164	0.783	215	1.184
12	0.714	63	1.010	114	0.902	165	1.651	216	1.249
13	1.311	64	1.360	115	1.215	166	2.052	217	2.030
14	2.692	65	1.093	116	0.727	167	1.397	218	2.807
15	2.859	66	0.179	117	0.193	168	1.432	219	2.797
16	1.601	67	0.326	118	0.676	169	1.337	220	2.783
17	2.298	68	0.691	119	0.859	170	1.405	221	1.753
18	2.729	69	1.115	120	0.518	171	1.907	222	1.199
19	2.027	70	1.110	121	0.234	172	2.455	223	1.308
20	1.357	71	2.970	122	1.444	173	1.993	224	1.128
21	0.704	72	2.035	123	0.917	174	1.541	225	1.179
22	1.196	73	2.074	124	1.191	175	0.940	226	1.119
23	0.887	74	1.277	125	1.261	176	0.870	227	1.377
24	0.867	75	0.988	126	--	177	1.913	228	1.645
25	0.863	76	1.680	127	0.815	178	2.248	229	1.877
26	0.979	77	1.119	128	1.716	179	2.953	230	2.277
27	0.526	78	1.135	129	1.079	180	2.718	231	1.993
28	1.017	79	1.097	130	0.690	181	2.680	232	1.615
29	1.323	80	1.386	131	2.773	182	2.517	233	3.762
30	0.588	81	1.270	132	1.644	183	1.473	234	5.312
31	0.234	82	1.686	133	1.181	184	2.014	235	3.194
32	0.870	83	1.704	134	2.107	185	1.247	236	0.651
33	1.301	84	1.482	135	2.683	186	1.973	237	3.819
34	1.279	85	2.598	136	1.177	187	2.620	238	1.524
35	1.107	86	1.310	137	0.952	188	2.145	239	1.949
36	1.818	87	1.728	138	1.189	189	1.173	240	1.972
37	1.127	88	1.045	139	0.127	190	0.617	241	1.277
38	1.408	89	0.794	140	0.839	191	0.640	242	1.462
39	0.805	90	0.733	141	0.692	192	0.972	243	1.332
40	2.039	91	0.657	142	1.578	193	0.759	244	0.617
41	1.926	92	1.768	143	0.906	194	1.081	245	1.047
42	2.600	93	1.312	144	1.668	195	0.278	246	1.080
43	3.563	94	1.143	145	1.578	196	0.935	247	0.887
44	1.872	95	1.150	146	1.620	197	0.954	248	0.856
45	2.409	96	1.114	147	1.395	198	1.738	249	0.646
46	0.593	97	1.021	148	2.099	199	1.218	250	2.379
47	0.817	98	1.554	149	2.239	200	0.600	251	1.739
48	1.360	99	2.070	150	2.148	201	0.835	252	1.247
49	0.443	100	1.471	151	2.257	202	2.491	253	1.990
50	1.588	101	3.008	152	2.180	203	1.670	254	1.461
51	1.129	102	2.736	153	1.100	204	0.585	255	2.074
52	2.955	103	1.674	154	1.453	205	0.683	256	1.462
53	5.131	104	1.663	155	1.928	206	1.235	257	1.330
54	4.156	105	1.317	156	0.467	207	0.212	258	1.240

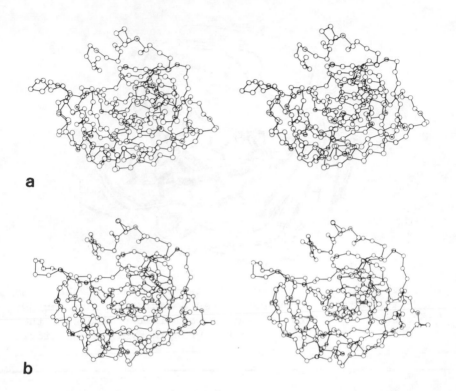

a

b

Fig. 1. Stereoscopic α - carbon diagrams of a) HCAB and
b) HCAC. Dr. Carrol N. Johnson's ORTEP program has been used
to produce the computer drawings in this article.

by 10 chain segments. Only two pairs of strands are parallel
to each other. There are a number of smaller antiparallel
β-structures located elsewhere. In all about 35% of the residues are involved in all β-structures.

The active site cavity is formed by chain segments 3-6 of
the large β-structure together with loops formed by residues
122-135 and 196-204. The essential zinc ion is located at
the bottom of this conical cavity and is bound to the protein
through three histidyl residues (Fig. 2).

Secondary structures in HCAB and HCAC: There are 6 helical
segments in HCAB and 7 in HCAC. Helices B, E, and F are of
the 3_{10} type. Helix E is rather distorted and could be considered as two helices as there is no continuity in the

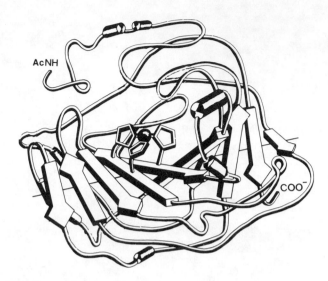

Fig. 2. A schematic drawing of· the main chain of the carbonic anhydrase molecule. Cylinders represent helices and arrows represent β-structure. The dark ball in the middle is the essential zinc ion bonded to the protein by three histidyl ligands.

---

hydrogen bonding and the helix is listed as E_1 and E_2 in Table 3, where all the secondary structures of the two isozymes are listed. Helix D is almost in the regular α - helical conformation. Helix G is the longest helix closest to the α - helical conformation.

Fig. 3 gives the general hydrogen bond pattern of HCAC. The nonhomologous residues of the HCAB enzyme are given in the one letter code.

There are 10 bends occuring at the same points in the three dimensional structures. In HCAB 8 of these form hydrogen bonds and are of the 3_{10} bend type (Venkatachalam, 1971; Crawford et al., 1973; Lewis et al., 1973), whereas there are only 5 of this type in HCAC.

The β-structures are located identically in the two enzymes (Figs. 4 a and 4 b) even though there are local variations in the hydrogen bond pattern especially in the smaller pleated sheets. The large β-structure is almost identical in the two structures. Out of the 70 residues located in this β-structure 28 differ in the two isozymes and all but 4 of these are

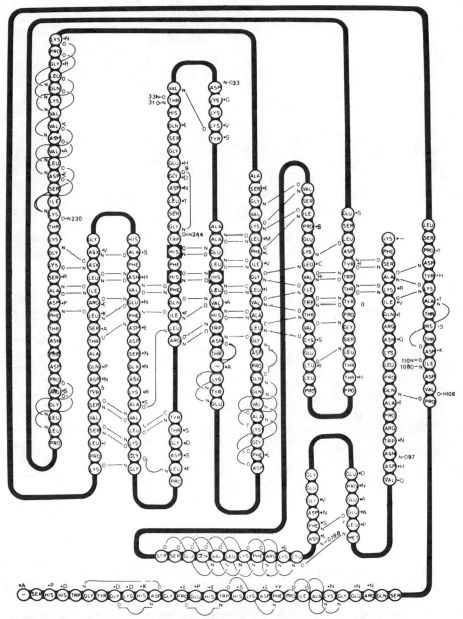

Fig. 3. Hydrogen bond pattern observed in the carbonic anhydrase isozymes. The amino acid sequence of HCAC is given in the 3 letter code and the non homologous residues of HCAB are given in the single letter code. The zinc ligands are marked by a square instead of ring.

TABLE 3 a
SECONDARY STRUCTURES

Helix		Turns		β -	Chain No. in pleated	Residues
HCAC	HCAB	HCAC	HCAB	structure	sheet	involved
		3_{10} I	Reverse			8- 11
		3_{10} I	3_{10} I			13- 16
Aα	Aα					16- 20
B3_{10}	B3_{10}					20- 25
Cα	-		Reverse			34- 38
				X	10	39- 41
		Open	3_{10} I			51- 54
				X	2	55- 62
		Reverse	Reverse			62- 65
				X	3	65- 71
		3_{10} I	3_{10} I			81- 84
				X	4	87- 96
		Reverse	3_{10} II			109-112
				X	5	116-124
		3_{10} I	3_{10} I			124-128
Dα,3_{10}	Dα,3_{10}					130-137
		3_{10} I	3_{10} I			137-140
				X	6	140-151
E$_1$	E$_1$					155-162
E$_2$	E$_2$					162-168
				X	1	171-176
F3_{10}	F3_{10}					181-185
				X	8	191-196
				X	7	206-212
Gα	Gα					219-229
		Open	3_{10} I			233-236
		Open	3_{10} I			251-254
				X	9	256-259

conservative changes. Most of these differences occur in the outermost pleated sheet segments. About 30 residues constitute the pleated sheet segments 3-6 which form part of the active site cavity. There are only 8 amino acid changes in this region and all are conservative replacements. Furthermore, there are no deletions in this region. One might expect that this part of the pleated sheet would be preserved to a great extent even in other carbonic anhydrases.

TABLE 3 b

Other β-structures; segments A_1 and A_2 etc. are hydrogen bonded to each other.

Segments	Amino acid involved in HCAC	in HCAB if different
A 1	30	
A 2	249	
B 1	31- 33	
B 2	108-110	
C 1	45- 53	
C 2	82- 76	
D 1	87- 90	
D 2	79- 77	
E 1	87- 88	
E 2	126-124	
F 1	215-218	
F 2	147-150	
G 1	229-231	
G 2	241-239	

Folding of the molecule: It has been observed that the enzymes whose molecular structures have been established are folded from a number of small globular domains and the active site cavity is formed at the crevice formed between the domains (Wetlaufer, 1973; Liljas and Rossmann, 1974). This is not the case in the carbonic anhydrases. Residues 44-150, which can be considered to form a domain, form the core of the molecule around which the rest of the polypeptide is folded (Fig. 2). This unique folding is achieved by a process of wrapping long chain segments about this central core of the molecule and stabilizing the structure by the different secondary structures, hydrophobic interactions, and the side chain hydrogen bonds. A major part of the active site cavity is formed by parts of the twisted pleated sheet segments 3-6 in this central core.

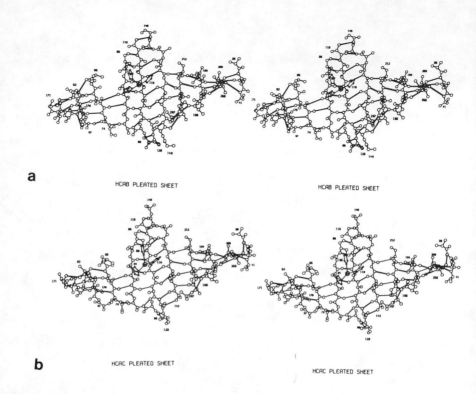

a

HCAB PLEATED SHEET   HCAB PLEATED SHEET

b

HCAC PLEATED SHEET   HCAC PLEATED SHEET

Fig. 4. Stereo diagram of the large pleated sheet structure in a) HCAB and b) HCAC.

Side chain packing in HCAB and HCAC: Hydrophobic side chains are tucked away from the solvent region and most of the hydrophilic side chains are on the surface of the molecule. There is a common pattern of packing of the aromatic side chains in the two isozymes (Figs. 5 a and 5 b). There are 7 phenylalanyl residues in an aromatic cluster which are homologous residues in the two enzymes. This cluster is found between the pleated sheet strands 2 to 6 and the chain segments that form the bottom plane of the molecule (Fig. 2). These residues together with a number of hydrophobic residues located between the twisted pleated sheet and the bottom plane of the molecule provide the main hydrophobic interaction and are very likely important in stabilizing the molecule. The residues found in this hydrophobic core are listed in Table 4. Of the 32 side chains that are packed in this region 24 are invariant in the two enzymes.

A second aromatic cluster is found in the amino-terminus

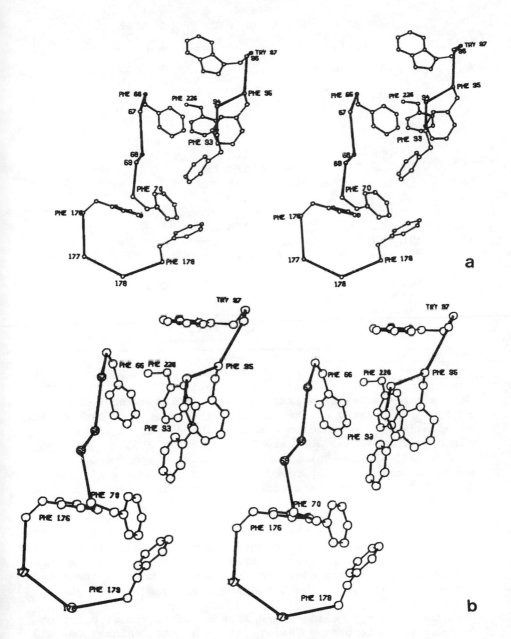

Fig. 5. Stereo pair of the aromatic cluster II in a) HCAB and b) HCAC.

TABLE 4
RESIDUES FORMING THE HYDROPHOBIC CORE

Residue in		Residue in	
HCAC	HCAB if different	HCAC	HCAB if different
59 Leu	Ile	157 Leu	
66 Phe		160 Val	
68 Val		161 Val	Leu
70 Phe		163 Val	Ala
79 Leu		164 Leu	
90 Leu	Phe	176 Phe	
93 Phe		179 Phe	
95 Phe		181 Pro	
97 Trp		184 Leu	
116 Ala		185 Leu	
118 Leu		186 Pro	
120 Leu		210 Ile	
122 His		212 Leu	Cys
144 Leu	Ile	216 Ile	
146 Ile	Val	218 Val	
148 Leu	Met	226 Phe	

region and involves mainly 2 tryptophanyl, 1 tyrosyl, and 1 histidyl residue, which are also invariant in the two enzymes. Here again the major interaction between the amino-terminus region and the rest of the molecule is achieved by hydrophobic side chains. The residues found in the different aromatic clusters in HCAB and HCAC are listed in Table 5.

The 240-310 nm region in the ORD and CD spectra of the mammalian carbonic anhydrases is unusually complex with multiple Cotton effects resulting from aromatic chromophores (Rosenberg, 1966). It has been proposed (Beychock et al., 1966; Rosenberg, 1966; Coleman, 1969) that the complex ORD and CD patterns arise from the presence of tightly packed aromatic residues, presumably with charged groups in their vicinity producing specific asymmetric environments. It is possible that the aromatic clusters observed in the structures of the two isozymes could also contribute towards their complex ORD and CD spectra.

The cysteinyl residues are located at different parts of the isozyme molecules. The cysteine in HCAB replaces Leu 212 in HCAC and is located slightly shielded inside the molecule. It is possible for the cysteine in HCAB to react with small mercurials like $HgCl_2$ but not with PCMBS or similar aromatic

TABLE 5
AROMATIC CLUSTERS

I		II		III	
Residues		Residues		Residues	
HCAC	HCAB if different	HCAC	HCAB if different	HCAC	HCAB if different
His 4	Asp	Phe 66		His 107	
Trp 5		Phe 70		Tyr 114	
Tyr 7		Phe 93		Phe 147	Leu
Trp 16		Phe 95		Trp 192	
Phe 20	Tyr	Trp 97		Tyr 194	
His 64		Phe 176		Trp 209	
		Phe 179		Tyr 191	Phe
		Phe 226		Phe 260	

mercurials presumably due to sterical hindrance. HCAB contains 2 methionyl residues, whereas there is only one in HCAC. Met 241, located homologously in the sequences (Table 1) and in the three-dimensional structures of the two enzymes is inaccessible to the solvent. The second methionine in HCAB is Met 152 and is replaced by Leu 152 in HCAC. This residue is also buried.

Structural stabilization: Comparison of the chemical sequences of the isozymes reveal two deletions and one insertion in HCAC with respect to HCAB. These changes occur on the surface of the molecule and the conformations in the vicinity of these changes are altered to a very small extent in HCAB as compared to HCAC (see Table 2). It may be unfavorable to delete more than two residues from the carboxy-terminus of the B-enzyme as this would disturb the pleated sheet formation at strands 8, 9, and 10 and, thus the stability of the structure, because strand 10 is mainly stabilized in this conformation by the hydrogen bonds formed with strand 9 (Fig. 3).

There are two buried histidines in HCAB and HCAC which are hydrogen bonded to two buried tyrosines. These four residues are all conserved in the two sequences. These are His 122 - Tyr 51 and His 107 - Tyr 194. His 107 is furthermore hydrogen bonded to Glu 117 in both the enzymes in accordance with the titration studies of Riddiford (1964).

Homologous arginyl residues 88, 225, and 244 have similar hydrogen bonding in the structures of HCAB and HCAC. Arg 88 is hydrogen bonded to Asp 75 in HCAB and HCAC. Thus many of the homologous residues have similar environments and inter-

actions in the two isozymes. It may be expected that this
pattern will be followed in other carbonic anhydrases and that
these interactions may have importance in the stabilization
and conservation of the three-dimensional structure.

The active site of HCAB and HCAC: The active site cavities of
human carbonic anhydrases B and C are formed mainly by the
twisted pleated sheet segments 3 - 6 and the loops formed by
residues 128 - 135 and 196 - 202. This dead end cavity is
about 12 Å deep and the essential zinc ion is located at the
bottom. The zinc ion is firmly bonded to the enzyme by homologous residues His 94, 96, and 119. His 94 and 96 ligand to
the zinc through the $N^{\varepsilon 2}$ nitrogen while the $N^{\delta 1}$ nitrogen of
His 119 is the third ligand. The fourth coordination site of
the distorted tetrahedron is occupied by a solvent molecule,
presumably a water molecule or a hydroxyl ion.
The conical active site cavity can principally be divided
into a hydrophobic half and a hydrophilic half (partitioned
through the zinc ion). The residues involved in these two
halves are listed in Table 6 for both the enzymes and shown
in Figs. 6 a and 6 b.

TABLE 6

EXPOSED SIDE CHAINS IN THE ACTIVE SITE OF HCAC AND HCAB

Polar residues		Nonpolar residues	
HCAC	HCAB if different	HCAC	HCAB if different
Tyr 7		Ala 65	Ser
Asn 61		Ile 91	Phe
His 64		Val 121	Ala
Asn 67	His	Phe 131	Leu
Glu 69	Asn	Leu 141	
Gln 92		Val 143	
His 94		Gly 145	
His 96		Pro 201	
His 119	(cis)	Pro 202	
Thr 199		Val 207	
Thr 200	His	Val 211	Ile
		Val 207	
		X) Leu 204	Tyr
		Cys 206	Ser

X) The cysteinyl residue is not exposed in HCAC but
   the corresponding serinyl residue in HCAB is exposed.

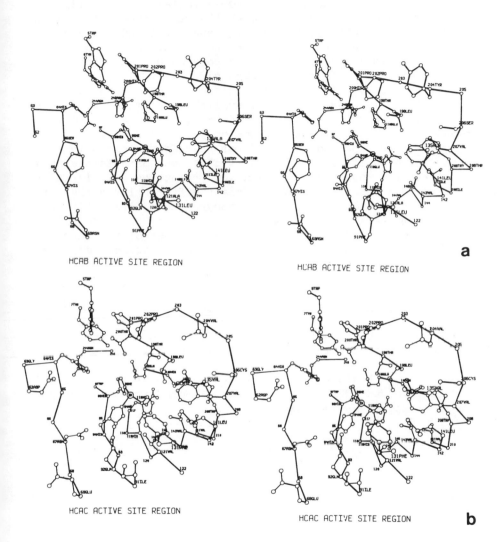

Fig. 6. The residues in the active site region of a) HCAB and b) HCAC.

The hydrophobic part of the cavity in HCAB contains a polar residue, Ser 206, which replaces Cys 206 in HCAC. Tyr 204 in HCAB which replaces Leu 204 is oriented away from the active site cavity and this makes this region somewhat more open in HCAB compared to HCAC. Ile 91 in HCAC is replaced by

Phe 91 in HCAB and Phe 131 in HCAC is replaced by Leu 131 in HCAB. These changes together with the replacement of valines 121 and 135 in HCAC by alanines 121 and 135 in HCAB change the content and character of the hydrophobic half of the active site in the isozymes. There are also a number of changes in the hydrophilic half of the active site cavity. Residues Asn 67 and Thr 200 in HCAC are replaced by His 67 and His 200 respectively in HCAB. Thus the available volume of the active site cavity in the neighborhood of the zinc ion would be reduced to certain extent in HCAB compared to HCAC.

It may be expected that some of these differences in the active site cavities of the two isozymes would influence their catalytic rate and inhibitor binding properties. It is known that certain modifying reagents react differently to the active sites of HCAB and HCAC. Thus bromopyruvate modifies His 64 in HCAC irreversibly whereas it modifies His 200 in HCAB (Göthe et al., 1972). In both instances most of the enzyme activity is destroyed. His 64 in HCAC has also been modified with loss of activity by bromoacetazolamide (Kandel et al., 1970) but this residue has not so far been modified in the B enzyme even though it is similarly situated. However, His 67 and 200 shield this residue to some extent from the active site region in the three-dimensional structure of HCAB. His 67 in HCAB has also been modified irreversibly by a chlorothiazide derivative with complete loss of activity (Whitney et al., 1967).

His 94 is hydrogen bonded to Gln 92 in both isozymes. His 119 is involved in a hydrogen bond network with a number of buried residues as indicated below and

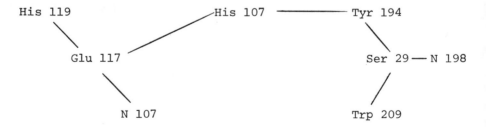

illustrated in Figs. 7 a and 7 b for the two isozymes. This hydrogen bond network occurs in an identical fashion in HCAB and HCAC. It is possible that this may have a functional significance in the form of a charge distribution system together with the zinc ion. However, it may only be stabilizing the structure in that region. All these residues depicted above are invariant in primate and equine enzymes (Tashian, 1974).

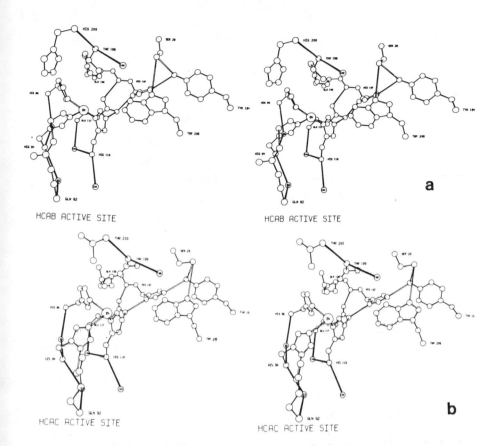

Fig. 7. Stereoscopic drawing of the hydrogen bond network observed in a) HCAB and b) HCAC.

Fig. 8 gives a schematic representation of some of the active site residues and the possible hydrogen bonding between them.
The solvent molecule at the fourth ligand position of the zinc ion is hydrogen bonded to Thr 199, which is hydrogen bonded to Glu 106 in HCAB and Gln 106 in HCAC. It is possible that Thr 199, the only residue in hydrogen bond contact to the zinc bound hydroxyl group may have a functional importance together with the activity linked zinc-hydroxyl (Lindskog et al. 1971). Thr 199 may facilitate the orientation of the hydroxyl ion and function as an intermediate proton acceptor for this ion. It is also conceivable that this zinc-hydroxyl-Thr 199 system may function as a proton pump in and out of the solvent. It is interesting in this regard that Thr 199

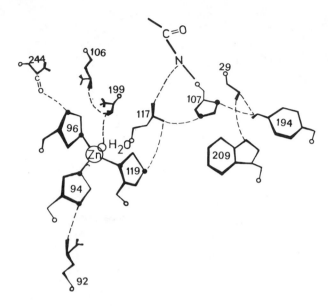

Fig. 8. A schematic drawing of the hydrogen bonding observed in the active site of carbonic anhydrase isozymes.

---

is invariant in primate and equine carbonic anhydrase isozymes (Tashian, 1974).

Inhibitor binding to HCAB and HCAC: Some of the heavy atom derivatives used in the crystal structure investigation of the carbonic anhydrases are inhibitors and bind in the active site cavity near the essential zinc ion. Thus the high resolution structure of the enzymes complexed with acetazolamide and gold cyanide, respectively, have been investigated for the two isozymes.

The sulfonamide inhibitors bind at the active site with the sulfonamide moiety replacing the fourth zinc ligand in the native enzyme. The sulfonamide moiety is also hydrogen bonded with Thr 199 in both isozymes. The heterocyclic group of the inhibitor is situated in the hydrophobic part of the active site cavity. There are small differences in the interaction between this part of the sulfonamide and the residues in the active site of the two enzymes because of the differences in these residues.

The anionic inhibitor gold cyanide binds at the active site cavity in an orientation very similar to that of the sulfonamide inhibitors. However, this inhibitor does not replace the solvent molecule at the fourth ligand site of zinc in either of the isozymes.

Structure of other carbonic anhydrases: The physicochemical properties of mammalian and bacterial carbonic anhydrases reveal a number of similarities (Lindskog et al., 1971). The sequence of bovine, sheep, and other mammalian carbonic anhydrases reveal a large amount of homology (Tashian et al., 1971 and 1974). The spectroscopic properties especially the ORD and CD spectra and denaturation studies of mammalian carbonic anhydrases (e.g. see Lindskog et al., 1971; Edsall, 1968) indicate great similarity in these enzymes.

It can be inferred from the volume of evidence available now that the tertiary structure of carbonic anhydrases from other species will be very similar to that of the human carbonic anhydrases. It may be further expected that the zinc ligands and the phenylalanyl residues forming the aromatic cluster will be preserved in other carbonic anhydrases. It therefore seems likely that the structures of other carbonic anhydrases can be studied by fitting the chemical sequence to the tertiary structure of the human enzyme. Thus it is of special importance to have sequence information from other species and also from mutant carbonic anhydrases with reduced catalytic rates so that a better understanding of the catalytic mechanism of this "fast" enzyme can be achieved.

ACKNOWLEDGEMENTS

We would like to thank our coworkers S. Lövgren, K. Fridborg, A. Ohlsson and M. Petef for the continuous and valuable assistance. We would also like to thank Prof. B. Strandberg for fruitful discussions and support and encouragement throughout the work. We thank Drs. P. Lentz, A. Liljas, P. Nyman and S. Lindskog for stimulating discussions. We would like to thank Professors C. Kurland and I. Olovsson for interest shown in this work. We are very grateful to Professors J. T. Edsall and M. F. Perutz for valuable discussions. We are indebted to G. Johansson and G. Lindman for typing the manuscript and to H. Ukkonen for the photographical work and to the staff of the Uppsala Data Centre for helpful cooperation.

This work was supported by grants from the Faculty of Science, University of Uppsala; the National Institutes of Health, U. S. Public Health Service (Grant No. AI 07382); the Swedish Computing Authorities; the Swedish Medical Science Research Council (Grant No. 13X-26); the Swedish Natural

Science Research Council (Grant No. 2142); the Tricentennial Fund of the Bank of Sweden and the Knut and Alice Wallenberg Foundation.

REFERENCES

Andersson, B., P. O. Nyman, and L. Strid 1972. Amino acid sequence of human erythrocyte carbonic anhydrase B. *Biochem. Biophys. Res. Commun.* 48: 670-677.

Armstrong, J. McD., D. V. Myers, J. A. Verpoorte and J. T. Edsall 1966. Purification and properties of human erythrocyte carbonic anhydrases. *J. Biol. Chem.* 241: 5137-5149.

Beychock, S., J. McD. Armstrong, C. Lindblow and J. T. Edsall 1966. Optical rotatory dispersion and circular dichroism of human carbonic anhydrases B and C. *J. Biol. Chem.* 241: 5140-5160.

Blow, D. and F. H. C. Crick 1959. The treatment of errors in the isomorphous replacement method. *Acta Cryst.* 12: 794-802.

Boyer (Ed.) The Enzymes 1971. Hydrolysis of peptide bonds. *The Enzymes* 3d ed. Vol. III. 165-212, 323-376, 547-562.

Coleman, J..E. 1969. Carbonic anhydrase: Conformation of the active center and the mechanism of action. *NASA, (Natl. Aeron. Space Admin.). Special Publication* sp-188: 141-156.

Crawford, J. L., W. N. Lipscomb and C. G. Shellman 1973. The reverse turns as a polypeptide conformation in globular proteins. *Proc. Nat. Acad. Sci. USA.* 70: 538-542.

Edsall, J. T. 1968. The carbonic anhydrases of erythrocytes. *The Harvey Lectures,* Series 62: 191-230.

Göthe, P. O. and P. O. Nyman 1972. Inactivation of human carbonic anhydrases by bromopyruvate. *FEBS Letters,* 21: 159-164.

Henderson, L. E., D. Henriksson, P. O. Nyman and L. Strid 1971. On the primary structure of the B and C isozymes of human erythrocyte carbonic anhydrases. *Alfred Benzon Symposium* IV 341-352.

Henderson, L. E., D. Henriksson and P. O. Nyman 1973. Amino acid sequence of human erythrocyte carbonic anhydrase C. *Biochem. Biophys. Res. Commun.* 52: 1388-1394.

Kandel, M., A. G. Gornall, S. -C. C. Wong and S. I. Kandel 1970. Some characteristics of human, bovine and horse carbonic anhydrases as revealed by inactivation studies. *J. Biol. Chem.* 245: 2444-2450.

Kannan, K. K., A. Liljas, I. Waara, P. C. Bergsten, K. Fridborg, S. Lövgren, B. Strandberg, U. Bengtsson, U.

Carlbom, L. Järup and M. Petef 1971. Crystal structure of human erythrocyte carbonic anhydrase C. VI. The three-dimensional structure at high resolution in relation to other mammalian carbonic anhydrases. *Cold Spring Harbor Symposia* 36: 221-231.

Kannan, K. K., K. Fridborg, P. C. Bergstén, A. Liljas, S. Lövgren, M. Petef, B. Strandberg, I. Waara, L. Adler, S. O. Falkbring, P. O. Göthe and P. O. Nyman 1972. Structure of human carbonic anhydrase B. I. Crystallization and heavy atom modifications. *J. Mol. Biol.* 63: 601-604.

Kannan, K. K., B. Notstrand, K. Fridborg, S. Lövgren, A. Ohlsson and M. Petef 1973. Crystal structure of human carbonic anhydrase B. Abstracts. *The Stockholm Symposium on the structure of Biological Molecules*, Stockholm, Sweden.

Kuang-Tzu, D. L., and H. F. Deutsch 1973. Human carbonic anhydrases. XI. The complete primary structure of carbonic anhydrase B. *J. Biol. Chem.* 248: 1885-1893.

Lewis, P. N., F. A. Momany and H. A. Scheraga 1973. Chain reversals in proteins. *Biochim et Biophys. Acta* 303: 211-229.

Liljas, A., K. K. Kannan, P. C. Bergsten, K. Fridborg, L. Järup, S. Lövgren, H. Paradies, B. Strandberg and I. Waara 1969. X-ray diffraction studies of the structure of carbonic anhydrase. *In CO_2: Chemical Biochemical and Physiological aspects.* (R. E. Forster, J. T. Edsall, A. B. Otis and F. Jr. Roughton, Eds.) NASA, Washington, D. C. sp 188: 89-99.

Liljas, A., K. K. Kannan, P. C. Bergsten, I. Waara, K. Fridborg, B. Strandberg, U. Carlbom, L. Järup, S. Lövgren and M. Petef 1972. Crystal structure of human carbonic anhydrase C. *Nature New Biol.* 235: 131-137.

Liljas, A., and M. G. Rossmann 1974. X-ray studies of protein interactions, *Ann. Rev. Biochem.* 43: in press.

Lindskog, S., L. E. Henderson, K. K. Kannan, A. Liljas, P. O. Nyman and B. Strandberg 1971. Carbonic anhydrase. *The Enzymes* (P. D. Boyer, ed.)3d ed. 5: 587-665.

Maren, T. H. 1967. Carbonic anhydrase chemistry, physiology and inhibition. *Physiol. Rev.* 47: 595-838.

Marriq, C., M. Sciaky, N. Giraud, D. Foveau and G. Laurent-Tabusse 1973. Structure primaire de l´anhydrase carbonique érythrocytaire B humaine. II. Clivage par le bromure de cyanogène et séquence des résidus 1-148. *Biochemie.* 55: 1361-1379.

Nyman P. O. 1961. Purification and properties of carbonic anhydrase from human erythrocytes. *Biochem. Biophys. Acta.* 52: 1-12.

Nyman, P. O., L. Strid and G. Westermark 1968. Carboxy-terminal region of human and bovine carbonic anhydrases. I. Amino acid sequences of terminal cyanogen bromide fragments. *Euorp. J. Biochem.* 6: 172-189.

Pocker, Y., and J. E. Meany 1967. The catalytic versatility of erythrocyte carbonic anhydrase. II. Kinetic studies of the enzyme catalysed hydration of pyridine aldehydes. *Biochemistry.* 6: 239-246.

Pocker, Y., and J. T. Stone 1967. The catalytic versatility of erythrocyte carbonic anhydrase. III. Kinetic studies of the enzyme catalysed hydrolysis of p-nitrophenyl acetate. *Biochemistry.* 6: 668-678.

Rao, S. T., and M. G. Rossmann 1973. Comparison of super secondary structures in proteins. *J. Mol. Biol.* 76: 241-256.

Richards, F. M. 1968. The matching of physical models to three-dimensional electron density maps. A simple optical device. *J. Mol. Biol.* 37: 225-230.

Rickli, E. E., S. A. S. Ghazanfar, B. H. Gibbons and J. T. Edsall 1964. Carbonic anhydrase from human erythrocytes. Preparation and properties of two enzymes. *J. Biol. Chem.* 239: 1065-1078.

Riddiford, L. M. 1965. Acid difference spectra of human carbonic anhydrases. *J. Biol. Chem.* 240: 168-172.

Rosenberg, A. 1966. The optical rotary dispersion of bovine and human carbonic anhydrases in the ultraviolet region. *J. Biol. Chem.* 241: 5126-5136.

Strandberg, B., B. Tilander, K. Fridborg, S. Lindskog and P. O. Nyman 1962. The crystallization and x-ray investigation of one form of human carbonic anhydrase. *J. Mol. Biol.* 5: 583-584.

Tashian, R. E., D. P. Douglas and Y. S. L. Yu 1964. Esterase and hydrase activity of carbonic anhydrase. I. From primate erythrocytes. *Biochem. Biophys. Res. Commun.* 14: 256-261.

Tashian, R. E., R. J. Tanis and R. E. Ferrell 1971. Comparative aspects of the primary structures and activities of the carbonic anhydrase. *Alfred Benzon symposium.* IV: 353-362.

Tashian, R. E., (personal Commun.) The evolution of structure and function in the carbonic anhydrase isozymes. These Proceedings. 1974.

Tilander, B., B. Strandberg and K. Fridborg 1965. Crystal structure studies on human erythrocyte carbonic anhydrase C. (II). *J. Mol. Biol.* 12: 740-760.

Timasheff, S. 1970. Some physical probes of enzyme structure in solution. *The Enzymes.* 3d ed. 2: 371-443.

Venkatachalam, C. 1968. Stereochemical criteria for polypeptides and proteins. V. Conformation of a system of three linked peptide units. *Biopolymers.* 6: 1425-1436.

Verpoorte, J. A., S. Mehta and J. T. Edsall 1967. Esterase activities of human carbonic anhydrases B and C. *J. Biol. Chem.* 242: 4221-4229.

Wetlaufer, D. B., and Ristow 1973. Acquisition of three dimensional structures of proteins. *Proc. Nat. Acad. Sci. USA.* 70: 697-701.

Whitney, P. L., G. Fölsch, P. O. Nyman and B. G. Malmström 1967. Inhibition of human erythrocyte carbonic anhydrase B by chloroacetyl sulfonamides with labeling of the active site. *J. Biol. Chem.* 242: 4206-4211.

MULTIPLICITY OF ISOZYMES OF LIPOAMIDE DEHYDROGENASE AND PHOSPHOLIPASE A

WILLIAM C. KENNEY
Department of Biochemistry and Biophysics,
University of California, San Francisco,
California 94143, and Division of Molecular
Biology, Veterans Administration Hospital,
San Francisco, California 94121

ABSTRACT. Six major and two minor molecular forms of lipo-amide dehydrogenase are found in pig heart mitochondria. Three of the main components are associated with 2-oxo-glutarate dehydrogenase complex and the other three with the pyruvate dehydrogenase complex. Uncomplexed lipoamide dehydrogenase originates, in the main, from the latter com-plex. The number of electrophoretically distinct forms de-tected are, in part, due to the conditions of preparation and analysis of the enzyme. The chemical basis of the differences among the molecular forms is as yet uncertain, but similarities in electrophoretic patterns from several species, as well as lack of difference in catalytic proper-ties, amino acid composition, and immunochemical properties within a given species indicate conformational isomerism to be a possibility.

Phospholipase A has been found to exist in at least nine forms in *Naja naja* venom and seven forms in *Vipera russellii* venom. In the former enzyme the molecular weights range from 8,500 to 20,000 and in the latter from. < 15,000 to 24,000. The enzyme consists of a single polypeptide chain, but the lower molecular weight components do not arise by proteolysis or preparative modification. In those forms with different molecular weights, separate genes are proba-bly involved. Similarity in amino acid composition and other properties for other forms, however, implicate con-figurational isomerism as a likely reason for the multi-plicity.

Several reasons have been postulated to account for the presence of multiple molecular forms of enzymes (Markert, 1968; Kaplan, 1968: Vessell, 1968). In a general sense these can be divided into two categories based on whether genetic or epi-genetic factors are involved. The five isozymes of lactate dehydrogenase arising from random combination of two genetic-ally independent subunits into a tetrameric molecule is a classic example of the former category. Monomers, dimers, trimers, etc., of a single polypeptide chain, deamidation, and

601

conjugation with cofactors or other such groups are examples
of the latter category. Also included in this group are
enzymes existing in different conformational states. Since
the tertiary structure of an enzyme required for catalytic
function is dictated by the primary sequence, it has been
argued that only one unique conformational state exists for an
enzyme. It is possible, however, that a given primary sequence might exist in a limited number of conformational
states each capable of enzymatic activity (Epstein and Schechter, 1968). In this as well as in many other laboratories,
multiple forms of enzymes have been observed, the origin of
which cannot be explained by random combination of different
subunits, as in lactate dehydrogenase, nor as a result of preparative artifacts, but appear to be differences in the conformational states of the enzyme.

Lipoamide dehydrogenase, a component of the multienzyme
pyruvate dehydrogenase complex and 2-oxoglutarate dehydrogenase
complex, has been reported to exist in several electrophoretically distinct forms in mammalian tissue. Although Lusty (1963;
Lusty and Singer,1964) found six components of lipoamide dehydrogenase with identical catalytic properties in beef heart
and beef liver - results which were substantiated for the
corresponding enzyme from pig heart (Cohn et al., 1968; Cohn
and McManus, 1972) - controversies appeared in the literature
in that anywhere from 2 to 13 forms were detected (Atkinson et
al., 1962; Hayakawa et al., 1968; Sakurai et al., 1970; Stein
and Stein, 1965) and one report (Wilson, 1971) stated that the
molecular forms arise from proteolytic degradation and only
one form exists in the cell. Current knowledge indicates that
six major forms exist for this enzyme (Kenney et al., 1972;
Cohn and McManus, 1972; McManus, 1974) and other minor catalytically active components are probably a result of association
of lipoamide dehydrogenase with other constituents of the complex in which it is associated and of the sensitivity of the
detecting system. Of the six major forms, Cohn et al.(1968)
provided evidence that 3 of the 6 arise from the pyruvate
dehydrogenase complex, the other three from 2-oxoglutarate
dehydrogenase complex which subsequent findings supported
(Kenney et al., 1972). These components were detected under a
variety of conditions of isolation from a mild enzymatic treatment of mitochondria to urea treatment of the purified keto
acid dehydrogenase complexes, which results in dissociation of
the complexes into their component enzymes.

In addition to the question regarding the number of forms
of lipoamide dehydrogenase, also considered is whether free,
i.e. uncomplexed, lipoamide dehydrogenase, found in crude extracts of mitochondria, is of biological significance or

arises from one or both of the keto acid dehydrogenase complexes.

Phospholipase A, another enzyme containing multiple molecular forms investigated in this laboratory (Salach et al.,
1971a), is rather unique since several mechanisms need to be
postulated in order to account for its multiplicity found in
some snake venoms. Thus, the differences in molecular weights
of some of the monomeric molecular forms indicate that these
are under separate genetic control. In some molecular forms,
however, no differences in molecular weights or amino acid
compositions have been found (Currie et al., 1968; Wahlström,
1971; Wells and Hanahan, 1969), and the origin of these multiple forms is best interpreted by conformational isomerism.
Though the reason for the large number of molecular forms of
phospholipase A in snake venoms is not clearly understood,
differences in catalytic activity, especially with respect to
solubilization of enzymes from biological membranes have been
found for some of the purified forms (Salach et al., 1968,
1971b).

ISOZYMES OF LIPOAMIDE DEHYDROGENASE IN
PIG HEART

As well as being associated with the two keto acid dehydrogenase complexes, a considerable portion of lipoamide dehydrogenase occurs in free or uncomplexed form in mitrochondria.
This fraction can be readily obtained by ammonium sulfate
fractionation under conditions which remove the keto acid
dehydrogenase complexes. When mitochondria are treated with
phospholipase A, which disrupts the mitochondrial membrane,
uncomplexed lipoamide dehydrogenase, 2-oxoglutarate dehydrogenase complex, and part of the pyruvate dehydrogenase complex
can be extracted.

Multiple molecular forms of lipoamide dehydrogenase of pig
heart mitochondria are readily apparent after polyacrylamide
gel electrophoresis of various enzyme fractions (FIGURE 1).
The uncomplexed fraction, i.e., free form, of lipoamide dehydrogenase shows five electrophoretically distinct bands (gel
1). Upon treatment of the phospholipase digested mitochondria
with urea, which destroys the keto acid dehydrogenase complexes, three new molecular forms appear (gel 2), resulting
in six distinct major bands as detected by diaphorase of NADH
fluorescence assay and one to two minor, faster moving bands.
If urea treatment is omitted, the mitochondrial extract shows
bands corresponding to forms 4, 5, and 6 plus a band at the
origin representing the complexes (gel 3) which are too large
to penetrate the gel. Three of the six major bands (4, 5, and

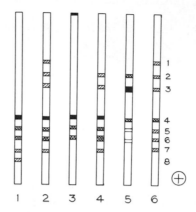

Fig. 1. Polyacrylamide gel electrophoresis of pig heart lipoamide dehydrogenase. Electrophoresis at pH 8.1 for 2 hr, as
previously described (Kenney et al., 1972). Origin is at the
top. Enzyme was visualized by the diaphorase reaction with
nitroblue tetrazolium (Cohn et al., 1968). Gel 1, uncomplexed
lipoamide dehydrogenase; gel 2, urea treated mitochondria
(total enzyme pattern); gel 3, phospholipase extract of mitochondria; gel 4, urea treated pyruvate dehydrogenase complex;
gel 5, urea treated 2-oxoglutarate dehydrogenase complex, after
elution from calcium phosphate gel with 19% saturated $(NH_4)_2SO_4$;
gel 6, same as 5, but after second elution of gel with 45%
saturated $(NH_4)_2SO_4$.

6) and one to two minor ones (7, and occasionally 8) are seen
in urea extracts of the pyruvate dehydrogenase complex (gel 4)
while urea treatment of 2-oxogluturate dehydrogenase complex
show the other three bands (1,2, and 3) with only minor cross-
contamination of forms 4, 5, and 6 (gels 5 and 6). Since the
preparation of pyruvate dehydrogenase complex used was free of
2-oxoglutarate dehydrogenase activity, but the latter complex
preparation contained 10-20% of the pyruvate dehydrogenase
activity, these results are in agreement with the conclusion
of Cohn et al. (1968) that three of the six components of lipoamide dehydrogenase arise from the 2-oxoglutarate dehydrogenase complex the other three from the pyruvate dehydrogenase
complex.

Comparison of the patterns of urea treated pyruvate dehydrogenase complex (gel 4) and of uncomplexed lipoamide dehydrogenase (gel 1) shows that they are identical, indicating
that the free or uncomplexed forms originate to a major extent
from this complex. One further point to be emphasized is that
when the enzymatic forms of lipoamide dehydrogenase originating

from 2-oxoglutarate dehydrogenase complex were eluted from calcium phosphate gel with 19% saturated solution of ammonium sulfate, component one was absent (gel 5) but was present on further elution of the gel with 45% saturated ammonium sulfate (gel 6) indicating that some molecular forms may be discarded in the course of isolation of the enzyme.

Additional evidence for the conclusion that free lipoamide dehydrogenase originates, in the main, from the pyruvate dehydrogenase complex came from studies on the lipoamide dehydrogenase activity on ammonium sulfate fractionation of phospholipase extracts of mitochondria of heart and liver from various mammaliam species (TABLE 1). The results show that 23% of the lipoamide dehydrogenase content (as activity) of phospholipase treated beef heart mitochondria and 55-60% of beef liver mitochondria appears to be uncomplexed. The much higher yield of free lipoamide dehydrogenase from liver than from heart is accompanied by a considerably lower activity of pyruvate dehydrogenase complex detectable in beef liver mito-chondria than in heart. This inverse correlation between pyruvate dehydrogenase activity and uncomplexed lipoamide de-hydrogenase is also seen in other species. For example, pig heart mitochondria have relatively high pyruvate dehydrogenase activity and low apparent free lipoamide dehydrogenase con-tent, whereas rat liver mitochondria have low pyruvate de-hydrogenase activity and relatively high uncomplexed lipo-amide dehydrogenase. The uncomplexed lipoamide dehydrogenase and pyruvate dehydrogenase activity of rabbit heart and liver mitochondria followed essentially the same relationship as in beef heart and liver mitochondria. Guinea pig liver mitochon-drial lipoamide dehydrogenases appear to be an exception, since the percentage of uncomplexed enzyme is relatively low, despite the low pyruvate dehydrogenase activity.

LIPOAMIDE DEHYDROGENASE ISOZYMES OF RAT LIVER

The isoelectric focusing pattern of lipoamide dehydrogenase from phospholipase A digested extracts of rat liver mitochon-dria and of the free forms after removal of high molecular weight complexes indicates that a number of forms of the enzyme are present with different isoelectric points (FIGURE 2). Also of significance is the relative ratio of components 4, 5, and 6 in both the crude extract and after isolation of the uncomplexed form of the enzyme. Although the components with high isoelectric points are removed upon purification, the remaining components migrate to the same isoelectric point, and appear to be in the same relative ratio as found in the crude extract. Assignment of these multiple forms to origi-

TABLE 1

RELATION OF FREE LIPOAMIDE DEHYDROGENASE TO PYRUVATE DEHYDROGENASE ACTIVITY OF MITOCHONDRIA IN DIFFERENT SPECIES

Tissue	Lipoamide dehydrogenase activity extracted	Uncomplexed lipoamide dehydrogenase	Activity in phospholipase A treated mitochondria (20 mg/ml)		
			Lipoamide dehydrogenase	2-Oxoglutarate dehydrogenase	Pyruvate dehydrogenase
	%	%	units/ml	units/ml	units/ml
Beef heart	61	23	3.35	1.26	1.53
Beef liver	87	55	1.20	0.27	0.034
Rabbit heart	58	32	1.10	0.82	0.60
Rabbit liver	66	53	0.69	0.20	0.12
Guinea pig heart	57	32	1.93	0.41	0.57
Guinea pig liver	79	29	0.56	0.29	0.17
Pig heart	49	9	2.39	1.63	1.24
Rat liver	87	66	1.31	0.30	0.02

Mitochondria were treated with phospholipase A as previously described (Kenney et al., 1972). Extracted activity is the fraction remaining in the 144,000 x g supernatant. Uncomplexed lipo-amide is that remaining soluble in 45% saturated ammonium sulfate.

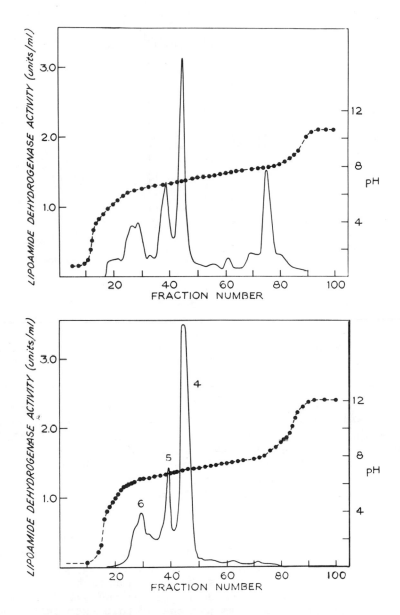

Fig. 2. Isoelectric focusing of lipoamide dehydrogenase from rat liver mitochondria. **Upper**, phospholipase extract; **lower**, uncomplexed lipoamide dehydrogenase. Two percent ampholine, pH 6 to 8, 90 hr at 0.5°C, 2.5 mA, and 1000 V. Fraction size 1.2 ml; lipoamide dehydrogenase activity (———); pH (·——·).

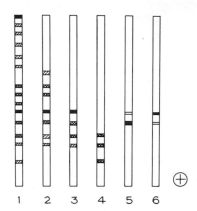

Fig. 3. Polyacrylamide gel electrophoresis of lipoamide dehydrogenase from rat liver mitochondria. Conditions as in
FIGURE 1, origin at top. Enzyme was visualized by the fluorescence of enzymatically produced NADH (Lusty, 1963). Gel 1,
phospholipase extract; gel 2, same after removal of keto acid
complexes by Sephadex G-200 chromatography; gel 3, uncomplexed
lipoamide dehydrogenase; gel 4, peak 6 of isoelectric focusing
experiment FIGURE 5 lower; gel 5, peak 5 of same; gel 6, peak
4 of same.

---

nate from either the pyruvate dehydrogenase complex or the
2-oxoglutarate dehydrogenase complex is made by analogy with
the multiple forms from pig heart mitochondria. The multiple
forms of lipoamide dehydrogenase with isoelectric points
>pH = 7.5 thus arise from the 2-oxoglutarate dehydrogenase
complex. This complex is removed in the course of isolation
of the uncomplexed lipoamide dehydrogenase, and consequently
the component lipoamide dehydrogenase forms do not appear upon
isoelectric focusing.

Evidence for the multiple enzyme forms is confirmed by polyacrylamide gel electrophoresis. In the phospholipase extract
as many as thirteen electrophoretically distinct forms have
been detected from this species, in addition to the keto acid
dehydrogenase complexes which do not enter the gel (Fig. 3, gel 1).
After mild treatment, such as passage of this extract through
Sephadex G-200, which removes the keto acid dehydrogenase
complexes, the number of detectable components decreases to
seven (gel 2). The same result was obtained if the rat liver
mitochondria extract was dialyzed extensively for 18 hr
against 0.1 mM EDTA, pH 7.5. Another instance in which mild

treatment results in a decrease in the number of forms is in ammonium sulfate fractionation. Four electrophoretically distinct forms are readily detected for uncomplexed lipoamide dehydrogenase obtained after fractionation of the crude extract with ammonium sulfate (gel 3). Isolation of free lipoamide dehydrogenase by ammonium sulfate fractionation removes the 2-oxoglutarate dehydrogenase complex, and, as in the iso-electric focusing experiment (FIGURE 2), the molecular forms of lipoamide dehydrogenase associated with this complex are removed. Results of electrophoresis of the enzyme in peaks 6, 5, and 4 of the uncomplexed forms separated by isoelectric focusing (FIGURE 2) are given in gels 4, 5, and 6 (FIGURE 3), respectively. Their mobility is in line with the isoelectric points found.

MULTIPLICITY AND ORIGIN OF ISOZYMES
OF LIPOAMIDE DEHYDROGENASE

Turning to the question of the reason for the number of forms of lipoamide dehydrogenase, Massey et al. (1962) has shown that this enzyme has a molecular weight of 100,000 and consists of two identical subunits. It is already clear that six to seven molecular forms cannot arise from random com-bination of two subunits, nor are preparative artifacts feasi-ble. Since conformational isomerism is a likely possibility several sources of the enzyme were compared as parallel changes in the relative mobilities of the various forms of the enzyme among different species would be consistent with this hypothesis. This would not be expected if the multiple forms of the enzyme were composed of nonidentical subunits (Kitto et al., 1966). Phospholipase A extracts and the uncomplexed fractions therefrom were examined by gel electrophoresis at pH 8.1 in several species. The results show (FIGURE 4) that within a given species essentially the same lipoamide dehydro-genase pattern is observed on polyacrylamide gel electro-phoresis between the phospholipase A extract (rat liver, gel 2; rabbit liver, gel 4) and the uncomplexed enzyme (rat liver, gel 3; rabbit liver, gel 5) and similar to the pattern observed for the uncomplexed forms from pig heart mitochondria (gel 1) which were shown to be identical to those forms arising from pyruvate dehydrogenase complex. One other point to be emphasized from this FIGURE is that although the distri-bution of intensities is similar, the migration of, for example, lipoamide dehydrogenase forms of pig heart (gel 1) and rat liver (gel 3) are not identical. Such results are consistent with configurational isomers or conformers.

Another observation concerning the number of molecular

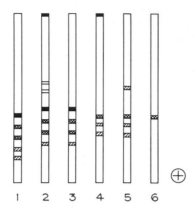

Fig. 4. Polyacrylamide gel electrophoresis of lipoamide dehydrogenase from various species. Conditions as in FIGURE 1.
Origin at top. Gel 1, uncomplexed lipoamide dehydrogenase
from pig heart mitochondria; gel 2, phospholipase A extract
of rat liver mitochondria; gel 3, uncomplexed rat liver
enzyme; gel 4, phospholipase A extract of rabbit liver mitochondria; gel 5, uncomplexed rabbit liver enzyme; gel 6, uncomplexed rabbit heart enzyme.

---

forms detected is the distance of migration. After 1 hr of
electrophoresis, corresponding to the time for the marker dye
to migrate through the gel, only two broad lipoamide dehydrogenase bands are detected in rat liver extracts, several bands
appear after 1½ hr electrophoresis, and six distinct bands are
readily seen after electrophoresis for 2 hr (Kenney et al.,
1972).

The present study explains many discrepant findings in the
literature regarding the origin and number of forms of lipoamide dehydrogenase. Uncomplexed forms of this enzyme appear
to originate from keto acid dehydrogenase complexes containing
this enzyme, particularly the pyruvate dehydrogenase complex.
Since this complex is relatively stable in heart mitochondria,
and shows no tendency to be inactivated or dissociated during
phospholipase treatment, either breakdown of the complex
occurs during preparation of the heart mitochondria, which
seems unlikely, or the uncomplexed forms may be normal metabolic products of the dissociation of the complex, a precursor thereof, or existing in equilibrium with the other
components of the complex.

In several organisms and mammalian species six major molecular forms occur. A greater number may be detected which can be a result of association of lipoamide dehydrogenase with other components of the complexes in which it is found.

The number of forms observed is in part a result of separation, detection, and degree of purification of the enzyme. For example, it was shown that one of the forms of the 2-oxoglutarate dehydrogenase complex is easily removed in the course of purification. Experimental conditions, e.g. time of electrophoresis, may dictate the number of forms seen. Since lipoamide dehydrogenase is associated with keto acid dehydrogenase complexes, the number of forms observed is also a function of the ease of extraction of these complexes from the mitochondria. Phospholipase A or sonication extracts only $\sim$ 20% of the pyruvate dehydrogenase complex from pig heart mitochondria, however, up to 90% of the 2-oxoglutarate dehydrogenase activity is solubilized by the former treatment (Kenney et al., 1972). The relative stabilities of these complexes both in vivo and after extraction also can give rise to variable numbers of molecular forms. In confirmation of reports from other laboratories, nevertheless, up to six forms of lipoamide dehydrogenase are detectable; three arising from the pyruvate dehydrogenase complex and three from the 2-oxogluturate dehydrogenase complex. In addition, minor components are detectable if sufficient quantities are analyzed.

That the number of forms of lipoamide dehydrogenase is not the result of preparative artifacts is evidenced from the similarity in relative distribution on isoelectric focusing of three of the forms in the uncomplexed fraction of lipoamide dehydrogenase with comparable forms found in the phospholipase extract of rat liver mitochondria (FIGURE 1). In addition, multiple forms have been observed in crude phospholipase extracts (Kenney et al., 1972; Lusty, 1963; Lusty and Singer, 1964), in purified preparations, and in preparations of the keto acid dehydrogenase complexes after urea cleavage (Cohn and McManus, 1972; Kenney et al., 1972; Sakurai et al., 1970).

The question concerning the chemical basis of the differences among the forms of lipoamide dehydrogenase still remains. No differences in catalytic properties, hydrodynamic parameters, amino acid compositions, end group analyses, or immunochemical properties have been observed in the molecular forms examined (Cohn and McManus, 1972; Lusty, 1963; Sakurai, et al., 1970; Stein et al., 1965). The possibility that the molecular forms are conformers is supported by the similarities in electrophoretic patterns from different species. Evidence of conformational isomers has also been supported by results of

Sakurai et al. (1970) based on the finding that the two forms
of lipoamide dehydrogenase, each derived from one of the keto
acid dehydrogenase complexes, which they studied, were inter-
changeable in function and ability to reconstitute complexes.

MULTIPLE MOLECULAR FORMS OF PHOSPHOLIPASE A

Phospholipase A is another instance in which an extensive
amount of published work leads to the conclusion that genetic
factors alone cannot explain the existence of the multiple
molecular forms of the enzyme, nor are they, in many cases,
a result of random combination of different subunits or polymerization of a single polypeptide chain. More importantly,
within this class of enzymes from snake venon, two diametric-
ally opposed situations exist. In *Crotalus adamanteus*, for
example, two forms of the enzyme are found. Each form is
dimeric, all subunits being identical, with a molecular weight
of 15,000. No difference is found in catalytic properties
(Wells, 1971), and the only tenable conclusion is that these
forms are a result of configurational isomerism. On the
other hand, *Naja naja* and *Vipera russellii* phospholipases A
contain from 7 to 14 molecular forms, differing not only in
charge and catalytic activity, but also in molecular weight
(Salach et al., 1971a, Shiloah et al., 1973). These multiple
forms do not contain subunits but are composed of single polypeptide chains, indicating them to be under separate genetic
control.

As is presented in previous sections, phospholipase A is
used as a mild enzymatic means to disrupt mitochondria in
order to liberate lipoamide dehydrogenase, and the multienzyme complexes with which it is associated. Since this
enzyme disrupts permeability barriers of mitochondria to
charged substrates (Arrigoni and Singer, 1962) it is often
employed in establishing enzyme levels and kinetic studies of
enzymes in this organelle.

A combination of isoelectric focusing, polyacrylamide gel
electrophoresis, and gel exclusion on Sephadex indicated the
presence of at least 9 molecular forms of phospholipase A in
Naja naja venom, and seven molecular forms have been found in
Vipera russellii venom (Salach et al., 1971a). Isoelectric
focusing of partially purified *Naja naja* phospholipase A is
presented in FIGURE 5. The presence of 5 components is readily seen in this experiment. On pooling the three fractions
as shown in FIGURE 5 and subjecting them to isoelectric focusing over narrower pH ranges, additional components are
detected (FIGURE 6). Two species are observed from fraction
I, the main component containing two forms detectable by

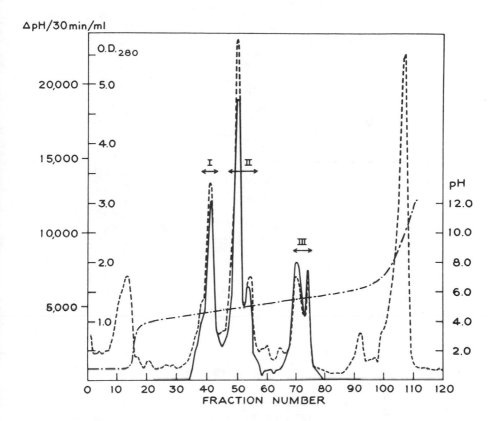

Fig. 5.   Isoelectric focusing of partially purified *Naja naja* venom phospholipase A.   Focusing in pH range 4 to 6 for 43 hours at 1000 volts and 0°, then collected in fractions of 1.0 to 1.2 ml.   pH (·-·-·), absorbance at 280 nm (- - -), and enzymatic activity (——).   Enzyme fractions were combined as indicated.   From Salach et al. (1971a).

polyacrylamide gel electrophoresis and Sephadex G-100 chromatography.   Three components are derived from the original two components of fraction II, and three forms are found for fraction III.

   The isoelectric points of the different molecular forms and their molecular weights as determined by gel chromatography on calibrated Sephadex G-100 columns is given in TABLE 2. Forms IA and IIA were also examined by the sedimentation equilibrium technique of Schachman (1957) and gave comparable values to those listed in this TABLE.   Results for Peak III forms are from a mixture of the three components III A, B, and

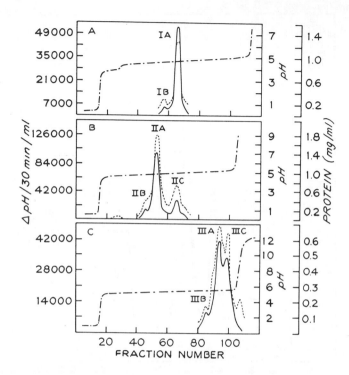

Fig. 6. Separation of isozyme peaks obtained from pH 4 to 6 isoelectric focusing (FIGURE 5) on narrow range ampholytes. A, peak 1 (FIGURE 5), pH gradient 4.2 to 5.2, 1500 volts for 65 hr at 0°. B, peak II (FIGURE 5) pH gradient 4.7 to 5.5, 1500 volts, 66 hr at 0°. C, peak III (FIGURE 5), pH gradient 4.8 to 5.8, 1500 volts, 96 hr at 0°. Phospholipase A activity (——), protein (- - -), pH (·-·-·). From Salach et al., (1971a).

C, since they are not well enough resolved from each other by isoelectric focusing, consequently, the assignment of a molecular weight value to an individual form cannot be made. It is apparent, nevertheless, that the molecular weights of the different forms are not the same, but range from 9,000 to 20,000.

The findings of these multiple molecular forms has recently been confirmed by Shiloah et al. (1973). They isolated six phospholipase A containing fractions from *Naja naja* venom by chromatography on CM- and DEAE-cellulose and on Sephadex G-50.

TABLE 2

ISOELECTRIC POINTS AND MOLECULAR WEIGHTS OF *NAJA NAJA*
VENOM PHOSPHOLIPASE A[a]

Form	Isoelectric point	Molecular weight
I A	4.63	10,800
		8,500
I B	4.60	
II A	4.95	20,200
II B	4.75	
II C	5.02	
III A	5.56	15,900
III B	5.51	14,400
III C	5.66	11,200

[a]Taken from Salach et al., 1971a.

Four fractions each contained two electrophoretically distinct
bands and two fractions each contained three bands yielding a
total of 14 forms. On sodium dodecyl sulfate (SDS) polyacrylamide gel electrophoresis minimum molecular weights
ranging from 11,000 to 24,000 were observed. Result of immunodiffusion and immunoelectrophoresis experiments indicated
antigenic identity amongst all six fractions.

In addition to *Naja naja* phospholipase A, the corresponding enzyme from *Vipera russellii* can be resolved into 7 forms
(FIGURE 7) with isoelectric points of 9.90, 9.75, 9.52, 9.29,
9.05, 8.83, and 4.62 (Salach et al., 1971a). Fraction 7 can
be isolated in homogeneous form by isoelectric focusing in
the pH range of 4 to 6. The molecular weight of phospholipase
A in peaks 3 and 7 were found to be 15,900 and 23,800 respectively, by Sephadex G-100 chromatography, and from the sedimentation coefficient of 1.88, a molecular weight of < 15,000 is
estimated for component 1. Forms 1 and 3, containing most of
the phospholipase A activity, are obtained in nearly 90% homogeneous states after a single isoelectric focusing experiment
of the crude venom.

The origin of the multiple forms of phospholipase A is of
interest since several different mechanisms may be involved.
The possibility of preparative artifacts seems very unlikely
as all of the purified forms of phospholipase A of *Naja naja*
are present in crude venom. Secondly, incubation of form

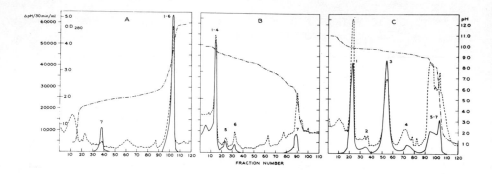

Fig. 7. Isoelectric focusing of *Vipera russellii* venom phospholipase A. **A**, pH gradient 4 to 6, 1000 volts for 46 hr at 0°. **B**, pH gradient 7 to 10, 700 volts, 144 hr at 0°. **C**, pH range 9.1 to 10.0, 1000 volts, 18 hr at 0°. Enzymatic activity (——), absorbance at 280 nm (— — —), pH (·—·—·). From Salach et al. (1971a).

IIA of phospholipase A, the highest molecular weight species, with crude venom did not reveal on polyacrylamide gel electrophoresis degradation to lower molecular weight components. Combination of identical or different subunits does not explain the origin of forms since polyacrylamide gel electrophoresis in 8.8 M urea of partially purified *Naja naja* venom or crude venom of *Vipera russellii* revealed the same bands corresponding to phospholipase A as without urea (Salach et al., 1971a) and in addition SDS-polyacrylamide gel electrophoresis of purified phospholipase A from *Naja naja* also indicated multiple molecular weights ranging from 11,000 to 24,000 for minimum values (Shiloah et al., 1973).

The phospholipases A of *Crotalus atrox* and *Crotalus adamanteus*, on the other hand, have molecular weights of 30,000, consisting of two identical 15,000 molecular weight subunits (Hachimori, et al., 1971; Wells, 1971). In the latter species, two forms of the dimeric enzyme are found. They have identical sedimentation and diffusion coefficients, amino acid composition, molecular weights, and catalytic characteristics. However, they are distinguishable by polyacrylamide gel electrophoresis. These phospholipases A appear to be two distinct dimeric forms, since one form is not generated from the other during the isolation procedure.

Although it might be expected that separate genes are involved in the synthesis of the different molecular weight species, the identity in amino acid composition for the major phospholipase A forms found in *Naja naja* and in *Crotalus adamanteus* implicate configurational isomerism as most probable for those forms having identical amino acid composition and molecular weight.

ACKNOWLEDGEMENT

The original data reported herein were obtained with the support of Program Project No. 1 PO 1 HL 16251-01 of the National Heart Institute and by a Grant-in-Aid (73 674) from the American Heart Association.

REFERENCES

Arrigoni, O., and T. P. Singer 1962. Limitations of the phenazine methosulphate assay for succinate dehydrogenase and related dehydrogenases. *Nature* 193: 1256-1258.

Atkinson, M. R., M. Dixon, and J. M. Thornber 1962. Multiple forms of flavoprotein oxidoreductases from heart-muscle particles. *Biochem. J.* 82: 29-30.

Cohn, Major I., and I. R. McManus 1972. Studies on the distribution and properties of the multiple forms of mammalian lipoamide dehydrogenase. *Biochim. Biophys. Acta* 276: 70-84.

Cohn, M. L., L. Wang, W. Scouten, and I. R. McManus 1968. Intramitochondrial distribution of multiple forms of pig heart lipoamide dehydrogenase. *Biochim. Biophys. Acta* 159: 182-185.

Currie, Byron T., D. E. Oakley, and C. A. Broomfield 1968. Crystalline phospholipase A associated with a cobra venom toxin. *Nature* 220: 371.

Epstein, Charles J., and A. N. Schechter 1968. An approach to the problem of conformational isozymes. *Ann. N.Y. Acad. Sci.* 151: 85-101.

Hachimori, Yutaka, M. A. Wells, and D. J. Hanahan 1971. Observations on the phospholipase A_2 of *Crotalus atrox*. Molecular weight and other properties. *Biochemistry* 10: 4084-4089.

Hayakawa, Taro, Y. Sakurai, T. Aikawa, Y. Fukuyoshi, and M. Koike 1968. On the interpretation of the multiple forms of mammalian lipoamide dehydrogenase. In: *Flavins and flavoproteins*. K. Yagi, Ed.: 99-106. Univ. of Tokyo Press, Tokyo.

Kaplan, N. O. 1968. Nature of multiple molecular forms of enzymes. *Ann. N.Y. Acad. Sci.* 151: 382-399.

Kenney, William C., D. Zakim, P. K. Hogue, and T. P. Singer
1972. Multiplicity and origin of isoenzymes of lipoyl
dehydrogenase. *Eur. J. Biochem.* 28: 253-260.

Kitto, G. B., P. M. Wassarman, and N. O. Kaplan 1966. Enzymatically active conformers of mitochondrial malate dehydrogenase. *Proc. Nat. Acad. Sci. U.S.* 56: 578-585.

Lusty, C. J. 1963. Lipoyl dehydrogenase from beef liver mitochondria. *J. Biol. Chem.* 238: 3443-3452.

Lusty, C. J., and T. P. Singer 1964. Lipoyl dehydrogenase.
Free and complexed forms in mammalian mitochondria. *J.
Biol. Chem.* 239: 3733-3742.

Markert, Clement L. 1968. The molecular basis for isozymes.
Ann. N.Y. Acad. Sci. 151: 14-40.

Massey, V., T. Hofmann, and G. Palmer 1962. The relation of
function and structure in lipoyl dehydrogenase. *J. Biol.
Chem.* 237: 3820-3828.

McManus, I. R. and M. L. Cohn 1974. Properties of multiple
molecular forms of lipoamide dehydrogenase. *I. Isozymes*:
Molecular Structure, C. L. Markert, ed., Acad. Press, N. Y.

Sakurai, Yukihiko, Y. Fukuyoshi, M. Hamada, T. Hayakawa, and
M. Koike 1970. Mammaliam α-keto acid dehydrogenase complexes. VI. Nature of the multiple forms of pig heart
lipoamide dehydrogenase. *J. Biol. Chem.* 245: 4453-4462.

Salach, J. I., P. Turini, J. Hauber, R. Seng, H. Tisdale, and
T. P. Singer 1968. Isolation of phospholipase A isoenzymes from *Naja naja* venom and their action on membrane
bound enzymes. *Biochem. Biophys. Res. Commun.* 33:
936-941.

Salach, James I., P. Turini, R. Seng, J. Hauber, and T. P.
Singer 1971a. Phospholipase A of snake venoms. I. Isolation and molecular properties of isoenzymes from *Naja
naja* and *Vipera russellii* venoms. *J. Biol. Chem.* 246:
331-339.

Salach, James I., R. Seng, H. Tisdale, and T. P. Singer 1971b.
Phospholipase A of snake venoms. II. Catalytic properties of the enzyme from *Naja naja.* *J. Biol. Chem.* 246:
340-347.

Schachman, H. K. 1957. Ultracentrifugation, diffusion and
viscometry. *Meth. in Enzymol.* 4: 32-103.

Shiloah, J., C. Klibansky, and A. deVries 1973. Phospholipase
isozymes from *Naja naja* venom. I. Purification and
partial characterization. *Toxicol* 11: 481-490.

Stein, Abraham M., and J. H. Stein 1965. Studies on the
Straub diaphorase. I. Isolation of multiple forms.
Biochemsitry 4: 1491-1500.

Stein, Abraham M., B. Wolf, and J. H. Stein 1965. Studies on
the Straub diaphorase. II. Properties of an antibody to

the Straub diaphorase. *Biochemistry* 4: 1500-1505.

Vessell, Elliot S. 1968. Introduction. In: Multiple molecular forms of enzymes. *Ann. N.Y. Acad. Sci.* 151: 5-13.

Wahlström, Agneta 1971. Purification and characterization of phospholipase A from the venom of *Naja nigricollis*. *Toxicol* 9: 45-56.

Wells, Michael A. 1971. Evidence that the phospholipases A_2 of *Crotalus adamanteus* venom are dimers. *Biochemistry* 10: 4074-4078.

Wells, Michael A., and D. J. Hanahan 1969. Studies on phospholipase A. I. Isolation and characterization of two enzymes from *Crotalus adamanteus* venom. *Biochemistry* 8: 414-424.

Wilson, John E. 1971. A comparative study of the multiple forms of pig heart lipoyl dehydrogenase. *Arch. Biochem. Biophys.* 144: 216-223.

PROPERTIES OF MULTIPLE MOLECULAR FORMS OF LIPOAMIDE DEHYDROGENASE[1]

IVY R. McMANUS and MAJOR L. COHN
Department of Biochemistry
Faculty of Arts and Sciences
and
Department of Anesthesiology
School of Medicine
University of Pittsburgh
Pittsburgh, Pennsylvania 15261

ABSTRACT. Multiple forms of lipoamide dehydrogenase ($NAD^+ \cdot NADH$: lipoamide oxidoreductase, EC 1.6.4.3) have been isolated from a variety of mammalian tissues. Six main enzymically active anodal species can be separated of which three are associated with the pyruvate dehydrogenase complex and three with the α-ketoglutarate dehydrogenase complex. Whether or not these isozymes exist in vivo or arise as a consequence of the experimental conditions employed in the isolation of lipoamide dehydrogenase has been the subject of several previous investigations. Further experimental efforts, including the use of the proteolytic enzyme inhibitor, pepstatin, during enzyme purification and isolation of the isozymes from sonicated fresh pig heart mitochondria, provide evidence against a solely artifactual origin.

Species from heart, skeletal muscle, and brain are immunochemically equivalent and pH optima for lipoamide dehydrogenase and transhydrogenase activities of the pig heart enzymes are the same. Activities of isozymes derived from the pyruvate dehydrogenase complex are inhibited 47.5% to 54% by 1.0×10^{-4} M DTNB in the presence of NADH while those derived from the α-ketoglutarate dehydrogenase complex are inhibited 16.6% to 18.7% under the same conditions. Both groups are equally sensitive to cupric ions and to p-chloromercuribenzoate.

The turnover of the major species of lipoamide dehydrogenase from heart is being studied, using double isotope radiolabeling techniques with L-leucine-U-^{14}C and L-leucine-4,5-^{3}H. Preliminary results suggest that these species differ in their in vivo rates of degradation and provide an impetus for investigation of the turnover and interrelationships of the multiple forms of lipoamide dehydrogenase in the cell.

Lipoamide dehydrogenase ($NAD^+ \cdot NADH$:lipoamide oxidoreduc-

tase, EC 1.6.4.3) is a flavoenzyme associated with the multienzyme pyruvate and α-ketoglutarate dehydrogenase complexes
as an essential component required for the oxidation of protein-
bound dihydrolipoic acid. Mammalian tissues and yeast contain
several forms of lipoamide dehydrogenase and the enzyme from
pig heart (Atkinson et al., 1962; Stein and Stein, 1965; Cohn
et al., 1968), beef heart (Lusty and Singer, 1964), beef liver
(Lusty,1963), and rabbit and rat liver (Kenney et al., 1972)
may be resolved into at least six species having identical
catalytic activities but differing in electrophoretic mobilities
and behavior on ion exchange columns. The isozymes, as isolated
from pig heart by the Massey procedure (Massey, 1960) or following urea treatment of purified pyruvate and α-ketoglutarate
dehydrogenase complexes (Hayakawa et al., 1966; Hirashima et al.,
1967), exist as charge isomers of the same molecular size with
three higher mobility anodal species derived from the pyruvate
dehydrogenase complex and three lower mobility anodal species
derived from the α-ketoglutarate dehydrogenase complex (Cohn
et al., 1968; Cohn and McManus, 1972; Kenney et al., 1972).
These six forms may be resolved partially by electrofocusing
in free solution and the distinctive charge behavior of the
forms associated with the two α-keto acid dehydrogenase complexes
permits their separation by isoelectric focusing.

In contrast to these results, Hayakawa et al. (1968) and
Sakurai et al. (1970) detected three forms of pig heart lipoamide
dehydrogenase, one associated with the α-ketoglutarate dehydrogenase complex, one with the pyruvate dehydrogenase complex,
and a third form which they designated as a free form of lipoamide dehydrogenase. Wilson (1971a) investigated the influence
of mode of enzyme isolation and the effects of digestion of the
enzyme with added trypsin and carboxypeptidase on the appearance of the multiple forms of the enzyme and concluded that
lipoamide dehydrogenase exists in the intact cell as a single
component and that the multiple forms arise as products of
proteolytic enzyme attack on the enzyme.

Efforts to resolve the problem of the origin and multiplicity
of the forms of lipoamide dehydrogenase have been reported from
this laboratory (Cohn and McManus, 1972) and by Kenney et al
(1972) and the present investigation summarizes further work
on some properties of the multiple forms, on the role of endogenous proteases in the genesis of these forms, and some preliminary experiments relating to the synthesis and degradation
of the enzyme in rat heart.

Typical results following electrofocusing of lipoamide
dehydrogenase prepared from pig heart according to the Massey
procedure (Massey, 1960) are shown in Fig. 1. Isoelectric
focusing in free solution in a sucrose gradient containing

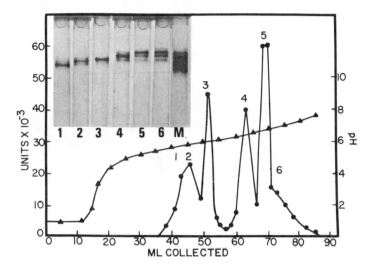

Fig. 1: Isoelectric focusing of purified pig heart lipoamide dehydrogenase. 14.7 mg enzyme were electrofocused in a LKB Model 8102 unit in pH 5-8 carrier ampholyte for 65 hr at 700 v at 4°C. Insert shows the patterns obtained after gel electrophoresis of 10 µl aliquots of numbered fractions (1.5 ml/fraction) collected after electrofocusing. Samples were electrophoresed for 1.5 hr at 1.5 mA per tube. ● , Lipoamide dehydrogenase activity; ▲ , pH (Cohn, M.L. and I. R. McManus, 1972).

pH 5-8 carrier ampholyte was performed in an LKB 8102 unit following the procedure developed by Vesterberg and Svensson (1966) for 65 hr at 700 volts with the column cooled to 4° C and the anode solution at the bottom of the column. Fractions of 1.5 ml were then collected from the bottom at a flow rate of 18 drops per min and these fractions were assayed for lipoamide dehydrogenase activity, protein concentration as indicated by absorbance at 280 nm, and flavin content as shown by absorbance at 455 nm. Enzyme activity was assayed using DL-lipoamide as substrate (Massey, 1960). The reaction was initiated by addition of lipoamide and the rate of lipoamide-dependent NADH oxidation was measured over a 3 minute period at 340 nm at room temperature using a recording spectrophotometer. Enzyme activity is expressed as units of activity per ml and a unit is defined as the amount of enzyme giving a change of 0.001 absorbance unit per min. Four enzymically active regions were resolved, indicated as 2 - 5 as well as distinct shoulders labeled as areas 1 and 6. These regions showed complete correspondence between enzyme activity, 280 nm absorbance and 455 nm absorbance, and a flavoprotein absorption spectrum typical of oxidized lipoamide dehydrogenase (Massey, 1963) was obtained when fractions indicated as peaks 3 and 5 were scanned between 325 and 550 nm. The spectral ratio of absorbance at 280 nm and 455 nm for peaks 3 and 5 was 5.3 indicative of pure flavoenzyme (Massey, 1963). The insert shows the electrophoretic patterns of these species on 4% polyacrylamide gels following disc electrophoresis (Clark, 1964) in 0.005M Tris buffer, pH 8.1 cooled between 3-5° with a current of 1.5 mA per tube for 1.5 hours. The gels were reacted with nitro-blue tetrazolium chloride, 0.3 mg/ml, and NADH, 0.6 mg/ml, in 0.1M phosphate buffer, pH 7.5 to detect diaphorase-lipoamide dehydrogenase activity. The patterns showed that isoelectric focusing had separated effectively species 1, 2, and 3, representing the highest mobility forms, and had partially resolved the lower mobility forms.

A similar analysis of lipoamide dehydrogenase obtained by resolution of purified pig heart pyruvate dehydrogenase complex in 6M urea buffered with 0.01M phosphate buffer, pH 7.5 showed that the higher mobility species 1, 2, and 3, corresponding to enzyme electrofocusing between pH 5.6 - 6.0 were derived from the pyruvate dehydrogenase complex. The lower mobility species 4, 5, and 6 having isoelectric points between pH 6.5 and 6.8 were shown to correspond to lipoamide dehydrogenase derived from the α-ketoglutarate dehydrogenase complex (Cohn and McManus, 1972).

Similar isozyme patterns are demonstrable when the enzyme is isolated from other ovine tissues, including skeletal muscle, kidney, and brain (Cohn and McManus, 1972; Millard et al., 1969). Figure 2 presents evidence for the immunochemical equivalence of lipoamide dehydrogenases from heart, skeletal muscle, and brain ovine tissues. Antiserum to pig heart lipoamide dehydrogenase was prepared as described by Stein et al (1965). Enzyme was preincubated for 10 minutes at 25° with increasing amounts of antiserum and then assayed for lipoamide-dependent oxidation of NADH. A progressive inhibition of the activities of heart, skeletal muscle and brain enzymes was observed and 90-98% inhibition resulted in the presence of 80 to 100 μl antiserum. Normal rabbit serum obtained prior to sensitization with lipoamide dehydrogenase failed to inhibit the activities of the several enzymes.

In contrast to the heterogeneity of the mammalian mitochondria lipoamide dehydrogenases, the enzyme purified from *E. coli*, including *E. coli B*, *E.coli*, Crookes strain, and *E. coli* K12, (Williams 1965; Reed and Willms,1966) is electrophoretically homogeneous

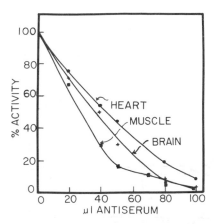

Fig. 2: Titration of lipoamide-NADH activity of lipoamide dehydrogenase purified from pig heart, skeletal muscle, and brain with antiserum to pig heart enzyme. Each enzyme was preincubated for 10 min at 25°C with indicated volume of antiserum prior to assay.

and lipoamide dehydrogenases derived from the pyruvate and α-ketoglutarate dehydrogenase complexes are indistinguishable both by electrophoretic and immunological criteria (Pettit and Reed, 1967; Scouten and McManus, 1971). Similarly, a single electrophoretic species with an isoelectric point of about pH 5.5 is found when the enzyme purified from *E. coli* K12 is electrofocused in a pH 5 - 8 sucrose gradient (Fig. 3A and B). Absence of isozymes of lipoamide dehydrogenases in several other prokaryotes including *S. marcescens*, *Ps. flourescens*, *A. agilis*, and *B. subtilis* has been noted previously (Scouten and McManus, 1971).

Lack of agreement in the literature on the degree of heterogeneity and the number of isozymes of lipoamide dehydrogenase in mammalian tissues has generated investigations designed to examine the contribution of endogenous proteolytic attack, particularly during the initial stages of isolation of the enzyme, to the observed multiplicity. In addition, no evidence exists that unequivocally rules out the possibility that variable deamidation of glutamine and asparagine residues occurring during enzyme purification may contribute to the formation of these isozymes. This latter possibility was examined by Wilson (1971b) who compared the free carboxyl group content of the isozymes from pig heart with the number of residues of glutamate and aspartate found after strong acid hydrolysis of two of the forms. He estimated that the multiple forms contained 34 to 39 amidated residues per FAD and failed to obtain evidence for

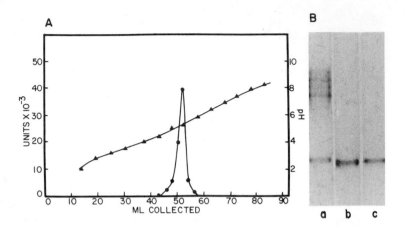

Fig. 3: A. Isoelectric focusing of purified *E. coli* lipoamide dehydrogenase. 4.2 mg enzyme were electrofocused in pH
5 - 8 carrier ampholyte for 60 hr at 700 v at 4°C. ● ,
Lipoamide dehydrogenase activity; ▲ , pH. B. Gel electrophoresis of *E. coli* lipoamide dehydrogenase. (a) 10 μg *E. coli*
enzyme prepared by Williams' (1965) method was mixed with 10
μg pig heart enzyme. (b) 10 μg *E. coli* enzyme derived from the
pyruvate dehydrogenase complex. (c) 10 μg *E. coli* enzyme
derived from α-ketoglutarate dehydrogenase complex. All gels
were electrophoresed for 1 hr, 15 min at 2 mA per tube in
0.005M Tris-glycine buffer, pH 8.5.

significant differences among the forms. Efforts in this
laboratory to detect differences in the primary structure of
the isozymes from analysis of amino acid composition of hy-
drolysates obtained by strong acid hydrolysis (Cohn and
McManus, 1972) and by total enzymic hydrolysis and by mapping
of tryptic peptides, have failed to provide evidence for
significant modification of the primary structure such as
might be predicted if the forms were the result of modification by limited proteolysis.

Experiments have been reported (Cohn and McManus, 1972)
showing that electrofocusing of a crude preparation of enzyme
prepared under conditions designed to minimize proteolytic
enzyme attack yielded a pattern of multiple forms essentially
identical to those seen using a conventionally purified preparation. Additional efforts to ascertain the possible role
of limited proteolysis in generating the multiple forms have
included the use of the proteolytic enzyme inhibitor, pepstatin[2], and the preparation of the enzyme from a sonicate of

freshly prepared purified mitochondria. Pepstatin is a potent
inhibitor of Cathepsins D and E (Barrett and Dingle, 1972) and
it has been used as an effective agent for suppressing proteolytic attack on skeletal muscle troponin (Hartshorne and Dreizen,
1972). At a ratio of 1:5,000, pepstatin to protein concentration, cathepsin activity was totally suppressed and pepstatin
was added to the homogenizing medium and included in all steps
in purification of the enzyme. Enzyme prepared under these
conditions was applied to an electrofocusing column in a pH
5 - 8 gradient as described before and the results are illustrated in Fig. 4A and B. The results are qualitatively similar
to those observed using the conventional enzyme preparation,
with the exception of some displacement of enzyme into the
peak 2 region. The significance of this alteration is difficult to assess but, as seen in Fig 4B, disc gel electrophoresis
has resolved a total of at least 6 isozymes.

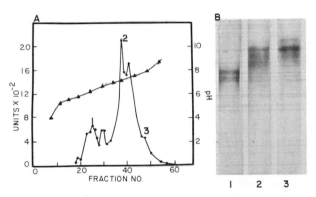

Fig. 4: A. Isoelectric focusing of pig heart lipoamide dehydrogenase purified in the presence of pepstatin. 2.4 mg
enzyme were electrofocused in pH 5-8 carrier ampholyte for 41
hr at 700 v at 4°C. ● , Lipoamide dehydrogenase activity;
▲ pH. B. Gel patterns obtained after electrophoresis of
30 µl aliquots from numbered areas.

The usual procedure for isolation of lipoamide dehydrogenase
yields a Keilin-Hartree acid precipitated mitochondrial fraction which serves as the effective fraction for isolation of
the enzyme. This fraction contains, in addition to mitochron-
dria, a complex mixture which is derived from much of the sedi-
mentable material of the heart muscle. To test the effect of
varying this crucial step on the appearance of the isozyme,
intact heart mitochondria were prepared by the method of Smith
(1967) and a suspension of the middle layer was sonicated for

15 seconds at 3°, using a Branson W-185 C Sonifier at a setting
of 6. The sonicate, obtained after centrifugation at 6500
x g for 15 min, was centrifuged for 30 minutes at 144,000 x g
to sediment the α-keto acid dehydrogenase complexes. Under
these conditions, the α-ketoglutarate dehydrogenase complex
is recovered, and approximately 20% of the pyruvate dehydrogenase complex is reported to be recovered (Kenney et al., 1972).
The sedimented pellet was incubated in 10 mM phosphate-1 mM EDTA-
6M urea for 2 hours at 4° and the resolved lipoamide dehydrogenase was recovered by chromatography on a calcium phosphate-gel-
cellulose column.

Fig. 5 shows a photograph of the diaphorase-reactive species
obtained by electrophoresis of 25 µg of enzyme obtained from
this sonicated preparation (Gel A). Comparison of this pattern
with that obtained using 30 µg of enzyme obtained by the standard Massey procedure (Gel B) shows a qualitatively identical
pattern. The preparation derived from sonicated mitochondria
shows a less well defined, more diffuse species 1 and 2 which
may reflect the lower recovery of the pyruvate dehydrogenase
complex in this procedure, but no evidence of a major alteration
in the pattern is apparent. These results, in addition to the
earlier reported experiments (Cohn and McManus, 1972) and those
of Kenney et al (1972) in which the enzyme was derived from
phospholipase A extracts of mitochondria from several species
show that a variety of methods of preparation fail to alter
the number or relative relationships of the isozymes as might
be expected if the forms were in fact derived from a single
species by the attack of endogenous proteolytic enzymes.

Using isozyme fractions recovered from resolved pig heart
lipoamide dehydrogenase, some catalytic properties of the
isozymes were examined. These included comparison of the pH
optima for lipoamide dehydrogenase and transhydrogenase
activities and sensitivity to the thiol group reactive agents,
cupric ions and dithiobis (2-nitrobenzoic acid) (DTNB).
Lipoamide dehydrogenase activity was measured as described
above and transhydrogenase was assayed essentially as described by Weber and Kaplan (1957) in which reduction of 3-
acetylpyridine-NAD by NADH is followed spectrophotometrically
as the change in absorbance at 395 nm.

The pH activity curves for isozymes representative of the
major species derived from the pyruvate and α-ketoglutarate
dehydrogenase complexes are shown in Fig. 6. Acetate buffer
was used for the pH 4 to 5.5 range, phosphate buffer from pH
5.5 to 7.5, and barbital buffer from pH 7.0 to 8.8. The pH
optima for either lipoamide-dependent NADH oxidation or for
transhydrogenase activity were between pH 6 and pH 6.6 for
all species.

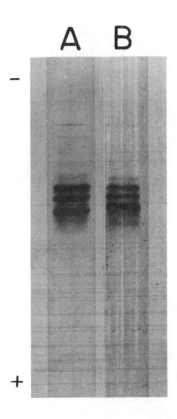

Fig. 5: Polyacrylamide gel electrophoresis of pig heart lipoamide dehydrogenase. A. 35 µg enzyme prepared from sonicated fresh mitochondria were applied to gel. B. 30 µg enzyme purified by Massey procedure were applied to gel. Electrophoresed for 1.5 hr at 2 mA per tube.

No significant differences in activities of the various isozymes were noted when cupric chloride was added to the incubation medium. The isozymes associated with the pyruvate dehydrogenase complex were inhibited 55% by 1.9×10^{-4}M $CuCl_2$ and 83% at a concentration of 6.4×10^{-4}M $CuCl_2$. Addition of 6.4×10^{-4}M $CuCl_2$ to the assay mixtures containing the isozymes derived from the α-ketoglutarate dehydrogenase complex inhibited activities over 91%.

In contrast to the uniform effects observed with added Cu++ ions, the lipoamide dehydrogenase activity of the isozymes showed differences in sensitivity to DTNB in the presence of NADH. Fractions representing areas 1 plus 2, 3, 4 and 5 from

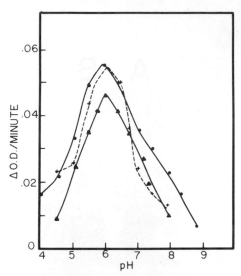

Fig. 6: Effect of pH on the lipoamide-NADH activity and
transhydrogenase activity of lipoamide dehydrogenase derived
from the pyruvate and α-ketoglutarate dehydrogenase complexes.
● , Lipoamide-NADH activity of enzyme derived from α-ketoglutarate dehydrogenase complex; dashed line, lipoamide-NADH activity of enzyme derived from pyruvate dehydrogenase complex;
▲ , transhydrogenase assay of each of the α-keto acid dehydrogenase complexes. Conditions for the assays are described
in the text.

the electrofocusing column, as designated in Fig. 1, were preincubated with 1.0 and 2.0 x 10^{-4}M DTNB in the presence and
absence of equimolar concentrations of NADH for 0 to 30 minutes.
The initial concentration of each of the enzyme containing
fractions was adjusted by suitable dilution to give an activity
catalysing a change in absorbance at 340 nm of 0.24 per minute
per 0.1 ml enzyme solution. Maximal inhibition occurred after
preincubation of the fractions with either 1 or 2 x 10^{-4}M
DTNB and NADH for 10 minutes. Fractions 1 plus 2 was inhibited
47.5% and Fraction 3, 54%, while fractions 4 and 5 were inhibited only 18.7% and 16.6% respectively. None of the enzyme fractions was inhibited by DTNB in the absence of NADH. The thiol-
reactive reagent, p-chloromercuribenzoate, strongly inhibited
all fractions in the presence of reducing agents. These results are consistent with the findings of Sakurai et al (1970)
who found that the form designated FP-II was inhibited to a
much greater degree than form FP-I in the presence of DTNB and
NADH. These results taken in conjunction with measurements of
the number of DTNB-reactive groups in FP-I and FP-II and comparison of optical rotatory dispersion and circular dichroism
spectra, suggested to Sakurai et al. (1970) that the two forms

might be distinguished by conformational differences in the
active site region. The view that the multiple forms might
be "conformers" of lipoamide dehydrogenase has also been expressed by Wilson (1971), but he failed to detect significant
differences in circular dichroism spectra or in fluorescence
properties of the isozymes such as would be expected if conformational differences were causative factors in the existence of the isozymes. The effects of DTNB reported here
suggest that differences in the environment of the reactive
disulfide linkage may distinguish the pyruvate and α-ketoglutarate dehydrogenase derived lipoamide dehydrogenases,
providing a chemical probe for further characterization of
the multiple forms derived from the respective α-keto acid
dehydrogenase complexes.

As an alternative to the structural and enzymatic approaches
that have been widely used in studying the origin and properties of the lipoamide dehydrogenase isozymes, consideration was
given to exploring isotope tracer techniques as a tool for a
study of the cellular relationships of the isozymes with the
aim of examining the rates of synthesis and degradation of the
major species. The enzyme may be isolated in good yield from
heart and it can be highly purified and characterized by a
combination of electrofocusing and immunochemical techniques.
The design of these preliminary studies was based on double
isotope radio-labeling techniques developed and applied by
Arias et al. (1969) and Glass and Doyle (1972) to measurement of protein turnover in animal cells. Male white rats
weighing between 190 and 210 g were fasted for 16 hr prior to
intraperitoneal injection of 25 μc L-leucine-U-^{14}C (348 mc/
mmole). Six days and ten days after the initial injection the
animals were again fasted and 150 μc L-leucine-4,5-^{3}H (1 c/
mmole) were given intraperitoneally to groups of 7 rats. Each
of the groups was sacrificed 4 hours after receiving the labeled leucine and hearts were removed, washed with 0.9% NaCl-10
mM leucine, and lipoamide dehydrogenase was isolated and purified. The ratio of ^{3}H to ^{14}C in the isolated protein was
determined. ^{3}H and ^{14}C were counted in a Packard Tri-Carb
scintillation spectrometer using channels in which the ^{3}H
window contained 12.8% of ^{14}C counts and the ^{14}C window contained 1.4% of the ^{3}H counts. Crossover corrections were
applied and counting efficiencies were about 26% for ^{3}H and 50%
for ^{14}C. The ^{3}H counts represent the initial time point on the
decay curve for the protein and the ^{14}C counts represent the
amount of activity remaining in the protein after the time
between injections and the ratio is independent of specific
radioactivity or absolute yield of labeled protein. The
protein with more rapid rates of degradation will have higher
^{3}H/^{14}C ratios.

Material used for preliminary measurement of $^3H/^{14}C$ ratios was obtained following isoelectric focusing of partially purified labeled rat heart lipoamide dehydrogenase (Fig. 7). The insert shows a comparison of patterns obtained following electrophoresis of 30 μg of the rat heart enzyme (A) and a similar amount of pig heart enzyme. The lower mobility forms were well resolved and are significantly retarded when compared to the pig heart enzyme. A similar pattern was also obtained when the enzyme was purified from rat liver. Electrofocusing of the enzyme separated it into two major enzymically active regions having isoelectric points of about pH 6 and 8.2. Under the conditions employed here, the individual species failed to resolve and fractions 1 and 2 represent the mixed higher and lower mobility species respectively. Fractions 28-30 and fraction 39 were recovered as representatives of the two groups and these were used to determine $^3H/^{14}C$ ratios. Table 1 summarizes the results of this experiment. After six days, the pyruvate and α-ketoglutarate dehydrogenase-derived enzymes have $^3H/^{14}C$ ratios and 2.7 and 2.3 respectively which increase to 4.4 and 3.7 after 10 days. These may be compared with control ratios of 2.07 and 1.97. Although preliminary and subject to the limitation that the differences may be due to the presence of a small amount of contaminating protein, the results suggest that the rates of degradation for the two groups of isozymes are different with a faster rate of degradation observed for the pyruvate dehydrogenase-complex-derived enzyme. Further work is in progress to establish the experimental validity of these results and to apply this approach to a study of the physiological relationships and significance of the lipoamide dehydrogenase isozymes in the cell.

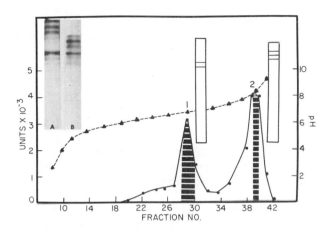

Fig. 7:   Isoelectric focusing of purified rat heart lipoamide
dehydrogenase 3.2 mg enzyme were electrofocused in a sucrose
gradient with pH 5-8 carrier ampholyte for 65 hr at 700 v at
4°C. 2.3 ml fractions were collected.   Gel electrophoresis
patterns drawn diagrammatically show the enzyme species obtained in peaks 1 and 2.   The hatched areas of peaks 1 and 2
represent the fractions used for determination of $^3H/^{14}C$ ratios
in radioactive leucine incorporation experiments. Insert shows:
A.   30 µg of purified rat heart lipoamide dehydrogenase.
B.   30 µg of purified pig heart lipoamide dehydrogenase.
Both samples were electrophoresed for 1.5 hr at 1.5 mA per tube.

TABLE I
$^3H{:}^{14}C$ Ratios of Rat Heart Lipoamide Dehydrogenase

Time of injection with ^3H-Leucine	PDC-derived Enzyme	KDC-derived Enzyme
	$^3H{:}^{14}C$ Ratios	
0 days	2.07	1.97
6 days	2.70	2.30
10 days	4.40	3.70

Each of seven 190-210 g male white rats in a group
were injected IP with 25 µc L-leucine-U-^{14}C (348 mc/
mmole).   Six and ten days later, each rat received 150
µc L-leucine-4,5-^3H (1 c/mmole).   Four hours after injection, the rats were sacrificed and enzyme was isolated
from rat hearts. A control (0 days) was done in which
L-leucine-U-^{14}C and L-leucine-4,5-^3H were administered
at the same time and the rats then were sacrificed after
4 hours. PDC = pyruvate dehydrogenase complex; KDC =
α-ketoglutarate dehydrogenase complex.

ACKNOWLEDGEMENTS

This work was supported by Research Grants No. AM-02914 and 11147 from the National Institute of Arthritis and Metabolic Diseases, National Institutes of Health.

The authors wish to thank Dr. David Hartshorne of Carnegie-Mellon University for the gift of pepstatin.

REFERENCES

Arias, Irwin M., D. Doyle, and R. T. Schimke 1969. Studies on the synthesis and degradation of proteins of the endo-plasmic reticulum of rat liver. *J. Biol. Chem.* 244: 3303-3315.

Atkinson, M. R., M. Dixon, and J. M. Thornber 1962. Multiple forms of flavoprotein oxidoreductases from heart muscle particles. *Biochem. J.* 82: 29P-30P.

Barrett, A. J. and J. T. Dingle 1972. The inhibition of tissue acid proteinases by pepstatin. *Biochem. J.* 127: 439-441.

Clarke, John T. 1964. Simplified "disc" (polyacrylamide gel) electrophoresis. *Ann. N. Y. Acad. Sci.* 121: 428-436.

Cohn, Major L., L. Wang, W. Scouten, and I. R. McManus 1968. Intramitochondrial distribution of multiple forms of pig heart lipoamide dehydrogenase. *Biochim. Biophys Acta* 159: 182-185.

Cohn, Major L. and I. R. McManus 1972. Studies on the distribution and properties of the multiple forms of mammalian lipoamide dehydrogenase. *Biochim. Biophys. Acta* 276: 70-84.

Glass, Richard D. and D. Doyle 1972. Measurement of protein turnover in animal cells. *J. Biol. Chem.* 247: 5234-5242.

Hartshorne, David J. and P. Dreizen, 1972. Studies on the subunit composition of troponin. *Cold Spring Harbor Symposia on Quant. Biol.* 37: 225-234.

Hayakawa, Taro, M. Hirashima, S. Ide, M. Hamada, K. Okabe, and M. Koike 1966. Mammalian α-keto acid dehydrogenase complexes. I. Isolation, purification and properties of pyruvate dehydrogenase complex of pig heart muscle. *J. Biol. Chem.* 241: 4694-4699.

Hayakawa, Taro, Y. Sakurai, T. Aikawa, Y. Fukuyoshi, and M. Koike 1968. On the interpretation of the multiple forms of mammalian lipoamide dehydrogenase in *Flavins and Flavoproteins*, Yagi, K. ed., University of Tokyo Press, Tokyo, pp. 99-105.

Hirashima, Masahiio, T. Hayakawa, and M. Koike 1967. Mammalian α-keto acid dehydrogenase complexes. II. An improved procedure for the preparation of 2-oxoglutarate dehydrogenase complex from pig heart muscle. *J. Biol. Chem.* 242: 902-907.

Kenney, William C., D. Zakim, P. K. Hogue, and T. P. Singer 1972. Multiplicity and origin of isozymes of lipoyl dehydrogenase. *Eur. J. Biochem.* 28: 253-260.

Lusty, C. J. 1963. Lipoyl dehydrogenase from beef liver mitochondria. *J. Biol. Chem.* 238: 3443-3452.

Lusty, C. J. and T. P. Singer 1964. Lipoyl dehydrogenase. Free and complexed forms in mammalian mitochondria. *J. Biol. Chem.* 239: 3733-3742.

Massey, Vincent 1960. The identity of diaphorase and lipoyl dehydrogenase. *Biochim. Biophys. Acta* 37: 314-322.

Massey, Vincent 1963. Lipoyl dehydrogenase in *The Enzymes* 7: 275-306, 2nd ed., Boyer, Paul D., H. Lardy, and K. Myrback, eds., Academic Press, New York.

Millard, Sara, A., A. Kubose, and E. M. Gal 1969. Brain lipoyl dehydrogenase. Purification, properties,and inhibitors. *J. Biol. Chem.* 244: 2511-2515.

Pettit, Flora H. and L. J. Reed 1967. α-Keto acid dehydrogenase complexes. VIII. Comparison of dihydrolipoyl dehydrogenases from pyruvate and α-ketoglutarate dehydrogenase complexes of *Escherichia coli*. *Proc. Natl. Acad. Sci. U. S.* 58: 1126-1130.

Reed, Lester J. and C. R. Willms 1966. Purification and resolution of the pyruvate dehydrogenase complex *(Escherichia coli)*, in *Methods in Enzymol.* 9: 247-265, Wood, W. A., ed., Academic Press, New York.

Sakurai, Yukihiko, Y. Fukuyoshi, M. Hamada, T. Hayakawa, and M. Koike 1970. Mammalian α-keto acid dehydrogenase complexes. VI. Nature of the multiple forms of pig heart lipoamide dehydrogenase. *J. Biol. Chem.* 245: 4453-4462.

Scouten, William H. and I. R. McManus 1971. Microbial lipoamide dehydrogenase. Purification and some characteristics of the enzyme derived from selected microorganisms. *Biochim. Biophys. Acta* 227: 248-263.

Smith, Archie. L. 1967. Preparation, properties, and conditions for assay of mitochondria: slaughterhouse material small scale, in *Methods in Enzymol* 10: 81-86, Estabrook, R. W. and M. E. Pullman, eds., Academic Press, New York.

Stein, Abraham M. and J. H. Stein 1965. Studies on the Straub diaphorase. I. Isolation of multiple forms. *Biochem.* 4: 1491-1500.

Stein, Abraham M., B. Wolf, and J. H. Stein 1965. Studies on the Straub diaphorase. II. Properties of an antibody to the Straub diaphorase. *Biochem.* 4: 1500-1505.

Vesterberg, O. and H. Svensson 1966. Isoelectric fractionation, analysis, and characterization of ampholytes in natural pH gradients. IV. Further studies on the resolving power in connection with separation of myoglobins. *Acta Chem. Scand.* 20: 820-834.

Weber, Morton M. and N. O. Kaplan 1957. Flavoprotein-catalyzed pyridine nucleotide transfer reactions. *J. Biol. Chem.* 225: 909-920.

Williams, Charles H., Jr. 1965. Studies on lipoyl dehydrogenase from *Escherichia coli*. *J. Biol. Chem.* 240: 4793-4800.

Wilson, John E. 1971a. The origin of the multiple forms of pig heart lipoyl dehydrogenase in *Proc. 3rd Int. Symp. on Flavins and Flavoproteins,* Kamin, H., ed., University Park Press, Baltimore, Md., pp. 313-318.

Wilson, John E. 1971b. A comparative study of the multiple forms of pig heart lipoyl dehydrogenase. *Arch. Biochem. Biophys.* 144: 216-223.

MULTIPLE FORMS OF β-GLUCURONIDASE: MOLECULAR NATURE, TRANSFORMATION, AND SUBCELLULAR TRANSLOCATION

CHI-WEI LIN
Tufts Cancer Research Center
and Department of Pathology
Tufts University School of Medicine
Boston, Massachusetts 02111

ABSTRACT. β-Glucuronidase in lysosomes and microsomes exhibits different electrophoretic forms. The microsomal enzyme is associated with the membrane while the lysosomal enzyme is mainly soluble. It has also been established that in the microsomes at least four isozymic forms exist all of which differ from the lysosomal form by containing one to four accessory protein chains.

Characterization of a highly purified β-glucuronidase of mouse kidney which is electrophoretically identical to the lysosomal form revealed that the enzyme is a tetramer containing four subunits each having a molecular weight of 74,000 daltons. All the isozymes of β-glucuronidase have similar antigenicity since the antiserum prepared from the purified enzyme cross-reacted with all the isozymic forms.

Result of isoelectric focusing in polyacrylamide gel demonstrated that the lysosomal enzyme, even though it is homogeneous by criteria of electrophoresis and chromatography, contains multiple components, possibly differing in molecular charge. Furthermore, the microsomal enzyme can be readily converted to the lysosomal forms by heat, urea or acid treatments, an event presumably resulting from the dissociation of the accessory protein chains from the microsomal enzyme molecules. The conversion at 37° occurs more readily at pH 5 than at a neutral pH and does not cause the release of the microsomal enzyme from the membrane, thus suggesting that the accessory protein chain does not serve as a ligand between the enzyme and the microsomal membrane.

Finally, a time-course study on the induction of β-glucuronidase in lysosomal, microsomal, and Golgi fractions by a single dose of gonadotrophin showed that the Golgi apparatus may play a role in the intracellular translocation of β-glucuronidase.

INTRODUCTION

β-Glucuronidase is one of the acid hydrolases generally identified with the lysosome. Presumably, this enzyme participates in the catabolic functions of the lysosome in the breakdown of mucopolysaccharides (Levvy and Conchie, 1966).

637

Also, it is believed that this enzyme is involved in the metabolic hydrolysis of conjugated compounds (Wakabayashi, 1972).

However, in the cell about 30 to 40% of the enzyme is
located in sites outside the lysosome. In this article the
various forms of β-glucuronidase present in the lysosome and
extralysosomal sites, mainly in the microsome, will be discussed. Specifically, the molecular basis for the multiple
isozymic forms, the characterization of a purified β-glucuronidase, and the possible involvement of Golgi apparatus in the
intracellular translocation of this enzyme are the main con-
cerns of this paper.

The experimental system for this study is the induction in
mouse kidney of β-glucuronidase by androgenic hormones
(Fishman, 1965). This induction is specific to kidney β-glucuronidase; the enzyme in other organs or other kidney acid
hydrolases such as acid phosphatase are not induced (Ide and
Fishman, 1969). The induction has also been shown to be due
to an increased de novo synthesis of the enzyme (Frieden et
al, 1964; Swank et al, 1973).

The separation of lysosomal and microsomal β-glucuronidase
from mouse liver by electrophoresis has been reported by
Ganschow and Bunker (1970) and from rat liver by Mameli et al
(1972). However, microsomal β-glucuronidase was shown by
Swank and Paigen (1973) to consist of multiple forms. We have
confirmed this finding in which mouse kidney homogenate, extracted by 5% Triton X-100, is resolved into five components
in polyacrylamide gel electrophoresis (Fig. 1). These components are designated as L, M1, M2, M3, and M4 according to
Swank and Paigen.

The microsomal enzyme is membrane-associated and cannot be
released by freezing and thawing although this treatment rup-
tures the lysosomes and releases the lysosomal β-glucuronidase
(Paigen, 1961; Ide and Fishman, 1969). Such a selective release of the enzyme can be used to remove contaminations in
the microsomal and lysosomal fractions. Thus, when isolated
fractions are frozen and thawed, and centrifuged, a relatively
uncontaminated lysosomal enzyme preparation can be recovered
from the supernatant and a microsomal enzyme preparation from
the precipitate. The isozyme patterns of microsomal and lysosomal β-glucuronidase so prepared are also shown in Fig. 1.
The lysosomes contain essentially a single component, L, while
the microsomes contain all the M components with a significant
amount of the L component present. Therefore, the L component
is essentially lysosomal while the M forms are mainly microsomal.

Genetic studies by Paigen (1961) have shown that a single
gene codes for the enzyme in both lysosomes and microsomes.
Recently, Swank and Paigen (1973) further showed that the

638

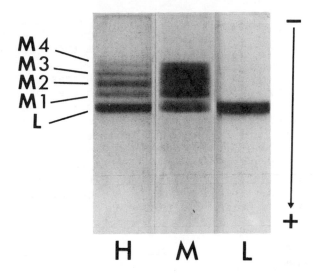

Fig. 1. β-Glucuronidase zymograms of kidney homogenate (H), microsomal (M) and lysosomal (L) fractions. The electrophoresis was carried out in polyacrylamide gel system described by Clarke (1964). The enzyme staining procedure was adapted from the histochemical method of Hayashi et al (1964) which used naphthol AS-BI glucuronide as substrate and coupled simultaneously with hexazonium pararosanilin. The designation of isozymes was that of Swank and Paigen (1973).

multiple forms of β-glucuronidase, separated by electrophoresis, resulted from differences in molecular weight of the components. That is, all the enzyme forms contain L as the basic structure. The M forms, from M1 to M4, contain one to four extra protein chains respectively, each contributing 50,000 daltons of molecular weight to the enzyme molecule.

CHARACTERIZATION OF A PURIFIED MOUSE KIDNEY β-GLUCURONIDASE

β-Glucuronidase in the mouse kidney has been actively studied in relation to its biosynthesis, intracellular distribution, and translocation, and for its significance in metabolic hydrolysis of conjugated components. However, it has not been highly purified. The highest specific activity reported was that by Pettengill and Fishman (1962) who obtained a preparation of 81,000 Fishman units/mg protein. The preparation was not shown to be homogeneous. The preparation of this enzyme from mouse kidney in a high degree of purity is hindered

by the unavailability of large quantities of this tissue and
by the relatively low level of the enzyme in the tissue.

By taking advantage of the fact that this enzyme in mouse
kidney is specifically induced by androgenic hormone to a much
higher level, we were able to purify the enzyme to a high degree of purity[1]. The purified enzyme has a specific activity
of 284,000 Fishman units per mg of protein which represents
a 1090-fold purification from the crude homogenate of the
induced kidney and is about 17,000 times higher than the enzyme
level of the non-induced kidney (16.7 units/mg of protein).
The final preparation showed a single protein peak in isoelectric focusing and in gel filtration with Sephadex G-200. It
also appeared as a single band in SDS-gel electrophoresis and
formed a single precipitin line when its antibody prepared in
rabbits was reacted against enzyme preparations of various
purities.

As shown in Fig. 2, the purified enzyme is electrophoretically identical to the L-component. This is also seen along
with M2 and M3 in the isozyme pattern of the hormone-induced
kidney.

The antibody against the purified enzyme, however, does
cross-react with all the isozyme components present in the

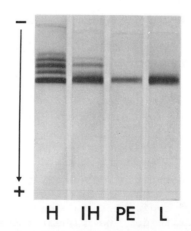

Fig. 2.   Isozyme patterns of β-glucuronidase of the gonado-
trophin-induced kidney homogenate (IH) and the purified enzyme
(PE).   H and L are non-induced kidney homogenate and lysosomal
enzyme respectively.

homogenate. Thus, as demonstrated in Fig. 3, when the same amount of enzyme extract from non-induced kidney was mixed with the antiserum at different dilutions and followed by electrophoresis in polyacrylamide gel, all the components formed a complex with the antiserum, which is visible at the origin in the gel. The result also showed that the enzyme of the induced kidney is not antigenically different from the non-induced enzyme. Also, the antigen-antibody complex is enzymatically active.

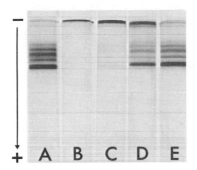

Fig. 3. Cross-reactivity of the antibody to the purified enzyme to all the isozyme components. Diluted antiserum was mixed with enzyme extracts (by 5% Triton X-100) of uninduced kidney homogenate before being applied to electrophoresis in polyacrylamide gel. A. Kidney extract with no antiserum; B. to E. with diluted antiserum at 1:200, 1:500, 1:1000, and 1:5000 respectively.

The purified enzyme has a subunit molecular weight of 74,000 daltons as determined by SDS-gel electrophoresis. The molecular weight of the native enzyme is around 300,000 daltons determined by electrophoresis as a gel gradient of 4 to 30% polyacrylamide. This result indicates that the molecule, which is electrophoretically identical to the L-isozyme, has a tetrameric structure with four subunits, each with a molecular weight of 74,000.

MULTIPLE LYSOSOMAL FORMS AND CONVERSION OF MICROSOMAL INTO LYSOSOMAL FORMS

Even though the purified enzyme from mouse kidney, electrophoretically identical to the L-form, is homogeneous by criteria of electrophoresis, chromatography, and immunodiffusion, it was earlier reported by Plapp and Cole (1967) that the purified enzyme from bovine liver can be separated into five components by ion-exchange chromatography. In their study, all components have similar molecular weights, around 280,000, and carbohydrate analysis revealed that the multiple isozymes might result from differences in carbohydrate composition.

A similar result was reported recently by Potier and Gianetto (1973) from their study on the lysosomal β-glucuronidase of rat liver. In this study, five active components of β-glucuronidase were fractionated by DEAE ion-exchange chromatography from purified lysosomal enzyme. All the components were also shown to be similar in molecular weight by electrophoresis and by ultracentrifugation. Therefore, the possibility that multiple lysosomal forms exist in the purified enzyme from mouse kidney was examined.

As shown in Fig. 4, the purified enzyme from mouse kidney can be resolved into at least four components by isoelectric

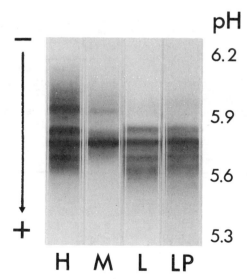

Fig. 4. Enzyme patterns of β-glucuronidase in isoelectric focusing in polyacrylamide gel. 1% ampholine, pH 5-8 was mixed in 5% polyacrylamide gel to provide the pH gradient. The isoelectric focusing was carried out for 16 hours at 4° with a current of 6mA maintained by a pulse power supply at 40 pulses per second. The gels were incubated in 0.2 M acetate buffer, pH 5.0 for 20 min before stained for enzyme activity. The sections shown are areas where enzyme activity is located. H. Kidney homogenate; M. microsomal fraction; L. lysosomal fraction; and LP. purified enzyme which had been shown to be identical to the L-form by electrophoresis.

focusing in polyacrylamide gel, which separates molecules by their differences in charge. The enzyme bands are located at a pH between 5.6 to 6.0 with the major component at pH 5.8. The purified enzyme exhibited a similar isozyme pattern to that of the lysosomal fraction while the kidney homogenate contained three extra bands in the region of higher pH. These three extra components are presumably of microsomal origin because they match bands in the homogenate sample in the non-lysosomal positions. Some of the microsomal isozymes might have been converted to the lysosomal forms by the increase in temperature during the 16-hour run. As will be explained in greater detail in the following sections, the M-forms readily convert to the L-form at low pH or at high temperature. The observation that the purified enzyme can be resolved into multiple components by isoelectric focusing is consistent with the belief that lysosomal β-glucuronidase is heterogenous, perhaps due to differences in carbohydrate content of the molecule.

According to Swank and Paigen (1973), the microsomal forms of β-glucuronidase can be converted into a lower molecular weight form X by incubation in 6 M urea at 0° C. The X-component has a similar molecular weight to the L-component and therefore the conversion is attributed to the dissociation of the accessory protein chain from the microsomal isozymes.

As shown in Fig. 5, the conversion occurs more readily at pH 5 than at a neutral pH. When the microsomal enzyme, solubilized by 5% Triton X-100, was incubated at pH 5.0 in 0.1 M

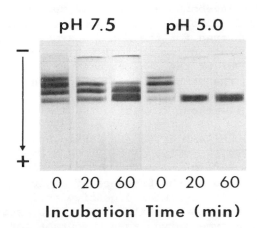

Fig. 5. Transformation of β-glucuronidase isozymes. The microsomal enzyme extract (by 5% Triton X-100) was incubated at 37° at pH 7.5 or 5.0 for different lengths of time before electrophoresis was performed.

acetate buffer at 37° all the M-forms converted to the L-form in less than 20 min. At pH 7.5 complete conversion did not occur even after 60 min of incubation.

It is interesting to note that the pH at which microsomal isozymes are rapidly converted to the lysosomal form is close to the biological pH of the lysosome. This is also the enzyme's optimal pH at which the activity of the enzyme is generally measured.

All the biochemical data available so far have yet to show any difference in the enzymatic and catalytic properties between the lysosomal and the microsomal enzymes even though we now know that the microsomal enzymes have one to four extra protein chains attached to them (Swank and Paigen, 1973). The failure to demonstrate the difference in catalytic properties between the two might have been due to the pH of the enzyme assay, a pH at which the microsomal isozymes are readily converted into the lysosomal isozymes.

The function of the accessory protein chain of the microsomal isozymes is not known. One of the possibilities considered is its function in the attachment of the microsomal β-glucuronidase to the microsomal membrane. If this is the case, the conversion of the M-forms into the L-form which resulted in the dissociation of the accessory protein chain from the M-forms would also release the enzyme from the microsomal membrane. However, the experimental result shown in Fig. 6 does not support this hypothesis.

In this experiment, microsomal enzyme was incubated at pH 5 with and without being previously extracted with Triton X-100 to compare membrane-bound and free enzyme. Conversion of the M-forms occurred in both cases although the conversion was less readily achieved for the bound enzyme. When the incubated microsomal preparation (no Triton) was centrifuged it was found that more than 90% of the β-glucuronidase activity remained associated with the membrane. It is apparent from this result that the dissociation of the accessory protein chain from the enzyme can occur without releasing the enzyme from the membrane.

GOLGI-APPARATUS AND INTRACELLULAR TRANSLOCATION OF β-GLUCURONIDASE

The significance of the presence of β-glucuronidase in the microsome where intracellular digestive activities are usually not evident is not known. Kato et al (1970) suggested that the microsomal enzyme serves as a precursor for the lysosomal enzyme. In a temporal sequential study of induction of the enzyme activity in these organelles, Kato et al have shown that within 10 hr after a single hormone injection an increase of the enzyme activity occurs in the microsomal fraction, while

Solubilized   Bound

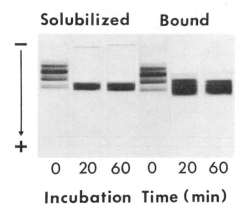

0   20   60   0   20   60

Incubation Time (min)

Fig. 6. Effect of solubilization of the enzyme from microsomal membrane on the transformation of isozyme. The transformation of the microsomal enzyme at 37° C, pH 5.0 is compared
between Triton X solubilized enzyme and the membrane-bound
enzyme. Result of incubation did convert most of the bound
microsomal enzyme into the lysosomal form without releasing
the enzyme from membrane. More than 90% of the enzyme activity remained with the precipitate when the incubated enzyme
was centrifuged.

---

the lysosomal enzyme did not respond until 40 hr after the
injection. The sequential pattern of induction has been confirmed by Ganschow (1973) and the concept of the precursor-
product relationship between the two is also supported by the
study of Van Lancker and Lentz (1970) on the biosynthesis of
β-glucuronidase and its appearance in lysosomes in the hypoxic
rat liver. A further study by Kato et al (1972) suggested
that the route of intracellular transport of this enzyme is
from the membrane-bound ribosomes across the membrane into the
cisternae of the rough endoplasmic reticulum and then into the
lysosome via the smooth endoplasmic reticulum. However, this
concept is disputed by the genetic and biochemical studies of
Swank and Paigen (1973), who hold that the majority of the microsomal enzyme is not a precursor of the lysosomal enzyme.

Working on the assumption that the microsomal enzyme is
modified and then transported into various cellular sites,
including lysosomes, a study was undertaken to investigate
whether the Golgi apparatus is involved in these processes
(Lin et al, 1973; Marsh et al, 1974).

A Golgi-enriched fraction was isolated from mouse kidney by

the method described by Leelavathi et al (1970). This fraction was considerably enriched in the Golgi marker enzymes galactosyl transferase and thiamine pyrophosphatase. Microsomal and lysosomal fractions were also prepared. The response of β-glucuronidase in the Golgi, lysosomal, and microsomal fractions to a single dose of gonadotrophin injection were followed in a time-course study, and the results are shown in Fig. 7.

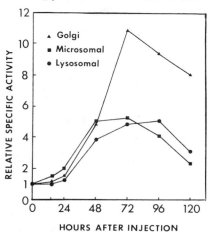

HOURS AFTER INJECTION

Fig. 7. Response of β-glucuronidase in mouse kidney lysosomal (●), microsomal (■), and Golgi (▲) fractions to gonadotrophin injection (2.5 i.u./mouse). Relative specific activities were calculated with respect to the non-injected controls. Enzyme activity was assayed by the method described by Fishman (1973), using phenolphthalein glucuronic acid (1 mM) as substrate in 0.1 M acetate buffer, pH 4.5. Each subcellular preparation was made from the pooled kidneys of six mice (with permission from *Biochemical Journal*).

In this experiment the increases in β-glucuronidase activity in the lysosomal and microsomal fractions were similar in temporal sequence to those reported by Kato et al (1970). The enzyme activity in the Golgi fraction was stimulated to a much higher level than those of the microsomal and lysosomal fractions and was about eleven-fold that of the non-induced level. Furthermore, the enzyme level of the Golgi fraction remained elevated throughout the entire period of the experiment. These results suggest that the Golgi complex is involved in the induction of this enzyme by androgen. The prolonged elevation of the enzyme activity in the Golgi fraction also suggests that a portion of the Golgi β-glucuronidase is retained in the Golgi apparatus.

The electrophoretic patterns of β-glucuronidase in the

Golgi fraction obtained from hormone-induced and non-induced kidney are shown in Fig. 8. It is important to note that the Golgi fraction contains both the lysosomal and the microsomal isozymes.

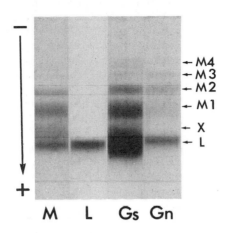

Fig. 8. Electrophoretic patterns of β-glucuronidase in microsomal (M), lysosomal (L), and hormone stimulated (Gs) and non-stimulated (Gn) Golgi fractions (with permission from *Biochemical Journal*).

DISCUSSION

It is apparent from the above information that β-glucuronidase of both lysosomes and microsomes exhibits multiple enzyme forms. The microsomal isozymes are different in molecular weights due to the attachment of different numbers of the accessory protein chain to the molecule. The lysosomal isozymes are different in molecular charge possibly due to difference in carbohydrate content. Since all the isozymic forms originated from the same structural gene, it is apparent that modifications of the enzymatic protein must have taken place after initial synthesis. What is not apparent at this time is the mechanism involved in this transformation of β-glucuronidase isozymes.

The pathway of translocation proposed by Kato et al (1970) for β-glucuronidase implies that the transformation could occur during the transport of enzyme from endoplasmic reticulum to lysosomes. Goldstone and Koenig (1973) suggested that acid hydrolases, including β-glucuronidase, are modified by the addition of acid carbohydrate molecules to the enzymatic protein during transport from endoplasmic reticulum to lysosomes. Evidence supporting this idea were their observations of a progressive increase in solubility and electrophoretic mobility of these hydrolases in subcellular fractions in the order from rough microsomes, smooth microsomes, Golgi, and lysosomes. However, the difference in electrophoretic mobility could be due to difference in molecular weight as Swank and Paigen (1973) have demonstrated. The difference in solubility produced by freezing and thawing of various fractions could be due to lysosomal contamination in these fractions, since lysosomal enzyme is soluble and the microsomal enzyme is membrane bound.

Swank and Paigen (1973) suggested that the transformation of the microsomal enzymatic forms does not involve the lysosomal enzyme, the latter being formed directly from the structural gene product. However, the result presented earlier in this article (Fig. 4) which showed the existence of multiple forms of lysosomal enzyme suggests that modification of molecular charge must have occurred to the lysosomal enzyme.

It is significant that the Golgi apparatus is shown to be involved in the transport of β-glucuronidase. This organelle has been postulated to be the site of biosynthesis of glycoproteins as well as their assembly, transport, and intracellular distribution (Novikoff et al, 1971; Whaley et al, 1972; Cook, 1973). It is possible that it may play a key role in the transformation and distribution of β-glucuronidase isozymes, especially in androgen induction of this enzyme.

Adding to the complexity of the problem is the realization that the microsomal fraction contains cytomembrane systems derived from various organelles such as rough and smooth endoplasmic reticulum, Golgi-apparatus, and plasma membrane, all of which are known to be sites of β-glucuronidase localization. (The localization of β-glucuronidase in basal infolding plasma membrane of the kidney distal tubule cells is shown in Fig. 9.) The elucidation of the isozymic forms in these organelles therefore would be a significant step in the understanding of the intracellular transport and distribution of β-glucuronidase.

ACKNOWLEDGMENT

The author is indebted to Dr. W. H. Fishman for his suggestions and encouragement in the course of this investigation. Part of the work presented was done in collaboration with Dr. C. A. Marsh. We also thank Dr. M. Sasaki for providing the electron micrograph of the β-glucuronidase localization in androgen-induced mouse kidney. The technical assistance of Mrs. M. L. Orcutt is also acknowledged.

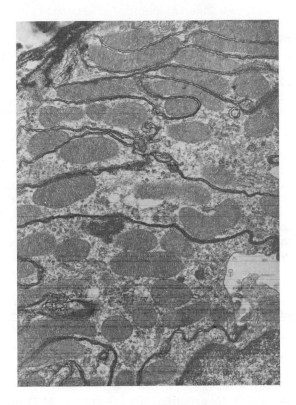

Fig. 9. Localization of β-glucuronidase in the epithelial cells of distal tubules of mouse kidney. Strong enzyme activity is located in the basal unfolding membranes. Smith and Fishman (1969) staining techniques were used to visualize the β-glucuronidase activity. Magnification 5,500 X.

FOOTNOTE

1. Lin, C. W., Orcutt, M. L., and W. H. Fishman. β-glucuronidase of mouse kidney; purification and characterization. In preparation.

REFERENCES

Clarke, J. T. 1964. Simplified "disc" (polyacrylamide gel) electrophoresis. *Ann. N. Y. Acad. Sci.* 121: 404-427.
Cook, G. M. W. 1973. The Golgi apparatus: Form and function. in *Lysosomes in Biology and Pathology* (Dingle, ed.). North-Holland, Amsterdam, 3: 237-277.

Fishman, W. H. 1965. The influence of steroids on β-glucuron-
    idase of mouse kidneys. in *Method in Hormone Research*
    (Dorfman, ed.) Academic Press, New York, 4: 273-326.
Fishman, W. H. 1973. β-Glucuronidase. in *Methods of Enzymatic
    Analysis* (Bergmeyer, ed.) Academic Press, 1393-1406.
Frieden, E. H., A. A. Harper, F. Chin, and W. H. Fishman 1964.
    Dissociation of androgen-induced enzyme synthesis and
    amino acid incorporation in mouse kidney by actinomycin
    D. *Steroids* 4: 777-786.
Ganschow, R. E. 1973. The genetic control of acid hydrolases.
    in *Metabolic Conjugation and Metabolic Hydrolysis*
    (Fishman, ed.) Academic Press, New York, 3: 189-207.
Ganschow, R. E., and B. G. Bunker 1970. Genetic control of
    glucuronidase in mice. *Biochem. Genet.* 4: 127-133.
Goldstone, A., and H. Koenig 1973. Physiochemical modifications
    of lysosomal hydrolases during intracellular transport.
    Biochem. J. 132: 267-282.
Hayashi, M., Y. Nakajima, and W. H. Fishman 1964. The cytologic
    demonstration of β-glucuronidase employing naphthol AS-BI
    glucuronide and hexazonium pararosanilin; A preliminary
    report. *J. Histochem. Cytochem.* 12: 293-297.
Ide, H., and W. H. Fishman 1969. Dual localization of β-glu-
    curonidase and acid phosphatase in lysosomes and in
    microsomes. II. Membrane-associated enzymes. *Histochemie*
    20: 300-321.
Kato, K., I. Hirohata, W. H. Fishman, and H. Tsukamota 1972.
    Intracellular transport of mouse kidney β-glucuronidase
    induced by gonadotrophin. *Biochem. J.* 127: 425-535.
Kato, K., H. Ide, T. Shirahama, and W. H. Fishman 1970. Incor-
    poration of [^{14}C] - glucosamine and [^{14}C] - leucine into
    mouse kidney β-glucuronidase induced by gonadotrophin.
    Biochem. J. 117: 161-167.
Leelavathi, D. E., L. W. Estes, D. S. Feingold, and B. Lombardi
    1970. Isolation of a Golgi-rich fraction from rat liver.
    Biochim. Biophys. Acta 211: 124-138.
Levvy, G. A., and J. Conchie 1966. β-Glucuronidase and the
    hydrolysis of glucuronides. in *Glucuronic Acid* (Dutton,
    ed.) Academic Press, New York, 301-364.
Lin, C. W., C. Marsh, M. Sasaki, and W. H. Fishman 1973. Golgi
    β-glucuronidase of androgen-stimulated mouse kidney.
    J. Histochem. Cytochem. 21: 409-410.
Mameli, L., M. Potier, and R. Gianetto 1972. Difference in
    electrophoretic mobility between the lysosomal and the
    microsomal β-glucuronidase of rat liver. *Biochem. Biophys.
    Res. Commun.* 46: 560-563.
Marsh, C. A., C. W. Lin, and W. H. Fishman 1974. Golgi β-glu-
    curonidase of androgen-stimulated mouse kidney. *Biochem.
    J.* in press.

Novikoff, P. M., A. B. Novikoff, N. Quintana, and J. J. Hauw 1971. Golgi apparatus, GERL, and lysosomes of neurons in rat dorsal root ganglia, studied by thick section and thin section cytochemistry. *J. Cell Biol.* 50: 859-886.

Paigen, K. 1961. The effect of mutation on the intracellular location of β-glucuronidase. *Exp. Cell Res.* 25: 286-301.

Pettengill, O. S., and W. H. Fishman 1962. The preparation and purification of β-glucuronidase from mouse liver, kidney, and urine. *J. Biol. Chem.* 237: 24-28.

Plapp, B. V., and R. D. Cole 1967. Demonstration and partial characterization of multiple forms of bovine liver β-glucuronidase. *Biochem.* 6: 3676-3681.

Potier, M., and R. Gianetto 1973. β-Glucuronidase isoenzymes of rat-liver lysosomes. *Can. J. Biochem.* 51: 973-979.

Smith, R. E., and W. H. Fishman 1969. p-(Acetoxymercuric) aniline diazotate, a reagent for visualizing the naphthol AS-BI product of acid hydrolase action at the level of the light and electron microscope. *J. Histochem. Cytochem.* 17: 1-22.

Swank, R. T., and K. Paigen 1973. Biochemical and genetic evidence for a macromolecular β-glucuronidase complex in microsomal membranes. *J. Mol. Biol.* 77: 371-389.

Swank, R. T., K. Paigen, and R. E. Ganschow 1973. Genetic control of glucuronidase induction in mice. *J. Mol. Biol.* 81: 225-243.

Van Laneker, J. L., and P. L. Lentz 1970. Study on the site of biosynthesis of β-glucuronidase and its appearance in lysosome in normal and hypoxic rats. *J. Histochem. Cytochem.* 18: 529-541.

Wakabayashi, M. 1972. β-Glucuronidase in metabolic hydrolysis. in *Metabolic Conjugation and Metabolic Hydrolysis* (Fishman, ed.) Academic Press, New York, 2: 520-602.

Whaley, W. G., M. Dauwalder, and J. E. Kephart 1972. Golgi apparatus; Influence on cell surfaces. *Science* 75: 596-599.

GEL ISOELECTRIC FOCUSING AND THE MICROHETEROGENEITY
OF HUMAN GLUCOSE-6-PHOSPHATE DEHYDROGENASE

BRUCE J. TURNER[1] and STEPHEN D. CEDERBAUM
Division of Medical Genetics, Mental Retardation Unit
UCLA Neuropsychiatric Institute, Los Angeles
California 90024

ABSTRACT. Human erythrocyte glucose-6-phosphate dehydrogenase (G6PD) was analyzed by an improved method of polyacrylamide gel isoelectric focusing. The patterns resolved from crude hemolysates and partially purified preparations were indistinguishable and consisted of 7 to 11 or
more bands, depending upon sample concentration. The patterns were stable to aging and treatment with thioglycolate
and mercuric ion, and apparently were not artifacts induced by residual persulfate. The physicochemical properties and interrelationships of the multiple forms of G6PD
resolved by this technique await elucidation, but the
method shows great promise for studies of the microheterogeneity of the enzyme in human and other mammalian tissues.

INTRODUCTION

The discovery and elucidation of multiple molecular forms
of enzymes has been primarily dependent on advances in the
techniques of analytical protein chemistry. The successive
introductions of ion exchange and gel permeation-chromatogra-
phy, and especially starch and polyacrylamide gel electrophoresis have each led to fundamental insights into the genetic
control, structure, and function of a vast number of enzymes.
Analytical scale isoelectric focusing in polyacrylamide gels
was developed in 1968 shortly after the introduction of preparative isoelectric focusing in sucrose density gradients by
Svensson and Vesterberg (reviewed by Haglund, 1971). The technique has since been modified or refined by other workers (e.
g. Haglund, 1971; Finlayson and Chrambach, 1971; and Vesterberg, 1972). Among these modifications, the technique of
Righetti and Drysdale (1971) appears to provide the most linear
and stable pH gradients; both properties are prerequisites for
the general applicability of gel isoelectric focusing to enzyme studies.
In this paper we describe the results of the application of

---

[1]Present Address: The Rockefeller University, New York, New
York 10021

the method of Righetti and Drysdale, with slight modifications, to glucose-6-phosphate dehydrogenase (G6PD) from human erythrocytes, a well-characterized enzyme of considerable genetic and clinical importance.

MATERIALS AND METHODS

Hemolysate preparation: Blood for the entire series of experiments was taken from a single male Caucasian subject; glass beads were routinely used to remove clotting factors. Erythrocytes were separated from plasma by centrifugation and washed three times with 20 volumes of 0.9% NaCl containing 1mM beta-mercaptoethanol, 1mM EDTA, and 50μM NADP, pH 7.0 at 4°C (1mM DFP was also included in some experiments). Concentrated hemolysates were prepared by adding an equal volume of cold extracting buffer (1mM Pi, 1mM mercaptoethanol, 1mM EDTA, and 50μM NADP, pH 6.4) and allowing the mixture to stand at 4°C for 30 to 40 min with periodic vigorous shaking. Stroma was removed by centrifugation at 49,500 x g for 2 hrs at 4°C.

Partial purification of enzyme and experimental treatments: G6PD was partially purified from 50 cc of whole blood by batch-process DEAE-cellulose chromatography followed by ammonium sulfate precipitation (Motulsky and Yoshida, 1969). These procedures result in the virtual complete elimination of hemoglobin and, in our hands, an 85-90% elimination of 6-phosphogluconate dehydrogenase activity; yields were always better than 80% of initial total activity. The resultant enzyme was stored for up to 14 days at 5 to 7°C in extracting buffer containing 0.5mM DFP and 1 mM epsilon-amino-n-caproate; aliquots were dialyzed overnight against extracting buffer prior to electrofocusing experiments. In some experiments, 25mM thioglycolate was included in the buffer used for dialysis. Experimental treatments of the enzyme with Hg^{2+} ion were in the manner of Taketa and Watanabe (1971); in these cases, mercaptoethanol was omitted from the dialysis buffer.

Gel isoelectric focusing: Ampholines isoelectric at pH 5 to 7 were used at 2% final concentration. Acrylamide was recrystallized once from 1:1 benzene-ethyl acetate and twice from chloroform and N, N'-methylenebisacrylamide ("Bis") was recrystallized twice from acetone; to assure efficient polymerization, stock solutions of both were prepared fresh daily. Ammonium persulfate and TEMED (purified by distillation) were used at final concentrations of 0.015%; polymerization was complete in three hours at 24°C.

Significant modifications of the technique of Righetti and Drysdale (1971) were as follows:

1. The gel solution was 4.5% in total monomer concentration of which 4.5% was "Bis," and contained 10% v/v glycerol, 5%

sucrose, and 50μM NADP; NADP was added to the catholyte at the same concentration.

2. Gels 10.5 cm high were prepared in 12 x 0.5 cm (I.D.) pyrex glass tubes. The top 1.5 cm of the tubes were treated with 1% Photoflo to facilitate water-layering. Prior to loading with gel mixture, the bottoms of the glass tubes were capped with a piece of dialysis membrane previously soaked in 0.2% Gelamide (American Cynamid Corporation); this assures an even bottom to the gel. The dialysis membrane, held by a 5 mm length of Tygon tubing, was left in place throughout the entire experiment.

A water-cooled apparatus similar to that of Righetti and Drysdale (1971) was employed with coolant at 1°C. Gels were allowed to cool to 1°C and "pre-run" for 45-60 min at a constant potential of 200 V/12 tubes. The pre-run tends to eliminate residual persulfate. In some experiments, 2-5μl of 20% thoiglycolate were layered on top of the gel before the pre-run as a further precaution against persulfate-induced artifacts (Maurer, 1970:60); at the end of the experiment, the thioglycolate occupies the lowest 1 cm of the gel and if this is removed, no interference with the tetrazolium-based assay occurs. After the pre-run, the sample was layered on the tops of the gels and focusing continued at constant potentials of 300 V for 12 hrs and then 400 V for 6 hrs.

Sample solutions were 1:1 dilutions of concentrated hemolysate or an equivalent amount of dialyzed enzyme preparation with a "layering-solution" containing 8% ampholine (pH 5-8) and 10% sucrose and adjusted to a final pH of 7.0 with NaOH. Samples contained 50μl of the hemolysate-layering solution mixture per tube, corresponding to an activity of 0.008 I.U. as defined by Motulsky and Yoshida (1969); 0.016 I.U. of enzyme were used in certain experiments.

Enzyme assays: G6PD activity in hemolysates and partially purified preparations was assayed by the method of Motulsky and Yoshida (1969); concentrated hemolysates were diluted at least five-fold prior to assay. The chromogenic assay used after electrofocusing contained (per 100 ml of 0.3M Tris-HCl, pH 8.35) 25 mg glucose-6-phosphate (monosodium salt), 25 mg nitroblue tetrazolium, 10 mg NADP+, and 10 mg phenazine methosulfate; gels were incubated in this mixture for 60 min at 37°C. In some experiments, equimolar amounts of galactose-6-phosphate or 6-phosphogluconate were substituted for the glucose-6-phosphate. After staining, gels were fixed in 7.5% acetic acid and stored overnight at 4°C prior to photography. The tetrazolium-based chromogenic assay appeared entirely compatible with the ampholines in the gels.

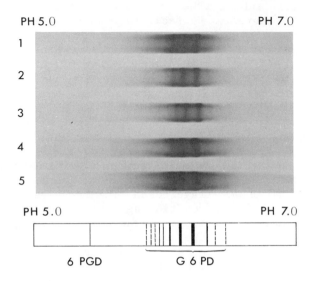

PH 5.0   PH 7.0

PH 5.0   PH 7.0

6 PGD   G 6 PD

Fig. 1. Gel isoelectric focusing of human erythrocyte G6PD. Samples were as follows: 1) 0.012 I.U., crude hemolysate; 2) 0.008 I.U., crude hemolysate; 3) 0.008 I.U., crude hemolysate treated with Hg^{2+} (see text); 4) 0.016 I.U., crude hemolysate; 5) 0.016 I.U., partially purified enzyme. The diagram below shows the average relative positions and intensities of the 11 discernible bands of enzyme activity and the position of 6PGD.

RESULTS

Specific staining of the gels for G6PD activity after 17 to 21 hrs of isoelectric focusing revealed a pattern of 7 to 11 bands (Fig. 1). Of these, three central bands were of greatest prominence; the intensity of the remaining 4 to 8 bands was dependent on the amount of enzyme activity placed on the gel. In terms of the number of bands, the heterogeneity detected in G6PD with gel isoelectric focusing appears to be considerably greater than that obtained by other methods.

The banding patterns of hemolysates and partially purified preparations were virtually identical. The hemolysate pattern was unchanged after storage at 4°C for at least 36 hrs; that of partially purified enzyme was stable for at least 14 days. Dialysis of both hemolysate and purified enzyme against 25mM thioglycolate or treatment of each with Hg^{2+} ion had no disernable effect on the banding pattern. Mercuric ion tends to increase the electrophoretic heterogeneity of human hemolysate and rat liver G6PDs (Taketa and Watanabe, 1971; Watanabe et al., 1972).

The direct application of thioglycolate to the gels prior to the prerun also had no discernable effect on the banding pattern; this is reasonably convincing evidence that the heterogeneity obtained was not the result of some oxidizing effect of residual persulfate or its by-products.

No bands were detected when galactose-6-phosphate was substituted for glucose-6-phosphate in the chromogenic assay. A single band was detected with 6-phosphogluconate; this band corresponded to a light-staining band of very low isoelectric point that was also detected with glucose-6-phosphate, and is probably the enzyme 6-phosphogluconate dehydrogenase.

G6PD from crude human leukocyte preparations could not be successfully resolved by gel isoelectric focusing. The pattern was always a broad smear over about 40% of the length of the gel, sometimes with indefinite bands. Incorporation of up to 3mM DFP into the gel or application of as much as seven-fold more enzyme did not increase resolution. Other workers (e. g. Yoshida et al., 1968) have noted the instability of crude leukocyte G6PD and have suggested that the enzyme is attacked by a cathepsin or pepsin-like protease in these preparations.

The prolonged stability of the focused pH gradients of the Righetti and Drysdale technique was a necessary feature of our experiments with G6PD. Small changes in the banding pattern of the enzyme were detected until about 17 hrs after the start of the experiment; banding patterns of gels analyzed at 17.5 to about 21 hrs showed no detectable differences.

DISCUSSION

There have been several types of apparently non-genetic heterogeneity of G6PD reported in the literature. Of these, two apparently have no relationship to the heterogeneity that we have noted. These are the "stromal effects" discussed by Carson et al. (1966) and the "aging effect" studied by Walter and Caccam (1966). The first is a partial inactivation of the enzyme apparently caused by stromal NADPase; it is prevented by the presence of NADP in concentrations lower than that used in our experiments. The second effect seems to involve binding of oxidized glutathione to some sulfhydryl site on the enzyme with a resultant increase in electrophoretic mobility; it occurs in vitro in stored erythrocytes and hemolysates, and is largely prevented by the presence of mercaptoethanol in the storage medium. The hemolysates used in our experi-

ments contained mercaptoethanol and were routinely prepared
less than 6 hrs prior to electrofocusing.

Another potential source of heterogeneity is the G6PD-pro-
tein (hemoglobin) complex demonstrated by Kirkman and Hanna
(1968). This complex is apparently disassociated by 14mM mercaptoethanol. The hemolysates used in our experiments were
of high protein concentration and contained only 1mM mercaptoethanol. It is, therefore, conceivable that the G6PD-protein
complex might provide a partial explanation of some of the
isoelectric heterogeneity. However, the enzyme-protein complex might be expected to be disassociated with overnight
dialysis against 25mM thioglycolate (a sulfhydryl agent very
similar to mercaptoethanol); this treatment failed to influence the banding pattern obtained in our experiments.

Human G6PD is heterogeneous under certain conditions of
polyacrylamide "disc" electrophoresis. Bakay and Nyhan (1969)
noted a three-banded pattern from hemolysates and suggested
the existence of two genetically distinct subunits. However,
Watanabe et al. (1972) have shown that the bands resolved by
disc electrophoresis have similar molecular weights and are
interconvertible with certain sulfhydryl reagents. By treatment with Hg^{2+} ion, a two-banded pattern from fresh hemolysates
displayed a third component without change of overall enzyme
activity; the conversion was reversible with mercaptoethanol.
The completely negative results we obtained with Hg^{2+} treatment, even at ionic concentrations twice that used by the
Japanese group, at temperatures up to 37°C and at treatment
times of up to 2 hrs, fails to confirm the hypothesis that the
isoelectric h eterogeneity of the enzyme and the electrophoretic studied by Watanabe et al. (1972) have a similar molecular
basis.

The demonstration of enzyme microheterogeneity with a relatively new analytical tool always raises the question of possible experimental artifact. The most obvious potential
source of artifacts in our system of gel isoelectric focusing
is residual persulfate or its by-products (Maurer, 1971).
However, pretreatment of the gels with thioglycolate, which
destroys persulfate, had no detectable effect on the banding
pattern of the enzyme. Although not tested directly in this
work, the investigation of Hayes and Wellner (1969) using
tritiated ampholines suggests that ampholine-protein binding
is not a significant problem. Moreover, the finding of similar microheterogeneity by several groups, using relatively
diverse methods, suggests that the phenomenon is real.

Considerable physicochemical heterogeneity has been demonstrated for human G6PD. The enzyme has been shown to exist
as a tetramer near its isoelectric point and predominantly as

a dimer at its maximally active pH of 8.0 (Yoshida and Hoagland, 1970). Cancedda et al. (1973) presented evidence which suggests that the dimeric form of the enzyme may exist in at least two distinct conformational states mediated by NADP. This oligomeric and conformational heterogeneity may provide a theoretical basis for at least some of the microheterogeneity detected by gel isoelectric focusing.

We have no direct evidence that the isoelectric heterogeeity of human G6PD is of any physiological significance. However, the stability of the observed pattern to aging, partial purification, and treatment with various sulfhydryl reagents suggests the possibility of the existence of the multiple forms in vivo. The microheterogeneity of human G6PD might ultimately be analogous to that of erythrocyte pyruvate kinase where interconvertible isozymes with different physical and kinetic properties have been demonstrated (Ibsen et al., 1971).

ACKNOWLEDGEMENTS

This research was supported in part by: University of California at Los Angeles; Mental Retardation Program, Neuropsychiatric Institute and NIH grants HD-04612, HD-00345, and HD-05615.

REFERENCES

Bakay, B. and W. L. Nyhan 1969. An improved technique for the separation of glucose 6-phosphate dehydrogenase isoenzymes by disc electrophoresis on polyacrylamide gel. *Biochem. Genet.* 3: 571-582.

Cancedda, R., G. Ogunmola, and L. Luzzatto 1973. Genetic variants of human erythrocyte glucose-6-phosphate dehydrogenase; discrete conformational states stabilized by NADP+ and NADPH. *Eur. J. Biochem.* 34: 199-204.

Carson, P. E., F. Ajmar, F. Hashimoto, and J. E. Bowman 1966. Electrophoretic demonstration of stromal effects on haemolysate glucose-6-phosphate dehydrogenase and 6-phosphogluconic dehydrogenase. *Nature* 210: 813-815.

Finlayson, G. R. and A. Chrombach 1971. Isoelectric focusing in polyacrylamide gel and its preparative application. *Anal. Biochem.* 40: 292-311.

Haglund, H. 1971. Isoelectric focusing in pH gradients - a technique for fractionation and characterization of empholytes; pp. 1-104 in: D. Glick (ed), *Methods of Biochemical Analysis,* Vol. 19.

Hayes, M. B. and D. Wellner 1969. Microheterogeneity of L-amino acid oxidase; separation of multiple components by

polyacrylamide gel electrofocusing. *J. Biol. Chem.*
244; 6636-6644.

Ibsen, K. H., K. W. Schiller, and T. A. Haas 1971. Interconvertible kinetic and physical forms of human erythrocyte pyruvate kinase. *J. Biol. Chem.* 246: 1233-1240.

Kirkman, H. N. and J. E. Hanna 1968. Isozymes of human red cell glucose-6-phosphate dehydrogenase. *Ann. N.Y. Acad. Sci.* 151: 133-144.

Maurer, H. R. 1971. *Disc Electrophoresis and Related Techniques of Polyacrylamide Gel Electrophoresis*, 2nd ed. Walter de Gruyter, Berlin and New York, xvi +222 pp.

Motulsky, A. G. and A. Yoshida 1969. Methods for the study of red cell glucose-6-phosphate dehydrogenase; pp. 51-93 in: J. J. Yunis (ed), *Biochemical Methods in Red Cell Genetics*. Academic Press, New York.

Righetti, P. and J. W. Drysdale 1971. Isoelectric focusing in polyacrylamide gels. *Biochim. Biophys. Acta* 236: 17-38.

Taketa, K. and A. Watanabe 1971. Interconvertible microheterogeneity of glucose-6-phosphate dehydrogenase in rat liver. *Biochim. Biophys. Acta* 235: 19-26.

Vesterberg, O. 1972. Isoelectric focusing of proteins in polyacrylamide gels. *Biochim. Biophys. Acta* 257: 11-19.

Walter, H. and J. F. Caccam 1966. Effect of oxidized glutathione on some enzymes of erythrocytes and its relation to erythrocyte enzyme activity and electrophoretic mobility. *Biochem. J.* 100: 274-277.

Watanabe, A., K. Taketa, and K. Kosaka 1972. Three interconvertible forms of glucose-6-phosphate dehydrogenase in human liver and erythrocyte. *Enzyme* 13: 203-210.

Yoshida, A. and V. D. Hoagland 1970. Active molecular unit and NADP content of human glucose-6-phosphate dehydrogenase. *Biochem. Biophys. Res. Commun.* 40: 1167-1172.

Yoshida, A., G. Stamatoyannopoulos, and A. G. Motulsky 1968. Biochemical genetics of glucose-6-phosphate dehydrogenase variation. *Ann. N.Y. Acad. Sci.* 155: 868-879.

INVESTIGATIONS ON THE QUATERNARY STRUCTURE AND FUNCTION OF OLIGOMERIC PROTEINS BY HYBRIDIZATION WITH VARIANTS PRODUCED BY CHEMICAL MODIFICATION*

E. A. MEIGHEN

Department of Biochemistry, McGill University
Montreal, Quebec, Canada

*Supported by a grant from the Medical Research Council of Canada (MA-4314).

ABSTRACT. A relatively homogeneous electrophoretic variant with the same quaternary structure as the native enzyme can be prepared by reaction of limited amounts of succinic anhydride with an oligomeric protein. The modification of approximately 40% of the lysyl residues of *E. coli* alkaline phosphatase or rabbit muscle aldolase will give such variants. These variants are useful for hybridization experiments in order to investigate subunit structure and subunit interactions. Hybridization of native and succinyl aldolase results in three hybrid species consistent with a tetrameric structure. The succinyl derivative of alkaline phosphatase was hybridized with another chemical variant produced by chemical modification of the tyrosine residues of alkaline phosphatase with tetranitromethane. Since both the succinyl and nitrotyrosyl derivatives have dimeric structures and differ in charge, one hybrid species was detected on electrophoresis. The succinyl nitrotyrosyl hybrid was isolated by anion exchange chromatography and then converted to a succinyl aminotyrosyl hybrid by reduction of the nitrotyrosyl residues with sodium dithionite. The activities of the two hybrids agreed with values predicted, assuming that the subunits of alkaline phosphatase function independently. These results are not consistent with mechanisms involving negative homotropic interactions during the rate limiting step of the alkaline phosphatase reaction.

INTRODUCTION

A variant of an oligomeric protein can be prepared by reaction of the native protein with succinic anhydride (Meighen and Schachman, '70a; Klapper and Klotz, '72). This reaction results primarily in the conversion of the positively charged lysyl residues of the protein to negatively charged succinyl lysyl residues as shown below.

$$-NH_3^+ \; + \; \begin{matrix} CH_2-C \overset{O}{\overset{\|}{}} \\ | \rangle O \\ CH_2-C \underset{\|}{\underset{O}{}} \end{matrix} \longrightarrow \; -NH-\overset{O}{\overset{\|}{C}}-(CH_2)_2-\overset{O}{\overset{\|}{C}}-O^- \; + \; 2H^+$$

At appropriate extents of modification, dependent on the particular enzyme, a succinylated variant is produced that (i) can be easily resolved from the native enzyme by electrophoresis or chromatography, (ii) is sufficiently homogeneous to give a relatively sharp band on electrophoresis, (iii) maintains the same quaternary structure as the native enzyme, and (iv) can be dissociated into subunits and then can be reassociated either with succinyl or native subunits. Such a variant satisfies the necessary criteria for hybridization experiments with the native enzyme in order to elucidate the subunit structure (Meighen and Schachman, '70a). Furthermore, these hybrids can be separated and purified and the function of the hybrid species containing two types of subunits investigated. If the kinetic properties of the two types of subunits are sufficiently different then it is possible to determine whether the subunits function independently in the hybrid species. This type of experiment is particularly useful to study enzymes that exhibit negative or positive homotropic interactions in which the subunits should not function independently.

PRODUCTION OF CHEMICAL VARIANTS BY REACTION WITH SUCCINIC ANHYDRIDE

The extent of succinylation of an oligomeric protein that is necessary to give a suitable variant for hybridization experiments can be elucidated by reaction of different amounts of succinic anhydride with the native protein followed by analysis of the properties of the succinyl derivatives by electrophoresis (Figs. 1 and 2). Such experiments are extremely rapid since preparation of the variants and analysis by cellulose acetate electrophoresis can be accomplished in a few hours.

The addition of a low molar excess of succinic anhydride (one to five molecules per lysyl residue) to *E. coli* alkaline phosphatase results in the production of succinyl derivatives that migrate upon electrophoresis as relatively wide bands of higher mobility than the native enzyme (Fig. 1). However, at a higher extent of modification (7-fold molar excess), a sharper and more homogeneous band is obtained. This succinyl

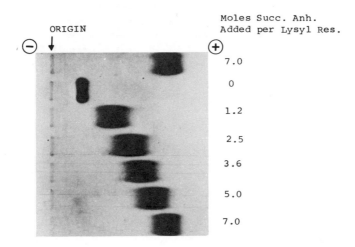

Fig. 1. Cellulose acetate electrophoresis of *E. coli* alkaline phosphatase modified with different amounts of succinic
anhydride. The derivatives were prepared by addition of succinic anhydride to a 0.6% solution of alkaline phosphatase
in 0.05 M Tris-chloride, pH 8.0, and the pH maintained at
8.0 by the addition of NaOH. Electrophoresis was conducted
for 15 minutes at 200 volts in 0.04 M Na/K phosphate, pH 7.0.
The membrane was fixed and stained for 7 minutes with Ponceau
S in trichloroacetic acid and sulfosalicyclic acid and then
stained further in 0.002% nigrosin in 2% acetic acid.

---

variant has 40% of its lysyl residues modified and is ideal
for hybridization experiments since it not only migrates on
electrophoresis as a relatively sharp band with a different
mobility than the native enzyme but sedimentation velocity
experiments have shown that the modified alkaline phosphatase
has the same quaternary structure as the native enzyme.

The results of some earlier work on aldolase (Meighen and
Schachman, '70a) are perhaps more representative of the effects of succinylation on oligomeric proteins since extensive
modification results in the dissociation of aldolase (Fig. 2).
If 24% of the lysyl residues of aldolase are succinylated,
a relatively wide band is obtained on electrophoresis, however, if the extent of modification is increased to 46% the
electrophoretic mobility is not only increased but the band
becomes considerably sharper. These results are similar to
the results obtained for alkaline phosphatase. Further modification of aldolase results in the appearance of a new band
of lower mobility whose relative proportion increases as the

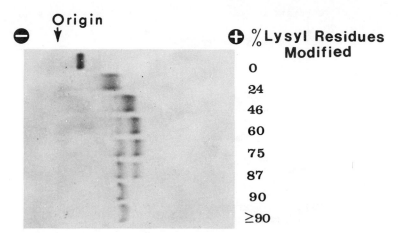

Fig. 2. Cellulose acetate electrophoresis of rabbit muscle aldolase succinylated to different extents. Electrophoresis was conducted for 12 minutes at 250 volts in 0.02 M potassium phosphate, pH 6.5, and the membrane stained for protein as described in Fig. 1. From Meighen and Schachman ('70a).

TABLE I

SUCCINYLATED PROTEINS SUITABLE FOR HYBRIDIZATION EXPERIMENTS[a]

Enzyme	%Lysyl Residues Succinylated	Subunit Structure
Aldolase[b]	45	Tetramer
Glyceraldehyde-3-phos. Dehydrogenase[c]	27	Tetramer
Aspartate Transcarbamylase[d] (catalytic subunit)	58	Trimer
Bacterial luciferase[e]	45	Dimer
Asparaginase (E. coli B)[f]	22	Tetramer
Asparaginase (Erwinia carotovora)[g]	85	Tetramer
Alkaline phosphatase[h]	40	Dimer

[a]Only those enzymes for which it has been possible to elucidate the subunit structure by hybridization of the native and succinylated enzyme are listed. [b]Meighen and Schachman, '70a; [c]Meighen and Schachman, '70b; [d]Meighen et al, '70; [e]Meighen et al, '71; [f]Shifrin and Grochowski, '72; [g]Shifrin et al, '73; [h]Meighen and Yue, unpublished experiments.

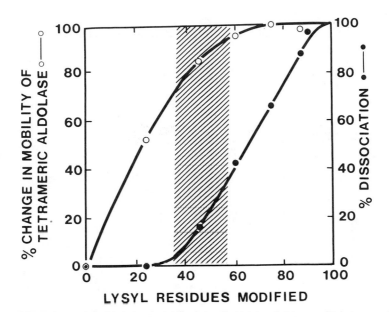

Fig. 3. The change in electrophoretic mobility on cellulose acetate and the extent of dissociation of tetrameric succinyl aldolase as a function of the degree of succinylation of the lysyl residues (Meighen and Schachman, '70a).

extent of modification is increased. Sedimentation velocity experiments on the different succinyl preparations have shown that this species arises from the dissociation of aldolase into subunits. It has been observed with other proteins, as well as aldolase, that the succinyl species with the same quaternary structure as the native enzyme has a higher mobility on cellulose acetate electrophoresis than other succinyl species present in the same preparation due to the dissociation and/or aggregation of extensively modified molecules (Meighen and Schachman, '70; Meighen et al, '71). If this conclusion is applicable to most proteins, it is therefore possible to elucidate by cellulose acetate electrophoresis, the extent of succinylation that will alter the quaternary structure of a specific oligomeric protein.

For enzymes that are dissociated by high extents of succinylation, the selection of a suitable variant is a compromise between obtaining a sample with the minimum amount of dissociated material and yet having the intact succinylated derivative give a sharp band on electrophoresis with a sufficiently different mobility than the native enzyme. Fig. 3

Electrophoresis of Succinyl—aldolase and Aldolase

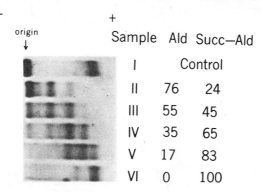

Fig. 4. Hybridization of different percentages of native and succinyl aldolase by dissociation in 4 M urea and reconstitution by dialysis. The control is a mixture of native and succinyl aldolase that has not been dissociated and reconstituted. Cellulose acetate electrophoresis was conducted for 12 minutes at 250 volts in 0.02 M potassium phosphate, pH 6.5, and the membrane stained for protein as described in Fig. 1. From Meighen and Schachman ('70a).

gives a plot of the percentage of dissociated molecules and the percentage change in electrophoretic mobility of tetrameric aldolase as a function of the degree of succinylation of the lysyl residues. A suitable compromise for the degree of modification to give a variant for hybridization experiments is indicated by the slashed area. This data suggests that there is a class of lysyl residues (~40%) on the "exterior" of the protein that can be modified without dissociation of aldolase. If additional residues are modified, the molecule dissociates and the exposed lysyl residues react at a faster rate. A similar conclusion has been reached by Shifrin and Grochowski ('72) for *E. coli* B asparaginase. These workers have directly confirmed such a hypothesis by showing that the degree of modification of the dissociated molecules is greater than that for intact succinyl asparaginase.

The exact extent of succinylation of lysyl residues required to give the most suitable variant for hybridization experiments is dependent on the particular protein (Table I). Other amino acids are also irreversibly modified by succinic anhydride, particularly serine and threonine residues (Gournaris and Perlmann, '67), however, at lower extents of modi-

fication the reaction appears to be reasonably specific for
lysyl residues (Shifrin and Grochowski, '72).

ELUCIDATION OF SUBUNIT STRUCTURE BY THE HYBRIDIZATION OF NATIVE AND SUCCINYLATED PROTEINS

The dissociation and reconstitution of a mixture of a native and succinyl oligomeric enzyme will result in hybrid
species providing that after dissociation, the succinyl subunits can then associate with native as well as succinyl
subunits. Although it has been observed in some cases that
the renaturation of the succinyl derivative is not as efficient nor as rapid as the native enzyme, high ionic strengths
in the renaturation media can increase both the rate and efficiency of refolding of the succinyl enzyme. High salt concentrations presumably eliminate intramolecular and intermo-
lecular charge repulsions due to the negatively charged succinyl residues. It is necessary in these experiments not
only that the succinyl derivatives can be reconstituted but
that the rate of renaturation of the native and succinyl enzymes be of the same order of magnitude or otherwise no
hybrids will be produced.

The results of hybridization of different percentages of
native and succinyl aldolase is given in Fig. 4. Dissociation and reconstitution of succinyl aldolase alone results
in the production of a species with an electrophoretic mobility identical to the undissociated succinyl derivative as
well as a minor component that does not refold. Hybridization of the native and succinyl enzymes gives back not only
the initial species but three hybrid or intermediate species
as would be expected for a tetrameric molecule. These species do not correspond to the minor component in the inactive
succinyl aldolase sample since activity can be detected in
all three intermediate bands.

From the number of hybrid species, the subunit structure
of a homopolymer can be determined; n-1 hybrid species will
be produced for a protein that contains n subunits. If the
subunits are not identical, the number of hybrids observed
will depend upon the relative extent of succinylation of the
different subunits. For example, only one hybrid species was
detected on hybridization of succinyl and native bacterial
luciferase, a heteropolymer containing two nonidentical subunits (α β), since the electrophoretic mobilities of the hybrids, $\alpha_{succ.}$ β and α $\beta_{succ.}$, were identical (Meighen et
al, '71).

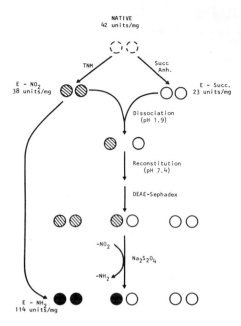

Fig. 5. Experimental protocol for hybridization of chemical variants of alkaline phosphatase. The native enzyme was prepared from *E. coli*, strain C-90, by a procedure similar to that of Malamy and Horecker ('64). The specific activity is given as the units of activity (μmoles/min) at 24° in 10^{-3} M p-nitrophenyl phosphate, 1.0 M Tris-chloride, pH 8.0, per milligram of enzyme. The hybridization experiments and the preparation of the succinyl (E-succ.), nitrotyrosyl (E-NO_2), and aminotyrosyl (E-NH_2) derivatives are described in the text.

INVESTIGATION OF SUBUNIT EXCHANGE

The mixing of two variants under conditions where the enzyme is not dissociated is a useful technique for investigating association-dissociation equilibria and subunit exchange. The earliest work using succinyl derivatives to study subunit exchange (Keresztes-Nagy et al, '65) showed that hybrid species of intermediate electrophoretic mobility were obtained on mixing native and succinyl hemerythrin. Although it was not necessary for the succinyl derivative to give a sharp band on electrophoresis, it is advantageous to have a relatively homogeneous variant for subunit exchange experiments. The hybridization of native and succinyl glyceraldehyde-3-phosphate dehydrogenase was conducted under conditions where

668

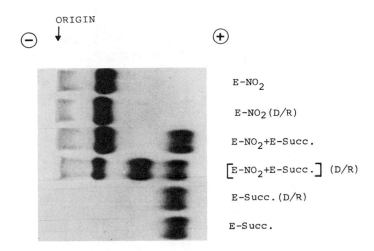

Fig. 6. Hybridization of the nitrotyrosol and succinyl derivatives of alkaline phosphatase. The derivatives were dissociated at pH 1.9 and then reconstituted at neutral pH. The
mixture of nitrotyrosyl and succinyl derivatives that has
been hybridized, [E-NO$_2$ + E-Succ.](D/R), contains 40% succinyl and 60% nitrotyrosyl alkaline phosphatase. Cellulose
acetate electrophoresis was conducted for 15 minutes at 200
volts in 0.04 M Na/K phosphate, pH 7.0, and the membrane
stained for enzyme activity with α-napthyl acid phosphate and
fast red violet salts (Hoffman and Wilhelm, '70).

the enzyme was in a tetrameric structure (Meighen and Schachman, '70b). These experiments showed that a hybrid species
containing two native and two succinyl subunits was produced
before hybrids containing three of one type of subunit and
one of the other. The results indicate a dimer-tetramer and
a monomer-dimer association dissociation equilibria and thus
provide evidence for isologous interactions between the subunits of glyceraldehyde-3-phosphate dehydrogenase. Furthermore, such studies are useful for studying the relative
strengths of different subunit interactions within an oligomeric protein (Smith and Schachman, '71).

INVESTIGATION OF SUBUNIT FUNCTION IN HYBRIDS

The hybrid tetramers containing inactive succinyl and
native aldolase subunits (Fig. 4) have specific activities
proportional to the number of native chains thus indicating
that the subunits function independently (Meighen and Schachman, '70a). Such a result would not necessarily be predicted

for oligomeric proteins with allosteric properties. For example, if a dimeric enzyme could only bind substrate to one
subunit as a result of negative homotropic interactions, then
a hybrid species with one blocked binding site and one native
binding site would be expected to have the same activity as
the native enzyme. This conclusion assumes that blocking the
one active site does not have the same effect as binding substrate; otherwise, the hybrid would be inactive. However, if
the experimental observation of only one binding site in the
native enzyme was due to the presence of two (or more) enzyme
forms, one inactive and one active, then the hybrid species
would be expected to have 50% activity. The investigation of
a hybrid species containing subunits with different functional or kinetic properties should be useful therefore to test
such possibilities and determine whether the subunits function independently.

Alkaline phosphatase from $E.$ $coli$ is a dimeric enzyme for
which negative allosteric behavior has been proposed as part
of the mechanism (Ladzunski et al, '71). This enzyme catalyzes the hydrolysis of phosphate esters as shown below for
p-nitrophenyl phosphate. A covalent bond is formed
initially

$$E + \text{p-nitrophenyl} - \circled{P} \xrightarrow{\text{p-nitrophenol}} E \; \circled{P} \xrightarrow[\text{H}_2\text{O}]{\text{Tris}} P_i + \text{Tris} - \circled{P}$$

between the enzyme and phosphate, resulting in the release of
p-nitrophenol. This bond is then either hydrolyzed to give
inorganic phosphate (P_i) or the phosphate is transferred to
an acceptor alcohol (e.g. Tris buffer).

A number of studies have shown that only one mole of alcohol is released in a presteady state burst under conditions
where the rate limiting step occurs after formation of the
phosphorylated enzyme (Fernley and Walker, '66; Trentham and
Gutfreund, '68). In addition, binding studies have shown
two sites for phosphate with different affinities (Simpson
and Vallee, '70). Ladzunski et al ('71) has proposed a
flip-flop or reciprocating subunit model to explain this behavior. Recent evidence, however, has seriously questioned
the proposal of negative allosteric behavior and indicated
that the subunits of alkaline phosphatase function independently at least in the presteady state (Bloch and Schlesinger, '73). Since both mechanisms (independent and reciprocating subunits) give Michaelis-Menten kinetics, it is not
possible to distinguish between these mechanisms by steady
state kinetics on the native enzyme.

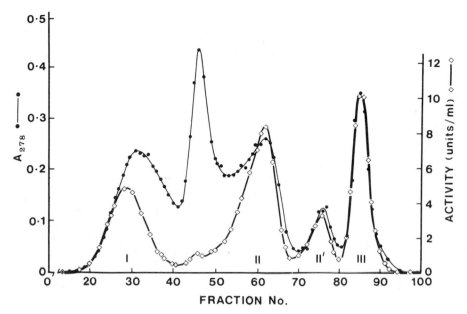

Fig. 7. DEAE-Sephadex chromatography of the hybrid set from
Fig. 6. The sample was applied to a 1.5 x 30 cm DEAE-
Sephadex A-50 column and eluted with a linear gradient (140
ml) of 0.15 to 0.75 M NaCl, in 0.01 M Tris-chloride, 3×10^{-3}
M mercaptoethanol, 4×10^{-3} M $MgCl_2$, 1×10^{-5} M $ZnCl_2$, pH
7.4, at 4°.   Fractions were analyzed for absorbance at 278 nm
and activity (μmoles/min) in 1.0 M Tris-chloride, pH 8.0,
containing 10^{-3} M p-nitrophenyl phosphate.

If , however, we form a hybrid species containing two alkaline phosphatase subunits with different activities, then
steady state kinetics predicts different activities for the
hybrid in the reciprocating and independent subunit models.
Such a hybrid is represented below in the flip-flop
mechanism,

where the open and filled symbols represent alkaline phosphatase subunits with different activities. In this simplified representation, the circle represents an "active" subunit and the square represents a subunit incapable of functioning until the other subunit has turned over in the rate limiting step of the reaction. The enzyme thus reciprocates
between alternate states with both subunits being required
to function. The activity predicted by steady state kine-

TABLE II
THEORETICAL ACTIVITIES OF ALKALINE
PHOSPHATASE HYBRIDS

Predicted activity[a]

Hybrid	Independent subunits	Reciprocating subunits
Succinyl nitrotyrosyl	$v(-) = 30.5 (E_t)$	$v(-) = 29 (E_t)$
Succinyl aminotyrosyl	$v(+) = 68.5 (E_t)$	$v(+) = 38 (E_t)$
	$v(+)/v(-) = 2.25$	$v(+)/v(-) = 1.32$

[a]The predicted activities for independent and reciprocating subunits are given by $v = (k_1 + k_2) (E_t)/2$ and $v = 2k_1k_2 (E_t)/(k_1 + k_2)$ where $k_2 = 23$ units/mg and $k_1 = 38$ units/mg and 114 units/mg for the nitrotyrosyl and aminotyrosyl hybrids, respectively. The activities are given in terms of the amount (mg) of hybrid assayed (E_t) before (-) and after (+) reduction of the nitrotyrosyl residues in the sample with $Na_2S_2O_4$.

tics for a hybrid species in the reciprocating subunit mechanism is given by $v = 2k_1k_2 (E_t/(k_1 + k_2))$, where E_t is the milligrams of hybrid assayed, and k_1 and k_2 have been equated to the specific activity (units/mg) of a dimer containing open or filled subunits respectively. In contrast, the specific activity predicted for a hybrid species with independent subunits is simply the average of the two specific activities ($v = (k_1 + k_2) (E_t)/2$). Consequently it is possible to distinguish between the reciprocating and independent subunit models by steady state kinetics on hybrids.

Such studies require that we have two alkaline phosphatase variants that not only have the same dimeric structure and are electrophoretically distinct but have a sufficient difference in activity so these two models can be distinguished. Fig. 5 describes hybridization experiments between two such chemical variants of alkaline phosphatase.

In this experiment, one variant (E-Succ.) is produced by reaction of alkaline phosphatase with succinic anhydride and the other variant (E-NO$_2$) is produced by nitration of six tyrosine residues of alkaline phosphatase with tetranitromethane (Christen et al., 1971). A mixture of these two variants upon dissociation at acid pH and then reconstitution at neutral pH results in two species with the same electrophoretic mobility as E-NO2 and E-Succ. as well as a band with intermediate electrophoretic mobility corresponding to a hybrid containing one succinyl and one nitrotyrosyl subunit (Fig. 6). The members of the

TABLE III
EFFECT OF $Na_2S_2O_4$ ON ACTIVITIES OF THE
MEMBERS OF SUCCINYL NITROTYROSYL HYBRID SET[a]

Sample	v (+) / v (-)
$E-NO_2$	3.0
I	2.95
II	2.35
II'	2.25
III	1.0
E-Succ.	1.0

[a]The activities were measured in 1.0M Tris-chloride, pH 8.0, either before (-) or after (+) reduction of the nitrotyrosyl residues with $Na_2S_2O_4$. The activity ratio (v (+)/v (-)) is given for a constant volume of sample.

succinyl nitrotyrosyl hybrid set can then be resolved by DEAE-Sephadex chormatography as shown in Fig. 7. Three major activity peaks, I, II, and III, are resolved, corresponding to $E-NO_2$, the hybrid and E-Succ. In addition, a minor activity peak, II', can be observed. No kinetic difference has been observed between this peak (II') and the hybrid species (II). The optical density does not follow the activity at all positions, particularly between peaks I and II, apparently due to inactive protein arising from incomplete renaturation in the hybridization experiment.

Each of the members of the succinyl nitrotyrosyl hybrid set can be investigated kinetically or the nitrotyrosyl residues can be reduced to aminotyrosyl residues with $Na_2S_2O_4$. It may be noted that the activity is greatly stimulated on conversion of the nitrotyrosyl derivative to an aminotyrosyl derivative (Fig. 5) and thus the kinetic properties of the subunits in a succinyl aminotyrosyl hybrid will be substantially different.

The reciprocating and independent subunit models can be distinguished by investigation of the activities of the succinyl nitrotyrosyl and the succinyl aminotyrosyl hybrids. Table II gives the predicted activities for the hybrids in terms of the amount of hybrid assayed. Since the hybrid species cannot be completely resolved experimentally from inactive material (Fig. 7), direct measurement of the exact amount of hybrid assayed was not possible. However, this problem can be eliminated if we measure the increase in activity on conversion of the nitrotyrosyl to aminotyrosyl residues for a constant volume of hybrid (v(+)/v(-)). For the independent subunit model, a 2.25-fold increase in activity is predicted whereas only a 1.32-fold increase in activity is predicted for the reciprocating subunit mechanism.

The effect of treatment with $Na_2S_2O_4$ on the activity of the members of the hybrid set is given in Table III. The three fold stimulation of activity for I corresponds with that observed for the nitrotyrosyl derivative ($E-NO_2$) whereas III is not affected by the addition of $Na_2S_2O_4$ as expected for a molecule containing only succinyl subunits. The hybrid species, II (and II') is stimulated to intermediate extents on reduction of the nitrotyrosyl residues. This result for the hybrid agrees with the predicted activities for the independent subunit model and not with the activities predicted for the reciprocating subunit mechanism. Similar hybridization experiments should be applicable to investigate other enzymes for which negative homotropic interactions have been proposed as part of the mechanism.

REFERENCES

Bloch, W. and M. J. Schlesinger 1973. The phosphate content of *Escherichia coli* alkaline phosphatase and its effect on stopped flow kinetic studies. *J. Biol Chem.* 248: 5794-5805.

Christen, P., B. L. Vallee, and R. T. Simpson 1971. Sequential chemical modifications of tyrosyl residues in alkaline phosphatase of *Escherichia coli*. *Biochemistry,* 10: 1377-1384.

Fernley, N. H. and P. G. Walker 1966. Phosphorylation of *Escherichia coli* alkaline phosphatase by substrate. *Nature* 212: 1435-1437.

Gournaris, A. D. and G. E. Perlman 1967. Succinylation of pepsinogen. *J. Biol. Chem.* 242: 2739-2745.

Hoffman, E. P. and R. C. Wilhelm 1970. Genetic mapping and dominance of the amber suppressor, Sul (supD), in *Escherichia coli* K 12. *J. Bacteriol.* 103: 32-36.

Keresztes-Nagy, S., L. Lazer, M. H. Klapper, and I. M. Klotz 1965. Hybridization experiments: Evidence of dissociation equilibrium in hemerythrin. *Science* 150: 357-359.

Klapper, M. H. and I. M. Klotz 1972. Hybridization of chemically modified proteins. *Methods in Enzymology* 25: 536-540.

Ladzunski, M., C. PetitClerc, D. Chappelet, and C. Ladzunski 1971. Flip-flop mechanisms in enzymology, a model: the alkaline phosphatase of *Escherichia coli*. *Eur. J. Biochem.* 20: 124-139.

Malamy, M. H. and B. L. Horecker 1964. Purification and crystallization of alkaline phosphatase of *Escherichia coli*. *Biochemistry* 3: 1893-1897.

Meighen, E. A. and H. K. Schachman 1970a. Hybridization of native and chemically modified enzymes, I. Development of a general method and its application to the study of the subunit structure of aldolase. *Biochemistry* 9: 1163-1176.

Meighen, E. A. and H. K. Schachman 1970b. Hybridization of native and chemically modified enzymes, II. Hybridization of native and succinylated glyceraldehyde 3 - phosphate dehydrogenase. *Biochemistry* 9: 1177-1184.

Meighen, E. A., V. Piqiet, and H. K. Schachman 1970. Hybridization of native and chemically modified enzymes, III. The catalytic subunits of aspartate transcarbamylase. *Proc. Natl. Acad. Sci. USA* 65: 234-241.

Meighen, E. A., M. Z. Nicoli, and J. W. Hastings 1971. Hybridization of bacterial luciferase with a variant produced by chemical modification. *Biochemistry* 10: 4062-4068.

Shifrin, S. and B. J. Grochowski 1972. L-asparaginase from *Escherichia coli* B succinylation and subunit interactions. *J. Biol. Chem.* 247: 1048-1054.

Shifrin, S., B. G. Solis, and I. M. Chaiken 1973. L-asparaginase from *Erwinia carotovara*. Physiochemical properties of the native and succinylated enzyme. *J. Biol Chem.* 248: 3464-3469.

Simpson, R. T. and B. L. Vallee 1970. Negative homotropic interactions in binding of substrate to alkaline phosphatase of *Escherichia coli*. *Biochemistry* 9: 953-958.

Smith, G. D. and H. K. Schachman 1971. A disproportionation mechanism for the all-or-none dissociation of mercurial-treated glyceraldehyde phosphate dehydrogenase. *Biochemistry*. 10: 4576-4588.

Trentham, D. R. and H. Gutfreund 1968. The kinetics of the reaction of nitrophenyl phosphates with alkaline phosphatase from *Escherichia coli*. *Biochem. J.* 106: 455-460.

RADIOCHEMICAL ASSAY OF ELECTROPHORETICALLY SEPARATED PHOSPHOTRANSFERASES USING LANTHANUM PRECIPITATION

BOHDAN BAKAY
Department of Pediatrics, School of Medicine
University of California at San Diego
La Jolla, California 92037

ABSTRACT. The precipitation by lanthanum chloride of products generated by an enzyme reaction from radioactive
substrates has been used for the detection and quantitation
of adenine phosphoribosyltransferase (APRT), hypoxanthine
guanine phosphoribosyl transferase (HGPRT), xanthine oxidase (XO), adenosine kinase (AK), inosine kinase (IK),
thymidine kinase (TK), galactokinase (GalK), glycerol kinase (GlyK), hexokinase (HexK), and lactate dehydrogenase
(LDH). In each case the isozymes have been separated by
polyacrylamide gel disc electrophoresis. It is clear that
this method has considerable generality, particularly for
reactions in which a labeled substrate is phosphorylated.

It has become recognized that elucidation of gene chromosome, and gene-gene linkage relationships, and gene expression
depend in great part on the availability of specific genetic
markers. For this reason, differences between the various
specialized products arising from the facultative functions of
differentiated cells, such as serum proteins and hormones and
the constitutive (housekeeping) products such as some enzymes,
are constantly sought and explored. Cell products with specific biological functions, which differ electrophoretically
from those of the other cell types and species are especially
valuable.

Mutation in the genome may cause a substitution of an amino
acid in a protein polypeptide chain. This may give rise to
change in the net charge of the protein, it may alter the conformation of the protein, and further it may change the biological properties of the protein. The protein may have a
different electrophoretic migration. Changes arising from
nongenetic processes such as post-translational molecular
reassociations, end group modifications, or binding of cofactors are equally responsible for the formation of protein
markers.

Combined, these processes give rise to multiple molecular
forms of enzymes and proteins with normal and abnormal activities which prevail in various populations and species. For
the most part they represent the accumulation of what may be
termed "benign" mutations that were preserved in populations

and species by the selective processes of evolution. Numerous mutations, however, cause a decrease or a total loss of the biological activity of enzyme proteins.

For a number of years in the Department of Pediatrics at the University of California, San Diego we have, in collaboration with Dr. William L. Nyhan, been involved in the study of inborn errors of metabolism. Some of our efforts have been directed to the study of deficiencies of the X-linked enzymes glucose-6-phosphate dehydrogenase (G6PD; E.C:1.1.1.49) and hypoxanthine guanine phosphoribosyl transferase (HGPRT; E.C:2.4.2.8). Deficiency of G6PD gives rise to hemolytic anemia (Beutler, 1972), while lack of HGPRT activity gives rise to two distinctly different types of disease. An essentially complete deficiency of HGPRT activity is responsible for the Lesch-Nyhan syndrome (Lesch and Nyhan, 1964; Seegmiller et al, 1967). This disorder is characterized by an overproduction of uric acid and hyperuricemia, cerebral dysfunction, mental retardation, and self-mutilation. On the other hand, partial deficiency of HGPRT activity also results in an overproduction of uric acid and hyperuricemia and may cause renal stones and gout, but it does not cause cerebral or behavioral abnormalities (Kogut et al, 1970). Both deficiencies are due to mutation in the structural gene for HGPRT, and are inherited as X-linked recessive traits (Nyhan et al, 1967).

ANALYSIS OF PHOSPHORIBOSYL TRANSFERASES

Hypoxanthine guanine phosphoribosyl transferase catalyzes the transfer of the ribose phosphate moiety of phosphoribosyl pyrophosphate (PRPP) to hypoxanthine and guanine and converts these purine bases to inosine monophosphate (IMP) and guanosine monophosphate (GMP), respectively. In the absence of HGPRT, hypoxanthine is converted by xanthine oxidase (E.C. 1.2.3.3) to xanthine and uric acid.

Another closely related, autosomally linked enzyme which utilizes PRPP converts adenine to adenosine monophosphate (AMP) is called adenine phosphoribosyl transferase (APRT; E.C:2.4.2.7).

In addition to its important metabolic role, and its role as an X-linked genetic marker, HGPRT is of interest because on the one hand it controls the survival of cells grown in vitro in selective media containing hypoxanthine aminopterin and thymidine (HAT) (Szybalski et al, 1962), and on the other hand it causes death of cells in media containing 8-azaguanine (Szybalski et al, 1959). The selection of cells based on the presence or absence of APRT in the media containing adenine and alanosine or 2-fluoroadenine has also been reported (Tischfield and Ruddle, 1974).

As there were no simple tests for HGPRT activity we first developed a radiochemical test tube assay for this enzyme (Bakay et al, 1969). This method which is also applicable to APRT is based on a unique reaction of lanthanum chloride with compounds containing a phosphate group. Products of the HGPRT and APRT reactions are respectively the mononucleotides of hypoxanthine, guanine, and adenine, and therefore they can be precipitated by lanthanum chloride. If the reaction is carried out with radioactively labeled purines, the activity of these enzymes can be established by either analyzing the substrate for the residual radioactivity or by analyzing the insoluble lanthanum nucleotide complexes.

This method was then adapted for the assay of HGPRT and APRT isozymes (Bakay and Nyhan, 1971). Following separation by polyacrylamide disc gel electrophoresis, gels are incubated in solutions containing ^{14}C or ^{3}H labeled substrates hypoxanthine or adenine, or both, and PRPP. Next gels are placed in a solution of lanthanum chloride and washed in deionized water. Since hypoxanthine and adenine are soluble in water, while lanthanum complexes with the IMP and AMP are not, the unutilized free purine bases are washed out and precipitated nucleotides remain fixed in the gel. Gels are then examined for radioactivity by segmentation and analysis in vials (Bakay and Nyhan, 1971), or by automated fractionation and scintillation counting in a hollow tube flow cell (Bakay, 1971).

The apparatus used for the automated determination of radioactivity in the gel is shown schematically in Fig. 1. It consists of a one-channel Beckman β-Mate II Scintillation Spectrometer, a hollow tube flow cell, a Duplex Milton Roy, 31 rpm (160 ml/hr/side) Mini-Pump, a 10 inch linear chart recorder, a six digit printer, a mixer, and a motor-driven gel fractionator (Bakay, 1971). In this system the fractionated gel is picked up by 0.1 $\underline{M}$ Tris-EDTA pH 9 buffer, pumped by piston B of the Mini-Pump, and transported to the mixer, where it is mixed with a water miscible scintillation fluid which is pumped by the piston A. The combined fluids are directed into the flow cell monitored by the Scintillation Spectrometer and out to the waste reservoir.

The profiles of the HGPRT and APRT activities of human erythrocytes are shown in Fig. 2. The top profile was obtained by transverse slicing of the gel into 80 slices, solubilization in hydrogen peroxide, and counting in vials as described before (Bakay and Nyhan, 1971). The tracings shown in the middle and on the bottom represent distribution of radioactivity in the gel as recorded by the chart recorder

679

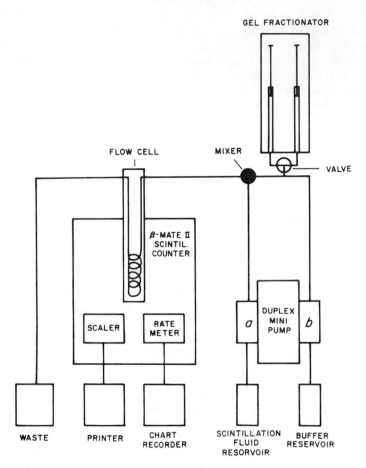

Fig. 1. Interphasing of apparatus for the automated detection of radioactivity in polyacrylamide gel radioelectropherograms.

when gels were analyzed automatically.

Electrophoretically faster migrating APRT produced a symmetrical peak of activity on the right, while slower migrating HGPRT produced a broad zone in the middle of the profile. The APRT enzyme consisted of a single protein, and HGPRT consisted of four distinct isozymes designated here by the letters A, B, C, and D.

The profile on the bottom was produced by hemolysate of a patient displaying about 50 percent of normal HGPRT activity. This variant had a different distribution of isozymes than the normal enzyme above. Furthermore, it migrated at a slower rate than the normal enzyme. Most of the other deficient variants which we have encountered so far migrated 12-15 per-

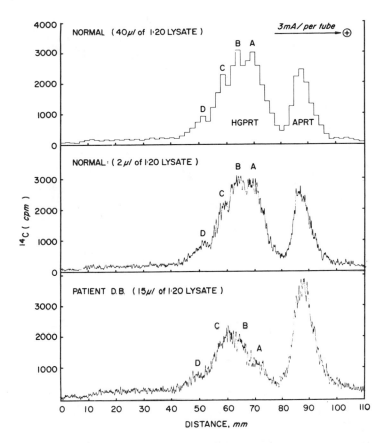

Fig. 2. Radioelectropherograms of HGPRT and APRT of human red cells.

cent faster than normal enzymes (Bakay and Nyhan, 1972a; Bakay et al, 1972).

These profiles also show that the automated method of analysis required 20 times less sample to generate a profile which was comparable to that obtained by the slicing and digestion method of analysis.

By connecting the scaler of the scintillation counter to a digital printer, it was possible to document the accumulated counts generated by HGPRT and APRT and evaluate enzymatic activity of these two enzymes quantitatively (Borden et al, 1974) (Fig. 3). Assay of different aliquots of hemolysates produced values which were linearly proportional to the size of aliquots. When the assay was done under precisely con-

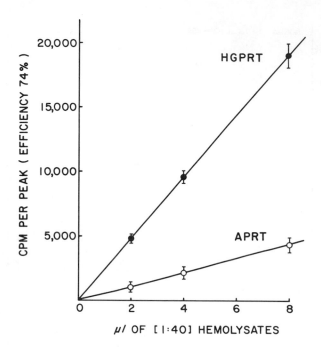

Fig. 3. Quantitation of HGPRT and APRT by automated system of assay in gel.

---

trolled conditions, ^{14}C was counted at 74% and ^{3}H at 13% efficiency with an accuracy of 1 and 2 percent.

We have now used this method for the analysis of a great number of various types of samples. Thus, we found that there are two electrophoretically distinct human variants of APRT in which the enzyme activity is normal (Bakay and Nyhan, 1971). In contrast, so far we have found no variants of HGPRT with normal activity. There are, however, a number of variants with decreased activity among the patients with the Lesch-Nyhan syndrome and those with partial deficiency of HGPRT. This subject was discussed by Dr. W. L. Nyhan and this author at a different session of this conference.

One particularly interesting case was a heterozygous female who had two brothers with hyperuricemia and kidney stone disease (Bakay et al, 1972; Nyhan and Bakay, this volume). Hemolysate of S.F. yielded profile with two HGPRT enzyme variants. (See Nyhan and Bakay, this monograph.) One which migrated like the normal human enzyme designated as HGPRT-M, and the other, which migrated like the partially active

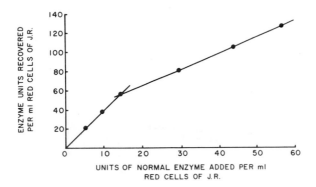

Fig. 4. Activation of HGPRT-LN variant of hemolysates from a patient with the Lesch-Nyhan syndrome by normal enzyme. Samples containing mixture of hemolysates from patient J.R. and a normal individual were separated by electrophoresis and assayed by automated method of analysis.

HGPRT-L⁻ variant of her brother R.L. This indicated that she carries two populations of red cells, one which has normal HGPRT and the other which has deficient enzyme. In this respect, female carriers of the gene for total deficiency of HGPRT differ from some of the female carriers of partial deficiency in that they express only the normal HGPRT in their red cells (Nyhan et al, 1970). When we compared the activity of the HGPRT-L⁻ variant present in the profile of S.F. with that of her brother R.L., it became evident that her variant enzyme was much more active than that of R.L. As we found later this increase of her HGPRT activity in the variant area was due to an interaction of the variant enzyme with the normal enzyme. (Bakay and Nyhan, 1972b).

A similar activation was also observed with hemolysates of patients with the Lesch-Nyhan syndrome. This is shown in Fig. 4. By plotting enzyme units of HGPRT activity recovered against enzyme units added we obtained a kinetic curve whose slope was different as low concentrations of added normal enzyme from that obtained at high concentrations. Up to 15 units of added normal enzymes, the recovery of HGPRT activity was about 400 percent; at higher concentration it was about 200-300 percent. An exchange of subunits may be responsible for this activation (Bakay and Nyhan, 1972b).

The electrophoretic method of HGPRT assay was also adapted for the rapid detection of heterozygous carriers of the Lesch-Nyhan disease by analyzing their hair roots (Francke et al,

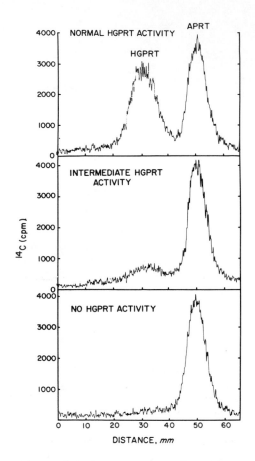

Fig. 5. Detection of heterozygotes for HGPRT activity by analyzing hair roots.

1973) (Fig. 5). As first observed by Gartler and colleagues (1971a, 1971b) many hair follicles develop from a single cell. Therefore, hair roots of heterozygous females express genes of one or the other X chromosome in accordance with the X-inactivation hypothesis of Lyon. Some roots from hetero-zygotes have normal HGPRT and others completely lack activity and produce profiles identical to those of patients with the Lesch-Nyhan syndrome. Some hairs produce intermediate values, indicating that they contain a mixed population of cells. In this diagnostic test HGPRT activity is related to APRT activity which serves as an internal reference standard for each root and an index of its viability. In order to achieve the necessary margin of confidence, we usually analyze 30 hair roots, randomly pulled from the scalp of each female.

The determination of HGPRT activity in polyacrylamide gel has also been utilized for the prenatal diagnosis of the Lesch-Nyhan syndrome. Samples of fetal cells are obtained by amniocentesis and propagated in tissue culture. Along with normal controls they are analyzed for HGPRT and APRT on poly-acrylamide gel. As in the hair root analysis, APRT serves as an internal reference. Cultured normal amniotic fluid cells produced a profile identical to that observed with other normal tissues. Amniotic fluid cells were first tested using this methodology from two concurrent pregnancies one in the mother of two known patients with the Lesch-Nyhan syndrome and the other in her younger sister whose carrier status was not known. The profiles obtained indicated that the first mother was carrying another male fetus with no HGPRT activity. On the basis of this and other tests she decided to terminate this pregnancy. In contrast, the amniotic cells of her sis-ter had normal HGPRT activity, and later she delivered a nor-mal child. We have subsequently had an opportunity to ana-lyze the hair roots of this lady and she is not a heterozy-gote. The first mother is now pregnant for the fourth time. Amniocentesis has been performed and it is clear that this time she is carrying a male fetus with normal HGPRT activity.

Because of the excellent resolving power of polyacrylamide gel electrophoresis it is also possible to separate and dis-tinguish the HGPRT enzymes of cells derived from different species. As shown in Fig. 6 human HGPRT can be distinguished from that of chick, muntjak (a Himalyan deer), and rat, but not from that of mouse. Similarly, human APRT can be sep-arated from that of muntjak, mouse, and rat, but not from that of chick.

ELECTROPHORETIC ANALYSIS OF OTHER ENZYMES

The combination of polyacrylamide gel electrophoresis and precipitation of the products of enzyme reaction with lantha-num has been now used for detection of a variety of other enzymes. Not all phosphorylated compounds are precipitated by lanthanum chloride. Therefore, we first tested the pre-cipitation reaction with various nonlabeled compounds. In this way we accumulated a list of products which could be fixed in the gel by the lanthanum reaction. With this infor-mation we were able to determine which enzymes could be detected by this system of analysis.

In order to establish whether the polyacrylamide gel assay procedure can discriminate between the isozymes of different species, we experimented with the 151,000 x g supernatant

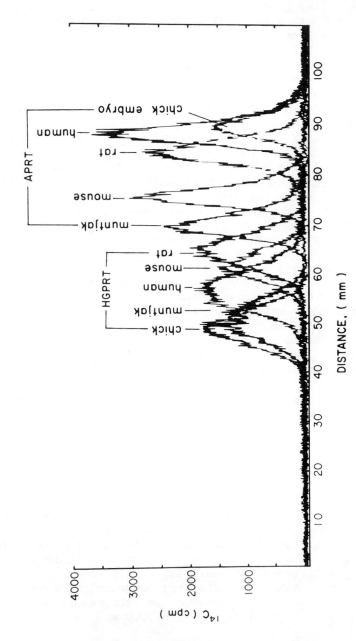

Fig. 6. Radioelectropherograms of HGPRT and APRT isozymes of various species.

fluid prepared from extracts of livers of mouse, rat, and guinea pig. Tissues were homogenized in buffered saline in the presence of dithiothreitol (DTT). Samples of these extracts were separated by electrophoresis in 6.5 or 8 percent polyacrylamide gel and incubated in solutions containing appropriate ^{14}C-labeled substrates and cofactors.

In some instances analysis of liver extracts produced unexpected results (Fig. 7). When gels were developed in sub-

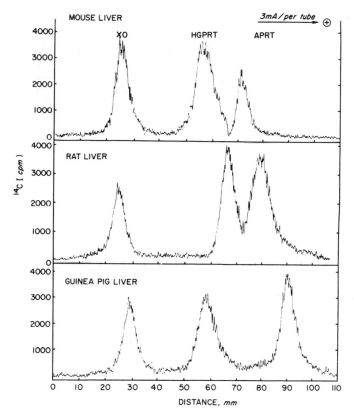

Fig. 7. Radioelectropherograms of xanthine oxidase, HGPRT, and APRT produced by liver extracts of mouse, rat, and guinea pig.

strates containing hypoxanthine and adenine they produced three zones of activity instead of the two we had seen in erythrocytes, fibroblasts, and amniotic fluid cells. The zone on the right represented APRT activity, and the one in the middle, HGPRT. The third zone, on the left, turned out

to have been generated by the xanthine oxidase reactions from
hypoxanthine. Uric acid is formed in this reaction, and uric
acid reacts with lanthanum chloride to form a water insoluble
complex which is fixed in the gel. In 8% polyacrylamide gel,
the APRT enzymes of the mouse, rat, and guinea pig migrated
at different rates and were easy to distinguish. HGPRT of
the rat could be distinguished from that of the mouse and
guinea pig, and the xanthine oxidase of the guinea pig could
be separated from that of the mouse and rat. This experiment
demonstrated also that in this system of enzyme analysis, it
is possible to study side by side two or even three enzyme
markers on the same gel.

When the analysis of liver extracts was carried out in
6.5% gel, and developed in substrate containing xanthine-8-
^{14}C, we obtained better resolved xanthine oxidase profiles in
which the three species were more readily distinguished. In
all three profiles (Fig. 8) this enzyme produced broad zones
of activity, each consisting of at least three components.
As in 8% gel, the xanthine oxidase (XO) of guinea pig migrated
faster than that of mouse and rat. However, the separation
obtained in 120 min of electrophoresis would be insufficient
to distinguish those enzymes in a mixture, and longer runs
are needed to achieve better separations.

Because of our interest in the purine metabolizing enzymes
we proceeded to assess the separation and detection of adenosine kinase (AK: E.C:2.7.1.20) and inosine kinase (IK: E.C:
2.7.1.73). Electrophoretic separation of the proteins of
mouse, rat and guinea pig liver was carried out in 6.5% gel.
The gels were incubated in substrates containing either adeno-
sine-8-^{14}C and ATP, or inosine-8-^{14}C and ATP, washed in water
and analyzed for radioactivity (Fig. 8). After two hours of
electrophoresis, the adenosine kinase of all three species
migrated only a short distance and formed a broad zone of
activity, each consisting of at least two components. Because
their rates of migration were similar the adenosine kinase
enzymes of these three species could not be distinguished
from each other.

Inosine kinase of the same three species migrated also a
short distance (Fig. 8). This enzyme also formed broad peaks
of activity consisting of three distinct isozymes which have
been designated here according to their rate of electrophoretic migration with Roman numerals I, II, and III. Three
isozymes were particularly clearly evident in the profiles
of the guinea pig. A longer run may have resolved these isozymes further, but it would not separate the enzymes of the
three species from each other. Perhaps other electrophoretic
systems could achieve this.

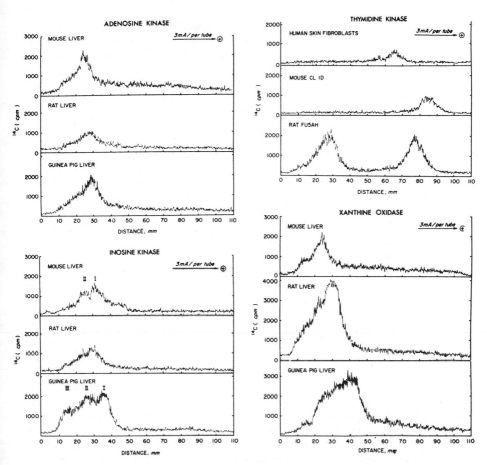

Fig. 8. Radioelectropherograms of adenosine kinase, inosine kinase, thymidine kinase and xanthine oxidase produced by liver extracts of mouse, rat, and guinea pig and human, mouse, and rat fibroblasts, respectively.

Analysis of thymidine kinase (E.C.:2.7.1.21) was carried out on extracts of human, mouse and rat fibroblasts in 8% gel (Fig. 8). After electrophoresis, the gels were developed in a solution containing thymidine-2-^{14}C and ATP and processed in the usual manner. Extracts of human and mouse cells generated well separated single peaks, while extracts of rat FU5AH cells generated two peaks (Fig. 8). The fast and slow migrating isozymes of the rat cell appear to be of different cellular origins. According to reports in the literature the origin of slowly migrating isozymes is in the cytoplasm while that of the faster migrating one is that of mitochondria

(Taylor et al, 1972). Using liver extracts of mouse and rat we also examined the isozymes of galactokinase (E.C.:2.7.1.6) and glycerol kinase (E.C.:2.7.1.30). In this test samples were separated on 6.5 and 8% gel, respectively. The 6.5% gels were incubated in a solution containing D-galactose-1-^{14}C and ATP, and the 8% gels in a solution containing glycerol-^{14}C and ATP. The results are shown in Fig. 9. Galactokinase of mouse and rat migrated at different distances and formed heterogeneous peaks suggesting that this enzyme may consist of more than one molecular form. The separation achieved in 120 min of electrophoresis indicates that these enzymes can be separated from each other well enough to be distinguished in mixtures.

In contrast, glycerol kinase of mouse and rat liver extracts generated single, sharp peaks which were located at about the same distance along the gel, and were indistinguishable from each other.

Separation of mouse, rat, and guinea pig liver extracts on 6.5% gel and incubation in solutions containing D-glucose-^{14}C and ATP were used for the detection of hexokinase (E.C.:2.7.1.1) (Fig. 9). At least four components were evident at widely different distances along the gel in each profile. Furthermore, some components yielded peaks of different size. These differences are important as they might help in establishing the identity of mouse, rat, and guinea pig isozymes.

As tradition goes, each new isozyme system of analysis should include profiles of lactic dehydrogenase (E.C.:1.1.1.27) (LDH) isozymes. With this in mind we charged 8% gels with extracts of normal amniotic fluid cells grown in tissue culture and after electrophoresis, we incubated them in solutions containing lactate and ^{14}C-labeled NAD. As it turned out the generated NADH is precipitable by lanthanum chloride while NAD is not. After washing in water the gels were analyzed in the automated isotope detection system. They produced the profile shown in Fig. 10 consisting of five peaks, representing the five isozymes of LDH. They are marked according to their rate of migration with Roman numerals. This experiment is particularly interesting as it shows that radioactive assays of the enzymes in gel could be used also for the detection of enzymes which generate NADH. We have found that this is also applicable to NADPH.

The question arises of course, as to why one would want to run a radiochemical assay on an enzyme which can be detected by histochemical stains. The main reason is the sensitivity of radiochemical analysis.

In conclusion, I would like to mention that lanthanum precipitation has been used for the detection of HGPRT and other

690

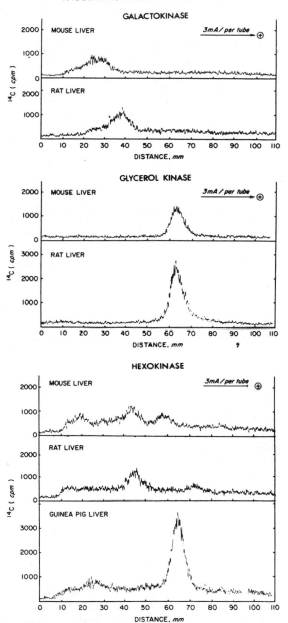

Fig. 9. Radioelectropherograms of galactokinase, glycerol kinase and hexokinase of liver extracts of mouse, rat, and guinea pig, respectively.

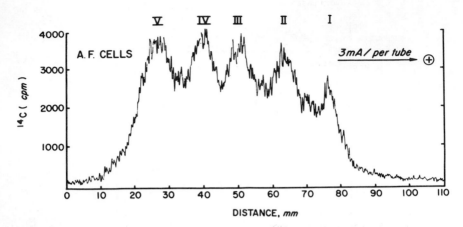

Fig. 10. Radioelectropherograms of LDH of cultured human amniotic fluid cells.

---

phosphotransferases in flat acrylamide gels with the detection of enzymes by radioautography. These methods of analysis have been reported by Tischfield et al (1973) and by McBride and Ozer (1973).

ACKNOWLEDGMENTS

The technical assistance of Marsha Graf and Roswitha Enright is gratefully acknowledged. This work was supported by grants No. GM11702 from the National Institutes of General Medical Sciences, National Institutes of Health, U.S. Public Health Service; and No. CRBS 242 from the National Research Foundation.

REFERENCES

Bakay, B. 1971. Detection of radioactive components in poly-acrylamide gel disc electropherograms by automated mechanical fractionation. *Anal. Biochem.* 40: 429–439.

Bakay, B. and W. L. Nyhan 1971. The separation of adenine and hypoxanthine-guanine phosphoribosyltransferase isoenzymes by disc gel electrophoresis. *Biochem. Genet.* 5: 81–90.

Bakay, B. and W. L. Nyhan 1972a. Electrophoretic properties of hypoxanthine-guanine phosphoribosyl transferase in erythrocytes of subjects with the Lesch-Nyhan syndrome. *Biochem. Genet.* 6: 139–147.

Bakay, B. and W. L. Nyhan 1972b. Activation of variants of hypoxanthine-guanine phosphoribosyl transferase by the normal enzyme. *Proc. Natl. Acad. Sci.* U.S.A. 69: 2523-2527.

Bakay, B., W. L. Nyhan, N. Fawcett, and M. D. Kogut 1972. Isoenzymes of hypoxanthine-guanine phosphoribosyl transferase in a family with partial deficiency of the enzyme. *Biochem. Genet.* 7: 73-85.

Bakay, B., M. A. Telfer, and W. L. Nyhan 1969. Assay of hypoxanthine-guanine and adenine phosphoribosyl transferases. a simple screening test for the Lesch-Nyhan syndrome and related disorders of purine metabolism. *Biochem. Med.* 3: 230-243.

Beutler, E. 1972. Glucose-6-phosphate dehydrogenase deficiency. in *The Metabolic Basis of Inherited Disease* (Stanbury, Wyngarden, and Fredrickson, eds). McGraw-Hill Book Co., 1358-1388.

Borden, M., W. L. Nyhan, and B. Bakay 1974. Increased activity of adenine phosphoribosyl transferase in erythrocytes of normal newborn infants. *Pediat. Res.* 8: 31-36.

Francke, U., B. Bakay, and W. L. Nyhan 1973. Detection of heterozygous carriers of the Lesch-Nyhan syndrome by electrophoresis of hair root lysates. *J. Pediat.* 82: 472-478.

Gartler, S. M., E. Gandini, H. T. Hutchison, B. Campbell, and G. Zechhi 1971. Glucose-6-phosphate dehydrogenase mosaicism: utilization in the study of hair follicle variegation. *Ann. Hum. Genet. Lond.* 35: 1-7.

Gartler, S. M., R. C. Scott, J. L. Goldstein, B. Campbell, and R. Sparkes 1971. Lesch-Nyhan syndrome: Rapid detection of heterozygotes by use of hair follicles. *Science* 172: 572-573.

Kogut, M. D., G. N. Donnell, W. L. Nyhan, and L. Sweetman 1970. Disorder of purine metabolism due to partial deficiency of hypoxanthine-guanine phosphoribosyl transferase. *Amer. J. Med.* 48: 148-161.

Lesch, M. and W. L. Nyhan 1964. A familial disorder of uric acid metabolism and central nervous system function. *Amer. J. Med.* 36: 561-570.

McBride, O. W. and H. L. Ozer 1973. Transferase of genetic information by purified metaphase chromosomes. *Proc. Natl. Acad. Sci.* U.S.A. 70: 1258-1262.

Nyhan, W. L., B. Bakay, J. D. Connor, J. F. Marks, and D. K. Keele 1970. Hemizygous expression of glucose-6-phosphate dehydrogenase in erythrocytes of heterozygotes for the Lesch-Nyhan syndrome. *Proc. Natl. Acad. Sci.* U.S.A. 65: 214-218.

Nyhan, W. L., J. Pesek, L. Sweetman, D. G. Carpenter, and C. H. Carter 1967. Genetics of an X-linked disorder of uric acid metabolism and cerebral function. *Pediat. Res.* 1: 5-13.

Seegmiller, J. E., F. M. Rosenbloom, and W. N. Kelley 1967. Enzyme defect associated with a sex-linked human neurological disorder and excessive purine synthesis. *Science* 155: 1682-1684.

Szybalski, W. and M. J. Smith 1959. Genetics of human cell lines. I. 8-azaguanine resistance, a selective "single step" marker. *Proc. Soc. Exp. Biol. and Med.* 101: 662-666.

Szybalski, W., E. H. Szybalski, and F. Rangi 1962. Genetic studies with human cell lines. *Nat. Cancer Inst. Mono.* 7: 75-88.

Taylor, A. T., M. A. Stafford, and O. W. Jones 1972. Properties of thymidine kinase partially purified from human fetal and adult tissue. *J. Biol. Chem.* 247: 1930-1935.

Tischfield, J. A., H. P. Bernhard, and F. H. Ruddle 1973. A new electrophoretic autoradiographic method for the visual detection of phosphotransferases. *Anal. Biochem.* 53: 545-554.

Tischfield, J. A. and F. H. Ruddle 1974. Assignment of the gene for adenine phosphoribosyltransferase to human chromosome 16 by mouse human somatic cell hybridization. *Proc. Natl. Acad. Sci.* U.S.A. 71: 45-59.

PRESENCE OF A DIPHOSPHOGLYCERATE PHOSPHATASE ACTIVITY IN THE MULTIPLE MOLECULAR FORMS OF 3-PHOSPHOGLYCERATE MUTASE.

RAYMONDE ROSA, ISABELLE AUDIT, AND JEAN ROSA
Unite 91 de l'N.S.E.R.M. - Hôpital Henri Mondor
94010   Creteil   France

ABSTRACT. The electrophoretic patterns of 3-phospho-glycerate mutase (3-PG mutase) and 2,3-diphosphoglycerate phosphatase (DPG phosphatase) have been studied in human tissue extracts. At pH 7.5, 3-PG mutase activity was revealed as three bands in heart tissue, one band in liver, and one band in skeletal muscle. On the other hand, two bands were detected in white blood cells and three bands were apparent in red blood cells.

DPG phosphatase activity was detected in all of the electrophoretic bands of 3-PG mutase from heart, liver, and skeletal muscle. Only one of the three bands of red cell 3-PG mutase was shown to possess DPG phosphatase activity and no activity was detected in white blood cells.

The electrophoretic pattern of red cell 3-PG mutase has been studied in several animal species. In every case the DPG phosphatase activity, when it is present, has been found in one band of 3-PG mutase. Chromatographic separation of the three forms of the human red cell 3-PG mutase has been accomplished.

We consider here a mechanism for the simultaneous presence of 3-PG mutase and DPG phosphatase activities.

2,3-Diphosphoglycerate (2,3-DPG) has been shown to play a very important role in the oxygen affinity of hemoglobin by Benesch and Benesch (1967). Many papers have been published about the two enzymatic activities implicated in its metabolism: 3-phosphoglycerate mutase (3-PG mutase, EC 2.7.5.3.), which catalyzes the reversible transformation of 2-phospho-glycerate to 3-phosphoglycerate in the presence of 2,3-DPG, and 2,3-diphosphoglycerate phosphatase (DPG phosphatase, EC 3.1.3.13), which catalyzes the hydrolysis of 2,3-DPG to 3-phosphoglycerate and inorganic phosphate, and many attempts have been made to separate these two activities by column chromatography. The red cell 3-phosphoglycerate mutase could be obtained free of phosphatase activity, but in most cases the peak of DPG phosphatase contained a slight 3-PG mutase activity. We tried to approach the problem by an electrophoretic study similar to that employed in the comparative studies of DPG phosphatase and DPG mutase, and this proved to be successful (Rosa, Gaillardon, and Rosa, 1973). We report here an electrophoretic study of 3-PG mutase and DPG phosphatase activities from human tissue extracts. Extracts of human heart, liver and skeletal muscle, obtained from autopsy, were prepared by homogenizing in saline solution and toluene, and centrifuging at 30,000 x g for 60 min. The clear supernatant fluid was used for electrophoresis. Human white blood cells were prepared according to the method of Evans and Kaplan (1966). Hemolysates were prepared from washed red cells by freezing and thawing three times and lysis with two volumes of distilled water.

Electrophoresis was performed at 4°C for three hours at 220 V on cellulose acetate strips (Cellogel from Chemetron - Italy). Electrode compartments contained 75mM Tris-citric acid buffer, pH 7.5, with 4mM EDTA, according to the technique of Meera Khan (1971).

The enzymatic activities were revealed by observing the fluorescence decrease resulting from the oxidation of NADH to NAD.

For the detection of 3-PG mutase we applied the following sequence of reactions:

$$\begin{array}{c} (2,3\text{-diphosphoglycerate}) \\ \text{2-phosphoglycerate} \xrightarrow{} \text{3-phosphoglycerate} \\ \textit{3-phosphoglycerate mutase} \end{array}$$

$$\begin{array}{c} \text{3-phosphoglycerate + ATP} \xrightarrow{} \text{1,3-phosphoglycerate + ADP} \\ \textit{3-phosphoglycerate kinase} \end{array}$$

$$\begin{array}{c} \text{1,3-phosphoglycerate + NADH} \xrightarrow{} \text{3-phosphoglyceraldehyde + NAD} \\ \textit{3-phosphoglyceraldehyde dehydrogenase} \end{array}$$

The DPG phosphatase activity was revealed according to the technique previously described by us (Rosa, Gaillardon, and Rosa, 1973). In the presence of 2-phosphoglycolate added as as activator according to Rose and Liebowitz (1970), the following sequence of reactions was followed:

$$\begin{array}{c} (2\text{-phosphoglycolate}) \\ \text{2,3-diphosphoglycerate} \xrightarrow{} \text{3-phosphoglycerate + Pi} \\ \textit{2,3-DPG phosphatase} \end{array}$$

$$\begin{array}{c} \text{3-phosphoglycerate + ATP} \xrightarrow{} \text{1,3-phosphoglycerate + ADP} \\ \textit{3-phosphoglycerate kinase} \end{array}$$

$$\begin{array}{c} \text{1,3-phosphoglycerate + NADH} \xrightarrow{} \text{3-phosphoglyceraldehyde + NAD} \\ \textit{3-phosphoglyceraldehyde dehydrogenase} \end{array}$$

Figure 1. shows the electrophoretic pattern of 3-PG mutase from human tissues, at pH 7.5. In liver extracts, one band

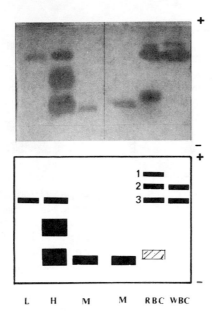

L   H   M   M   RBC WBC

Fig. 1.   Electrophoresis of 3-PG mutase activity from human
tissues: From left to right: Liver (L), Heart (H), Muscle
(M), Muscle (M), red blood cells (RBC), and white blood cells
(WBC).   The hatched zone indicates the location of hemoglobin.
The staining mixture contained: 0.05 M triethanolamine-HCl
buffer pH 7.5, 0.8 mM 2-phosphoglycerate, 0.08 mM 2,3-diphos-
phoglycerate, 3 mM ATP, 1 mM $MgCl_2$; 0.25 mM NADH, 4.8 units
of phosphoglycerate kinase and 4.8 units of 3-phosphoglyce-
raldehyde dehydrogenase.   Five ml of the staining mixture
previously heated to 45°C, 1.5 ml of liquid 3% agarose were
added.   After the solution had cooled to 45°C, the mixture was
poured on the strip which was submitted to U.V. light.   All
the substrates and enzymes came from Boehringer when not
otherwise specified.

of 3-PG mutase activity is visible, migrating at the same
level as the fastest of the three bands appearing in heart
tissue, while the band of the skeletal muscle has the same
electrophoretic mobility as the slowest band of heart tissue.
Two additional minor bands, not represented in this figure,
appeared in liver and skeletal muscle when a more concentrated extract was applied.   These bands migrated at the same
level as the other two bands of heart tissue.
    The electrophoretic pattern of red blood cells also shows
three bands of 3-PG mutase activity, one of which migrated
like the band from liver.   The other bands migrated more
rapidly.   In white blood cells, the two bands migrated with

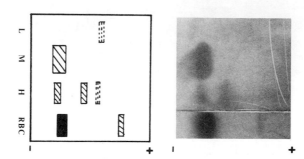

Fig. 2. Electrophoresis of 2-phosphoglycolate enhanced-DPG phosphatase activity of human tissues. From top to bottom: liver (L), muscle (M), heart (H), and red blood cells (RBC). The dark zone represents hemoglobin. The staining mixture contained: 0.05 M triethanolamine-HCl buffer pH 7.5, 0.8 mM 2,3-diphosphoglycerate, 3mM ATP, 1mM $MgCl_2$, 0.25 mM NADH, 4.8 units of phosphoglycerate kinase and 4.8 units of 3-phosphoglyceraldehyde dehydrogenase. 1 mM phosphoglycolate was added to the mixture as an activator.

the two major bands of red cell 3-PG mutase.

Figure 2 shows the results of staining for the 2-phosphoglycolate enhanced-DPG phosphatase activity of the same human tissues. A very weak activity appeared in liver extracts and in the fastest band of heart, the two bands migrating at the same level. Two other major bands were revealed in heart tissue and one pronounced zone of activity appeared in skeletal muscle, migrating at the level of the slowest band in heart extracts.

On the other hand, the red blood cells showed a single band, migrating more rapidly than the other bands. No phosphatase activity was detected in white blood cells.

The pattern of the two enzymatic activities, 3-PG mutase and DPG phosphatase are compared in Figure 3. In liver, heart muscle, and skeletal muscle extracts, DPG phosphatase activity is revealed at the same level as all the bands of 3-PG mutase whereas, in red cells, the band of DPG phosphatase activity appeared at the same level as the band 1 of 3-PG mutase.

In order to eliminate artifacts arising from the use of sequential reactions, several controls were run: 1. omission of each reagent from the staining mixture; 2. staining with the LDH system; 3. staining for the second product of the DPG phosphatase reaction, i.e. inorganic phosphate. These controls allow us to conclude that the zones of flourescence that decrease are indeed due to the 3-PG mutase and the DPG phosphatase activities.

RED BLOOD CELL 3-PGM AND DPGP :

ELECTROPHORETIC PATTERN FROM DIFFERENT MAMMALS

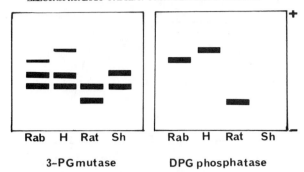

3-PG mutase   DPG phosphatase

Fig. 3.   Comparative electrophoretic pattern of 3-PG mutase
and DPG phosphatase activities of human tissues.   From left
to right, the applied samples are extracts of: liver (L),
heart (H), muscle (M), and red blood cells (RBC).   The left
side of the figure represents the staining for 3-PG mutase,
the right side, staining for DPG phosphatase.

   The results obtained with human hemolysates led us to test
the electrophoretic pattern of 3-PG mutase and DPG phosphatase activity in hemolysates from several animal species
(Fig. 4). We detected three bands of mutase activity appearing in hemolysates of rabbit and horse blood and two bands in
hemolysates of rat and sheep blood.
   On the other hand, the DPG phosphatase activity appears as
one band in hemolysates of rabbit, horse, and rat, at the

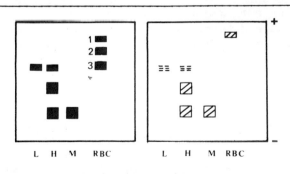

3-PGM   DPG Phosphatase

Fig. 4.   Electrophoretic pattern of red blood cell 3-PG
mutase and DPG phosphatase from different mammals.   From left
to right the samples are hemolysates of: rabbit, horse, rat,
and sheep.

699

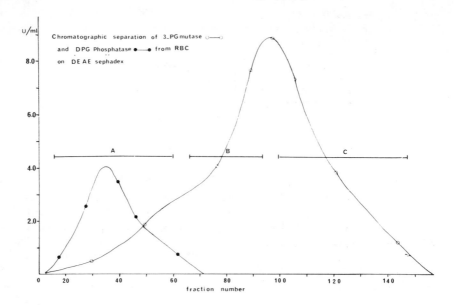

Fig. 5. Chromatography of human red cell 3-PG mutase and DPG phosphatase on DEAE Sephadex. A crude preparation was obtained free of hemoglobin by passage through a CM Sephadex column in 0.01 M sodium phosphate buffer pH 6.4. This sample was applied on a DEAE Sephadex column in a 0.05 M potassium phosphate buffer pH 6.2. Elution of 3-PG mutase and DPG phosphatase was obtained by a gradient of 0.05 M to 0.1 M potassium phosphate buffer pH 6.2. All the buffers contained 20% glycerol, 1mM EDTA and 1mM β-mercaptoethanol. The enzymatic activities were determined, using the same principle and the same mixture as used for the electrophoretic staining (see legend Fig. 1 and 2). The fractions were collected in three pools, as represented on the figure by A, B, C.

---

level of the band 1 of the 3-phosphoglycerate mutase. No phosphatase activity was detected in hemolysates of sheep blood. Several buffers were tested at different pH values between 6.5 and 9.0. In every case the band of DPG phosphatase activity, when present, migrated at the same level as one of the bands of 3-PG mutase.

In order to confirm the existence of three forms of 3-PG mutase in human erythrocytes, we attempted the chromatographic separation of these three forms.

In chromatography on DEAE Sephadex (Fig. 5), the DPG phosphatase activity was detected in the early fractions,

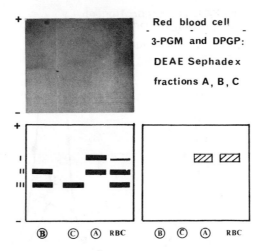

Red blood cell
3-PGM and DPGP:
DEAE Sephadex
fractions A, B, C

Fig. 6. Electrophoresis of red cell 3-PG mutase and DPG phosphatase activities eluted from DEAE Sephadex. Elution fractions A, B, C are compared to a crude extract of RBC.

and 3-PG mutase activity began to appear in the later fractions with the main portion of this activity appearing after the mutase was eluted. After completion of chromatography, these fractions were collected in three pools A, B, and C with pool A containing the main portion of DPG phosphatase activity. Each pool was concentrated by ultrafiltration before being analyzed by electrophoresis. As shown in Fig. 6, fraction A containing some 3-PG mutase activity shows two bands of activity migrating at the same level as bands 1 and 2 of red blood cells, whereas fraction B contains two bands migrating at the level of the bands 2 and 3 of red blood cells. Fraction C, which contains the bulk of the 3-PG mutase activity peak, yields a single band migrating at the level of band 3 of red cell 3-PG mutase. On the right side of the figure, the DPG phosphatase activity is revealed in fraction A as a single band migrating at the same level as the band 1 of the mutase. No DPG phosphatase activity was detected in the fractions B and C. These results furnish evidence for a possible separation of the band 3 of the red cell 3-PG mutase from the other two bands. In order to separate the band 1 from band 2 we have submitted the fraction A to chromatography on hydroxylapatite.

Fig. 7 represents the elution profile of this chromatography, showing two distinct peaks. The first peak contained most of the 3-PG mutase activity, the second peak contained

701

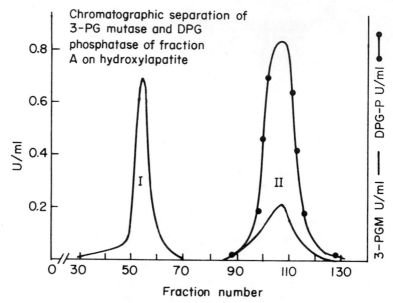

Fig. 7. Chromatographic separation of 3-PG mutase from DPG phosphatase on hydroxylapatite. Fraction A eluted from DEAE Sephadex was applied to a hydroxylapatite column, in 0.01 M potassium phosphate buffer, pH 7.2. Elution of peak I was with 0.05 M potassium phosphate buffer pH 7.2. Peak II was eluted with 0.075 M potassium phosphate buffer at the same pH. All the buffers contained 20% glycerol and 1mM β−mercaptoethanol.

all of the DPG phosphatase activity. In addition, a trace of 3-PG mutase activity was detected in the second peak. These two peaks were concentrated and submitted to electrophoresis. Fig. 8 shows the two enzymatic activities revealed from the two peaks after electrophoresis. A single band of 3-PG mutase activity is visible in each peak. The band of the peak I migrates at the level of the slower band of the fraction A eluted from DEAE Sephadex, which has the same mobility as the band 2 of RBC.

On the right side of Fig. 8, the DPG phosphatase activity appears in the band of the peak II of hydroxylapatite, migrating at the same level as the band 3-PG mutase of this peak. No DPG phosphatase activity was detected in the first peak.

In summary, the electrophoretic pattern of 3-PG mutase from red blood cells showed three bands. Band 1 carried a 2-phosphoglycolic acid-enhanced-DPG phosphatase activity. Bands 2

ELECTROPHORESIS OF PEAKS I AND II ELUTED
FROM HYDROXYLAPATITE

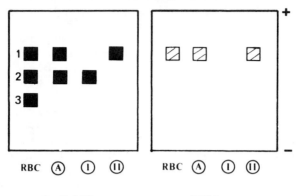

3-PGM   **DPGP**

Fig. 8. Electrophoretic pattern of the peaks I and II obtained by chromatography on hydroxylapatite of the fraction
A from the DEAE Sephadex column, and for comparison a crude
preparation of RBC. The left side of the figure represents
the 3-PG mutase activity, whereas the DPG phosphatase activity is shown on the right side.

---

and 3 were devoid of DPG phosphatase activity. These three
bands have been separated by chromatography. Band 3 has been
isolated on DEAE Sephadex, bands 2 and 1 on hydroxylapatite.

On the other hand, our electrophoretic technique allowed
us to detect three different bands of 3-PG mutase activity in
skeletal muscle, heart, and liver, and DPG phosphatase activity was associated with each band.

These results confirm those of other authors who have
attempted chromatographic separation of DPG phosphatase
activity from the 3-PG mutase peak. Usually, their preparations contained 3-PG mutase activity. Data of Joyce and
Grisolia (1958) on animal muscles are consistent with ours.
De Verdier and Groth (1973) working on human red cells have
demonstrated by another technique using ^{32}P labelled substrate that, in spite of several steps of purification, the
peak of DPG phosphatase activity contained a 3-PG mutase
activity.

On the other hand we could not detect the two DPG phosphatase activities in red cells as reported by Harkness,
Thompson, Roth, and Grayson (1970). These authors used bisulfite and pyrophosphate as activators of the DPG phosphatase. When we employed these reagents with our technique,
we observed the same bands as with the 2-phosphoglycolic acid

in skeletal muscle, heart, and liver tissues. But we did not
observe any band of DPG phosphatase activity in red cell
preparations. The failure to reveal these two activities in
red cells appears to be due to the fact that the stimulation
of DPG phosphatase by bisulfite and pyrophosphate is not
sufficient to allow the enzymatic detection by our technique.
Nevertheless this does not exclude the possible existence in
red blood cells of another DPG phosphatase activity not stimulated by 2-phosphoglycolic acid.

In this report we have demonstrated the existence of multiple molecular forms of 3-PG mutase. Until now, there has
been little information concerning the electrophoretic pattern
of 3-PG mutase (Omenn and Wade Cohen, 1971). Therefore, the
comparison of the electrophoretic patterns we have obtained
with extracts from human heart muscle, skeletal muscle, and
liver are pertinent to an interpretation of the three bands
observed in the heart. In all of the various electrophoretic
conditions tested, one band of the heart showed the same
electrophoretic mobility as that of the muscle, and another
migrated with the band from the liver. The existence of one
isozyme of liver type and one isozyme of muscle type is suggested. In addition, the presence of a third band occupying
an intermediate position suggests that it may be a liver-
muscle hybrid isozyme.

The interpretation of the electrophoretic pattern of
RBC 3-PG mutase is much more difficult. While band 3 has the
same electrophoretic mobility as that of the "liver" isozyme,
the two other bands have different mobilities from those of
the "muscle" isozyme. These facts suggest that there is
another isozyme type which is present in red and white blood
cells.

These multiple molecular forms of 3-PG mutase are not
peculiar to man since we have found an identical isozymic
pattern in rabbit tissues. However, no definitive conclusions
can be drawn at the present time and further research is needed to interpret these results.

Among the different electrophoretic and chromatographic
forms of 3-PG mutase, certain ones carry a DPG phosphatase
activity, while in other forms this activity is very slight
or absent. These data allow us to suggest several hypotheses
as to the structure of these molecules: one is that DPG
phosphatase and 3-PG mutase represent two distinct molecules
having the same isoelectric point, which are not separated,
at least by our electrophoretic techniques. However, it was
found that the electrophoretic mobilities of both enzymes
differed in RBC from one species to another (Fig. 4); such
differences in electrophoretic mobility are proof of several

mutations that led to some modifications of the structure of these enzymes during phylogeny. It is improbable that such mutations exhibit the same modifications of structure leading to a similar charge carried by both enzymes in all species tested.

A second hypothesis is that DPG phosphatase and 3-PG mutase are two distinct proteins which are united in a multienzyme complex. But, this does not explain why, in red cells, only one of the three bands contains DPG phosphatase activity whereas this activity is found an all three bands of heart muscle. Until now this possibility cannot be excluded.

A third hypothesis is that the two enzymatic acitivities could be carried by a single molecule. This type of enzyme should be represented by band 1 of red blood cells. In a previous article (Rosa, Gaillardon, and Rosa, 1973) we have provided some evidence that this band of DPG phosphatase contained a DPG mutase activity. In addition, Rose and Whalen (1973), studying a purified preparation of red cell DPG mutase, had demonstrated that this preparation contained both DPG phosphatase and 3-PG mutase activities. On the other hand, we have studied some properties of the 3-PG mutase of the two peaks eluted from hydroxylapatite. There is evidence for kinetic differences between the two activities. The 3-PG mutase of peak II containing the whole DPG phosphatase activity (Fig. 7) is stimulated by the addition of 2,3-DPG to the standard assay. However, peak I which contains the main portion of 3-PG mutase activity is not stimulated by this addition of 2,3-DPG. These results are consistent with those of Rose and Whalen (1973) who have obtained the same kind of data by another technique using ^{32}P labelled substrates.

Further experiments are necessary to elucidate the molecular basis for the presence of 2,3-diphosphoglycerate phosphatase activity in certain electrophoretic and chromatographic forms of 3-PG mutase.

ACKNOWLEDGEMENT

The authors wish to acknowledge the technical assistance of Josette Leheurteux.

REFERENCES

Benesch, R. and R. E. Benesch 1967. The effect of organic phosphates from the human erythrocyte on the allosteric properties of hemoglobin. *Biochem. Biophys. Res. Commun.* 26: 162-167.

De Verdier, C. H. and T. L. Groth 1973. Phosphate transfer between 2,3-bisphosphoglycerate and monophosphoglycerate catalysed by bisphosphoglycerate phosphatase. *Eur. J. Biochem.* 32: 188-196.

Evans, A. and N. O. Kaplan 1966. Pyridine nucleotide transhydrogenase in normal human and leukemic leucocytes. *J. Clin. Invest.* 45: 1268-1272.

Harkness, D. R., W. Thompson, S. Roth, and V. Grayson 1970. The 2,3-diphosphoglyceric acid phosphatase activity of phosphoglyceric acid mutase purified from human erythrocytes. *Arch. Biochem. Biophys.* 138: 208-219.

Joyce, B. K. and S. Grisolia 1958. Studies on glycerate-2,3-diphosphatase. *J. Biol. Chem.* 233: 350-354.

Meera Khan, P. 1971. Enzyme electrophoresis on cellulose acetate gel. *Arch. Biochem. Biophys.* 145: 470-483.

Omenn, G. S. and P. T. Wade Cohen 1971. Electrophoretic methods for differentiating glycolytic enzymes of mouse and human origin. *In Vitro* 7: 132-139.

Rosa, R., J. Gaillardon, and J. Rosa 1973. Characterization of 2,3-diphosphoglycerate phosphatase activity: electrophoretic study. *Biochim. Biophs. Acta.* 293: 285-289.

Rosa, R., J. Gaillardon, and J. Rosa 1973. Diphosphoglycerate mutase and 2,3-diphosphoglycerate phosphatase activities of red cells: comparative electrophoretic study. *Biochem. Biophs. Res Commun.* 51: 536-541.

Rose, Z. and J. Liebowitz 1970. 2,3-Diphosphoglycerate phosphatase from human erythrocytes. *J. Biol. Chem.* 245: 3232-3241.

Rose, Z. B. and R. G. Whalen 1973. The phosphorylation of diphosphoglycerate mutase. *J. Biol. Chem.* 248: 1513-1519.

HETEROGENEITY, POLYMORPHISM, AND SUBSTRATE SPECIFICITY OF ALCOHOL DEHYDROGENASE FROM HORSE LIVER

REGINA PIETRUSZKO

Center of Alcohol Studies, Rutgers University
New Brunswick, New Jersey 08903

ABSTRACT: Electrophoresis on starch gel demonstrates the presence of multiple (9-12) components in horse liver alcohol dehydrogenase (HLADH). Some of the cathodal components are active with steroids. The subunit composition of only three of these components can be accounted for in terms of genetically distinct subunits E and S (occurring in all horse livers) and subunit A (occurring in some horse livers). On starch gels the AA isozyme is electrophoretically superimposable with the SS isozyme and cannot be distinguished from it before purification. The A subunit can hybridize with the S subunit to form AS isozyme and can probably also hybridize with the E subunit. Consequently, an additional complexity in the postulated (Pietruszko and Theorell) subunit composition of the electrophoretically separable isozymes of HLADH is introduced.

The sites located on E, S, and A subunits are catalytically distinct. Only the "S site" is active with steroids, which readily distinguishes it from either the "E site" and the "A site" both of which are steroid inactive. In addition to its distinct substrate specificity the "S site" has different nucleotide specificity; its affinity for NADPH being increased in comparison with that of the classical HLADH. With some substrates turnover rates are greater with NADPH than with NADH as coenzyme. The physiological significance of steroid activity or the altered nucleotide specificity is unknown.

Horse liver alcohol dehydrogenase (HLADH) was first crystallized by Bonnichsen and Wassen (1948) and has been available commercially for a long time. As a result, its properties have been extensively investigated, perhaps most extensively of all the enzymes known. The molecule of HLADH of 80,000 MW consists of two subunits of 40,000 MW (Jörnvall, 1970a). In the presence of substrate-competitive inhibitors such as pyrazole (Theorell and Yonetani, 1964) or isobutyramide (Winer and Theorell, 1960) the enzyme can be titrated with NAD or NADH employing spectrophotometric and fluorometric procedures, and the presence of two active sites per molecule (one per subunit) can be easily demonstrated. HLADH is a metalloenzyme containing four zinc atoms per molecule (Åkeson, 1964). The enzyme

catalyzes the reversible interconversion of a large variety of
aliphatic and aromatic, primary and secondary but not tertiary,
saturated and unsaturated alcohols and the corresponding aldehydes and ketones utilizing NAD(H) as coenzyme (Sund and
Theorell, 1963).

Alcohol dehydrogenase is universally distributed and occurs
consistently in mammalian livers, but its physiological role
is unknown. After the discovery of ethanol (15μM) in the
hepatic portal vein of germ-free rats, Krebs and Perkins (1970)
suggested an ethanol metabolizing role for hepatic alcohol dehydrogenase although in mammalian organisms ethanol is not a
quantitatively important metabolite. The following compounds
have also been suggested as possible physiological substrates:
glycerol (Holzer, Schneider, and Lange, 1955), isoprenoid
alcohols such as dimethylallyl alcohol, farnesol and geraniol
(Christophe and Popjack, 1961), retinol (Bliss, 1951), pan-
thothenyl alcohol (Abiko, Tomikawa and Shimizu, 1969), 5β-
cholestane-3α,7α,12α, 26-tetrol, an intermediate in cholesterol biosynthesis (Okuda and Takagawa, 1970), and ω-hydroxy-
fatty acids (Björkhem, 1972).

In 1961 Ungar observed that commercial preparations of
HLADH had small but significant activity with 3β-hydroxyster-
oids and the corresponding 3-ketones of A/B *cis* configuration.
The subject was investigated further by Graves, Clark and
Ringold (1965a and b) who noted that although the enzyme exhibited preference for A/B *cis* configuration, steroid ketones of
A/B trans configuration were reduced at the rate of about
1/10th of A/B *cis* structures. In 1965 Waller, Theorell and
Sjövall described the substrate specificity of HLADH with respect to bile acids. Their preparation of HLADH catalyzed
dehydrogenation of 3β-hydroxyl groups of bile acids of A/B *cis*
configuration by a mechanism distinct from that of Theorell
and Chance (1951), applicable to dehydrogenation of ethanol by
HLADH. The steroid activity was, however, ascribed to HLADH
as a whole and not to the separate enzyme.

Despite observations already reported by Dalziel (1958)
that HLADH preparations may not be homogeneous the enzyme was
treated as a single catalytically homogeneous component during
the sixties. Even, when McKinley-McKee and Moss (1964) published starch gel zymograms of commercial and laboratory
preparations of HLADH, clearly showing five distinct HLADH
zones, the explanation for the apparent heterogeneity was provided in terms of artefact formation with buffer ions or complexes with coenzymes.

Graves et al. (1965a) also observed that steroid activity of
commercial Boehringer and Worthington preparations of HLADH
varied and concluded that these preparations were contaminated
with a small amount of a separate steroid enzyme. Since the

reduction of certain steroidal ketones had an unusual steric course attempts were made to isolate this enzyme from commercial Worthington preparations in order to investigate its stereospecificity. This led to the unravelling of the quaternary structure of HLADH isozymes (Pietruszko, Clark, Graves, and Ringold, 1966), which was subsequently explained in terms of two subunits E and S, subunit E active with classical substrates of HLADH, and subunit S active also with steroids (Pietruszko and Theorell, 1969).

Current evidence indicates, however, that in some horse livers the most cathodal component is heterogeneous and contains in addition to the steroid active HLADH SS a new, hitherto unreported polymorphic form AA, devoid of activity with 3-keto steroids of A/B *cis* configuration. (Ryzewski and Pietruszko, 1974; Pietruszko, 1974; Pietruszko and Ryzewski, manuscript in preparation.)

The electrophoretic separation pattern of HLADH on starch gel in tris-HCl buffer at pH 8.5 is shown schematically in Table 1. The separated zones are assigned a nomenclature in terms of E, S, and A subunits representing the dimeric molecules of the enzyme. The isozymes containing subunit S are steroid active, those devoid of the S subunit lack steroid activity. In some horse livers the SS isozyme appears to be present in pure form. In other horse livers, however, it occurs electrophoretically superimposed with a polymorphic form AA. Subunit S can hybridize with the A subunit to form the hybrid AS. Therefore, in these livers the fastest migrating starch gel zone represents three isozymes: SS, AS, and AA, as shown in Table 1. In the livers containing AA isozyme the zone marked ES probably represents ES and EA isozymes.

The subunit composition of the isozymes denoted by ' and " symbols is unknown. The assigned structure (Pietruszko and Theorell, 1969) is a postulate based on certain common features with the main isozymes in the group whose quaternary structure is expressed in terms of E and S subunits.

The relative instability and high solubility of the S subunit accounts for the absence of the most cathodal isozyme zones from commercial preparations of HLADH. It can be seen from Table 1 that in both Boehringer and Worthington preparations isozyme EE is the major component comprising 77% and 64% respectively of the total protein. The steroid active hybrid ES is a minor contaminant of the commercial preparations. The second major component of both Boehringer and Worthington preparations is EE' isozyme. The ES (or ES + EA) appears to be the major component in crude extracts of horse liver.

In our attempts to isolate the separate steroid enzyme from

TABLE 1

Isozymes of HLADH: Subunit composition, occurrence and distribution

electrophoresis separation pattern	nomen-clature*	present nomen-clature and occurrence		distribution		nomen-clature**	relative proportions present in commercial preparations	
		some horse livers	other horse livers	liver	commer-cial prep-arations		Boehringer	Worthington
	EE"			+	+	5	5.6	8.2
	EE'			+	+	4	15.1	22.6
	EE	EE	EE	+	+	3	77	64.0
	ES"			+	+	2A	Trace	Trace
	ES'			+	+	2	1.3	1.2
	ES	ES	ES + EA	+	+	1	1.0	4.1
	SS"			+	−		0	0
	SS'			+	−		0	0
	SS	SS	SS,AS,AA	+	−		0	0

* Pietruszko and Theorell (1969)

** Pietruszko, Clark, Graves, and Ringold (1966)

710

the commercial preparations of HLADH it was readily established
that steroid activity is essentially confined to a relatively
minor electrophoretically separable zone of HLADH which still
possesses a high activity with the classical HLADH substrates.
When this zone was isolated by chromatography on CM-cellulose
a majority of the initial activity with steroids was associated with the isolated component. Activity with classical
substrates of HLADH was inseparable from steroid activity by
physical means or by heat treatment or decomposition on storage.

The steroid active component of commercial HLADH (now known
as ES) was subjected to kinetic analysis. While the Km for
5β-dihydrotestosterone (5βDHT) was found to be 5 x 10^{-5}M the
Vmax was only 1/55th of that of cyclohexanone or acetaldehyde,
yet at a concentration of 1 x 10^{-4}M the steroid ketone failed
to inhibit the reduction of either cyclohexanone (Km = 1.5 x
10^{-2}M) or acetaldehyde. Further, n-butanol, another classical
substrate of HLADH, at 1 x 10^{-2}M exhibited virtually complete
inhibition of the reduction of cyclohexanone and acetaldehyde
but failed to inhibit the reduction of 5βDHT by more than 10%.
*These results indicated strongly that steroid and cyclohexa-
none-acetaldehyde activities were located at distinct catalytic sites.* The catalytic site concerned with interconversion
of steroids will henceforward be referred to as the steroid
site or "S site" and the site concerned with interconversion
of the classical substrates of ADH such as acetaldehyde, cyclohexanone or ethanol as the alcohol site or "E site".

Theorell, Taniguchi, Åkeson, and Skursky (1966), after
receiving a letter from Ringold describing the above experiments, crystallized ES isozyme (HLADH$_S$) and confirmed the
distinctness of the active sites by employing lithocholic acid
(3α-hydroxycholanoic acid) which inhibited the reaction with
steroids but not the oxidation of ethanol.

In order to find out whether the "E site" and the "S site"
are located on the same protein molecule of ES, the EE isozyme
was purified from commercial preparations of HLADH, and rabbit
antibody to EE was developed for crossreaction with the ES
isozyme. The antibody reacted with both EE and ES, inhibiting
the activity of the "E site" while only slightly inhibiting
the activity of the "S site". Antigen-antibody precipitate
was formed with either antigen (Pietruszko and Ringold, 1968).
Fig. 1 illustrates that when antibody to EE is reacted with ES
and the precipitate is removed by centrifugation, both the
cyclohexanone activity and the steroid activity of ES undergo
progressive decrease in a manner identical to the decrease of
cyclohexanone activity of EE isozyme (homologous antigen)
included for comparison. The same equivalent of antibody

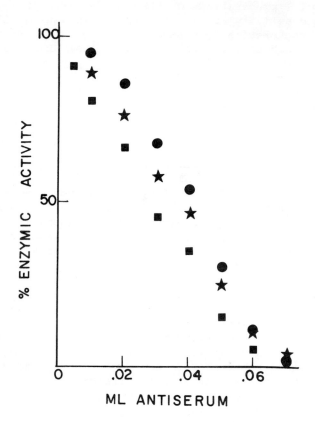

Fig. 1. Enzymic activity of ES and EE after crossreaction
with the increasing amounts of antiserum to EE and removal of
the precipitate by centrifugation; ● - 5βDHT as substrate,
ES isozyme; ■ - cyclohexanone as substrate, ES isozyme; ★ -
cyclohexanone as substrate, EE isozyme.

(ca 0.07 ml) precipitates the homologous antigen EE as the
heterologous antigen ES. The fact that both the steroid activity and the cyclohexanone activity of ES are precipitated by
the antibody to EE, which is itself devoid of steroid activity,
at essentially the same equivalence point, strongly argues
that the "E site" and "S site" of ES are located on the same
protein molecule and establishes structural relationship between EE and ES.

After several attempts it was also possible to obtain an
antibody to ES isozyme, specific for the "S site" (Pietruszko,
Ringold,Kaplan, Everse, 1968). When this antibody was cross-
reacted with the homologous antigen ES the steroid activity

but not the cyclohexanone activity, was inhibited. The catalytic activity of EE was not inhibited by the antibody to "S
site" but the precipitate was formed and could be removed by
centrifugation. Fig. 2 shows that the same equivalent of anti-

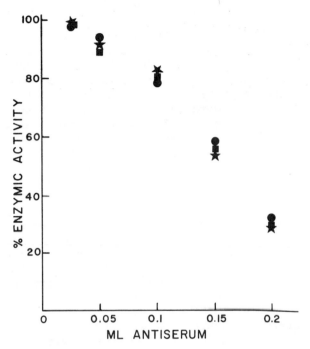

Fig. 2. Enzymic activity of ES and EE after crossreaction
with the increasing amounts of antiserum to ES and removal of
the precipitate by centrifugation; ● - 5βDHT as substrate,
ES isozyme; ■ - cyclohexanone as substrate, ES isozyme; ★ -
cyclohexanone as substrate, EE isozyme.

body precipitated the heterologous antigen EE as the homologous
antigen ES. This demonstrates that the area of the ES molecule
bearing the "S site" is structurally related to EE isozyme.
    In the meantime a method for dissociation and reconstitu-
tion of HLADH was developed by Vallee and his group at Harvard
(Drum, Harrison, Li, Bethune and Vallee, 1966). Employing
their method the EE isozyme was dissociated, reconstituted and
analyzed by electrophoresis and kinetics. These experiments
showed that the dissociated EE isozyme reconstituted to form
only EE suggesting that subunits composing EE isozyme were
structurally identical. However, when isozyme ES was disso-

ciated and reconstituted it yielded isozyme ES, isozyme EE and an isozyme migrating on starch gels faster than ES. This isozyme (SS) was absent from the commercial preparations of HLADH. Detailed examination of the isozyme composition of fresh horse livers demonstrated that SS was present in some horse livers in appreciable amounts. The SS was subsequently isolated from horse liver by chromatography on DEAE cellulose and like ES was found to be active with steroids. With 5βDHT as substrate, SS was shown to have a specific activity about twice as high as that of the ES isozyme, at the same time showing an appreciable activity with ethanol as substrate. The electrophoretic mobility characteristics of this newly isolated (SS) isozyme were the same as those of the fastest migrating component obtained from dissociation and reconstitution of ES isozyme. On dissociation and reconstitution SS produced only itself and no other isozymes indicating that its subunits were identical and distinct from those of EE isozyme. The results of this investigation are summarized in Fig. 3 and have been published (Pietruszko et al, 1968; Pietruszko and Theorell, 1969).

The isolated ADH isozymes shown to be related by dissociation and reconstitution were electrophoresed in a dissociated state. It can be seen (Fig. 4) that two polypeptides, differing in charge, are present in ES isozyme. The more negatively charged of these polypeptides occurs exclusively in EE isozyme while the more positively charged polypeptide occurs exclusively in the newly isolated SS isozyme. The "E site" of HLADH is therefore associated with the more negatively charged subunit (E) and the "S site" is associated with the more positively charged subunit (S) (Pietruszko and Theorell, 1969).

Both E and S subunits have now been fully sequenced (Jörnvall, 1970a and b), and have been shown to consist of 374 aminoacid residues with acetylated serine at the aminoterminal and phenylalanine at the carboxyterminal end of the subunit polypeptide. There are six aminoacids in the polypeptide composing the S subunit which differ from the aminoacids composing the E subunit (Jörnvall, 1970b). This difference in aminoacid sequence shows that E and S subunits are of distinct genetic origin.

The most cathodal isozyme-zone (visualized on starch gel) of HLADH present in the crude liver homogenates (see Table 1) has been for some time considered to be the pure SS homodimer. However, current investigations show that it can also represent another isozyme with which SS is super imposable.

In Table 2, specific activities of the fastest migrating zone of HLADH, isolated from four different horse livers, are shown with acetaldehyde, 5βDHT, cyclohexanone and ethanol as

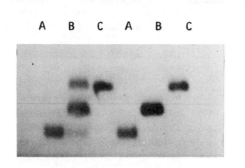

Fig. 3. EE, ES, and SS isozymes before and after dissociation
and reconstitution. A - SS; B - ES; C - EE. The left side of
the photograph represents electrophoretic separation after
dissociation in urea followed by a reconstitution procedure,
the right side the controls before dissociation and reconstitution. Electrophoresis on cellulose acetate in 0.025M tris-
HCl buffer pH 8.5; anode and cathode are marked with + and -
signs respectively.

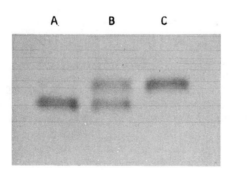

Fig. 4. Electrophoresis of EE, ES, and SS in the dissociated
state on cellulose acetate in glycine buffer, 0.025M pH 9.6
containing 8M urea and 0.1M 2-mercaptoethanol. A - SS; B - ES;
C - EE. Anode and cathode are marked with + and - signs
respectively.

substrates. These results show that two kinds of preparations
of the fastest migrating component of HLADH have been obtained:
preparations of A-type which have high specific activity with

TABLE 2

Specific activities with ethanol, acetaldehyde, 5βDHT and cyclohexanone of HLADH "SS" prepared from different horse livers.

liver no.	turnover no./active site/min				acetaldehyde
	ethanol	acetalde-hyde	5βDHT	cyclo-hexanone	5βDHT
1	93	1139	54	969	21.1
2	45	162	121	173	1.3
3	–	1352	58	–	23.3
4	–	110	79	–	1.4

The assay system contained: acetaldehyde (freshly distilled), 1.2mM; or 5βDHT, 0.115µM (added in 10µl dioxane to 3ml volume); or cyclohexanone, 12.8mM; at NADH, 0.17mM; in 0.1M phosphate buffer pH 7.0. For oxidation of ethanol the reaction was carried out in glycine buffer, 0.062M pH 10.0, at 500µM NAD and 10.8mM ethanol at 25°. The specific activities are based on active site concentration determined by titration with NADH in the presence of excess isobutyramide (Winer and Theorell, 1960).

acetaldehyde or cyclohexanone and preparations of S-type with specific activity with these substrates of about 1/10th of the preparations of A-type. The specific activity of the A-type preparations with ethanol as substrate is only two times more than that of the S-type preparations. The differences between these preparations are most apparent when the ratio of acetaldehyde (or cyclohexanone) to steroid activity is calculated. This ratio is about 20 for the preparations of the A-type and only about 1 for the preparations of the S-type. Further investigations have shown that the preparations of A-type can be separated by electrophoresis on ionagar into three zones, while the preparations of S-type migrates as one zone in the same conditions. In Fig. 5 a schematic representation of the electrophoretic separation of the A-type and the S-type preparations on ionagar gel is given. There are three HLADH components present in the A-type preparations. The fastest migrating component of the A-type preparation is electrophoretically identical with the single component present in the S-type preparations. The fastest migrating component as well as the intermediate migrating component of the A-type preparations are active with 3β-hydroxy steroids of A/B *cis* config-

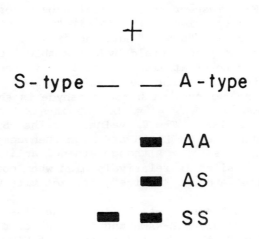

Fig. 5. Schematic representation of electrophoresis of the
S-type and A-type preparations in 5mM phosphate buffer pH 7.0
on ionagar. Anode and cathode are marked with + and - signs
respectively.

uration while the slowest migrating component is *inactive* with
steroids but active with the classical substrates of HLADH.
Preliminary experiments show that the latter component can
hybridize with the enzyme present in the S-type preparations.
On the basis of some preliminary hybridization studies and
some kinetic data involving the use of different HLADH inhibitors a subunit composition of AA for the slowest migrating
component of the A-type preparations has been postulated.
The fastest migrating component is the SS isozyme and the
intermediate migrating component is, therefore, a hybrid AS
(see Fig. 5). Since AA isozyme occurs only in certain horse
livers, and is not universally distributed like EE and SS
isozymes, it has been concluded that AA is a polymorphic form
of HLADH. The high acetaldehyde activity of the A-type preparations is due to the catalytic site located on the A subunit
which shall be referred to as the "A site".
The catalytic properties of the sites located on EE, SS,

and AA homodimers are listed in Table 3. The classical substrates of HLADH serve as substrates for all three homodimers but the catalytic constants vary. Steroid activity is confined to SS homodimer (the E and A subunits being inactive with steroids). With respect to classical substrates of HLADH the catalytic properties of the "A site" are intermediate between those of the "E site" and the "S site". While Km values for ethanol and acetaldehyde resemble those of the "E site" the turnover at Vmax for ethanol is the same as that of the "S site" (this value being about 1/3 of the "E site" turnover). The turnover with acetaldehyde is about 8 times more than that of the "S site" but about six times less than that of the "E site". The Km values for the "S site" with the classical substrates of HLADH are high when compared with those for the "E site". Although ethanol activity (turnover no.) of the "S site" is relatively high when compared with that of the "E site" and the "A site", its activity with acetaldehyde is low.

According to Theorell et al (1969) the dissociation constants of the binary (EE-NADH and SS-NADH) complexes are somewhat lower for SS than EE, indicating that NADH is more tightly bound to SS. The "S site" also appears to have a greater affinity for NADPH. While the Km value for NADPH is still a relatively large number (140µM, Pietruszko, 1973) it is considerably less than that determined for the classical HLADH (3,000µM, Dalziel and Dickinson, 1965). The reduction of 5βDHT is faster with NADPH than with NADH as coenzyme (Pietruszko, 1973).

The electrophoretic mobility, stability and solubility of the EE and the SS isozymes are distinct (Table 4). The EE is more stable than the SS, but the SS is more soluble than the EE and for this reason more difficult to crystallize. Preliminary experiments with the AA homodimer show that it is also more stable than the SS on heat treatment and on storage. Presence of the A subunit in the horse liver may render the S subunit more stable; the AS heterodimer is more resistant to heat treatment than the SS isozyme (Pietruszko and Ryzewski, unpublished).

The results discussed so far account for the subunit structure and some of the catalytic characteristics of only three electrophoretically separable zones of HLADH. Seven or nine zones can be routinely observed following electrophoresis on starch (Pietruszko and Theorell, 1969). On the basis of results from experiments not involving hybridization resulting in the conversion of EE' isozyme into a gel zone corresponding electrophoretically to the EE isozyme it has been suggested (Lutstorf and von Wartburg, 1969) that EE'

TABLE 3
Some catalytic constants of EE, SS, and AA isozymes

Substrate	EE		SS		AA	
	K_m	V_{max}	K_m	V_{max}	K_m	V_{max}
Ethanol	1.2	314	11.2	93	0.7	92
Acetaldehyde	0.2	5618	1.4	360	0.08	1210
5β-androstan-3β-ol-17-one	–	–	0.003	70	–	–
5βDHT	–	–	0.03	150	–	–

With ethanol and steroidal alcohol as substrates the constants were determined in glycine buffer 0.062M pH 9.5 at 500μM NAD. With 5βDHT and acetaldehyde as substrates the constants were determined in 0.1M phosphate buffer pH 7.0 at 170μM NADH. The reactions were carried out in a Beckman Spectrophotometer at 25°. K_m = mM, V_{max} = turnover no./active site/minute. Active site concentration determined as in Table 2.

and other zones marked with ' and " symbols (see Table 1) are conformational isomers of EE, ES and SS isozymes. The existence of such isomers has been discussed by Kaplan (1965). In order to provide proof for the conformational isozymes, the proof of the identity of aminoacid sequences of the suspected isomers is necessary. Even peptide maps of ' and " isozymes of HLADH are not yet available. In the light of the currently available evidence for the third subunit of HLADH the possible involvement of this subunit or of its isomer in the structure of the isozymes denoted by ' and " will have to be considered.

The biological significance of the steroid active alcohol dehydrogenase is uncertain although it also occurs in species other than the horse (Raynier, Theorell, Sjövall, 1969); 3β-hydroxysteroids which are good substrates for the enzyme do not occur naturally at significant concentrations (Dorfman and Ungar, 1961). When 3-keto-5β-cholanoic acid is administered to rats (Cronholm, Makino and Sjövall, 1972), all metabolites carry 3α-hydroxyl groups indicating that it is not metabolized via alcohol dehydrogenase. Dehydrogenation of 3β-hydroxysteroids is, therefore, an unlikely physiological

TABLE 4

Physicochemical characteristics of EE, SS, and AA homodimers

isozyme	solubility	heat treat- ment 70° 5 min % initial activity	storage .05M phos- phate pH 7.5 4°	electropho- retic migra- tion, starch gel	primary structure	coenzyme specificity
EE	crystal- lizable from 8% ethanol 4°		stable	separated from AA and SS	EE = primary standard	NAD
SS	does not crystallize from 30% ethanol - 18°	10	unstable	super- imposable with AA	different from EE 12 places (Jörnvall, 1970)	NAD (increased affinity for NADP)
AA	?	90	stable	super- imposable with SS	?	NAD

role for the enzyme. It has been suggested that steroids may exert some regulatory function (Cronholm et al, 1972). Lithocholic (3α-hydroxy-5β-cholanoic) acid is a good inhibitor of steroid active alcohol dehydrogenase and is formed naturally from chenodeoxycholic acid (a major bile acid metabolite) by bacterial action in the intestine (Norman and Sjövall, 1960). Steroids of the 3α-, 5β-configuration may, therefore control the activity of the steroid active alcohol dehydrogenase. In man it has been observed that different isozymes appear at different developmental stages (Pikkarainen and Raiha, 1969; Murray and Motulsky, 1971; Murray and Price, 1972) which suggests that alcohol dehydrogenase isozymes may play some specific yet distinct roles in development. The appearance of the last isozymes of human alcohol dehydrogenase between 11 and 14 years of age (puberty) may in some way be connected with steroid metabolism.

ACKNOWLEDGEMENTS

The work described was carried out in three laboratories. The part comprising E and S subunits was started and almost completed with Dr. Howard J. Ringold, at the Worcester Foundation of Experimental Biology, Shrewsbury, Mass., and completed during the first six months of my stay with Professor Hugo Theorell, at the Nobel Institute, Karolinska, Sweden. The polymorphic subunit A has been discovered only recently at the address given above. Supported by NIMH - NIAAA AA00186 Grant and Charles and Johanna Busch Memorial Fund.

REFERENCES

Abiko, Y., M. Tomikawa, and M. Shimizu 1969. Enzymatic conversion of pantothenyl-alcohol to pantothenic acid. *J. Vitaminology* 15: 59-69.

Åkeson, Å. 1964. On the zinc content of horse liver alcohol dehydrogenase. *Biochem. Biophys. Res. Commun.* 17: 211-214.

Björkhem, I. 1972. On the role of alcohol dehydrogenase in ω-oxidation of fatty acids. *Eur. J. Biochem.* 30: 441-451.

Bliss, A. F. 1951. The equilibrium between vitamin A alcohol and aldehyde in the presence of alcohol dehydrogenase. *Arch. Biochem. Biophys.* 31: 197-204.

Bonnichsen, R. K. and A. M. Wassén 1948. Crystalline alcohol dehydrogenase from horse liver. *Arch. Biochem. Biophys.* 18: 361-363.

Christophe, J. and G. Popjack 1961. Studies on the biosynthesis of cholesterol. XIV. The origin of prenoic acids from allyl pyrophosphates in liver enzyme systems. *J. Lipid Res.* 2: 244-257.

Cronholm, T., I. Makino, and J. Sjövall 1972. Steroid metabolism in rats given $\begin{bmatrix} 2 \\ 1-H_2 \end{bmatrix}$ ethanol. Biosynthesis of bile acids and reduction of 3-keto-5β cholanoic acid. *Eur. J. Biochem.* 24: 507-519.

Dalziel, K. 1958. On the purification of liver alcohol dehydrogenase. *Acta Chem. Scand.* 12: 459-464.

Dorfman, R. I. and F. Ungar 1965. *Metabolism of steroid hormones.* Academic Press, N. Y.

Drum, D. E., J. H. Harrison, IV, T-K. Li, J. L Bethune, and B. L. Vallee 1967. Structural and functional zinc in horse liver alcohol dehydrogenase. *Proc. Natl. Acad. Sci.* 57: 1434-1440.

Graves, J. M., A. Clark, and H. J. Ringold, 1965a. The 3-hydroxysteroid dehydrogenase associated with crystalline horse liver alcohol dehydrogenase (Lad). *Sixth Pan American Congress of Endocrinology* Mexico, D. F., E86.

Graves, J. M. H., A. Clark, and H. J. Ringold 1965b. Stereochemical aspects of the substrate specificity of horse liver alcohol dehydrogenase. *Biochem.* 4: 2655-2671.

Holzer, H., S. Schneider, and K. Lange 1955. Funktion der Leber-Alcoholdehydrase. *Angew. Chem.* 67: 276-277.

Jörnvall, H. 1970a. Horse liver alcohol dehydrogenase: The primary structure of the ethanol-active isoenzyme. *Eur. J. Biochem.* 16: 25-40.

Jörnvall, H. 1970. Horse liver alcohol dehydrogenase, on the primary structure of the isoenzymes. *Eur. J. Biochem.* 16: 41-49.

Kaplan, N. O. 1968. Nature of multiple molecular forms of enzymes. *Ann. N. Y. Acad. Sci.* 151. Art. 1: 382-399.

Krebs, H. A. and J. R. Perkins 1970. The physiological role of liver alcohol dehydrogenase. *Biochem. J.* 118: 635-644.

Lutstorf, U. M. and J. P. von Wartburg 1969. Subunit composition of horse liver alcohol dehydrogenase isoenzymes. *FEBS Letters* 5: 202-206.

McKinley-McKee, J. S. and D. W. Moss 1965. Heterogeneity of liver alcohol dehydrogenase on starch-gel electrophoresis. *Biochem. J.* 96: 583-587.

Murray, R. F., Jr. and A. G. Motulsky 1971. Developmental variation in the isoenzymes of human liver and gastric alcohol dehydrogenase. *Science* 171: 71-73.

Murray, R. R., Jr. and P. H. Price 1972. Ontogenetic, polymorphic and interethnic variation in the isoenzymes of human alcohol dehydrogenase. *Ann. N. Y. Acad. Sci.* 197: 68-72.

Norman, A. and J. Sjövall 1960. Formation of lithocholic acid from chenodeoxycholic acid in the rat. *Acta Chem. Scand.* 14: 1815-1818.

Okuda, K. and N. Takagawa 1970. Rat liver 5β-cholestane-3α, 7α, 12α,26-tetrol dehydrogenase as a liver alcohol dehydrogenase. *Biochem. Biophys. Acta* 220: 141-148.

Pietruszko, R., A. F. Clark, J. Graves, and H. J. Ringold 1966. Steroid activity and multiplicity of crystalline horse liver alcohol dehydrogenase. *Biochem. Biophys. Res. Commun.* 23: 526-533.

Pietruszko, R. and H. J. Ringold 1968. Antibody studies with multiple forms of horse liver alcohol dehydrogenase. I. *Biochem. Biophys. Res. Commun.* 33: 497-502.

Pietruszko, R., H. J. Ringold, N. O. Kaplan, and J. Everse 1968. Antibody studies with multiple forms of horse liver alcohol dehydrogenase. II. *Biochem. Biophys. Res. Commun.* 33: 503-507.

Pietruszko, R., H. J. Ringold, T. K. Li, B. L. Vallee, Å. Åkeson, and H. Theorell 1969. Structure and function relationships of isoenzymes of horse liver alcohol dehydrogenase. *Nature* 221: 440-443.

Pietruszko, R. and H. Theorell 1969. Subunit composition of horse liver alcohol dehydrogenase. *Arch. Biochem. Biophys.* 131: 288-298.

Pietruszko, R. 1973. Activity of horse liver alcohol dehydrogenase SS with NADP (H) as coenzyme and its sensitivity to barbiturates. *Biochem. Biophys. Res. Commun.* 54: 1046-1052.

Pietruszko, R. 1974. Polymorphism of horse liver alcohol dehydrogenase. *Biochem. Biophys. Res. Commun.* In press.

Pikkarainen, P. and N. C. R. Raiha 1969. Change in alcohol dehydrogenase isoenzyme pattern during development of human liver. *Nature* 222: 563-564.

Reynier, M., H. Theorell, and J. Sjövall 1969. Studies on the stereo-specificity of lever alcohol dehydrogenase (LADH) for 3β-hydroxy- 5βsteroids, inhibition effects of pyrazole and 3α-hydroxycholanoic acid. *Acta Chem. Scand.* 23: 1130-1136.

Ryzewski, C. and R. Pietruszko 1974. A third subunit of horse liver alcohol dehydrogenase. *Fed. Proc.* 32: 2093 **Abs.**

Sund, H. and H. Theorell 1963. Alcohol dehydrogenase. In: *The Enzymes* (Boyer, Lardy, and Myrback, eds). Vol. 7, Academic Press, New York, 25-83.

Theorell, H., Å. Åkeson, B. Liszka-Kopec, and C. deZalenski 1970. Equilibrium and rate constants for the reaction between NADH and horse liver alcohol dehydrogenases "EE," "ES," and "SS." *Arch. Biochem. Biophys.* 139: 241-247.

Theorell, H. and B. Chance 1951. Studies on liver alcohol dehydrogenase. II. The kinetics of the compound of horse liver alcohol dehydrogenase and reduced diphosphopyridine nucleotide. *Acta Chem. Scand.* 5: 1127-1144.

Theorell, H., S. Taniguchi, Å. Åkeson, and L. Skursky 1966. Crystallization of a separate steroid-active liver alcohol dehydrogenase. *Biochem. Biophys. Res. Commun.* 24: 603-610.

Theorell, H. and T. Yonetani 1964. Studies on liver alcohol dehydrogenase complexes. IV. Spectrophotometric observation of the enzyme complexes. *Arch. Biochem. Biophys.* 106: 252-258.

Ungar, F. 1960. 3-Hydroxysteroid dehydrogenase activity in mammalian liver. *Univ. Minnesota, Med. Bul.* 31: 226-242.

Waller, G., H. Theorell, and J. Sjövall 1965. Liver alcohol dehydrogenase as a 3β-hydroxy-5β-cholanic acid dehydrogenase. *Arch. Biochem. Biophys.* 111: 671-684.

Winer, A. and H. Theorell 1960. Dissociation constants of ternary complexes of fatty acids and fatty acid amides with horse liver alcohol dehydrogenase coenzyme complexes. *Acta Chem. Scand.* 14: 1729-1742.

DROSOPHILA ALCOHOL DEHYDROGENASE:
ORIGIN OF THE MULTIPLE FORMS

MARCIA SCHWARTZ, LARRY GERACE,[1] JANIS O'DONNELL, and
WILLIAM SOFER
Department of Biology, The Johns Hopkins University,
Baltimore, Maryland 21218

ABSTRACT. Alcohol dehydrogenase in some homozygous strains
of *Drosophila* consists of three forms separable by electrophoresis or chromatography-ADH 5, 3, and 1. All three
have similar, if not identical, primary structures as
shown by amino acid analysis and peptide fingerprinting.
After growing flies in medium containing labeled nicotinamide, radioactivity is associated with ADH 3 and ADH 1 but
not with ADH 5. In addition, a substance can be released
from mixtures containing ADH 3 and ADH 1, but not from
ADH 5, which is capable of converting ADH 5 into a form
whose electrophoretic migration is identical to ADH 3. We
interpret these data to mean that ADH 3 and ADH 1 are composed of the same peptides as ADH 5, but differ in charge,
activity, and stability because of the presence of a
nicotinamide containing molecule which is non-covalently
bound.

Strains of *Drosophila melanogaster* homozygous for one
electrophoretic variant of alcohol dehydrogenase show three
zones of enzyme activity after electrophoresis, called ADH 5,
ADH 3, and ADH 1 (Ursprung and Leone, 1965). These three
forms differ not only in charge but also in heat stability;
ADH 1 being the most stable, and ADH 5 the least stable of
the multiple activities (Grell et al., 1968).
  There are two fundamentally different mechanisms by
which these multiple activities might be generated. One
possibility is that like lactate dehydrogenase and similar
enzymes (Masters and Holmes, 1972), alcohol dehydrogenase is
·composed of unlike polypeptide chains which combine in varying proportions to give an array of proteins. The alternative possibility is that all forms of alcohol dehydrogenase
consist of the same peptide chains. The multiple zones of
activity might then result from different degrees of polymerization, binding of varying quantities of a ligand to the
different forms, or to the existence of a variety of conformational states. Some combination of these alternatives is

---

[1]Current address: Rockefeller University, New York, New York.

also possible.

A number of studies have dealt with these possibilities. For example, it has been shown that adding high concentrations of NAD^+ to ADH 5, results in the formation of zones of enzyme activity whose electrophoretic mobilities and heat stabilities are similar to ADH 3 and ADH 1 (Ursprung and Carlin, 1968; Jacobson, 1968). These data were taken to mean that the multiple forms of alcohol dehydrogenase were due to differential binding of NAD^+. Recently, however, Jacobson et al. (1972) found that acetone was also effective in converting ADH 5 to ADH 3 and ADH 1, and therefore concluded that the multiple forms were due to conformational differences.

While these studies are important in indicating potential mechanisms for generating the multiple forms, a number of direct experiments have not been performed. These include analysis of the primary structure of the various forms of alcohol dehydrogenase, analysis of the naturally occurring multiple species to determine whether any contain bound molecules, and finally, a determination as to whether any of the forms contains a factor which can cause mutual interconversion. These are the experiments which we have undertaken.

MATERIALS AND METHODS

NCS reagent was obtained from Nuclear Chicago. Aquasol was from New England Nuclear. Trypsin previously treated with L-(1-tosylamido-2-phenyl) ethyl chloromethyl ketone was purchased from Worthington Biochemical Corp. The dried medium used in radioactive labelling experiments was Formula 4-24 Blue Instant Drosophila Medium from Carolina Biological Co., to which we added [7-^{14}C]nicotinamide, 2-10 mCi/mM, from New England Nuclear. NAD^+ was Type III from Sigma Chemical Co. Marker proteins for sodium dodecyl sulfate (SDS) electrophoresis were also from Sigma. Hydroxylapatite was from Bio-Rad Labotatories, Bio Gel HTP. Alcohol dehydrogenase was purified from a white-eyed strain of *Drosophila melanogaster*, which is homozygous for the ADH^F or ADH 5, 3, 1 variant.

AMINO ACID ANALYSIS

Protein samples were dried in acid washed ignition tubes (14 x 100mm) in a vacuum desiccator. After the addition of 1 ml of constant boiling HCl and 10 µl of 6% phenol (DeLange et al., 1969), the tubes were evacuated to a pressure of less than 20µM of mercury and sealed. Hydrolysis was carried out at 110° for either 24, 48, or 72 hours. After hydrolysis, the tubes were opened, the HCl removed in a vacuum desiccator over pellets of NaOH, and the samples dissolved in sodium citrate buffer, pH2.2. Amino acid analyses were carried out on a Technicon TSM amino acid analyzer. Yields of serine and threonine were computed by extrapolation to zero time. Cystine and cysteine were measured as cysteic acid after performic acid oxidation by the method of Hirs (1967).

PEPTIDE FINGERPRINTING

Trypsin (250 µg) was added to between 5 and 10 mg of alcohol dehydrogenase dissolved in 0.05 M NH_4HCO_3, pH 8.3. Preliminary experiments using an automatic titrator showed that digestion was essentially completed after 40-50 min at 37°. In the experiments reported here we carried out the digestions for 5-6 hours at room temperature. The reaction was stopped by bringing the solution to pH 6.5. Samples were then frozen, thawed, centrifuged (to remove the very small amount of undigested protein), and lyophilized. The digest was then dissolved in 0.01 M NH_4HCO_3, pH 8.3, just before chromatography.

The tryptic peptides were analyzed by the procedure of Bennett (1967). Approximately 0.05 µmoles of protein were applied to a 44 x 57 cm sheet of Whatman 3MM paper for descending chromatography in n-butanol:acetic acid:water (4:1:5). After chromatography for 13 hours, the paper was dried, turned through 90°, and electrophoresis was carried out at 2000 volts for 75 min at 25° in pyridine:acetic acid:water (1:10:89).

Peptides and amino acid markers were visualized by dipping the papers in buffered ninhydrin (Easley, 1965). Tryptophan containing peptides were stained by the Ehrlich reaction (Easley, 1965).

GEL ELECTROPHORESIS

Agar and acrylamide gel electrophoresis were carried out as previously described (Ursprung and Leone, 1965; Sofer and Ursprung, 1968). Electrophoresis in SDS was done by the technique of Weber and Osborne (1969) except that alcohol dehydrogenase and marker proteins were incubated under a variety of conditions prior to electrophoresis. Electrophoresis in SDS-urea was performed according to the procedure of Swank and Munkres (1971).

For the determination of radioactivity in gels, alcohol dehydrogenase activity was first localized by lightly staining the gels in an histochemical staining mix (Ursprung et al., 1970). Then, the gels were sliced into 0.5 cm sections and each section was incubated at 40° in 0.5 ml of NCS reagent: water (9:1) for 5 hours. Ten ml of toluene-based fluor and one ml of aquasol were added before counting. We estimate counting efficiency at about 40% under these conditions.

ADH PURIFICATION

ADH was purified according to procedures previously published (Sofer and Ursprung, 1968; Borack and Sofer, 1971) except for the following modifications. Instead of passing the 40-60% ammonium sulfate fraction through Sephadex G-150, we precipitated much extraneous protein by dialyzing this fraction overnight against 0.01 M sodium phosphate buffer, pH 6.0. The dialyzate was centrifuged at 27,000 x g for 30 minutes. The resultant supernatant fluid was dialyzed overnight against 10 mM Tris-HCl, 0.1 mM EDTA, pH 8.3, 5 mM NaCl, and then applied to a 5.0 x 60 cm column of Sephadex-QAE A-50 previously equilibrated with the same buffer. The column was first thoroughly washed with this buffer, then 10 mM Tris-HCl, 0.1 mM EDTA, 25 mM NaCl, pH 8.3, was used to elute ADH 5. Subsequent elution with 10 mM Tris-HCl, 0.1 mM EDTA, 100 mM NaCl, pH 8.3, resulted in the release of a mixture of ADH 5, 3, and 1, although ADH 3 and ADH 1 usually predominated.

ADH 5 and the mixture of forms were then separately dialyzed against 0.01 M sodium phosphate buffer, pH 7.5, and each was applied to a 2.5 x 25 cm column of hydroxylapatite equilibrated in the same phosphate buffer. We find large differences in the ability of various lots of hydroxylapatite to bind alcohol dehydrogenase, but the enzyme usually will elute in 0.01 M sodium phosphate.

In the experiment where we purified alcohol dehydrogenase after feeding of radioactive nicotinamide, we omitted the QAE-Sephadex step in order to increase the yield. Here, the supernatant after pH 6.0 dialysis was dialyzed against 0.01 M phosphate buffer, pH 7.5, and applied directly to hydroxylapatite.

RADIOACTIVE LABELLING

Approximately 500 eggs were placed in each of four 1/2 pint milk bottles containing 50 ml of dried medium, 50 ml of water, and 20 μCi of [7-^{14}C]nicotinamide. A small amount of live, dried yeast was added, and the cultures were incubated at 22-25°.

Radioactivity was monitored during the purification of alcohol dehydrogenase by pipetting a known volume of solution onto a 2.5 cm Whatman GF-B filter and counting the dried filter in toluene-based fluor in a Nuclear Chicago series 720 scintillation counter.

728

ADH ASSAYS

One unit of alcohol dehydrogenase activity is defined as the amount catalyzing the reduction of 1 nmole of NAD^+ per minute. The assay has been described previously (Sofer and Ursprung, 1968).

COMPLEMENTATION

All pairwise matings of ten alcohol dehydrogenase null mutants (Adh^{n1}-ADH^{n10}) were set up in 2.5 x 8 cm vials. The resultant F_1 adults were allowed to age for more than 3 days, and then two or three were ground up in a few drops of histo-chemical staining mixture (Ursprung et al., 1970) modified to contain 0.05 M sodium phosphate buffer, pH 7.5. The nearly neutral pH of this staining solution enables us to carry out the incubation for two or more hours without spontaneous deposition of formazan. For comparison, a fly known to show wild-type levels of alcohol dehydrogenase and several null mutants are ground up at the same time. We estimate that we are able to detect about 10% of wild-type activity with this assay.

RESULTS

The direct way of determining whether all forms of alcohol dehydrogenase contain the same polypeptides would be to iso-late each of them and compare their primary structures. Un-fortunately, while we can readily prepare pure ADH 5, we always obtain ADH 3 and ADH 1 contaminated with ADH 5 and each other. All of our data therefore rests on comparisons of pure ADH 5 and a mixture enriched with varying amounts of ADH 3 and ADH 1.

AMINO ACID AND PEPTIDE ANALYSIS

Amino acid analyses of ADH 5 and a mixture of ADH 5, 3, and 1 are shown in Table I. (A densitometric tracing of the mix-ture after acrylamide gel electrophoresis is shown in Figure 1A). The two preparations have very similar amino acid com-positions.

A more sensitive method for distinguishing between two proteins is "fingerprinting" (Bennett, 1967). Figure 2 compares the pattern of tryptic peptides of ADH 5 with that of a mixture of forms. (A densitometric tracing of this mixture after acrylamide electrophoresis is shown in Figure 1B.) The patterns are very much alike. We conclude from these experi-ments that ADH 5, ADH 3, and ADH 1 all contain similar, if

TABLE I

AMINO ACID COMPOSITION OF ALCOHOL DEHYDROGENASE

Amino Acid	Mixture of ADH 5,3, and 1 [a] Number of residues [b]		ADH 5 Number of residues [b]	
Lysine	16.0	($\pm$1.0) [c]	16.7	($\pm$0.8)
Histidine	4.4	($\pm$0.4)	5.6	($\pm$0.8)
Arginine	5.9	($\pm$0.5)	4.9	($\pm$0.3)
Aspartate	24.6	($\pm$0.9)	24.1	($\pm$3.0)
Threonine	27.2	($\pm$0.3)	24.3	($\pm$1.0)
Serine	9.4	($\pm$0.8)	10.5	($\pm$0.6)
Glutamate	17.5	($\pm$0.6)	17.4	($\pm$0.8)
Proline	10.4	($\pm$0.5)	9.7	($\pm$0.6)
Glycine	18.6	($\pm$0.8)	18.9	($\pm$0.6)
Alanine	20.9	($\pm$1.8)	21.7	($\pm$0.5)
Half-cystine	—		2.1 [d]	
Valine	18.4	($\pm$0.8)	19.8	($\pm$1.5)
Methionine	0 [e]		0 [e]	
Isoleucine	18.6	($\pm$0.7)	19.8	($\pm$0.9)
Leucine	23.1	($\pm$1.1)	25.6	($\pm$0.6)
Tyrosine	5.9	($\pm$0.6)	6.0	($\pm$0.1)
Phenylalanine	8.7	($\pm$0.3)	8.8	($\pm$0.2)
Tryptophan	0 [f]		0 [f]	

[a]For the composition of this mixture see Figure 1A.

[b]Values are normalized to a subunit molecular weight of 25,000. Samples of alcohol dehydrogenase were subjected to acid hydrolysis for 24, 48, and 72 hours. The values reported are averages of 2-5 experiments for each time point. Yields of serine and threonine were extrapolated to zero time.

[c]Values in parentheses are standard deviations.

[d]Determined as cysteic acid after performic acid oxidation according to Hirs (1967). This experiment was performed only once.

[e]No methionine was detected after acid hydrolysis or acid hydrolysis preceded by performic acid oxidation.

[f]We have been unable to detect tryptophan by basic hydrolysis with Ba(OH)$_2$ (Noltmann et al., 1962) or hydrolysis with p-Toluene sulfonic acid (Liu, 1972). However, staining of our peptide maps with the Ehrlich reaction (Easley, 1965) shows

TABLE 1 (cont.)

two tryptophan containing peptides. Alcohol dehydrogenase has
been reported to contain tryptophan by Jacobson et al. (1970a.)
If Jacobson's data is renormalized to our subunit molecular
weight, we predict 2 tryptophans per subunit.

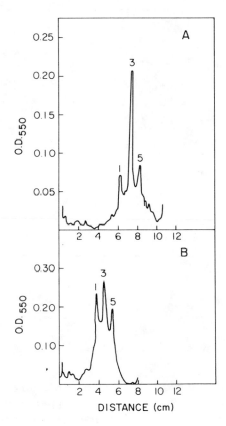

Figure 1. Composition of enzyme mixture enriched for ADH 3
and ADH 1. After hydroxylapatite chromatography the mixture
was electrophoresed on 5% polyacrylamide gels for 2 hr at 3
mA per tube in Tris-glycine buffer at pH 9.3. Gels were
stained with Coomassie Blue and scanned at 550 nm in a Gilford
gel scanner attached to a Beckman DU spectrophotometer
equipped with a Gilford model 2000 recorder. A, composition
of the mixture used for the amino acid analysis reported in
Table 1. B, composition of the mixture used for the peptide
map in Figure 2.

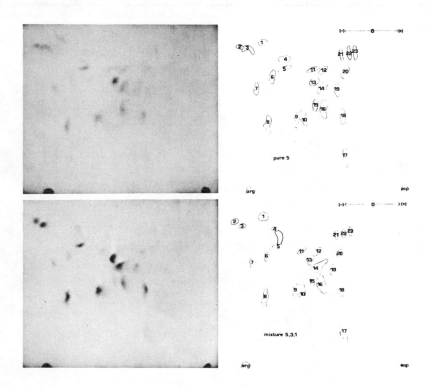

Figure 2. Peptide maps of ADH 5 and a mixture of forms. The tryptic peptides were spotted in the upper right hand corner at 0. After chromatography, the paper was dried and amino acid markers (aspartic acid and arginine) were spotted just below the solvent front. The upper map and tracing are of pure ADH 5, the lower map and tracing are of the mixture shown in Figure 1B.

not identical, polypeptide chains. However, it is clear that neither amino acid analysis nor fingerprinting can rule out small differences in primary structure.

Another piece of information can be derived from the data in Table 1 and Figure 2. We have previously measured the molecular weight of native alcohol dehydrogenase to be

48,000 (Sofer and Ursprung, 1968, see Discussion). From the
number of lysine (16-17) and arginine (5-6) residues, we calculate that if alcohol dehydrogenase is composed of one polypeptide chain, we should find 43-47 spots after fingerprinting. On the other hand, if ADH consists of two indentical
polypeptides, each of approximately 24,000 molecular weight,
then there should be between 22 and 24 spots. Since we find
23 spots, our data is consistent with a model of two identical subunits for the enzyme.

SDS-ELECTROPHORESIS

In contrast to the above data, ADH 5 and the mixture of
ADH 5, 3, and 1 show marked differences in migration in SDS-
electrophoresis after incubation in SDS for 2 hrs at 37°.
ADH 5 behaves as a single polypeptide with an apparent molecular weight of 2.3×10^4. On the other hand, electrophoresis
of the mixture of forms resulted in the appearance of three
rapidly migrating bands (Fig.3) with apparent molecular
weights below 1.5×10^4 (Table 2 and Fig. 4A). We have not
calculated the molecular weights of these rapidly migrating
species because it is not proper to extrapolate the plot of
molecular weight versus distance of migration into the low
molecular weight region (Weber and Osborn, 1969), hence the
dotted line in Fig. 4A. Notice that no band appeared at a
position similar to that shown by ADH 5 alone, although this
enzyme preparation contained 55% ADH 5.

In order to determine what caused the difference in behavior of the mixture of forms compared to pure ADH 5, we
varied the conditions of incubation and electrophoresis (Table
2). The omission of β-mercaptoethanol or substitution of
0.014 M dithiothreitol was without effect. Decreasing the
time of incubation of the mixture in SDS resulted in the
appearance of a small amount of material whose migration rate
was similar to that of ADH 5. Increasing the temperature of
incubation to 60° for 1 hr, or 100° for 5 min (then 37° for
2 hrs) resulted in the formation of a single protein zone
with an apparent molecular weight of 2.3×10^4 (Table 2 and
Fig. 4B). However, if the mixture was first incubated at 37°
for 2 hours and then boiled for 5 min, three low molecular
weight zones appeared.

Finally we modified both the incubation and electrophoretic
conditions by including urea in the gels and in the incubation
solution (Swank and Munkres, 1971). This treatment resulted
in the appearance of a single band with an apparent molecular
weight of 2.7×10^4 (Table 2, Figs. 5 and 6). We emphasize
that pure ADH 5 behaves as a single peptide chain with an

(−)

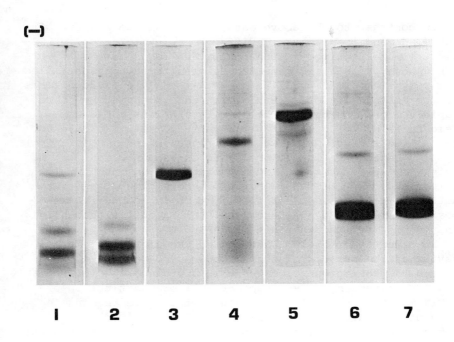

1	2	3	4	5	6	7

Figure 3. SDS electrophoresis of alcohol dehydrogenase and marker proteins. Samples were incubated in 0.01 $\underline{M}$ Na phosphate buffer, pH 7.5, 1% SDS, 1% β-mercaptoethanol for the times and temperatures given. Electrophoresis and staining were done by the procedure of Weber and Osborn (1969). Samples are: 1. ADH 5,3, and 1 incubated 1 hr, 37°C; 2. ADH 5,3,and 1 incubated 2 hr, 37°C; 3. ADH 5,3, and 1 incubated 1 hr, 60°C; 4. pepsin incubated 2 hr, 37°C; 5. ovalbumin incubated 2 hr, 37°C; 6. cytochrome c incubated 2 hr, 37°C; 7. ribonuclease incubated 2 hr, 37°C.

TABLE 2

INCUBATION AND SDS-ELECTROPHORESIS OF ALCOHOL DEHYDROGENASE UNDER A VARIETY OF CONDITIONS

Enzyme preparation	Incubation condition	Electrophoretic conditions	# of protein bands	Apparent molecular weight(s)
ADH 5	Std.[a], 37°, 2 hr.	Std.[a]	one	2.3×10^4
ADH 5,3,1	Std., 37°, 2 hr.	Std.	three	$<1.5 \times 10^4$
ADH 5,3,1	Std., 37°, 1 hr.	Std.	five	2.3×10^4 $<1.5 \times 10^4$ $<1.5 \times 10^4$
ADH 5,3,1	no reducing agent, 37°, 2 hr.	std.	three	$<1.5 \times 10^4$
ADH 5,3,1	0.014M DTT, 37°, 2 hr.	Std.	three	$<1.5 \times 10^4$
ADH 5,3,1	Std., 60°, 1 hr.	Std.	one	2.3×10^4
ADH 5,3,1	Std., 100°, 5 min; followed by 37°, 2 hr.	Std.	one	2.3×10^4
ADH 5,3,1	Std., 37°, 2 hrs; followed by 100°, 5 min.	Std.	three	$<1.5 \times 10^4$
ADH 5,3,1	urea + SDS[b]	urea + SDS[b]	one	2.7×10^4

[a] Weber and Osborn (1969)

[b] Swank and Munkres (1971)

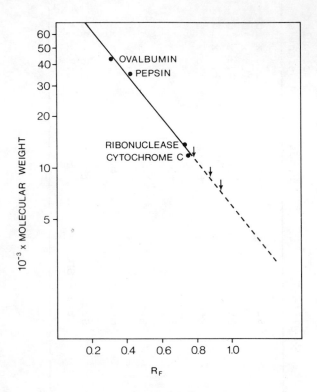

Figure 4A. Molecular weight of alcohol dehydrogenase subunits found by SDS electrophoresis. Migration relative to bromphenol blue is plotted against log of molecular weight. Molecular weights of marker proteins are from Weber and Osborn (1969). A mixture of ADH 5,3, and 1 and marker proteins were incubated for 2 hours at 37° in 1% SDS, 1% β-mercaptoethanol, 0.01 $\underline{M}$ sodium phosphate buffer, pH 7.5. Electrophoresis and staining were according to Weber and Osborn (1969). Gels were cut at the tracking dye prior to staining. The arrows indicate the positions of bands corresponding to alcohol dehydrogenase.

---

apparent molecular weight of between 2.3×10^4 and 2.7×10^4 under all of these conditions.

Our interpretation of these data is that ADH 5, ADH 3, and ADH 1 are dimers, each composed of two polypeptides with apparent molecular weight of approximately 2.5×10^4, but that the mixture of forms is contaminated with a very small quantity of proteolytic enzyme which is active in SDS at 37°. This results in the production of a variety of low molecular weight

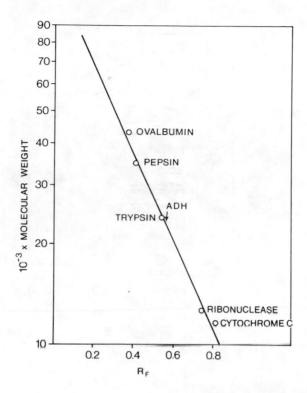

Figure 4B. Molecular weight of alcohol dehydrogenase subunits found by SDS electrophoresis. A mixture of ADH 5,3, and 1 and marker proteins treated as in Fig. 4A except that incubation was at 60° for 1 hour.

peptides. When the temperature is raised to 60° or 100°, or when urea is included during incubation, the proteolytic activity is destroyed. Under these conditions only a single protein band (whose apparent molecular weight is approximately 2.5×10^4) is discernible.

To confirm this interpretation, we incubated ADH 5 and an enzyme preparation containing ADH 5, 3, and 1 with ovalbumin at 37° for 2 hours in SDS. The samples were then boiled for 5 min, cooled, and subjected to electrophoresis. Control tubes first boiled for 5 min and then incubated in SDS for 2 hours at 37° were run at the same time. The data are given in Table 3. They show that our preparation of the mixture of ADH 5, 3, and 1, but not ADH 5 alone, contains a heat labile component capable of increasing the mobility of ovalbumin

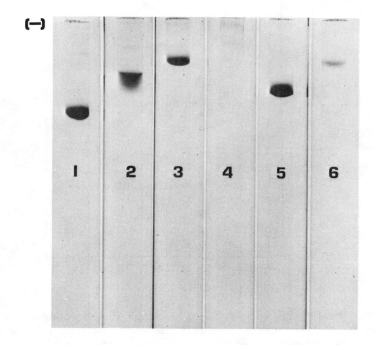

Figure 5. SDS-urea electrophoresis of alcohol dehydrogenase and marker proteins. The proteins were incubated for 2 hours at 37° in 0.01 $\underline{M}$ sodium phosphate buffer, pH 7.5, 1% SDS, 1% β-mercaptoethanol, 8 $\underline{M}$ urea. Electrophoresis was in gels containing 0.1% SDS and 8 $\underline{M}$ urea according to the procedure of Swank and Munkres (1971). Gels were cut off at the position of the bromphenol blue marker before staining. Samples are: 1. cytochrome c, 2. myoglobin, 3. Drosophila alcohol dehydrogenase mixture, 4. pepsin, 5. ribonuclease, 6. trypsin.

(and also hemoglobin, in subsequent experiments) in SDS electrophoresis. We interpret this increased mobility to be due to proteolysis.

In conclusion, our electrophoresis data strongly suggests that the three forms of alcohol dehydrogenase are all dimers.

TABLE 3

SDS-ELECTROPHORESIS OF ALCOHOL DEHYDROGENASE AND OVALBUMIN

Proteins	Incubation conditions	Number of Bands	Apparent Molecular weight
ADH 5	A	one	2.4×10^4
ADH 5 + ovalbumin	A	two	4.3×10^4, 2.4×10^4
ADH 5	B	one	2.4×10^4
ADH 5 + ovalbumin	B	two	4.3×10^4, 2.4×10^4
ADH 5,3,1	A	one	2.4×10^4
ADH 5,3,1 + ovalbumin	A	two	2.4×10^4, 4.3×10^4
ADH 5,3,1	B	three	$<1.5 \times 10^4$
ADH 5,3,1 + ovalbumin	B	four	$<1.5 \times 10^4$
ovalbumin	A	one	4.3×10^4
ovalbumin	B	one	4.3×10^4

Incubation condition A: the proteins were brought to 100° for 5 min, cooled, and then incubated at 37° for 2 hr.
Incubation condition B: the proteins were incubated at 37° for 2 hr and then brought to 100° for 5 min.
The incubation solution contained 1% SDS, 1% 2-mercaptoethanol, and protein in 0.01 M sodium phosphate buffer, pH 7.5.

COMPLEMENTATION ANALYSIS

Further confirmation that alcohol dehydrogenase consists of only one type of peptide chain comes from genetic analysis of ADH-null mutants. Ten ADH-null mutants were mated in all pairwise combinations. After eclosion and aging for 3-5 days, the flies were homogenized in a small volume of histochemical reaction mix. None of the mutations showed complementation, suggesting that all of these mutations are within a single polypeptide chain.

[14C]NICOTINAMIDE LABELING

If all species of alcohol dehydrogenase have the same primary structure, something other than differences in amino acid sequence must be causing the appearance of multiple forms. Several laboratories (Ursprung and Carlin, 1968; Jacobson, 1968) had previously shown that the addition of NAD[+] to ADH 5 resulted in the generation of forms whose heat stabilities and electrophoretic migrations approximated those

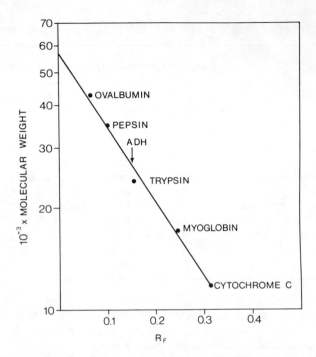

Figure 6. Molecular weight of alcohol dehydrogenase subunits. Plot of logarithm of molecular weight versus mobilities relative to bromphenol blue. Conditions are given in the legend to Figure 5.

---

of ADH 3 and ADH 1. We guessed, therefore, that a similar mechanism might be operating in vivo and that the substance bound to alcohol dehydrogenase might be NAD^+ or an analogous compound.

Accordingly we grew flies in medium containing [^{14}C]nicotinamide, and purified alcohol dehydrogenase by a modified procedure. Table 4 compares the yield of radioactivity with that of the enzyme. We calculate that between 0.1% and 0.4% of the total counts taken up by the flies is bound to alcohol dehydrogenase. When the enzyme is passed over a hydroxylapatite column it separates into two peaks (Fig. 7). The major peak, a, is composed mostly of ADH 5; the minor peak, b, of all three forms (see the densitometer tracings in Fig 8). Notice that peak b contains nearly three times the radioactivity of peak a although there is about 25 times more enzyme activity in a than b. When material from these peaks is pooled, concentrated and then electrophoresed in acrylamide

TABLE 4

PURIFICATION OF ALCOHOL DEHYDROGENASE FROM FLIES
LABELLED WITH [^{14}C]NICOTINAMIDE

	Volume (ml)	Total enzyme units	Total CPM
Crude homogenate	23.0	73,600	4.3×10^6
Protamine sulfate supernatant	24.8	62,000	4.4×10^6
40-60% ammonium sulfate cut	3.5	55,300	1.7×10^5
60% ammonium sulfate supernatant	28	28,600	3.5×10^6
pH 6.0 supernatant, dialyzed against 0.01M Na phosphate buffer, pH 7.5	4.0	56,000	3.1×10^4
Hydroxylapatite			
Peak b	17.0	873	4200 [a]
Peak a	47.0	22,250	1500 [a]

[a]Approximately 33% more protein was found in peak a than in peak b.

gels, most of the radioactivity is associated with ADH 3 and ADH 1; virtually none with ADH 5 (Figure 8). (Some of the radioactivity migrates with the tracking dye. This probably represents material which was released from denatured protein.) An analogous experiment where the multiple forms were separated by chromatography on DEAE-cellulose resulted in the same observation. That is, most of the radioactivity was associated with ADH 3 and ADH 1, and less than 5% with ADH 5. In both these instances, when the mixture of ADH 5, 3, and 1 labeled with nicotinamide was heated to 90° for 5 min, more that 90% of the radioactivity was released from the precipitated protein.

These labeling experiments indicate that some substance(s) derived from [^{14}C]nicotinamide is bound to ADH 3 and ADH 1, but not to ADH 5, and that it can be released by denaturation.

INTERCONVERSION OF FORMS

If the substance(s) bound to ADH 3 and ADH 1 is responsible for the appearance of multiple forms, then we expected that treating ADH 5 with this material might generate ADH 3 and ADH 1. To test this hypothesis, we heated a preparation of ADH 5, 3, and 1 to 90° for 5 min, centrifuged down the

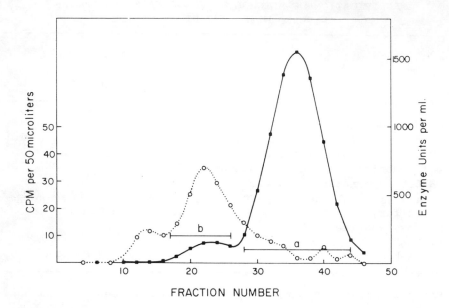

Figure 7. Hydroxylapatite chromatography of alcohol dehydrogenase from Drosophila raised on medium containing [^{14}C]nico-
tinamide.   The early steps of purification followed published
procedures (Sofer and Ursprung, 1968).   Four grams of radioactive flies were used.   The 40-60% ammonium sulfate fraction
was dialyzed overnight against 0.01 $\underline{M}$ sodium phosphate buffer,
pH 6.0, and then centrifuged at 27,000 x g for 20 minutes.
The supernatant was dialyzed against 0.01 $\underline{M}$ sodium phosphate
buffer, pH 7.5, for several hours and applied to a 1.5 x 20
cm column of hydroxylapatite in 0.01 $\underline{M}$ sodium phosphate buffer,
pH 7.5.   The same buffer was used to elute alcohol dehydrogenase activity.   ······ ^{14}C counts per minute per 50 μl.
————, alcohol dehydrogenase activity, enzyme units per ml.

denatured protein, and then incubated the supernatant solution
with pure ADH 5.   As a control, pure ADH 5 was also heat denatured and a supernatant fluid derived.   As shown in Figure
9, the supernatant from the enzyme mixture was effective in
generating a zone of alcohol dehydrogenase activity whose
migration matched that of ADH 3.   Supernatant derived from
ADH 5 was ineffective and neither of the supernatant fluids
possessed any intrinsic enzyme activity.

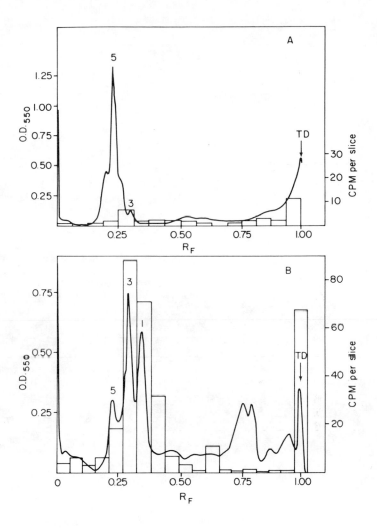

Figure 8. Polyacrylamide gel electrophoresis of labelled alcohol dehydrogenase. Peaks a and b from hydroxylapatite chromatography (Figure 7) were pooled and concentrated before electrophoresis on 5% gels. Staining, slicing and counting procedures are described under Methods. In each case, a parallel gel was stained for protein with Coomassie Blue and scanned at 550 nm with a Guilford gel scanner. A, electrophoresis of peak a from hydroxylapatite. B, electrophoresis of peak b from hydroxylapatite, ————, O.D. 550. Bars indicate cpm per slice. T.D. indicates position of tracking dye (bromphenol blue).

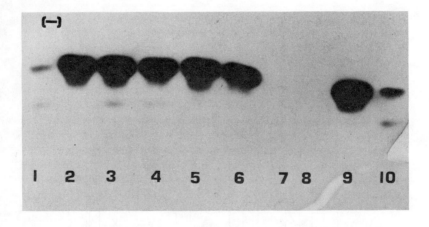

Figure 9. Agar gel electrophoresis of ADH 5 and derived forms.
Agar gel electrophoresis was carried out as previously described (Ursprung and Leone, 1965) and the gel was stained in
the histochemical stain for alcohol dehydrogenase activity
(Ursprung et al., 1970). Pure ADH 5 was incubated at room
temperature with supernatant solution derived from heat treatment (90°, 5 minutes) of ADH 5 or a mixture of ADH 5,3, and 1.
Samples are: 1. crude extract of an ADH[F] strain, used as
electrophoretic marker; 2. ADH 5; 3. ADH 5 incubated with
supernatant from mixture of ADH 5,3, and 1, 30 hrs; 4. ADH 5
incubated with supernatant from mixture of ADH 5, 3, and 1,
18 hrs; 5. ADH 5 incubated with supernatant from ADH 5, 30 hrs;

6. ADH 5 incubated with supernatant from ADH 5, 18 hrs; 7. supernatant from mixture of ADH 5, 3, and 1; 8. supernatant from ADH 5; 9. ADH 5; 10. crude extract of an ADH^F strain.

DISCUSSION

A model for the structure of alcohol dehydrogenase consistent with the data presented in this paper is shown in Figure 11B. Before discussing this model in detail, it would be well to review some additional experiments which bear on the interpretations we will be making.

1. There are three electrophoretic variants of alcohol dehydrogenase, each of which displays three zones of activity after starch, agar or acrylamide electrophoresis (Fig. 10, ADH^F or Fast, ADH^S or Slow, and ADH^D). ADH^F and ADH^S are naturally occurring variants; ADH^D was derived from ADH^F after mutagenesis with ethyl methanesulfonate by Grell et al. (1968).

One striking fact about these variants is that in each, the most cathodally migrating zone of activity exhibits both the highest activity (Fig. 10) and the lowest heat stability (Ursprung and Leone, 1965; Grell et al., 1968; Jacobson, 1968). These data indicate that a single mutation at the alcohol dehydrogenase structural locus (ADH$^F \longrightarrow$ ADH^D) can result in a simultaneous change in migration of all three isozymes; this in turn, we feel, means that at least one subunit is common to all three enzyme forms in any one variant.

2. ADH^F/ADH^S and ADH^F/ADH^D heterozygotes exhibit bands of intermediate mobility in addition to parental bands after electrophoresis (Fig. 10). Such intermediate bands are not produced by simply mixing extracts of the parental types (Ursprung and Carlin, 1968). We feel that these "hybrid" zones are caused by the formation of heteropolymers, and that, therefore alcohol dehydrogenase is a polymer composed of at least two subunits.

3. After electrophoresis in SDS (with incubation at 60°C), and in SDS-urea, a single protein band with an apparent molecular weight of approximately 1/2 of the native molecule is observed.

If one accepts these three observations and their interpretations (that at least one subunit is held in common among the three forms, that the intermediate forms represent heteropolymers, and that alcohol dehydrogenase is a dimer), then one can rule out on theoretical grounds a model in which alcohol dehydrogenase is composed of two unlike polypeptide chains. Consider the model shown in Figure 11A. Columns 1 and 2 are two representations wherein dimeric molecules share one subunit "A" in common. In these models, polypeptides B, C, and

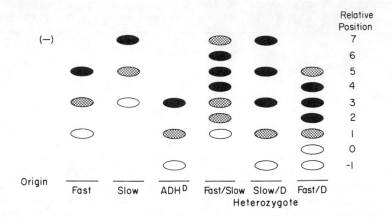

Figure 10. Electrophoretic variants of alcohol dehydrogenase.
Schematic representation of the patterns seen after agar gel
electrophoresis of crude homogenates and staining for alcohol
dehydrogenase activity. Note the presence of bands of intermediate mobility in the Fast/Slow and Fast/D heterozygotes.
We interpret these to be heterodimers. We believe that heterodimers are present in the Slow/D heterozygotes as well and
that their mobility coincides with that of the homodimers.
Solid circles represent the most intense staining; cross-
hatched circles, medium intensity; and open circles, the least
intense staining.

D impart the differential mobilities to the different forms.
If a mutation now occurs which changes the mobility of subunit
"A" to "A'" (column 2) an electrophoretic variant would be
produced. Consistent with our observations, all three forms
would be affected. But this model cannot account for all the
hybrid bands observed in heterozygotes (Fig. 10). In fact,
if alcohol dehydrogenase is considered a dimer, we feel forced
to conclude that it is composed of two <u>identical</u> subunits.

   Additional data bearing on this point come from genetic and
biochemical experiments.

   4. Complementation data support the conclusion that alcohol dehydrogenase consists of only one kind of polypeptide
chain.

   5. Amino acid analysis and peptide fingerprinting imply
that all three forms of the enzyme have the same primary
structure. In addition, they suggest that all the alcohol dehydrogenases are composed of two identical polypeptide chains
of approximately 25,000 molecular weight.

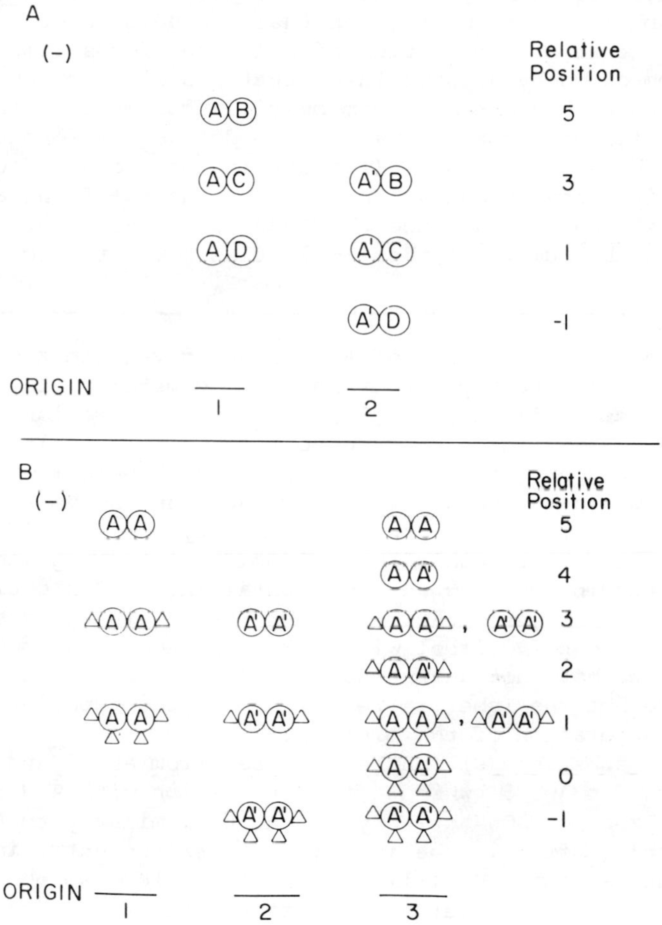

Figure 11. Models for the structure of alcohol dehydrogenase.
A. In this model the difference in electrophoretic migration of the three forms of alcohol dehydrogenase in any
one strain is due to the association of protein A with three
different proteins; B, C, and D. The observation that a single
genetic event causes a change in migration of all three forms
simultaneously is accommodated by hypothesizing that protein
A changes to A' by a change in amino acid composition (column
2). This model cannot readily account for the presence of
intermediate forms ("hybrids") in heterozygous individuals.
B. In this model the difference in electrophoretic

747

migration of the three forms of alcohol dehydrogenase in any
one strain is due to the association of protein A with a non-
protein, heat stable, low molecular weight molecule (triangles
in column 1). The observation that a single genetic event
causes a change in migration of all three forms simultaneously
is accomodated by hypothesizing that protein A changes to A'
by a change in amino acid composition (column 2). Intermediate forms ("hybrid enzyme") are due to the formation of
heterodimers (column 3). (The binding of two and four triangles is formally equivalent to the binding of one and two
triangles, each with twice the charge. We have not, as yet,
measured the number of these molecules bound to alcohol dehydrogenase.)

If the multiple forms of ADH in any given homozygous strain
have the same primary structure, then something else must be
causing the difference in electrophoretic migration. (One
mechanism which has been ruled out is that the multiple forms
are due to aggregation (Jacobson, 1970a; Imberski et al.,
1968).) Our data support the view that the isozymes differ
in charge, activity and thermal lability due to differential
binding of a heat stable, nicotinamide containing substance.

6. Feeding *Drosophila* food containing [^{14}C]nicotinamide
and subsequent purification of ADH results in the association
of the radioactive label with ADH 3 and ADH 1 but not with
ADH 5. We have not shown that the label remains in nicotinamide but the label is released into the supernatant after
heat denaturation of the protein.

7. A substance(s) can be released from ADH 3 and ADH 1,
but not from ADH 5, which after incubation with ADH 5 produces a zone of ADH activity migrating similarly to ADH 3.
Presumably, after incubation with higher concentrations of the
substance(s) some ADH 1-like activity would also be produced.

8. Studies to determine the structure of the factor released from ADH 3 and ADH 1 by heat or ethanol denaturation
are underway. Preliminary results indicate that the factor
is flourescent and probably represents an NAD^+-carbonyl compound addition complex of the general type first described by
Kaplan (Everse et al., 1972). It exhibits the pH dependent
shift in flourescence excitation and emission maxima which
is characterictic of such complexes. The factor migrates similarly, but apparently not identically, to NAD^+-acetone on thin
layer chromatography on cellulose plates in several solvent
systems. The products of snake venom phosphodiesterase digestion of both the factor and NAD^+-acetone also show similar
migrations (O'Donnell et al., unpublished data).

Some of our data conflicts with results published previously.   For example, we estimate the molecular weight of alcohol dehydrogenase at 5.0 x 10^4, based on SDS electrophoresis and amino acid analysis.   Jacobson et al.(1970a) had calculated a molecular weight of 6.0 x 10^4 and we had reported a value of 4.4 x 10^4 (Sofer and Ursprung, 1968).   Our earlier report was based on an assumed partial specific volume for ADH. From the amino acid composition reported in this paper we calculate a partial specific volume of 0.749 by the method of Cohn and Edsall (1943).   This changes our previously calculated molecular weight to 4.8 x 10^4.   We cannot resolve the discrepancy between this molecular weight and that reported by Jacobson et al.

Another apparent discrepancy between our data and those of Jacobson et al. (1970a) is in the measurement of the molecular weight of the subunits.   They used SDS electrophoresis and found, as we do, a variety of low molecular weight forms.   We present data here which attribute these low molecular weight forms to the action of contaminating proteolytic enzymes.   In our hands, the presence of urea in the incubation mixture inhibits proteolysis and, in fact, when Jacobson et al. used urea in the presence of SDS to dissociate ADH they found larger subunits than in the absence of urea.

The physiological significance of the multiple forms of *Drosophila* ADH is unknown. ADH 3 and ADH 1 might represent either newly synthesized enzyme or old enzyme on its way to degradation.

In summary, we believe that *Drosophila* alcohol dehydrogenase (like liver alcohol dehydrogenase (Masters and Holmes, 1972)) is a dimer composed of two identical polypeptide chains. Multiple electrophoretic species are formed from this protein by differential binding of a heat stable substance probably containing nicotinamide.   All the biochemical analyses reported in this paper have been done on the ADH^F variant.   We believe that the same model applies to the ADH^S and ADH^D variants. We postulate that each is a dimer of two identical polypeptide chains and that the ADH^D polypeptide differs from the ADH^F polypeptide by a single amino acid and that the ADH^S polypeptide differs from the ADH^F polypeptide by one or more amino acids.   As in the case of ADH^F, we believe that the three forms in the ADH^D and ADH^S sets are formed by differential binding of a heat stable ligand (Fig. 11B).

ACKNOWLEDGEMENTS

This work was supported by N.I.H. Grant GM18254-03.   Publication number 778 from the Department of Biology, The Johns Hopkins University.

ACKNOWLEDGEMENTS (CONT.)

M. Schwartz and J. O'Donnell were predoctoral fellows supported by N.I.H. Training Grant GM-57.

We thank Dr. Allen Shearn for many informative discussions and Dr. Dennis Powers for help with the amino acid analyses.

REFERENCES

Bennett, J. C. 1967. Paper chromatography and electrophoresis; special procedure for peptide maps. In *Methods in Enzymology*. (S. P. Colowick and N. O. Kaplan, Eds.) Vol. XI, pp. 330-339, Academic press, New York.

Borack, L. I., and W. Sofer 1971. *Drosophila* β-L-Hydroxy acid dehydrogenase: purification and properties. *J. Biol. Chem.* 246, 5345-5350.

Cohn, E. J. and J. T. Edsall 1943. *Proteins, Amino Acids and Peptides*, pp. 370-381, Reinhold, New York.

DeLange, R. J., D. M. Fambrough, E. L. Smith, and J. Bonner 1969. Calf and pea histone IV: the complete amino acid sequence of calf thymus histone IV; presence of ε-N-Acetyllysine. *J. Biol. Chem.* 244, 319-334.

Easley, C. W. 1965. Combinations of specific color reactions useful in the peptide mapping technique. *Biochim. Biophy. Acta* 107, 386-388.

Everse, J., R. L. Berger, and N. O. Kaplan 1972. Complexes of pyridine nucleotides and their function. In *Structure and Function of Oxidation-Reduction Enzymes* (A. Akeson and A. Ehrenberg, Eds.) Pergamon Press, Oxford.

Grell, E. H., K. B. Jacobson, and J. B. Murphy 1968. Alterations of genetic material for analysis of alcohol dehydrogenase isozymes of *Drosophila melanogaster*. *Ann. N. Y. Acad. Sci.* 151, 441-445.

Hirs. C. H. W. 1967. Determination of cystine as cysteic acid. In *Methods in Enzymology* (S. P. Colowick and N. O. Kaplan, Eds.) Vol. XI, pp. 59-62 Academic Press, New York.

Imberski, R. B., W. H. Sofer, and H. Ursprung 1968. *Experientia* 24, 504-505.

Jacobson, K. B. 1968. Alcohol dehydrogenase of *Drosophila*: interconversion of isoenzymes. *Science* 159, 324-325.

Jacobson, K. B., J. B. Murphy, and F. C. Hartman 1970a. Isoenzymes of *Drosophila* Alcohol dehydrogenase: isolation and interconversion of different forms. *J. Biol. Chem.* 245, 1075-1083.

Jacobson, K. B. and P. Pfuderer 1970b. Interconversion of isoenzymes of *Drosophila* alcohol dehydrogenase: physical characterization of the enzyme and its subunits. *J. Biol. Chem.* 245, 3938-3944.

Jacobson K. B., , J. B. Murphy, J. A. Knopp, and J. R. Ortiz 1972. Multiple forms of *Drosophila* alcohol dehydrogenase: conversion of one form to another by nicotinamide adenine dinucleotide or acetone. *Arch. Biochem. Biophys.* 149, 22-35.

Liu, T. Y. 1972. Determination of tryptophan. In *Methods in Enzymology* (S. P. Colowick and N. O. Kaplan, Eds.) Vol. XXV, pp. 44-55, Academic Press, New York.

Masters, C. J. and R. S. Holmes 1972. Isoenzymes and ontogeny. *Biol. Rev.* 47, 309-361.

Noltmann, E. A., T. A. Mahowald, and S. A. Kuby 1962. Studies on adenosine triphosphate transphosphorylases: amino acid composition of adenosine triphosphate-adenosine 5'-phosphate transphorylase (Myokinase). *J. Biol. Chem.* 237, 1146-1154.

Sofer, W. and H. Ursprung 1968. *Drosophila* alcohol dehydrogenase: purification and partial characterization. *J. Biol. Chem.* 243, 3110-3115.

Swank, R. T. and K. D. Munkres 1971. Molecular weight analysis of oligopeptides by electrophoresis in polyacrylamide gel with sodium dodecyl sulfate. *Anal. Biochem.* 39, 462-477.

Ursprung, H., and J. Leone 1965. Alcohol Dehydrogenases: a polymorphism in *Drosophila melanogaster*. *J. Exp. Zool.* 160, 147-154.

Ursprung, H. and L. Carlin 1968. *Drosophila* alcohol dehydrogenase: in vitro changes of isozyme patterns. *Ann. N.Y. Acad. Sci.* 151, 456-475.

Ursprung, H., W. H. Sofer, and N. Burroughs 1970. Ontogeny and tissue distribution of alcohol dehydrogenase in *Drosophila melanogaster*. *Wilhelm Roux' Arch. Entw. Mech.* 164, 201-208.

Weber, K. and M. Osborn 1969. The reliability of molecular weight determinations by dodecyl sulfate-polyacrylamide gel electrophoresis. *J. Biol. Chem.* 244, 4406-4412.

MULTIPLE MOLECULAR FORMS OF PNEUMOCOCCAL NEURAMINIDASE

TOVA FRANCUS, STUART W. TANENBAUM,
HELGA LUNDGREN AND MICHAEL FLASHNER
SUNY College of Environmental Science & Forestry,
Syracuse, New York 13210 and
Department of Microbiology, Columbia University
New York, New York 10032

ABSTRACT. The isozymes earlier described for neuraminidase from *D. pneumoniae* can be reduced in number by carrying out isolation procedures in the presence of $5.7 \times 10^{-4}M$ phenylmethylsulfonyl fluoride and by using cells rather than growth filtrates as the enzyme source. Although glycohydrolase activity has not been ruled out, the multiple molecular neuraminidase species appear in large measure to result from serine-esterase proteolysis. The molecular weight of the presumptive "parental" cell-bound enzyme is of the order of 90-95,000 daltons, in contrast to the 70,000 size reported for the numerous filtrate-derived multiple components. The larger enzyme species is a glycoprotein containing approximately 15 residues per mole of glucosamine but not galactosamine, and appears to lack cysteine residues. The latter finding accords with prior circumstantial evidence which indicated that the isozymes are single-chain proteins. Following a study of the effect of various synthetic inhibitors on neuraminidase activity, an alternative, rapid method employing affinity chromatography was devised for isozyme purification. The procedure involves recycling chromatography on a support of Sepharosyl-hexamethylene-amidobenzene-azo-p-β-hydroxyphenyl-α-mercaptoacrylic acid. This approach obviates the use of inhibitors during purification and provides enzyme with properties like those of cellular neuraminidase obtained by conventional procedures.

Neuraminidases are ubiquitous enzymes of more than intrinsic importance. From the historical aspect, myxoviral neuraminidase was the first component of a virion to exhibit enzymatic function (Hirst, 1942). Subsequently, this enzyme was detected among such diverse microorganisms as *Cl. welchii*, *C. diphtheriae*, *Klebsiella aerogenes*, *Bacteroides spp.*, *Streptomyces albus*, and among pneumococci and streptococci. Neuraminidase activity has also been shown to be present in many organs and tissues of mammalian origin. The occurrence and properties of neuraminidases and their substrate specificities have been the subject of recent reviews by Drzeniek (1972, 1973).

753

The selection of neuraminidase as a subject for further investigation of structure-function parameters and isozyme genesis relates to its presumptive role in the molecular pathology of viral and bacterial infections, to its widespread use as a biological tool in the study of the architecture of the surface of cells in culture, and because neuraminidases of differing linkage specificity may be useful in the structural analysis and modification of glycoproteins. For these applications and investigations, the availability of a bacterial preparation that has been separated from adventitious proteases, glycohydrolases, phospholipases, and other macromolecular contaminants is mandatory.

Our own studies followed the work of Hughes and Jeanloz (1964) on pneumococcal enzyme purification and of Lee and Howe (1966) on enzyme production. As a result of these investigations, a purification procedure was evolved for neuraminidase in the filtrate from an overnight culture of pneumococcus. This resulted in an enzyme that crystallized from ammonium sulfate but which was shown by disc gel electrophoresis to contain numerous isozymes as well as enzymatically inactive protein (Tanenbaum et al., 1970). The appearance of these multiple components was reminiscent of the work of Hayes and Wellner (1969) with L-amino acid oxidase, in which it was shown that crystalline preparations exhibited some 18 peaks of enzymatic activity upon isoelectric focusing. These analytical findings caused us to reexamine the various steps of the enrichment sequence, including especially one that took advantage of the relative heat stability of neuraminidase. It was found that deletion of the heat-treatment stage, though diminishing the number of multiple forms that could be observed on ion-exchange chromatography, still provided highly purified preparations which exhibited a multiplicity of clearly separable activities following either CM- or DEAE-cellulose chromatography.

To account for the appearance of these growth filtrate isozymes, we considered aggregation-disaggregation, subunit dissociation and recombination, differences in conformational states, or differences in primary sequence. For each of four pneumococcal neuraminidase isozymes, an estimated molecular weight of $70,600 \pm 1000$ was determined by gel filtration chromatography. An identical value was estimated for the most homogeneous of these preparations, following its treatment with $8M$ urea and $0.01M$ β-mercaptoethanol and subsequent chromatography on a molecular sieve equilibrated with the denaturant (Tanenbaum and Sun, 1971). These results restricted the choice of alternatives to these involving conformational or charge differences. Application of a slightly modified purification procedure to growth filtrate enzyme derived from the same isolate

of pneumococcus by Stahl and O'Toole (1972) substantiated both
the chromatographic profiles of multiple activity and the
molecular weight reported from our laboratory.

In order to minimize the possibility that population heterogeneity was responsible for the foregoing observations on
multiple molecular forms, the original *D. pneumoniae* strain 70
was cloned. Three serial single colony isolates were grown,
and the enzymes in the filtrates were concentrated, purified,
and each was subjected to ion-exchange chromatography. The
elution-activity profiles were identical to those found for
the parental strain. To eliminate other factors that might have
caused isozyme formation *via* subtle chemical changes such as
deamidation (Funakoshi et al., 1969) during the course of
conventional salt fractionation, a large bulk fermentation
medium was divided into two portions. One of these was purified by the usual procedure (Tanenbaum et al., 1970), while
the other was concentrated according to the simpler procedures
outlined in the left-hand side of the flow sheet (Scheme I).
Again, the ion-exchange chromatographic elution pattern was
identical with the pattern of multiple activity peaks obtained
from the salt-fractionated portion (Scheme I, right side).

Our attention was next directed toward proteolysis as an
alternative explanation for neuraminidase isozyme generation.
Pneumococcus is not notable for the production of proteolytic
activity and indeed several tests for proteolytic activity in
the growth filtrate, such as azacoll digestion (Moore, 1960)
and casein hydrolysis (Subramanian and Kalnitsky, 1964), were
negative. Nevertheless, we observed that the degree of microheterogeneity was reduced in proportion to the rapidity with
which the earlier-devised purification procedures were carried
out. It was also considered possible that the extracellular
neuraminidase might have been cleaved from a larger polypeptide prior to secretion or as a concomitant enzymatic step
during the autolysis of the pneumococci. Following the work
of Clark and Jacoby (1970) with yeast in which esterase inhibitors were shown to enhance enzyme yields and restrict heterogeneity, a series of experiments was carried out in which cell-
bound enzyme was obtained by disruption at pH 7.2 in a hydraulic
press and in the presence of the serine-protease inhibitor
phenylmethylsulfonyl flouride (PMSF) at $1.5 \times 10^{-3}M$ concentration (Gold and Fahrney, 1964). The cell lysate was then
dialyzed against buffer which contained PMSF at a concentration
of $5.7 \times 10^{-4}M$. The inhibitor at this final concentration
was included in all further operations. Chromatography on DEAE-
cellulose of the enzyme purified by ammonium sulfate fractionation now yielded one enzymatic peak that coincided with the
major isozyme fraction in enzyme preparations purified in the

SCHEME I

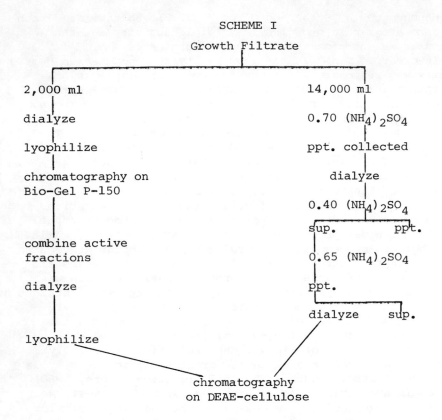

Growth Filtrate

2,000 ml

dialyze

lyophilize

chromatography on
Bio-Gel P-150

combine active
fractions

dialyze

lyophilize

14,000 ml

0.70 $(NH_4)_2SO_4$

ppt. collected

dialyze

0.40 $(NH_4)_2SO_4$

sup.   ppt.

0.65 $(NH_4)_2SO_4$

ppt.

dialyze   sup.

chromatography
on DEAE-cellulose

absence of the inhibitor. Furthermore, the conversion of this relatively homogeneous enzyme into the other isozyme forms, even in the presence of dilute PMSF, was demonstrated by incubation *in vitro* at 37° for 18 hours.

Since the initial specific activity of pneumococcal neuraminidase from ruptured cells was 25-50 fold higher than enzyme obtained from the growth filtrate, a more convenient method was developed for enzyme purification. This procedure, detailed in Scheme II, involved collection of the cells, liberation of enzyme by the use of a French press and chromatography on DEAE-cellulose followed by resolution on a CM-ion exchanger with 5.7 x $10^{-4}M$ PMSF included in all the operations. Adherence to these experimental protocols resulted in 112-fold purification and specific activity of ∿6000 units/mg protein (Table I). The apparent increase in enzyme activity at Step 2 deserves comment. This phenomenon was reproducible, having also been observed with filtrate-derived enzyme, and reported

SCHEME II
FLOWSHEET FOR PURIFICATION OF CELL-BOUND
PNEUMOCOCCAL NEURAMINIDASE

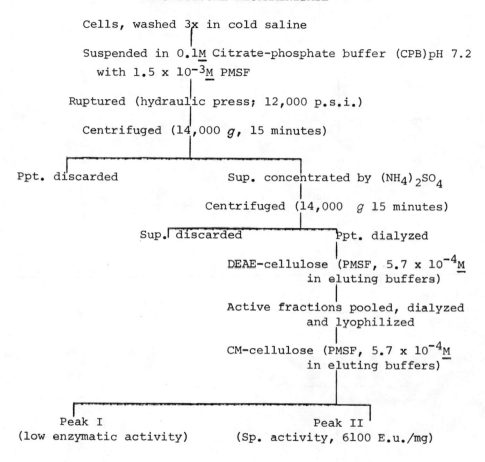

Cells, washed 3x in cold saline

Suspended in 0.1M Citrate-phosphate buffer (CPB)pH 7.2
with 1.5 x 10^{-3}M PMSF

Ruptured (hydraulic press; 12,000 p.s.i.)

Centrifuged (14,000 g, 15 minutes)

Ppt. discarded   Sup. concentrated by $(NH_4)_2SO_4$

Centrifuged (14,000 g 15 minutes)

Sup. discarded   Ppt. dialyzed

DEAE-cellulose (PMSF, 5.7 x 10^{-4}M
in eluting buffers)

Active fractions pooled, dialyzed
and lyophilized

CM-cellulose (PMSF, 5.7 x 10^{-4}M
in eluting buffers)

Peak I   Peak II
(low enzymatic activity)   (Sp. activity, 6100 E.u./mg)

earlier for Clostridial (Cassidy et al., 1965) and diphtherial
neuraminidases. In this connection, it has been suggested that
a natural inhibitor accompanies these neuraminidases (Warren
and Spearing, 1963), but this possibility has not been further
investigated.

The pneumococcal enzyme so purified appeared homogeneous
when examined by polyacrylamide disc gel electrophoresis, though
an enzymatically active and putative "aggregated" species that
did not migrate into the separating gel remained in the stacking
gel. The enzyme preparation had a K_m value of 5.8 x $10^{-4}M$ with
N-acetylneuraminyl lactose as substrate. Of perhaps greater
interest was the finding that the pneumococcal neuraminidase

TABLE I

PURIFICATION OF INTRACELLULAR PNEUMOCOCCAL NEURAMINIDASE

Step	Fraction	Final volume (ml)	Total protein (mg)	Total activity (E.U.)*	Specific activity (E.U./mg protein)	Recovery %
1	Supernatant of broken cells	100	1,050	58,000	55	100
2	0.40 (NH$_4$)$_2$SO$_4$ (sup.)	100	725	89,500	123	(154)
3	0.70 (NH$_4$)$_2$SO$_4$ (ppt.)	10	210	38,250	182	66
4	DEAE-cellulose	5	14.5	10,000	690	17.2
5	CM-cellulose Peak II	1	0.41	2,625	6,158	4.5

*An enzyme unit is defined as one nanomole of N-acetylneuraminic acid released per min. from glycoprotein or other substrates.

obtained in this fashion was estimated by gel exclusion chromatography to have a molecular weight of 88-90,000 and by SDS-
polyacrylamide electrophoresis, $95,000 \pm 3,000$ daltons. These
data are consistent with the value of 88,000 given by Drzeniek
(1972) as an unpublished observation and support the conclusion
that the numerous isozyme species of 70,000 daltons isolated
earlier from culture filtrates represent a homologous series
of proteolytic degradation products.

Amino acid analysis was carried out with this higher molecular weight "parental" enzyme. These data (Table II) revealed two salient features. First, this enzyme was devoid of
cysteine residues in accord with prior evidence found with
70,000 dalton isozymes for a single polypeptide chain; and
second, it is a glycoprotein containing approximately 15 moles
of glucosamine but no galactosamine. The latter observation
relates the pneumococcal neuraminidase to the viral enzymes
(Compans and Choppin, 1971), although it should be pointed out
that the latter, with a subunit molecular weight of $\sim 50,000$,
contain cysteine and are multimers (Kendal and Eckert, 1972).
Several proteins and bacterial enzymes lack cysteine, including the α-amylase of *B. subtilis* (Junge et al., 1959) and
penicillinases from various sources (Kirschenbaum, 1971).

We sought a still simpler, efficient isolation procedure
to obtain sufficient material for several biochemical studies,
including C- and N-terminal amino acid analyses, investigation
of the precise structural differences among the lower molecular weight isozymes, and analysis of the nature of the catalytic site. During the course of our investigations, Cuatre-
cases and Illiano (1971) published an affinity chromatographic
procedure for purification of neuraminidases from *Cl. perfringens* and *Vibrio cholerae* based upon the use of p-aminophenylox-
amate as the affinity ligand, but this procedure was not applicable to the cell-bound enzyme from *D. pneumoniae*, for which
the compound is a poor enzyme inhibitor (Table III). Of greater potential application was the finding that a number of compounds in the β-aryl-α-mercaptoacrylic acid series, as well as
the o-aminophenylbenzimidazoles, as reported by Haskell et al.
(1970), were effective inhibitors of pneumococcal neuraminidase. We prepared β-(p-hydroxy-m-[4']-carboxyphenylazo)-α-
mercaptoacrylic acid, which was found to have a K_i of 1.1 x
$10^{-4}M$, and which was shown to inhibit the enzyme competitively.
The aza-arabino-octonic lactone derivative claimed by Khorlin
et al. (1970) to be a potent (and perhaps irreversible) inhibitor of a variety of neuraminidases was, on the other hand,
found to be inactive even at $10^{-3}M$.

The relative effectiveness with pneumococcal neuraminidase
of a number of affinity chromatography matrixes, based on the

TABLE II

PROXIMATE AMINO ACID ANALYSIS* OF PNEUMOCOCCAL NEURAMINIDASE

Amino Acid	nmoles/28μg	μg/28μg	Residues per 95,000 daltons
Lysine	22	2.820	79
Histidine	7	0.963	25
Arginine	14.47	2.261	52
Aspartic Acid	25.98	2.990	93
Threonine	17	1.719	61
Serine	13.39	1.165	48
Glutamic Acid	30.53	3.759	110
Proline	7.74	0.752	28
Glycine	22.47	1.282	81
Alanine	31.46	2.236	113
Half-Cystine	---	---	---
Valine	16.23	1.609	58
Methionine	1.54	0.202	6
Isoleucine	9.68	1.095	35
Leucine	22	2.489	79
Tyrosine	3.7	0.605	13
Phenylalanine	7.77	1.144	28
Glucosamine	4.22	0.680	15
Galactosamine	---	---	---

* Specific activity 6,100. Trypotophan not assayed.

foregoing inhibition experiments was then tested (Table IV).
Under a variety of conditions, and with differing degrees of
ligand substitution, and lengths of polypeptide chain extensions,
the oxamate gels yielded inconsistent results and showed lack
of specificity for neuraminidase. In the interim, Rood and
Wilkinson (1974) reported that at pH 5.5 Sepharosyl-
glycyltyrosyl-N-p-aminophenyloxamate retarded not only
neuraminidase but also *Clostridial* hemolysin and hemagglutin,
and at pH 7.5 only partial resolution of neuraminidase was
obtained. We observed that the most promising purification
and resolution of the enzyme was accomplished on columns that
were composed of β-$(p$-hydroxy-m-⌈4'⌉-carboxyphenylazo)-α-
mercaptoacrylate attached by a peptide bond through a hexyl
extender to Sepharose (number 4, Table IV). Reproducibly good
recoveries with high specific activity enzyme were obtained with
this matrix. The relative capacity for enzyme of gels numbered 1-4 was 1:0.25:1:4. When gel #4 was eluted with 0.2M borate
buffer at pH 9.3, only 6% of the applied enzyme units were removed; the remainder was eluted with 0.01M NaOH - 1.0M NaCl. In

TABLE III

INHIBITION OF PNEUMOCOCCAL NEURAMINIDASE BY SYNTHETIC COMPOUNDS

Protocol: Crude enzyme (sp. act. 310 E.u./mg protein); 1.0 mg
Collocalia mucoid substrate; CPB buffer, pH 6.0;
30 mins. at 37°

Inhibitor	Conc'n mM	% Inhibition
N-(p-aminophenyl)oxamic acid	1.1	0
β-(p-nitrophenyl)-α-mercapto-acrylic acid	0.90	95
β-(p-hydroxyphenyl)-α-mercapto-acrylic acid	1.0	10
p-hydroxyphenylpyruvic acid	1.1	5
β-(4-HO-3-MeO-5-NO$_2$)-2-mercapto-acrylic acid	0.70	1
β-(p-HO-m-[4']-carboxyphenylazo)-α-mercaptoacrylic acid	1.0	61
3-Aza-2,3,4-trideoxy-4-oxo-D-arabino-octonic acid δ-lactone	1.0	0
2-(o-aminophenyl)-benzimidazole	0.80	90

contrast, approximately 100% of applied neuraminidase was
removed from the "control gel" (#7) by use of the first buffer
system. These findings are interpreted as indicating specific
adsorption of enzyme by affinity columns of type #4.

Application of these results let to the development of
a rapid procedure for purification of neuraminidase based upon
several cycles of chromatography with matrix #4. The experimental details are outlined in Scheme III. The enzyme that
resulted from application of these steps was almost identical
in terms of specific activity and its disc gel electrophoretic
pattern to that obtained by the lengthier procedures given in
Table I. In addition, it was demonstrated that the protein
band within the gel that exhibited enzymatic activity upon
slicing and Warren assay was coincident with activity as
revealed by the chromogenic Tuppy and Palese (1969) reagent
MPN. It is felt that this procedure will provide, with a
minimum of artifacts, the "parental" as well as the modified

TABLE IV
AFFINITY CHROMATOGRAPHY GEL MATRIXES TESTED

(1) 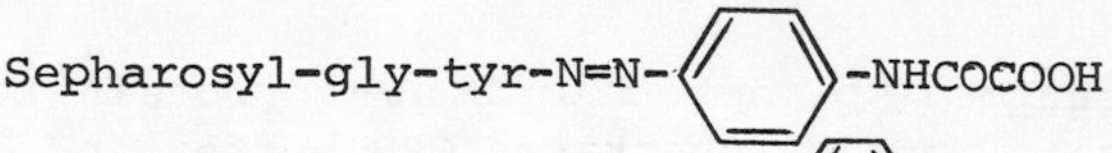 Sepharosyl-gly-tyr-N=N-⟨benzene⟩-NHCOCOOH

(2) 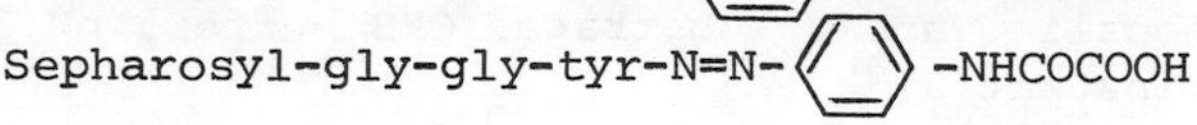 Sepharosyl-gly-gly-tyr-N=N-⟨benzene⟩-NHCOCOOH

recoveries (10 runs), 5-66%; sp. activity average;
1000 E.u./mg disc gel examination: (+) isozymes
(+) aggregation

(3) 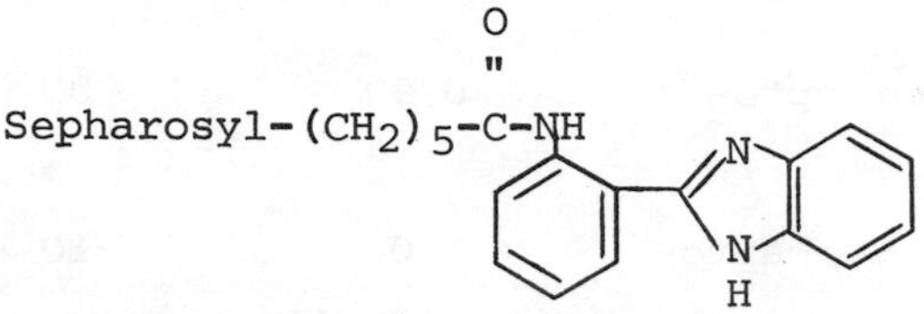 Sepharosyl-(CH$_2$)$_5$-C(=O)-NH ⟨benzimidazole structure⟩
      |
      H

recovery 90%; sp. activity ∿700 E.u./mg

(4) 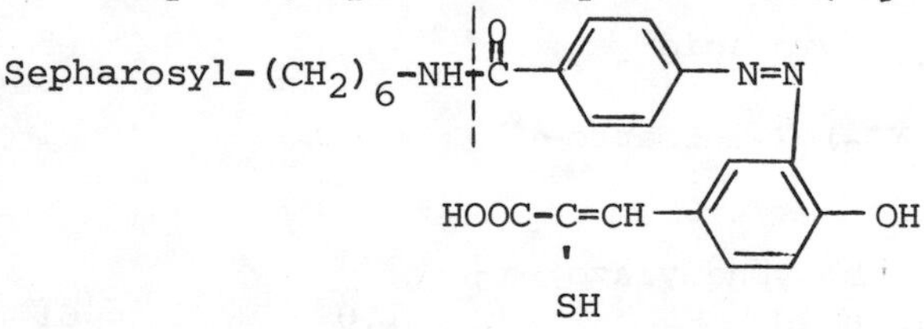 Sepharosyl-(CH$_2$)$_6$-NH-C(=O)-⟨benzene⟩-N=N

HOOC-C=CH-⟨benzene⟩-OH
      |
      SH

recovery 35-92%; sp. act. 1000-3000 E.u./mg

(5) Recycling columns (1) then (4): recovery ∿30%; sp. act.
     800 E.u/mg

(6) Recycling column (4) and repeat: recovery ∿60%; sp. act.
     average 6000 E.u./mg. Disc gel:No isozymes (+) aggre-
     gation; MPN activity ≡ gel slice activity

(7) 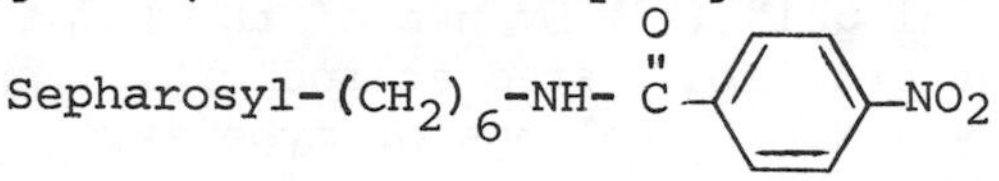 Sepharosyl-(CH$_2$)$_6$-NH-C(=O)-⟨benzene⟩-NO$_2$

"Gel control," complete recovery, sp. act. ∿470 E.u/mg

---

isozymes in good yields for further structural determinations.
    The biochemical and genetic parameters responsible for
the occurrence of pleiotropic forms of pneumococcal neuraminidase are far from clear. Even enzymes of the higher molecular
weight class, isolated from cells by procedures that incorpor-
ate PMSF, exhibit isozymes after ion-exchange chromatography.
Thus, the relatively homogeneous enzyme that served for the
amino acid analyses was selected from among two peaks present

SCHEME III

RAPID NEURAMINIDASE PURIFICATION: RECYCLING WITH LIGAND (4)

Step I (a) Column equilibrated with 0.1M citrate-
phosphate buffer, pH 7.2

(b) Crude enzyme sample applied in same buffer

(c) Flow rate 25 ml/hr, V_t = 8 ml, 0.9 cm diameter

(d) Wash with 0.2M borate buffer, pH 9.3, using
3-4x V_t (removes almost all protein but
retains enzyme)

(e) Elute enzyme with 0.01M NaOH - 1M NaCl, collect
in buffer pH 5-7. Sp. activity, 1000-3000
E.u./mg; yield 75-100%

Step II (a) Adjust pH to 7.2 (desalt; optional), and add
mercaptoethanol to 0.01M

(b) Apply to Step I column and repeat (c) - (e)
as above.
Enzyme recovery 80-100%, sp. activity $\cong$
6000 E.u./mg.

---

at the final CM-cellulose separation step. It is conceivable
that certain forms of the enzyme are the result of glycohydrolase activity as well as of proteolysis. Post-translational
proteolysis of enzymes, related either to genetic and hence to
physiological processes, or as a result of the biochemical
events necessary for their release from the cells, is becoming
of increasing interest. Among several instances of proteolysis
as the responsible factor in isozyme generation is the recent
analysis of the situation for the *E. coli* alkaline phosphatases
(Kelley et al., 1973). Here, evidence for a small, discrete
difference in a single N-terminal amino acid residue, occurring
after maturation, has been shown for two of these isozymes.
Perhaps more analogous to the neuraminidase isozymes is the
occurrence of galactosyltransferase of milk in 58,000 and 42,000
dalton forms (Magee et al., 1974); the larger convertible to
the smaller by trypsinolysis. From our own work with neuraminidase (unpublished observations) and the findings with
galactosyltransferase, it would appear that the components of
different size posess almost identical catalytic and related
physio-chemical properties. The identification of the truly
physiologically significant isozyme forms of these biocatalysts,

therefore, awaits further investigations. That selective proteolytic degradation of microbial enzymes, often in response to nutritional alternations in the growth medium, may be more widespread than heretofore thought, has been further underscored by the recent studies of Goldberg et al. (1974) on inhibition of protein turnover in *E. coli* by toluenesulfonyl fluoride or pentamidine. Thus the pneumococcal isozymes apparently constitute examples of epigenetic or post-translational modifications of a single primary protein structure encoded in a single gene.

ACKNOWLEDGEMENT

Submitted in partial fulfillment of the requirements for the degree of Doctor of Philosophy, Columbia University, 1972, by T. Francus. One of us (H. L.) was on leave from the Research Dept., Pharmacia Fine Chemicals, Uppsala, Sweden. We are grateful for materials supplied by this firm which aided the research. We thank Dr. Allen Gold for help with amino acid analyses, and also acknowledge the expert technical assistance provided by Ms. Denise Wheeler.

REFERENCES

Cassidy, J. T., G. W. Jourdian and S. Roseman 1965. The sialic acids VI. Purification and properties of sialidase from *Cl. perfringens*. *J. Biol Chem*. 240: 3501-3506

Clark, Julia F. and W. B. Jakoby 1970. Yeast aldehyde dehydrogenase III. Preparation of three homogeneous species. *J. Biol. Chem*. 245: 6065-6071

Compans, Richard W. and P. Choppin 1971. The structure and assembly of influenza and parainfluenza viruses. *Comparative Virology* PP. 407-432. Academic Press, N.Y.

Cuatrecasas, P. and G. Illiano 1971. Purification of neuraminidases from *V. cholerae*, *Cl. perfingens* and influenza virus by affinity chromatography. *Biochem. Biophys. Res. Communs*. 44: 178-184

Drzeniek, R. 1972. Viral and bacterial neuraminidases. *Current Topics in Microbiol. and Immunol*. 59: 35-73.

Drzeniek, R. 1973. Substrate specificity of neuraminidases. *Histochem. J*. 5: 271-290.

Funakoshi, S. and H. E. Deutsch 1969. Human carbonic anhydrases II. Some physicochemical properties of native isoenzymes and of similar isoenzymes generated *in vitro*. *J. Biol. Chem*. 244: 3438-3446.

Gold, Allen M. and D. Fahrney 1964. Sulfonyl fluorides as

inhibitors of esterases II. Formation and reactions of phenylmethyl-sulfonyl-α-chymotrypsin. *Biochemistry* 3: 783-790.

Goldberg, Alfred L., E. M. Howell, J. B. Li, S. B. Martel and W. F. Prouty 1974. Physiological significance of protein degradation in animal and bacterial cells. *Fed. Proc.* 33: 1112-1120.

Haskell, Theodore H., F. E. Peterson, D. Watson, N. R. Plessas and T. Culbertson 1970. Neuraminidase inhibition and viral chemotherapy. *J. Med. Chem.* 13: 697-704.

Hayes, Melvin B. and D. Wellner 1969. Microheterogeneity of $\underline{\underline{L}}$-amino acid oxidase. Separation of multiple components by polyacrylamide gel electrofocusing. *J. Biol. Chem.* 244: 6636-6644.

Hirst, G. K. 1942. Adsorption of influenza hemagglutinins and virus by red blood cells. *J. Exper. Med.* 76: 195-206.

Hughes, Colin R. and R. W. Jeanloz 1964. The extracellular glycosidases of *D. pneumoniae*. Purification and properties of a neuraminidase and a β-galactosidase. *Biochemistry* 3: 1535-1544.

Junge, Josephine M., E. A. Stein, H. Neurath and E. H. Fisher 1959. The amino acid composition of α-amylase from *B. subtilis*. *J. Biol. Chem.* 234: 556-561.

Kelley, Philip M., P. A. Neumann, K. Schriefer, F. Cancedda, M. J. Schlesinger and R. A. Bradshaw 1973. Amino acid sequence of *E. coli* alkaline phosphatase. Amino- and carboxyl-terminal sequences and variations between two isoenzymes. *J. Biol. Chem.* 12: 3499-3503.

Kendal, A. P. and E. A. Eckert 1972. The preparation and properties of ^{14}C-carboxyamidomethylated subunits from A/2-1957 influenza neuraminidase. *Biochim. Biophys. Acta* 258: 448-459.

Kirschenbaum, Donald M. 1971. A compilation of amino acid analyses of proteins I. *Anal. Biochem.* 44: 159-173.

Khorlin, A. Y., I. M. Privalova, L. Y. Zakstelskaya, E. V. Molibog and E. V. Evstigneeva 1970. Synthetic inhibitors of *V. cholerae* neuraminidase and neuraminidases of some influenza virus strains. *FEBS Letters* 8: 17-19.

Lee, Lucile T. and C. Howe 1966. Pneumococcal neuraminidase. *J. Bact.* 91: 1418-1426.

Magee, Steve C., R. Marval and K. E. Ebner 1974. Multiple forms of galactosyl transferase from bovine milk. *Biochemistry* 13: 99-102.

Moore, Gerald L. 1969. Use of azo-dye-bound collagen to measure reaction velocities of proteolytic enzymes. *Anal. Biochem.* 32: 122-127.

Rood, Julian I. and R. G. Wilkinson 1974. Affinity chromatography of *Cl. perfringens* sialidase: non-specific adsorption of haemagglutinin, hemolysin, and phospholipase C to Sepharosyl-glycyltyrosyl-N-ρ-aminophenyloxamic acid. *Biochim. Biophys. Acta* 334: 168-178.

Stahl, W. L. and R. D. O'Toole 1972. Pneumococcal neuraminidase: purification and properties. *Biochim. Biophys. Acta* 268: 480-487.

Subramanian, A. R. and G. Kalnitsky 1964. The major alkaline proteinase of *A. oryzae*. Aspergillopeptidase B. Isolation in homogeneous form. *Biochemistry* 3: 1861-1869.

Tanenbaum, S. W. and S. C. Sun 1971. Some molecular properties of pneumococcal neuraminidase. *Biochim. Biophys. Acta* 229: 824-828.

Tanenbaum, S. W., J. Gulbinsky, M. Katz and S. C. Sun 1970. Separation, purification and some properties of pneumococcal neuraminidase isoenzymes. *Biochim. Biophys. Acta* 198: 242-254.

Tuppy, Hans and P. Palese 1969. A chromogenic substrate for the investigation of neuraminidases. *FEBS Letters* 3: 72-76.

Warren, Leonard and C. W. Spearing 1963. Sialidase (neuraminidase) of *C. diphtheriae* *J. Bact.* 86: 950-955.

ISOZYMES OF THE PENTOSE PHOSPHATE PATHWAY

ORESTES TSOLAS

Roche Institute of Molecular Biology
Nutley, New Jersey   07110

ABSTRACT.   There are two primary forms of the enzymes of
the pentose phosphate pathway in *Candida utilis*, a yeast
that can grow on pentoses as well as on hexoses.   Transaldolase, one of the enzymes in this pathway, has an additional form which is a hybrid formed by exchange of subunits of the other two isozymes.   There are no differences
in the kinetic properties of these isozymes; their only
difference lies in their structure, indicating different
genetic origin as shown by the sequence of a peptide from
the active site of transaldolase I and III.   In this peptide, substitution of a tyrosine residue by a histidine
residue is observed.   Even though there are two identical
chains in isozymes I or III, a single active site can be
detected.   The participation of a lysine and a histidine
and the role of a neighboring cysteine in the mechanism of
action of the enzyme are discussed.

The yeast *Candida utilis* can grow well on pentoses as well
as on hexoses (Horecker, 1962).   This is in contrast to Saccharomyces which grows very poorly on five-carbon sugars.   The
vigorous growth of Candida on pentoses is utilized commercially
for the waste treatment of sulfite liquors in paper manufactur-
ing and the yeast obtained from the processing of these paper
mill wastes is used in animal feeds.

The utilization of pentoses in Candida follows the pathway
shown in Fig. 1.   Xylose is reduced to xylitol, which in turn
is converted to xylulose and then phosphorylated, thus entering
the main pathways of sugar metabolism via the pentose phosphate
cycle.   Glucose is catabolized mainly by the oxidative pathway,
with NADPH utilized in the biosynthesis of lipids.   Thus in
this organism the pentose phosphate pathway plays a predominant
role in the metabolism of carbohydrate and the enzymes participating in this pathway are found in abundance.   As a consequence
Candida is an unusually rich source of the enzymes of this pathway.

All the enzymes investigated in detail for this pathway are
present in at least two molecular forms (Table I).   These enzymes include glucose-6-phosphate dehydrogenase, 6-phosphoglu-
conic dehydrogenase, transaldolase, and transketolase.   A third
isozyme of transaldolase found in the extracts is a hybrid

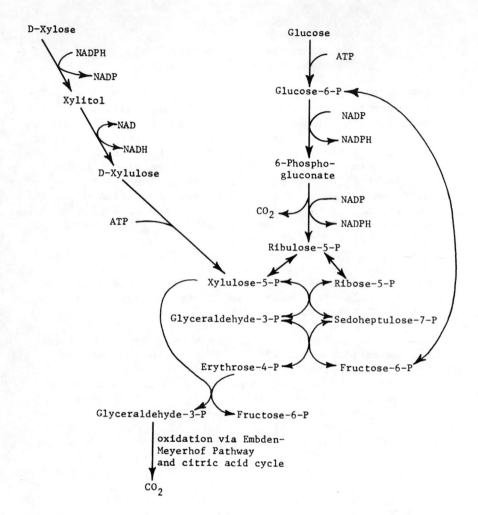

Fig. 1. Pathways of D-glucose and D-xylose metabolism in *Candida utilis*.

formed by exchange of subunits of the other two isozymes (Tsolas and Horecker, 1970b). This isozyme is, in all probability, formed by exchange of subunits of the two primary isozymes during the extraction procedure. It may also be that the third isozyme of transketolase is a derivative of the other two. The enzymes of pentose isomerization and epimerization have not been studied, and it would not be surprising if indeed

these enzymes are also found to exist in two molecular forms.
There are few if any differences in the enzymatic properties in these isozymes. Their main differences seem to lie
in their molecular nature. This would be a reflection in
differences in the primary sequence, indicating independent
genetic origin. It is possible that there are two sets of
genes for the pathway, one set activated by hexoses and the
other by pentoses. These observations raise interesting questions which deserve further investigation.

Table I

Isozymes of the Pentose Phosphate
Pathway in *Candida utilis*

Enzyme	Number of Isozymes	Reference
Glucose 6-phosphate dehydrogenase	2	Domagk et al. (1969)
6-Phosphogluconate dehydrogenase	2	Rippa et al. (1967)
Transaldolase	3 (One Hybrid)	Tchola and Horecker (1966)
Transketolase	3 (One Hybrid?)	Kiely et al. (1969)
R-5-P Isomerase	Not Known	
Xu-5-P Epimerase	Not Known	

Transaldolase

The reaction of transaldolase and the molecular properties
of the isozymes will now be discussed in more detail. The
partial reactions and the overall reaction catalyzed by transaldolase are shown in equations 1-3.

Fructose 6-P + Transaldolase $\rightleftharpoons$ Dihydroxyacetone-
transaldolase + Glyceraldehyde 3-P (1)

Dihydroxyacetone-transaldolase + Erythrose 4-P $\rightleftharpoons$
Sedoheptulose 7-P + Transaldolase (2)

Fructose 6-P + Erythrose 4-P $\rightleftharpoons$ Sedoheptulose 7-P
+ Glyceraldehyde 3-P (3)

This enzyme transfers a three carbon fragment from fructose 6-phosphate to erythrose 4-phosphate giving sedoheptulose
7-phosphate and glyceraldehyde 3-phosphate. A dihydroxyacetone-

769

transaldolase intermediate can be identified and the overall reaction can be written in two steps. Fructose 6-phosphate combines with the enzyme to form a dihydroxyacetone-transaldolase complex (eq. 1), which in turn transfers the dihydroxyacetone moiety to erythrose 4-phosphate to give sedoheptulose 7-phosphate and to regenerate the free enzyme (eq. 2).

The three transaldolase isozymes found in extracts of Candida yeast have been named I, II, and III, based on their order of elution from DEAE-Sephadex columns (Fig. 2) (Tchola and Horecker 1966). Isozymes I and III have been purified and crystallized (Tsolas and Horecker 1970a, Tsolas and Joris, in press). As mentioned earlier, isozyme II is a hybrid form, composed of subunits of the other two and no further work has been done on this isozyme.

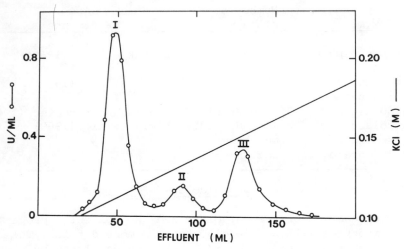

Fig. 2. Analytical chromatography of the isozymes of trans-aldolase. (From Tsolas and Horecker 1970a).

The primary isozymes I and III have similar catalytic properties but their molecular properties differ, indicating different primary structures. The molecular weights of all three isozymes are in the range of 65,000 to 68,000 (Table II). Isozymes I and III differ in their C-terminal amino acids; isozyme I has two phenylalanines, whereas isozyme III has two alanines. No free amino terminals are detected in either isozyme. On dissociation in guanidinium hydrochloride each species gives subunits of half the original molecular weight. The two primary isozymes show distinctive patterns on chromatography of their peptic digests (Fig. 3). These differences

Table II

Molecular Properties of Transaldolases [1]

Property	Transaldolase Isozyme		
	I	II	III
Molecular Weight	68,000 [2]	68,100	65,200
Carboxyl-Terminals	2 Phe	-	2 Ala
Amino-Terminals	None Detectable		None Detectable
Dissociation	One-Half M.W.		One-Half M.W.
Subunit Composition	$\alpha\alpha$	$\alpha\beta$	$\beta\beta$
Number of Active Sites	1	1	1

[1] From Tsolas and Horecker (1970b), and Sia (1970).
[2] D. Luk and O. Tsolas, unpublished observations.

are pronounced in the middle of the elution profile and they can be taken as a good indication of different internal amino acid sequences. The amino acid composition of isozymes I and III is very similar (Table III). However there are significant differences in amino acids found in least amounts. There are four residues of histidine, tryptophan, and cysteine in isozyme I. Significantly, the number of histidine and tryptophan residues is reduced by two, an even number, in isozyme III. This could indicate two identical subunits for each species. Even though isozyme I has a slightly higher molecular weight, it contains less valine and methionine, thus eliminating the possibility that isozyme III, the smaller species, is derived from I.

The evidence on C-terminals, and on the amino acid and subunit composition indicates two identical subunits in the molecule, and leads to a subunit composition of $\alpha\alpha$ and $\beta\beta$ (Table II). Even though the evidence is compatible with identical subunits, each isozyme contains only one active site per molecule. This observation has recently been confirmed with burst experiments performed with fructose 6-phosphate (Fig. 4). When fructose 6-phosphate is added to micromolar quantities of the enzyme, an initial burst of glyceraldehyde 3-phosphate is observed (eq. 1). The amount of product formed in this very fast reaction is equal to the amount of enzyme present on a molar basis. This equivalence is observed over a three-fold range of enzyme concentration.

The Active Site of Transaldolase

771

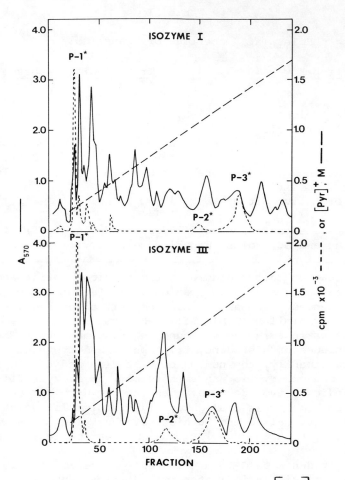

Fig. 3 Chromatography of peptic digests of $\begin{bmatrix} ^{14}C \end{bmatrix}$ carboxymethyl transaldolase isozymes I and III on Dowex 50. (From Tsolas and Sun, manuscript in preparation).

The dihydroxyacetone in the dihydroxyacetone-transaldolase intermediate is bound as a Schiff's base to the ε-amino group of lysine, in the ratio of one mole per mole of enzyme (Horecker et al. 1961). The complex can be stabilized for acid hydrolysis by reduction with sodium borohydride to the β-glyceryl derivative. The sequence around this lysine residue has already been determined (Lai et al. 1967): Ile-βGLys-Ile-Ala-Ser-Thr-Tyr.

Table III

The Amino Acid Composition of Transaldolase
Isozymes I and III[1]

Amino acid	Isozyme I (residues per mole)	Isozyme III (residues per mole)
Lysine	53.8	51.3
Histidine	4.0	2.0
Arginine	24.0	18.5
Aspartic acid	57.6	49.4
Threonine	55.3	42.8
Serine	39.5	31.9
Glutamic acid	76.0	68.8
Proline	20.0	18.8
Glycine	37.1	33.0
Alanine	73.9	73.4
Valine	41.3	48.1
Methionine	4.0	6.6
Isoleucine	48.2	34.6
Leucine	62.0	54.0
Tyrosine	25.4	20.1
Phenylalanine	28.0	23.3
Tryptophan	4.2	2.3
Cysteine	4.2	3.6
Total	658.5	582.5

[1] From Tsolas and Horecker (1970b).

There is also evidence for a functional histidine residue
(Brand et al. 1969). On photooxidation in the presence of the
dye rose Bengal, one of the two histidines in isozymes III
is destroyed and the enzymatic activity is completely lost.
On longer exposure to light the second histidine is also
destroyed. Loss of catalytic activity and destruction of the
two histidines is prevented when photooxidation is carried
out in the presence of substrate. This suggests that the
histidine residues are functional at the active site.

Each isozyme contains four cysteine residues, and all
four residues are reactive with chlorodinitrobenzene (Tsolas
et al. 1971). Dinitrophenylation of these residues is
associated with complete loss of activity. Catalytic activity
is again protected by either fructose 6-phosphate or erythrose
4-phosphate. It is remarkable that the enzyme in its native

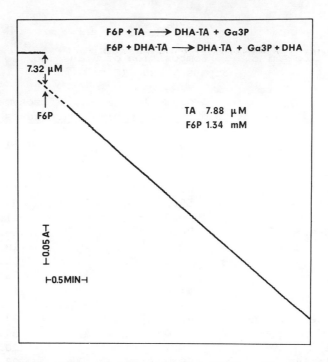

Fig. 4. Burst value of transaldolase III. The enzyme (7.88 μM), 1.0 ml, was incubated at pH 7.8, 25°, in a cuvette with 1 cm light-path in a Gilford spectrophotometer. At the time shown fructose 6-phosphate (2 λ, 1.35 mM final concentration) was added with rapid stirring. At 15 seconds the recorder was started and the reaction shown in the lower equation in the figure was followed for a few minutes by measuring the glyceraldehyde 3-phosphate generated with excess triose isomerase and α-glycerophosphate dehydrogenase in the presence of NADH at 340 nm. This reaction was extrapolated to the time of addition of fructose 6-phosphate and the difference in absorbance before and after addition of the substrate was taken to represent the amount of glyceraldehyde 3-phosphate, 7.32 μM, produced in the very fast reaction shown in the upper equation. (From Tsolas and Horecker, manuscript in preparation).

---

form is not affected by any other sulfhydryl reagent, such as silver salts, iodoacetic acid, p-mercuribenzoate, or Ellman's reagent. Only the hydrophobic reagent chlorodinitrobenzene will react with the native enzyme.

Since isozyme III contains only one histidine per subunit, it was of interest to isolate the peptide containing this histidine, since our evidence suggests that it is a functional

residue in catalysis. This peptide was obtained by peptic
digestion and was purified by conventional procedures. Further
cleavage of this peptide with trypsin gave a nonapeptide with
the following sequence:

Tyr-Gly-Ile-His-Cys-Asn-Thr-Leu-Leu

(Tsolas and Sun, manuscript in preparation). Interestingly,
in this peptide the histidine and cysteine residues are next
to each other. A similar peptide has been isolated from
isozyme I:

His-Gly-Ile-His-Cys-Asn-Thr-Leu-Leu

(Tsolas and Sun, manuscript in preparation).

The only difference between the two peptides is the replacement of tyrosine with histidine. This is the additional histidine found in this isozyme. Since the two isozymes have identical enzymatic properties, the proximity of the second histidine to the one associated with the activity does not seem to
affect appreciably the catalytic properties of the enzyme.
On the other hand, this internal substitution of amino acids
indicates conclusively that the two isozymes have different
genetic origin. The sequence bears no resemblance to that of
the lysine peptide binding the dihydroxyacetone, indicating
that the two peptides would be close to each other in the
active site only after extensive folding of the subunit chain.

We can summarize the role of these functional groups in
the mechanism of action of transaldolase as follows (Tsolas
and Horecker 1973) (Fig. 5): Fructose 6-phosphate reacts with
the ε-amino group of lysine to form a Schiff's base intermediate, which then with the simultaneous liberation of glyceraldehyde 3-phosphate gives the dihydroxyacetone-transaldolase
complex shown. The dealdolization reaction involves the
abstraction of a proton from the substrate, and from the
photooxidation experiments we believe this proton is stabilized by the imidazole nucleus of histidine, as shown in the
lower left part of the figure. We know this proton is held
by the enzyme because it does not exchange with the protons
of the medium (Brand et al. 1969). Also the protonated histidine becomes resistant to photooxidation. In the presence
of an acceptor aldehyde like erythrose 4-phosphate the dihydroxyacetone will be transferred to this acceptor and free
enzyme will be generated, thus completing the sequence of
events in the reaction of transaldolase.

The histidine-cysteine peptide described earlier is shown
in the space-filling model in Fig. 6. In this model the
proximity of the imidazole to the sulphur of cysteine can
be seen. This arrangement makes it possible for the protonated histidine in the dihydroxyacetone transaldolase complex
to be stabilized by hydrogen bonding to the sulphur of cysteine.

775

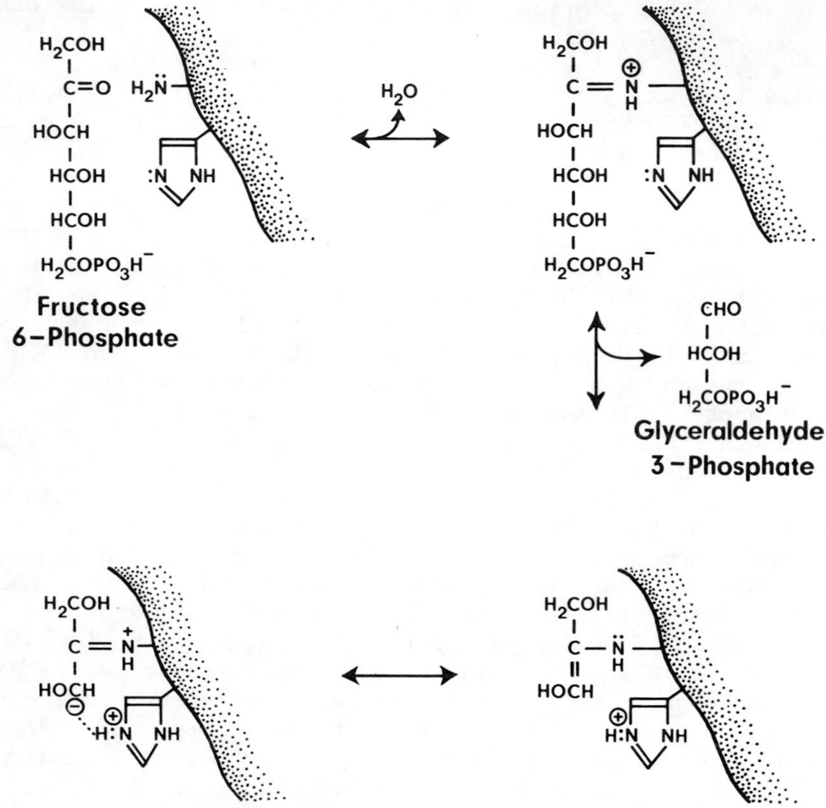

Fig. 5. Mechanism of action of transaldolase. (From Tsolas and Horecker 1973).

One of the two leucines in the conformation shown would create a hydrophobic pocket in which the cysteine residue can be partly shielded from the medium, and this should further help the stabilization of the shared proton. Evidence for this hydrophobic pocket was presented earlier when discussing the reaction of cysteine with chlorodinitrobenzene.

Possible Functions for the Isozymes of the Pentose Phosphate Pathway

Although essentially the same histidine-cysteine peptide has been isolated from both isozymes and no enzymatic differences have been observed in the two isozymes, we can speculate that the different genetic origin in this case, and probably in the case of the other pairs of enzymes in the hexose monophosphate pathway indicated earlier, may serve distinct physiological functions. Evidence to support this claim comes from work in another system. In Neurospora, Tatum and his co-workers have shown that there are three glucose 6-phosphate dehydrogenases, each coded by an independent gene (Brody and Tatum, 1966; Scott and Tatum, 1970). Mutation in any one of the three loci results in morphologically defective growth. Normal appearance can be restored when heterokaryons between any pair of these mutants are made. Similar observations have been made on the next enzyme in the pathway, 6-phosphogluconate dehydrogenase (Lechner et al. 1971, Scott and Abramsky 1973). It is apparent, therefore, that these isozymes are necessary for the normal growth of the organism and defects in one cannot be compensated by the presence of the other isozymes.

Similarly, Tolbert and his co-workers (Schnarrenberger et al. 1973) have shown that in spinach leaves there are two isozymes of glucose 6-phosphate dehydrogenase and two of 6-phosphogluconate dehydrogenase, one set in the chloroplast and the other in the cytosol. An analogous situation may exist in Candida. This organism, as has already been indicated, can grow on pentoses as well as hexoses; it could be that one set of isozymes serves in the metabolism of pentoses and the other in the metabolism of hexoses. Further experiments should clarify this question.

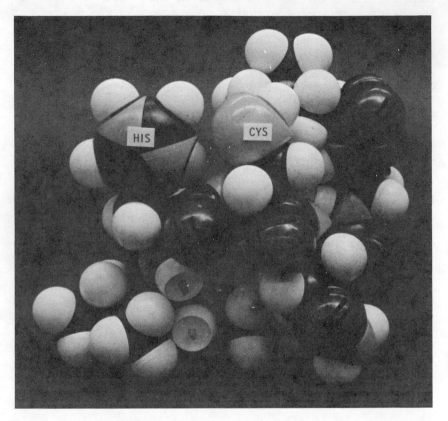

Fig. 6. The histidine-cysteine diad in the Gly-Ile-His-Cys-Asn-Thr-Leu-Leu peptide. (From Tsolas and Sun, manuscript in preparation).

REFERENCES

Brand, K., O. Tsolas, and B. L. Horecker 1969. Evidence for a specific function for histidine residues in transaldolase. *Arch. Biochem. Biophys*. 130: 521-529.

Brody, S., and E. L. Tatum 1966. The primary biochemical effect of a morphological mutation in *Neurospora crassa*. *Proc. Nat. Acad. Sci*. 56: 1290-1297.

Domagk, G. F., W. Domschke, H. J. Engel 1969. Hinweise auf eine unterchiedliche Proteinstruktur der beiden aus *Candida utilis* kristallisierten Glucose-6-phosphat-dehydrogenasen. *Hoppe-Seyler's Z. Physiol. Chem*. 350: 1242-1246.

Horecker, B. L. 1962. *Pentose Metabolism in Bacteria*. John Wiley and Sons, New York, N. Y.

Horecker, B. L., S. Pontremoli, C. Ricci, and T. Cheng 1961. On the nature of the transaldolase-dihydroxyacetone complex. *Proc. Nat. Acad. Sci.* 47: 1949-1955.

Kiely, M. E., E. L. Tan, and T. Wood 1969. The purification of transketolase from *Candida utilis*. *Can. J. Biochem.* 47: 455-460.

Lai, C. Y., C. Chen, and O. Tsolas 1967. Isolation and sequence analysis of a peptide from the active site of transaldolase. *Arch. Biochem. Biophys.* 121: 790-797.

Lechner, J. F., K. E. Fuscaldo, and G. Bazinet 1971. Genetic and biochemical studies of the hexose monophosphate shunt in *Neurospora crassa* II. Characterization of the biochemical defects of the morphological mutants colonial 2 and colonial 3. *Can. J. Microbiol.* 17: 789-794.

Rippa, M., M. Signorini, and S. Pontremoli 1967. Purification and properties of two forms of 6-phosphogluconate dehydrogenase from *Candida utilis*. *Europ. J. Biochem.* 1: 170-178.

Schnarrenberger, C., A. Oeser, and N. E. Tolbert 1973. Two isoenzymes each of glucose 6-phosphate dehydrogenase and 6-phosphogluconate dehydrogenase in spinach leaves. *Arch. Biochem. Biophys.* 154: 438-448.

Scott, W. A., and T. Abramsky, 1973. Neurospora 6-phosphogluconate dehydrogenase II. Properties of two purified mutant enzymes. *J. Biol. Chem.* 248: 3542-3545.

Scott, W. A., and E. L. Tatum 1970. Glucose 6-phosphate dehydrogenase and Neurospora morphology. *Proc. Nat. Acad. Sci.* 66: 515-522.

Sia, C. L. 1970. Subunits of transaldolase III - ultracentrifugation studies. *Arch. Biochem. Biophys.* 136: 318-319.

Tchola, O., and B. L. Horecker 1966. Transaldolase. *In Methods in Enzymology*. S. Colowick and N. O. Kaplan, Eds. 9: 499-505. Academic Press, Inc., New York, N. Y.

Tsolas, O., J. de Castro, and B. L. Horecker 1971. The reaction of the isoenzymes of transaldolase with chlorodinitrobenzene. *Arch. Biochem. Biophys.* 143: 516-525.

Tsolas, O., and B. L. Horecker 1970a. Isoenzymes of transaldolase in *Candida utilis* I. Isolation of three isoenzymes from yeast extracts. *Arch. Biochem. Biophys.* 136: 287-302.

Tsolas, O., and B. L. Horecker 1970b. Isoenzymes of transaldolase in *Candida utilis* II. The molecular properties of transaldolase isoenzymes I and III and the conditions for hybridization to form isoenzyme II. *Arch. Biochem.*

Biophys. 136: 303-319.

Tsolas, O., and B. L. Horecker 1973. Transaldolase: A model for studies of isoenzymes and half-site enzymes. *Mol. Cell. Biochem.* 1: 3-13.

Tsolas, O., and L. Joris. Transaldolase. *In Methods in Enzymology.* (in press).

POST-TRANSLATIONAL ALTERATIONS
OF HUMAN ERYTHROCYTE ENZYMES

BRYAN M. TURNER[1], RACHEL A. FISHER, and HARRY HARRIS
MRC Human Biochemical Genetics Unit,
University College, London, England

ABSTRACT. A description is given of the electrophoretic
properties of 16 enzymes (the products of 18 gene loci) in
erythrocyte fractions of differing mean age prepared by
density gradient centrifugation. Of these 16 enzymes, 5
(the products of 6 gene loci) were found to show pronounced electrophoretic changes with increasing red cell
age. These were purine-nucleoside phosphorylase, phosphoglucomutase (from both the PGM_1 and PGM_2 loci), triosephosphate isomerase, inorganic pyrophosphatase, and hypoxanthine-guanine phosphoribosyltransferase. The electro-
phortic changes were shown to be due in each case to the
progressive generation of more anodal, 'secondary' isozymes
at the expense of the 'primary' enzyme form. The electrophoretic change shown by purine-nucleoside phosphorylase
with aging of the red cell could be reproduced in vitro
using enzyme from cultured lymphoid cell lines. The rate
of isozyme formation has been shown to be temperature
dependent and to be increased at high pH.

It has become apparent during recent years that the occurrence of multiple electrophoretic forms of enzymes is the
rule rather than the exception. In view of the diversity of
isozyme systems it has proved convenient to classify isozymes
in terms of their underlying causes and three categories have
been suggested (Harris 1969). These are: 1. isozymes due
to multiple gene loci, 2. isozymes due to multiple alleles
at a single locus and 3. 'secondary' isozymes. Isozymes
within the third category, with which this report is primarily
concerned, are those thought to arise through post-translational modifications of protein structure.

Where an enzyme shows genetic variation the extent to which
multiple loci or multiple alleles are responsible for the
isozyme pattern may be determined by genetic studies. Where
genetic variation has not been found other criteria must be
applied to determine the factors responsible for isozyme

---

[1]Present Address: Division of Medical Genetics, Department
of Pediatrics, Mount Sinai School of Medicine of the City
University of New York, New York, N. Y. 10029

formation. The study of electrophoretic changes in erythrocyte enzymes with increasing cell age is particularly useful in this respect. The mature human erythrocyte is incapable of protein synthesis and therefore any isozymes which appear during the 120 day life-span of this cell must arise through modification of pre-existing enzyme protein. They can therefore be classified as secondary isozymes.

Various methods have been described for the separation of human and rabbit red blood cells by age (e. g. Lief and Vinograd, 1964, Prentice and Bishop, 1965, Sass et al., 1963). During these studies we used a density gradient technique employing a layered gradient made up of mixtures of Ficoll (Pharmacia) and Triosil (Nyegaard) as described by Turner et al. (1971, 1974). We found the method convenient and easily reproducible. The decline in reticulocyte count and G6PD activity were used to monitor the degree of age separation achieved (Turner et al., 1974). This compared favorably with that obtained by other density gradient techniques (Piomelli et al., 1968). Six, seven, or eight cell fractions were obtained from a single gradient depending on the number of layers used. Separations were usually carried out using erythrocyte preparations from normal, healthy donors. Reticulocyte counts in the lightest fraction were in the 5 to 10% range.

It has been established that the density of human and rabbit erythrocytes increases in direct proportion to their age (Bishop and Prentice, 1966, Lief and Vinograd, 1964, Piomelli et al., 1967). It is therefore possible to estimate the mean age of a red cell fraction from its known position in a density gradient. A method by which this may be done has been described in detail elsewhere (Piomelli et al., 1968, Turner et al., 1974). Using this method and assuming a red cell survival time of 120 days, we have been able to calculate the age range covered by the red cell fractions obtained during the present experiments. The mean cell age of the lightest fraction obtained after centrifugation of normal red cells always fell within the range 25-40 days. That of the densest fraction was always within the range 80-95 days. The use of this information in the interpretation of some of the electrophoretic findings is discussed further below.

Sixteen enzymes were examined electrophoretically in density gradient fractions. They are listed in Table 1. These enzymes were selected only in so far as they could all be examined by well established methods for electrophoresis and specific detection. Horizontal starch gel electrophoresis was used for all the enzymes studied. Enzyme detection was usually by visible staining using MTT and PMS. The methods used for specific enzymes were the same or similar to those summarized by Povey et al. (1973). Inorganic pyrophosphatase

782

TABLE I

Enzyme	Rate of Activity Loss	References
Purine-nucleoside phosphorylase	Negligible*	Turner et al. 1974
Phosphoglucomutase	-	
Inorganic pyrophosphatase	70 days	R. A. Fisher (unpublished)
Triosephosphate isomerase	135 days	S. Piomelli (personal comm.)
Hypoxanthine-guanine phosphoribosyltransferase	Very slow	Fox et al. 1971
Adenosine deaminase	-	
Adenylate kinase	-	
Glucose-6-phosphate dehyd.	42, 53, 62 days	Lohr and Waller, 1962; Piomelli et al. 1968; Turner et al. 1974
Hexokinase	23 days	Chapman and Schaumberg 1967
Inosine triphosphatase	-	
Lactate dehyd.	165 days	Lohr and Waller 1962
6-phosphogluconate dehyd.	Very slow*	Turner et al. 1974
Phosphoglucose isomerase	151 days	S. Piomelli (personal comm.)
s-Glutamic-oxaloacetic transaminase	Rapid	Sass et al. 1964
s-Isocitrate dehyd. (NADP)	28 days	Turner et al. 1974
s-Malate dehyd. (NAD)	-	

*Signigicant loss of activity occurs very early in the life of the circulating cell, but not subsequently.

Table 1. A list of the 16 enzymes examined electrophoretically in human red cell fractions of differing mean age together with information on their rates of activity loss. Half-life estimates are shown where possible. Dashes indicate that no information on rates of activity loss is available.

was detected as described by Fisher et al. (1974). Hypoxanthine-guanine phosphoribosyltransferase was detected by the
autoradiographic method described by Watson et al. (1972).

ENZYMES SHOWING AGE-RELATED ELECTROPHORETIC ALTERATION

Five of the sixteen enzymes examined were found to undergo
pronounced electrophoretic changes with increasing red cell
age. These were purine-nucleoside phosphorylase (PNP), phosphoglucomutase (PGM), inorganic pyrophosphatase (PPase), triosephosphate isomerase (TPI), and hypoxanthine-guanine phosphoribosyltransferase (HGPRT). The changes in electrophoretic
pattern shown by these enzymes are illustrated in Figures 1-6.
All photographs are of single gels, but in some cases the
prints have been cut and rearranged for greater clarity.

1. Purine-nucleoside phosphorylase: Of all the enzymes
studied PNP showed the most striking age-dependent alteration.
In mature human erythrocytes PNP consists of at least seven
electrophoretically distinct isozymes. As shown in Figure 1,
young red cells contain mainly the least anodal of these isozymes (1-3). With aging of the cell a progressive shift in
the isozyme pattern occurred until, in the oldest cells, the
most anodal isozymes (4-7) were the most active components.

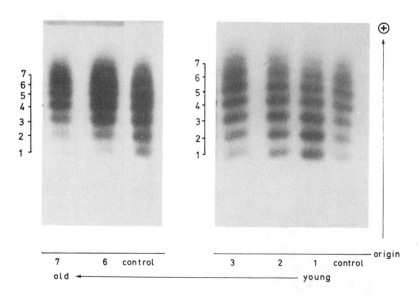

Fig. 1. Electrophoretic patterns of purine-nucleoside phosphorylase in red cell fractions of differing mean age.

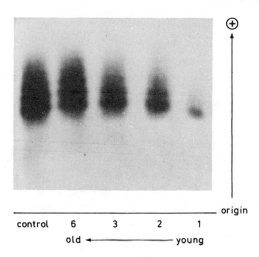

Fig. 2. Electrophoretic patterns of purine-nucleoside phosphorylase in red cell fractions of differing mean age prepared from a red cell sample from an individual with hereditary
spherocytosis. The control sample consisted of unfractionated cells from this individual.

The PNP isozyme pattern seen in density gradient fractions
containing predominantly young cells was also seen in unfractionated hemolysates from individuals with a shortened red cell
survival time. The isozyme pattern of PNP in red cells from
an individual with hereditary spherocytosis and a whole blood
reticulocyte count of 8%, is shown in Figure 2 (control).
The density gradient method was used to prepare fractions
from the red cells of this individual and the electrophoretic
patterns of PNP in these fractions are also shown in Figure
2. The enzyme in the lighest fractions (fractions 1 and 2)
consisted primarily of the least anodal isozyme. In this
respect it corresponded to the enzyme from rapidly dividing
cells such as cultured fibroblasts (Edwards et al. 1971) and
lymphoid cell lines (Turner et al. 1971). Fifty-one percent
of the cells in fraction 1 were reticulocytes and had therefore been in the circulation for less than 2 days.
    2. Phosphoglucomutase: Both the PGM_1 and PGM_2 isozymes
were found to change electrophoretically with increasing red
cell age. In the case of PGM_1, phenotype 1 (Spencer et al.
1964) there was a progressive increase with red cell age in
the staining intensity of band c relative to band a (Figure
3). The 2-1 phenotype of PGM_1 showed a similar progressive
increase in bands c and d relative to a and b. In most

785

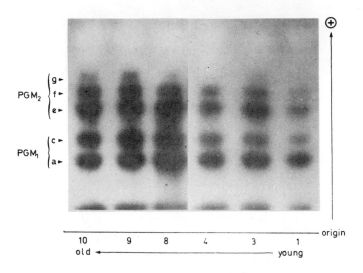

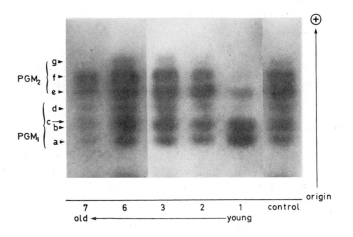

Fig. 3. Electrophoretic patterns of phosphoglucomutase in red cell fractions of differing mean age. a) PGM₁ phenotype 1, PGM₂ phenotype 1; b) PGM₁ phenotype 2-1, PGM₂ phenotype 1.

experiments there was a similar, though less pronounced, change in the PGM₂ isozymes (e, f, and g) with bands f and g becoming increasingly more active relative to band e.

These results confirm previous observations on age-related changes in PGM isozymes made using different methods (Monn 1969).

A third change becomes apparent when one examines the relative staining intensities of the PGM_1 and PGM_2 isozymes. In the younger cells the PGM_1 isozymes are the more active while in older cells the activities of the two sets of isozymes are approximately the same. This can be most simply accounted for by postulating a more rapid loss of activity of the PGM_1 isozymes with increasing red cell age. This explanation is supported by the finding that the PGM_2 isozymes (e, f, and g) are more thermostable than the PGM_1 isozymes a to d (McAlpine et al., 1970).

3. Inorganic Pyrophosphatase: Inorganic pyrophosphatase in the human erythrocyte consists of three isozymes (Fisher et al., 1974). In the youngest red cell fractions examined the three isozymes were of approximately equal staining intensity. With increasing cell age the two less anodal bands underwent a progressive loss of activity until in the oldest fractions the most anodal band was by far the most prominent (Figure 4.)

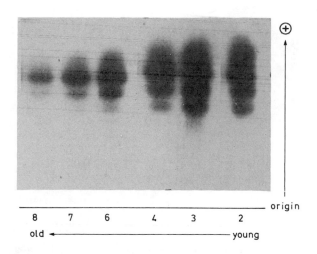

Fig. 4. Electrophoretic patterns of erythrocyte inorganic pyrophosphatase in red cell fractions of differing mean age. The enzyme from a normal, unfractionated red cell sample has an electrophoretic pattern similar to fraction 4.

4. Triosephosphate Isomerase: Like pyrophosphatase, TPI from human red cells exhibits three major electrophoretic bands (Peters et al., 1973). In young cells band a, the least anodal, was the most active. With aging of the cell the activities of bands b and c showed a marked increase relative to band a (Figure 5).

5. Hypoxanthine-guanine phosphoribosyltransferase: In mature human erythrocytes HGPRT consists of at least seven isozymes. The electrophoretic pattern resembles that of purine-nucleoside phosphorylase. With increasing red cell age there was a progressive increase in the relative staining intensities of the more anodal HGPRT isozymes (Figure 6).

INTERPRETATION OF ELECTROPHORETIC CHANGES

In the case of the enzymes PNP and PGM (both PGM_1 and PGM_2) the oldest red cell fractions were repeatedly found to contain anodal isozymes not present in the youngest cells. This can only have occurred through the formation of these isozymes during aging of the cell, presumably by successive modifications of the primary form of the enzyme. However, for the other three enzymes HGPRT, TPI and PPase the interpretation of the results is less clear-cut. On the basis of the electrophoretic results alone, it is possible that changes in band

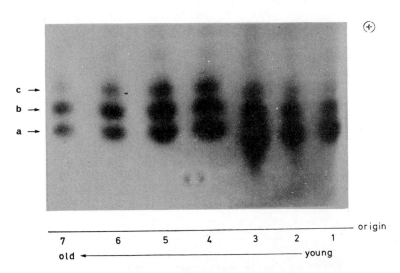

Figure 5. Electrophoretic patterns of triosephosphate isomerase in red cell fractions of differing mean age. The enzyme from a normal, unfractionated red cell sample has an electrophoretic pattern similar to fraction 4.

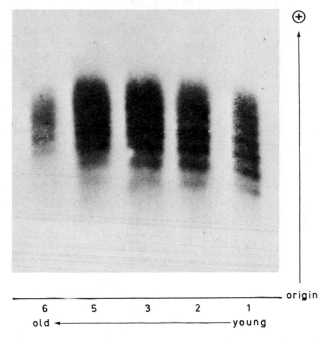

⊕

origin

| 6 | 5 | 3 | 2 | 1 |

old ◄─────────────────► young

Figure 6. Electrophoretic patterns of hypoxanthine-guanine phosphoribosyltransferase in red cell fractions of differing mean age. The enzyme from a normal, unfractionated red cell sample has an electrophoretic pattern similar to fraction 3.

intensity are due to differences in the rates of activity loss of different isozymes as the cell ages. For example, the apparent increase in activity of TPI isozyme b, relative to a, with aging of the cell (Figure 5) could be due to the pre-ferential inactivation of band a rather than a conversion of a to b. In general terms, this explanation requires that the young erythrocyte should have a full complement of isozymes and that some of these should then be lost more rapidly than others, leading to age-related changes in the electrophoretic pattern. All enzymes for which this explanation is correct should therefore lose activity with increasing erythrocyte age. Information on rates of enzyme activity loss with aging of the erythrocyte is therefore useful in the interpretation of electrophoretic changes.

Available data on the rates of activity loss of the sixteen enzymes studied is summarized in Table 1. Half-life estimates have been included where available. The information shown in

Table 1 suggests that electrophoretic changes and rates of activity loss are not closely related.

HGPRT has been reported to show virtually no loss of activity with aging of the human erythrocyte (Fox et al., 1971). Isozyme activity loss cannot therefore contribute to the electrophoretic change shown by this enzyme. TPI also loses activity relatively slowly with age, having a half-life in the red cell of 135 days (S. Piomelli, personal communication). As mentioned previously the density gradient fractions used in the present series of experiments had mean cell ages ranging from 25–40 days in the top fraction to 80–95 days in the bottom. The ages of the top and bottom red cell fractions from most gradients therefore differed by about 60 days. With a half-life of 135 days the TPI activity of the cells would fall by less than 25% over this time period. Such a small loss of activity is clearly insufficient to account for the magnitude of the electrophoretic change.

Inorganic pyrophosphatase shows appreciable age-related activity loss (R. A. Fisher, unpublished results). The PPase activity of the densest fractions was 50–60% of the activity of the lightest. It can be seen from Figure 4 that aging of this enzyme involves the virtual disappearance of two of the original three, equally active bands. If these bands were simply inactivated, an activity loss of 65–70% would be expected (i.e. the densest fraction should have about one-third the activity of the lightest). This is assuming that the third, most anodal isozyme is completely stable. It is therefore likely that the electrophoretic change is due, at least in part, to conversion of the least anodal isozyme into the more anodal forms.

In summary, the results show that the five enzymes PGM, TPI, PPase, HGPRT, and PNP all form more anodal isozymes during aging of the erythrocyte in vivo. These isozymes must arise by post-translational modification of the enzyme. Examination of the electrophoretic changes shown in Figures 1–6 shows that the formation of secondary isozymes takes place at an apparently steady rate over the approximate age range 30–90 days. Furthermore, isozyme formation seems to occur in sequential fashion leading to the generation of progressively more anodal bands. For example, the PNP isozymes are apparently formed in the sequence 1→2→3→... and those of TPI in the sequence a→b→c.

ENZYMES IN WHICH LITTLE ELECTROPHORETIC ALTERATION WAS SEEN

Certain minor electrophoretic changes were found among the 11 enzymes in this group. For example the single glucose-6-

phosphate dehydrogenase isozyme had a mobility 2-3% greater in the oldest cells than in the younger cells. Also LDH 5 was consistently seen in the youngest red cell fractions but not in older cells. In general, however, these 11 enzymes showed no major electrophoretic alteration in the red cell fractions examined.

Of these 'unaltered' enzymes, seven consistently show multiple isozymes after starch gel electrophoresis. These are lactate dehydrogenase (LDH), adenosine deaminase (ADA), adenylate kinase (AK), red cell hexokinase (HK), phosphoglucose isomerase (PGI), soluble glutamicoxaloacetic transaminase (sGOT), and soluble malate dehydrogenase (sMDH). The major isozymes of LDH seen in the red cell are known to be the products of two gene loci, but genetic evidence suggests that each of the remaining six enzymes is coded by only a single locus (for a review see Harris and Hopkinson 1972). If this is the case then the occurrence of multiple isozymes is due in all probability to post-translational modification of the primary gene product.

Two explanations may be offered for the fact that no age related alteration in these isozyme systems was seen. One is that these isozymes arise only as a result of lysis of the cell or during electrophoresis and do not occur in vivo. Such an explanation has been proposed for the PGI isozymes (Payne et al., 1972). Alternatively, modification of these enzymes may occur only during the early life of the red cell, possibly even before entry into the circulation.

We have examined the PGI isozymes in red cell fractions prepared from an individual with thalassemia minor and a low mean red cell age due to premature hemolysis. The lightest fraction obtained from this individual had a reticulocyte count of more than 20%. In the lightest fractions the minor bands of PGI were far less prominent, relative to the major band, than in older cells. This indicates that the minor isozymes of PGI do arise within the red cell and are not simply experimental artifacts. The observation also suggests that, for certain enzymes, secondary isozyme formation can occur only very early in the life of the red cell and not subsequently. Further studies on very young red cell populations are necessary to confirm this.

SECONDARY ISOZYME FORMATION IN VITRO

In order to investigate further the processes involved in secondary isozyme generation, we have attempted to reproduce in vitro the electrophoretic changes seen during in vivo aging of the red cell. For these experiments we used the

enzyme purine-nucleoside phosphorylase from cultured human
lymphoid cells. In these cells the enzyme consists almost
exclusively of the least anodal electrophoretic band (Turner
et al., 1971). Storage of extracts of these cells resulted in
the gradual appearance of more anodal isozymes corresponding
in mobility to those seen in erythrocytes. The rate of isozyme generation was both temperature and pH dependent, being
considerably faster at pH 9-10 (using carbonate/bicarbonate
buffer) than at neutrality and some five times faster at room
temperature (in the presence of antibiotics) than at +4°C.
At pH 10.0 isozymes 1-4 were clearly visible after 10 days at
+4°C (Figure 7) and 2 days at +20°C. Identical results were
obtained using PNP partially purified from lymphoid cells by
ion-exchange chromatography. Electrophoretic mobility is not,
of course, conclusive evidence that isozymes generated in
vitro are identical to those formed in vivo. However isozymes
of PNP generated in vitro from lymphoid cell extracts have
been shown to have the same catalytic properties as those
formed within the red cell (Turner et al., 1971). It therefore seems likely that the same enzyme modification is responsible for the generation of PNP isozymes both in vivo and
in vitro.

The finding that partially purified PNP generated secondary isozymes in vitro at a rate identical to that of cell extracts shows that a complete intracellular environment is not

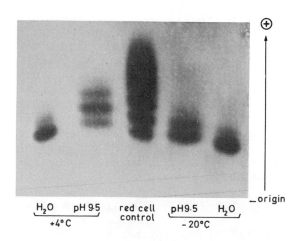

Figure 7. Electrophoretic patterns of purine-nucleoside phos-
phosylase in a red cell sample and in extracts of a lymphoid
cell line stored for 10 days at +4° and -20°C.

necessary for isozyme formation. This suggests that the process of isozyme formation involves a relatively simple chemical or structural alteration rather than a complex interaction between intracellular components.

The results also establish that the formation of secondary
isozymes of PNP is not peculiar to the erythrocyte, thus confirming the conclusions drawn from a study of the enzyme in
various human tissues (Edwards et al., 1971). PGM, PPase,
TPI, and HGPRT have also been examined in a variety of human
tissues (Fisher and Harris, 1972; Fisher et al., 1974; Peters
et al., 1973; Watson et al., 1972; and unpublished results).
At least some of the secondary isozymes seen in erythrocytes
were also seen in tissues. However, the secondary isozymes
were always found to be less prominent, relative to the primary form, than in the erythrocyte.

For each of these enzymes the prominence of their secondary
isozymes seems to be a function of the rate of protein turnover and cell division. In rapidly dividing cultured cells
such as fibroblasts or lymphoid lines the secondary isozymes
are weak or absent. In tissues secondary isozymes are rather
more prominent and in the erythrocyte, where protein synthesis does not occur, the activity of the secondary isozymes
becomes equal to or greater than that of the primary isozyme.

The fact that all the secondary isozymes formed during
aging of the red cell were found to have a more anodal electrophoretic mobility than the primary isozyme raises the possibility that a common mechanism is involved in their formation.
The results on the in vitro 'aging' of PNP suggests that this
may be a relatively simple chemical or physical change. The
enhancement of isozyme generation by high pH is consistent
with a process of deamidation. However, there is at present
little firm evidence on which to base a detailed chemical explanation of secondary isozyme formation. The present results
do enable us to eliminate certain mechanisms. For example,
the formation of mixed disulphides between red cell glutathione
and enzyme sulphydryl groups can be ruled out as none of the
five enzymes showing age-related isozyme formation was affected by mercaptoethanol. It is interesting that adenosine
deaminase and phosphoglucose isomerase, two enzymes known to
be susceptible to disulphide formation (Spencer et al., 1968,
Payne et al., 1972) showed little or no electrophoretic alteration over the major part of the red cell's life span.

REFERENCES

Bishop, C. and T. C. Prentice 1966. Separation of rabbit red
cells by density in a bovine serum albumin gradient and
correlation of red cell density with cell age after in

vivo labeling with [59]Fe. *J. Cell. Physiol.* 67: 197–207.

Chapman, R. G. and L. Schaumberg 1967. Glycolysis and glycolytic enzyme activity of aging red cells in man. *Brit. J. Haemat.* 13: 665–678.

Edwards, Y. H., D. A. Hopkinson, and H. Harris 1971. Inherited variants of human nucleoside phosphorylase. *Ann. Hum. Genet.*, Lond. 34: 395–408.

Fisher, R. A. and H. Harris 1972. 'Secondary' isozymes derived from the three PGM loci. *Ann. Hum. Genet.*, Lond. 36: 68–77.

Fisher, R. A., B. M. Turner, H. L. Dorkin, and H. Harris 1974. Studies on human erythrocyte inorganic pyrophosphatase. *Ann. Hum. Genet.*, Lond. 37: 341-353.

Fox, R. M., M. H. Wood, and W. J. O'Sullivan 1971. Studies on the coordinate activity and lability of orotidylate phosphoribosyltransferase and decarboxylase in human erythrocytes and the effects of allopurinol administration. *J. Clin. Invest.* 50: 1050–1060.

Harris, H. 1969. Genes and isozymes; *Proc. Roy. Soc. B.* 174: 1–31.

Harris, H. and D. A. Hopkinson 1972. Average heterozygosity per locus in man: an estimate based on the incidence of enzyme polymorphisms. *Ann. Hum. Genet.*, Lond. 36: 9–20.

Leif, R. C. and J. Vinograd 1964. The distribution of buoyant density of human erythrocytes in bovine albumin solutions. *Proc. Nat. Acad. Sci.* 51: 520–528.

Lohr, G. W. and H. D. Waller 1962. Zur biochemie der erythrocytenalterung. *Folia Haematologia* 78: 385–402.

McAlpine, P. J., D. A. Hopkinson, and H. Harris 1970. Thermostability studies on the isozymes of human phosphoglucomutase. *Ann. Hum. Genet.*, Lond. 34: 61–71.

Monn, E. 1969. Relation between blood cell phosphoglucomutase isoenzymes and age of cell population. *Scand. J. Haemat.* 6: 133–138.

Payne, D. M., D. W. Porter, and R. W. Gracy 1972. Evidence against the occurrence of tissue-specific variants and isoenzymes of phosphoglucose isomerase. *Arch. Biochem. Biophys.* 151: 122–127.

Peters, J., D. A. Hopkinson, and H. Harris 1973. Genetic and non-genetic variation of triose phosphate isomerase isozymes in human tissues. *Ann. Hum. Genet.*, Lond. 36: 297–312.

Piomelli, S., G. Lurinsky, and L. R. Wasserman 1967. The mechanism of red cell aging 1. Relationship between cell age and specific gravity evaluated by ultracentrifugation in a discontinuous density gradient. *J. Lab. Clin. Med.* 69: 659–674.

Piomelli, S., L. M. Corash, D. D. Davenport, J. Miraglia, and E. L. Amorosi 1968. In vivo lability of glucose-6-phosphate dehydrogenase in Gd^A-and $Gd^{Mediterranean}$ deficiency. *J. Clin. Invest.* 47: 940-948.

Povey, S., S. E. Gardiner, B. Watson, S. Mowbray, H. Harris, E. Arthur, C. M. Steel, C. Blenkinsop, and H. J. Evans 1973. Genetic studies on human lymphoblastoid lines. *Ann. Hum. Genet.*, Lond. 36: 247-266.

Prentice, T. C. and C. Bishop 1965. Separation of rabbit red cells by density methods and characterictics of separated layers. *J. Cell and Comp. Physiol.* 65: 113-125.

Sass, M. D., L. M. Levy, and H. Walter 1963. Characteristics of erythrocytes of different ages II. Enzyme activity and osmotic fragility. *Can. J. Biochem. Physiol.* 41: 2287-2296.

Sass. M. D., E. Vorsanger, and P. W. Spear 1964. Enzyme activity as an indicator of red cell age. *Clin. Chim. Acta* 10: 21-26.

Spencer, N., D. A. Hopkinson, and H. Harris 1964. Phosphoglucomutase polymorphism in man. *Nature*, Lond. 204: 742-745.

Spencer, N., D. A. Hopkinson, and H. Harris 1968. Adenosine deaminase polymorphism in man. *Ann. Hum. Genet.*, Lond. 32: 9-14.

Turner, B. M., R. A. Fisher, and H. Harris 1971. An association between the kinetic and electrophoretic properties of human purine-nucleoside phosphorylase isozymes. *Eur. J. Biochem.* 24: 288-295.

Turner, B. M., R. A. Fisher, and H. Harris 1974. The age related loss of activity of four enzymes in the human erythrocyte. *Clin. Chim. Acta* 50: 85-95.

Watson, B., I. P. Gormley, S. E. Gardiner, H. J. Evans, and H. Harris 1972. Reappearance of murine hypoxanthine guanine phosphoribosyl transferase activity in mouse A9 cells after attempted hybridization with human cell lines. *Exp. Cell Res.* 75: 401-409.

A NEW PURIFICATION PROCEDURE FOR PIG HEART FUMARASE

BEULAH M. WOODFIN
Department of Biochemistry, School of Medicine
University of New Mexico
Albuquerque, New Mexico 87131

ABSTRACT. Pig heart fumarase was thought for some years to be a homopolymer with four subunits. Heterogeneity was demonstrated by electrofocusing, and polydispersity by chromatography, both in phosphate systems. In this paper, evidence is presented that the heterogeneity may be a function of the anionic environment. Since the established purification procedures for pig heart fumarase expose the protein to high concentrations of phosphate and sulfate, both inhibitors of the enzymic activity, a new purification procedure has been undertaken.

Heterogeneity has been demonstrated in the new procedure. One form has been purified to a large extent, and exhibits some kinetic properties which differ from those reported earlier. The other form appears to be still complexed with other mitochondrial proteins and may or may not be a distinct species.

Electrofocusing data were presented three years ago by Cohen's group (Lin et al., 1971; Penner and Cohen, 1971) indicating that fumarase from pig heart was heterogeneous, composed of possibly ten forms. Under dissociating conditions, electrofocusing showed as many as six subunit forms. These data contrasted with the structural studies of Hill's group (Kanarek et al., 1964; Hill and Kanarek, 1964; Teipel and Hill, 1971), who concluded that fumarase was a polymer of four essentially identical subunits. Even allowing for four different, independently-coded subunits, only four forms should have been observed under dissociating conditions.

We had been interested in fumarase earlier because of anomalous Arrhenius plots reported originally by Massey (1953). The reports by Cohen's group revived our interest, and the present report describes experiments suggesting that the heterogeneity observed with fumarase may be due to the formation of forms described by Dr. Horecker as "metazymes". We also report the development of a new purification procedure which may prevent the production of such derived isozymes.

In particular, we were interested in the paper from Cohen's group (Lin et al., 1971) showing unresolved heterogeneity on DEAE-Sephadex, using a phosphate buffer gradient as the eluting agent. Since phosphate is a competitive inhibitor of the

enzyme, as well as the buffer of choice in most of the purification and experimental procedures reported for fumarase, we
decided to try a modified "substrate elution" procedure as a
possible preparative-scale demonstration of heterogeneity. As
reported earlier (Woodfin, 1973) and shown in Figure 1, the
chromatography of 16 mg of commercial crystalline fumarase on
DEAE-cellulose, using L-malate as the eluting agent, yielded
the profile shown. The isoelectric point reported for fumarase
has ranged from pH 6.3 to pH 7.6, depending upon the nature and
concentration of the buffer anion. We have chosen to work at

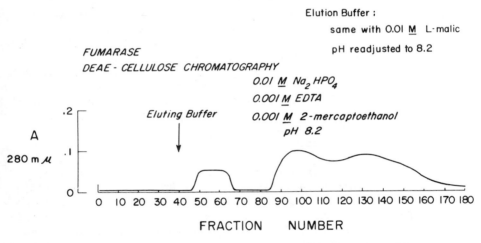

Fig. 1. Chromatography of a commercial crystalline fumarase
preparation (Woodfin, 1973). Room temperature.

pH 7.8 primarily because of the limited solubility of the
enzyme at low pH (less than 0.5 mg/ml at pH 6.5). Some of
the absorbance at 280 nm shown for the second, non-symmetrical
peak eluted from the column is due to the fumarate produced
from malate by the action of the enzyme. The activity profile
roughly followed the A_{280} profile, tapering off about five fractions ahead of the A_{280}.

With this evidence of heterogeneity, we felt that a more
specific chromatographic reagent might improve resolution. The
reagent shown in Figure 2 was prepared and attached to cellulose by the method of Cuatrecasas (1970), as adapted by Lowe
and Dean (1971). Titration of the product indicated the incorporation of five μmoles of the dicarboxylic acid (9.8 μequiv)
per gram of cellulose.

798

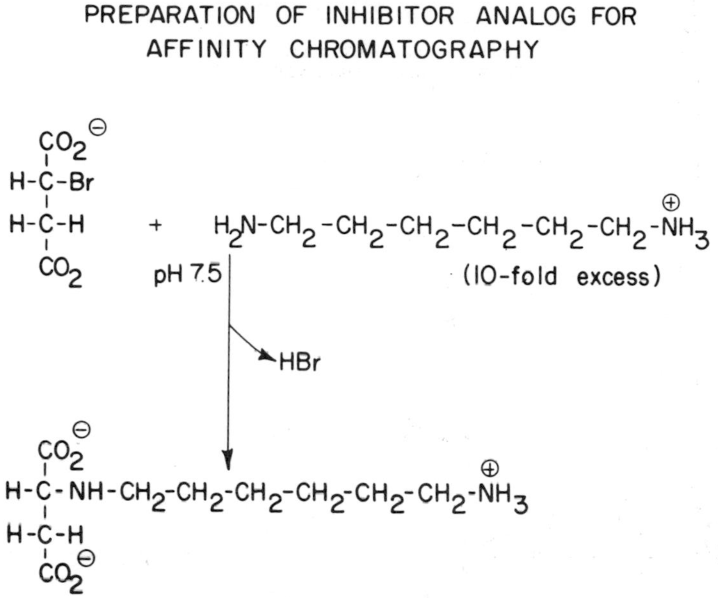

PREPARATION OF INHIBITOR ANALOG FOR
AFFINITY CHROMATOGRAPHY

$$\begin{array}{c} CO_2^{\ominus} \\ | \\ H-C-Br \\ | \\ H-C-H \\ | \\ CO_2 \end{array} \quad + \quad H_2N-CH_2-CH_2-CH_2-CH_2-CH_2-CH_2-\overset{\oplus}{N}H_3$$

pH 7.5   (10-fold excess)

→ HBr

$$\begin{array}{c} CO_2^{\ominus} \\ | \\ H-C-NH-CH_2-CH_2-CH_2-CH_2-CH_2-CH_2-\overset{\oplus}{N}H_3 \\ | \\ H-C-H \\ | \\ CO_2^{\ominus} \end{array}$$

Fig. 2. Preparation of the affinity reagent. The product was separated from the excess starting material by chromatography on Dowex 50 (H+)

Commercial crystalline fumarase was chromatographed on this medium, in phosphate buffer containing EDTA and 2-mercaptoethanol, using the competitive inhibitor, succinate, as the eluting agent in order to avoid the interfering ultraviolet absorbance of fumarate. This is shown in Figure 3a. Note the appearance of some inactive protein in fraction 2; bovine serum albumin, chromatographed on this column as a control, appears in fractions 2, 3, and 4. Fractions 38 through 55 were pooled, precipitated by 60% saturation with ammonium sulfate, redissolved and dialyzed. Rechromatography yielded the profile shown in Figure 3b (inset).

Figure 4 shows an identical column chromatographic procedure using barbital buffer at pH 7.8. The commercial fumarase preparation was exhaustively dialyzed to remove as much sulfate and phosphate as possible. Although the general form of the elution profile is similar to that in phosphate buffer, this shows somewhat less heterogeneity, and a significant retardation of the fumarase activity on the column. The activity does not appear until fraction 6.

The possible origins of this heterogeneity are not clear. The established purification procedures of Massey (1952), of

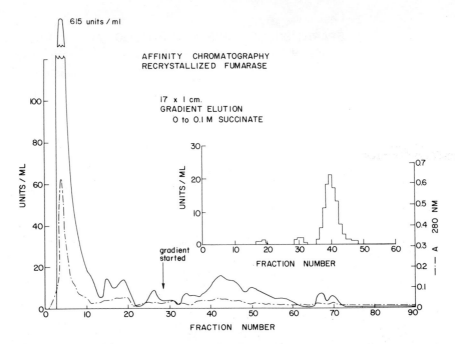

Fig. 3. Affinity chromatography of a commercial crystalline fumarase preparation. Buffer: 0.01 M sodium phosphate, 0.001 M EDTA, 0.001 M 2-mercaptoethanol; room temperature. (a) Initial column. (b) Inset – Rechromatography of fractions 38-55.

Frieden et al., (1954), and of Hill's group (Kanarek and Hill, 1964) all involved homogenization in phosphate buffer, ammonium sulfate fractionation, and precipitation at pH 5.0 to 5.4. Phosphate and sulfate (as well as other polyvalent anions) are inhibitors of the enzyme; low pH at early stages of purification may result in modification of the protein by non-specific proteases. The affinity chromatography experiments suggested that the presence of phosphate and/or sulfate changed the binding of the protein to the anionic reagent. Therefore, we undertook the development of a new purification procedure in order to rule out the possibility of artifacts of isolation caused by the binding of polyvalent anions.

The procedure now involves homogenization in barbital buffer, and chromatography on DEAE-cellulose and cellulose phosphate, using sodium chloride gradients for elution. We have succeeded in purifying one fraction some 500-fold (to approximately 30% pure, based on previously reported extinction coefficients), but the bulk of the activity remains associated

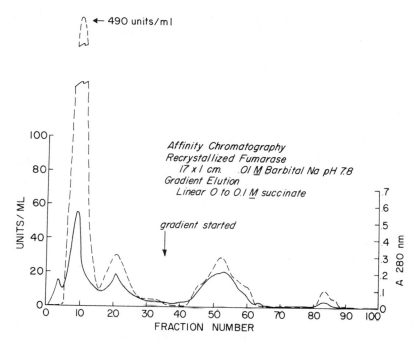

Fig. 4. Affinity chromatography of a commercial crystalline fumarase preparation. Buffer: 0.01 M̲ barbital, 0.001 M̲ EDTA, 0.001 M̲ 2-mercaptoethanol; room temperature.

with a red protein at about 80-fold purification.

Purification Procedure. A single pig heart (obtained from Schwartzman Packing Co., Albuquerque, N.M., when not undergoing labor difficulties) is trimmed of connective tissue, valves, and fat, and is minced, weighed, and homogenized in two volumes of 0.01 M̲ barbital, 0.001 M̲ EDTA, 0.001 M̲ 2-mercaptoethanol, pH 8.2. This particular mixture will be referred to hereafter as "the standard buffer". After homogenization for two minutes in a Waring blender, the pH of the mixture, between 6.6 and 6.8, is adjusted to 8.2 by the addition of approximately 12 ml of 1 N̲ NaOH, added dropwise with stirring. The homogenate is then centrifuged at 1300 x g in the IEC PR-2 centrifuge for 30 minutes; the supernatant liquid is centrifuged at 27,000 x g in a Sorvall RC2-B (GSA rotor) for 90 minutes. With the exception of this final centrifugation at 5-6°, the procedure is carried out near room temperature. Assay at 295 nm (Massey, 1952) shows 18,000 to 22,000 units at 27°C.

The supernatant liquid from the 27,000 x g centrifugation is applied to a large DEAE-cellulose column equilibrated with the standard buffer, at room temperature. The size of this column has been adjusted to maximize separation, and a typical experiment is shown in Figure 5. The second peak showing fumarase activity is intensely red in color, perhaps containing other mitochondrial proteins.

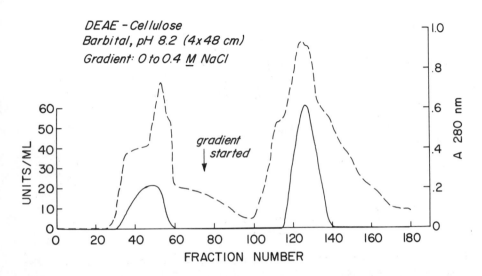

Fig. 5. Chromatography of heart muscle extract. Buffer: 0.01 $\underline{M}$ barbital, 0.001 $\underline{M}$ EDTA, 0.001 $\underline{M}$ 2-mercaptoethanol; room temperature.

At pH 8.2, fumarase should be negatively charged, assuming that the previously reported information on isoelectric points is accurate. Similarly, most of the other proteins in the extract should be negatively charged. Thus, there will be considerable competition for binding to the positively charged chromatographic medium. The appearance of fumarase activity in what is essentially the breakthrough region of the effluent suggests that it is a protein either below or at its isoelectric point, or a protein only moderately negatively charged, such that it is displaced in interaction with the column by a more strongly negatively charged protein. The fairly tight binding exhibited by the remaining fumarase activity, and its association with a red-colored protein, suggest that this portion of the enzyme may still be associated with other mitochondrial proteins. The report of Hamm and co-workers (1972) indicates that fumarase is relatively tightly associated with the mitochondrial membrane. It is possible that the relatively mild extraction conditions used here have only partially solubilized

the proteins, while the much stronger buffers (which also contained polyvalent anions) used in earlier purification procedures more readily solubilized at least fumarase.

The fumarase activity that chromatographs in the breakthrough
area was pooled and chromatographed on cellulose phosphate, in
Tris containing EDTA and 2-mercaptoethanol, as shown in Figure
6. A high concentration of sodium chloride was necessary to
elute this activity from the column. It should be noted that,
since phosphate is a competitive inhibitor of the enzyme,
cellulose phosphate itself functions as an affinity chromatography reagent. This fumarase activity removed from cellulose
phosphate is nearly pure. Based on previously-reported extinction coefficients for fumarase (Kanarek and Hill, 1964), and
assuming the maximal activity unchanged, this enzyme is about
30% pure. It is quite unstable, and on several occasions the
activity has been lost within 48 hours. In order to avoid this
loss, we proceeded to use this fraction immediately, with high
sodium chloride, in order to obtain some comparative kinetic
data. The pH activity profile for this fraction is shown in
Figure 7. The values plotted are for V_{max} at each pH. Ionic
strength was maintained using sodium chloride; the assay was
that of Massey (1952). The K_m's determined in this experiment

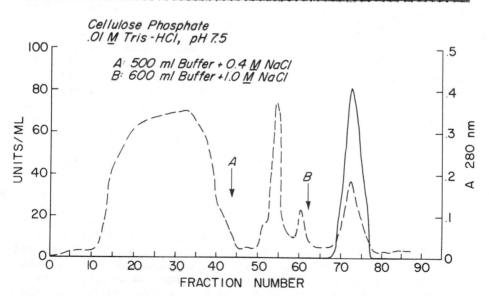

Fig. 6. Cellulose phosphate chromatography of first fumarase
fraction from DEAE-cellulose (fumarase "A"). Buffer: 0.01 M
Tris-HCl, 0.001 M EDTA, 0.001 M 2-mercaptoethanol; room temperature.

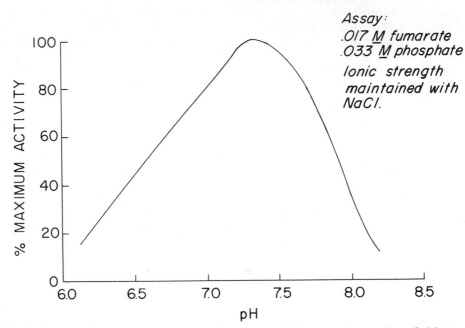

Fig. 7. The pH profile of the first fumarase fraction following cellulose phosphate chromagraphy.

were similar to those reported previously (Massey, 1953; Kanarek
and Hill, 1964). The major difference to be noted is the pH
optimum at approximately 7.3, which is a marked shift from the
6.7 reported by the same workers. The relatively narrow limits
of this profile are also of interest. Massey showed that the
enzyme is active from about pH 5 to pH 9. Isotopic exchange
studies by Berman et al. (1971), also showed an optimum exchange
at pH 7.3, but they observed exchange between pH 5 and pH 9.6.

When the second DEAE-cellulose fraction was rechromatographed, the fractions containing fumarase activity (coinciding
with a red peak) were combined and dialyzed against Tris buffer,
pH 7.5; the fraction was applied to a cellulose phosphate column under the same conditions as with the first fraction. This
activity was also bound by cellulose phosphate, although it
remained associated with some red color. The purification was
approximately eight-fold. Thus, on cellulose phosphate, both
fractions appear to have similar chromatographic properties.

The crucial question as to the nature of fumarase heterogeneity is impossible to settle at this time. Certainly, the
physical studies of Hill's group suggest a relatively homogeneous protein. Separation of isozymes by electrophoresis has

been unsuccessful with this protein, and attempts to demonstrate heterogeneity by disc gel electrophoresis and other standard methodologies have also been fruitless. The only demonstrations of heterogeneity have been the electrofocusing studies of Cohen's group, and the anion-mediated chromatographic experiments reported here. The large number of species appears to argue for a basis other than the independently-coded sub-unit explanation of classic isozymes.

The demonstration of differences in chromatographic behavior with the affinity reagent for crystalline enzyme exposed to different ionic environments suggests that the exposure of the protein to anions known to bind may in part account for the heterogeneity. The apparent heterogeneity demonstrated so far in this purification procedure may be real, or it may be due to the incomplete solubilization of the protein from the mitochondrial matrix under the mild extraction conditions used. In any case, the pH profile demonstrated for the only highly purified fraction indicates that, to some extent, the properties of the enzyme purified by this modified procedure, in the absence of polyvalent anions, have been somewhat altered, when compared with the data for the enzyme purified by classical procedures. The evidence is most supportive of the conclusion that these multiple forms of fumarase are "metazymes", as defined by Dr. Horecker, that is, heterogeneous forms resulting from events occurring after the synthesis of otherwise identical sub-units.

ACKNOWLEDGEMENT

Technical assistance for these studies has been provided by two undergraduate research students at the University of New Mexico, Mr. Malcolm Ennis, senior chemistry major, and Mr. Robert Bussey, senior biology major.

REFERENCES

Berman, K., E.C. Di Novo, and P.D. Boyer 1971. Relationships of pH to exchange rates and deuterium isotope effects in the fumarase reaction. *Bioorg. Chem.* 1: 234-242.

Cuatrecasas, P. 1970. Protein purification by affinity chromatography. *J. Biol. Chem.* 245: 3059-3065.

Frieden, C., R.M. Bock, and R.A. Alberty 1954. Studies of the enzyme fumarase II. Isolation and physical properties of crystalline enzyme. *J. Am. Chem. Soc.* 76: 2482-2484.

Hamm, R., A.A. El-Badawi, and L. Tetzlaff 1972. Lokalisierung einiger enzyme in den skeletmuskel-mitochondrien des schweines. *Z. Lebensm. Unters Forsch.* 149: 7-17.

Hill, R.L. and L. Kanarek 1964. The subunits of fumarase. *Brookhaven Symposia in Biology.* 17: 80-97.

Kanarek, L. and R.L. Hill 1964. The preparation and characterization of fumarase from swine heart muscle. *J. Biol. Chem.* 239: 4202-4206.

Kanarek, L., E. Marler, R.A. Bradshaw, R.E. Fellows, and R.L. Hill 1964. The subunits of fumarase. *J. Biol. Chem.* 239: 4207-4211.

Lin, y.-C., C.F. Scott, and L.H. Cohen 1971. Multiple forms of fumarase of pig heart. *Arch. Biochem. Biophys.* 144: 741-748.

Lowe, C.R. and P.D.G. Dean 1971. Affinity chromatography of enzymes on insolubilized cofactors. *FEBS Letters* 14: 313-316.

Massey, V. 1952. The crystallization of fumarase. *Biochem. J.* 51: 490-494.

Massey, V. 1953. Studies on fumarase. 3. The effect of temperature. *Biochem. J.* 53: 72-79.

Penner, P.E. and L.H. Cohen 1971. Fumarase: demonstration, separation, and hybridization of different subunit types. *J. Biol. Chem.* 246: 4261-4265.

Teipel, J.W. and R.L. Hill 1971. Subunit interactions of fumarase. *J. Biol. Chem.* 246: 4859-4865.

Woodfin, B.M. 1973. Separation and characterization of multiple forms of fumarase. *Fed. Proc.* Abst. No. 2002, 57th Meeting.

MULTIPLE FORMS OF PIG HEART AND *Herellea vaginicola* SUCCINATE THIOKINASE, EFFECTS OF SULFHYDRAL GROUPS AND LIGANDS

DAVID P. BACCANARI, CECIL J. KELLY, and SUNGMAN CHA
Division of Biological and Medical Sciences
Brown University, Providence, Rhode Island 02912

ABSTRACT. Multiplicity of pig heart succinate thiokinase has been observed in isoelectric focusing studies. The pH values of the isozymic forms are 6.4, 6.2, 6.0, 5.9, and 5.8. All are kinetically indistinguishable and have similar molecular weights, heat stabilities, and pH optimum.

The pH 6.2, 6.0, 5.9, and 5.8 enzymes are phosphorylated and contain the same amount of bound phosphate per mole of enzyme. The pH 6.4 enzyme, which lacks exchangeable phosphate or CoA, is called the free enzyme. It is unstable but can be phosphorylated to the stable pH 6.2 and 6.0 phosphoenzymes. The pH 6.2 and 6.0 forms can each be dephosphorylated to the pH 6.4 enzyme.

The pH 6.2 species is converted to the pH 6.0 form by dithiothreitol. Titration of the pH 6.2 and 6.0 enzymes with DTNB has indicated that possibly only one or two disulfides are reduced during the conversion of the pH 6.2 to the 6.0 enzyme.

Modified enzyme was prepared by incubating succinate thiokinase with oxidized [^{3}H]CoA. Two species with pH values of 5.3 and 5.2 were observed; each had the same amount of CoA bound per mole of enzyme. The CoA was discharged by dithiothreitol indicating a disulfide compound was formed between succinate thiokinase and oxidized CoA.

Herellea vaginicola contains two succinate thiokinases, one specific for guanine nucleotide and the other non-specific with regard to ATP or GTP. These enzymes also exhibit polymorphism.

Succinate thiokinases (E.C. 6.2.1.4 and E.C. 6.2.1.5) catalyze the phosphorylation of GDP or ADP by inorganic phosphate utilizing succinyl CoA as an energy source. In the past few years some of the structural properties of the enzyme have been determined. It has been shown that succinate thiokinase can exist in either a phosphorylated or a non-phosphorylated form (Kreil and Boyer, '64; Cha et al., '65). The pig heart enzyme (M.W. 77,000) consists of a dimer of two dissimilar subunits; only one is capable of being phosphorylated (Brownie and Bridger, '72). Antibody to succinate thiokinase has been prepared, and the reactivity of sulfhy-

dryl groups and immunochemical properties of the enzyme have
also been examined (Murakami and Nishimura, '74).

This report summarizes our observations on the occurrence
of multiple forms of pig heart succinate thiokinase, their
characteristics, and their interconvertibility (Baccanari
and Cha, '73, Baccanari and Cha, '74). Revised values for
the sulfhydryl group content and the amount of various
ligands bound per mole of enzyme are also presented.

Isoelectric focusing

Either partially or highly purified succinate thiokinase
isolated from a single pig heart or a mixture of hearts
showed electrofocusing patterns similar to Fig. 1. Although
the pH of each individual peak varied up to ± 0.05 pH unit
between experiments, the enzymes are called the pH 6.4, 6.2,
6.0, 5.9, and 5.8 forms. Since electrofocusing was conducted
at 5° and the pH was measured at room temperature, the pH
values differ from the true isoelectric points. In most
cases the pH 6.2 and 6.0 enzymes were the major species; however, when the enzyme purification procedure was modified

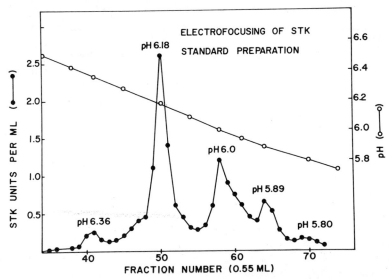

Fig. 1. Isoelectric focusing (pH 5 to 7) of succinate thiokinase. The enzyme (30 units, 10.5 units per mg) was added
to the "light" solution before gradient formation. Electrofocusing was continued for 66 hours with a final power consumption of 0.54 watt at 600 volts.

TABLE I

Enzyme species (pI)	Apparent Michaelis Constants			Ratios of reaction rates		
	GTP	Succinate	CoA	dGTP:GTP	ITP:GTP	Itaconate: Succinate
6.4	0.009	0.39	0.004	0.85	0.88	0.38
6.2	0.011	0.46	0.006	0.84	0.85	0.39
6.0	0.010	0.45	0.006	0.87	0.80	0.39
5.9	0.011	0.43	0.007	0.88	0.84	0.37
5.8	0.011	0.43	0.005	0.84	0.86	0.42

Apparent Michaelis constants (mM) and ratios of reaction rates between substrate analogs and substrate.

(see below), the pH 6.4 form became a major species. The pH 5.9 and 5.8 enzymes were present in only small amounts.

Physical and kinetic properties

Primary considerations in characterizing enzyme multiplicity are the physical and kinetic properties of the various forms. Our first indication that the multiple forms of heart succinate thiokinase were very similar was obtained by kinetic measurements. All enzyme forms were indistinguishable with respect to K_m for succinic acid, GTP or CoA. Another example of similarities in the kinetic properties among the electrofocused enzymes was seen when the reaction rates with substrate analogs were compared to that of the standard enzyme assay; the ratios with dGTP and ITP as GTP analogs or itaconate as a succinate analog were similar for each enzyme form (Table I). It was also observed that the pH 6.2, 6.0, 5.9, and 5.8 enzymes had similar heat stabilities and pH optimum.

Multiplicity could be caused by protein aggregation during electrofocusing in a low ionic strength medium. This appeared unlikely for several reasons. If aggregation was caused by low ionic strength in the electrofocusing column, then dialysis against buffer should cause a dissociation. However, when the isolated pH 6.4, 6.2, and 6.0 peaks were dialyzed against 100 mM Tris-acetate buffer (pH 7.2) and refocused, only one peak of activity at the original pH was observed. Secondly, each enzyme has an apparent molecular weight of approximately 78,000 as determined by Sephadex G-100 gel filtration at pH 8 and at the pH value for each isolated peak (Table II). There was no measureable activity in tubes corresponding to multiples or fractions of the molecular weight. Also, the observed heterogeneity was not due to artifacts unique to electrofocusing technique since

TABLE II

Enzyme species (pI)	Molecular Weight	
	At pH 8.0	At pI
6.4	82,250	80,000
6.2	78,000	78,000
6.0	79,250	74,000
5.9	77,000	

Molecular weights of electrofocused succinate thiokinase by Sephadex G-100 gel filtration.

The molecular weights were determined at pH 8 and at a pH corresponding to the isoelectric point of each species. The Sephadex G-100 column (2.5 cm x 50 cm) was calibrated at both pH 8.0 and 6.0; essentially identical plots of elution volume versus log molecular weight were obtained.

a good correlation between electrofocusing and polyacrylamide gel electrophoresis patterns was observed (Baccanari and Cha, '73).

Identification of phosphorylated forms

E. coli succinate thiokinase is isolated from the cell as a phosphorylated protein (Kreil and Boyer, '64); the pig heart enzyme can also be phosphorylated and, in the presence of CoA, inorganic phosphate exchanges with the enzyme bound phosphate (Cha, et al., '65). Therefore to determine whether or not phosphorylation was responsible for the multiple forms of succinate thiokinase, the enzyme was incubated with $^{32}P_i$, CoA and MgCl$_2$ and subjected to electrofocusing. Fig. 2 shows that all the enzymes, except the pH 6.4 species, were phosphorylated. Since no new enzyme-phosphate bonds could be formed, these enzyme species were already phosphorylated in the native state. In calculating the moles of various ligands bound per mole of succinate thiokinase, the value 8.4 units per nmole of enzyme has been used in the past (Cha, et al., '67). The uncertainties associated with these calculations were discussed by Baccanari and Cha ('73). Since that time, Murakami and Nishimura ('74) found the $E_{280}^{1\%}$ of pig heart succinate thiokinase to be 3.5 ± 0.5. Also numerous determinations have shown the molecular weight of the enzyme to be 77,000 rather than 72,000 (Brownie and Bridger, '72, Baccanari and Cha, '73, Murakami and Nishimura, '74). Therefore, a value of 3.6 units/nmole is now being used in calculating the molar concentration of succinate thiokinase, and the appropriate values for the amounts of various ligands bound per mole of enzyme are presented. For example in

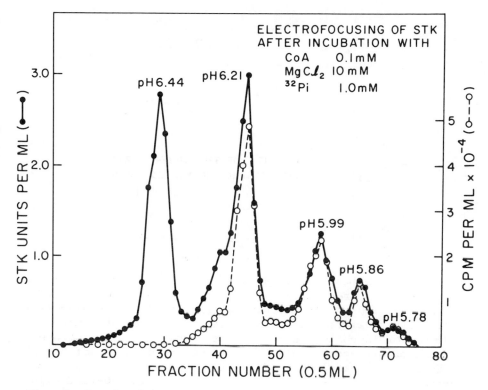

Fig. 2. Isoelectric focusing pattern of succinate thiokinase phosphorylated by $^{32}P_i$. The reaction mixture contained, in micromoles: Tris-acetate (pH 7.4), 70; $MgCl_2$, 10; CoA, 0.1; $K_2H^{32}PO_4$, 1 (4×10^7 cpm); and 3.9 mg of succinate thiokinase (40.5 units) in a final volume of 1.0 ml.

Fig. 2 the pH 6.2, 6.0, 5.9, and 5.8 forms each contained approximately 1 mole of phosphate per mole of enzyme.

Stability of the pH 6.4 enzyme

In conducting the preceding experiment it was found that the nonphosphorylated pH 6.4 enzyme was unstable in Tris-acetate buffer. Glycerol and ampholine carrier ampholytes stabilized the enzyme significantly while succinic acid, GDP or dithiothreitol did not. The pH 6.4 enzyme stored in Tris-acetate buffer containing 30% glycerol retained 70% of its activity over a two month period. When buffer containing 20% glycerol was used in the enzyme purification, the total yield was doubled, while the specific activity was increased three-fold over a comparable step without glycerol. Also,

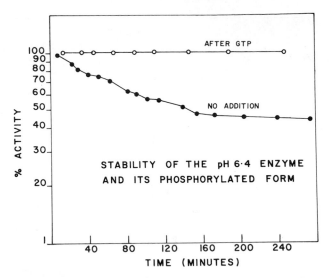

Fig. 3. Stability of the pI 6.4 enzyme and its phosphorylated form at 0°. Previously isolated pI 6.4 enzyme (0.8 unit in 0.1 ml of 100 mM Tris-acetate buffer, pH 7.4, 20% glycerol) was either diluted with 4.9 ml of buffer without glycerol, (●—●) or incubated at room temperature for 15 min with 0.05 μmole of GTP and 1 μmole of $MgCl_2$ before dilution of 5.0 ml (O—O). The samples were kept in ice and assayed for enzymic activity at various times. In both cases the 50-fold dilution with buffer (100 mM Tris-acetate, pH 7.4) reduced the concentration of glycerol from 20% to a value which was no longer protective.

the pH 6.4 enzyme was then seen to be a major form upon electrofocusing.

Phosphorylation of the pH 6.4 enzyme by GTP also affords protection. Fig. 3 illustrates the stabilities of the pH 6.4 enzyme with and without prior incubation with GTP. At zero time buffer was added to dilute each sample 50-fold so that the concentration of glycerol was no longer protective. It can be seen that the phosphorylated enzyme was stable over the four hour duration of the experiment, whereas the non-phosphorylated enzyme lost 50% activity in 140 minutes.

Interconversion of phosphorylated and nonphosphorylated enzymes

Since there is only one nonphosphorylated enzyme and four phosphorylated forms, it was of obvious interest to examine the electrofocusing pattern of the pH 6.4 enzyme after phos-

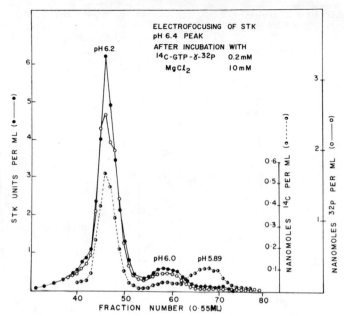

Fig. 4. Electrofocusing pattern of the pI 6.4 enzyme after
phosphorylation by GTP. The 2.1 ml reaction mixture (in 20%
glycerol) contained, in micromoles: $[8-^{14}C, \gamma-^{32}P]$GTP,
0.4 (6 x 10^6 cpm of ^{14}C, 7.2 x 10^6 cpm of ^{32}P); $MgCl_2$, 20;
Tris-acetate buffer, pH 7.4, 200; and 12.5 units of the pI
6.4 enzyme.

TABLE III

Enzyme species	Moles of sulfhydryl per mole enzyme	
	Without SDS	With SDS
Unresolved mixture	5.3	15.6
pH 6.2	4.7	14.0
pH 6.0	5.2	15.9

Accessible and total sulfhydryl groups of succinate thiokinase.

The reaction with DTNB was monitored by measuring the absorbance at 410 nm using an E_m value of 13,700 M^{-1} cm^{-1} at
pH 8.0. The value 3.6 enzyme units/nmole was used to calculate succinate thiokinase concentration. The specific activities of the unresolved mixture, the pH 6.2 enzyme and the
pH 6.0 enzyme were 40.1, 47.1, and 34.2 units/mg, respectively.

phorylation by GTP. As shown in Fig. 4, the pH 6.4 enzyme is
converted to the pH 6.2 and 6.0 forms by $[\gamma-^{32}P, 8-^{14}C]$GTP.
Each enzyme contains approximately 1 mole of phosphate and
0.4 mole of guanine nucleotide per mole of enzyme, showing

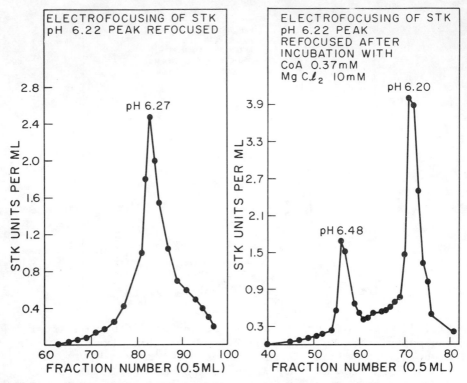

Fig. 5. Electrofocusing the pI 6.2 enzyme before and after incubation with CoA. The pI 6.2 enzyme (18 units) was isolated and dialyzed against 100 mM Tris-acetate buffer, pH 7.4. (A) An aliquot of the enzyme was refocused. (B) Another aliquot of enzyme incubated in a reaction mixture containing 20% glycerol, 1 μmole of CoA and 27 μmoles of $MgCl_2$ in a final volume of 2.7 ml. After a 30-min incubation at room temperature, the solution was refocused.

---

that the multiplicity is not due to differences in bound phosphate or nucleotide.

It has been determined that the enzyme can be dephosphorylated by CoA (Cha, et al., '65). Fig. 5 shows that the pH 6.2 enzyme can be isolated and refocused to the same pH value, but upon incubation with CoA the pH 6.4 enzyme is formed. Similar results were seen with the pH 6.0 enzyme (Fig. 6). Experiments with [^{3}H]CoA have shown that there is no significant amount of radioactivity associated with the newly formed pH 6.4 peak. Thus, the pH 6.4 form has been termed the free enzyme, containing neither exchangeable phosphates nor CoA.

The fact that there are two phosphoenzymes capable of

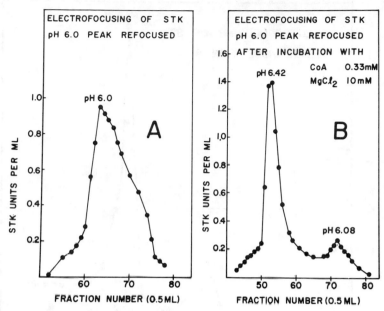

Fig. 6. Electrofocusing the pI 6.0 enzyme before and after incubation with CoA. The pI enzyme (12 units) was isolated and dialyzed against 100 mM Tris-acetate buffer (pH 7.4). (A) An aliquot of the enzyme was refocused. (B) Another aliquot of enzyme was incubated in a reaction mixture containing 20% glycerol, 1 μmole of CoA and 28 μmoles of $MgCl_2$ in a final volume of 2.8 ml.

forming the same free enzyme suggested that the pH 6.2 and 6.0 forms may be interconvertible (Baccanari and Cha, '73). More convincing evidence was obtained by investigating the effects of thiols on succinate thiokinase. When an enzyme preparation with an electrofocusing pattern similar to Fig. 1 was incubated with 10 mM dithiothreitol, the pH 6.2 enzyme was almost completely converted to the pH 6.0 form (Fig. 7). Mercaptoethanol gave qualitatively similar but less dramatic effects. Since dithiothreitol serves as a reducing agent for disulfide groups, it appears that the sulfhydryl group oxidation state of the pH 6.2 and 6.0 enzymes may be responsible for the differences in the isoelectric points.

Titration of the sulfhydryl groups of the pH 6.2 and 6.0 forms by DTNB shows that they contain 4.7 and 5.2 moles of sulfhydryl group per mole of enzyme, respectively (Table III). When the sulfhydryl group titration was performed in the presence of 1% SDS, the pH 6.2 and 6.0 forms contained 14.0 and 15.9 moles of sulfhydryl group per mole of enzyme,

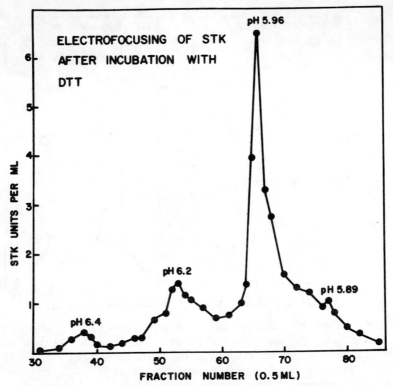

Fig. 7. Electrofocusing succinate thiokinase after incubation with dithiothreitol. The enzyme (35 units, 31 units/
mg) in 100 mM Tris-acetate buffer, pH 7.4, containing 25%
glycerol, was incubated for 1.5 hr at room temperature with
10 mM dithiothreitol, then electrofocused.

respectively. These data are in fair agreement with the
value of 12 moles of sulfhydryl group per mole of enzyme reported by Murakami and Nishimura ('74) for an unresolved
preparation of succinate thiokinase. Thus it appears that
the pH 6.2 and 6.0 enzymes do not differ greatly in total
sulfhydryl group content and that their interconversion by
dithiothreitol may involve only one or two disulfides.

Modified succinate thiokinase

In further investigating the reactivity of succinate thiokinase sulfhydryl groups, we found that it was possible to
generate an altered enzyme form when the heterogeneous enzyme was incubated with oxidized [^{3}H]CoA, GTP, and $MgCl_2$.
The electrofocusing pattern (Fig. 8) shows two peaks of en-

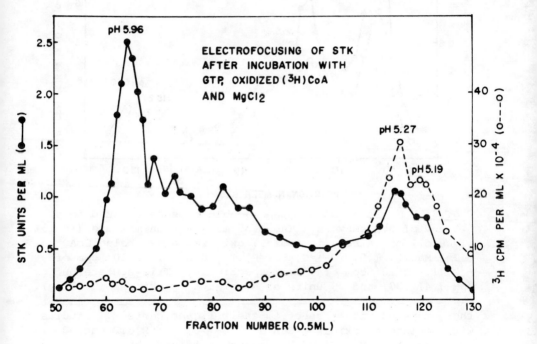

Fig. 8. Isolation of two species containing enzyme bound [³H]CoA. The 2.25 ml reaction mixture (in 20% glycerol) contained, in μmoles: oxidized [³H]CoA, 0.15 (3.3 x 10⁸ cpm/μmole); MgCl₂, 2.5; GTP, 0.25; Tris-acetate buffer (pH 7.4) 200; and 32 units of succinate thiokinase (2.3 mg).

zyme activity, which are not natural constituents of succinate thiokinase, one at pH 5.3 and the other at pH 5.2. The [³H] radioactivity profile indicates that 2.7 moles of CoA were bound per mole of each enzyme. Since it has been shown that the pH 6.2 and 6.0 phosphoenzymes can bind guanine nucleotide (cf. Fig. 4), the pH 5.3 and 5.2 species were fur-

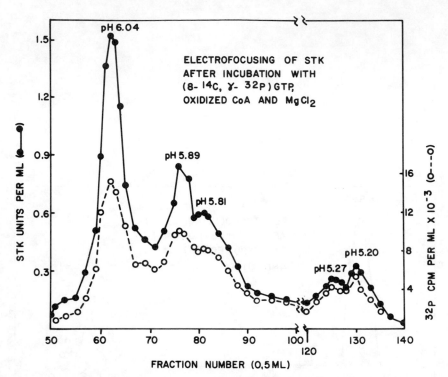

Fig. 9. Demonstration of exchangeable phosphate bound to pI 5.3 and pI 5.2 enzymes. The 1.35 ml reaction mixture (in 15% glycerol) contained, in μmoles: oxidized non-labeled CoA, 0.25; MgCl$_2$, 1.5; [8-^{14}C, γ-^{32}P]GTP, 0.3 (1.2 x 10^7 cpm of ^{14}C, 1.5 x 10^7 cpm of ^{32}P, respectively); Tris-acetate buffer (pH 7.4) 100; and 29 units of succinate thiokinase (0.77 mg).

---

ther examined with respect to their ligand content. Studies with nonlabeled oxidized CoA and [8-^{14}C, γ-^{32}P]GTP showed that both the pH 5.3 and 5.2 forms contained about 1 mole of phosphate per mole of enzyme but no bound guanine nucleotide (Fig. 9). Other experiments have shown that these forms could be generated in the absence of GTP, indicating that phosphate was not discharged from the naturally occurring phosphoenzymes during the formation of the modified enzyme. Both the pH 5.3 and 5.2 species contain the same number of phosphate and CoA per mole of enzyme, but it is not known why they differ in isoelectric point.

If disulfide compounds were being formed between succinate thiokinase and oxidized CoA, then a reducing agent such as dithiothreitol should discharge the CoA. Indeed, when dithiothreitol was added to a mixture of succinate thiokinase and

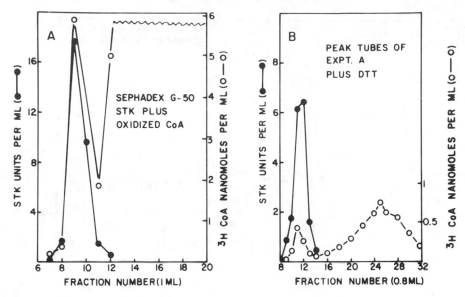

Fig. 10. Effects of dithiothreitol on the enzyme CoA-compound. In Experiment A, the enzyme-CoA compound was prepared by mixing succinate thiokinase with, in μmoles (in 0.66 ml of 15% glycerol): oxidized [^{3}H]CoA, 0.15 (3.3 x 10^8 cpm/μmole); GTP, 0.1; MgCl$_2$, 5; Tris-acetate buffer (pH 8) 50; and 41 units of succinyl-CoA synthetase (31 unit/mg). The mixture was incubated for 1 hr at room temperature and filtered through a Sephadex G-50 column. In Experiment B, the enzyme from the peak tube in Experiment A (17 units, 1.6 nmoles [^{3}H]CoA/3.6 units enzyme) was incubated for 45 min with 5.3 mM dithiothreitol then applied to another Sephadex G-50 column. In each case, the Sephadex G-50 column (1.2 x 20 cm) was eluted with 100 mM Tris-acetate buffer (pH 8) containing 20% glycerol.

oxidized CoA, the modified enzyme forms were not seen upon electrofocusing. The discharge of CoA from the modified enzyme by a reducing agent was also demonstrated in another experiment. In Fig. 10A, succinate thiokinase was incubated with [^{3}H] oxidized CoA and filtered through a Sephadex G-50 column; the enzyme in the peak tube contained 1.2 moles of CoA per mole enzyme. When the enzyme from this tube was incubated with dithiothreitol and applied to another Sephadex G-50 column (Fig. 10B), greater than 80% of the bound [^{3}H]CoA was discharged.

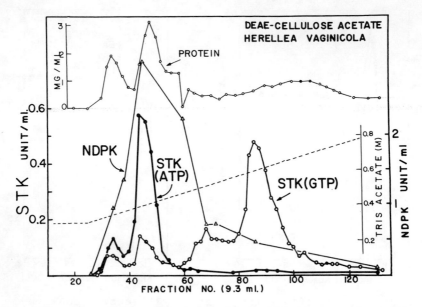

Fig. 11. Separation of two succinate thiokinases in crude
extract of *Herellea vaginicola*. The crude extract (600 mg
protein in 30 ml of 50 mM potassium phosphate buffer, pH 7.4)
was loaded on a DEAE-cellulose acetate column (5 x 26 cm)
equilibrated with 0.3 M Tris-acetate buffer, pH 7.4. After
washing with 300 ml of the acetate buffer, the column was
eluted with a linear gradient of the acetate buffer from 0.3
to 0.8 M in a total volume of 1 liter.

Succinate thiokinase from *Herellea vaginicola*

While the succinate thiokinases from mammalian sources are
specific for guanine nucleotides, it has been reported that
the enzyme from *Rhodopseudomonas spheroides* (Burnham, '63) is
nonspecific for nucleotide substrates and that GTP can serve
as an alternative substrate for the *E. coli* enzyme (Murakami,
et al., '72). As can be seen upon DEAE-cellulose column
chromatography (Fig. 11), an extract of *Herellea vaginicola*
grown in a medium containing succinate contains two succinate
thiokinases. One is specific for GTP and the other can utilize either ATP of GTP. The GTP specific succinate thiokinase
from this organism also exhibits multiple forms on isoelectric focusing (Fig. 12). The properties of the nonspecific
enzyme and the nature of the polymorphic GTP specific enzymes
are currently being investigated.

In conclusion, the multiplicity of pig heart succinate thioki-
has been examined. The pH 6.4 form is the unstable free enzyme,

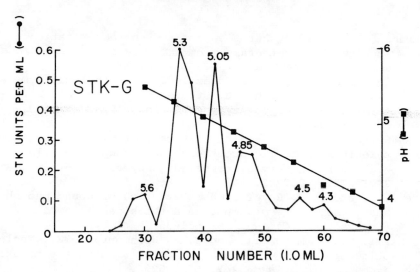

Fig. 12. Isoelectric focusing of GTP-specific succinate thiokinase from *Herellea vaginicola*. The enzyme (26 units, 8.3 units/mg) partially purified by chromatography steps on DEAE-cellulose (acetate) and calcium phosphate gel-cellulose columns was subjected to isoelectric focusing in 1% ampholine carrier ampholyte, pH 4-6 for 56 hrs with a final power consumption of 0.7 watt at 800 volts.

---

containing neither exchangeable phosphate nor CoA. Phosphorylation of the free enzyme by GTP results in the formation of the pH 6.2 amd 6.0 phosphoenzymes. Dephosphorylation of the pH 6.2 and 6.0 enzymes by CoA leads to a regeneration of the pH 6.4 enzyme. The pH 6.2 and 6.0 forms can be directly interconverted by sulfhydryl reagents. Therefore, all evidence obtained in this laboratory indicate that the three major forms of pig heart succinate thiokinase do not qualify as genetically determined isozymes.

(Figs. 1-6 were reproduced from *J. Biol. Chem.* 248: 15-24 (1973) and Figs. 7-10 from *Biochim. Biophys. Acta* 334: 226-234 (1974), with permission of the respective journals.

REFERENCES

Baccanari, D. P. and S. Cha 1973. Succinate thiokinase VI. Multiple interconvertible forms of the enzyme. *J. Biol. Chem.* 248: 15-24.
Baccanari, D. P. and S. Cha 1974, Sulfhydryl groups, a factor in the polymorphism of succinyl-CoA synthetase (GDP Forming). *Biochim. Biophys. Acta* 334: 226-234.

Burnham, B. F. 1963. Purification and characterization of succinyl CoA synthetase from *Rhodopseudomonas spheroides*. *Acta. Chem. Scand.* 17: S123-S128.

Brownie, E. R. and W. A. Bridger 1972. Succinyl coenzyme A synthetase of pig heart: studies of the subunit structure and of phosphoenzyme formation. *Can. J. Biochem.* 50: 718-724.

Cha, S., C.-J. M. Cha, and R. E. Parks, Jr. 1965. Succinic thiokinase III. The occurrence of a nonphosphorylated high energy intermediate of the enzyme. *J. Biol. Chem.* 240: PC3700-PC3702.

Kreil, G. and P. D. Boyer 1964. Detection of bound phosphohistidine in *E. coli* succinate thiokinase. *Biochem. Biophys. Res. Commun.* 16: 551-555.

Murakami, Y., T. Mitchell, and J. S. Nishimura 1972. Nucleotide specificity of *Escherichia coli* succinate thiokinase. *J. Biol. Chem.* 247: 6247-6252.

Murakami, Y. and J. S. Nishimura 1974. Porcine heart succinate thiokinase; reactivity of sulfhydryl groups, cross-linking and immunochemical properties of the enzyme. *Biochim. Biophys. Acta* 336: 252-263.

SUBTILISIN-LIKE ISOZYME IN PRONASE

WILLIAM M. AWAD, JR.
KLAUS D. VOSBECK
STEVEN SIEGEL
MARIA S. OCHOA
Departments of Medicine and Biochemistry
University of Miami School of Medicine
P. O. Box 520875 Biscayne Annex
Miami, Florida 33152

ABSTRACT. The subtilisin-like enzyme of the K-1 strain of *Streptomyces griseus* undergoes rapid irreversible denaturation in 6 M guanidine·HCl if EDTA is also present. In the absence of EDTA the protease is both active and stable in either 8 M urea or 6 M guanidine·HCl. The native enzyme is stable to 70°; if EDTA is added, inactivation begins above 40°. Calcium ion was found to be the specific metal which protects this protein. This enzyme shows maximal stability in the pH range of 6 to 9; it is acid labile, like the well-studied subtilisins, in contrast to the chymotrypsin-like endopeptidases in Pronase. Enhanced rates and extents of hydrolysis of casein by this protease are found if 8 M urea or 6 M guanidine·HCl is present. Native ovalbumin appears to be resistant to hydrolysis by this enzyme. In 8 M urea, ovalbumin is very slowly hydrolyzed whereas the rate is very rapid in 6 M guanidine·HCl. These findings reflect the selective unfolding of substrate proteins by denaturants. The Michaëlis constant against Nα-acetyl-L-tyrosine ethyl ester is similar to those noted for the two companion chymotrypsin-like isozymes; however, the maximal velocity against this substrate is about ten-fold greater. As yet no good explanation can be offered as to why two different classes of serine enzymes have been generated and conserved by this microorganism.

INTRODUCTION

The commercial protease preparation, Pronase, consists of a mixture of extracellular enzymes synthesized by the K-1 strain of *Streptomyces griseus*. Pronase was first studied and partially characterized by Nomoto and his colleagues (1959 a, b, c), who demonstrated its broad specificity of proteolytic action (1960 b). Initial studies revealed a narrow range of molecular weights for the components and it was believed that most of the activity could be attributed to one protein. Nomoto and his colleagues (Nomoto et al,

1964; Hiramatsu and Ouchi, 1963; cited in Awad and Wilcox, 1963) later corrected this misinterpretation, pointing out that there were different properties and specificities associated with different components. The multi-component nature of Pronase remains unknown to a surprisingly large number of investigators who presently use this agent.

Most of the more recent studies related to Pronase have been directed towards the purification of endopeptidases which also possess esterase activity. Wählby and Engström (1968) showed that three of these enzymes demonstrated homology with bovine chymotrypsin. The only other microbial enzyme with the chymotrypsin sequence, Asp-Ser-Gly, at the active site is the α-lytic protease of a Sorangium species (Olson et al, 1970). One of the three Pronase chymotrypsin-like enzymes demonstrates marked homology with bovine trypsin (Olson et al, 1970) and activates bovine chymotrypsinogen in the same manner as bovine trypsin (Awad and Wilcox, 1963).

Following these early results a series of studies was begun to purify to homogeneity several of the Pronase components. It was noted that, in addition to the chymotrypsin-like enzymes, other components with esterase activity were present (Awad et al, 1971, 1972 b; Gertler and Trop, 1971). One of these latter proteins was treated with [^{32}P] diisopropylfluorophosphate and thereafter treated by partial acid hydrolysis. The pattern of electrophoretic migration of labelled peptides was identical to that of similarly derived peptides from the subtilisins but distinctly different from the pattern of derived peptides of chymotrypsin (Awad et al, 1972 b). Therefore this Pronase component has been tentatively designated a subtilisin-like protease and presumably has the sequence, Thr-Ser-Met, around the reactive serine residue.

Initial studies did not allow complete purification of the subtilisin-like enzyme. However, our earlier studies had revealed marked stability in denaturant of one chymotrypsin-like Pronase component (Siegel et al, 1972). This led to the procedure of permitting that enzyme to undergo self-purification by incubation of crude Pronase in denaturant where the stable enzyme could hydrolyze all companion enzymes to low-molecular-weight products. Surprisingly, it was found that the subtilisin-like enzyme also was stable in 6 M guanidinium chloride and could be readily purified after chromatography through CM-cellulose (Awad et al, 1972 a). All of our studies have been based upon this preparation of the enzyme. It can not be excluded that a small peptide fragment was lost during purification in denaturant. However, a very large peptide fragment is unlikely to have been removed by

824

proteolysis since the apparent molecular weight of this protein, as determined by gel filtration, does not change after exposure to denaturant.

The finding of both chymotrypsin-like and subtilisin-like enzymes in Pronase raises several interesting questions. Since the K-1 strain of *S. griseus* is not available for definitive studies, it has not been clearly ruled out that Pronase may be the product of a mixed culture with enzyme components from different microbial sources. Because of differences in primary (Hartley, 1964; Smith et al, 1968) and tertiary structures (Matthews et al, 1967; Wright et al, 1969), chymotrypsin and subtilisin have been considered to have evolved from different ancestral genes. Since the charge relay system of the two active sites are identical, the two enzymes have been thought to be molecular examples of convergent evolution (Smith et al, 1970; Neurath and Bradshaw, 1970). If these enzymes are produced by one organism, no good explanation can be offered yet as to why both families of genes are retained to carry out very similar functions. These Pronase endopeptidases represent the first examples of isozymes which are not products of gene duplication. The present report describes some of the properties of the subtilisin-like protease.

EXPERIMENTAL RESULTS

The conformation in 6 M guanidine·HCl of guanidine-stable chymoelastase, one of the chymotrypsin-like endopeptidases in Pronase, was found to be susceptible to the effects of EDTA (Siegel et al, 1972). The specific requirement of calcium ion to protect that protein against heat denaturation was later demonstrated (Siegel and Awad, 1973). This led to our examination of the possible requirement of a cation for stabilization of the subtilisin-like enzyme. Fig. 1 reveals that this enzyme as expected is completely stable in 6 M guanidine·HCl when 125 mM $CaCl_2$ is also present. In the presence of EDTA, a very rapid loss of activity occurs with the process being complete within 25 minutes. This rate of inactivation is faster than that noted under similar conditions with the chymotrypsin-like enzyme (Siegel et al, 1972).

Following the lead of our earlier studies (Siegel and Awad, 1973), the effect of EDTA upon the heat stability of the subtilisin-like enzyme was examined. Fig. 2 reveals that the native enzyme is stable to about 70°. No effect of EDTA can be seen at 40° or lower temperatures but at higher temperatures a dramatic drop in activity is noted. As observed in the studies with guanidine, the effect of EDTA on the sub-

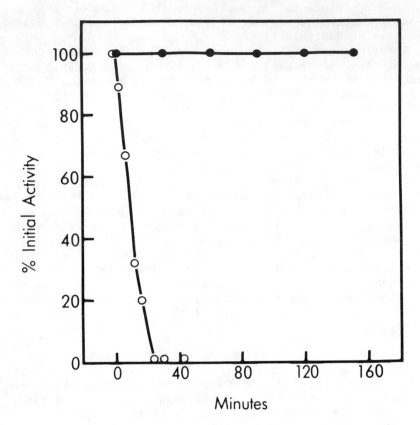

Fig. 1. Stability of subtilisin-like enzyme in 6 M guanidine·HCl. The activities against Ac-Tyr-OEt[1] are demonstrated at the indicated time intervals after incubation in 6 M guanidine·HCl - 8 mM Tris·HCl (pH 8.0) with either 125 mM CaCl$_2$ (●——●) or with 12.5 mM EDTA (O——O); enzyme concentration was 0.85 mg/ml. Aliquots of 50 μl were removed to determine the activities against Ac-Tyr-OEt according to the method of Awad and Wilcox (1963).

tilisin-like enzyme is more pronounced than with the chymotrypsin-like enzyme (Siegel and Awad, 1973). The present enzyme in the native state is more heat stable than the protein studied earlier. However, in the presence of EDTA, the latter protein is more heat stable than the former.

Cations could be removed reversibly from the chymotrypsin-like components by protonation of liganding groups at pH 3.5 because of the stability of these proteins in the acid

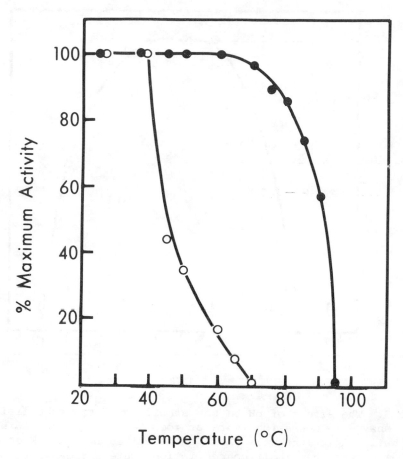

Fig. 2. Temperature stability of subtilisin-like enzyme.
The activities against Ac-Tyr-OEt are demonstrated after 30
min incubations at the indicated temperatures in 8 mM Tris·HCl
(pH 8.0) with either 75 mM $CaCl_2$ (●——●) or 7.5 mM EDTA
(○——○). Each incubation mixture was cooled for one hour in
an ice-bath before assay. The enzyme concentration was 0.43
mg/ml; aliquots of 50 µl were removed to determine the activities against Ac-Tyr-OEt according to the method of Awad and
Wilcox (1963).

---

pH range. In contrast to the known acid stability of the
chymotrypsin family of enzymes, the subtilisins are characterized as being acid labile (Ottesen et al, 1970). Fig. 3
depicts the stability of the subtilisin-like enzyme as a function of pH. No loss in activity is noted at pH values between

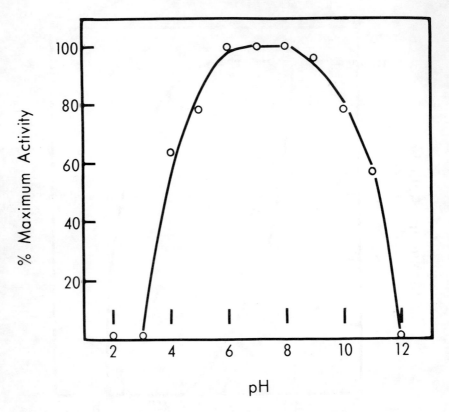

Fig. 3. The effect of pH on the stability of the subtilisin-like enzyme. Incubations were at room temperature for 16 hrs. The buffer solutions were those as described by Vosbeck et al (1973) except that their concentrations were increased to 50 mM. The enzyme concentration was 0.43 mg/ml; aliquots of 50 μl were removed to determine the activities against Ac-Tyr-OEt (Awad and Wilcox, 1963).

6 and 9; however, incubation at lower pH values yields progressively greater degrees of denaturation. Therefore, to determine the specific cation requirement for stability of this enzyme, apparent metal-free protein was prepared at pH 8 by repeated suspension and concentration in an EDTA-containing solution. Following the principles applied earlier (Siegel and Awad, 1973) calcium ion was found to be the only metal to protect to any extent against denaturation at 75°. The observation that only 83% of the activity is recovered with calcium can readily be explained by the data plotted in Fig. 3: after incubation at 75° some loss of activity of the

native enzyme occurs. Our results indicate that calcium is
not required for enzyme activity. The finding that calcium
helps to keep this protein in the native conformation is
not surprising since calcium has been found to stabilize
subtilisin BPN' (Matsubara et al, 1958).

In an earlier report guanidine-stable chymoelastase was
demonstrated to be active in 6 M guanidine·HCl with the property of enhanced digestion of substrate proteins in this
denaturant. We carried out a similar study with the subtilisin-like enzyme. Fig. 4 depicts the results. With casein

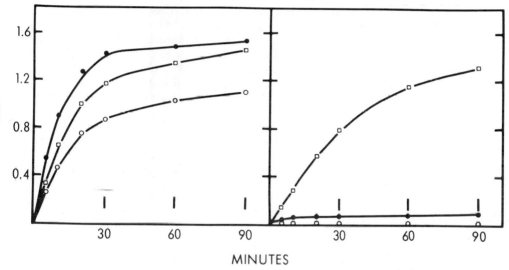

Fig. 4. Hydrolysis of casein (left panel) and ovalbumin
(right panel) by the subtilisin-like protease in 50 mM Tris·
HCl - 50 mM KCl - 5 mM CaCl$_2$ (pH 8.0) in the absence of
denaturant (O——O) or with 6 M guanidine·HCl (□——□) or
with 8 M urea (●——●). Casein and ovalbumin concentrations
were 9.5 mg/ml; the protease concentration was 0.43 mg/ml.
All incubations were at 37°. The assay procedure was that
as described by Siegel et al (1972).

as the substrate, extensive hydrolysis is seen in the absence
of denaturant. However, it can be seen clearly that, within
a 90 minute period, the hydrolysis proceeds more rapidly and
to a greater extent if either 8 M urea or 6 M guanidine·HCl
is also present. In contrast to these results, native ovalbumin appears to be resistant to the action of this Pronase
enzyme. Though limited cleavage of the protein has not been
entirely excluded, no soluble peptides (with aromatic resi-

dues) are noted after trichloroacetic acid treatment. The presence of 8 M urea in the hydrolysis reaction allows the very slow generation of soluble peptides. However, if 6 M guanidine·HCl is present a rapid extensive release of soluble peptides occurs. It is known that ovalbumin unfolds rather slowly in 8 M urea at pH 8 at 20° (Simpson and Kauzmann, 1953) whereas very rapid unfolding occurs in guanidine concentrations of 4.9 M or greater (Schellman et al., 1953). The present studies were carried out after substrate proteins had been incubated in denaturant for well over two hours. After this time interval, the changes in specific rotation for ovalbumin are almost the same whether in 6 M guanidine or 8 M urea. The present findings suggest, therefore, that despite the similar spectral data in the different denaturant solutions, the conformational distribution of ovalbumin may be different in 8 M urea when compared to that in 6 M guanidine. It is unlikely that the effects of urea are substantially different from those of guanidine on the subtilisin-like enzyme in view of the almost similar rates of hydrolysis of casein by the enzyme whether in urea or guanidine.

The kinetic constants of the present enzyme were determined against Ac-Tyr-OEt (Table 2); the values are very close to those determined for the subtilisins (Smith et al, 1970). Though the Michaelis constant is not greatly different from those determined for the two Pronase chymoelastases, the maximal velocity is approximately one order of magnitude greater.

DISCUSSION

This study demonstrates the specific requirement of calcium ion for stabilization of the subtilisin-like component of Pronase; thus, all six Pronase components purified to homogeneity in this laboratory are stabilized by this metal ion (Awad et al, 1972 a, b; Vosbeck et al, 1973). The two Pronase aminopeptidases are the only components which have been demonstrated to require calcium also for activity. The technique of determining cation specificity by analysis for stability at an appropriate temperature may be applicable to many proteins. More definitive studies are needed to determine the stoichiometry of calcium binding to the subtilisin-like endopeptidase. The basis for the remarkable stability of this protein in denaturant remains unknown. The two stable proteins in Pronase are, to our knowledge, the only enzymes found to be active and stable in 6 M guanidine·HCl. Since they are not homologous a study of their structures may reveal some common features which may relate

TABLE 1
EFFECT OF CATIONS ON THERMAL STABILITY

Added cation	Preincubation at 75° for 30 min	Percentage of maximum activity
None	No	100
None	Yes	0
Ca^{2+} [a]	Yes	83
Other [b]	Yes	0

Pronase subtilisin was treated with repeated dilution in 9 mM Tris·HCl (pH 8.0) with 0.5 mM sodium ethylenediaminetetraacetate alternating with concentration by ultrafiltration in order to prepare metal-free protein for these studies. For each assay against Nα-acetyl-L-tyrosine ethyl ester, 15 μg of enzyme was used (Awad et al, 1972 b).

[a] $CaCl_2$ at 10 mM concentration.

[b] The chloride salt of Ba^{2+}, Cd^{2+}, Co^{2+}, Cu^{2+}, Mg^{2+}, Mn^{2+}, Ni^{2+}, Sr^{2+}, or Zn^{2+} at 10 mM concentration.

TABLE 2
KINETIC CONSTANTS FOR HYDROLYSIS OF Ac-Tyr-OEt

Protein	K_m	V_{max}
Guanidine-stable chymoelastase	26mM	12 sec^{-1}
Lysine-free chymoelastase	47	35
Subtilisin-like endopeptidase	67	280

Assays were carried out at pH 8.0 in the pH stat at 24° as described before (Awad and Wilcox, 1963). The range of Ac-Tyr-OEt concentrations used was from 5 to 30 mM in 3% (v/v) dioxane. The nomenclature of the chymoelastases follows that of Siegel and Awad (1973).

the convergent development of not only the catalytic centers but also of some specific conformations which confer resistance to denaturant effects. The present demonstration of acid lability confirms our earlier explanation for the progressive loss of the subtilisin-like protein during purification procedures carried out at pH 5 (Awad et al, 1972 b). This protease may be useful for structural studies of proteins which are resistant to proteolysis in the native state. Furthermore, it may provide a unique method of preparing protein-free polynucleotides or polysaccharides by hydrolysing any associated proteins to small peptide fragments in denaturant solution. The pH range for maximal activity of this serine protease is very similar to that of two of its companion chymotrypsin-like enzymes. Preliminary studies have revealed that the subtilisin-like endopeptidase has no marked variance in substrate specificity from these two chymotrypsin-like enzymes. Thus any explanation for conservation of the two families of serine enzymes in this microorganism must also describe the necessity of retaining three different enzymes carrying out similar functions. These enzymes are considered to have primarily extracellular functions. However, each protein may have unknown specific and unique intracellular functions; this may be the reason for retention of so many similar endopeptidases.

ACKNOWLEDGMENTS

This investigation was supported by United States Public Health Service grants (NIH-AM-09001, NIH-AM-05472, and NIH-GM-02011). This is Paper no. VIII in the series, "The Proteolytic Enzymes of the K-1 Strain of *Streptomyces griseus* Obtained from a Commercial Preparation (Pronase)".

We wish to express our appreciation for the continued encouragement and support extended by Mr. B. B. Sigelbaum to the efforts of this laboratory.

FOOTNOTE

[1] The abbreviation used is: Ac-Tyr-OEt, N^α-acetyl-L-tyrosine ethyl ester.

REFERENCES

Awad, W. M., Jr., M. S. Ochoa, and T. P. Toomey 1972a. Self Purification of Two Serine Endopeptidases. *Proc. Natl. Acad. Sci. USA* 69: 2561-2565.

Awad, W. M., Jr., W. E. Skiba, G. G. Bernstrom, and M. S.

Ochoa 1971. Pronase Contains an Enzyme Homologous with Subtilisin. *Fed. Proc.* 30, Abstracts: 1296.

Awad, W. M., Jr., A. R. Soto, S. Siegel, W. E. Skiba, G. G. Bernstrom, and M. S. Ochoa 1972b. The Proteolytic Enzymes of the K-1 Strain of *Streptomyces griseus* Obtained from a Commercial Preparation (Pronase). I. Purification of Four Serine Endopeptidases. *J. Biol. Chem.* 247: 4144-4154.

Awad, W. M., Jr., and P. E. Wilcox 1963. The Activation of Chymotrypsinogen-A by a Protease from *Streptomyces griseus*. *Biochim. Biophys. Acta* 73: 285-292.

Blow, D. M., J. J. Birktoft, and B. S. Hartley 1969. Role of a Buried Acid Group in the Mechanism of Action of Chymotrypsin. *Nature* 221: 337-340.

Gertler, A. and M. Trop 1971. The Elastase-Like Enzymes from *Streptomyces griseus* (Pronase). Isolation and Partial Characterization. *Eur. J. Biochem.* 19: 90-96.

Hartley, B. S. 1964. Amino-Acid Sequence of Bovine Chymotrypsinogen-A. *Nature* 201: 1284-1287.

Hiramatsu, A. and T. Ouchi 1963. On the Proteolytic Enzymes from the Commercial Preparation of *Streptomyces griseus* (Pronase P). *J. Biochem.* (Tokyo) 54: 462-464.

Markland, F. S., Jr. and E. L. Smith 1971. Subtilisins: Primary Structure, Chemical and Physical Properties, in *The Enzymes*, 3rd Edition (P. D. Boyer, ed) Academic Press, New York, 561-608.

Matsubara, H., B. Hagihara, M. Nakai, T. Komaki, T. Yonetani, and K. Okunuki 1958. Crystalline Bacterial Proteinase. II. General Properties of Crystalline Proteinase of Bacillus Subtilis N'. *J. Biochem.* (Tokyo) 45: 251-258.

Matthews, B. W., P. B. Sigler, R. Henderson, and D. M. Blow 1967. Three-dimensional Structure of Tosyl-α-chymotrypsin. *Nature* 214: 652-656.

Neurath, H. and R. A. Bradshaw 1970. Evolution of Proteolytic Function. *Accts. Chem. Res.* 3: 249-257.

Nomoto, M. and Y. Narahashi 1959a. A Proteolytic Enzyme of *Streptomyces griseus*. I. Purification of a Protease of *Streptomyces griseus*. *J. Biochem.* (Tokyo). 46: 653-667.

Nomoto, M. and Y. Narahashi 1959b. A Proteolytic Enzyme of *Streptomyces griseus*. II. An Improved Method for Purification of a Protease of *Streptomyces griseus*. *J. Biochem.* (Tokyo). 46: 839-847.

Nomoto, M. and Y. Narahashi 1959c. A Proteolytic Enzyme of *Streptomyces griseus*. III. Homogeneity of the Purified Enzyme Preparation. *J. Biochem.* (Tokyo). 46: 1481-1487.

Nomoto, M., Y. Narahashi, and M. Murakami 1960a. A Proteolytic Enzyme of *Streptomyces griseus*. V. Protective

Effect of Calcium Ion on the Stability of Protease. *J. Biochem.* (Tokyo). 48: 453-463.

Nomoto, M., Y. Narahasi, and M. Murakami 1960b. A Proteolytic Enzyme of *Streptomyces griseus*. VII. Substrate Specificity of *Streptomyces griseus* Protease. *J. Biochem.* (Tokyo). 48: 906-918.

Nomoto, M., Y. Narahashi, T. Ouchi, and A. Hiramatsu 1964. Isolation and Properties of *Streptomyces griseus* Proteases from a Commercial Protease Powder, Pronase. *Proc. Int. Congr. Biochem*, Abstracts, 325.

Olson, M. O. J., N. Nagabhushan, M. Dzwiniel, L. B. Smillie, and D. R. Whitaker 1970. Primary Structure of α-Lytic Protease: a Bacterial Homologue of the Pancreatic Serine Proteases. *Nature* 228: 438-442.

Ottesen, M., J. T. Johansen, and I. Svendsen 1970. Subtilisin: Stability Properties and Secondary Binding Sites, in *Structure-Function Relationships of Proteolytic Enzymes* (P. Desnuelle, H. Neurath, and M. Ottesen, eds.) Academic Press, New York, 175-186.

Schellman, J. A., R. B. Simpson, and W. Kauzmann 1953. The Kinetics of Protein Denaturation. II. The Optical Rotation of Ovalbumin in Solutions of Guanidinium Salts. *J. Am. Chem. Soc.* 75: 5152-5154.

Siegel, S. and W. M. Awad, Jr. 1973. The Proteolytic Enzymes of the K-1 Strain of *Streptomyces griseus* Obtained from a Commercial Preparation (Pronase). IV. Structure-Function Studies of the Two Smallest Serine Endopeptidases; Stabilization by Glycerol During Reaction with Acetic Anhydride. *J. Biol. Chem.* 248: 3233-3240.

Siegel, S., A. H. Brady, and W. M. Awad, Jr. 1972. The Proteolytic Enzymes of the K-1 Strain of *Streptomyces griseus* Obtained from a Commercial Preparation (Pronase). II. The Activity of a Serine Enzyme in 6 M Guanidinium Chloride. *J. Biol. Chem.* 247: 4155-4159.

Simpson, R. B. and W. Kauzmann 1953. The Kinetics of Protein Denaturation. I. The Behavior of the Optical Rotation of Ovalbumin in Urea Solutions. *J. Am. Chem. Soc.* 75: 5139-5152.

Smith, E. L., R. J. DeLange, W. H. Evans, M. Landon, and F. S. Markland 1968. Subtilisin Carlsberg. V. The Complete Sequence; Comparison with Subtilisin BPN'; Evolutionary Relationships. *J. Biol. Chem.* 243: 2184-2191.

Smith, E. L., F. S. Markland, and A. N. Glazer 1970. Some Structure-Function Relationships in the Subtilisins, in *Structure-Function Relationships of Proteolytic Enzymes* (P. Desnuelle, H. Neurath, and M. Ottesen, eds.) Academic Press, New York, 160-172.

Vosbeck, K. D., K.-F. Chow, and W. M. Awad, Jr. 1973. The Proteolytic Enzymes of the K-1 Strain of *Streptomyces griseus* Obtained from a Commercial Preparation (Pronase). Purification and Characterization of the Aminopeptidases. *J. Biol. Chem.* 248: 6029-6034.

Wählby, S. and L. Engström 1968. Studies on *Streptomyces griseus* Protease. II. The Amino Acid Sequence Around the Reactive Serine Residue of DFP-Sensitive Components with Esterase Activity. *Biochim. Biophys. Acta* 151: 402-408.

Wright, C. S., R. A. Alden, and J. Kraut 1969. Structure of Subtilisin BPN' at 2.5 Å Resolution. *Nature* 221: 235-242.

IMPLICATIONS OF MACROMOLECULE-SMALL MOLECULE INTERACTIONS AND KINETICALLY CONTROLLED REACTIONS FOR SEPARATION OF PROTEINS BY MASS TRANSPORT

JOHN R. CANN

Department of Biophysics and Genetics
University of Colorado Medical Center
Denver, Colorado 80220

ABSTRACT. The implications of macromolecular interactions for the separation and purification of proteins by electrophoresis, sedimentation, and chromatography are discussed with emphasis on a variety of mechanistically different, rapidly equilibrating ligand-mediated interactions and several representative kinetically-controlled, nonmediated reactions. With rare exception each of these interactions has the potentiality for showing bimodal reaction boundaries or zones, and a physical explanation for resolution into two peaks is given. Such transport patterns could easily be misinterpreted as indicative of inherent heterogeneity, and unequivocal proof of heterogeneity is afforded only by analysis of fractions to see if they run true. Moreover, it is essential that several independent physiochemical methods be brought to bear in order to establish the nature of the interaction. Multizoning due to interaction with constituents of the solvent medium may often be avoided by judicious choice of buffer.

INTRODUCTION

The pioneering work of Gilbert (Gilbert, 1955 and 1959) about two decades ago on the theory of sedimentation of reversibly associating macromolecules has had a profound influence on our concepts concerning the electrophoretic, sedimentation, and chromatographic behavior of interacting systems, particularly with respect to the nature of reaction boundaries and zones. Like most advances of this kind, Gilbert's work has stimulated an ever increasing interest in the theory and practice of the mass transport of interacting macromolecules (Cann, 1970; Nichol and Winzor, 1972). The most provocative of the new insights provided by these investigations is the realization that the electrophoretic, sedimentation, and chromatographic patterns of homogeneous, but interacting, macromolecules can and do show multiple peaks and zones despite instantaneous establishment of equilibrium. This is of con-

837

siderable practical significance for it is a disturbing fact
that many such patterns could easily be misinterpreted as
indicating inherent heterogeneity. It cannot be overemphasized
that unequivocal proof of inherent heterogeneity is afforded
only by isolation of the various components. In this review
of our work in the area we shall address ourselves primarily
to the implications of macromolecular interactions for the
separation and purification of proteins by the several zonal
modes of mass transport, but sight must not be lost of the
applications of both zonal and moving-boundary methods for
quantitative characterization of biologically important interactions.

RESULTS AND DISCUSSION

Our interest in the mass transport of interacting systems
originally stemmed from observations on the moving-boundary
electrophoresis of a variety of highly purified proteins which
display nonenantiographic, bimodal electrophoretic patterns
in acidic media containing acetate or other carboxylic acid
buffers (Cann, 1970). Experimental evidence was advanced to
support interpretation of the patterns in terms of reversible
complexing of the protein with undissociated buffer acid, with
concomitant increase in the net positive charge on the protein
molecule without significant change in its fractional coefficient. Subsequently it was shown that the change in net positive charge on the protein is most probably due to binding
of the undissociated buffer acid to side-change carboxyl
groups with concomitant increase in their pK (Cann, 1971a).
It was also found that under appropriate conditions of pH
and buffer concentration interaction of aliphatic acids with
ribonuclease A is a cooperative phenomenon associated with a
change in tertiary structure without substantial alteration
in secondary structure (Cann, 1971b), but that binding of
undissociated acid to a protein need not necessarily be cooperative for resolution of bimodal reaction boundaries (Cann,
1971a and 1973).

A theoretical basis for interpretation of the electrophoretic patterns was provided (Cann and Goad, 1965a) by a
theory of electrophoresis of reversibly interacting systems
of the type
$$P + nHA \rightleftharpoons P(HA)_n \tag{I}$$
where P represents a protein molecule or other macromolecular
ion in solution and $P(HA)_n$ its complex formed by binding n
moles of a small, uncharged constituent, HA, of the solvent
medium, e.g. undissociated buffer acid. It is assumed that

P and P(HA)$_n$ possess different electrophoretic mobilities and that equilibrium is established instantaneously. These computations account for the essential features of the moving-boundary electrophoretic behavior of proteins in acidic media containing varying concentration of carboxylic acid buffer.

Upon realization that a single macromolecule can give moving-boundary patterns showing two peaks, the computations were extended to include zone electrophoresis on a solid support or in a density gradient. It is evident from Figs. 1, 2, and 4 in Cann and Goad (1965b) that the whole spectrum of experimentally recognized types of zone electrophoretic patterns, may in principle arise from a rapidly equilibrating interaction of a single macromolecule with a small neutral molecule in accordance with Reaction I. In particular, the patterns can show two well resolved and stable peaks (Fig. 4 in Cann and Goad, 1965b). This, despite instantaneous reestablishment of equilibrium during differential transport of P and P(HA)$_n$. Resolution is dependent upon the changes in concentration of HA which accompany reequilibration and maintenance of the resulting concentration gradients of the electrically neutral molecule along the electrophoresis column, the peaks in the pattern corresponding to different equilibrium compositions and not simply to P and P(HA)$_n$. Although the protein concentration eventually becomes very small between the peaks, it never vanishes. In other words, the patterns display bimodal reaction zones; but in practical terms it may be said that a single macromolecule can give two zones due to reversible and rapid interaction with the solvent medium.

This important prediction was verified experimentally (Cann, 1970), and it now appears that multiplicity of zones due to interaction is of rather frequent occurrence. However, with due precaution fractionation provides an unambiguous method for distinguishing between interaction and heterogeneity. The protein eluted from each unstained zone is subject to zone electrophoresis in the same buffer that was used for the original separation. For interactions the fractions will behave like the unfractionated material and show multiple zones, while for heterogeneity a single zone will be obtained.

Since the generation of multiple zones due to interaction will in general be sensitive to protein concentration, the fractions should be analyzed at about the same concentration as used in the original separation. Alternatively, the method can be calibrated by analyzing unfractionated material at several different concentrations. Fractionation is then carried out at a concentration sufficiently high that the concentration of the fractions falls within the range over which the original material exhibits multiple zones.

In conventional analytical and preparative applications of electrophoresis it is obviously desirable to avoid conditions conducive to the production of transport patterns which do not faithfully reflect the inherent state of homogeneity of the protein. Multizoning due to protein-buffer interactions have been encountered with carboxylate, phosphate-borate, borate, Tris-borate, amino acid and even phosphate buffers. Choice of a buffer in which a particular protein at specified pH and ionic strength will not show multiple zones due to interaction is largely empirical, although there are some guide lines. For example, electrophoresis in barbital buffer appears to be free of these complications. If strong interaction with the solvent is unavoidable, one can attempt to minimize interpretative difficulties by lowering the protein concentration and/or increasing the buffer concentration. At a sufficiently low protein to buffer concentration ratio a truly homogeneous but interacting protein will in principle show a single zone. But such conditions may be difficult to realize in practice, and failure to do so must not be accepted in lieu of fractionation as indicative of inherent heterogeneity.

The foregoing electrophoretic considerations also apply to ultracentrifugation. Thus, it is conceivable that binding of small molecules or ions by a protein could cause a significant change in its sedimentation coefficient due to alteration in macromolecular conformation. Given the appropriate conditions such a system might show a bimodal reaction zone upon sedimentation through a density gradient containing the ligand. Another possibility is that binding of a small ligand molecule mediates molecular association or dissociation of the protein; and we have developed the theory of sedimentation for ligand-mediated association-dissociation reactions (Cann, 1970 and 1973; Cann and Goad, 1970 and 1972; Cann and Kegeles, 1974).

Consider, for example, the reversible reaction

$$mM + nX \rightleftharpoons M_mX_n \tag{II}$$

in which a macromolecule M associates into a polymer containing m monomeric units (m-mer), with the mediation of a small ligand molecule or ion X, of which a fixed number n are bound into the complex. While moving-boundary sedimentation patterns have been calculated for both dimerization and tetramerization, the results for dimerization (Reaction II with m = 2 and n = 1 or 6) are, in certain respects, the more revealing with regards to fundamental principles. A particularly provocative result is that ligand-mediated dimerization can give rise to a well-resolved bimodal reaction boundary (Fig. 1A) despite instantaneous establishment of equilibrium. As in the case of electrophoresis, resolution of the two sedimenting peaks depends upon the production and maintenance of concentration gradients of unbound ligand along the centrifuge cell by reequilibration during differential transport of macromonomer and dimer. This behavior is in contradistinction to that predicted by the Gilbert theory (Gilbert, 1955 and 1959) and observed experimentally (Field and O'Brien, 1955; Field and Ogston, 1955) for nonmediated dimerization,

$$2 \text{ M} \rightleftharpoons \text{M}_2 \qquad\qquad (III)$$

Systems of this kind always give sedimentation patterns showing a single peak. Of course, at sufficiently high ligand concentration, ligand-mediated dimerization will also give patterns that show a single peak. The set of patterns presented in Fig. 1 illustrate how the shape of the reaction boundary depends upon total ligand concentration at constant macromolecule concentration. Increasing the initial ligand concentration, holding the degree of dimerization constant, results in progressive coalescence of the two peaks in the reaction boundary with concomitant drift in their velocities toward the weight-average value until resolution disappears entirely at the highest concentration. In the limit where for any reason the concentration of unbound ligand along the centrifuge cell is not significantly perturbed by the reaction during differential sedimentation of the macromolecular species, the system effectively approaches the case of nonmediated dimerization considered by Gilbert.

The theoretical prediction of bimodal reaction boundaries has found experimental verification in the reversible dimerization of New England lobster hemocyanin mediated by the binding of 4-6 Ca++ and 2-4 H+ which occurs when the pH is lowered from above pH 9.6 to below 9.2. Kegeles and his coworkers (Morimoto and Kegeles, 1971; Tai and Kegeles, 1971; Kegeles and Tai, 1973) have examined this system in some detail by the combined application of fractionation experiments, Archibald molecular weight determinations, and rapid kinetic measurements. Theirs is an exemplary study of an interacting system which in addition demonstrates that sedimentation patterns showing two or more peaks must be interpreted with caution.

The same consideration also applies to band sedimentation in the analytical ultracentrifuge and to zone sedimentation through a preformed density gradient in the preparative ultracentrifuge. Consider, for example, the band sedimentation patterns computed for ligand-mediated dimerization (Reaction II with m = 2 and n = 6). The results displayed in Fig. 2 are in certain respects analogous to those displayed in Fig. 1 for moving-boundary sedimentation. Thus, for appropriate

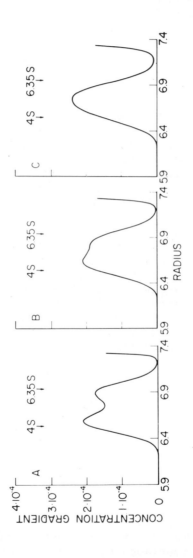

Fig. 1. Theoretical moving-boundary sedimentation patterns for the ligand-mediated dimerization reaction, $2M + 6X \rightleftharpoons M_2X_6$. Dependence of boundary shape at 50% dimerization upon the initial concentration of unbound small ligand, C_{30}: A, $C_{30} = 5 \times 10^{-5}M$; B, $10^{-4}M$; C, $2 \times 10^{-4}M$. Macromolecular concentrations and parameters approximate the sedimentation of a 1% solution of a protein of molecular weight about 60,000; sedimentation coefficient of monomer, 4S; dimer, 6.35S; ligand, 0.15S. From Cann (1970).

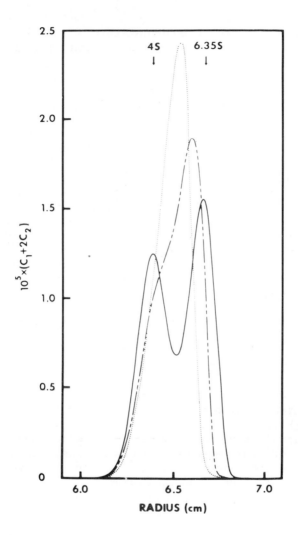

Fig. 2.   Theoretical band sedimentation patterns for the
ligand-mediated dimerization reaction, $2M + 6X \rightleftharpoons M_2X_6$.
Dependence of zone shape at 50% dimerization upon the initial
concentration of unbound small ligand, C_{30}:——— , $C^{30} = 10^{-6}M$;
——— — — ——— , $10^{-5}M$; · · · · · · , $5 \times 10^{-5}M$ Parameters are
the same as in Fig. 2.   From Cann and Goad (1970).   Copyright
1970 by the American Association for the Advancement of
Science.

choice of ligand concentration, the computed band pattern shows two well resolved peaks which constitute a bimodal reaction zone. Increasing the initial ligand concentration, holding the percent dimerization constant, results in progressive coalescence of the two peaks with concomitant drift in the sedimentation velocities toward a value close to the weight average until resolution disappears entirely at the highest concentration.

Once again we see that ligand-mediated interactions can give well resolved bimodal transport patterns even for rapid equilibration. The frequent occurrence of this type of pattern begs a physical explanation. Scrutiny of the calculated concentration profiles for macromonomer, dimer, and ligand as a function of time of sedimentation reveals the following mechanism for resolution: Because of the coupling between reequilibration during differential sedimentation of macromonomer and dimer and the transport of ligand bound into dimer, the back half of the sedimenting zone is almost depleted of ligand. Consequently, the remaining monomer in this region is deprived of the dimerization-mediating small molecule and thus lags behind to form a second peak. At the same time the fast peak is enriched in ligand to an extent which more than compensates for dilution of the macromolecule in the spreading zone. As a result, the macromolecule in this peak is actually more highly dimerized than in the starting zone. The two peaks never completely resolve, however, because the association reaction is reversible.

Because zone sedimentation is so widely employed for separation of proteins and other macromolecules, we extended the foregoing calculations to encompass the sedimentation behavior of a variety of ligand-mediated interactions. In addition to Reaction II and the ligand-mediated dissociation reaction

$$M_m + mnX \rightleftharpoons mMX_n \qquad \text{(IV)}$$

these include representative noncooperative interactions such as the set of sequential and simultaneous reactions (Reaction set V)

$$M + X \rightleftharpoons MX, \ K_1$$
$$MX + M \rightleftharpoons M_2X, \ K_2$$
$$2 \ MX \rightleftharpoons M_2X_2, \ K_3 \qquad \text{(V)}$$

the sequential monomer-tetramer reaction (Reaction set VI)

$$M + X \rightleftharpoons MX, \ K_1$$
$$4MX \rightleftharpoons M_4X_4, \ K_2 \qquad \text{(VI)}$$

sequential and progressive tetramerization (Reaction set VII)

$$M + X \rightleftharpoons MX, \ K_1$$
$$2 \ MX \rightleftharpoons M_2X_2, \ K_2$$
$$M_2X_2 + MX \rightleftharpoons M_3X_3, \ K_3$$
$$M_3X_3 + MX \rightleftharpoons M_4X_4, \ K_4 \tag{VII}$$
$$K_2 = K_3 = K_4 \equiv K$$

and dissociation of a dimer driven by ligand binding to the monomer (Reaction set VIII)

$$M_2 \rightleftharpoons 2 \ M, \ K_1$$
$$M + X \rightleftharpoons MX, \ K_2 \tag{VIII}$$

Except for one case (tetramerization mediated by cooperative interaction with ligand, Reaction II with m = 4 and n = 4), each of these interactions can, under appropriate conditions, give zone patterns showing two peaks. Thus, with only a single exception, the generalization seems justified that rapidly equilibrating ligand-mediated interactions in general have the potentiality for showing bimodal reactions zones irrespective of reaction mechanism subject to the provisos that overall ligand-binding is sufficiently strong to generate large gradients of unbound ligand along the centrifuge tube but that binding to the reactant itself is not so strong that in effect the system behaves like the analogous non-mediated interaction. Since the generation of bimodal zones is not unique for any one reaction mechanism, it is imperative that in addition to fractionation experiments appeal be made to the combined application of one or more independent physical methods in order to elucidate the nature of the interaction.

The results described above for zone electrophoresis and zone sedimentation can be applied qualitatively to chromatography. This is illustrated quantitatively in Fig. 3 by theoretical gel-permeation chromatographic profiles calculated for the sequential monomer-tetramer Reaction set VI. For purposes of discussion it is convenient to characterize Reaction set VI in terms of the ration, R_{10}, of the initial equilibrium concentration of MX to that of M so that

$$K_1 = R_{10}/C'_{50} \tag{1}$$

$$K_2 = \frac{C'_{40}}{C'^4_{10}} \cdot \frac{(1 + R_{10})^4}{R^4_{10}} \tag{2}$$

in which C'_{50} is the initial equilibrium concentration of X; C'_{40}, initial equilibrium concentration of M_4X_4, and C'_{10},

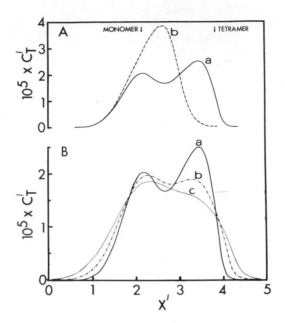

Fig. 3. Factors governing the shape of the gel-permeation chromatographic profile computed for Reaction set VI on sepha-dex G200R, 50% tetramerization. A - Dependence of shape on R_{10} for flow rate F = 1.2ml/hr: a, R_{10} = 0.10; b, 10.00. B - Dependence of shape on F for R_{10} = 0.10: a, 1.2ml/hr; b, 5ml/hr; c, 9.6ml/hr. The times are such that the volume passed (V = Ft) is 5ml for all curves. $C'_{10} + 4C'_{40}$ = 14 x 10^{-5}M; C'_{50} = 5 x 10^{-7}M. Monomer, 17,000 daltons. From Cann (1973).

initial equilibrium concentration of total monomer both uncom-plexed, M, and complexed, MX, with ligand. The concentrations are primed to indicate the total column frame of reference. For appropriate choice of R_{10} = 0.10 and C'_{50} = 5 x 10^{-7}M at 50% tetramerization, overall ligand-binding is sufficiently strong to generate large gradients of unbound ligand along the chromatographic column. Consequently, the chromatographic profile (curve a, Fig. 3A) shows a well-resolved bimodal re-action zone. Increasing the value of R_{10} to 10.0 holding other things constant leads to loss of resolution (curve b, Fig. 3A) because ligand-binding is now so strong (Equation 1) that the limit is approached in which all of the monomer is complexed with ligand. In this limit the system in effect behaves like the nonmediated tetramerization reaction

$$4M \rightleftharpoons M_4 \qquad (IX)$$

and shows a somewhat skewed unimodal zone which migrates at a velocity less than the initial weight average velocity due to progressive dissociation of the tetramer in the spreading zone. Referring back to those conditions conducive to bimodal reaction zones, it is interesting that increasing the flow rate (Fig. 3B) causes progressive coalescence of the two peaks in the broadening zone. This behavior is due to progressive increase in the axial dispersion of the ligand which tends to smooth its concentration gradients through the zone.

Although the chromatographic profiles were computed for the total column frame of reference and are thus directly related to the column-scanning mode of data acquisition (Ackers, 1969), the elution profile will show the same features. Accordingly, the same precaution that was emphasized above for electrophoresis and sedimentation also applies to gel filtration; namely, fractions must be rechromatographed to see if they run true.

The foregoing discussion has focused on ligand-mediated interactions for which reequilibration is instantaneous, but other classes of interaction also have important implications for separation of proteins by mass transport. In particular, proteins may undergo slow and irreversible changes during the separation process; for example, irreversible isomerization

$$A \xrightarrow{k} B \tag{X}$$

where k is the specific rate constant, and the isomer B has a different electrophoretic mobility or sedimentation coefficient from isomer A; irreversible dimerization

$$2M \xrightarrow{k} M_2 \tag{XI}$$

and irreversible dissociation into identical subunits

$$M_m \xrightarrow{k} mM \tag{XII}$$

The question posed is whether the electrophoretic, sedimentation or chromatographic pattern of a protein undergoing such a reaction during the time course of the experiment will show two peaks or a single skewed peak with a long trailing or leading edge. As illustrated by the representative results displayed in Fig. 4, theoretical moving-boundary and zone patterns typically show two well-resolved peaks for half-times of reaction ranging from about 0.3 to 2.5 times the duration of the transport experiment depending upon the difference in velocity between product and reactant. The greater the difference, the better is the resolution for longer half-times. There is no indication of a single skewed peak with long trailing or leading edges as long as the reaction does not go to completion during the course of the experiment. It is characteristic of the patterns (Figs. 4B and C) that the peak corresponding to a mixture of reactant and some product is sharp, while the one corresponding largely to product is broad and skewed in the direction from whence it was formed. The product peak grows

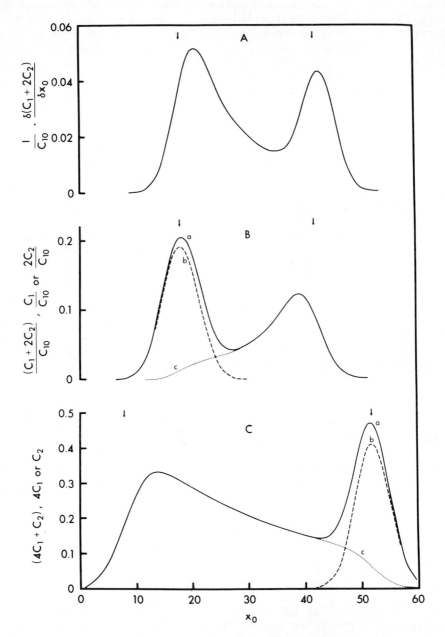

Fig. 4. Theoretical transport patterns. A – Moving-boundary
sedimentation pattern for irreversible dimerization, Reaction
XI; sedimentation from right to left. B – Zone electrophoretic

Fig. 4 (con't). or sedimentation pattern for irreversible dimerization, Reaction XI: a, total concentration normalized to initial concentration of monomer, $(C_1 + 2C_2)/ C_{10}$; b, normalized concentration of monomer, C_1/ C_{10}; c, normalized concentration of dimer, $2C_2/ C_{10}$; migration from left to right. C – Zone electrophoretic and sedimentation pattern for irreversible dissociation into four identical subunits, Reaction XII with m = 4: a, total concentration, $4C_1 + C_2$; b, concentration of tetramer, $4C_1$; c, concentration of subunit, C_2; pattern normalized by assigning unit initial concentration to M_4; essentially the same pattern is shown for dissociation into half-mers, Reaction XII with m = 2; migration from left to right. In each case the protein exists entirely as reactant at the start of the experiment. From Cann and Oates (1973).

---

with time at the expense of reactant, and the pattern never reaches the baseline between the peaks. In other words, the two peaks constitute a single reaction boundary or zone whose interpretation is subject to the same precautions as in the case of rapidly equilibrating, ligand-mediated interactions.

The physical explanation for resolution into two peaks is rather different from the one given above for rapidly equilibrating, ligand-mediated association-dissociation reactions. Consider, for example, a protein which dissociates slowly and irreversibly into subunits during zone sedimentation. As the zone of reactant departs from the meniscus and begins to sediment down the centrifuge tube, the reaction is occurring at its maximum rate determined by the initial concentration of reactant. Because of the difference in sedimentation velocity between product and reactant, the product lags behind the advancing peak of reactant and would form a long trailing plateau if the reaction were to proceed at a constant rate. But, in fact, the rate of reaction decreases progressively with time as reactant is consumed. Since the amount of product formed per unit time decreases as differential sedimentation of product and reactant proceeds, the concentration profile of product must pass through a maximum and skew forward to blend into the peak of reactant. For the special case in which the rate of sedimentation of the product is negligibly small compared to reactant, the maximum in its concentration profile will be located at the meniscus.

Finally, mention should be made of kinetically-controlled macromolecular isomerization,

$$A \underset{k_2}{\overset{k_1}{\rightleftharpoons}} B$$

(XIII)

where k_1 and k_2 are the specific rates of interconversion between two isomers with different mobilities. The zone electrophoretic, sedimentation, and chromatographic patterns of such a system can show one, two, or three peaks depending upon the half-times of reaction relative to the duration of the transport experiment (Cann 1970). Dissociation of a macromolecular complex, C, into its components, A and B,

$$C \rightleftharpoons A + B \qquad\qquad (XIV)$$

is also a source of multizoning even when the reactions are rapid (Bethune and Kegeles, 1961; Cann and Oates, 1973).

ACKNOWLEDGMENT

Supported in part by Research Grant 5R01 HL 13909-23 from the National Heart and Lung Institute, National Institutes of Health, United States Public Health Service. This publication is No. 605 from the Department of Biophysics and Genetics, University of Colorado Medical Center, Denver, Colorado 80220.

REFERENCES

Ackers, Gary K. 1969. Analytical gel chromatography of proteins. *Adv. Protein Chem.* 24: 343-446.

Bethune, J. L., and G. Kegeles 1961. Countercurrent distribution of chemically reacting systems. II. Reactions of the type A + B $\rightleftharpoons$ C. *J. Phys. Chem.* 65: 1755-1760.

Cann, John R. 1970. *Interacting Macromolecules. The Theory and Practice of Their Electrophoresis, Ultracentrifugation and Chromatography*, Academic Press, N. Y.

Cann, John. R. 1971a. Interaction of acetic acid with poly-L-glutamic acid and serum albumin. *Biochem.* 10: 3707-3712.

Cann, John R. 1971b. Cooperative interaction of aliphatic acids with ribonuclease A. *Biochem.* 10: 3713-3722.

Cann, John R. 1973. Theory of zone sedimentation for non-cooperative ligand-mediated interactions. *Biophys. Chem.* 1: 1-10.

Cann, John R., and W. B. Goad 1965a. Theory of moving boundary electrophoresis of reversibly interacting systems. Reaction of proteins with small uncharged molecules such as undissociated acid. *J. Biol. Chem.* 240: 148-155.

Cann, John R. and W. B. Goad 1965b. Theory of zone electrophoresis of reversibly interacting systems. Two zones from a single macromolecule. *J. Biol. Chem.* 240: 1162-1164.

Cann, John R., and W. B. Goad 1970. Bimodal sedimenting zones due to ligand-mediated interactions. *Science* 170: 441-445.

Cann, John R., and W. B. Goad 1972. Theory of sedimentation
    for ligand-mediated dimerization. *Arch. Biochem. Biophys.*
    153: 603-609.
Cann, John R. and G. Kegeles 1974. Theory of sedimentation
    for kinetically controlled dimerization reactions. *Biochem.*
    13: 1868-1874.
Cann, John R., and D. C. Oates 1973. Theory of electrophor-
    esis and sedimentation for some kinetically controlled
    interactions. *Biochem.* 12: 1112-1119.
Field, E. O. and J. R. P. O'Brien 1955. Dissociation of hu-
    man hemoglobin at low pH. *Biochem. J.* 60: 656-661.
Field, E. O., and A. G. Ogston 1955. Boundary spreading in
    the migration of a solute in rapid dissociation equili-
    brium. Theory and its application in the case of human
    haemoglobin. *Biochem. J.* 60: 661-665.
Gilbert, G. A. 1955. Discuss. Faraday Soc. 20: 68-71.
Gilbert, G. A. 1959. Sedimentation and electrophoresis of
    interacting substances. *Proc. Roy. Soc. London* 250A:
    377-388.
Kegeles, Gerston, and M. Tai 1973. Rate constants for the
    hexamerdodecamer reaction of lobster hemocyanin. *Biophys.*
    Chem. 1: 46-50.
Morimoto, Keisuke, and G. Kegeles 1971. Subunit interactions
    of lobster hemocyanin. I. Ultracentrifuge studies. *Arch.*
    Biochem. Biophys. 142: 247-257.
Nichol, L. W., and D. J. Winzor 1972. *Migration of Inter-*
    acting Systems. Clarendon Press, Oxford.
Tai, Mei-Sheng, and G. Kegeles 1971. Subunit interactions
    of hemocyanin. II. Temperature-jump kinetic studies.
    Arch. Biochem. Biophys. 142: 258-267.

Index